GENERALIZATIONS OF CYCLOSTATIONARY SIGNAL PROCESSING

GENERALIZATIONS OF CYCLOSTATIONARY SIGNAL PROCESSING

SPECTRAL ANALYSIS AND APPLICATIONS

Antonio Napolitano
University of Napoli Parthenope, Italy

WILEY
A John Wiley & Sons, Ltd., Publication

IEEE
IEEE PRESS

This edition first published 2012

Registered office
John Wiley & Sons Ltd, The Atrium, Southern Gate, Chichester, West Sussex, PO19 8SQ, United Kingdom

For details of our global editorial offices, for customer services and for information about how to apply for permission to reuse the copyright material in this book please see our website at www.wiley.com.

Library of Congress Cataloging-in-Publication Data

Napolitano, Antonio, 1964–
Generalizations of cyclostationary signal processing : spectral analysis and applications / Antonio Napolitano.
pages cm
Includes bibliographical references and index.
ISBN 978-1-119-97335-5
1. Signal processing. 2. Spectrum analysis. 3. Cyclostationary waves. I. Title.
TK5102.9.N37 2012
621.382′2–dc23
2012015740

A catalogue record for this book is available from the British Library.
Print ISBN:9781119973355

Set in Times New Roman 10/12 by Thomson Digital, Noida, India
Printed and bound in Malaysia by Vivar Printing Sdn Bhd

1 2012

To Ubalda and Asha

Contents

About the Author

Antonio Napolitano was born in Naples, Italy, in 1964. He received his Ph.D. in Electronic Engineering and Computer Science in 1994 from the University of Napoli Federico II, Italy. Since 2005 he has been Professor of Telecommunications at the University of Napoli "Parthenope", Italy. He is an IEEE Senior Member. From 2006 to 2009, and since 2011 he has been an Associate Editor of the IEEE Transactions on Signal Processing; and since 2008 he has been on the Editorial Board of Signal Processing (Elsevier). He is a Member of the Signal Processing Theory and Method Technical Committee (SPTM-TC) of the IEEE Signal Processing Society. In 1995 and 2006, he received the Best Paper Award from the European Association for Signal Processing (EURASIP) for articles on higher order cyclostationarity and the functional approach for signal analysis, respectively. In 2008, he received from Elsevier the Most Cited Paper Award for a review article on cyclostationarity.

Acknowledgements

Parts of these IEEE materials are reprinted with permission:

- A. Napolitano, "Uncertainty in measurements on spectrally correlated stochastic processes," *IEEE Transactions on Information Theory*, vol. 49, no. 9, pp. 2172–2191, September 2003.
- A. Napolitano, "Estimation of second-order cross-moments of generalized almost-cyclostationary processes," *IEEE Transactions on Information Theory*, vol. 53, no. 6, pp. 2204–2228, June 2007.
- A. Napolitano, "Discrete-time estimation of second-order statistics of generalized almost-cyclostationary processes," *IEEE Transactions on Signal Processing*, vol. 57, no. 5, pp. 1670–1688, May 2009.
- A. Napolitano, "Sampling of spectrally correlated processes," *IEEE Transactions on Signal Processing*, vol. 59, no. 2, pp. 525–539, February 2011.

Parts of this material from Elsevier are reprinted with permission: A. Napolitano, "Sampling theorems for Doppler-stretched wide-band signals," *Signal Processing*, vol. 90, no. 7, pp. 2276–2287, July 2010, Elsevier.

A special thank you is due to my parents. I also gratefully acknowledge Dominique Dehay, William A. Gardner, Harry L. Hurd, Luciano Izzo, Giacomo Lauridia, Jacek Leśkow, Luigi Paura and Chad M. Spooner for support, guidance, discussions and friendship.

AN

Preface

Many processes in nature arise from the interaction of periodic phenomena with random phenomena. The results are processes which are not periodic, but whose statistical functions are periodic functions of time. These processes are called cyclostationary and are an appropriate mathematical model for signals encountered in telecommunications, radar, sonar, telemetry, astronomy, mechanics, econometric, biology. In contrast, the classical model of stationary processes considers statistical functions which do not depend on time. More generally, if different periodicities are present in the generation mechanism of the process, the process is called almost cyclostationary (ACS). Almost all modulated signals adopted in communications, radar, and sonar can be modeled as ACS. Thus, in the past twenty years the exploitation of almost-cyclostationarity properties in communications and radar has allowed the design of signal processing algorithms for detection, estimation, and classification that significantly outperform classical algorithms based on a stationary description of signals. The gain in performance is due to a proper description of the nonstationarity of the signals, that is, the time variability of their statistical functions.

In this book, mathematical models for two general classes of nonstationary processes are presented: generalized almost-cyclostationary (GACS) processes and spectrally correlated (SC) processes. Both classes of processes include cyclostationary and ACS processes as special cases. SC and GACS processes are appropriate models for the received signal in mobile communications or radar scenarios when the transmitted signal is ACS and the propagation channel is a (possibly multipath) Doppler channel due to the relative motion between transmitter, receiver, and/or surrounding scatters or targets. SC processes are shown to be useful in the description of processes encountered in multirate systems and spectral analysis with nonuniform frequency spacing. GACS processes find application in the description of communications signals with slowly varying parameters such as carrier frequency, baud rate, etc.

The problem of statistical function estimation is addressed for both GACS and SC processes. This problem is challenging and of great interest at the applications level. In fact, once the nonstationary behavior of the observed signal has been characterized, statistical functions need to be estimated to be exploited in applications. The existence of reliable statistical function estimators for ACS processes is one of the main motivations for the success of signal processing algorithms based on this model. The results presented in this book extend most of the techniques used for ACS signals to the more general classes of GACS and SC signals. Mean-square consistency and asymptotic Normality properties are proved for the considered statistical function estimators. Both continuous- and discrete-time cases are considered and the problem of sampling and aliasing is addressed. Extensive simulation results are presented to corroborate the theoretical results.

How to use this book

The book is organized so that it can be used by readers with different requirements. Chapter 1 contains background material for easy reference in the subsequent chapters. Chapters 2 and 4 contain the main results presented in the form of theorems with sketches of proofs and illustrative examples. Thus, these chapters can be used by the non-specialist who is only interested in recipes or results and wants to grasp the main ideas. Each of these two chapters is followed by a chapter containing complements and proofs (Chapters 3 and 5). Each proof is divided into two parts. The first part consists of the formal manipulations to find the result. This part is aimed at advanced readers with a background of graduate students in engineering. The second part of the proof consists of the justification of the formal manipulations and is therefore aimed at specialists (e.g., mathematicians).

Book outline

In **Chapter 1**, the statistical characterization of persistent (finite-power) nonstationary stochastic processes is presented. Both strict-sense and wide-sense characterizations are considered. Harmonizability and time-frequency representations are treated. Definition and properties of almost periodic functions are provided. A brief review of ACS processes is also presented. The chapter ends with some properties of cumulants.

In **Chapter 2**, GACS processes are presented and characterized. GACS processes have multivariate statistical functions that are almost-periodic function of time. The (generalized) Fourier series of these functions have both coefficients and frequencies, named lag-dependent cycle frequencies, which depend on the lag shifts of the processes. ACS processes are obtained as special case when the frequencies do not depend on the lag parameters. The problems of linear filtering and sampling of GACS processes are addressed. The cyclic correlogram is shown to be, under mild conditions, a mean-square consistent and asymptotically Normal estimator of the cyclic autocorrelation function. Such a function allows a complete second-order characterization in the wide-sense of GACS processes. Numerical examples of communications through Doppler channels due to relative motion between transmitter and receiver with constant relative radial acceleration are considered. Simulation results on statistical function estimation are carried out to illustrate the theoretical results. Proofs of the results in Chapter 2 are reported in Chapter 3.

In **Chapter 3**, complements and proofs for the results presented in Chapter 2 are reported. Each proof consists of two parts. The first part contains formal manipulations that lead to the result. The second part contains the justifications of the mathematical manipulations of the first part. Thus, proofs can be followed with two different levels of rigor, depending on the background and interest of the reader.

In **Chapter 4**, SC processes are presented and characterized. SC processes have the Loève bifrequency spectrum with spectral masses concentrated on a countable set of support curves in the bifrequency plane. ACS processes are obtained as a special case when the curves are lines with a unit slope. The problems of linear filtering and sampling of SC processes are addressed. The time-smoothed and the frequency-smoothed cross-periodograms are considered as estimators of the spectral correlation density. Consistency and asymptotic Normality properties are analyzed. Illustrative examples and simulation results are presented. Proofs of the results in Chapter 4 are reported in Chapter 5.

In **Chapter 5**, complements and proofs for the results presented in Chapter 4 are reported. The system used is the same as in Chapter 3.

In **Chapter 6**, the problem of signal modeling and statistical function estimation is addressed in the functional or fraction-of-time (FOT) approach. Such an approach is an alternative to the classical one where signals are modeled as sample paths or realizations of a stochastic process. In the FOT approach, a signal is modeled as a single function of time and a probabilistic model is constructed by this sole function of time. Nonstationary models that can be treated in this approach are discussed.

In **Chapter 7**, applications in mobile communications and radar/sonar systems are presented. A model for the wireless channel is developed. It is shown how, in the case of relative motion between transmitter and receiver or between radar and target, the ACS transmitted signal is modified into a received signal with a different kind of nonstationarity. Conditions under which the GACS or SC model are appropriate for the received signal are derived.

In **Chapter 8**, citations are classified into categories and listed in chronological order.

List of Abbreviations

a.e. = almost everywhere
ACS = almost cyclostationary
AP = almost periodic
CDMA = code-division multiple access
DFS = discrete Fourier series
DFT = discrete Fourier transform
DSSS = direct-sequence spread-spectrum
DT-CCC = discrete-time cyclic cross-correlogram
fBm = fractional Brownian motion
FOT = fraction-of-time
FRESH = frequency shift
GACS = generalized almost-cyclostationary
H-CCC = hybrid cyclic cross-correlogram
LAPTV = linear almost-periodically time-variant
lhs = left-hand side
LTI = linear time-invariant
LTV = linear time-variant
MIMO = multi-input multi-output
MMSE = minimum mean-square error
PAM = pulse amplitude modulated
pdf = probability density function
QTI = quadratic time-invariant
RCS = radar cross-section
rhs = right-hand side
RX = receiver
SC = spectrally correlated
SCD = spectral correlation density
STFT = short-time Fourier transform
t.m.s.s. = temporal mean-square sense
TX = transmitter
w.a.p. = weakly almost periodic
w.p.1 = with probability 1
w.r.t. = with respect to
WSS = wide-sense stationary

1

Background

In this chapter, background material that will be referred to in the subsequent chapters is reviewed. In Section 1.1, the statistical characterization of persistent (finite-power) nonstationary stochastic processes is presented. Second-order statistics in both time, frequency, and time-frequency domains are considered. In Section 1.2, definitions of almost-periodic functions and their generalizations (Besicovitch 1932) and related results are reviewed. Almost-cyclostationary (ACS) processes (Gardner 1985, 1987d) are treated in Section 1.3. Finally, in Section 1.4, some results on cumulants are reviewed.

1.1 Second-Order Characterization of Stochastic Processes

1.1.1 Time-Domain Characterization

In the classical stochastic-process framework, statistical functions are defined in terms of ensemble averages of functions of the process and its time-shifted versions. Nonstationary processes have these statistical functions that depend on time.

Let us consider a continuous-time real-valued process $\{x(t, \omega),\ t \in \mathbb{R},\ \omega \in \Omega\}$, with abbreviate notation $x(t)$ when it does not create ambiguity, where Ω is a sample space equipped with a σ-field $\mathcal{F}$ and a probability measure P defined on the elements of $\mathcal{F}$. The cumulative distribution function of $x(t)$ is defined as (Doob 1953)

$$F_x(\xi; t) \triangleq P[x(t, \omega) \leqslant \xi] = \int_\Omega \mathbf{1}_{\{\omega\,:\,x(t,\omega)\leqslant\xi\}}\,\mathrm{d}P(\omega) \triangleq \mathrm{E}\left\{\mathbf{1}_{\{\omega\,:\,x(t,\omega)\leqslant\xi\}}\right\} \tag{1.1}$$

where

$$\mathbf{1}_{\{\omega\,:\,x(t,\omega)\leqslant\xi\}} \triangleq \begin{cases} 1, & \omega\,:\,x(t,\omega) \leqslant \xi, \\ 0, & \omega\,:\,x(t,\omega) > \xi \end{cases} \tag{1.2}$$

is the indicator of the set $\{\omega \in \Omega : x(t, \omega) \leqslant \xi\}$ and $\mathrm{E}\{\cdot\}$ denotes statistical expectation (ensemble average). The expected value corresponding to the distribution $F_x(\xi; t)$ is the statistical mean

$$\int_{\mathbb{R}} \xi \, \mathrm{d}F_x(\xi; t) = \int_{\Omega} x(t, \omega) \, \mathrm{d}P(\omega) = \mathrm{E}\{x(t, \omega)\}. \tag{1.3}$$

Analogously, at second-order, the process is characterized by the second-order joint distribution function (Doob 1953)

$$\begin{aligned} F_x(\xi_1, \xi_2; t, \tau) &\triangleq P[x(t+\tau, \omega) \leqslant \xi_1, x(t, \omega) \leqslant \xi_2] \\ &= \mathrm{E}\left\{\mathbf{1}_{\{\omega : x(t+\tau,\omega) \leqslant \xi_1\}} \mathbf{1}_{\{\omega : x(t,\omega) \leqslant \xi_2\}}\right\} \end{aligned} \tag{1.4}$$

and the autocorrelation function

$$\mathrm{E}\{x(t+\tau, \omega) \, x(t, \omega)\} = \int_{\mathbb{R}^2} \xi_1 \xi_2 \, \mathrm{d}F_x(\xi_1, \xi_2; t, \tau). \tag{1.5}$$

If $F_x(\xi; t)$ and $F_x(\xi_1, \xi_2; t, \tau)$ depend on t, the process is said to be nonstationary in the strict sense. If $F_x(\xi; t)$ $[F_x(\xi_1, \xi_2; t, \tau)]$ does not depend on t, the process $x(t)$ is said to be 1st-order [2nd-order] stationary in the strict sense. If both mean and autocorrelation function do not depend on t, the process is said to be wide-sense stationary (WSS) (Doob 1953).

In the following, we will focus on the second-order statistics of complex-valued nonstationarity processes.

The complex-valued stochastic process $x(t)$ is said to be a *second-order process* if the second-order moments

$$\mathcal{R}_{\boldsymbol{x}}(t, \tau) \triangleq \mathrm{E}\left\{x(t+\tau) \, x^{(*)}(t)\right\} \tag{1.6}$$

exist $\forall t$ and $\forall \tau$. In Equation (1.6), superscript $(*)$ denotes optional complex conjugation, and subscript $\boldsymbol{x} \triangleq [x, x^{(*)}]$. That is, $\mathcal{R}_{\boldsymbol{x}}(t, \tau)$ denotes one of two different functions depending if the complex conjugation is considered or not in subscript $\boldsymbol{x}$. If conjugation is present, then (1.6) is the *autocorrelation function*. If the conjugation is absent, then (1.6) is the *conjugate autocorrelation function* also referred to as *relation function* (Picinbono and Bondon 1997) or *complementary correlation* (Schreier and Scharf 2003a). Note that, in the complex case the order of the distribution functions turns out to be doubled with respect to the real case. For example, the joint distribution function of $x(t)$ and $x(t+\tau)$ is a fourth-order joint distribution of the real and imaginary parts of $x(t)$ and $x(t+\tau)$.

The *(conjugate) autocovariance* is the (conjugate) autocorrelation of the process reduced to be zero mean by subtracting its mean value

$$\mathcal{C}_{\boldsymbol{x}}(t, \tau) \triangleq \mathrm{E}\Big\{\big[x(t+\tau) - \mathrm{E}\{x(t+\tau)\}\big]\,\big[x(t) - \mathrm{E}\{x(t)\}\big]^{(*)}\Big\}. \tag{1.7}$$

Even if $\mathcal{C}_{\boldsymbol{x}}(t, \tau) = \mathcal{R}_{\boldsymbol{x}}(t, \tau)$ only for zero-mean processes, in some cases the terms autocorrelation, autocovariance, and covariance are used interchangeably. When the terms autocovariance or covariance are adopted, from the context it is understood if the mean value is subtracted or not. In statistics, the definition of autocorrelation includes in (1.6) also a normalization by the standard deviations of $x(t)$ and $x(t+\tau)$.

1.1.2 Spectral-Domain Characterization

The characterization of stochastic processes in the spectral domain can be made by resorting to the concept of harmonizability (Loève 1963). A second-order stochastic process $x(t)$ is said to be *harmonizable* if its (conjugate) autocorrelation function can be expressed by the Fourier-Stieltjes integral

$$\mathrm{E}\left\{x(t_1)\, x^{(*)}(t_2)\right\} = \int_{\mathbb{R}^2} e^{j2\pi[f_1 t_1 + (-) f_2 t_2]}\, \mathrm{d}\gamma_{\boldsymbol{x}}(f_1, f_2) \tag{1.8}$$

where $\gamma_{\boldsymbol{x}}(f_1, f_2)$ is a spectral correlation function of bounded variation (Loève 1963):

$$\int_{\mathbb{R}^2} |\,\mathrm{d}\gamma_{\boldsymbol{x}}(f_1, f_2)| < \infty. \tag{1.9}$$

In (1.8), $(-)$ is an optional minus sign that is linked to $(*)$. $\gamma_{\boldsymbol{x}}(f_1, f_2)$ denotes one of two different functions depending if the complex conjugation is considered or not in subscript $\boldsymbol{x}$.

Under the harmonizability condition, $x(t)$ is said to be *(strongly) harmonizable* and can be expressed by the Cramér representation (Cramér 1940)

$$x(t) = \int_{\mathbb{R}} e^{j2\pi f t}\, \mathrm{d}\chi(f) \tag{1.10}$$

where $\chi(f)$ is the *integrated spectrum* of $x(t)$.

In (Loève 1963), it is shown that a necessary condition for a stochastic process to be harmonizable is that it is second-order continuous (or mean-square continuous) (Definition 2.2.11, Theorem 2.2.12). Moreover, it is shown that a stochastic process is harmonizable if and only if its covariance function is harmonizable. In fact, convergence of integrals in (1.8) and (1.10) is in the mean-square sense. In (Hurd 1973), the harmonizability of processes obtained by some processing of other harmonizable processes is studied.

If the absolutely continuous and the discrete component of $\chi(f)$ are (possibly) nonzero and the singular component of $\chi(f)$ is zero with probability 1 (w.p.1) (Cramér 1940), we can formally write $\mathrm{d}\chi(f) = X(f)\,\mathrm{d}f$ (w.p.1) (Gardner 1985, Chapter 10.1.2), where

$$X(f) = \int_{\mathbb{R}} x(t)\, e^{-j2\pi f t}\, \mathrm{d}t \tag{1.11}$$

is the Fourier transform of $x(t)$ which possibly contains Dirac deltas in correspondence of the jumps of the discrete component of $\chi(f)$. For *finite-power processes*, that is such that the time-averaged power

$$P_x \triangleq \lim_{T\to\infty} \frac{1}{T} \int_{-T/2}^{T/2} \mathrm{E}\left\{|x(t)|^2\right\} \mathrm{d}t \tag{1.12}$$

exists and is finite, relation (1.11) is intended in the sense of distributions (Gelfand and Vilenkin 1964, Chapter 3), (Henniger 1970).

Let $x(t)$ be an harmonizable stochastic process. Its *bifrequency spectral correlation function* or *Loève bifrequency spectrum* (Loève 1963; Thomson 1982), also called *generalized spectrum* in (Gerr and Allen 1994), *cointensity spectrum* in (Middleton 1967), or *dual frequency spectral correlation* in (Hanssen and Scharf 2003), is defined as

$$\mathcal{S}_x(f_1, f_2) \triangleq \mathrm{E}\left\{X(f_1)\, X^{(*)}(f_2)\right\} \tag{1.13}$$

and if $\chi(f)$ and $\gamma_x(f_1, f_2)$ do not contain singular components w.p.1, in the sense of distributions the result is that

$$\mathrm{d}\gamma_x(f_1, f_2) = \mathrm{E}\left\{\mathrm{d}\chi(f_1)\, \mathrm{d}\chi^{(*)}(f_2)\right\} \tag{1.14a}$$

$$= \mathrm{E}\left\{X(f_1)\, X^{(*)}(f_2)\right\} \mathrm{d}f_1\, \mathrm{d}f_2 \tag{1.14b}$$

and, accordingly with (1.8), we can formally write

$$\mathrm{E}\left\{x(t_1)\, x^{(*)}(t_2)\right\} = \int_{\mathbb{R}^2} \mathrm{E}\left\{X(f_1)\, X^{(*)}(f_2)\right\} e^{j2\pi[f_1 t_1 + (-) f_2 t_2]}\, \mathrm{d}f_1\, \mathrm{d}f_2 \tag{1.15}$$

$$\mathrm{E}\left\{X(f_1)\, X^{(*)}(f_2)\right\} = \int_{\mathbb{R}^2} \mathrm{E}\left\{x(t_1)\, x^{(*)}(t_2)\right\} e^{-j2\pi[f_1 t_1 + (-) f_2 t_2]}\, \mathrm{d}t_1\, \mathrm{d}t_2 \tag{1.16}$$

A spectral characterization for nonstationary processes that resembles that for WSS processes (Section 1.1.4) can be obtained starting from the *time-averaged (conjugate) autocorrelation function*

$$R_x(\tau) \triangleq \lim_{T\to\infty} \frac{1}{T} \int_{-T/2}^{T/2} \mathrm{E}\left\{x(t+\tau)\, x^{(*)}(t)\right\} \mathrm{d}t$$

$$\equiv \left\langle \mathrm{E}\left\{x(t+\tau)\, x^{(*)}(t)\right\}\right\rangle_t \tag{1.17}$$

when the limit exists. Its Fourier transform is called the *power spectrum*, is denoted by $S_x(f)$, and represents the spectral density of the time-averaged power $R_x(0)$ of the process. The time-averaged autocorrelation function and the power spectrum defined here for nonstationary processes exhibit the same properties of the autocorrelation function and power spectrum defined for wide-sense stationary processes (Wu and Lev-Ari 1997).

1.1.3 Time-Frequency Characterization

The Loève bifrequency spectrum (1.13) provides a description of the nonstationary behavior of the process $x(t)$ in the frequency domain. A description in terms of functions of time and frequency can be obtained by resorting to the time-variant spectrum, the Rihaczek distribution, and the Wigner-Ville spectrum.

The Fourier transform of the second-order moment (1.6) with respect to (w.r.t.) the lag parameter τ is the *time-variant spectrum*

$$\mathcal{S}_x(t, f) \triangleq \int_{\mathbb{R}} \mathcal{R}_x(t, \tau)\, e^{-j2\pi f\tau}\, \mathrm{d}\tau. \tag{1.18}$$

By substituting (1.6) into (1.18), interchanging the order of the expectation and Fourier-transform operators, and accounting for the formal relation $\mathrm{d}\chi(f) = X(f)\,\mathrm{d}f$, one obtains

$$\mathcal{S}_x(t, f)\,\mathrm{d}f = \mathrm{E}\left\{ \mathrm{d}\chi(f)\, x^{(*)}(t) \right\} e^{j2\pi f t} \tag{1.19}$$

where the right-hand-side is referred to as the *(conjugate) Rihaczek distribution* of $x(t)$ (Scharf *et al.* 2005).

By the variable change $t' = t + \tau/2$ in (1.6) and Fourier transforming w.r.t. τ, we obtain a time-frequency representation in terms of *Wigner-Ville spectrum* for stochastic processes (Martin and Flandrin 1985)

$$\mathcal{W}_x(t', f) \triangleq \int_{\mathbb{R}} \mathrm{E}\left\{ x(t' + \tau/2)\, x^{(*)}(t' - \tau/2) \right\} e^{-j2\pi f \tau}\, \mathrm{d}\tau \tag{1.20a}$$

$$= \int_{\mathbb{R}} \mathrm{E}\left\{ X(f + \nu/2)\, X^{(*)}(f - \nu/2) \right\} e^{j2\pi \nu t'}\, \mathrm{d}\nu \tag{1.20b}$$

where the second equality follows using (1.11).

Extensive treatments on time-frequency characterizations of nonstationary signals are given in (Amin 1992), (Boashash *et al.* 1995), (Cohen 1989, 1995), (Flandrin 1999), (Hlawatsch and Bourdeaux-Bartels 1992). Most of these references refer to finite-energy signals.

1.1.4 Wide-Sense Stationary Processes

Second-order nonstationary processes have (conjugate) autocorrelation function depending on both time t and lag parameter τ and the function defined in (1.6) is also called the time-lag (conjugate) autocorrelation function. Equivalently, their time-variant spectrum depends on both time t and frequency f. In contrast, second-order WSS processes are characterized by a (conjugate) autocorrelation and time-variant spectrum not depending on t. That is

$$\mathcal{R}_x(t, \tau) = R_x(\tau) \tag{1.21a}$$

$$\mathcal{S}_x(t, f) = S_x(f). \tag{1.21b}$$

In such a case, for $(*)$ present, the Fourier-transform (1.18) specializes into the Wiener-Khinchin relation that links the autocorrelation function and the *power spectrum* $S_x(f)$ (Gardner 1985)

$$S_x(f) = \int_{\mathbb{R}} R_x(\tau)\, e^{-j2\pi f \tau}\, \mathrm{d}\tau. \tag{1.22}$$

Condition (1.21a) is equivalent to the fact that the time–time (conjugate) autocorrelation function (1.8) depends only on the time difference $t_1 - t_2$. This time dependence in the spectral domain corresponds to the property that the Loève bifrequency spectrum (1.13) is nonzero only on the diagonal $f_2 = -_{(-)} f_1$. That is,

$$\mathcal{S}_x(f_1, f_2) = S_x(f_1)\, \delta(f_2 +_{(-)} f_1) \tag{1.23}$$

where $\delta(\cdot)$ denotes Dirac delta. When $(*)$ is present, $S_x(f_1)$ is the power spectrum of the process $x(t)$. From (1.23), it follows that for WSS processes distinct spectral component are uncorrelated. In contrast, the presence of spectral correlation outside the diagonal is evidence of nonstationarity in the process $x(t)$ (Loève 1963). Finally, for WSS processes the Wigner-Ville spectrum is independent of t' and is coincident with the power spectrum. That is, $\mathcal{W}_x(t', f) = S_x(f)$.

Extensive treatments on WSS processes are given in (Brillinger 1981), (Cramér 1940), (Doob 1953), (Grenander and Rosenblatt 1957), (Papoulis 1991), (Prohorov and Rozanov 1989), (Rosenblatt 1974, 1985).

1.1.5 *Evolutionary Spectral Analysis*

In (Priestley 1965), the class of zero-mean processes for which the autocovariance function admits the representation

$$\mathrm{E}\left\{x(t_1)\, x^*(t_2)\right\} = \int_{\mathbb{R}} \phi_{t_1}(\omega)\, \phi_{t_2}^*(\omega)\, \mathrm{d}\mu(\omega) \tag{1.24}$$

is considered, where $\{\phi_t(\omega)\}$ is a family of functions defined on the real line ($\omega \in \mathbb{R}$) indexed by the suffix t and $\mathrm{d}\mu(\omega)$ is a measure on the real line. In (Grenander and Rosenblatt 1957, paragraph 1.4), it is shown that if the autocovariance has the representation (1.24), then the process $x(t)$ admits the representation

$$x(t) = \int_{\mathbb{R}} \phi_t(\omega)\, \mathrm{d}Z(\omega) \tag{1.25}$$

where $Z(\omega)$ is an orthogonal process with

$$\mathrm{E}\left\{\, \mathrm{d}Z(\omega_1)\, \mathrm{d}Z^*(\omega_2)\right\} = \delta(\omega_1 - \omega_2)\, \mathrm{d}\mu(\omega_1). \tag{1.26}$$

In fact, we formally have

$$\begin{aligned}
\mathrm{E}\left\{x(t_1)\, x^*(t_2)\right\} &= \mathrm{E}\left\{\int_{\mathbb{R}} \phi_{t_1}(\omega_1)\, \mathrm{d}Z(\omega_1) \int_{\mathbb{R}} \phi_{t_2}^*(\omega_2)\, \mathrm{d}Z^*(\omega_2)\right\} \\
&= \int_{\mathbb{R}} \int_{\mathbb{R}} \phi_{t_1}(\omega_1)\, \phi_{t_2}^*(\omega_2) \mathrm{E}\left\{\, \mathrm{d}Z(\omega_1)\, \mathrm{d}Z^*(\omega_2)\right\} \\
&= \int_{\mathbb{R}} \int_{\mathbb{R}} \phi_{t_1}(\omega_1)\, \phi_{t_2}^*(\omega_2)\, \delta(\omega_1 - \omega_2)\, \mathrm{d}\mu(\omega_1) \\
&= \int_{\mathbb{R}} \phi_{t_1}(\omega_1)\, \phi_{t_2}^*(\omega_1)\, \mathrm{d}\mu(\omega_1)
\end{aligned} \tag{1.27}$$

where, in the last equality, the sampling property of the Dirac delta (Zemanian 1987, Section 1.7) is used.

When the process is second-order WSS, a valid choice for the family $\{\phi_t(\omega)\}$ is $\phi_t(\omega) = e^{j\omega t}$. The autocovariance is

$$\mathrm{E}\left\{x(t_1)\, x^*(t_2)\right\} = \int_{\mathbb{R}} e^{j\omega(t_1 - t_2)}\, \mathrm{d}\mu(\omega) \tag{1.28}$$

which is function of $t_1 - t_2$. The function $\mu(\omega)$ is the *integrated power spectrum*. If $\mu(\omega)$ is absolutely continuous or contains jumps and has zero singular component (Cramér 1940), then in the sense of distributions $\mathrm{d}\mu(\omega) = S(\omega)\, \mathrm{d}(\omega)$, where $S(\omega)$ is the power spectrum (with $\omega = 2\pi f$) which contains Dirac deltas in correspondence of the jumps in $\mu(\omega)$.

The function of t, $\phi_t(\omega)$ is said to be an *oscillatory function* if, for some real-valued function $\theta(\omega)$, it results in

$$\phi_t(\omega) = A_t(\omega)\, e^{j\theta(\omega)t} \tag{1.29}$$

where the modulating function $A_t(\omega)$, as a function of t, has a (generalized) Fourier transform with an absolute maximum in the origin (that is, as a function of t, it is a low-pass function) and can be seen as the "envelope" of $x(t)$. In addition, if the function $\theta(\cdot)$ is invertible with inverse $\theta^{-1}(\cdot)$, then by substituting (1.29) into (1.24) and making the variable change $\lambda = \theta(\omega)$ we have

$$\begin{aligned} \mathrm{E}\left\{x(t_1)\, x^*(t_2)\right\} &= \int_{\mathbb{R}} A_{t_1}(\omega)\, A^*_{t_2}(\omega)\, e^{j\theta(\omega)(t_1-t_2)}\, \mathrm{d}\mu(\omega) \\ &= \int_{\mathbb{R}} A_{t_1}(\theta^{-1}(\lambda))\, A^*_{t_2}(\theta^{-1}(\lambda))\, e^{j\lambda(t_1-t_2)}\, \mathrm{d}\mu(\theta^{-1}(\lambda)) \\ &= \int_{\mathbb{R}} \bar{A}_{t_1}(\lambda)\, \bar{A}^*_{t_2}(\lambda)\, e^{j\lambda(t_1-t_2)}\, \mathrm{d}\bar{\mu}(\lambda) \end{aligned} \tag{1.30}$$

where $\bar{A}_t(\cdot) \triangleq A_t(\theta^{-1}(\cdot))$ and $\mathrm{d}\bar{\mu}(\cdot) \triangleq \mathrm{d}\mu(\theta^{-1}(\cdot))$. The process is said to be an *oscillatory process* and admits the representation

$$x(t) = \int_{\mathbb{R}} \bar{A}_t(\lambda)\, e^{j\lambda t}\, \mathrm{d}\bar{Z}(\lambda) \tag{1.31}$$

with respect to the family of oscillatory functions

$$\{\bar{\phi}_t(\lambda)\} \equiv \{\bar{A}_t(\lambda)\, e^{j\lambda t}\} \tag{1.32}$$

where $\bar{Z}(\lambda)$ is an orthogonal process with

$$\mathrm{E}\left\{\mathrm{d}\bar{Z}(\lambda_1)\, \mathrm{d}\bar{Z}^*(\lambda_2)\right\} = \delta(\lambda_1 - \lambda_2)\, \mathrm{d}\bar{\mu}(\lambda_1). \tag{1.33}$$

In fact, by using (1.31) and (1.33) into the autocovariance definition leads to the rhs of (1.30).

Motivated by the fact that, according to (1.30),

$$\mathrm{E}\left\{|x(t)|^2\right\} = \int_{\mathbb{R}} |\bar{A}_t(\lambda)|^2\, \mathrm{d}\bar{\mu}(\lambda) \tag{1.34}$$

the *evolutionary spectrum* at time t with respect to the family (1.32) is defined as

$$\mathrm{d}F_t(\lambda) = |\bar{A}_t(\lambda)|^2\, \mathrm{d}\bar{\mu}(\lambda). \tag{1.35}$$

This definition is consistent with the interpretation of (1.31) as an expression of the process $x(t)$ as the superposition of complex sinewaves with orthogonal time-varying random amplitudes $\bar{A}_t(\lambda)\, \mathrm{d}\bar{Z}(\lambda)$.

The WSS processes are obtained as a special case of oscillatory processes if $A_t(\omega) = 1$ for all t and ω and $\theta(\omega) = \omega$, or, equivalently, $\bar{A}_t(\lambda) = 1$ for all t and λ. In such a case, the evolutionary spectrum is coincident with $\mathrm{d}\bar{\mu}(\lambda)$ and WSS processes are expressed as the superposition of complex sinewaves with orthogonal time-invariant random amplitudes (Cramér 1940).

For generalizations and applications, see (Matz *et al.* 1997), (Hopgood and Rayner 2003).

1.1.6 Discrete-Time Processes

The characterization of discrete-time nonstationary stochastic processes can be made similarly to that of continuous-time processes with the obvious modifications. The harmonizability condition for the discrete-time process $x_{\rm d}(n)$ is

$$\mathrm{E}\left\{x_{\rm d}(n_1)\, x_{\rm d}^{(*)}(n_2)\right\} = \int_{[-1/2,1/2]^2} e^{j2\pi[\nu_1 n_1 + (-)\nu_2 n_2]}\, \mathrm{d}\widetilde{\gamma}_{x_{\rm d}}(\nu_1, \nu_2) \tag{1.36}$$

with $\widetilde{\gamma}_{x_{\rm d}}(\nu_1, \nu_2)$ spectral correlation function of bounded variation when $(\nu_1, \nu_2) \in [-1/2, 1/2]^2$. Under the harmonizability condition, $x_{\rm d}(n)$ can be expressed as

$$x_{\rm d}(n) = \int_{[-1/2,1/2]} e^{j2\pi\nu n}\, \mathrm{d}\chi_{\rm d}(\nu) \tag{1.37}$$

where we can formally write $\mathrm{d}\chi_{\rm d}(\nu) = X_{\rm d}(\nu)\,\mathrm{d}\nu$ (Gardner 1985, Chapter 10.1.2) with

$$X_{\rm d}(\nu) = \sum_{n\in\mathbb{Z}} x_{\rm d}(n)\, e^{-j2\pi\nu n} \tag{1.38}$$

Fourier transform of $x_{\rm d}(n)$ to be intended in the sense of distributions (Gelfand and Vilenkin 1964, Chapter 3; Henniger 1970).

The possible presence of Dirac deltas on the edges of the integration domain in (1.36) or for $\nu = \pm 1/2$ in (1.38) must be managed, accounting for the periodicity with period 1 w.r.t. variables ν_1 and ν_2 in (1.36) and w.r.t. variable ν in (1.38). If one delta term is considered, then its replica must be neglected.

The Loève bifrequency spectrum of $x_{\rm d}(n)$ is defined as (Loève 1963)

$$\widetilde{\mathbb{S}}_{x_{\rm d}}(\nu_1, \nu_2) \triangleq \mathrm{E}\left\{X_{\rm d}(\nu_1)\, X_{\rm d}^{(*)}(\nu_2)\right\}. \tag{1.39}$$

Let $x_{\rm d}(n) \triangleq x(t)|_{t=nT_s}$ be the discrete-time process obtained by uniformly sampling with period $T_s = 1/f_s$ the continuous-time process $x(t)$. The (conjugate) autocorrelation function of $x_{\rm d}(n)$ turns out to be the sampled version of that of $x(t)$ at sampling instants $t_1 = n_1 T_s$ and $t_2 = n_2 T_s$. The Loève bifrequency spectrum of $x_{\rm d}(n)$ can be expressed as

$$\widetilde{\mathbb{S}}_{x_{\rm d}}(\nu_1, \nu_2) = f_s^2 \sum_{p_1\in\mathbb{Z}} \sum_{p_2\in\mathbb{Z}} \mathbb{S}_{x}((\nu_1 - p_1) f_s, (\nu_2 - p_2) f_s). \tag{1.40}$$

Uniform sampling is a linear periodically time-variant transformation of a continuous-time process into a discrete-time process. Since the transformation is time-variant, in general the nonstationary behavior of the discrete-time process can be different from that of the continuous-time one.

1.1.7 Linear Time-Variant Transformations

In this section, linear time-variant (LTV) transformations of stochastic processes are considered. Input/output relations are derived in both time and frequency domains with reference to processes and their second-order moments.

The input/output relationship of a LTV system is given by

$$y(t) = \int_{\mathbb{R}} h(t, u)x(u)\,\mathrm{d}u \tag{1.41}$$

where $h(t, u)$ is the system *impulse-response function*. That is,

$$x(u) = \delta(u - u_0) \quad \Rightarrow \quad y(t) = h(t, u_0) \tag{1.42}$$

where $\delta(\cdot)$ denotes Dirac delta. By Fourier transforming both sides of (1.41), one obtains the input/output relationship in the frequency domain

$$\begin{aligned} Y(f) &\triangleq \int_{\mathbb{R}} y(t)e^{-j2\pi ft}\,\mathrm{d}t \\ &= \int_{\mathbb{R}} H(f, \lambda)X(\lambda)\,\mathrm{d}\lambda \end{aligned} \tag{1.43}$$

where the *transmission function* $H(f, \lambda)$ (Claasen and Mecklenbräuker 1982) is the double Fourier transform of the impulse-response function:

$$H(f, \lambda) \triangleq \int_{\mathbb{R}^2} h(t, u)e^{-j2\pi(ft-\lambda u)}\,\mathrm{d}t\ \mathrm{d}u. \tag{1.44}$$

In (1.43), (1.44), and the following, Fourier transforms are assumed to exist at least in the sense of distributions (generalized functions) (Zemanian 1987).

1.1.7.1 Input/Output Relations in the Time Domain

Let us consider two LTV systems with impulse-response functions $h_1(t, u)$ and $h_2(t, u)$, excited by $x_1(t)$ and $x_2(t)$, respectively (Figure 1.1). The output signals are

$$y_i(t) = \int_{\mathbb{R}} h_i(t, u)\, x_i(u)\,\mathrm{d}u \qquad i = 1, 2. \tag{1.45}$$

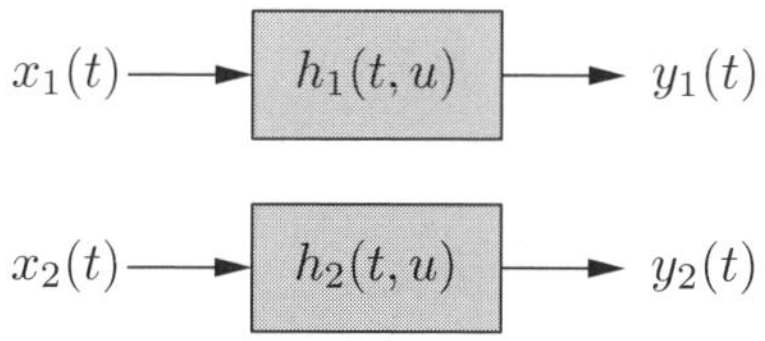

Figure 1.1 LTV systems – time domain

The (conjugate) cross-correlation of the outputs $y_1(t)$ and $y_2(t)$ can be expressed in terms of that of the input signals $x_1(t)$ and $x_2(t)$:

$$
\begin{aligned}
&\mathrm{E}\left\{y_1(t+\tau_1)\, y_2^{(*)}(t+\tau_2)\right\} \\
&= \mathrm{E}\left\{\int_{\mathbb{R}} h_1(t+\tau_1, u_1)\, x_1(u_1)\, \mathrm{d}u_1 \int_{\mathbb{R}} h_2^{(*)}(t+\tau_2, u_2)\, x_2^{(*)}(u_2)\, \mathrm{d}u_2\right\} \\
&= \mathrm{E}\left\{\int_{\mathbb{R}} h_1(t+\tau_1, t+s_1)\, x_1(t+s_1)\, \mathrm{d}s_1 \int_{\mathbb{R}} h_2^{(*)}(t+\tau_2, t+s_2)\, x_2^{(*)}(t+s_2)\, \mathrm{d}s_2\right\} \\
&= \int_{\mathbb{R}}\int_{\mathbb{R}} h_1(t+\tau_1, t+s_1)\, h_2^{(*)}(t+\tau_2, t+s_2) \\
&\quad\; \mathrm{E}\left\{x_1(t+s_1)\; x_2^{(*)}(t+s_2)\right\} \mathrm{d}s_1\, \mathrm{d}s_2
\end{aligned}
\tag{1.46}
$$

where, in the second equality, the variable changes $u_i = t + s_i$, $i = 1, 2$ are made in order to allow, in the third equality, the interchange of integral and expectation operators also when the proof is made in the functional approach (see Chapter 6).

A sufficient condition to allow the interchange of integral and expectation operators is given by the Fubini and Tonelli theorem (Champeney 1990, Chapter 3)

$$
\begin{aligned}
&\int_{\mathbb{R}}\int_{\mathbb{R}} \left|h_1(t+\tau_1, t+s_1)\, h_2^{(*)}(t+\tau_2, t+s_2)\right| \\
&\qquad \mathrm{E}\left\{\left|x_1(t+s_1)\; x_2^{(*)}(t+s_2)\right|\right\} \mathrm{d}s_1\, \mathrm{d}s_2 < \infty.
\end{aligned}
\tag{1.47}
$$

1.1.7.2 Input/Output Relations in the Frequency Domain

The frequency-domain counterpart of the input/output relation (1.45) is (Figure 1.2)

$$
Y_i(f) = \int_{\mathbb{R}} H_i(f, \lambda)\, X_i(\lambda)\, \mathrm{d}\lambda \qquad i = 1, 2. \tag{1.48}
$$

where $H_i(f, \lambda)$, $i = 1, 2$, are the transmission functions defined according to (1.44).

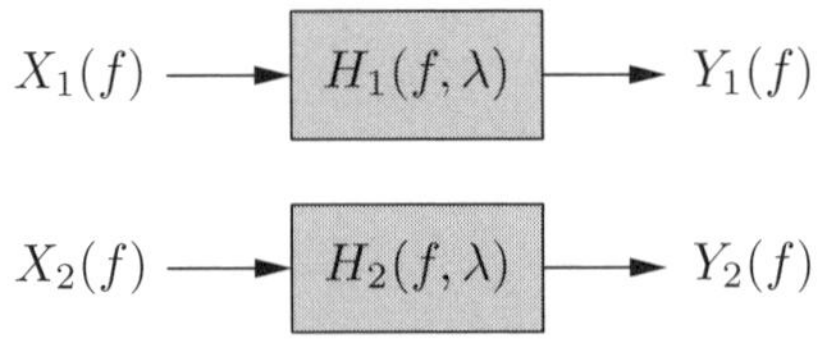

Figure 1.2 LTV systems – frequency domain

The Loève bifrequency cross-spectrum of the outputs $y_1(t)$ and $y_2(t)$ can be expressed in terms of that of the input signals $x_1(t)$ and $x_2(t)$ as follows

$$\begin{aligned} &\mathrm{E}\left\{Y_1(f_1)\, Y_2^{(*)}(f_2)\right\} \\ &\quad = \mathrm{E}\left\{\int_{\mathbb{R}} H_1(f_1, \lambda_1)\, X_1(\lambda_1)\, \mathrm{d}\lambda_1 \int_{\mathbb{R}} H_2^{(*)}(f_2, \lambda_2)\, X_2^{(*)}(\lambda_2)\, \mathrm{d}\lambda_2\right\} \\ &\quad = \int_{\mathbb{R}}\int_{\mathbb{R}} H_1(f_1, \lambda_1)\, H_2^{(*)}(f_2, \lambda_2)\, \mathrm{E}\left\{X_1(\lambda_1)\, X_2^{(*)}(\lambda_2)\right\} \mathrm{d}\lambda_1\, \mathrm{d}\lambda_2 \end{aligned} \tag{1.49}$$

provided that (Fubini and Tonelli theorem (Champeney 1990, Chapter 3))

$$\int_{\mathbb{R}}\int_{\mathbb{R}} \left|H_1(f_1, \lambda_1)\, H_2^{(*)}(f_2, \lambda_2)\right| \mathrm{E}\left\{\left|X_1(\lambda_1)\, X_2^{(*)}(\lambda_2)\right|\right\} \mathrm{d}\lambda_1\, \mathrm{d}\lambda_2 < \infty. \tag{1.50}$$

1.2 Almost-Periodic Functions

In this section, definitions and main results on almost-periodic (AP) functions and their generalizations are presented for both continuous- and discrete-time cases. For extensive treatments on almost-periodic functions, see (Besicovitch 1932), (Bohr 1933), and (Corduneanu 1989) for continuous-time, and (Corduneanu 1989, Chapter VII), (Jessen and Tornehave 1945), and (von Neumann 1934) for discrete-time.

1.2.1 Uniformly Almost-Periodic Functions

Definition 1.2.1 (Besicovitch 1932, Chapter 1). *A function $z(t)$, $t \in \mathbb{R}$, is said to be* uniformly almost-periodic *if $\forall \epsilon > 0\ \exists\, \ell_\epsilon > 0$ such that for any interval $I_\epsilon = (t_0, t_0 + \ell_\epsilon)\ \exists\, \tau_\epsilon \in I_\epsilon$ such that*

$$\sup_{t \in \mathbb{R}} |z(t + \tau_\epsilon) - z(t)| < \epsilon. \tag{1.51}$$

The quantity τ_ϵ is said translation number of $z(t)$ corresponding to ϵ. □

A set $D \subseteq \mathbb{R}$ is said to be *relatively dense* in $\mathbb{R}$ if $\exists\, \ell > 0$ such that $\forall\, I = (t_0, t_0 + \ell)$ the result is that $D \cap I \neq \emptyset$.

Thus, defined the set of the translation numbers of $z(t)$ corresponding to ϵ

$$E(z, \epsilon) \triangleq \left\{\tau \in \mathbb{R} \,:\, \sup_{t \in \mathbb{R}} |z(t + \tau) - z(t)| < \epsilon\right\} \tag{1.52}$$

according to Definition 1.2.1, the function $z(t)$ is uniformly almost periodic if and only if $\forall \epsilon > 0$ the set $E(z, \epsilon)$ is relatively dense in $\mathbb{R}$. That is, there are many translation numbers of $z(t)$ corresponding to ϵ.

Theorem 1.2.2 (Besicovitch 1932, Chapter 1). *Any uniformly AP function is the limit of a uniformly convergent sequence of trigonometric polynomials in t (generalized Fourier series):*

$$z(t) = \sum_{\alpha \in A} z_\alpha \, e^{j2\pi\alpha t} \tag{1.53}$$

where the frequencies $\alpha \in A$, with A countable set of possibly incommensurate reals and possibly containing cluster points, and

$$z_\alpha \triangleq \lim_{T\to\infty} \frac{1}{T} \int_{t_0-T/2}^{t_0+T/2} z(t)\, e^{-j2\pi\alpha t} \,\mathrm{d}t \tag{1.54}$$

with the limit independent of t_0. Thus, $z(t)$ is bounded and uniformly continuous. □

Theorem 1.2.3 (Besicovitch 1932, Chapter 1). *For any uniformly AP function the following Parseval's equality holds*

$$\lim_{T\to\infty} \frac{1}{T} \int_{-T/2}^{T/2} |z(t)|^2 \,\mathrm{d}t = \sum_{\alpha\in A} |z_\alpha|^2. \tag{1.55}$$

□

A function $z(t)$ is *periodic* with period $T_0 > 0$ if T_0 is the smallest nonzero value such that

$$z(t) = z(t + T_0) \qquad \forall t \in \mathbb{R}. \tag{1.56}$$

Thus, periodic functions are obtained as special case of almost-periodic functions with $\tau_\epsilon = kT_0$ independent of ϵ, $k \in \mathbb{Z}$. In such a case, the frequencies of the set A are all multiple of a fundamental frequency $1/T_0$, that is, $A = \{k/T_0\}_{k\in\mathbb{Z}}$ and (1.53) is the ordinary Fourier series expansion of the periodic function $z(t)$.

An example of uniformly AP function which is not periodic is

$$z(t) = \cos(2\pi t/T_0) + \cos(2\pi t/(\sqrt{2}T_0)). \tag{1.57}$$

Both cosines are periodic functions with periods T_0 and $\sqrt{2}T_0$, respectively. However, their sum is not periodic since the ratio of the two periods T_0 and $\sqrt{2}T_0$ is the irrational number $1/\sqrt{2}$.

The functions defined in Definition 1.2.1 and characterized in Theorem 1.2.2 are called almost-periodic in the sense of Bohr (Bohr 1933, paragraphs 84–92) or, equivalently, *uniformly almost periodic* in the sense of Besicovitch (Besicovitch 1932, Chapter 1), or, equivalently, *almost-periodic with respect to the sup norm.* More general classes of almost-periodic functions, including possibly discontinuous functions, are treated in (Besicovitch 1932, Chapter 2) and the following sections.

1.2.2 AP Functions in the Sense of Stepanov, Weyl, and Besicovitch

The almost-periodicity property can be defined with respect to the following norms or semi-norms, with $p \geqslant 1$, (Besicovitch 1932, Chapter 2):

1. *Stepanov S_T^p-norm:*

$$\|z\|_{S_T^p} \triangleq \left[\sup_{a\in\mathbb{R}} \frac{1}{T}\int_{a-T/2}^{a+T/2} |z(t)|^p \,\mathrm{d}t\right]^{1/p} \tag{1.58}$$

2. *Weyl W^p-norm:*

$$\|z\|_{W^p} \triangleq \lim_{T\to\infty} \|z\|_{S_T^p} = \left[\lim_{T\to\infty} \sup_{a\in\mathbb{R}} \frac{1}{T}\int_{a-T/2}^{a+T/2} |z(t)|^p \,\mathrm{d}t\right]^{1/p} \tag{1.59}$$

3. *Besicovitch B^p-seminorm:*

$$\|z\|_{B^p} \triangleq \left[\limsup_{T\to\infty} \frac{1}{T}\int_{-T/2}^{T/2} |z(t)|^p \,\mathrm{d}t\right]^{1/p}. \tag{1.60}$$

Functions belonging to $L^p_{\text{loc}}(\mathbb{R})$ with finite Besicovitch B^p-seminorm form a seminormed space called Marcinkiewicz space $\mathcal{M}^p$.

Let $\|\cdot\|_{G^p}$ denote any of the above defined (semi)norms, that is, $\|\cdot\|_{S_T^p}$, $\|\cdot\|_{W^p}$, or $\|\cdot\|_{B^p}$. For each (semi)norm, a definition of almost-periodicity can be given.

Definition 1.2.4 S_T^p, W^p, and B^p Almost Periodicity (Besicovitch 1932, Chapter 2). *A function $z(t)$ is said to be G^p* almost-periodic (G^p-AP), $p \geqslant 1$, *if $\forall\epsilon > 0 \;\exists\, \ell_\epsilon > 0$ such that for any interval $I_\epsilon = (t_0, t_0 + \ell_\epsilon) \;\exists\, \tau_\epsilon \in I_\epsilon$ such that*

$$\|z(t+\tau_\epsilon) - z(t)\|_{G^p} < \epsilon. \tag{1.61}$$

Specifically, if $G^p = S_T^p$, then $z(t)$ is said S_T^p-AP; If $G^p = W^p$, then $z(t)$ is said W^p-AP; If $G^p = B^p$, then $z(t)$ is said B^p-AP. □

Theorem 1.2.5 (Besicovitch 1932, Chapter 2). *Any G^p-AP function is G^p-bounded ($\|z\|_{G^p} < \infty$) and is the G^p-limit of a sequence of trigonometric polynomials in t:*

$$\lim_n \left\| z(t) - \sum_{\alpha\in A_n} z_\alpha\, e^{j2\pi\alpha t} \right\|_{G^p} = 0 \tag{1.62}$$

where A_n is an increasing sequence of countable sets such that $m < n \Rightarrow A_m \subseteq A_n$ and $\lim_n A_n \triangleq \cup_{n\in\mathbb{N}} A_n = A$. *The coefficients z_α of the generalized Fourier series are given by (1.54).* □

In (Besicovitch 1932, p. 74) it is shown that: If $\|z_1(t) - z_2(t)\|_{S_T^p} = 0$, then $z_1(t) = z_2(t)$ a.e.; If $\|z_1(t) - z_2(t)\|_{W^p} = 0$ or $\|z_1(t) - z_2(t)\|_{B^p} = 0$, then $z_1(t)$ and $z_2(t)$ may differ at a set of points of finite and even of infinite measure. In addition, for $p \geqslant 1$ the result is that (Besicovitch 1932, p. 73) $\sup_{t\in\mathbb{R}} |z(t)| \geqslant \|z(t)\|_{S_T^p} \geqslant \|z(t)\|_{W^p} \geqslant \|z(t)\|_{B^p}$.

Theorem 1.2.6 (Besicovitch 1932, Chapter 2). *For any G^2-AP function the following Parseval's equality holds*

$$\|z(t)\|_{G^2}^2 = \sum_{\alpha \in A} |z_\alpha|^2. \tag{1.63}$$

□

Further generalizations of almost-periodic functions can be found in (Besicovitch 1932, Chapter 2), (Bohr 1933, paragraphs 94–102), and (Corduneanu 1989, Chapter VI).

1.2.3 Weakly AP Functions in the Sense of Eberlein

Definition 1.2.7 Weakly Almost-Periodic Functions (Eberlein 1949, 1956). *A continuous and bounded function $z(t)$ is said to be* weakly almost-periodic (w.a.p.) *(in the sense of Eberlein) if the set of translates $z(t+\tau)$, $\tau \in \mathbb{R}$, is (conditionally) weakly compact in the set of continuous and bounded functions $C^0(\mathbb{R}) \cap L^\infty(\mathbb{R})$.* □

Examples of w.a.p. functions are the uniformly almost periodic functions (in the sense of Definition 1.2.1), the positive definite functions (hence Fourier-Stieltjes transforms), and functions vanishing at infinity (Eberlein 1949, Theorems 11.1 and 11.2). A w.a.p. function is uniformly continuous (Eberlein 1949, Theorem 13.1).

Theorem 1.2.8 (Eberlein 1956). *Every w.a.p. function $z(t)$ admits a unique decomposition*

$$z(t) = z_{\text{uap}}(t) + z_0(t) \tag{1.64}$$

where $z_{\text{uap}}(t)$ is a uniformly almost-periodic function in the sense of Definition 1.2.1 and $z_0(t)$ is a zero-power function

$$\lim_{T\to\infty} \frac{1}{T} \int_{-T/2}^{T/2} |z_0(t)|^2 \, \mathrm{d}t = 0. \tag{1.65}$$

Moreover, it results that

$$z_\alpha = \lim_{T\to\infty} \frac{1}{T} \int_{t_0-T/2}^{t_0+T/2} z_{\text{uap}}(t)\, e^{-j2\pi\alpha t} \, \mathrm{d}t \tag{1.66a}$$

$$= \lim_{T\to\infty} \frac{1}{T} \int_{t_0-T/2}^{t_0+T/2} z(t)\, e^{-j2\pi\alpha t} \, \mathrm{d}t \tag{1.66b}$$

and, accordingly with (1.65) and using the notation of Theorem 1.2.5, the result is that

$$\lim_{n} \lim_{T\to\infty} \frac{1}{T} \int_{t_0-T/2}^{t_0+T/2} \left| z(t) - \sum_{\alpha \in A_n} z_\alpha\, e^{j2\pi\alpha t} \right|^2 \mathrm{d}t = 0. \tag{1.67}$$

□

Theorem 1.2.9 (Eberlein 1956). *For any w.a.p. function the following Parseval's equality holds*

$$\lim_{T\to\infty}\frac{1}{T}\int_{-T/2}^{T/2}|z(t)|^2\,\mathrm{d}t=\lim_{T\to\infty}\frac{1}{T}\int_{-T/2}^{T/2}|z_{\mathrm{uap}}(t)|^2\,\mathrm{d}t=\sum_{\alpha\in A}|z_\alpha|^2. \tag{1.68}$$

□

It is worthwhile emphasizing that the set of w.a.p. functions, unlike other classes of generalized AP functions, is closed under multiplication (Eberlein 1949). That is, the product of two w.a.p. functions is in turn a w.a.p. function.

Other definitions of w.a.p. functions different from Definition 1.2.7 are given in (Amerio and Prouse 1971, Chapter 3), (Corduneanu 1989, Section VI.5), (Zhang and Liu 2010).

Theorem 1.2.10 (Eberlein 1949, Theorem 15.1). *Let $z_1(t)$ and $z_2(t)$ be w.a.p. functions. Then*

$$z(t)=\lim_{T\to\infty}\frac{1}{T}\int_{-T/2}^{T/2}z_1(t+s)\,z_2(s)\,\mathrm{d}s \tag{1.69}$$

exists and is a uniformly almost periodic function of t. A similar result also holds with different definitions of w.a.p. functions (Zhang and Liu 2010). □

1.2.4 Pseudo AP Functions

Definition 1.2.11 (Ait Dads and Arino 1996). *The function $z(t)$ is said to be* pseudo almost-periodic in the sense of Ait Dads and Arino, *shortly $z(t)\in\widetilde{\mathcal{PAP}}(\mathbb{R})$, if it admits the (unique) decomposition*

$$z(t)=z_{\mathrm{uap}}(t)+z_0(t) \tag{1.70}$$

where $z_{\mathrm{uap}}(t)$ is a uniformly almost-periodic function in the sense of Definition 1.2.1 and $z_0(t)$, referred to as the ergodic perturbation, is a Lebesgue measurable function such that

$$\lim_{T\to\infty}\frac{1}{T}\int_{-T/2}^{T/2}|z_0(t)|\,\mathrm{d}t=0 \tag{1.71}$$

shortly $z_0(t)\in\widetilde{\mathcal{PAP}}_0(\mathbb{R})$. □

The classes $\widetilde{\mathcal{PAP}}(\mathbb{R})$ and $\widetilde{\mathcal{PAP}}_0(\mathbb{R})$ are slight generalizations of the classes $\mathcal{PAP}(\mathbb{R})$ and $\mathcal{PAP}_0(\mathbb{R})$, respectively, of the pseudo almost-periodic functions in the sense of Zhang (Zhang 1994, 1995), where $z(t)$ and $z_0(t)$ are assumed to be continuous and bounded.

Theorem 1.2.12 (Ait Dads and Arino 1996). *Let be $z(t) \in \widetilde{\mathcal{P}}\mathcal{AP}(\mathbb{R})$. It results that*

$$z_\alpha = \lim_{T\to\infty} \frac{1}{T} \int_{t_0-T/2}^{t_0+T/2} z_{\text{uap}}(t)\, e^{-j2\pi\alpha t}\, \mathrm{d}t \tag{1.72a}$$

$$= \lim_{T\to\infty} \frac{1}{T} \int_{t_0-T/2}^{t_0+T/2} z(t)\, e^{-j2\pi\alpha t}\, \mathrm{d}t \tag{1.72b}$$

□

Proposition 1.2.13 (Ait Dads and Arino 1996). *If* $\lim_{|t|\to\infty} z_0(t)$ *exists, then* $\lim_{|t|\to\infty} z_0(t) = 0$ *and $z(t)$ belong to the class of the* asymptotically almost-periodic functions *in the sense of Frechet.*

Properties of asymptotically almost-periodic functions are given in (Leśkow and Napolitano 2006, Section 6.2).

1.2.5 AP Functions in the Sense of Hartman and Ryll-Nardzewski

Definition 1.2.14 (Kahane 1962), (Andreas *et al.* 2006, Definition 7.1). *The function $z(t) \in L^1_{\text{loc}}(\mathbb{R})$ is said to be* almost periodic in the sense of Hartman, *shortly $z(t) \in H^1_{\text{ap}}$, if, $\forall \alpha \in \mathbb{R}$*

$$z_\alpha \triangleq \lim_{T\to\infty} \frac{1}{T} \int_{-T/2}^{T/2} z(t)\, e^{-j2\pi\alpha t}\, \mathrm{d}t \tag{1.73}$$

exists and is finite. □

Definition 1.2.15 (Kahane 1962), (Andreas *et al.* 2006, Definition 7.2). *The function $z(t) \in L^1_{\text{loc}}(\mathbb{R})$ is said to be* almost periodic in the sense of Ryll-Nardzewski, *shortly $z(t) \in R^1_{\text{ap}}$, if, $\forall \alpha \in \mathbb{R}$*

$$z'_\alpha \triangleq \lim_{T\to\infty} \frac{1}{T} \int_{t_0}^{t_0+T} z(t)\, e^{-j2\pi\alpha t}\, \mathrm{d}t \tag{1.74}$$

exists uniformly with respect to $t_0 \in \mathbb{R}$ and is finite. □

Obviously, if $z(t) \in R^1_{\text{ap}}$, then $z(t) \in H^1_{\text{ap}}$ and $z'_\alpha = z_\alpha\ \forall \alpha \in \mathbb{R}$, but the converse is not true. That is, $R^1_{\text{ap}} \subset H^1_{\text{ap}}$.

If $z(t) \in S^p_T(\mathbb{R})$ or $z(t) \in W^p(\mathbb{R})$, then $z(t) \in R^1_{\text{ap}}$ and the Fourier coefficients of the (generalized) Fourier series in (1.62) are coincident with those in (1.74). If $z(t) \in B^p(\mathbb{R})$, then $z(t) \in H^1_{\text{ap}}$ (but not necessarily $z(t) \in R^1_{\text{ap}}$) and the Fourier coefficients of the (generalized) Fourier series in (1.62) are (obviously) those in (1.73).

Theorem 1.2.16 (Kahane 1961), (Andreas *et al.* 2006, Theorem 7.5). *The* spectrum *of $z(t) \in H^1_{\text{ap}}$, that is the set $\{\alpha \in \mathbb{R} : z_\alpha \neq 0\}$, is at most countable. Consequently, also the spectrum of $z(t) \in R^1_{\text{ap}}$ is at most countable.* □

Theorem 1.2.17 (Urbanik 1962), (Kahane 1961). *Let $z(t) \in H^1_{\mathrm{ap}} \cap \mathcal{M}^p$, $p > 1$. Then, the following unique decomposition holds*

$$z(t) = z_{\mathrm{Bap}}(t) + z_0(t) \tag{1.75}$$

where $z_{\mathrm{Bap}}(t)$ is a B^p-AP function and $z_0(t) \in H^1_{\mathrm{ap}}$ with empty spectrum, that is $\forall \alpha \in \mathbb{R}$

$$\lim_{T\to\infty} \frac{1}{T} \int_{-T/2}^{T/2} z_0(t)\, e^{-j2\pi\alpha t}\, \mathrm{d}t = 0. \tag{1.76}$$

□

In particular, since uniformly AP functions in the sense of Definition 1.2.1 are special cases of B^p-AP functions, the function $z_{\mathrm{Bap}}(t)$ in decomposition (1.75) can reduce to a uniformly AP function.

Let us define the sets

$$H_0 \triangleq \{z(t) \in H^1_{\mathrm{ap}} \ : \ z(t) \text{ has empty spectrum}\} \tag{1.77}$$

$$R_0 \triangleq \{z(t) \in R^1_{\mathrm{ap}} \ : \ z(t) \text{ has empty spectrum}\}. \tag{1.78}$$

Obviously $R_0 \subset H_0$.

Theorem 1.2.18 (Kahane 1962). *Let be $x(t) \in L^p_{\mathrm{loc}}(\mathbb{R})$. Then, there exists $z(t) \in R_0 \cap L^p_{\mathrm{loc}}(\mathbb{R})$ such that $|z(t)| = |x(t)|$. Let be $x(t) \in C^0(\mathbb{R})$. Then, there exists $z(t) \in R_0 \cap C^0(\mathbb{R})$ such that $|z(t)| = |x(t)|$.* □

Theorem 1.2.19 (Kahane 1962). *Let be $x(t) \in C^0(\mathbb{R})$. Then, there exist $z_1, z_2 \in R_0$ such that $x(t) = z_1(t)\, z_2(t)$.* □

Theorem 1.2.20 (Kahane 1962). *Let $x(t)$ be uniformly continuous and bounded. Then, there exist $z_1, z_2 \in R_0$ uniformly continuous and bounded such that $x(t) = z_1(t)\, z_2(t)$. In particular, $x(t)$ can be uniformly almost periodic.* □

1.2.6 AP Functions Defined on Groups and with Values in Banach and Hilbert Spaces

Almost-periodic functions and their generalizations on groups are treated in (Corduneanu 1989, Chapter VII), (von Neumann 1934), (Casinovi 2009).

Almost-periodic functions with values in Hilbert spaces are treated in (Phong 2007). Further classes of AP functions with values in Banach spaces and a survey of their properties are presented in (Andreas *et al.* 2006), (Chérif 2011a,b).

1.2.7 AP Functions in Probability

Let $\{z(t, \omega), t \in \mathbb{R}, \omega \in \Omega\}$, denoted shortly by $z(t)$, be a random process defined on a probability space $(\Omega, \mathcal{F}, P)$.

Definition 1.2.21 Random Functions Almost-Periodic in Probability (Corduneanu 1989, Sect. II.3). *A random process $z(t)$, $t \in \mathbb{R}$, is called* almost-periodic in probability *if* $\forall \epsilon > 0$, $\eta > 0$, *there exists* $\ell_{\epsilon,\eta} > 0$ *such that for every set of length* $\ell_{\epsilon,\eta}$, *say* $I_{\epsilon,\eta} = (t_0, t_0 + \ell_{\epsilon,\eta})$, *there exists at least one number* $\tau_{\epsilon,\eta} \in I_{\epsilon,\eta}$ *such that*

$$\sup_{t \in \mathbb{R}} P\Big\{ \Big| z(t + \tau_{\epsilon,\eta}, \omega) - z(t, \omega) \Big| \geqslant \eta \Big\} < \epsilon. \tag{1.79}$$

The real number $\tau_{\epsilon,\eta}$ is said (ϵ, η)-almost period in probability. □

Theorem 1.2.22 (Corduneanu 1989, Sect. II.3). *Any random process AP in probability is bounded in probability and is the limit in probability of a sequence of random trigonometric polynomials in t. That is,* $\forall \eta > 0$,

$$\lim_{n} \sup_{t \in \mathbb{R}} P\Big\{ \Big| z(t, \omega) - \sum_{\alpha \in A_n} z_\alpha(\omega)\, e^{j2\pi\alpha t} \Big| \geqslant \eta \Big\} = 0 \tag{1.80}$$

where A_n is an increasing sequence of countable sets of real numbers α and $z_\alpha(\omega)$ are random variables. □

1.2.8 AP Sequences

Definition 1.2.23 Discrete-Time Almost-Periodic Functions (Corduneanu 1989, Chapter VII), (Jessen and Tornehave 1945), (von Neumann 1934). *A sequence $z(n)$, $n \in \mathbb{Z}$, is said to be* almost-periodic *if* $\forall \epsilon > 0 \ \exists \ \ell_\epsilon \in \mathbb{N}$ *such that for any set* $I_\epsilon \triangleq \{n_0, n_0 + 1, \dots, n_0 + \ell_\epsilon\}$ $\exists \ m_\epsilon \in I_\epsilon$ *such that*

$$\sup_{n \in \mathbb{Z}} |z(n + m_\epsilon) - z(n)| < \epsilon \tag{1.81}$$

The integer m_ϵ is said to be the translation number of $z(n)$ corresponding to ϵ. □

Theorem 1.2.24 (Corduneanu 1989, Chapter VII), (von Neumann 1934). *Every AP sequence $\{z(n)\}_{n \in \mathbb{Z}}$ is the limit of a sequence of trigonometric polynomials in n:*

$$z(n) = \sum_{\widetilde{\alpha} \in \widetilde{A}} z_{\widetilde{\alpha}}\, e^{j2\pi\widetilde{\alpha} n} \tag{1.82}$$

where the frequencies $\widetilde{\alpha} \in \widetilde{A}$, with $\widetilde{A}$ countable set with possibly incommensurate elements in $[-1/2, 1/2)$ and possibly containing cluster points,

$$z_{\widetilde{\alpha}} \triangleq \lim_{N \to \infty} \frac{1}{2N + 1} \sum_{n = n_0 - N}^{n_0 + N} z(n)\, e^{-j2\pi\widetilde{\alpha} n} \tag{1.83}$$

with the limit independent of n_0 *and*

$$\sum_{\widetilde{\alpha}\in\widetilde{A}} |z_{\widetilde{\alpha}}| < \infty. \tag{1.84}$$

Thus, $z(n)$ *is bounded.* □

A sequence $z(n)$ is *periodic* with period N_0 if N_0 is the smallest non-zero integer such that

$$z(n) = z(n + N_0) \qquad \forall n \in \mathbb{Z}. \tag{1.85}$$

Thus, periodic sequences are obtained as a special case of almost-periodic sequences with $m_\epsilon = kN_0$ independent of ϵ, $k \in \mathbb{Z}$. In such a case, $\widetilde{A}$ is the finite set $\{0, 1/N_0, \ldots, (N_0 - 1)/N_0\}$ or any equivalent set $k_0 + A \triangleq \{k_0, k_0 + 1/N_0, \ldots, k_0 + (N_0 - 1)/N_0\}$ with k_0 integer, and (1.82) is the discrete Fourier series (DFS) of $z(n)$.

In continuous-time, the complex sinewave $z(t) = e^{j2\pi f_0 t}$ is periodic with period $T_0 = 1/f_0$ and the polynomial phase signal $z(t) = e^{j2\pi f_0 t^\gamma}$, $\gamma > 1$, is not almost periodic. Complex discrete-time sinewaves and polynomial phase sequences require more attention. The sequence $z(n) = e^{j2\pi\nu_0 n}$ with $\nu_0 = p/q \in \mathbb{Q}$, p, q relative prime integers or co-prime (that is, they have no common positive divisor other than 1 or, equivalently, their greatest common divisor is 1 or, equivalently, p/q is an irreducible fraction), is periodic with period q. In contrast, the sequence $z(n) = e^{j2\pi\nu_0 n}$ with $\nu_0 \notin \mathbb{Q}$, is almost periodic (not periodic). For every positive integer $L \geqslant 2$, the sequence $z(n) = e^{j2\pi\nu_0 n^L}$ with $\nu_0 = p/q \in \mathbb{Q}$, p, q relative prime integers, is periodic with period q whereas it is not almost periodic for $\nu_0 \notin \mathbb{Q}$.

1.2.9 AP Sequences in Probability

Definition 1.2.25 Almost-Periodic Random Sequences in Probability (Han and Hong 2007). *A random sequence* $z(n)$, $n \in \mathbb{Z}$, *is called* almost-periodic in probability *if* $\forall\epsilon > 0$, $\eta > 0$, *there exists* $\ell_{\epsilon,\eta} \in \mathbb{N}$ *such that for every set of length* $\ell_{\epsilon,\eta}$, *say* $I_{\epsilon,\eta} \triangleq \{n_0, n_0 + 1, \ldots, n_0 + \ell_{\epsilon,\eta}\}$, *there exists at least one number* $m_{\epsilon,\eta} \in I_{\epsilon,\eta}$ *such that*

$$\sup_{n\in\mathbb{Z}} P\left\{\left|z(n + m_{\epsilon,\eta}) - z(n)\right| \geqslant \eta\right\} < \epsilon. \tag{1.86}$$

The integer $m_{\epsilon,\eta}$ *is said to be* (ϵ, η)*-almost period in probability.* □

1.3 Almost-Cyclostationary Processes

Many processes encountered in telecommunications, radar, mechanics, radio astronomy, biology, atmospheric science, and econometrics, are generated by underlying periodic phenomena. These processes, even if not periodic, give rise to random data whose statistical functions vary periodically with time and are called cyclostationary processes. In this section, the properties of cyclostationary, or more generally, of almost-cyclostationary processes are briefly reviewed. For extensive treatments, see (Gardner 1985, Chapter 12, 1987d), (Gardner *et al.* 2006), (Giannakis 1998), (Hurd and Miamee 2007).

1.3.1 Second-Order Wide-Sense Statistical Characterization

The (finite-power) process $x(t)$ is said to be second-order *cyclostationary* in the wide sense or *periodically correlated* with period T_0 if its first- and second-order moments are periodic functions of time with period T_0. More generally, first- and second-order moments can be almost-periodic functions of time (in one of the senses considered in Section 1.2) and the process is said to be second-order *almost-cyclostationary* (ACS) in the wide sense or *almost-periodically correlated* (Gardner 1985, Chapter 12, 1987d). In such a case, its (conjugate) autocorrelation function (1.6) under mild regularity conditions can be expressed by the (generalized) Fourier series expansion

$$\mathcal{R}_{\boldsymbol{x}}(t,\tau) = \sum_{\alpha \in A} R_{\boldsymbol{x}}^{\alpha}(\tau)\, e^{j2\pi\alpha t} \tag{1.87}$$

where subscript $\boldsymbol{x} \triangleq [x x^{(*)}]$, A is the countable set (depending on $(*)$) of possibly incommensurate *cycle frequencies* α, and the Fourier coefficients

$$R_{\boldsymbol{x}}^{\alpha}(\tau) \triangleq \lim_{T\to\infty} \frac{1}{T} \int_{-T/2}^{T/2} \mathcal{R}_{\boldsymbol{x}}(t,\tau)\, e^{-j2\pi\alpha t}\, \mathrm{d}t$$

referred to as the *(conjugate) cyclic autocorrelation functions*, are complex-valued functions whose magnitude and phase represent the amplitude and phase of the finite-strength additive sinewave component at frequency α contained in $\mathcal{R}_{\boldsymbol{x}}(t,\tau)$. Thus, $R_{\boldsymbol{x}}^{\alpha}(\tau) \not\equiv 0$ if $\alpha \in A$ and $R_{\boldsymbol{x}}^{\alpha}(\tau) \equiv 0$ if $\alpha \notin A$. Cyclostationary processes are obtained as a special case when $A = \{k/T_0\}_{k\in\mathbb{Z}}$. WSS processes are a further specialization when A contains the only element $\alpha = 0$.

The function (1.87) is the *time-lag* (conjugate) autocorrelation function of ACS processes. By substituting $t = t_2$ and $\tau = t_1 - t_2$ into (1.87), one obtains the *time-time* (conjugate) autocorrelation function of ACS processes

$$\mathrm{E}\{x(t_1)\, x^{(*)}(t_2)\} = \sum_{\alpha \in A} R_{\boldsymbol{x}}^{\alpha}(t_1 - t_2)\, e^{j2\pi\alpha t_2}. \tag{1.88}$$

For a zero-mean process $x(t)$ with finite or practically finite memory the result is that $|R_{\boldsymbol{x}}^{\alpha}(\tau)| \to 0$ as $|\tau| \to \infty$. In contrast, if the process has non-zero almost-periodic expectation $\mathrm{E}\{x(t)\}$, then some $R_{\boldsymbol{x}}^{\alpha}(\tau)$ contain additive sinusoidal functions of τ which arise from products of finite-strength sinusoidal terms contained in $\mathrm{E}\{x(t)\}$ (Gardner and Spooner 1994). ACS signals in communications are zero mean unless pilot tones are present for synchronization purposes. In mechanical applications, ACS signals generally have an almost-periodic expected value (Antoni 2009). For the following, let us assume that

$$\varrho(\tau) \triangleq \sum_{\alpha \in A} \left| R_{\boldsymbol{x}}^{\alpha}(\tau) \right| \in L^1(\mathbb{R}) \cap L^{\infty}(\mathbb{R}). \tag{1.89}$$

Such an assumption is verified by finite-memory or practically finite-memory signals. From (1.89) it follows that $\varrho(\tau) \in L^p(\mathbb{R})$ for $p \geqslant 1$ and $R_{\boldsymbol{x}}^{\alpha}(\tau) \in L^p(\mathbb{R})$ for $p \geqslant 1$.

By Fourier transforming with respect to τ both sides of (1.87) we obtain the almost-periodically time-variant spectrum

$$\mathcal{S}_x(t, f) = \sum_{\alpha \in A} S_x^{\alpha}(f)\, e^{j2\pi\alpha t} \tag{1.90}$$

where

$$S_x^{\alpha}(f) = \int_{\mathbb{R}} R_x^{\alpha}(\tau)\, e^{-j2\pi f\tau}\, \mathrm{d}\tau \tag{1.91}$$

is the *(conjugate) cyclic spectrum* which can be expressed as

$$S_x^{\alpha}(f) = \lim_{\Delta f \to 0} \lim_{T \to \infty} \frac{1}{T} \int_{-T/2}^{T/2} \Delta f\, \mathrm{E}\left\{ X_{1/\Delta f}(t, f)\, X_{1/\Delta f}^{(*)}(t, {}_{(-)}(\alpha - f)) \right\} \mathrm{d}t \tag{1.92}$$

where the order of the two limits cannot be reversed and

$$X_Z(t, f) \triangleq \int_{t-Z/2}^{t+Z/2} x(s)\, e^{-j2\pi f s}\, \mathrm{d}s \tag{1.93}$$

is the short-time Fourier transform (STFT) of $x(t)$. Therefore, the (conjugate) cyclic spectrum is also called the (conjugate) *spectral correlation function* and (1.91) is the *Gardner Relation* (Gardner 1985) (also called *Cyclic Wiener-Khinchin Relation*). Under assumption (1.89) the Fourier transform $S_x^{\alpha}(f)$ exists in the ordinary sense. Accordingly with (1.90), the Wigner-Ville spectrum (1.20a) of an ACS signal is an almost-periodic function of t' with frequencies $\alpha \in A$ and Fourier-series coefficients $S_x^{\alpha}(f + \alpha/2)$.

From (1.8), (1.14b), and (1.87), in the case of ACS processes, one obtains the following Loève bifrequency spectrum

$$\mathcal{S}_x(f_1, f_2) = \sum_{\alpha \in A} S_x^{\alpha}(f_1)\, \delta(f_2 + {}_{(-)}(f_1 - \alpha)). \tag{1.94}$$

That is, ACS processes have Loève bifrequency spectrum with spectral masses concentrated on a countable set of lines with slope ± 1. Equivalently, ACS processes have distinct spectral components that are correlated only if the spectral separation belongs to a countable set which is the set A of the cycle frequencies.

In Figures 1.3 and 1.4, the magnitude of the cyclic autocorrelation as a function of α and τ and the magnitude of the bifrequency spectral correlation density $\bar{\mathcal{S}}_x(f_1, f_2)$ (obtained by replacing in (1.23) and (1.94) Dirac deltas with Kronecker ones) as a function of f_1 and f_2 are reported for a WSS and an ACS signal. The ACS signal is a cyclostationary PAM with a rectangular pulse $q(t)$. The considered WSS signal has the same autocorrelation function and power spectrum of the PAM signal. The WSS signal has a nonzero cyclic autocorrelation function only for $\alpha = 0$ (Figure 1.3 (top)). Equivalently, the bifrequency spectral correlation density in nonzero is only on the main diagonal of the bifrequency plane (Figure 1.4 (top)). In contrast, the ACS signal has a nonzero cyclic autocorrelation function in correspondence of the cycle frequencies which are multiples of a fundamental one (Figure 1.3 (bottom)). Equivalently, the bifrequency spectral correlation density is nonzero on lines which are parallel to the main diagonal so that spectral components which are correlated are separated by quantities equal to the cycle frequencies (Figure 1.4 (bottom)).

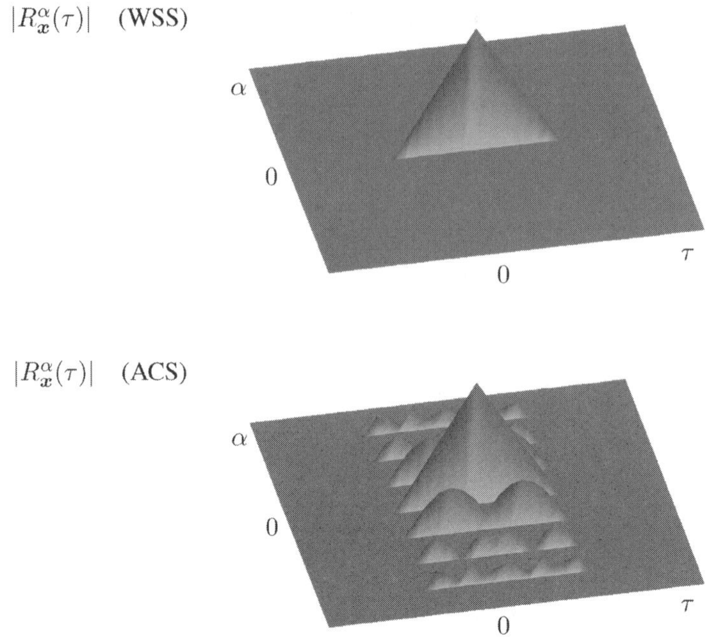

Figure 1.3 Magnitude of the cyclic autocorrelation function of a signal $x(t)$ as functions of α and τ. Top: WSS signal. Bottom: ACS signal

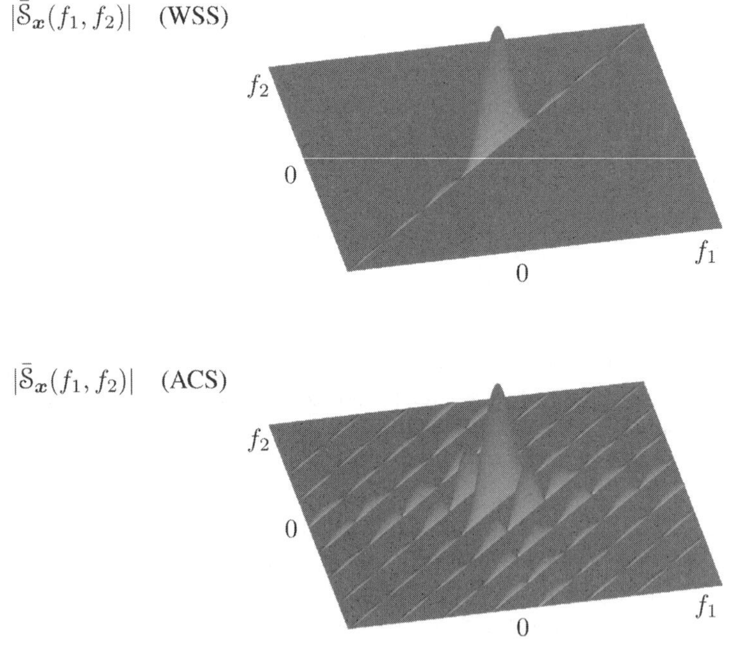

Figure 1.4 Magnitude of the bifrequency spectral correlation density of a signal $x(t)$ as functions of f_1 and f_2. Top: WSS signal. Bottom: ACS signal

1.3.2 Jointly ACS Signals

Let $x_i(t)$, $i = 1, 2$, $t \in \mathbb{R}$, be two complex-valued almost-cyclostationary (ACS) continuous-time processes with second-order (conjugate) cross-correlation function

$$R_{x_1 x_2^{(*)}}(t, \tau) \triangleq \mathrm{E}\left\{x_1(t + \tau)\, x_2^{(*)}(t)\right\} = \sum_{\alpha \in A_{12}} R^{\alpha}_{x_1 x_2^{(*)}}(\tau)\, e^{j2\pi\alpha t} \tag{1.95}$$

where

$$R^{\alpha}_{x_1 x_2^{(*)}}(\tau) \triangleq \lim_{T\to\infty} \frac{1}{T} \int_{-T/2}^{T/2} \mathrm{E}\left\{x_1(t+\tau)\, x_2^{(*)}(t)\right\} e^{-j2\pi\alpha t}\, \mathrm{d}t \tag{1.96}$$

is the *(conjugate) cyclic cross-correlation function* of x_1 and x_2 at (conjugate) cycle frequency α and

$$A_{12} \triangleq \{\alpha \in \mathbb{R} \,:\, R^{\alpha}_{x_1 x_2^{(*)}}(\tau) \not\equiv 0\} \tag{1.97}$$

is a countable set. If the set A_{12} contains at least one nonzero element, then the time series $x_1(t)$ and $x_2(t)$ are said to be *jointly ACS*. Note that, in general, the set A_{12} depends on whether $(*)$ is a conjugation or not and can be different from the sets A_{11} and A_{22} (both defined according to (1.97)).

The *(conjugate) cyclic cross-spectrum* of x_1 and x_2 at (conjugate) cycle frequency α is defined as

$$S^{\alpha}_{x_1 x_2^{(*)}}(f) \triangleq \lim_{\Delta f\to 0} \lim_{T\to\infty} \frac{1}{T} \int_{-T/2}^{T/2} \Delta f\, \mathrm{E}\left\{X_{1,1/\Delta f}(t, f)\, X^{(*)}_{2,1/\Delta f}(t, (-)(\alpha - f))\right\} \mathrm{d}t \tag{1.98}$$

and is related to the (conjugate) cyclic cross-correlation function by the *Gardner relation*

$$S^{\alpha}_{x_1 x_2^{(*)}}(f) = \int_{\mathbb{R}} R^{\alpha}_{x_1 x_2^{(*)}}(\tau)\, e^{-j2\pi f\tau}\, \mathrm{d}\tau. \tag{1.99}$$

In (1.98), $X_{i,1/\Delta f}(t, f)$ is the STFT of $x_i(t)$ defined according to (1.93).

In the case of distinct processes (or processes denoted as distinct), the use of optional conjugation can be avoided without a lack of generality. In fact, if $x_i(t) = x_0^*(t)$, then $X_i(f) = X_0^*(-f)$. However, the following results are useful for future reference. Let $\boldsymbol{x} \triangleq [x_1^{(*)_1}\; x_2^{(*)_2}]$, where superscript $(*)_i$ denotes ith optional complex conjugation.

If

$$\mathrm{E}\left\{x_1^{(*)_1}(t+\tau)\, x_2^{(*)_2}(t)\right\} = \sum_{\alpha \in A_{\boldsymbol{x}}} R^{\alpha}_{\boldsymbol{x}}(\tau)\, e^{j2\pi\alpha t} \tag{1.100}$$

then

$$\mathrm{E}\left\{X_1^{(*)_1}(f_1)\, X_2^{(*)_2}(f_2)\right\} = \sum_{\alpha \in A_{\boldsymbol{x}}} S^{\alpha}_{\boldsymbol{x}}((-)_1 f_1)\, \delta(f_2 - (-)_2(\alpha - (-)_1 f_1)) \tag{1.101}$$

where $(-)_i$ is an optional minus sign linked to $(*)_i$.

In fact, by making the variable change $t_1 = t + \tau$, $t_2 = t$ into (1.100) the result is that

$$\mathrm{E}\left\{x_1^{(*)_1}(t_1)\, x_2^{(*)_2}(t_2)\right\} = \sum_{\alpha \in A_{\boldsymbol{x}}} R_{\boldsymbol{x}}^{\alpha}(t_1 - t_2)\, e^{j2\pi\alpha t_2} \tag{1.102}$$

and in the sense of distributions we formally have

$$\begin{aligned}
&\mathrm{E}\left\{X_1^{(*)_1}(f_1)\, X_2^{(*)_2}(f_2)\right\} \\
&= \mathrm{E}\left\{\left[\int_{\mathbb{R}} x_1(t_1)\, e^{-j2\pi f_1 t_1}\, \mathrm{d}t_1\right]^{(*)_1} \left[\int_{\mathbb{R}} x_2(t_2)\, e^{-j2\pi f_2 t_2}\, \mathrm{d}t_2\right]^{(*)_2}\right\} \\
&= \int_{\mathbb{R}}\int_{\mathbb{R}} \mathrm{E}\left\{x_1^{(*)_1}(t_1)\, x_2^{(*)_2}(t_2)\right\} e^{-j2\pi[(-)_1 f_1 t_1 + (-)_2 f_2 t_2]}\, \mathrm{d}t_1\, \mathrm{d}t_2 \\
&= \int_{\mathbb{R}}\int_{\mathbb{R}} \sum_{\alpha \in A_{\boldsymbol{x}}} R_{\boldsymbol{x}}^{\alpha}(t_1 - t_2)\, e^{j2\pi\alpha t_2}\, e^{-j2\pi[(-)_1 f_1 t_1 + (-)_2 f_2 t_2]}\, \mathrm{d}t_1\, \mathrm{d}t_2 \\
&= \sum_{\alpha \in A_{\boldsymbol{x}}} \int_{\mathbb{R}} R_{\boldsymbol{x}}^{\alpha}(\tau)\, e^{-(-)_1 j2\pi f_1 \tau}\, \mathrm{d}\tau \int_{\mathbb{R}} e^{j2\pi[\alpha - ((-)_1 f_1 + (-)_2 f_2)]t}\, \mathrm{d}t
\end{aligned} \tag{1.103}$$

from which (1.101) follows since

$$\begin{aligned}
\int_{\mathbb{R}} e^{j2\pi[\alpha - ((-)_1 f_1 + (-)_2 f_2)]t}\, \mathrm{d}t &= \delta(\alpha - ((-)_1 f_1 + (-)_2 f_2)) \\
&= \delta(f_2 - (-)_2(\alpha - (-)_1 f_1)).
\end{aligned} \tag{1.104}$$

By specializing (1.100) and (1.101) we have

$$\mathrm{E}\left\{y(t+\tau)\, x^{(*)}(t)\right\} = \sum_{\alpha \in A_{yx^{(*)}}} R_{yx^{(*)}}^{\alpha}(\tau)\, e^{j2\pi\alpha t} \tag{1.105}$$

$$\mathrm{E}\left\{Y(f_1)\, X^{(*)}(f_2)\right\} = \sum_{\alpha \in A_{yx^{(*)}}} S_{yx^{(*)}}^{\alpha}(f_1)\, \delta(f_2 - (-)(\alpha - f_1)). \tag{1.106}$$

1.3.3 LAPTV Systems

Linear almost-periodically time-varying (LAPTV) systems have an impulse-response function that can be expressed by the (generalized) Fourier series expansion

$$h(t, u) = \sum_{\sigma \in J} h_{\sigma}(t - u) e^{j2\pi\sigma u} \tag{1.107}$$

or, equivalently,

$$h(t + \tau, t) = \sum_{\sigma \in J} h_{\sigma}(\tau) e^{j2\pi\sigma t}. \tag{1.108}$$

where J is a countable set of frequency shifts.

By substituting (1.107) into (1.41), the output $y(t)$ can be expressed in the two equivalent forms (Franks 1994), (Gardner 1987d)

$$y(t) = \sum_{\sigma \in J} h_\sigma(t) \otimes [x(t)\, e^{j2\pi\sigma t}] \tag{1.109a}$$

$$= \sum_{\sigma \in J} [g_\sigma(t) \otimes x(t)]\, e^{j2\pi\sigma t} \tag{1.109b}$$

where

$$g_\sigma(t) \triangleq h_\sigma(t)\, e^{-j2\pi\sigma t}. \tag{1.110}$$

Equations (1.109a) and (1.109b) can be re-expressed in the frequency domain:

$$Y(f) = \sum_{\sigma \in J} H_\sigma(f)\, X(f - \sigma) \tag{1.111a}$$

$$= \sum_{\sigma \in J} G_\sigma(f - \sigma)\, X(f - \sigma) \tag{1.111b}$$

where $H_\sigma(f)$ and $G_\sigma(f)$ are the Fourier transforms of $h_\sigma(t)$ and $g_\sigma(t)$, respectively.

From (1.109a) it follows that the LAPTV systems combine linear time-invariant filtered versions of frequency-shifted versions of the input signal. For this reason, LAPTV filtering is also referred to as *frequency-shift* (FRESH) filtering (Gardner 1993). Equivalently, from (1.109b) the result is that LAPTV systems can be realized by combining frequency shifted versions of linear time-invariant filtered versions of the input (Figure 1.5).

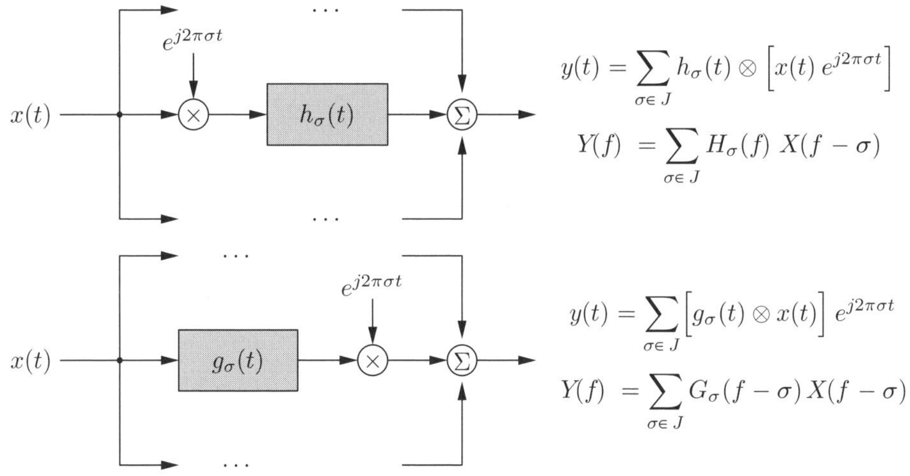

Figure 1.5 LAPTV systems: two equivalent realizations and corresponding input/output relations in time and frequency domains (see (1.109a)–(1.111b))

In the special case for which $J \equiv \{k/T_0\}_{k\in\mathbb{Z}}$ for some period T_0, the system is linear periodically time-variant (LPTV). If J contains only the element $\sigma = 0$, then the system is LTI and $h(t+\tau, t) = h_0(\tau)$.

1.3.3.1 Input/Output Relations

Let us consider two LAPTV systems with impulse-response functions

$$h_i(t,u) = \sum_{\sigma_i \in J_i} h_{\sigma_i}(t-u)\, e^{j2\pi\sigma_i u} \qquad i = 1, 2. \tag{1.112}$$

The (conjugate) cross-correlation of the outputs

$$y_i(t) = \int_{\mathbb{R}} h_i(t,u)\, x_i(u)\, \mathrm{d}u \qquad i = 1, 2 \tag{1.113}$$

is given by (see (Gardner *et al.* 2006, eq. (3.78)))

$$\begin{aligned} R_{y_1 y_2^{(*)}}(t,\tau) &\triangleq \mathrm{E}\left\{ y_1(t+\tau)\, y_2^{(*)}(t) \right\} \\ &= \sum_{\alpha\in A_{12}} \sum_{\sigma_1\in J_1} \sum_{\sigma_2 \in J_2} \left[R^{\alpha}_{x_1 x_2^{(*)}}(\tau)\, e^{j2\pi\sigma_1\tau} \right] \\ &\qquad \underset{\tau}{\otimes}\, r^{\alpha+\sigma_1+(-)\sigma_2}_{\sigma_1\sigma_2(*)}(\tau)\, e^{j2\pi(\alpha+\sigma_1+(-)\sigma_2)t} \end{aligned} \tag{1.114}$$

where $\underset{\tau}{\otimes}$ denotes convolution with respect to τ, $(-)$ is an optional minus sign that is linked to $(*)$, and

$$\begin{aligned} r^{\gamma}_{\sigma_1\sigma_2(*)}(\tau) &\triangleq \int_{\mathbb{R}} h_{\sigma_1}(s+\tau)\, h^{(*)}_{\sigma_2}(s)\, e^{-j2\pi\gamma s}\, \mathrm{d}s \\ &= h_{\sigma_1}(\tau) \otimes \left[h^{(*)}_{\sigma_2}(-\tau)\, e^{j2\pi\gamma\tau} \right] \end{aligned} \tag{1.115}$$

Thus (see (Napolitano 1995, eqs. (32) and (34)) and (Gardner *et al.* 2006, eqs. (3.80) and (3.81))), the (conjugate) cyclic cross-correlation function and the (conjugate) cyclic cross-spectrum of the outputs $y_1(t)$ and $y_2(t)$ are given by

$$\begin{aligned} R^{\beta}_{y_1 y_2^{(*)}}(\tau) &\triangleq \lim_{T\to\infty} \frac{1}{T} \int_{-T/2}^{T/2} \mathrm{E}\left\{ y_1(t+\tau)\, y_2^{(*)}(t) \right\} e^{-j2\pi\beta t}\, \mathrm{d}t \\ &= \sum_{\sigma_1\in J_1} \sum_{\sigma_2\in J_2} \left[R^{\beta-\sigma_1-(-)\sigma_2}_{x_1 x_2^{(*)}}(\tau)\, e^{j2\pi\sigma_1\tau} \right] \underset{\tau}{\otimes}\, r^{\beta}_{\sigma_1\sigma_2(*)}(\tau) \end{aligned} \tag{1.116}$$

$$\begin{aligned} S^{\beta}_{y_1 y_2^{(*)}}(f) &\triangleq \int_{\mathbb{R}} R^{\beta}_{y_1 y_2^{(*)}}(\tau)\, e^{-j2\pi f\tau}\, \mathrm{d}\tau \\ &= \sum_{\sigma_1\in J_1} \sum_{\sigma_2\in J_2} S^{\beta-\sigma_1-(-)\sigma_2}_{x_1 x_2^{(*)}}(f-\sigma_1)\, H_{\sigma_1}(f)\, H^{(*)}_{\sigma_2}((-)(\beta-f)) \end{aligned} \tag{1.117}$$

where

$$H_{\sigma_i}(f) \triangleq \int_{\mathbb{R}} h_{\sigma_i}(\tau)\, e^{-j2\pi f\tau}\, \mathrm{d}\tau \tag{1.118}$$

and, in the sums in (1.116) and (1.117), only those $\sigma_1 \in J_1$ and $\sigma_2 \in J_2$ such that $\beta - \sigma_1 - {}_{(-)}\sigma_2 \in A_{12}$ give nonzero contribution. In the derivation of (1.117), the Fourier transform pair

$$r^{\gamma}_{\sigma_1\sigma_2(*)}(\tau) = h_{\sigma_1}(\tau) \otimes \left[h^{(*)}_{\sigma_2}(-\tau)\, e^{j2\pi\gamma\tau}\right] \overset{\mathcal{F}}{\longleftrightarrow} H_{\sigma_1}(f)\, H^{(*)}_{\sigma_2}({}_{(-)}(\gamma - f)) \tag{1.119}$$

is used.

Equations (1.116) and (1.117) can be specialized to several cases of interest. For example, if $x_1 = x_2 = x$, $h_1 = h_2 = h$, $y_1 = y_2 = y$, and $(*)$ is conjugation, then we obtain the input-output relations for LAPTV systems in terms of cyclic autocorrelation functions and cyclic spectra:

$$R^{\beta}_{yy^*}(\tau) = \sum_{\sigma_1 \in J} \sum_{\sigma_2 \in J} \left[R^{\beta-\sigma_1+\sigma_2}_{xx^*}(\tau)\, e^{j2\pi\sigma_1\tau}\right] \underset{\tau}{\otimes} r^{\beta}_{12}(\tau) \tag{1.120}$$

$$S^{\beta}_{yy^*}(f) = \sum_{\sigma_1 \in J} \sum_{\sigma_2 \in J} S^{\beta-\sigma_1+\sigma_2}_{xx^*}(f - \sigma_1)\, H_{\sigma_1}(f)\, H^*_{\sigma_2}(f - \beta) \tag{1.121}$$

where

$$r^{\beta}_{12}(\tau) \triangleq \int_{\mathbb{R}} h_{\sigma_1}(\tau + s)\, h^*_{\sigma_2}(s)\, e^{-j2\pi\beta s}\, \mathrm{d}s. \tag{1.122}$$

By specializing (1.117) to the case $x_1 = x_2 = x$, $h_1 = h_2 = h$ LTI system, and $y_1 = y_2 = y$ we obtain

$$S^{\alpha}_{yy^{(*)}}(f) = S^{\alpha}_{xx^{(*)}}(f)\, H(f)\, H^{(*)}({}_{(-)}(\alpha - f)). \tag{1.123}$$

Analogously, by specializing (1.116), we obtain the inverse Fourier transform of both sides of (1.123)

$$R^{\alpha}_{yy^{(*)}}(\tau) = R^{\alpha}_{xx^{(*)}}(\tau) \otimes r^{\alpha}_{hh^{(*)}}(\tau) \tag{1.124}$$

where, accounting for the Fourier pairs

$$h(-\tau) \overset{\mathcal{F}}{\longleftrightarrow} H(-f) \qquad h^{(*)}(-\tau) \overset{\mathcal{F}}{\longleftrightarrow} H^{(*)}({}_{(-)}(-f))$$

$$h^{(*)}(-\tau)\, e^{j2\pi\alpha\tau} \overset{\mathcal{F}}{\longleftrightarrow} H^{(*)}({}_{(-)}(-(f - \alpha))) = H^{(*)}({}_{(-)}(\alpha - f))$$

the *(conjugate) ambiguity function* $r^{\alpha}_{hh^{(*)}}(\tau)$ can be written in one of the following equivalent forms

$$\begin{aligned} r^{\alpha}_{hh^{(*)}}(\tau) &\triangleq \int_{\mathbb{R}} h(\tau+s)\, h^{(*)}(s)\, e^{-j2\pi\alpha s}\, \mathrm{d}s = \int_{\mathbb{R}} h(\tau-s')\, h^{(*)}(-s')\, e^{j2\pi\alpha s'}\, \mathrm{d}s' \\ &= h(\tau) \otimes \left[h^{(*)}(-\tau)\, e^{j2\pi\alpha\tau}\right] \end{aligned} \tag{1.125a}$$

$$\begin{aligned} r^{\alpha}_{hh^{(*)}}(\tau) &\triangleq \int_{\mathbb{R}} h(\tau+s)\, h^{(*)}(s)\, e^{-j2\pi\alpha s}\, \mathrm{d}s = \underset{s\to\alpha}{\mathfrak{F}} \left[h(\tau+s)\, h^{(*)}(s)\right] \\ &= \left[H(\alpha)\, e^{j2\pi\alpha\tau}\right] \underset{\alpha}{\otimes} H^{(*)}((-)\alpha) = \int_{\mathbb{R}} H(\lambda)\, H^{(*)}((-)(\alpha-\lambda))\, e^{j2\pi\lambda\tau}\, \mathrm{d}\lambda \end{aligned} \tag{1.125b}$$

By specializing (1.116) and (1.117) to the case $x_1 = x_2 = x$, $h_2(t+\tau, t) = \delta(\tau)$, $y_1 = y$, $y_2 = x$, so that $J_2 = \{0\}$ and

$$r^{\beta}_{\sigma_1\sigma_2(*)}(\tau) = \int_{\mathbb{R}} h_{\sigma_1}(\tau+s)\, \delta(s)\, e^{-j2\pi\beta s}\, \mathrm{d}s = h_{\sigma_1}(\tau) \tag{1.126}$$

we obtain the cyclic cross-statistics of the output and input signals

$$R^{\beta}_{yx^{(*)}}(\tau) = \sum_{\sigma_1\in J_1} \left[R^{\beta-\sigma_1}_{xx^{(*)}}(\tau)\, e^{j2\pi\sigma_1\tau}\right] \underset{\tau}{\otimes} h_{\sigma_1}(\tau) \tag{1.127}$$

$$S^{\beta}_{yx^{(*)}}(f) = \sum_{\sigma_1\in J_1} S^{\beta-\sigma_1}_{xx^{(*)}}(f-\sigma_1)\, H_{\sigma_1}(f). \tag{1.128}$$

By specializing (1.117) to the case h_1 and h_2 LTI and $(*)$ absent, for $\alpha = 0$ we obtain

$$S_{y_1y_2}(f) = S_{x_1x_2}(f)\, H_1(f)\, H_2(-f) \tag{1.129}$$

$$\overset{\mathfrak{F}}{\longleftrightarrow} \quad R_{y_1y_2}(\tau) = R_{x_1x_2}(\tau) \otimes h_1(\tau) \otimes h_2(-\tau) \tag{1.130}$$

1.3.4 Products of ACS Signals

Let $x_1(t)$, $x_2(t)$, $c_1(t)$, and $c_2(t)$ be ACS signals with (conjugate) cross-correlation functions

$$R_{x_1x_2^{(*)}}(t,\tau) = \sum_{\alpha_x\in A_{x_1x_2^{(*)}}} R^{\alpha_x}_{x_1x_2^{(*)}}(\tau)\, e^{j2\pi\alpha_x t} \tag{1.131}$$

$$R_{c_1c_2^{(*)}}(t,\tau) = \sum_{\alpha_c\in A_{c_1c_2^{(*)}}} R^{\alpha_c}_{c_1c_2^{(*)}}(\tau)\, e^{j2\pi\alpha_c t}. \tag{1.132}$$

If $x_1(t)$ and $x_2(t)$ are statistically independent of $c_1(t)$ and $c_2(t)$, then their fourth-order joint probability density function factors into the product of the two second-order joint probability densities of x_1, x_2 and of c_1, c_2 (Gardner 1987d, 1994), (Brown 1987), and the (conjugate) cross-correlation functions of the product waveforms

$$y_i(t) = c_i(t)\, x_i(t) \qquad i = 1, 2 \tag{1.133}$$

also factors,

$$R_{y_1 y_2^{(*)}}(t, \tau) = R_{c_1 c_2^{(*)}}(t, \tau)\, R_{x_1 x_2^{(*)}}(t, \tau). \tag{1.134}$$

Therefore, the (conjugate) cyclic cross-correlation function and the (conjugate) cyclic cross-spectrum of $y_1(t)$ and $y_2(t)$ are

$$R^{\alpha}_{y_1 y_2^{(*)}}(\tau) = \sum_{\alpha_x \in A_{x_1 x_2^{(*)}}} R^{\alpha_x}_{x_1 x_2^{(*)}}(\tau)\, R^{\alpha-\alpha_x}_{c_1 c_2^{(*)}}(\tau) \tag{1.135}$$

$$S^{\alpha}_{y_1 y_2^{(*)}}(f) = \sum_{\alpha_x \in A_{x_1 x_2^{(*)}}} \int_{\mathbb{R}} S^{\alpha_x}_{x_1 x_2^{(*)}}(\lambda)\, S^{\alpha-\alpha_x}_{c_1 c_2^{(*)}}(f - \lambda)\, \mathrm{d}\lambda \tag{1.136}$$

where, in the sums, only those (conjugate) cycle frequencies α_x such that $\alpha - \alpha_x \in A_{c_1 c_2^{(*)}}$ give nonzero contribution. From (1.135) and (1.136), it follows that the set of (conjugate) cycle-frequencies of the (conjugate) cross-correlation of $y_1(t)$ and $y_2(t)$ is

$$\{\alpha\} = \left\{\alpha = \alpha_c + \alpha_x,\ \alpha_c \in A_{c_1 c_2^{(*)}}, \alpha_x \in A_{x_1 x_2^{(*)}}\right\}. \tag{1.137}$$

In the special case where $c_1(t)$ and $c_2(t)$ are almost-periodic functions with (generalized) Fourier series

$$c_i(t) = \sum_{\gamma_i \in G_i} c_{i,\gamma_i}\, e^{j2\pi\gamma_i t} \qquad i = 1, 2 \tag{1.138}$$

then

$$\begin{aligned} R_{c_1 c_2^{(*)}}(t, \tau) &= c_1(t + \tau)\, c_2^{(*)}(t) \\ &= \sum_{\gamma_1 \in G_1} \sum_{\gamma_2 \in G_2} c_{1,\gamma_1}\, c^{(*)}_{2,\gamma_2}\, e^{j2\pi\gamma_1\tau}\, e^{j2\pi(\gamma_1 + (-)\gamma_2)t} \end{aligned} \tag{1.139}$$

and

$$R^{\alpha}_{c_1 c_2^{(*)}}(\tau) = \sum_{\gamma_1 \in G_1} c_{1,\gamma_1}\, c^{(*)}_{2,(-)(\alpha-\gamma_1)}\, e^{j2\pi\gamma_1\tau} \tag{1.140}$$

$$S^{\alpha}_{c_1 c_2^{(*)}}(f) = \sum_{\gamma_1 \in G_1} c_{1,\gamma_1}\, c^{(*)}_{2,(-)(\alpha-\gamma_1)}\, \delta(f - \gamma_1) \tag{1.141}$$

where, in the sum, only those frequencies γ_1 such that $(-)(\alpha - \gamma_1) \in G_2$ give nonzero contribution. From (1.140) and (1.141) it follows that the set of the (conjugate) cycle-frequencies of the (conjugate) cross-correlation of $y_1(t)$ and $y_2(t)$ is

$$\{\alpha\} = \{\alpha = \gamma_1 + (-)\gamma_2,\ \gamma_1 \in G_1, \gamma_2 \in G_2\}. \tag{1.142}$$

In the special case $x_1 \equiv x_2$ and $c_1 \equiv c_2$, (1.139)–(1.141) reduce to (corrected versions of) (3.93)–(3.95) in (Gardner *et al.* 2006)). By substituting (1.140) and (1.141) into (1.135) and (1.136), respectively, and making the variable change $\gamma_2 = {}_{(-)}(\alpha - \alpha_x - \gamma_1)$, we get (see (Napolitano 1995, eqs. (46)–(48)))

$$R^{\alpha}_{y_1 y_2^{(*)}}(\tau) = \sum_{\gamma_1 \in G_1} \sum_{\gamma_2 \in G_2} c_{1,\gamma_1}\, c^{(*)}_{2,\gamma_2}\, R^{\alpha-(\gamma_1+{}_{(-)}\gamma_2)}_{x_1 x_2^{(*)}}(\tau)\, e^{j2\pi\gamma_1\tau} \tag{1.143}$$

$$S^{\alpha}_{y_1 y_2^{(*)}}(f) = \sum_{\gamma_1 \in G_1} \sum_{\gamma_2 \in G_2} c_{1,\gamma_1}\, c^{(*)}_{2,\gamma_2}\, S^{\alpha-(\gamma_1+{}_{(-)}\gamma_2)}_{x_1 x_2^{(*)}}(f - \gamma_1). \tag{1.144}$$

where, in the sums, only those values of γ_1 and γ_2 such that $\alpha - (\gamma_1 + {}_{(-)}\gamma_2) \in A_{x_1 x_2^{(*)}}$ give nonzero contribution. From (1.143) and (1.144), it follows that the set of the (conjugate) cycle-frequencies of the (conjugate) cross-correlation of $y_1(t)$ and $y_2(t)$ is

$$\{\alpha\} = \left\{\alpha = \gamma_1 + {}_{(-)}\gamma_2 + \alpha_x,\ \alpha_x \in A_{x_1 x_2^{(*)}},\ \gamma_1 \in G_1, \gamma_2 \in G_2\right\}. \tag{1.145}$$

1.3.5 Cyclic Statistics of Communications Signals

Cyclostationarity in man-made communications signals is due to signal processing operations used in the construction and/or subsequent processing of signals, such as modulation, sampling, scanning, multiplexing, and coding. Consequently, cycle frequencies are related to parameters such as sinewave carrier frequency, sampling rate, etc. In the following, the cyclic autocorrelation functions and cyclic spectra of two basic communications signals are reported. Further examples can be found in (Gardner 1985, Chapter 12, 1987d), (Gardner *et al.* 2006) and references therein.

The first signal is the double side-band amplitude-modulated (DSB-AM) signal

$$x(t) \triangleq s(t)\cos(2\pi f_0 t + \phi_0). \tag{1.146}$$

If the modulating process $s(t)$ is WSS, then $x(t)$ is cyclostationary with period $1/(2f_0)$, cyclic autocorrelation functions (Gardner 1985, Chapter 12, 1987d), (Gardner *et al.* 2006)

$$R^{\alpha}_{x}(\tau) = \begin{cases} \frac{1}{2} R^{0}_{s}(\tau)\, \cos(2\pi f_0 \tau) & \alpha = 0 \\ \frac{1}{4} R^{0}_{s}(\tau)\, e^{\pm j2\pi f_0 \tau}\, e^{\pm j2\phi_0} & \alpha = \pm 2 f_0 \\ 0 & \text{otherwise} \end{cases} \tag{1.147}$$

and cyclic spectra

$$S^{\alpha}_{x}(f) = \begin{cases} \frac{1}{4}\left\{S^{0}_{s}(f - f_0) + S^{0}_{s}(f + f_0)\right\} & \alpha = 0 \\ \frac{1}{4} S^{0}_{s}(f \mp f_0)\, e^{\pm j2\phi_0} & \alpha = \pm 2 f_0 \\ 0 & \text{otherwise.} \end{cases} \tag{1.148}$$

The second signal is the pulse-amplitude-modulated (PAM) signal

$$x(t) \triangleq \sum_{k \in \mathbb{Z}} a_k\, q(t - kT_0) \tag{1.149}$$

where $\{a_k\}_{k\in\mathbb{Z}}$ is a WSS sequence and $q(t)$ a finite-energy pulse. This PAM signal is cyclostationary with period T_0 and (conjugate) cyclic autocorrelation functions and cyclic spectra (Gardner 1985, Chapter 12, 1987d), (Gardner *et al.* 2006)

$$R_{\boldsymbol{x}}^{\alpha}(\tau) = \frac{\mathrm{E}\{a_k a_k^{(*)}\}}{T_0}\, r_{\boldsymbol{q}}^{\alpha}(\tau) \tag{1.150}$$

$$S_{\boldsymbol{x}}^{\alpha}(f) = \frac{\mathrm{E}\{a_k a_k^{(*)}\}}{T_0}\, Q(f)\, Q^{(*)}((-)(\alpha - f)) \tag{1.151}$$

for $\alpha = m/T_0$, $m \in \mathbb{Z}$, and zero otherwise. In (1.150),

$$r_{\boldsymbol{q}}^{\alpha}(\tau) \triangleq \int_{\mathbb{R}} q(t+\tau)\, q^{(*)}(t)\, e^{-j2\pi\alpha t}\, \mathrm{d}t \tag{1.152}$$

and in (1.151) $Q(f)$ is the Fourier transform of $q(t)$.

1.3.6 Higher-Order Statistics

A more complete characterization of stochastic processes can be obtained by considering higher-order statistics. Processes that have almost-periodically time-variant moment and cumulant functions are called higher-order almost-cyclostationary. They have been characterized in the time and frequency domains in (Dandawaté and Giannakis 1994, 1995), (Gardner and Spooner 1994), (Napolitano 1995), (Spooner and Gardner 1994).

A continuous-time complex-valued process $x(t)$ is said to *exhibit Nth-order wide-sense cyclostationarity* with cycle frequency $\alpha \neq 0$, for a given conjugation configuration, if the Nth-order *cyclic temporal moment function* (CTMF)

$$\begin{aligned}\mathcal{R}_{\boldsymbol{x}}^{\alpha}(\boldsymbol{\tau}) &\triangleq \mathcal{R}_{x^{(*)_1},\dots,x^{(*)_N}}^{\alpha}(\boldsymbol{\tau}) \\ &\triangleq \lim_{T\to\infty} \frac{1}{T}\int_{-T/2}^{T/2} \mathrm{E}\Big\{\prod_{n=1}^{N} x^{(*)_n}(t+\tau_n)\Big\}\, e^{-j2\pi\alpha t}\, \mathrm{d}t\end{aligned} \tag{1.153}$$

exists and is not zero for some delay vector $\boldsymbol{\tau}$ (Gardner and Spooner 1994). In (1.153), $\boldsymbol{\tau} \triangleq [\tau_1, \dots, \tau_N]^{\mathrm{T}}$ and $\boldsymbol{x} \triangleq [x^{(*)_1}, \dots, x^{(*)_N}]^{\mathrm{T}}$ are column vectors, and $(*)_n$ represents the nth optional complex conjugation. The process is said to be *Nth-order wide-sense almost-cyclostationary* if its Nth-order temporal moment function (TMF) is an almost-periodic function of t. Under mild regularity conditions the TMF can be expressed by the (generalized) Fourier series expansion

$$\begin{aligned}\mathcal{R}_{\boldsymbol{x}}(t, \boldsymbol{\tau}) &\triangleq \mathrm{E}\Big\{\prod_{n=1}^{N} x^{(*)_n}(t+\tau_n)\Big\} \\ &= \sum_{\alpha\in A_{\boldsymbol{x}}} \mathcal{R}_{\boldsymbol{x}}^{\alpha}(\boldsymbol{\tau})\, e^{j2\pi\alpha t}\end{aligned} \tag{1.154}$$

where $A_{\boldsymbol{x}}$ is a countable set and the sense of equality in (1.154) is one of those discussed in Section 1.2.

The N-fold Fourier transform of the CTMF

$$\mathcal{S}_{\boldsymbol{x}}^{\alpha}(\boldsymbol{f}) = \int_{\mathbb{R}^N} \mathcal{R}_{\boldsymbol{x}}^{\alpha}(\boldsymbol{\tau})\, e^{-j2\pi \boldsymbol{f}^{\mathrm{T}}\boldsymbol{\tau}}\, \mathrm{d}\boldsymbol{\tau} \tag{1.155}$$

which is called the Nth-order *cyclic spectral moment function* (CSMF), can be written as

$$\mathcal{S}_{\boldsymbol{x}}^{\alpha}(\boldsymbol{f}) = S_{\boldsymbol{x}}^{\alpha}(\boldsymbol{f}')\, \delta(\boldsymbol{f}^{\mathrm{T}}\mathbf{1} - \alpha) \tag{1.156}$$

where $\delta(\cdot)$ is the Dirac delta function, $\mathbf{1} \triangleq [1, ..., 1]^{\mathrm{T}}$, and primes denote the operator that transforms $\boldsymbol{v} = [v_1, \ldots, v_N]^{\mathrm{T}}$ into $\boldsymbol{v}' = [v_1, \ldots, v_{N-1}]^{\mathrm{T}}$. The function $S_{\boldsymbol{x}}^{\alpha}(\boldsymbol{f}')$, referred to as the Nth-order reduced-dimension CSMF (RD-CSMF), can be expressed as the $(N-1)$-fold Fourier transform of the Nth-order reduced-dimension CTMF (RD-CTMF) defined as

$$R_{\boldsymbol{x}}^{\alpha}(\boldsymbol{\tau}') \triangleq \mathcal{R}_{\boldsymbol{x}}^{\alpha}(\boldsymbol{\tau})\big|_{\tau_N=0}\,. \tag{1.157}$$

For $N = 2$ and conjugation configuration $[xx^*]$, the RD-CTMF is coincident with the *cyclic autocorrelation function* $R_{xx^*}^{\alpha}(\tau_1)$, whereas, for conjugation configuration $[xx]$ it is coincident with the *conjugate cyclic autocorrelation function* $R_{xx}^{\alpha}(\tau_1)$ that can be nonzero only if the signal is noncircular (Picinbono and Bondon 1997), (Schreier and Scharf 2003a,b). It is shown in (Gardner and Spooner 1994) that the RD-CSMF $S_{\boldsymbol{x}}^{\beta}(\boldsymbol{f}')$ can contain impulsive terms if the vector $\boldsymbol{f}$ with $f_N = \beta - \sum_{n=1}^{N-1} f_n$ lies on the β-submanifold, i.e., if there exists at least one partition $\{\mu_1, \cdots, \mu_p\}$ of $\{1, \cdots, N\}$ with $p > 1$ such that each sum $\alpha_{\mu_i} = \sum_{n\in\mu_i} f_n$ is an $|\mu_i|$th-order cycle frequency of $x(t)$, where $|\mu_i|$ is the number of elements in μ_i.

Note that for second-order statistics, unlike in (1.155), in the complex exponential of double Fourier transforms, an optional minus sign $(-)$ linked to the optional complex conjugation $(*)$ is accounted for (see (1.15), (1.16), and (1.101)).

A well-behaved frequency-domain function that characterizes a signal's higher-order cyclostationarity can be obtained starting from the Nth-order *cyclic temporal cumulant function* (CTCF), that is, the coefficient

$$\mathcal{C}_{\boldsymbol{x}}^{\beta}(\boldsymbol{\tau}) \triangleq \lim_{T\to\infty} \frac{1}{T} \int_{-T/2}^{T/2} \mathrm{cum}\Big\{x^{(*)_n}(t+\tau_n),\ n = 1, ..., N\Big\}\, e^{-j2\pi\beta t}\, \mathrm{d}t \tag{1.158}$$

of the (generalized) Fourier-series expansion of the Nth-order temporal cumulant function (Gardner and Spooner 1994)

$$\mathcal{C}_{\boldsymbol{x}}(t, \boldsymbol{\tau}) \equiv \mathrm{cum}\Big\{x^{(*)_n}(t+\tau_n),\ n = 1, \ldots, N\Big\}$$

$$\triangleq (-j)^N \frac{\partial^N}{\partial\omega_1 \cdots \partial\omega_N} \log_e \mathrm{E}\Big\{\exp\Big[j\sum_{n=1}^{N} \omega_n x^{(*)_n}(t+\tau_n)\Big]\Big\}\Big|_{\boldsymbol{\omega}=\mathbf{0}} \tag{1.159a}$$

$$= \sum_{\mathrm{P}} \Big[(-1)^{p-1}(p-1)! \prod_{i=1}^{p} \mathcal{R}_{\boldsymbol{x}_{\mu_i}}(t, \boldsymbol{\tau}_{\mu_i})\Big] \tag{1.159b}$$

where, $\boldsymbol{\omega} \triangleq [\omega_1, ..., \omega_N]$; P is the set of distinct partitions of $\{1, ..., N\}$, each constituted by the subsets $\{\mu_i,\ i = 1, ..., p\}$; $\boldsymbol{x}_{\mu_i}$ is the $|\mu_i|$-dimensional vector whose components are those of $\boldsymbol{x}$ having indices in μ_i, with $|\mu_i|$ the number of elements in μ_i. See Section 1.4.2 for a discussion of the definition of (cross-) cumulant of complex random variables and processes.

The N-fold Fourier transform of $\mathcal{C}_{\boldsymbol{x}}^{\beta}(\boldsymbol{\tau})$ is the Nth-order *cyclic spectral cumulant function* $\mathcal{P}_{\boldsymbol{x}}^{\beta}(\boldsymbol{f})$, which can be written as $\mathcal{P}_{\boldsymbol{x}}^{\beta}(\boldsymbol{f}) = P_{\boldsymbol{x}}^{\beta}(\boldsymbol{f}')\,\delta(\boldsymbol{f}^{\mathrm{T}}\mathbf{1} - \beta)$, where the Nth-order *cyclic polyspectrum* (CP) $P_{\boldsymbol{x}}^{\beta}(\boldsymbol{f}')$ is the $(N-1)$-fold Fourier transform of the reduced-dimension CTCF (RD-CTCF) $C_{\boldsymbol{x}}^{\beta}(\boldsymbol{\tau}')$ obtained by setting $\tau_N = 0$ into (1.158). The CP turns out to be a well-behaved function (i.e., it does not contain impulsive terms) under the mild assumption that the time series $x(t)$ and $x(t+\tau)$ are asymptotically $(\tau \to \infty)$ independent (Section 1.4.1). Moreover, except on a β-submanifold, it is coincident with the RD-CSMF $S_{\boldsymbol{x}}^{\beta}(\boldsymbol{f}')$.

Higher-order cyclostationarity can be exploited when the second-order cyclostationarity features of the signal of interest are zero or weak. Moreover, since they provide a more complete characterization of signals, higher-order cyclostationarity is suitable to be used in modulation format classification and cognitive radio (Spooner and Nicholls 2009).

1.3.7 Cyclic Statistic Estimators

Consistent estimates of second-order statistical functions of an ACS stochastic process can be obtained provided that the stochastic process has finite or "effectively finite" memory. Such a property is generally expressed in terms of mixing conditions or summability of second- and fourth-order cumulants. Under such mixing conditions, the cyclic correlogram is a consistent estimator of the cyclic autocorrelation function. Moreover, a properly normalized version of the cyclic correlogram is asymptotically zero-mean complex Normal as the observation interval approaches infinity. In the frequency domain, the cyclic periodogram is an asymptotically unbiased but not consistent estimator of the cyclic spectrum. However, under appropriate conditions, the frequency-smoothed cyclic periodogram is a consistent estimator of the cyclic spectrum. Moreover, a properly normalized version of the frequency-smoothed cyclic periodogram is asymptotically zero-mean complex Normal as the observation interval approaches infinity and the width of the frequency-smoothing window approaches zero. In addition, the time-smoothed cyclic periodogram is asymptotically equivalent to the frequency-smoothed cyclic periodogram.

Consistent estimators of (conjugate) cyclic autocorrelation function and cyclic spectrum are proposed and analyzed in (Hurd and Leśkow 1992a,b), (Dandawaté and Giannakis 1994, 1995), (Dehay 1994), (Dehay and Hurd 1994), (Hurd and Miamee 2007), and references therein. These results can be obtained by specializing to ACS processes the results presented in Chapters 2 and 4. Higher-order cyclic statistic estimators are presented in (Dandawaté and Giannakis 1994, 1995), and (Napolitano and Spooner 2000).

1.3.8 Discrete-Time ACS Signals

ACS processes, such as WSS processes, can also be defined in discrete time. The definitions are similar to those in continuous time with the obvious modifications (Gladyshev 1961), (Gardner 1985, 1994), (Dandawaté and Giannakis 1994), (Napolitano 1995), (Giannakis 1998), (Gardner *et al.* 2006).

Let us consider a discrete-time complex-valued stochastic process $\{x_{\mathrm{d}}(n, \omega), n \in \mathbb{Z}, \omega \in \Omega\}$, with abbreviated notation $x_{\mathrm{d}}(n)$ when this does not create ambiguity. The stochastic process

$x_{\mathrm{d}}(n)$ is said to be *second-order almost-cyclostationary in the wide sense* (Gardner 1985) if its mean value $\mathrm{E}\{x_{\mathrm{d}}(n)\}$ and its (conjugate) autocorrelation function

$$\widetilde{R}_{\boldsymbol{x}_{\mathrm{d}}}(n, m) \triangleq \mathrm{E}\{x_{\mathrm{d}}(n+m)\, x_{\mathrm{d}}^{(*)}(n)\} \tag{1.160}$$

with subscript $\boldsymbol{x}_{\mathrm{d}} \triangleq [x_{\mathrm{d}}\; x_{\mathrm{d}}^{(*)}]$, are almost-periodic functions of the discrete-time parameter n. Thus, under mild regularity assumptions the (conjugate) autocorrelation function can be expressed by the (generalized) Fourier series

$$\widetilde{R}_{\boldsymbol{x}_{\mathrm{d}}}(n, m) = \sum_{\widetilde{\alpha} \in \widetilde{A}} \widetilde{R}_{\boldsymbol{x}_{\mathrm{d}}}^{\widetilde{\alpha}}(m)\, e^{j2\pi\widetilde{\alpha}n} \tag{1.161}$$

where

$$\widetilde{R}_{\boldsymbol{x}_{\mathrm{d}}}^{\widetilde{\alpha}}(m) \triangleq \lim_{N\to\infty} \frac{1}{2N+1} \sum_{n=-N}^{N} \widetilde{R}_{\boldsymbol{x}_{\mathrm{d}}}(n, m)\, e^{-j2\pi\widetilde{\alpha}n} \tag{1.162}$$

is the *(conjugate) cyclic autocorrelation function* at *(conjugate) cycle frequency* $\widetilde{\alpha}$ and

$$\widetilde{A} \triangleq \left\{\widetilde{\alpha} \in [-1/2, 1/2) \;:\; \widetilde{R}_{\boldsymbol{x}_{\mathrm{d}}}^{\widetilde{\alpha}}(m) \not\equiv 0\right\} \tag{1.163}$$

is a countable set. Note that the (conjugate) cyclic autocorrelation function $\widetilde{R}_{\boldsymbol{x}_{\mathrm{d}}}^{\widetilde{\alpha}}(m)$ is periodic in $\widetilde{\alpha}$ with period 1. Thus, the sum in (1.161) can be equivalently extended to the set $\widetilde{A}_1 \triangleq \{\widetilde{\alpha} \in [0, 1) \;:\; \widetilde{R}_{\boldsymbol{x}_{\mathrm{d}}}^{\widetilde{\alpha}}(m) \not\equiv 0\}$.

In general, the set $\widetilde{A}$ (or $\widetilde{A_1}$) depends on the optional complex conjugation $(*)$ and possibly contains incommensurate cycle frequencies $\widetilde{\alpha}$ and cluster points. In the special case where $\widetilde{A}_1 \equiv \{0, 1/N_0, \ldots, (N_0-1)/N_0\}$ for some integer N_0, the (conjugate) autocorrelation function $\widetilde{R}_{\boldsymbol{x}_{\mathrm{d}}}(n, m)$ is periodic in n with period N_0. If also $\mathrm{E}\{x_{\mathrm{d}}(n)\}$ is periodic with period N_0, the process $x(n)$ is said to be *cyclostationary in the wide sense*. If $N_0 = 1$ then $x(n)$ is *wide-sense stationary*.

Let

$$\widetilde{X}_{\mathrm{d}}(\nu) \triangleq \sum_{n\in\mathbb{Z}} x_{\mathrm{d}}(n)\, e^{-j2\pi\nu n} \tag{1.164}$$

be the stochastic process obtained from Fourier transformation (in a generalized sense (Gelfand and Vilenkin 1964, Chapter 3)) of the ACS process $x(n)$. By using (1.161), in the sense of distributions, it can be shown that (Hurd 1991)

$$\mathrm{E}\left\{\widetilde{X}_{\mathrm{d}}(\nu_1)\, \widetilde{X}_{\mathrm{d}}^{(*)}(\nu_2)\right\} = \sum_{\widetilde{\alpha}\in\widetilde{A}} \widetilde{S}_{\boldsymbol{x}_{\mathrm{d}}}^{\widetilde{\alpha}}(\nu_1) \sum_{\ell\in\mathbb{Z}} \delta(\nu_2 + {\scriptstyle(-)}(\nu_1 - \widetilde{\alpha}) - \ell) \tag{1.165}$$

where

$$\widetilde{S}_{\boldsymbol{x}_{\mathrm{d}}}^{\widetilde{\alpha}}(\nu) \triangleq \sum_{m\in\mathbb{Z}} \widetilde{R}_{\boldsymbol{x}_{\mathrm{d}}}^{\widetilde{\alpha}}(m)\, e^{-j2\pi\nu m} \tag{1.166}$$

is the *(conjugate) cyclic spectrum* at (conjugate) cycle frequency $\widetilde{\alpha}$. The cyclic spectrum $\widetilde{\widetilde{S}}^{\alpha}_{\boldsymbol{x}_{\mathrm{d}}}(\nu)$ is periodic in both ν and $\widetilde{\alpha}$ with period 1.

In (Gladyshev 1961) and (Dehay 1994), it is shown that discrete-time cyclostationary processes are harmonizable.

1.3.9 Sampling of ACS Signals

Let

$$x(n) \triangleq x_a(t)|_{t=nT_s} \qquad n \in \mathbb{Z} \tag{1.167}$$

be the discrete-time process obtained by uniformly sampling with period $T_s = 1/f_s$ the continuous-time ACS signal $x_a(t)$ with Loève bifrequency spectrum (see (1.94))

$$\mathrm{E}\left\{X_a(f_1)\, X_a^{(*)}(f_2)\right\} = \sum_{\alpha \in A_{\boldsymbol{x}_a}} S^{\alpha}_{\boldsymbol{x}_a}(f_1)\, \delta(f_2 - {}_{(-)}(\alpha - f_1)). \tag{1.168}$$

where subscript $\boldsymbol{x}_a \triangleq [x_a\ x_a^{(*)}]$. Accounting for the relationship between the Fourier transform $X_a(f)$ of the continuous-time signal $x_a(t)$ and the Fourier transform $X(\nu)$ of the discrete-time sampled signal $x(n)$ (Lathi 2002, Section 9.5)

$$X(\nu) = \frac{1}{T_s} \sum_{k=-\infty}^{+\infty} X_a((\nu - k) f_s) \tag{1.169}$$

we have the following result.

Lemma 1.3.1 Aliasing Formula for the Loève Bifrequency Spectrum of a Sampled ACS Signal.

$$\begin{aligned}
\mathrm{E}\left\{X(\nu_1)\, X^{(*)}(\nu_2)\right\} &= \mathrm{E}\left\{\frac{1}{T_s}\sum_{n_1 \in \mathbb{Z}} X_a((\nu_1 - n_1) f_s)\, \frac{1}{T_s}\sum_{n_2 \in \mathbb{Z}} X_a^{(*)}((\nu_2 - n_2) f_s)\right\} \\
&= \frac{1}{T_s^2} \sum_{n_1 \in \mathbb{Z}} \sum_{n_2 \in \mathbb{Z}} \mathrm{E}\left\{X_a((\nu_1 - n_1) f_s)\, X_a^{(*)}((\nu_2 - n_2) f_s)\right\} \\
&= \frac{1}{T_s^2} \sum_{n_1 \in \mathbb{Z}} \sum_{n_2 \in \mathbb{Z}} \sum_{\alpha \in A_{\boldsymbol{x}_a}} S^{\alpha}_{\boldsymbol{x}_a}((\nu_1 - n_1) f_s) \\
&\quad \delta((\nu_2 - n_2) f_s - {}_{(-)}(\alpha - (\nu_1 - n_1) f_s)) \\
&= \frac{1}{T_s} \sum_{\alpha \in A_{\boldsymbol{x}_a}} \sum_{n_1 \in \mathbb{Z}} S^{\alpha}_{\boldsymbol{x}_a}((\nu_1 - n_1) f_s) \\
&\quad \sum_{n_2 \in \mathbb{Z}} \delta((\nu_2 - n_2) - {}_{(-)}(\alpha/f_s - (\nu_1 - n_1)))
\end{aligned} \tag{1.170}$$

where, in the last equality, the scaling property of the Dirac delta $f_s \delta(\nu f_s) = \delta(\nu)$ is used (Zemanian 1987, Section 1.7). □

Corollary 1.3.2 *From (1.170), it follows that the discrete-time signal* $x(n) \triangleq x_a(t)|_{t=nT_s}$ *obtained by uniformly sampling a continuous-time ACS signal* $x_a(t)$ *is a discrete-time ACS signal. In fact, its Loève bifrequency spectrum has support concentrated on lines with slope* ± 1 *in the bifrequency plane* (ν_1, ν_2) *and can be written in the form (1.165).*

The countable set $\widetilde{A}$ *in (1.165) is linked to* $A_{\boldsymbol{x}_a}$ *by the relationships*

$$\widetilde{A} = \bigcup_{\alpha \in A_{\boldsymbol{x}_a}} \left\{ \widetilde{\alpha} \in (-1/2, 1/2] \, : \, \widetilde{\alpha} = (\alpha/f_s) \bmod 1 \right\} \tag{1.171}$$

$$A_{\boldsymbol{x}_a} \subseteq \bigcup_{\widetilde{\alpha} \in \widetilde{A}} \bigcup_{p \in \mathbb{Z}} \left\{ \alpha \in \mathbb{R} \, : \, \alpha = \widetilde{\alpha} f_s - p f_s \right\} \tag{1.172}$$

with mod 1 *denoting the modulo 1 operation with values in* $(-1/2, 1/2]$ *and equality can hold in (1.172) only if* $x_a(t)$ *is not strictly bandlimited.* □

The (conjugate) cyclic spectrum and the (conjugate) cyclic autocorrelation function of $x(n)$ are given in the following result.

Lemma 1.3.3 Aliasing Formulas for Cyclic Statistics of ACS Signals (Izzo and Napolitano 1996), (Napolitano 1995).

$$\mathrm{E}\{x(n+m)\, x^{(*)}(n)\} = \mathrm{E}\{x_a(t+\tau)\, x_a(t)\}|_{t=nT_s, \tau=mT_s} \, . \tag{1.173}$$

$$\widetilde{R}_{\boldsymbol{x}}^{\widetilde{\alpha}}(m) = \sum_{p \in \mathbb{Z}} R_{\boldsymbol{x}_a}^{\alpha - p f_s}(\tau) \Big|_{\tau = m T_s, \alpha = \widetilde{\alpha} f_s} \tag{1.174}$$

$$\widetilde{S}_{\boldsymbol{x}}^{\widetilde{\alpha}}(\nu) = \frac{1}{T_s} \sum_{p \in \mathbb{Z}} \sum_{q \in \mathbb{Z}} S_{\boldsymbol{x}_a}^{\alpha - p f_s}(f - q f_s) \Big|_{f = \nu f_s, \alpha = \widetilde{\alpha} f_s} \, . \tag{1.175}$$

where $\boldsymbol{x} \triangleq [x \; x^{(*)}]$. □

Assumption 1.3.4 *The continuous-time process* $x_a(t)$ *is strictly band-limited, i.e.,* $S^0_{x_a x_a^*}(f) = 0 \quad |f| > B$ *(Gardner et al. 2006, Section 3.8).* □

Lemma 1.3.5 Support of the (Conjugate) Cyclic Spectrum. *Under Assumption 1.3.4 (*$x_a(t)$ *strictly band-limited), it results that (Gardner et al. 2006, eqs. (3.100) and (3.111), Figure 1):*

$$\operatorname{supp}\left\{ S_{\boldsymbol{x}_a}^{\alpha}(f) \right\} \subseteq \{(\alpha, f) \in \mathbb{R} \times \mathbb{R} \, : \, |f| \leqslant B, \; |\alpha - f| \leqslant B\} \tag{1.176a}$$

$$\subseteq \{(\alpha, f) \in \mathbb{R} \times \mathbb{R} \, : \, |f| \leqslant B, \; |\alpha| \leqslant 2B\} \, . \tag{1.176b}$$

Therefore,

$$S_{\boldsymbol{x}_a}^{\alpha}(f) = 0 \quad |f| > B, \; |\alpha| > 2B. \tag{1.177}$$

Moreover, in (1.168) we have

$$\begin{aligned}
&\text{supp}\left\{S^{\alpha}_{x_a}(f)\delta(f_2 - {}^{(-)}(\alpha - f_1))\right\} \\
&\quad \subseteq \{(f_1, f_2) \in \mathbb{R} \times \mathbb{R} \,:\, |f_1| \leqslant B,\ |\alpha - f_1| \leqslant B,\ f_2 = {}^{(-)}(\alpha - f_1)\} && (1.178a)\\
&\quad = \{(f_1, f_2) \in \mathbb{R} \times \mathbb{R} \,:\, |f_1| \leqslant B,\ |f_2| \leqslant B,\ f_2 = {}^{(-)}(\alpha - f_1)\}. && (1.178b)
\end{aligned}$$

□

Consequently, under Assumption 1.3.4 ($x_a(t)$ strictly band-limited), and accounting for (1.176a), the support of the replica with $n_1 = n_2 = 0$ in the aliasing formula (1.170) is given by

$$\begin{aligned}
&\text{supp}\left\{S^{\alpha}_{x_a}(\nu_1 f_s)\,\delta(\nu_2 - {}^{(-)}(\alpha/f_s - \nu_1))\right\} \\
&\quad \subseteq \{(\nu_1, \nu_2) \in \mathbb{R} \times \mathbb{R} \,:\, |\nu_1 f_s| \leqslant B,\ |\alpha - \nu_1 f_s| \leqslant B, \\
&\qquad \nu_2 - {}^{(-)}(\alpha/f_s - \nu_1) = 0\} && (1.179a)\\
&\quad = \{(\nu_1, \nu_2) \in \mathbb{R} \times \mathbb{R} \,:\, |\nu_1| \leqslant B/f_s,\ |\nu_2| \leqslant B/f_s,\ \alpha = (\nu_1 + {}^{(-)}\nu_2) f_s\} && (1.179b)
\end{aligned}$$

Theorem 1.3.6 ACS Signals: Sampling Theorem for the Discrete-Time Loève Bifrequency Spectrum. *Under Assumption 1.3.4 and for $f_s \geqslant 4B$, it results in*

$$\text{E}\left\{X(\nu_1)\,X^{(*)}(\nu_2)\right\} = \frac{1}{T_s} \sum_{\alpha \in A_{x_a}} S^{\alpha}_{x_a}(\nu_1 f_s)\,\delta\,(\nu_2 - {}^{(-)}(\alpha/f_s - \nu_1))$$
$$|\nu_1| \leqslant \frac{1}{2},\ |\nu_2| \leqslant \frac{1}{2}. \tag{1.180}$$

Moreover,

$$\text{E}\left\{X(\nu_1)\,X^{(*)}(\nu_2)\right\} = 0 \ \text{ for } \ 1/4 \leqslant |\nu_1| \leqslant 1/2 \ \text{ and } \ 1/4 \leqslant |\nu_2| \leqslant 1/2. \tag{1.181}$$

Proof: Replicas in (1.170) are separated by 1 in both ν_1 and ν_2 variables. Thus, from (1.179b) it follows that $B/f_s \leqslant 1/2$, that is,

$$f_s \geqslant 2B \tag{1.182}$$

is a *sufficient condition such that replicas do not overlap*. Note that, however, (1.182) does not assure that the mappings $f_1 = \nu_1 f_s$ and $\alpha = \widetilde{\alpha} f_s$ in

$$\widetilde{S}^{\widetilde{\alpha}}_{x}(\nu_1) = \frac{1}{T_s}\, S^{\alpha}_{x_a}(f_1)\Big|_{f_1 = \nu_1 f_s, \alpha = \widetilde{\alpha} f_s} \tag{1.183}$$

holds $\forall \widetilde{\alpha} \in [-1/2, 1/2)$ and $\forall \nu_1 \in [-1/2, 1/2)$. For example, for $(*)$ present and $\alpha = -B$, equality (1.183) holds only for $\nu_1 \in [-1/2, 0]$ and not for $\nu_1 \in [0, 1/2]$. In fact, for $\nu_1 \in [0, 1/2]$ the density of the replica with $n_1 = 0, n_2 = 1$ should be present in the right-hand side of (1.183).

In contrast, condition $B/f_s \leqslant 1/4$, that is,

$$f_s \geqslant 4B \tag{1.184}$$

assures in (1.179b)

$$|\nu_1| \leqslant 1/4, \quad |\nu_2| \leqslant 1/4, \quad |\alpha/f_s| = |\nu_1 +_{(-)} \nu_2| \leqslant |\nu_1| + |\nu_2| \leqslant 1/2 \tag{1.185}$$

and, consequently, the mappings $f_1 = \nu_1 f_s$ and $\alpha = \widetilde{\alpha} f_s$ in (1.183) hold $\forall \widetilde{\alpha} \in [-1/2, 1/2)$ and $\forall \nu_1 \in [-1/2, 1/2)$. Moreover, (1.180) holds. Finally, note that the effect of sampling at two times the Nyquist rate (see (1.184)) leads also to (1.181). □

Theorem 1.3.7 Sampling Theorem for Cyclic Statistics of ACS Signals (Napolitano 1995), (Izzo and Napolitano 1996). *Under Assumption 1.3.4 and for $f_s \geqslant 4B$, the result is*

$$\widetilde{\widetilde{R}}{}^{\widetilde{\alpha}}_{\boldsymbol{x}}(m) = R^{\alpha}_{\boldsymbol{x}_a}(\tau)\big|_{\tau = mT_s, \alpha = \widetilde{\alpha} f_s} \qquad |\widetilde{\alpha}| \leqslant \frac{1}{2} \tag{1.186a}$$

$$R^{\alpha}_{\boldsymbol{x}_a}(\tau)\big|_{\tau = mT_s} = \begin{cases} \widetilde{\widetilde{R}}{}^{\widetilde{\alpha}}_{\boldsymbol{x}}(m)\big|_{\widetilde{\alpha} = \alpha/f_s} & |\alpha| \leqslant \frac{f_s}{2} \\ 0 & \text{otherwise} \end{cases} \tag{1.186b}$$

$$R^{\alpha}_{\boldsymbol{x}_a}(\tau) = \sum_{m \in \mathbb{Z}} \widetilde{\widetilde{R}}{}^{\widetilde{\alpha}}_{\boldsymbol{x}}(m) \operatorname{sinc}\left(\frac{\tau}{T_s} - m\right) \qquad \forall \tau \in \mathbb{R}, \ \alpha = \widetilde{\alpha} f_s \tag{1.187}$$

$$\widetilde{\widetilde{S}}{}^{\widetilde{\alpha}}_{\boldsymbol{x}}(\nu) = \frac{1}{T_s} S^{\alpha}_{\boldsymbol{x}_a}(f)\big|_{f = \nu f_s, \alpha = \widetilde{\alpha} f_s} \qquad |\widetilde{\alpha}| \leqslant \frac{1}{2}, \ |\nu| \leqslant \frac{1}{2} \tag{1.188a}$$

$$S^{\alpha}_{\boldsymbol{x}_a}(f) = \begin{cases} T_s \, \widetilde{\widetilde{S}}{}^{\widetilde{\alpha}}_{\boldsymbol{x}}(\nu)\big|_{\nu = f/f_s, \widetilde{\alpha} = \alpha/f_s} & |\alpha| \leqslant \frac{f_s}{2}, \ |f| \leqslant \frac{f_s}{2} \\ 0 & \text{otherwise} \end{cases} \tag{1.188b}$$

where $\boldsymbol{x}_a = [x_a \, x_a^{(*)}]$ and $\boldsymbol{x} = [x \, x^{(*)}]$. By Fourier transforming both sides in (1.187) we get the following expression for the cyclic spectra of a strictly band-limited signal

$$S^{\alpha}_{\boldsymbol{x}_a}(f) = \sum_{m \in \mathbb{Z}} \widetilde{\widetilde{R}}{}^{\widetilde{\alpha}}_{\boldsymbol{x}}(m) \, e^{-j2\pi f m T_s} \, T_s \operatorname{rect}(f T_s) \, , \qquad \alpha = \widetilde{\alpha} f_s \tag{1.189}$$

which is equivalent to (1.188b). □

1.3.10 Multirate Processing of Discrete-Time ACS Signals

Expansion and decimation are linear time-variant transformations that are not almost-periodically time-variant. Consequently, the results of Section 1.3.3 cannot be used. In Section 4.3.1, it is observed that expansion and decimation belong to the class of linear time-variant systems that can be classified as deterministic in the FOT probability sense. These systems transform input ACS signals into output ACS signals with different almost-cyclostationarity characteristics (Sathe and Vaidyanathan 1993), (Izzo and Napolitano 1998b), (Akkarakaran and Vaidyanathan 2000). In contrast, in Section 4.10, it is shown that cross-statistics of signals generated with different rates by the same ACS signal give rise to signals that are jointly SC.

Results on multirate processing of discrete-time ACS processes are obtained as special cases of results on multirate processing of spectrally correlated processes in Section 4.10.

1.3.11 Applications

The existence of consistent estimators for the cyclic statistics is one of the main motivations for many applications of cyclostationarity in several fields such as communications, circuits, systems and control, acoustics, mechanics, econometrics, biology, and astronomy (Gardner *et al.* 2006). Applications in mechanics and vibroacoustic signal analysis are discussed in (Antoni 2009) and references therein.

In communications and radar/sonar applications, cyclostationarity properties can be used to design signal selective detection and estimation algorithms. In fact, in the case of additive noise uncorrelated with the signal of interest (SOI), if the SOI does not share at least one cycle frequency, say α_0, with the disturbance, then the cyclic autocorrelation function of the SOI plus noise at α_0 is coincident with the cyclic autocorrelation function of the SOI alone. Therefore, detection or estimation algorithms operating at α_0 are potentially immune to the effects of noise and interference, provided that a sufficiently long observation interval is adopted to estimate the cyclic statistics (Gardner 1985, 1987d), (Gardner and Chen 1992), (Chen and Gardner 1992), (Gardner and Spooner 1992, 1993), (Flagiello *et al.* 2000). In (Gardner *et al.* 2006) and references therein, applications of cyclostationarity for interference tolerant channel identification and equalization, signal detection and classification, source separation, periodic autoregressive (AR) and autoregressive moving-average (ARMA) modeling and prediction, are described. The spectral line regeneration property of ACS signal can be exploited for synchronization purposes (Gardner 1987d).

Cyclostationarity properties can be suitably exploited in minimum mean-square error (MMSE) linear filtering or Wiener filtering. If the useful signal, the data, and the disturbance signal are singularly and jointly ACS, then it can be shown that the optimum linear filter is linear almost-periodically time variant. By exploiting the spectral correlation property of ACS signals, the optimum filter cancels the spectral bands of the useful signal which are corrupted by noise and then reconstructs these bands exploiting the spectral components of the useful signal that are noise-free and correlated with the canceled bands (Gardner 1993). This MMSE filtering procedure is referred to as *cyclic Wiener filtering* or *frequency shift (FRESH) filtering*. It provides significant performance improvement with respect to the classical Wiener filtering that assumes a stationary model for the involved signals.

Second- and higher-order cyclostationarity properties can be exploited for modulation format classification. The classification is performed by comparing estimated second- and higher-order cyclic features of the received signal with those stored in a catalog (Spooner and Gardner 1994), (Spooner and Nicholls 2009).

1.4 Some Properties of Cumulants

In this section, some properties of cumulants that will be used in the subsequent chapters are reviewed. For comprehensive treatment of higher-order statistics of random variables and stationary and nonstationary signals see (Brillinger 1965, 1981), (Rosenblatt 1974), (Gardner and Spooner 1994), (Spooner and Gardner 1994), (Dandawaté and Giannakis 1994, 1995),

(Boashash *et al.* 1995), (Napolitano 1995), (Izzo and Napolitano 1998a), (Napolitano and Tesauro 2011).

1.4.1 Cumulants and Statistical Independence

Theorem 1.4.1 *If two real-valued random-variable vectors* $\boldsymbol{X}_1 \triangleq [X_1, \ldots, X_h]^{\mathrm{T}}$ *and* $\boldsymbol{X}_2 \triangleq [X_{h+1}, \ldots, X_k]^{\mathrm{T}}$ *are statistically independent, then*

$$\operatorname{cum}\{X_1, \ldots, X_k\} = 0. \tag{1.190}$$

Proof: Let $\boldsymbol{X} \triangleq [\boldsymbol{X}_1^{\mathrm{T}}\, \boldsymbol{X}_2^{\mathrm{T}}]^{\mathrm{T}}$, $\boldsymbol{\omega}_1 \triangleq [\omega_1, \ldots, \omega_h]^{\mathrm{T}}$, $\boldsymbol{\omega}_2 \triangleq [\omega_{h+1}, \ldots, \omega_k]^{\mathrm{T}}$, and $\boldsymbol{\omega} \triangleq [\boldsymbol{\omega}_1^{\mathrm{T}}\, \boldsymbol{\omega}_2^{\mathrm{T}}]^{\mathrm{T}}$. The *characteristic function* of the random vector $\boldsymbol{X}$ factorizes into the product of the characteristic functions of $\boldsymbol{X}_1$ and $\boldsymbol{X}_2$. That is,

$$\begin{aligned} \mathrm{E}\left\{e^{j\boldsymbol{\omega}^{\mathrm{T}}\boldsymbol{X}}\right\} &= \mathrm{E}\left\{e^{j\boldsymbol{\omega}_1^{\mathrm{T}}\boldsymbol{X}_1}\, e^{j\boldsymbol{\omega}_2^{\mathrm{T}}\boldsymbol{X}_2}\right\} \\ &= \mathrm{E}\left\{e^{j\boldsymbol{\omega}_1^{\mathrm{T}}\boldsymbol{X}_1}\right\} \mathrm{E}\left\{e^{j\boldsymbol{\omega}_2^{\mathrm{T}}\boldsymbol{X}_2}\right\} \end{aligned} \tag{1.191}$$

where the second equality is consequence of the statistical independence of the random vectors $\boldsymbol{X}_1$ and $\boldsymbol{X}_2$. Thus,

$$\begin{aligned} \operatorname{cum}\{X_1, \ldots, X_k\} &\triangleq (-j)^k \frac{\partial^k}{\partial\omega_1 \cdots \partial\omega_k} \log_e \mathrm{E}\left\{e^{j\boldsymbol{\omega}^{\mathrm{T}}\boldsymbol{X}}\right\}\Big|_{\boldsymbol{\omega}=\boldsymbol{0}} \\ &= (-j)^k \frac{\partial^k}{\partial\omega_1 \cdots \partial\omega_k} \left[\log_e \mathrm{E}\left\{e^{j\boldsymbol{\omega}_1^{\mathrm{T}}\boldsymbol{X}_1}\right\} + \log_e \mathrm{E}\left\{e^{j\boldsymbol{\omega}_2^{\mathrm{T}}\boldsymbol{X}_2}\right\}\right]_{\boldsymbol{\omega}=\boldsymbol{0}} \\ &= 0 \end{aligned} \tag{1.192}$$

where the kth-order derivative of each of the two terms in the square brackets is zero since the first one depends only on $\boldsymbol{\omega}_1 \triangleq [\omega_1, \ldots, \omega_h]^{\mathrm{T}}$, and the second one depends only on $\boldsymbol{\omega}_2 \triangleq [\omega_{h+1}, \ldots, \omega_k]^{\mathrm{T}}$.

This result can be extended with minor changes to the case of complex-valued random-variable vectors. □

As a consequence of Theorem 1.4.1, we have that if the stochastic process is asymptotically independent, that is, for every t the random variables $x(t)$ and $x(t+\tau)$ are asymptotically ($|\tau| \to \infty$) independent, then

$$\operatorname{cum}\{x(t),\ x(t+\tau_i), i = 1, \ldots, k-1\} \to 0 \quad \text{as } \|\boldsymbol{\tau}\| \to \infty \tag{1.193}$$

where $\|\boldsymbol{\tau}\|^2 \triangleq \tau_1^2 + \cdots + \tau_{k-1}^2$. In fact, if $\|\boldsymbol{\tau}\| \to \infty$, then at least one $\tau_i \to \infty$ and $x(t+\tau_i)$ becomes statistically independent of $x(t)$.

1.4.2 Cumulants of Complex Random Variables and Joint Complex Normality

In this section, the non-obvious result is proved that the Nth-order cumulant for complex random variables defined in (Spooner and Gardner 1994, App. A) (see also (1.209)) is zero for $N \geqslant 3$ when the random variables are jointly complex Normal (Napolitano 2007a, App. E).

Let $\boldsymbol{V} = [V_1, \ldots, V_N, V_{N+1}, \ldots, V_{2N}]^{\mathrm{T}} \triangleq [\boldsymbol{X}^{\mathrm{T}}, \boldsymbol{Y}^{\mathrm{T}}]^{\mathrm{T}}$ be the $2N$-dimensional column vector of real-valued random variables obtained by the N-dimensional column vectors $\boldsymbol{X} \triangleq [X_1, \ldots, X_N]^{\mathrm{T}}$, and $\boldsymbol{Y} \triangleq [Y_1, \ldots, Y_N]^{\mathrm{T}}$ of real-valued random variables. It is characterized by the $2N$th-order joint probability density function (pdf)

$$f_{\boldsymbol{V}}(\boldsymbol{v}) = f_{V_1 \cdots V_{2N}}(v_1, \ldots, v_{2N}) = f_{X_1 \ldots X_N Y_1 \ldots Y_N}(x_1, \ldots, x_N, y_1, \ldots, y_N). \tag{1.194}$$

Its *moment generating function* is the $2N$-dimensional Laplace transform

$$\Phi_{\boldsymbol{V}}(\boldsymbol{s}) \triangleq \mathrm{E}\{e^{\boldsymbol{s}^{\mathrm{T}}\boldsymbol{V}}\} = \int_{\mathbb{R}^{2N}} f_{\boldsymbol{V}}(\boldsymbol{v})\, e^{\boldsymbol{s}^{\mathrm{T}}\boldsymbol{v}}\, \mathrm{d}\boldsymbol{v} \tag{1.195}$$

with $\boldsymbol{s} \triangleq [\boldsymbol{s}_X^{\mathrm{T}}, \boldsymbol{s}_Y^{\mathrm{T}}]^{\mathrm{T}} \in \mathbb{C}^{2N}$, which is analytic in the region of convergence of the integral. The mean vector of $\boldsymbol{V}$ is

$$\boldsymbol{\mu}_V \triangleq \mathrm{E}\{\boldsymbol{V}\} = [\boldsymbol{\mu}_X^{\mathrm{T}}, \boldsymbol{\mu}_Y^{\mathrm{T}}]^{\mathrm{T}} \tag{1.196}$$

where $\boldsymbol{\mu}_X$ and $\boldsymbol{\mu}_Y$ are the mean vectors of $\boldsymbol{X}$ and $\boldsymbol{Y}$, respectively. The covariance matrix is

$$\boldsymbol{C}_{VV} \triangleq \mathrm{E}\{(\boldsymbol{V} - \mathrm{E}\{\boldsymbol{V}\})(\boldsymbol{V} - \mathrm{E}\{\boldsymbol{V}\})^{\mathrm{T}}\} = \begin{bmatrix} \boldsymbol{C}_{XX} & \boldsymbol{C}_{XY} \\ \boldsymbol{C}_{YX} & \boldsymbol{C}_{YY} \end{bmatrix} \tag{1.197}$$

where $\boldsymbol{C}_{XX}$ and $\boldsymbol{C}_{YY}$ are the covariance matrices of $\boldsymbol{X}$ and $\boldsymbol{Y}$, respectively, and

$$\boldsymbol{C}_{XY} \triangleq \mathrm{E}\{(\boldsymbol{X} - \mathrm{E}\{\boldsymbol{X}\})(\boldsymbol{Y} - \mathrm{E}\{\boldsymbol{Y}\})^{\mathrm{T}}\} = \boldsymbol{C}_{YX}^{\mathrm{T}} \tag{1.198}$$

is their cross-covariance matrix.

The $2N$th-order cumulant of the real random variables $V_1, \ldots, V_{2N}$ is defined as

$$\operatorname{cum}\{V_1, \ldots, V_{2N}\} \triangleq (-j)^{2N} \frac{\partial^{2N}}{\partial\omega_{X1} \cdots \partial\omega_{XN}\partial\omega_{Y1} \cdots \partial\omega_{YN}} \log_e \mathrm{E}\left\{\exp[j(\boldsymbol{\omega}_X^{\mathrm{T}}\boldsymbol{X} + \boldsymbol{\omega}_Y^{\mathrm{T}}\boldsymbol{Y})]\right\}\Big|_{\boldsymbol{\omega}_X=\boldsymbol{\omega}_Y=\boldsymbol{0}} \tag{1.199}$$

where $\boldsymbol{\omega}_X \triangleq [\omega_{X1}, \ldots, \omega_{XN}]^{\mathrm{T}}$ and $\boldsymbol{\omega}_Y \triangleq [\omega_{Y1}, \ldots, \omega_{YN}]^{\mathrm{T}}$.

Let us consider, now, the N-dimensional complex-valued column vector $\boldsymbol{Z} = [Z_1, \ldots, Z_N]^{\mathrm{T}} \triangleq \boldsymbol{X} + j\boldsymbol{Y}$. It is characterized by the same $2N$th-order joint pdf $f_{\boldsymbol{V}}(\boldsymbol{v})$ of the $2N$-dimensional real-valued vector $\boldsymbol{V}$ defined above.

Instead of considering the vector $\boldsymbol{V}$, the following complex augmented-dimension vector

$$\boldsymbol{\zeta} \triangleq \begin{bmatrix} \boldsymbol{Z} \\ \boldsymbol{Z}^* \end{bmatrix} \tag{1.200}$$

can be considered. Its mean vector and covariance matrix are given by

$$\boldsymbol{\mu}_{\zeta} \triangleq \begin{bmatrix} \mathrm{E}\{\boldsymbol{Z}\} \\ \mathrm{E}\{\boldsymbol{Z}^*\} \end{bmatrix} \tag{1.201}$$

$$\boldsymbol{\Gamma} \triangleq \mathrm{E}\left\{(\boldsymbol{\zeta} - \boldsymbol{\mu}_{\zeta})(\boldsymbol{\zeta} - \boldsymbol{\mu}_{\zeta})^{\mathrm{H}}\right\} = \begin{bmatrix} \boldsymbol{C}_{ZZ^*} & \boldsymbol{C}_{ZZ} \\ \boldsymbol{C}_{ZZ}^* & \boldsymbol{C}_{ZZ^*}^* \end{bmatrix} \tag{1.202}$$

where

$$\begin{aligned} \boldsymbol{C}_{ZZ^*} &\triangleq \mathrm{cov}\,\{\boldsymbol{Z}, \boldsymbol{Z}\} = \mathrm{E}\left\{(\boldsymbol{Z} - \mathrm{E}\{\boldsymbol{Z}\})(\boldsymbol{Z} - \mathrm{E}\{\boldsymbol{Z}\})^{\mathrm{H}}\right\} \\ &= \boldsymbol{C}_{XX} + \boldsymbol{C}_{YY} - j\boldsymbol{C}_{XY} + j\boldsymbol{C}_{YX} \end{aligned} \tag{1.203}$$

$$\begin{aligned} \boldsymbol{C}_{ZZ} &\triangleq \mathrm{cov}\left\{\boldsymbol{Z}, \boldsymbol{Z}^{\mathrm{H}}\right\} = \mathrm{E}\left\{(\boldsymbol{Z} - \mathrm{E}\{\boldsymbol{Z}\})(\boldsymbol{Z} - \mathrm{E}\{\boldsymbol{Z}\})^{\mathrm{T}}\right\} \\ &= \boldsymbol{C}_{XX} - \boldsymbol{C}_{YY} + j\boldsymbol{C}_{YX} + j\boldsymbol{C}_{XY} \end{aligned} \tag{1.204}$$

$$\boldsymbol{C}_{XX} = \frac{1}{2}\mathrm{Re}\,\{\boldsymbol{C}_{ZZ^*} + \boldsymbol{C}_{ZZ}\} \tag{1.205}$$

$$\boldsymbol{C}_{YY} = \frac{1}{2}\mathrm{Re}\,\{\boldsymbol{C}_{ZZ^*} - \boldsymbol{C}_{ZZ}\} \tag{1.206}$$

$$\boldsymbol{C}_{XY} = -\frac{1}{2}\mathrm{Im}\,\{\boldsymbol{C}_{ZZ^*} - \boldsymbol{C}_{ZZ}\} \tag{1.207}$$

$$\boldsymbol{C}_{YX} = \frac{1}{2}\mathrm{Im}\,\{\boldsymbol{C}_{ZZ^*} + \boldsymbol{C}_{ZZ}\} \tag{1.208}$$

with $\boldsymbol{C}_{ZZ^*}$ and $\boldsymbol{C}_{ZZ}$ referred to as *covariance matrix* and *conjugate covariance matrix*, respectively.

The Nth-order cumulant of the complex random variables $Z_1, \ldots, Z_N$ can be defined as (Spooner and Gardner 1994, App. A)

$$\begin{aligned} \mathrm{cum}\{Z_1, \ldots, Z_N\} &\triangleq (-j)^N \frac{\partial^N}{\partial\omega_1 \cdots \partial\omega_N} \log_e \mathrm{E}\left\{\exp[j\boldsymbol{\omega}^{\mathrm{T}}(\boldsymbol{X} + j\boldsymbol{Y})]\right\}\Big|_{\boldsymbol{\omega}=\boldsymbol{0}} \\ &= \sum_{\mathrm{P}} (-1)^{p-1}(p-1)! \prod_{i=1}^{p} \mathrm{E}\left\{\prod_{\ell\in\mu_i} Z_\ell\right\} \end{aligned} \tag{1.209}$$

where $\boldsymbol{\omega} \triangleq [\omega_1, \ldots, \omega_N]^{\mathrm{T}}$ and P are the set of distinct partitions of $\{1, \ldots, N\}$ each constituted by the subsets $\{\mu_i,\ i = 1, \ldots, p\}$. Note that since each complex variable Z_k is arbitrary, it can also be the complex conjugate of another complex variable. Such a definition turns out to be useful when applied to complex-valued stochastic processes or time series as shown in (Spooner and Gardner 1994, App. A), (Izzo and Napolitano 1998a), (Izzo and Napolitano 2002a). In particular, it is useful since it preserves the same relationship between moments and cumulants of real random variables (Leonov and Shiryaev 1959), (Brillinger 1965, 1981), (Brillinger and Rosenblatt 1967), as such a relationship is purely algebric. In addition, the characteristic function $\mathrm{E}\left\{\exp[j\boldsymbol{\omega}^{\mathrm{T}}(\boldsymbol{X} + j\boldsymbol{Y})]\right\}$ in (1.209) is an analytic function of the complex variables $X_k + jY_k$, $k = 1, \ldots, N$. In contrast, the characteristic function $\mathrm{E}\left\{\exp[j(\boldsymbol{\omega}_X^{\mathrm{T}}\boldsymbol{X} + \boldsymbol{\omega}_Y^{\mathrm{T}}\boldsymbol{Y})]\right\}$ in (1.199) depends separately on X_k and Y_k and hence, in general, is not an analytic function of the complex variables $X_k + jY_k$. Finally, note that, for deterministic liner time-variant systems, input/output relationships in terms of cumulants defined as in (1.209) have the same form as that in terms of moments (Napolitano 1995), (Izzo and Napolitano 2002a).

Let $\boldsymbol{Z}$ be a N-dimensional column vector of jointly complex Normal random variables. That is, $\boldsymbol{V}$ is a $2N$-dimensional column vector of jointly Normal real-valued random variables:

$$f_V(\boldsymbol{v}) = \frac{1}{(2\pi)^N |\det \boldsymbol{C}_{VV}|^{1/2}} \exp\left[-\frac{1}{2}(\boldsymbol{v}-\boldsymbol{\mu}_V)^{\mathrm{T}} \boldsymbol{C}_{VV}^{-1} (\boldsymbol{v}-\boldsymbol{\mu}_V)\right] \tag{1.210}$$

that can also be written in the complex form (Picinbono 1996), (van den Bos 1995)

$$f_{Z,Z^*}(z, z^*) = \frac{1}{\pi^N |\det \boldsymbol{\Gamma}|^{1/2}} \exp\left[-\frac{1}{2}(\boldsymbol{\zeta}_1-\boldsymbol{\mu}_\zeta)^{\mathrm{H}} \boldsymbol{\Gamma}^{-1} (\boldsymbol{\zeta}_1-\boldsymbol{\mu}_\zeta)\right] \tag{1.211}$$

where $\boldsymbol{\zeta}_1 \triangleq [z^{\mathrm{T}}, z^{\mathrm{H}}]^{\mathrm{T}}$, which is denoted as $\mathcal{N}(\boldsymbol{\mu}_\zeta, \boldsymbol{C}_{ZZ^*}, \boldsymbol{C}_{ZZ})$.

The moment-generating function (1.195) of $f_V(\boldsymbol{v})$ is the $2N$-dimensional Laplace transform

$$\Phi_V(\boldsymbol{s}) = \exp[\boldsymbol{s}^{\mathrm{T}}\boldsymbol{\mu}_V] \exp\left[\frac{1}{2}\boldsymbol{s}^{\mathrm{T}} \boldsymbol{C}_{VV} \boldsymbol{s}\right] \tag{1.212}$$

whose region of convergence is the whole complex space $\mathbb{C}^{2N}$. Therefore, both characteristic functions involved in the cumulant definitions (1.199) and (1.209) can be expressed as slices of the moment-generating function $\Phi_V(\boldsymbol{s})$ given in (1.212). Specifically, in (1.199) we have

$$\begin{aligned}
&\mathrm{E}\left\{\exp[j(\boldsymbol{\omega}_X^{\mathrm{T}}\boldsymbol{X} + \boldsymbol{\omega}_Y^{\mathrm{T}}\boldsymbol{Y})]\right\} \\
&\quad = \Phi_V(\boldsymbol{s})|_{\boldsymbol{s}_X = j\boldsymbol{\omega}_X, \boldsymbol{s}_Y = j\boldsymbol{\omega}_Y} \\
&\quad = \exp[j(\boldsymbol{\omega}_X^{\mathrm{T}}\boldsymbol{\mu}_X + \boldsymbol{\omega}_Y^{\mathrm{T}}\boldsymbol{\mu}_Y)] \\
&\quad\quad \exp\left[-\frac{1}{2}\left(\boldsymbol{\omega}_X^{\mathrm{T}}\boldsymbol{C}_{XX}\boldsymbol{\omega}_X + \boldsymbol{\omega}_X^{\mathrm{T}}\boldsymbol{C}_{XY}\boldsymbol{\omega}_Y + \boldsymbol{\omega}_Y^{\mathrm{T}}\boldsymbol{C}_{YX}\boldsymbol{\omega}_X + \boldsymbol{\omega}_Y^{\mathrm{T}}\boldsymbol{C}_{YY}\boldsymbol{\omega}_Y\right)\right]
\end{aligned} \tag{1.213}$$

from which it follows that $\log_e \mathrm{E}\left\{\exp[j(\boldsymbol{\omega}_X^{\mathrm{T}}\boldsymbol{X} + \boldsymbol{\omega}_Y^{\mathrm{T}}\boldsymbol{Y})]\right\}$ is a quadratic homogeneous polynomial in the real variables $\omega_{X1}, \ldots, \omega_{XN}, \omega_{Y1}, \ldots, \omega_{YN}$. As a consequence, we have the well-known result that jointly real-valued Normal random variables have kth-order cumulants of order $k \geqslant 3$ equal to zero. In addition, in (1.209) we have

$$\begin{aligned}
&\mathrm{E}\left\{\exp[j\boldsymbol{\omega}^{\mathrm{T}}(\boldsymbol{X} + j\boldsymbol{Y})]\right\} \\
&\quad = \Phi_V(\boldsymbol{s})|_{\boldsymbol{s}_X = j\boldsymbol{\omega}, \boldsymbol{s}_Y = -\boldsymbol{\omega}} \\
&\quad = \exp[j\boldsymbol{\omega}^{\mathrm{T}}\boldsymbol{\mu}_X - \boldsymbol{\omega}^{\mathrm{T}}\boldsymbol{\mu}_Y] \\
&\quad\quad \exp\left[-\frac{1}{2}\boldsymbol{\omega}^{\mathrm{T}}(\boldsymbol{C}_{XX} - \boldsymbol{C}_{YY})\boldsymbol{\omega} - j\frac{1}{2}\boldsymbol{\omega}^{\mathrm{T}}(\boldsymbol{C}_{XY} + \boldsymbol{C}_{YX})\boldsymbol{\omega}\right].
\end{aligned} \tag{1.214}$$

Hence, $\log_e \mathrm{E}\left\{\exp[j\boldsymbol{\omega}^{\mathrm{T}}(\boldsymbol{X} + j\boldsymbol{Y})]\right\}$ is a quadratic homogeneous polynomial in the real variables ω_i. Therefore, jointly *complex* Normal random variables are characterized by the fact that their cumulants of order $N \geqslant 3$ defined as in (1.209) are equal to zero.

2

Generalized Almost-Cyclostationary Processes

In this chapter, the statistical characterization of generalized almost-cyclostationary (GACS) processes is presented. Then the problem of estimating second-order cross-moments of GACS processes is addressed. GACS processes have statistical functions that are almost-periodic functions of time whose (generalized) Fourier series expansions have both frequencies and coefficients that depend on the lag shifts of the processes. The class of such nonstationary processes includes the almost-cyclostationary (ACS) processes which are obtained as a special case when the frequencies do not depend on the lag shifts. ACS processes filtered by Doppler channels and communications signals with time-varying parameters are further examples. The second-order cross-moment of two jointly GACS processes is shown to be completely characterized by the cyclic cross-correlation function. Moreover, the cyclic cross-correlogram is proved to be a mean-square consistent, asymptotically Normal estimator, of the cyclic cross-correlation function. It is shown that continuous-time GACS processes do not have a discrete-time counterpart. The discrete-time cyclic cross-correlogram of the discrete-time ACS process obtained by uniformly sampling a GACS process is considered as an estimator of samples of the continuous-time cyclic cross-correlation function. The asymptotic performance analysis is carried out by resorting to the hybrid cyclic cross-correlogram which is partially continuous-time and partially discrete-time. Its mean-square consistency and asymptotic complex Normality as the number of data-samples approaches infinity and the sampling period approaches zero are proved under mild conditions on the regularity of the Fourier series coefficients and the finite or practically finite memory of the processes expressed in terms of summability of cumulants. The asymptotic properties of the hybrid cyclic cross-correlogram are shown to be coincident with those of the continuous-time cyclic cross-correlogram. Hence, discrete-time estimation does not give rise to any loss in asymptotic performance with respect to continuous-time estimation. Well-known consistency and asymptotic Normality results for ACS processes are obtained by specializing the results for GACS processes.

Generalizations of Cyclostationary Signal Processing: Spectral Analysis and Applications, First Edition.

2.1 Introduction

In the past two decades, a huge effort has been devoted to analysis and exploitation of the properties of the almost-cyclostationary (ACS) processes. In fact, almost-all modulated signals adopted in communications can be modeled as ACS (Gardner 1994), (Gardner and Spooner 1994), (Spooner and Gardner 1994), (Gardner *et al.* 2006). For an ACS process, multivariate statistical functions are almost-periodic functions of time and can be expressed by (generalized) Fourier series expansions whose coefficients depend on the lag shifts of the processes and whose frequencies, referred to as cycle frequencies, do not depend on the lag shifts. Almost-cyclostationarity properties have been widely exploited for analysis and synthesis of communications systems. In particular, they have been exploited to develop signal selective detection and parameter estimation algorithms, blind-channel-identification and synchronization techniques, and so on (Gardner 1994), (Gardner *et al.* 2006). Moreover, ACS processes are encountered in econometrics, climatology, hydrology, biology, acoustics, and mechanics (Gardner 1994), (Gardner *et al.* 2006).

More recently, wider classes of nonstationary processes extending the class of the ACS processes have been considered in (Izzo and Napolitano 1998a), (Izzo and Napolitano 2002a,b, 2003, 2005), (Napolitano 2003, 2007a, 2009), (Napolitano and Tesauro 2011).

In (Izzo and Napolitano 1998a), the class of the generalized almost-cyclostationary (GACS) processes was introduced and characterized. Processes belonging to this class exhibit multivariate statistical functions that are almost-periodic functions of time whose (generalized) Fourier series expansions have coefficients and frequencies, referred to as lag-dependent cycle frequencies, that can depend on the lag shifts of the processes. The class of the GACS processes includes the class of the ACS processes as a special case, when the cycle frequencies do not depend on the lag shifts. Moreover, chirp signals and several angle-modulated and time-warped communication signals are GACS processes. In (Izzo and Napolitano 1998a), the higher-order characterization in the strict- and wide-sense of GACS signals is provided. Generalized cyclic moments and cumulants are introduced in both time and frequency domains. As examples of GACS signals, the chirp signal and a nonuniformly sampled signal are considered and their generalized cyclic statistics are derived. In (Izzo and Napolitano 2002a) and (Izzo and Napolitano 2002b), it is shown that several time-variant channels of interest in communications transform a transmitted ACS signal into a GACS one. In particular, in (Izzo and Napolitano 2002b) it is shown that the GACS model can be appropriate to describe the output signal of Doppler channels when the input signal is ACS and the channel introduces a quadratically time-variant delay. Therefore, the GACS model is appropriate to describe the received signal in presence of relative motion between transmitter and receiver with nonzero relative radial acceleration (Kelly and Wishner 1965), (Rihaczek 1967). In (Izzo and Napolitano 1998a), (Izzo and Napolitano 2002a), and (Izzo and Napolitano 2002b), it is also shown that communications signals with slowly time-varying parameters, such as carrier frequency or baud rate, should be modeled as GACS, rather than ACS, if the data-record length is such that the parameter time variations can be appreciated. In (Izzo and Napolitano 2003), the problem of sampling a GACS signal is addressed. It is shown that the discrete-time signal constituted by the samples of a continuous-time GACS signal is a discrete-time ACS signal. Moreover, it is shown that starting from the sampled signal, the ACS or GACS nature of the continuous-time signal can only be conjectured, provided that analysis parameters such as sampling period, padding factor, and data-record length are properly chosen. In (Izzo and Napolitano 1998a,

2002a,b, 2003), the signal analysis is carried out in the fraction-of-time probability or nonstochastic approach, where probabilistic parameters are defined through infinite-time averages of functions of a single time-series rather than through expected values (ensemble averages) of a stochastic process (Chapter 6), (Gardner 1987d), (Leśkow and Napolitano 2006). A survey of the GACS signals in the nonstochastic approach is provided in (Izzo and Napolitano 2005). In (Hanin and Schreiber 1998), the problem of consistently estimating the time-averaged autocorrelation function for a class of processes with periodically time-variant autocorrelation function is addressed, and it is shown that some results are still valid if the period depends on the lag parameter. This class of processes, therefore, is a subclass of the GACS processes.

The design of signal processing algorithms can require the estimation of second-order statistics. Statistical functions of GACS signals need to be estimated, for example, in channel identification and/or equalization problems if the channel model is linear time-variant but not almost-periodically time variant (e.g., when the channel introduces quadratically time-varying delays) (Izzo and Napolitano 2002a,b).

In this chapter, the problem of estimating second-order cross-moments of complex-valued jointly GACS processes is addressed. Estimators of autocorrelation function and conjugate autocorrelation function of a single process are obtained as special cases. The second-order cross-moment of jointly GACS processes can be expressed by a Fourier series expansion with lag-dependent cycle frequencies and lag-dependent coefficients referred to as generalized cyclic cross-correlation functions. Equivalently, it can be expressed by a (generalized) Fourier series expansion with constant cycle frequencies ranging in a countable set depending on the lag shift and lag-dependent coefficients referred to as cyclic cross-correlation functions. Thus, the second-order cross-moment is completely characterized by the cyclic cross-correlation function as a function of the two variables lag shift and cycle frequency. Such a function has support contained in a countable set of curves in the lag-shift cycle-frequency plane which are described by the lag-dependent cycle frequencies. For ACS processes these curves reduce to lines parallel to the lag-shift axis.

The cyclic cross-correlogram is proposed in (Napolitano 2007a) as an estimator of the cyclic cross-correlation function of jointly GACS processes. It is shown that, for GACS stochastic processes satisfying some mixing conditions expressed in terms of summability of cumulants, the cyclic cross-correlogram, as a function of the two variables lag shift and cycle frequency, is a mean-square consistent and asymptotically complex Normal estimator of the cyclic cross-correlation function. Furthermore, in the limit as the data-record length approaches infinity, the region of the cycle-frequency lag-shift plane where the cyclic cross-correlogram is significantly different from zero becomes a thin strip around the support curves of the cyclic cross-correlation function, that is, around the lag-dependent cycle frequency curves. Thus, the proved asymptotic complex Normality result can be used to establish statistical tests for presence of generalized almost-cyclostationarity.

The discrete-time cyclic cross-correlogram of the discrete-time jointly ACS processes obtained by uniformly sampling two continuous-time jointly GACS processes is considered in (Napolitano 2009) as an estimator for the continuous-time cyclic cross-correlation function. It is shown that for GACS processes no simple condition on the sampling frequency can be stated as for band-limited wide-sense stationary or ACS processes in order to avoid or limit aliasing. However, the discrete-time cyclic cross-correlogram is shown to be a mean-square consistent and asymptotically Normal estimator of the continuous-time cyclic cross-correlation function as the data-record length approaches infinity and the sampling period approaches zero, provided that some mild regularity conditions are satisfied. The well-known result for ACS processes that the cyclic correlogram is a mean-square consistent and asymptotically Normal estimator of the cyclic autocorrelation function (Hurd 1989a, 1991), (Hurd and Leśkow 1992b), (Dehay and Hurd 1994), (Genossar *et al.* 1994), (Dandawaté and Giannakis 1995) is obtained as a special case of the results established for GACS processes.

The discrete-time counterparts of the asymptotic results of (Napolitano 2007a) are not straightforward. In fact, in (Izzo and Napolitano 2003), it is shown that uniformly sampling a continuous-time GACS process gives rise to a discrete-time ACS process and a discrete-time counterpart of continuous-time GACS processes does not exist. Furthermore, the GACS or ACS nature of an underlying continuous-time process can only be conjectured starting from the analysis of the discrete-time ACS process. In addition, since GACS processes cannot be strictly band-limited (Izzo and Napolitano 1998a), (Napolitano 2007a), unlike the case of ACS or stationary processes, a minimum value of the sampling frequency to completely avoid aliasing in the discrete-time cyclic statistics does not exist. This constitutes a complication for the estimation of the cyclic autocorrelation function of a GACS process starting from the discrete-time process of its samples.

The asymptotic statistical analysis of the discrete-time cyclic correlogram of the ACS process obtained by uniformly sampling a continuous-time GACS process is carried out when the number of data samples approaches infinity (to get consistency) and the sampling period approaches zero (to counteract aliasing). It is shown that the discrete-time cyclic correlogram of the discrete-time process obtained by uniformly sampling a continuous-time GACS process is a mean-square consistent estimator of samples of the aliased continuous-time cyclic autocorrelation function of the GACS process as the number of data-samples approaches infinity. Moreover, it is pointed out that the discrete-time cyclic correlogram has the drawback that, when the sampling period approaches zero, the unnormalized lag parameter approaches zero and the unnormalized cycle frequency approaches infinity. It is shown that such a drawback is also present if the discrete-time cyclic correlogram is adopted to estimate the continuous-time cyclic autocorrelation function of ACS processes which are not strictly band-limited so that the sampling period should approach zero to reduce aliasing. Furthermore, the same problem is encountered with the correlogram estimate of the autocorrelation function of non strictly band-limited wide-sense stationary processes. This problem does not occur if continuous- and discrete-time estimation problems are treated separately, and the aliasing problem arising from sampling is not addressed (Brillinger and Rosenblatt 1967).

A procedure to carry out the asymptotic analysis of the discrete-time cyclic cross-correlogram as the number of data-samples approaches infinity and the sampling period approaches zero is proposed in (Napolitano 2009) by resorting to the hybrid cyclic correlogram. It is called "hybrid" since some parameters are continuous-time and others are discrete-time. Specifically, data-samples are discrete-time and lag parameter and cycle frequency are those of the continuous-time cyclic correlogram and are assumed to be constant with respect to the sampling period. It is shown that the hybrid cyclic correlogram is a mean-square consistent and asymptotically Normal estimator of the cyclic autocorrelation function when the number of samples approaches infinity and the sampling period approaches zero in such a way that the overall data-record length approaches infinity. Thus, it is shown that the mean-square error between the discrete-time cyclic correlogram and samples of the continuous-time cyclic autocorrelation function can be made arbitrarily small provided that the number of data samples is sufficiently large and the sampling period is sufficiently small. Moreover, it is shown that the asymptotic bias, covariance and distribution of the hybrid cyclic correlogram are coincident with those of the continuous-time cyclic correlogram. That is, the discrete-time analysis, asymptotically, does not introduce performance degradation with respect to continuous-time analysis. The proposed discrete-time asymptotic analysis can be applied also to the case of non strictly band-limited continuous-time ACS and stationary processes.

For continuous-time ACS processes continuous-time estimators are proposed in (Hurd 1989a, 1991), (Hurd and Leśkow 1992a,b), (Dehay 1994), and references therein, and for discrete-time ACS processes, discrete-time estimators are proposed in (Genossar *et al.* 1994), (Dandawaté and Giannakis 1995), (Schell 1995). Aliasing and estimation issues have not been previously considered together. The results in this chapter jointly treat the problem of reducing aliasing and getting consistency.

The chapter is organized as follows. In Section 2.2 the class of the GACS processes is characterized and motivations and examples are provided. In Section 2.3, the problem of linear time-variant filtering of GACS processes is addressed. In Section 2.4.1, the cyclic cross-correlogram is proposed as an estimator for the cyclic cross-correlation function of two jointly GACS processes. Moreover, its expected value and covariance are derived for finite data-record length. In Section 2.4.2, the mean-square consistency of the cyclic cross-correlogram is established and its asymptotic expected value and covariance are derived. In Section 2.4.3, the cyclic cross-correlogram is shown to be asymptotically complex Normal. In Section 2.5, the problem of sampling continuous-time GACS processes is treated. The discrete-time estimation of the cyclic cross-correlation function of continuous-time jointly GACS processes is addressed in Section 2.6. Specifically, in Section 2.6.1 the discrete-time cyclic cross-correlogram is defined and its mean and covariance for finite number N of data-samples and finite sampling period T_s are derived. Results of consistency and asymptotic complex Normality as $N \to \infty$ are derived in Section 2.6.2. Results for $N \to \infty$ and $T_s \to 0$ are derived in Section 2.6.3. Numerical results to corroborate the effectiveness of the theoretical results are reported in Section 2.7. A discussion on the stated results is made in Section 2.7.5. A Summary is given in Section 2.8. Proofs of the results presented in Sections 2.4.1, 2.4.2, 2.4.3, are reported in Sections 3.4, 3.5, and 3.6, in Chapter 3, respectively. Proofs of results of Sections 2.6.1, 2.6.2, and 2.6.3 are reported in Sections 3.9, 3.10, and 3.11, respectively. Some issues concerning complex processes are addressed in Sections 3.7 and 3.13.

2.2 Characterization of GACS Stochastic Processes

In this section, the strict-sense, second-order wide-sense, and higher-order characterizations of GACS stochastic processes are presented in the time domain. Moreover, the second-order spectral characterization is discussed. Examples of GACS processes and motivations to adopt such a model are provided. See (Izzo and Napolitano 1998a) and (Izzo and Napolitano 2002a) for a treatment in the nonstochastic or functional approach.

2.2.1 *Strict-Sense Statistical Characterization*

Definition 2.2.1 *The real-valued process* $\{x(t),\ t \in \mathbb{R}\}$, *is said to be* with almost-periodic structure in the strict sense *if its* N*th-order distribution function*

$$F_{x(t+\tau_1)\cdots x(t+\tau_{N-1})x(t)}(\xi_1, \ldots, \xi_{N-1}, \xi_N) \triangleq P\{x(t+\tau_1) \leqslant \xi_1, \ldots, x(t+\tau_{N-1}) \leqslant \xi_{N-1}, x(t) \leqslant \xi_N\} \tag{2.1}$$

is almost-periodic in t (in one of the senses considered in Section 1.2), for every fixed $\tau_1, \ldots, \tau_{N-1}$ *and* $\xi_1, \ldots, \xi_N$. □

Thus, for a process with almost-periodic structure in the strict sense, the Nth-order distribution function can be expressed by the (generalized) Fourier series

$$F_{x(t+\tau_1)\cdots x(t+\tau_{N-1})x(t)}(\xi_1, \ldots, \xi_{N-1}, \xi_N) = \sum_{\gamma \in \Gamma_{\boldsymbol{\tau},\boldsymbol{\xi}}} F_{\boldsymbol{x}}^{\gamma}(\boldsymbol{\xi}; \boldsymbol{\tau})\, e^{j2\pi\gamma t} \tag{2.2}$$

where $\Gamma_{\boldsymbol{\tau},\boldsymbol{\xi}}$ is a countable set depending on $\boldsymbol{\tau} \triangleq [\tau_1, \ldots, \tau_{N-1}]$ and $\boldsymbol{\xi} \triangleq [\xi_1, \ldots, \xi_N]$ and

$$F_{\boldsymbol{x}}^{\gamma}(\boldsymbol{\xi}; \boldsymbol{\tau}) \triangleq \lim_{T\to\infty} \frac{1}{T} \int_{-T/2}^{T/2} F_{x(t+\tau_1)\cdots x(t+\tau_{N-1})x(t)}(\xi_1, \ldots, \xi_{N-1}, \xi_N)\, e^{-j2\pi\gamma t}\, \mathrm{d}t \tag{2.3}$$

Definition 2.2.2 *The real-valued process* $\{x(t),\ t \in \mathbb{R}\}$, *is said to be* Nth-order generalized almost-cyclostationary in the strict sense *if for every* $\boldsymbol{\tau} \in \mathbb{R}^{N-1}$ *the set*

$$\Gamma_{\boldsymbol{\tau}} \triangleq \bigcup_{\boldsymbol{\xi} \in \mathbb{R}^N} \Gamma_{\boldsymbol{\tau},\boldsymbol{\xi}} \tag{2.4}$$

is countable (see (Izzo and Napolitano 2005) for the definition in the functional (or nonstochastic) approach). □

Definition 2.2.3 *The real-valued process* $\{x(t),\ t \in \mathbb{R}\}$, *is said to be* Nth-order almost-cyclostationary in the strict sense *if the set*

$$\Gamma \triangleq \bigcup_{\boldsymbol{\tau} \in \mathbb{R}^{N-1}} \Gamma_{\boldsymbol{\tau}} \tag{2.5}$$

is countable. □

Theorem 2.2.4 *For a* N*th-order generalized almost-cyclostationary process* ($\Gamma_{\boldsymbol{\tau}}$ *countable and* Γ *uncountable) the result is that*

$$F_{x(t+\tau_1)\cdots x(t+\tau_{N-1})x(t)}(\xi_1, \ldots, \xi_{N-1}, \xi_N) = \sum_{\gamma \in \Gamma_{\boldsymbol{\tau}}} F_{\boldsymbol{x}}^{\gamma}(\boldsymbol{\xi}; \boldsymbol{\tau})\, e^{j2\pi\gamma t} \tag{2.6a}$$

$$= \sum_{k \in \mathbb{K}} F_{\boldsymbol{x}}^{(k)}(\boldsymbol{\xi}; \boldsymbol{\tau})\, e^{j2\pi\gamma_{\boldsymbol{x}}^{(k)}(\boldsymbol{\xi};\boldsymbol{\tau})t} \tag{2.6b}$$

with $\mathbb{K}$ *countable and*

$$F_{\boldsymbol{x}}^{\gamma}(\boldsymbol{\xi}; \boldsymbol{\tau}) = \sum_{k \in \mathbb{K}} F_{\boldsymbol{x}}^{(k)}(\boldsymbol{\xi}; \boldsymbol{\tau})\, \delta_{\gamma - \gamma_{\boldsymbol{x}}^{(k)}(\boldsymbol{\xi};\boldsymbol{\tau})} \tag{2.7}$$

where δ_γ *denotes Kronecker delta, that is,* $\delta_\gamma = 1$ *for* $\gamma = 0$ *and* $\delta_\gamma = 0$ *for* $\gamma \neq 0$.

Proof: It is similar to the proof of Theorem 2.2.7. □

Theorem 2.2.5 *An Nth-order GACS process in the strict sense has the Nth-order temporal moment function which is an almost-periodic function of t with frequencies depending on* $\boldsymbol{\tau}$.

Proof: Accounting for (2.6a), the result is that

$$\begin{aligned}
&\mathrm{E}\{x(t+\tau_1)\cdots x(t+\tau_{N-1})x(t)\}\\
&\quad= \int_{\mathbb{R}^N} \xi_1\cdots\xi_N \,\mathrm{d}F_{x(t+\tau_1)\cdots x(t+\tau_{N-1})x(t)}(\xi_1,\ldots,\xi_{N-1},\xi_N)\\
&\quad= \sum_{\gamma\in\Gamma_{\boldsymbol{\tau}}} \int_{\mathbb{R}^N} \xi_1\cdots\xi_N \,\mathrm{d}_{\boldsymbol{\xi}} F_{\boldsymbol{x}}^{\gamma}(\boldsymbol{\xi};\boldsymbol{\tau})\, e^{j2\pi\gamma t}\\
&\quad= \sum_{\gamma\in\Gamma_{\boldsymbol{\tau}}} R_{\boldsymbol{x}}(\gamma,\boldsymbol{\tau})\, e^{j2\pi\gamma t}
\end{aligned} \tag{2.8}$$

where

$$\begin{aligned}
R_{\boldsymbol{x}}(\gamma,\boldsymbol{\tau}) &\triangleq \lim_{T\to\infty}\frac{1}{T}\int_{-T/2}^{T/2} \mathrm{E}\{x(t+\tau_1)\cdots x(t+\tau_{N-1})x(t)\}\, e^{-j2\pi\gamma t}\,\mathrm{d}t\\
&= \int_{\mathbb{R}^N} \xi_1\cdots\xi_N \,\mathrm{d}_{\boldsymbol{\xi}} F_{\boldsymbol{x}}^{\gamma}(\boldsymbol{\xi};\boldsymbol{\tau})
\end{aligned} \tag{2.9}$$

are the cyclic temporal moment functions.

Note that a similar result cannot be derived in the more general case of process with almost-periodic structure in the strict sense, that is, with the set $\Gamma_{\boldsymbol{\tau},\boldsymbol{\xi}}$ depending on both $\boldsymbol{\tau}$ and $\boldsymbol{\xi}$ and such that the set $\Gamma_{\boldsymbol{\tau}}$ defined in (2.4) is not countable. Digital signals $x(t)$ with values in a finite or countable set have $\Gamma_{\boldsymbol{\tau}}$ countable. □

2.2.2 *Second-Order Wide-Sense Statistical Characterization*

In this section, the second-order statistical characterization in the wide sense of GACS signals is provided. Emphasis is given to second-order (cross-) moment properties, while no emphasis is given to mean value properties.

Definition 2.2.6 *A finite-power complex-valued continuous-time stochastic process $x(t)$ is said to be* second-order GACS in the wide sense *if its mean value is an almost-periodic function and its autocorrelation function*

$$\mathcal{R}_{xx^*}(t,\tau) \triangleq \mathrm{E}\left\{x(t+\tau)\,x^*(t)\right\} \tag{2.10}$$

with $\mathrm{E}\{\cdot\}$ *denoting statistical expectation, for each $\tau\in\mathbb{R}$, is almost-periodic in t in the sense of Bohr (Bohr 1933, paragraphs 84–92) or, equivalently, uniformly almost periodic in sense of Besicovitch (Besicovitch 1932, Chapter 1), (Corduneanu 1989). That is, for each fixed τ, $\mathcal{R}_{xx^*}(t,\tau)$ is the limit of a uniformly convergent sequence of trigonometric polynomials in t*

$$\mathcal{R}_{xx^*}(t,\tau) = \sum_{\alpha\in A_\tau} R_{xx^*}(\alpha,\tau)\, e^{j2\pi\alpha t}. \tag{2.11}$$

In (2.11), the real numbers α and the complex-valued functions $R_{xx^*}(\alpha,\tau)$, referred to as cycle frequencies *and* cyclic autocorrelation functions, *are the frequencies and coefficients,* *respectively, of the (generalized) Fourier series expansion of* $\mathcal{R}_{xx^*}(t, \tau)$ *that is,*

$$R_{xx^*}(\alpha, \tau) \triangleq \lim_{T\to\infty} \frac{1}{T} \int_{t_0-T/2}^{t_0+T/2} \mathcal{R}_{xx^*}(t, \tau)\, e^{-j2\pi\alpha t}\, dt \tag{2.12}$$

with the limit independent of t_0. *Moreover,*

$$A_\tau \triangleq \{\alpha \in \mathbb{R} \, : \, R_{xx^*}(\alpha, \tau) \neq 0\} \tag{2.13}$$

is a countable set (of possibly incommensurate cycle frequencies) which, in general, depends on τ. □

Note that, even if the set A_τ is countable, the set

$$A \triangleq \bigcup_{\tau\in\mathbb{R}} A_\tau \tag{2.14}$$

is not necessarily countable. Thus, the class of the second-order wide-sense GACS processes extends that of the wide-sense ACS which are obtained as a special case of GACS processes when the set A is countable (Dehay and Hurd 1994), (Hanin and Schreiber 1998), (Hurd 1991).

A useful characterization of wide-sense GACS processes can be obtained by observing that the set A_τ can be expressed as

$$A_\tau = \bigcup_{n\in\mathbb{I}} \left\{\alpha \in \mathbb{R} \, : \, \alpha = \alpha_{xx^*}^{(n)}(\tau)\right\} \equiv \left\{\alpha_{xx^*}^{(n)}(\tau)\right\}_{n\in\mathbb{I}} \tag{2.15}$$

where $\mathbb{I}$ is a countable set and the functions $\alpha_{xx^*}^{(n)}(\tau)$, referred to as *lag-dependent cycle frequencies* and in the following denoted by $\alpha_n(\tau)$ for notation simplicity when this does not create ambiguity, are such that, for each α and τ, there exists at most one $n \in \mathbb{I}$ such that $\alpha = \alpha_n(\tau)$. Thus, accounting for the countability of A_τ for each τ, the support in the (α, τ)-plane of the cyclic autocorrelation function $R_{xx^*}(\alpha, \tau)$ is constituted by the closure of the set of curves defined by the explicit equations $\alpha = \alpha_n(\tau)$, $n \in \mathbb{I}$:

$$\begin{aligned} \operatorname{supp}\{R_{xx^*}(\alpha, \tau)\} &\triangleq \operatorname{cl}\{(\alpha, \tau) \in A_\tau \times \mathbb{R} \, : \, R_{xx^*}(\alpha, \tau) \neq 0\} \\ &= \operatorname{cl} \bigcup_{n\in\mathbb{I}} \left\{(\alpha, \tau) \in \mathbb{R} \times \mathcal{T}^{(n)} \, : \, \alpha = \alpha_n(\tau)\right\} \end{aligned} \tag{2.16}$$

where

$$\mathcal{T}^{(n)} \triangleq \{\tau \in \mathbb{R} \, : \, \alpha_n(\tau) \text{ is defined}\}\,. \tag{2.17}$$

The closure of a countable set of curves can be the whole plane $\mathbb{R}^2$ if infinities of clusters of curves exist. In such a case, the more appropriate concept of "being concentrated on" (Dehay 1994) can be used to describe the region of the (α, τ)-plane where $R_{xx^*}(\alpha, \tau) \neq 0$.

Starting from (2.16), the following result (Izzo and Napolitano 1998a, 2002a) provides an alternative representation for the autocorrelation function of GACS processes.

Theorem 2.2.7 *The autocorrelation function $\mathcal{R}_{xx^*}(t, \tau)$ of a second-order wide-sense GACS process can be expressed as*

$$\mathcal{R}_{xx^*}(t, \tau) = \sum_{n \in \mathbb{I}} R_{xx^*}^{(n)}(\tau)\, e^{j2\pi\alpha_n(\tau)t} \tag{2.18}$$

where the functions $R_{xx^*}^{(n)}(\tau)$, referred to as generalized cyclic autocorrelation functions, *are defined as*

$$R_{xx^*}^{(n)}(\tau) \triangleq \begin{cases} \lim_{T\to\infty} \dfrac{1}{T} \displaystyle\int_{t_0-T/2}^{t_0+T/2} \mathcal{R}_{xx^*}(t, \tau)\, e^{-j2\pi\alpha_n(\tau)t}\, \mathrm{d}t & \tau \in \mathcal{T}^{(n)} \\ 0 & \tau \in \mathbb{R} - \mathcal{T}^{(n)} \end{cases} \tag{2.19}$$

with the limit in (2.19) independent of t_0. □

Note that, in (2.18) the sum ranges over a set not depending on τ as, on the contrary, it occurs in (2.11). Moreover, unlike the case of second-order ACS processes, both coefficients and frequencies of the Fourier series in (2.18) depend on the lag parameter τ.

Fact 2.2.8 *The functions $\alpha_n(\tau)$ in (2.15)–(2.17) are such that, for each τ, $\alpha_n(\tau) \neq \alpha_m(\tau)$ for $n \neq m$. However, if more functions $\alpha_{n_1}(\tau), \cdots, \alpha_{n_K}(\tau)$ are defined in K (not necessarily coincident) neighborhoods of the same point τ_0, all have the same limit, say α_0, for $\tau \to \tau_0$, and only one of them is defined in τ_0, then it is convenient to* assume *all the functions $\alpha_{n_1}(\tau), \cdots, \alpha_{n_K}(\tau)$ defined in τ_0 with $\alpha_{n_1}(\tau_0) = \cdots = \alpha_{n_K}(\tau_0) = \alpha_0$ and, consequently,* define

$$R_{xx^*}^{(n_i)}(\tau_0) \triangleq \lim_{\tau\to\tau_0} R_{xx^*}^{(n_i)}(\tau) \quad i = 1, \ldots, K, \tag{2.20}$$

where, for each i, the limit is made with τ ranging in $\mathcal{T}^{(n_i)}$. With convention (2.20), by taking the coefficient of the complex sinewave at frequency α (see (2.12)) in both sides of (2.18), it follows that the cyclic autocorrelation function and the generalized cyclic autocorrelation functions are related by the relationship

$$R_{xx^*}(\alpha, \tau) = \sum_{n \in \mathbb{I}} R_{xx^*}^{(n)}(\tau)\, \delta_{\alpha-\alpha_n(\tau)} \tag{2.21}$$

where δ_γ denotes Kronecker delta, that is, $\delta_\gamma = 1$ for $\gamma = 0$ and $\delta_\gamma = 0$ for $\gamma \neq 0$. □

The wide-sense ACS processes are obtained as a special case of GACS processes when the lag-dependent cycle frequencies are constant with respect to τ and, hence, are coincident with the cycle frequencies (Izzo and Napolitano 1998a). In such a case,

$$\mathcal{R}_{xx^*}(t, \tau) = \sum_{n \in \mathbb{I}} R_{xx^*}^{\alpha_n}(\tau)\, e^{j2\pi\alpha_n t} \tag{2.22}$$

$$R_{xx^*}(\alpha, \tau) = \begin{cases} R_{xx^*}^{\alpha_n}(\tau) & \alpha = \alpha_n,\ \alpha_n \in A \\ 0 & \text{otherwise} \end{cases} \tag{2.23}$$

$$A = \{\alpha_n\}_{n \in \mathbb{I}}. \tag{2.24}$$

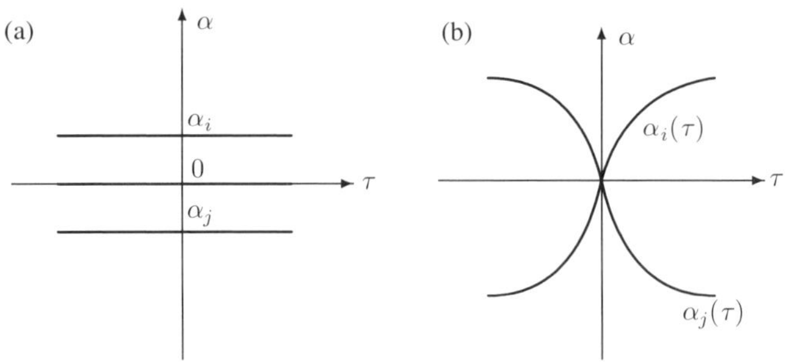

Figure 2.1 Support in the (α, τ)-plane of the cyclic autocorrelation function $R_{xx^*}(\alpha, \tau)$ of (a) an ACS process and (b) a GACS (not ACS) process

Moreover, for ACS processes only one term is present in the sum in (2.21) and, consequently, the generalized cyclic autocorrelation functions are coincident with the cyclic autocorrelation functions. Furthermore, the autocorrelation function $\mathcal{R}_{xx^*}(t, \tau)$ depends uniformly on the parameter τ (Corduneanu 1989, Chapter 2) and is uniformly continuous and the cyclic autocorrelation functions $R_{xx^*}(\alpha, \tau)$ are continuous in τ for each $\alpha \in A$ (which is countable) (Hurd 1991), (Dehay and Hurd 1994).

In Figure 2.1, the support in the (α, τ) plane of the cyclic autocorrelation function $R_{xx^*}(\alpha, \tau)$ is reported for (a) an ACS signal and (b) a GACS (not ACS) signal. For the ACS signal, such a support is contained in the lines $\alpha = \alpha_n$, $n \in \mathbb{I}$, that is, lines parallel to the τ axis corresponding to the cycle frequencies (see (2.23)). For the GACS signal, the support is constituted by the curves $\alpha = \alpha_n(\tau)$, $n \in \mathbb{I}$ (see (2.16)).

Lemma 2.2.9 *Second-order GACS processes in the wide sense have uniformly bounded second-order absolute moments.*

Proof: For every τ the autocorrelation function $\mathcal{R}_{xx^*}(t, \tau)$ is uniformly almost-periodic with respect to the variable t. Therefore, such a function is bounded and its (generalized) Fourier series is uniformly convergent (Definition 2.2.6). Thus, from $\mathrm{E}\{|x(t)|^2\} = \mathcal{R}_{xx^*}(t, 0)$ it follows that GACS processes have uniformly bounded second-order absolute moments (see (Dehay and Hurd 1994) for the case of ACS processes). □

For complex processes, two second-order moments need to be considered for a complete characterization in the wide sense: the autocorrelation function (2.10) and the *conjugate autocorrelation function* (Gardner 1987d) also called *relation function* (Picinbono and Bondon 1997) or *complementary correlation* (Schreier and Scharf 2003a)

$$\mathcal{R}_{xx}(t, \tau) \triangleq \mathrm{E}\{x(t+\tau)\, x(t)\}\,. \tag{2.25}$$

For complex-valued GACS processes, similar representations must hold for the autocorrelation and the conjugate autocorrelation functions and Definition 2.2.6, Theorem 2.2.7, and Fact 2.2.8 must be extended to $\mathcal{R}_{xx}(t, \tau)$. That is, $\mathcal{R}_{xx}(t, \tau)$ is the uniform limit of a sequence of trigonometric polynomials whose coefficients $R^{(n)}_{xx}(\tau)$ and frequencies $\beta_n(\tau)$ are referred to as *conjugate generalized cyclic autocorrelation functions* and *conjugate lag-dependent cycle frequencies*, respectively. The *conjugate cyclic autocorrelation function* at *conjugate cycle frequency* β is

$$R_{xx}(\beta, \tau) \triangleq \lim_{T\to\infty} \frac{1}{T} \int_{t_0-T/2}^{t_0+T/2} \mathcal{R}_{xx}(t, \tau)\, e^{-j2\pi\beta t}\, \mathrm{d}t. \tag{2.26}$$

The cyclic autocorrelation function (2.12) and the conjugate cyclic autocorrelation function (2.26) can both be represented by the concise notation

$$R_{xx^{(*)}}(\alpha, \tau) \triangleq \lim_{T\to\infty} \frac{1}{T} \int_{t_0-T/2}^{t_0+T/2} \mathrm{E}\{x(t+\tau)\, x^{(*)}(t)\}\, e^{-j2\pi\alpha t}\, \mathrm{d}t \tag{2.27}$$

where superscript $(*)$ represents an optional complex conjugation.

More general definitions of GACS processes can be given if the almost-periodicity property is considered in one of the generalized senses presented in Sections 1.2.2, 1.2.3, 1.2.4, and 1.2.5. In particular, a stochastic process $x(t)$ is said to *exhibit (conjugate) second-order cyclostationarity in the wide sense* with (conjugate) cycle frequency α if the (conjugate) cyclic autocorrelation function $R_{xx^{(*)}}(\alpha, \tau)$ is nonzero for some τ. Its (conjugate) autocorrelation function has the form

$$\mathcal{R}_{xx^{(*)}}(t, \tau) = \sum_{\alpha\in A_\tau} R_{xx^{(*)}}(\alpha, \tau)\, e^{j2\pi\alpha t} + r_{xx^{(*)}}(t, \tau) \tag{2.28a}$$

$$= \sum_{n\in\mathbb{I}} R^{(n)}_{xx^{(*)}}(\tau)\, e^{j2\pi\alpha^{(n)}_{xx^{(*)}}(\tau)t} + r_{xx^{(*)}}(t, \tau) \tag{2.28b}$$

with A_τ and $\mathbb{I}$ countable sets depending on $(*)$, and $r_{xx^{(*)}}(t, \tau)$ residual term not containing any finite-strength additive sinewave component

$$\lim_{T\to\infty} \frac{1}{T} \int_{t_0-T/2}^{t_0+T/2} r_{xx^{(*)}}(t, \tau)\, e^{-j2\pi\alpha t}\, \mathrm{d}t \equiv 0 \quad \forall\alpha \in \mathbb{R}. \tag{2.29}$$

For each fixed τ, the function $\mathcal{R}_{xx^{(*)}}(t, \tau)$ in (2.28b) is a generalized almost-periodic function of t in the sense of Hartman and Ryll-Nardzewski (Section 1.2.5). More generally, the function

$$\sum_{\alpha\in A_\tau} R_{xx^{(*)}}(\alpha, \tau)\, e^{j2\pi\alpha t} = \sum_{n\in\mathbb{I}} R^{(n)}_{xx^{(*)}}(\tau)\, e^{j2\pi\alpha^{(n)}_{xx^{(*)}}(\tau)t}$$

is not necessarily continuous in t and τ.

If condition (2.29) is assured by the asymptotic vanishing of the residual term

$$\lim_{|t|\to\infty} r_{xx^{(*)}}(t, \tau) = 0 \quad \forall\tau \in \mathbb{R} \tag{2.30}$$

and, in addition, the lag-dependent cycle frequencies do not depend on τ, then the stochastic process $x(t)$ is called asymptotically almost cyclostationary (Gardner 1978).

Definition 2.2.10 *Two finite-power complex-valued continuous-time stochastic process $y(t)$ and $x(t)$ are said to be* jointly GACS in the wide sense *if*

$$\mathcal{R}_{yx^{(*)}}(t, \tau) \triangleq \mathrm{E}\left\{y(t+\tau)\, x^{(*)}(t)\right\} \tag{2.31a}$$

$$= \sum_{\alpha \in A_\tau} R_{yx^{(*)}}(\alpha, \tau)\, e^{j2\pi\alpha t} \tag{2.31b}$$

$$= \sum_{n \in \mathbb{I}_{yx^{(*)}}} R^{(n)}_{yx^{(*)}}(\tau)\, e^{j2\pi\alpha_n(\tau)t}. \tag{2.31c}$$

In (2.31b) and (2.31c), A_τ and $\mathbb{I}_{yx^{(*)}}$ are countable sets and the lag-dependent cycle frequencies $\alpha_n(\tau) \equiv \alpha^{(n)}_{yx^{(*)}}(\tau)$ depend on the choice made for the optional complex conjugation $(*)$ and, in general, are not coincident with those of $x(t)$ or $y(t)$. The result is that

$$A_\tau \triangleq \left\{\alpha \in \mathbb{R} \,:\, R_{yx^{(*)}}(\alpha, \tau) \neq 0\right\} \tag{2.32a}$$

$$= \bigcup_{n \in \mathbb{I}_{yx^{(*)}}} \left\{\alpha \in \mathbb{R} \,:\, \alpha = \alpha^{(n)}_{yx^{(*)}}(\tau)\right\} \tag{2.32b}$$

$$\equiv \left\{\alpha^{(n)}_{yx^{(*)}}(\tau)\right\}_{n \in \mathbb{I}_{yx^{(*)}}}. \tag{2.32c}$$

The function

$$R_{yx^{(*)}}(\alpha, \tau) \triangleq \lim_{T\to\infty} \frac{1}{T} \int_{t_0-T/2}^{t_0+T/2} \mathcal{R}_{yx^{(*)}}(t, \tau)\, e^{-j2\pi\alpha t}\, \mathrm{d}t \tag{2.33}$$

is the cyclic cross-correlation function *and the functions $R^{(n)}_{yx^{(*)}}(\tau)$ are the* generalized cyclic cross-correlation functions *defined as*

$$R^{(n)}_{yx^{(*)}}(\tau) \triangleq \begin{cases} \lim\limits_{T\to\infty} \dfrac{1}{T} \displaystyle\int_{t_0-T/2}^{t_0+T/2} \mathcal{R}_{yx^{(*)}}(t, \tau)\, e^{-j2\pi\alpha_n(\tau)t}\, \mathrm{d}t & \tau \in \mathcal{T}^{(n)}_{yx^{(*)}} \\ 0 & \tau \in \mathbb{R} - \mathcal{T}^{(n)}_{yx^{(*)}} - D^{(n)} \end{cases} \tag{2.34}$$

$$R^{(n)}_{yx^{(*)}}(\tau_0) \triangleq \lim_{\tau\to\tau_0} R^{(n)}_{yx^{(*)}}(\tau) \quad \tau_0 \in D^{(n)} \tag{2.35}$$

where

$$\mathcal{T}^{(n)}_{yx^{(*)}} \triangleq \{\tau \in \mathbb{R} \,:\, \alpha_n(\tau) \text{ is defined},\ \alpha_n(\tau) \neq \alpha_m(\tau)\ \forall m \neq n\} \tag{2.36}$$

$$D^{(n)} \triangleq \{\tau \in \mathbb{R} \,:\, \alpha_n(\tau) = \alpha_m(\tau) \text{ for some } m \neq n\}\ , \tag{2.37}$$

and the limit in (2.35) is made with τ ranging in $\mathcal{T}^{(n)}_{yx^{(*)}}$. □

Moreover, from (2.34) for $\tau \in \mathcal{T}^{(n)}_{yx^{(*)}}$ one obtains

$$R^{(n)}_{yx^{(*)}}(\tau) = R_{yx^{(*)}}(\alpha, \tau)|_{\alpha=\alpha_n(\tau)} \tag{2.38}$$

and by reasoning as for a single process, it can be shown that for every τ

$$R_{yx^{(*)}}(\alpha, \tau) = \sum_{n \in \mathbb{I}_{yx^{(*)}}} R^{(n)}_{yx^{(*)}}(\tau)\, \delta_{\alpha - \alpha_n(\tau)}. \tag{2.39}$$

For GACS processes that are not ACS, even if the set A_τ defined in (2.15) and the (conjugate) autocorrelation function $\mathcal{R}_{xx^{(*)}}(t, \tau)$ are continuous functions of the lag parameter τ, the (conjugate) cyclic autocorrelation function $R_{xx^{(*)}}(\alpha, \tau)$ is not a continuous function of τ. Specifically, according to (2.39) with $y \equiv x$, $R_{xx^{(*)}}(\alpha, \tau)$ is constituted by sums of Kronecker deltas depending on τ.

It is well known that for every finite-power process $x(t)$ the conventional time-averaged autocorrelation function $R_{xx^*}(0, \tau)$ is continuous in $\tau = 0$ (and, hence, for any τ) if and only if the cross-correlation function $R_{xy^*}(0, \tau)$ is continuous for any $\tau \in \mathbb{R}$ and for any finite-power process $y(t)$ (Lee 1967, pp. 74–78). Therefore, if one defines the process

$$y_\alpha(t) \triangleq x^{(*)*}(t)\, e^{j2\pi\alpha t} \tag{2.40}$$

then the (conjugate) cyclic autocorrelation function of $x(t)$ can be written as

$$R_{xx^{(*)}}(\alpha, \tau) = R_{xy_\alpha^*}(0, \tau). \tag{2.41}$$

Consequently, the time-averaged autocorrelation function $R_{xx^*}(0, \tau)$ is continuous in $\tau = 0$ if and only if the (conjugate) cyclic autocorrelation function is continuous in $\tau \quad \forall \alpha \in \mathbb{R}$. Furthermore, such a property is still valid with reference to the continuity of all higher-order cyclic temporal moment functions (Izzo and Napolitano 1998a, 2005). Therefore, all GACS processes that are not ACS exhibit time-averaged autocorrelation functions discontinuous in $\tau = 0$. In particular, accounting for the symmetry property $R_{xx^*}(0, \tau) = R_{xx^*}(0, -\tau)^*$, it follows that $R_{xx^*}(0, \tau)$ contains the additive term $\overline{x^2}\, \delta_\tau$, where $\overline{x^2} = R_{xx^*}(0, 0)$ is the time-averaged power of $x(t)$. Consequently, as shown in Section 2.2.3, the power $\overline{x^2}$ (or quota of it) is uniformly spread over an infinite bandwidth and this constitutes a strong difficulty for the spectral characterization of GACS processes. Note that the above-discussed discontinuity property should not be confused with that examined in (Hurd 1974), where the discontinuity of the time-varying autocorrelation function is considered.

Definition 2.2.11 Mean-Square Continuity. *The stochastic process* $\{x(t),\ t \in \mathbb{R}\}$ *is said to be* mean-square continuous *in t if*

$$\lim_{\epsilon \to 0} \mathrm{E}\left\{|x(t+\epsilon) - x(t)|^2\right\} = 0. \tag{2.42}$$

□

Theorem 2.2.12 Necessary and Sufficient Condition for Mean-Square Continuity (Loève 1963, pp. 469–470). *The stochastic process* $\{x(t),\ t \in \mathbb{R}\}$ *is* mean-square continuous *in t if and only if*

$$\lim_{\substack{\epsilon_1 \to 0 \\ \epsilon_2 \to 0}} \left[\mathrm{E}\{x(t+\epsilon_1)\, x^*(t+\epsilon_2)\} - \mathrm{E}\{x(t)\, x^*(t)\}\right] = 0 \tag{2.43}$$

whatever way ϵ_1 *and* ϵ_2 *converge to zero. That is, if and only if* $\mathrm{E}\{x(t_1)\, x^*(t_2)\}$ *is continuous at* (t_1, t_1). □

From (2.43) it also follows that

$$\lim_{\epsilon \to 0} \mathrm{E}\{x(t_1 + \epsilon)\, x^*(t_2 + \epsilon)\} = \mathrm{E}\{x(t_1)\, x^*(t_2)\}. \tag{2.44}$$

Thus, as an immediate consequence of Theorem 2.2.12, we have that if $\mathrm{E}\{x(t_1)\, x^*(t_2)\}$ is continuous at every diagonal point (t_1, t_1), then it is continuous for every (t_1, t_2).

Theorem 2.2.13 Non Mean-Square Continuity of GACS Processes. *GACS processes which are not ACS are not mean-square continuous.*

Proof: From Theorem 2.2.7 it follows that (see (2.18))

$$\mathrm{E}\{x(t + \epsilon_1)\, x^*(t + \epsilon_2)\} = \sum_{n \in \mathbb{I}} R^{(n)}_{xx^*}(\epsilon_1 - \epsilon_2)\, e^{j2\pi\alpha_n(\epsilon_1 - \epsilon_2)\,(t+\epsilon_2)}. \tag{2.45}$$

The lack of mean-square continuity of GACS processes which are not ACS follows immediately from the fact that for the lag-dependent cycle frequency $\alpha_n(\tau) = 0$ we have $R^{(n)}_{xx^*}(\tau) = R_{xx^*}(0, \tau)$ discontinuous in $\tau = 0$. Therefore, the necessary condition of Theorem 2.2.12 is not satisfied. □

In contrast, ACS processes are mean-square continuous and are characterized by the following conditions (Hurd 1991), (Gardner *et al.* 2006):

1. The set $A \triangleq \bigcup_{\tau \in \mathbb{R}} A_\tau$ is countable.
2. The autocorrelation function $R_{xx^*}(t, \tau)$ is uniformly continuous in t and τ.
3. The time-averaged autocorrelation function $R^0_{xx^*}(\tau) \triangleq \langle R_{xx^*}(t, \tau)\rangle_t$ is continuous for $\tau = 0$ (and, hence, for every τ).
4. The process is mean-square continuous.

Definition 2.2.14 Mean-Square Integrability. *The stochastic process* $\{x(t),\ t \in \mathbb{R}\}$ *is said to be* (Riemann) mean-square integrable *in* $(-T/2, T/2)$ *if*

$$\lim_{T_s \to 0} \mathrm{E}\left\{\left| \sum_{n=-N}^{N} x(nT_s)\, T_s - \int_{-T/2}^{T/2} x(t)\, \mathrm{d}t \right|^2\right\} = 0 \tag{2.46}$$

where $N \triangleq \lfloor T/(2T_s) \rfloor$, *with* $\lfloor \cdot \rfloor$ *being the integer part.* □

Theorem 2.2.15 Necessary and Sufficient Condition for Mean-Square Integrability (Loève 1963, p. 472). *The zero-mean stochastic process* $\{x(t),\ t \in \mathbb{R}\}$ *is* mean-square integrable *in* $(-T/2, T/2)$ *if and only if*

$$\int_{-T/2}^{T/2} \int_{-T/2}^{T/2} \left|\mathrm{E}\{x(t_1)\, x^*(t_2)\}\right| \mathrm{d}t_1\, \mathrm{d}t_2 < \infty. \tag{2.47}$$

□

Assumption 2.2.16 Cesàro Summability of the Autocorrelation Function. *For every finite T it results that*

$$\sum_{n\in\mathbb{I}} \int_{-T}^{T} \left| R_{xx^*}^{(n)}(\tau) \right| \left(1 - \frac{|\tau|}{T}\right) \mathrm{d}\tau < \infty. \tag{2.48}$$

□

Since for $\tau \in (-T, T)$ it results $|1 - |\tau|/T| \leqslant 1$, a sufficient condition assuring that (2.48) holds is

$$\sum_{n\in\mathbb{I}} \int_{\mathbb{R}} \left| R_{xx^*}^{(n)}(\tau) \right| \mathrm{d}\tau < \infty. \tag{2.49}$$

Theorem 2.2.17 Mean-Square Integrability of GACS Processes. *Under Assumption 2.2.16, a GACS process is mean-square integrable in $(-T/2, T/2)$.*

Proof: See Section 3.1. □

In (Napolitano 2003) the class of the spectrally correlated (SC) processes is introduced, which is a further class that generalizes that of the ACS processes. SC processes have a Loève bifrequency spectrum (Loève 1963, Chapter X) with spectral masses concentrated on a countable set of curves in the bifrequency plane (Chapter 4). Thus, ACS processes are obtained as special case of SC processes when the support curves are lines with unit slope (Hurd 1989a, 1991), (Dehay and Hurd 1994). Specifically, SC processes exhibit spectral correlation between spectral components that are separated, and the separation between correlated spectral components depends on the shape of the support curves in the bifrequency plane. For ACS processes, correlation exists only between spectral components that are separated by quantities belonging to a countable set of values, the cycle frequencies, which are the frequencies of the (generalized) Fourier series expansion of the autocorrelation function which is an almost-periodic function of time. Therefore, ACS processes are obtained by the intersection of the class of the GACS processes and the class of the SC processes. That is, ACS processes are a subclass of GACS processes that exhibit the spectral correlation property (Napolitano 2007b) (Section 4.2.2).

Let $\mathcal{T}_{n,\alpha}$ be the set

$$\mathcal{T}_{n,\alpha} \triangleq \{\tau \in \mathbb{R} \; : \; \alpha_n(\tau) = \alpha\} \tag{2.50}$$

where $\alpha_n(\tau)$ in (2.50) are the lag-dependent cycle frequencies of $\mathcal{R}_{xx^{(*)}}(t, \tau)$. The GACS signal $x(t)$ is said *to contain an ACS component* (for the given conjugation configuration) if the Lebesgue measure of the set $\mathcal{T}_{n,\alpha}$ is positive for some $\alpha \in \mathbb{R}$ and some $n \in \mathbb{I}_{xx^{(*)}}$. The GACS signal $x(t)$ is said to be *purely GACS* or *not containing any ACS component* if the Lebesgue measure of the set $\mathcal{T}_{n,\alpha}$ is zero $\forall \alpha \in \mathbb{R}$ and $\forall n \in \mathbb{I}_{xx^{(*)}}$. GACS signals containing an ACS component have at least one lag-dependent cycle frequency $\alpha_n(\tau)$ which is constant with respect to τ in a set of values of τ with positive Lebesgue measure. A GACS signal containing an ACS component with cycle frequencies $\alpha_1, \ldots, \alpha_M$ exhibits spectral components separated by $\alpha_1, \ldots, \alpha_M$ that are correlated. Analogously, if $\alpha_n(\tau)$ in (2.50) are the lag-dependent cycle frequencies of the cross-moment $\mathcal{R}_{yx^{(*)}}(t, \tau)$, then the GACS signals $x(t)$ and $y(t)$ are said *to* *contain a joint ACS component* (for the given conjugation configuration) when the Lebesgue measure of the set $\mathcal{T}_{n,\alpha}$ is positive for some $\alpha \in \mathbb{R}$ and some $n \in \mathbb{I}_{yx^{(*)}}$. The GACS signals $x(t)$ and $y(t)$ are said to be *jointly purely GACS* or *not containing any joint ACS component* if the Lebesgue measure of the set $\mathcal{T}_{n,\alpha}$ is zero $\forall\alpha \in \mathbb{R}$ and $\forall n \in \mathbb{I}_{yx^{(*)}}$.

In the following, the relationship

$$R^{(n)}_{xx^{(*)}}(\tau) = R_{xx^{(*)}}(\alpha, \tau)\Big|_{\alpha=\alpha^{(n)}_{xx^{(*)}}(\tau)} \tag{2.51}$$

will be used for all values of τ such that there is no equality among different lag-dependent cycle frequencies (see (2.12), (2.19), and Fact 2.2.8). For these values of τ, symmetry relations involving (conjugate) cyclic autocorrelation functions, (conjugate) generalized cyclic autocorrelation functions, and (conjugate) lag-dependent cycle frequencies are derived.

Fact 2.2.18 *For the conjugation configuration xx^* (autocorrelation) we have*

$$R_{xx^*}(-\alpha, -\tau) = R^*_{xx^*}(\alpha, \tau)\, e^{j2\pi\alpha\tau}. \tag{2.52}$$

Moreover, the set (see (2.13) and (2.15))

$$A_\tau \triangleq \{\alpha \in \mathbb{R} \;:\; R_{xx^*}(\alpha, \tau)\} \tag{2.53a}$$

$$= \bigcup_{n\in\mathbb{I}_{xx^*}} \left\{\alpha \in \mathbb{R} \;:\; \alpha = \alpha^{(n)}_{xx^*}(\tau)\right\} \tag{2.53b}$$

is symmetric in the sense that

$$\alpha \in A_\tau \;\Rightarrow\; -\alpha \in A_{-\tau}. \tag{2.54}$$

Finally, $\exists n, n' \in \mathbb{I}_{xx^*}$ possibly coincident such that

$$\alpha^{(n)}_{xx^*}(\tau) = -\alpha^{(n')}_{xx^*}(-\tau) \tag{2.55}$$

$$R^{(n)}_{xx^*}(\tau) = R^{(n')*}_{xx^*}(-\tau)\, e^{-j2\pi\alpha^{(n')}_{xx^*}(-\tau)\tau}. \tag{2.56}$$

In the special case of ACS processes, we have, that for the conjugation configuration xx^* the countable set A defined in (2.14) is symmetric in the sense that $\alpha \in A \Rightarrow -\alpha \in A$.

Proof: See Section 3.1. □

Fact 2.2.19 *For the conjugation configurations xx^* and x^*x, we have*

$$R_{x^*x}(\alpha, \tau) = R^*_{xx^*}(-\alpha, \tau) \tag{2.57}$$

$\exists n' \in \mathbb{I}_{x^*x}$ and $n'' \in \mathbb{I}_{xx^*}$ such that

$$\alpha^{(n')}_{x^*x}(\tau) = -\alpha^{(n'')}_{xx^*}(\tau) \tag{2.58}$$

$$R^{(n')}_{x^*x}(\tau) = R^{(n'')*}_{xx^*}(\tau) \tag{2.59}$$

Proof: See Section 3.1. □

Fact 2.2.20 *For the conjugation configurations xx (conjugate correlation) and x^*x^*, we have*

$$R_{xx}(-\alpha, \tau) = R^*_{x^*x^*}(\alpha, \tau) \tag{2.60}$$

$\exists n' \in \mathbb{I}_{xx}$ *and* $n'' \in \mathbb{I}_{x^*x^*}$ *such that*

$$\alpha^{(n')}_{xx}(\tau) = -\alpha^{(n'')}_{x^*x^*}(\tau) \tag{2.61}$$

$$R^{(n')}_{xx}(\tau) = R^{(n'')*}_{x^*x^*}(\tau) \tag{2.62}$$

Proof: See Section 3.1. □

2.2.3 Second-Order Spectral Characterization

The presence of sums of Kronecker deltas depending on τ in the expressions of the (conjugate) cyclic autocorrelation function and of the cyclic cross-correlation function of (jointly) GACS processes (see (2.21) and (2.39)) when $\mathcal{T}_{n,\alpha}$ has zero Lebesgue measure, does not allow a spectral characterization of GACS processes in terms of ordinary Fourier transforms of these functions.

In Section 2.2.2, it is shown that the time-averaged autocorrelation function $R_{xx^*}(0, \tau)$ is discontinuous in $\tau = 0$ and contains the additive term $\overline{x^2}\,\delta_\tau$, where $\overline{x^2} = R_{xx^*}(0, 0)$ is the time-averaged power of $x(t)$. That is, the time-averaged autocorrelation function admits the decomposition

$$R_{xx^*}(0, \tau) = \overline{x_0^2}\,\delta_\tau + b(\tau) \tag{2.63}$$

with $b(\tau) = b^*(-\tau)$, $\overline{x_0^2} + b(0) = \overline{x^2}$, and $b(\tau)$ definite nonnegative and possibly discontinuous.

The Fourier transform of $\overline{x_0^2}\delta_\tau$, defined in terms of Lebesgue integral, provides an identically zero function. Consequently, the power spectrum for GACS processes (that are not ACS) cannot be defined in the ordinary sense. However, an heuristic approach to provide a spectral characterization of GACS processes can be based on the μ functional defined in Section 3.2. Specifically, the power spectrum of $x(t)$ can be obtained accounting for the Fourier pair in (3.30). We formally have

$$R_{xx^*}(0, \tau) = \overline{x_0^2}\,\delta_\tau + b(\tau) \overset{\mathcal{F}}{\longleftrightarrow} S_{xx^*}(f) = \overline{x_0^2}\,\mu(f) + B(f) \tag{2.64}$$

where $B(f)$ is the Fourier transform of $b(\tau)$. Due to the presence of the term $\overline{x_0^2}\mu(f)$, a quota of the power of $x(t)$ is spread over an infinite bandwidth. In fact, $\mu(f)$ can be interpreted as the limit of a very tiny and large rectangular window with unit area (Section 3.2). Consequently, low-pass or band-pass filtering (with finite or essentially finite bandwidth) of GACS signals cancels in the output signal the power contribution $\overline{x_0^2}$. In particular, if the process is purely GACS, the output signal is zero-power (Sections 2.3 and 2.7.7). This spectral characterization agrees with the observation in (Wiener 1949, pp. 39–40) that if the time-averaged autocorrelation function is discontinuous in the origin, then "there is a portion of the energy which does not belong to any finite frequencies, and which in a certain sense we must associate with infinite frequencies."

In (Izzo and Napolitano 1998a, 2002a, 2005), it is shown that a useful spectral characterization of (jointly) GACS processes can be obtained by considering the Fourier transforms of the generalized (conjugate) cyclic autocorrelation functions and of the generalized cyclic cross-correlation functions

$$S^{(n)}_{yx^{(*)}}(f) \triangleq \int_{\mathbb{R}} R^{(n)}_{yx^{(*)}}(\tau)\, e^{-j2\pi f\tau}\, \mathrm{d}\tau \tag{2.65}$$

and analogously for $y \equiv x$. As already observed, if $\mathcal{T}_{n,\alpha}$ has zero Lebesgue measure $\forall\alpha$, the function defined in (2.65) is not a spectral cross-correlation density function as it happens for ACS processes. In the following, the second-order spectral characterization is made resorting to the Loève bifrequency spectrum.

Definition 2.2.21 *Let $\{x(t),\ t \in \mathbb{R}\}$ be a complex-valued second-order harmonizable stochastic process. The* Loève bifrequency spectrum *(Loève 1963) or* spectral correlation function *is defined as (Definition 4.2.3 and (1.13))*

$$\mathcal{S}_{xx^{(*)}}(f_1, f_2) \triangleq \mathrm{E}\left\{X(f_1)\, X^{(*)}(f_2)\right\}. \tag{2.66}$$

In (2.66),

$$X(f) \triangleq \int_{\mathbb{R}} x(t)\, e^{-j2\pi f t}\, \mathrm{d}t \tag{2.67}$$

is the Fourier transform of $x(t)$ and is assumed to exist (at least) in the sense of distributions (Gelfand and Vilenkin 1964, Chapter 3), (Henniger 1970). Superscript $(*)$ denotes an optional complex conjugation. □

A stochastic process is said to be (strongly) harmonizable if it can be expressed as a Fourier-Stieltjes transform of a second-order spectral function with bounded-variation covariance (Loève 1963). A covariance function is said to be harmonizable if it can be expressed as a Fourier-Stieltjes transform of a (spectral) covariance function of bounded variation (Loève 1963). In (Loève 1963, p. 474), it is shown that a necessary condition for a stochastic process to be harmonizable is that it is second-order continuous. In fact, the bounded variation condition on the spectral covariance function implies that harmonizable stochastic processes are continuous in quadratic mean and harmonizable covariances are continuous and bounded. Moreover, since convergence of integrals is considered in the mean-square sense (Definition 2.2.14 and Theorem 2.2.15), in (Loève 1963, p. 476) it is shown that a stochastic process is harmonizable if and only if its covariance function is harmonizable. From Theorem 2.2.13 it follows that GACS processes are not mean-square continuous, and, hence, not (strongly) harmonizable. Consequently, for GACS processes the convergence of the integral

$$\mathrm{E}\left\{x(t_1)\, x^{(*)}(t_2)\right\} = \int_{\mathbb{R}^2} e^{j2\pi[f_1 t_1 + (-) f_2 t_2]}\, \mathrm{d}\gamma_{\boldsymbol{x}}(f_1, f_2) \tag{2.68}$$

with

$$\mathrm{d}\gamma_{\boldsymbol{x}}(f_1, f_2) = \mathrm{E}\left\{X(f_1)\, X^{(*)}(f_2)\right\} \mathrm{d}f_1\, \mathrm{d}f_2 \tag{2.69}$$

and with $(-)$ denoting an optional minus sign linked to the optional complex conjugation $(*)$, cannot be in the mean-square sense.

Convergence of the integral in (2.68) when $\gamma_{\boldsymbol{x}}(f_1, f_2)$ has no bounded variation can be considered in the sense of (Kolmogorov 1960) or in the sense of Morse-Transue (Morse and Transue 1956), (Rao 2008) and the process is said *weakly harmonizable*. Such an analysis, however, is beyond the scope of this book. In the following, a formal expression for the Loève bifrequency spectrum of a GACS process is derived, even if the process is not strongly harmonizable, without specifying the kind of convergence of the involved integrals.

Theorem 2.2.22 Loève Bifrequency Spectrum of GACS Processes. *Let $\{x(t),\ t \in \mathbb{R}\}$ be a complex-valued GACS stochastic process. Its Loève bifrequency spectrum is given by*

$$\begin{aligned}\mathrm{E}\left\{X(f_1)\, X^{(*)}(f_2)\right\} &= \sum_{n\in\mathbb{I}_d} S^{(n)}_{xx^{(*)}}(f_1)\, \delta(f_1 + (-)f_2 - \beta_n) \\ &\quad + \sum_{n\in\mathbb{I}_c} R^{(n)}_{xx^{(*)}}(\alpha_n^{-1}(f_1 + (-)f_2)) \\ &\quad \left|(\alpha_n^{-1})'(f_1 + (-)f_2)\right|\, e^{-j2\pi f_1 \alpha_n^{-1}(f_1+(-)f_2)}\end{aligned} \tag{2.70}$$

where the lag-dependent cycle frequencies curves have been chosen so that the set $\mathbb{I} \equiv \mathbb{I}_{xx^{(*)}}$ can be partitioned as $\mathbb{I} = \mathbb{I}_d \cup \mathbb{I}_c$, where

$$\mathbb{I}_d \triangleq \{n \in \mathbb{I} \,:\, \alpha_n(\tau) = \beta_n\} \qquad \text{(constant cycle frequencies)} \tag{2.71}$$

$$\mathbb{I}_c \triangleq \{n \in \mathbb{I} \,:\, \alpha_n(\tau) \text{ is invertible}\} \tag{2.72}$$

$\alpha_n^{-1}(\cdot)$ is the inverse function of $\alpha_n(\cdot)$ and

$$S^{(n)}_{xx^{(*)}}(f) \triangleq \int_{\mathbb{R}} R^{(n)}_{xx^{(*)}}(\tau)\, e^{-j2\pi f\tau}\, \mathrm{d}\tau \qquad n \in \mathbb{I}_d \tag{2.73}$$

are Fourier transforms of those (conjugate) generalized cyclic autocorrelation functions that are coincident with (conjugate) cyclic autocorrelation functions. That is, $S^{(n)}_{xx^{(*)}}(f)$, $n \in \mathbb{I}_d$, are the (conjugate) cyclic spectra of the ACS component.

Proof: See Section 3.2. □

Thus, the Loève bifrequency spectrum of a GACS process contains an impulsive part corresponding to the ACS component and a continuous part corresponding to the purely GACS component. In (Soedjack 2002) it is shown that the continuous term of the Loève bifrequency spectrum cannot be consistently estimated starting from a single realization or sample path, but only using several realizations.

2.2.4 Higher-Order Statistics

In this section, the higher-order characterization of GACS processes is provided in terms of moments and cumulants in the time domain. An extensive treatment on higher-order statistics of GACS signals in the functional (or nonstochastic) approach is made in (Izzo and Napolitano 1998a, 2002a, 2005).

Let $x_n(t)$, $n = 1, \ldots, N$, be N continuous-time complex-valued stochastic processes. Since the x_n are arbitrary, without lack of generality in this section we can avoid considering (optional) complex conjugations.

Definition 2.2.23 *The* Nth-order temporal cross-moment function (TCMF) *of the (complex-valued) processes* $x_1(t), \ldots, x_N(t)$ *is defined as*

$$\mathcal{R}_{\boldsymbol{x}}(t, \boldsymbol{\tau}) \triangleq \mathrm{E}\left\{\prod_{n=1}^{N} x_n(t + \tau_n)\right\} \tag{2.74}$$

where $\boldsymbol{x} \triangleq [x_1, \ldots, x_N]$ *and* $\boldsymbol{\tau} \triangleq [\tau_1, \ldots, \tau_{N-1}, \tau_N]$, *with* $\tau_N = 0$. □

Definition 2.2.24 *The processes* $x_1(t), \ldots, x_N(t)$ *are said to be* jointly GACS (for the moment) *if their* N*th-order TCMF is almost-periodic in* t *in the sense of Bohr (Bohr 1933, par. 84–92) or, equivalently, uniformly almost periodic in sense of Besicovitch (Besicovitch 1932, Chapter 1), (Corduneanu 1989). That is, for each fixed* $\boldsymbol{\tau}$, $\mathcal{R}_{\boldsymbol{x}}(t, \boldsymbol{\tau})$ *is the limit of a uniformly convergent sequence of trigonometric polynomials in* t *which can be written in the two following equivalent forms (Izzo and Napolitano 1998a, 2002a):*

$$\mathcal{R}_{\boldsymbol{x}}(t, \boldsymbol{\tau}) = \sum_{\alpha \in A_{\boldsymbol{\tau}}} R_{\boldsymbol{x}}(\alpha, \boldsymbol{\tau})\, e^{j2\pi\alpha t} \tag{2.75a}$$

$$= \sum_{k \in \mathbb{I}} R_{\boldsymbol{x}}^{(k)}(\boldsymbol{\tau})\, e^{j2\pi\alpha_k(\boldsymbol{\tau})t} \tag{2.75b}$$

In (2.75a), the real numbers α *and the complex-valued functions* $R_{\boldsymbol{x}}(\alpha, \boldsymbol{\tau})$, *referred to as* Nth-order (cross-moment) cycle frequencies *and* cyclic temporal cross-moment functions, *are the frequencies and coefficients, respectively, of the generalized Fourier series expansion of* $\mathcal{R}_{\boldsymbol{x}}(t, \boldsymbol{\tau})$ *that is,*

$$R_{\boldsymbol{x}}(\alpha, \boldsymbol{\tau}) \triangleq \lim_{T\to\infty} \frac{1}{T} \int_{-T/2}^{T/2} \mathcal{R}_{\boldsymbol{x}}(t, \boldsymbol{\tau})\, e^{-j2\pi\alpha t}\, \mathrm{d}t. \tag{2.76}$$

Furthermore, in (2.75a) and (2.75b),

$$A_{\boldsymbol{\tau}} \triangleq \{\alpha \in \mathbb{R} : R_{\boldsymbol{x}}(\alpha, \boldsymbol{\tau}) \neq 0\} \tag{2.77a}$$

$$= \bigcup_{k \in \mathbb{I}_{\mathrm{m}}} \{\alpha \in \mathbb{R} : \alpha = \alpha_k(\boldsymbol{\tau})\} \tag{2.77b}$$

is a countable set, $\mathbb{I}_{\mathrm{m}}$ *is also countable, the real-valued functions* $\alpha_k(\boldsymbol{\tau})$ *are referred to as* (cross-moment) lag-dependent cycle frequencies *and the complex-valued functions* $R_{\boldsymbol{x}}^{(k)}(\boldsymbol{\tau})$, *referred to as* generalized cyclic cross-moment functions, *are defined as*

$$R_{\boldsymbol{x}}^{(k)}(\boldsymbol{\tau}) \triangleq R_{\boldsymbol{x}}(\alpha, \boldsymbol{\tau})|_{\alpha=\alpha_k(\boldsymbol{\tau})} \tag{2.78}$$

for all values of $\boldsymbol{\tau}$ *such that two different N-dimensional varieties described by two different lag-dependent cycle frequencies do not intersect (Izzo and Napolitano 2002a), (Napolitano 2007a). It can be shown that (Izzo and Napolitano 1998a, 2002a)*

$$R_{\boldsymbol{x}}(\alpha, \boldsymbol{\tau}) = \sum_{k\in\mathbb{I}_{\mathrm{m}}} R_{\boldsymbol{x}}^{(k)}(\boldsymbol{\tau})\, \delta_{\alpha-\alpha_k(\boldsymbol{\tau})} \tag{2.79}$$

where δ_γ *denotes Kronecker delta, that is,* $\delta_\gamma = 1$ *for* $\gamma = 0$ *and* $\delta_\gamma = 0$ *for* $\gamma \neq 0$*. That is, the N-dimensional varieties* $\alpha = \alpha_k(\boldsymbol{\tau})$ *with* $\tau_N = 0$, $k \in \mathbb{I}_{\mathrm{m}}$*, describe the support of the cyclic cross-moment function* $R_{\boldsymbol{x}}(\alpha, \boldsymbol{\tau})$ *in the N-dimensional* $(\alpha, \tau_1, \ldots, \tau_{N-1})$*-space.* □

More generally, N processes $x_1(t), \ldots, x_N(t)$ are said to *exhibit joint Nth-order generalized almost-cyclostationarity (for the moment)* if the Nth-order TCMF is constituted by a (uniformly) almost-periodic component plus a residual term not containing any finite-strength additive sinewave component (Napolitano and Tesauro 2011):

$$\mathcal{R}_{\boldsymbol{x}}(t, \boldsymbol{\tau}) = \sum_{\alpha\in A_{\boldsymbol{\tau}}} R_{\boldsymbol{x}}(\alpha, \boldsymbol{\tau})\, e^{j2\pi\alpha t} + r_{\boldsymbol{x}}(t, \boldsymbol{\tau}) \tag{2.80a}$$

$$= \sum_{k\in\mathbb{I}_{\mathrm{m}}} R_{\boldsymbol{x}}^{(k)}(\boldsymbol{\tau})\, e^{j2\pi\alpha_k(\boldsymbol{\tau})t} + r_{\boldsymbol{x}}(t, \boldsymbol{\tau}) \tag{2.80b}$$

with

$$\lim_{T\to\infty} \frac{1}{T} \int_{-T/2}^{T/2} r_{\boldsymbol{x}}(t, \boldsymbol{\tau})\, e^{-j2\pi\alpha t}\, \mathrm{d}t = 0 \qquad \forall \alpha \in \mathbb{R}. \tag{2.81}$$

That is, $\mathcal{R}_{\boldsymbol{x}}(t, \boldsymbol{\tau})$ is an almost-periodic function in the sense of Hartman and Ryll-Nardzewski (Section 1.2.5).

Processes with statistical functions constituted by an almost-periodic component plus a residual term are appropriate models in mobile communications. For example, a multipath Doppler channel, that is a multipath channel introducing scaling amplitude, phase, time-delay, frequency shift, and time-scale factor for each path, excited by an ACS or GACS signal, has output with ACS- or GACS-kind statistical functions with residual terms, provided that the time-scale factors of at least two paths are different (Section 7.7.2) (Izzo and Napolitano 2002b).

In (Gardner and Spooner 1994), it is shown that for ACS signals, temporal cumulants, rather than temporal moments, properly describe the possible Nth-order almost-cyclostationarity of signals since computing cumulants is equivalent to removing all the sinewaves generated by beats of lower-order lag products whose orders sum to N. These considerations can easily be extended to GACS signals (Izzo and Napolitano 1998a, 2005).

Definition 2.2.25 *The* Nth-order temporal cross-cumulant function (TCCF) *of the (complex-valued) processes* $x_1(t), \ldots, x_N(t)$ *is defined as*

$$\mathcal{C}_{\boldsymbol{x}}(t, \boldsymbol{\tau}) \equiv \operatorname{cum}\{x_n(t+\tau_n),\ n = 1, \ldots, N\}$$

$$\triangleq (-j)^N \frac{\partial^N}{\partial\omega_1 \cdots \partial\omega_N} \log_e \mathrm{E}\left\{\exp\left[j\sum_{n=1}^{N} \omega_n x_n(t+\tau_n)\right]\right\}\Big|_{\boldsymbol{\omega}=\mathbf{0}} \tag{2.82a}$$

$$= \sum_{\mathrm{P}}\left[(-1)^{p-1}(p-1)! \prod_{i=1}^{p} \mathcal{R}_{\boldsymbol{x}_{\mu_i}}(t, \boldsymbol{\tau}_{\mu_i})\right] \tag{2.82b}$$

where $\boldsymbol{\tau} \triangleq [\tau_1, \ldots, \tau_N]$ *with* $\tau_N = 0$; $\boldsymbol{\omega} \triangleq [\omega_1, \ldots, \omega_N]$; P *is the set of distinct partitions of* $\{1, \ldots, N\}$, *each constituted by the subsets* $\{\mu_i,\ i = 1, \ldots, p\}$; $\boldsymbol{x}_{\mu_i}$ *is the* $|\mu_i|$*-dimensional vector whose components are those of* $\boldsymbol{x}$ *having indices in* μ_i, *with* $|\mu_i|$ *the number of elements in* μ_i. *See Section 1.4.2 for a discussion of the definition of (cross-) cumulant of complex random variables and processes.* □

Definition 2.2.26 *The processes* $x_1(t), \ldots, x_N(t)$ *are said to be* jointly GACS (for the cumulant) *if their* N*th-order TCCF is almost-periodic in* t *in the sense of Bohr (Bohr 1933, par. 84–92) or, equivalently, uniformly almost periodic in sense of Besicovitch (Besicovitch 1932, Chap. 1), (Corduneanu 1989). That is, for each fixed* $\boldsymbol{\tau}$, $\mathcal{C}_{\boldsymbol{x}}(t, \boldsymbol{\tau})$ *is the limit of a uniformly convergent sequence of trigonometric polynomials in* t *which can be written in the two following equivalent forms (Izzo and Napolitano 1998a, 2002a):*

$$\mathcal{C}_{\boldsymbol{x}}(t, \boldsymbol{\tau}) = \sum_{\beta \in B_{\boldsymbol{\tau}}} C_{\boldsymbol{x}}(\beta, \boldsymbol{\tau})\, e^{j2\pi\beta t} \tag{2.83a}$$

$$= \sum_{k \in \mathbb{I}_c} C_{\boldsymbol{x}}^{(k)}(\boldsymbol{\tau})\, e^{j2\pi\beta_k(\boldsymbol{\tau})t} \tag{2.83b}$$

In (2.83a), the real numbers β *and the complex-valued functions* $C_{\boldsymbol{x}}(\beta, \boldsymbol{\tau})$, *referred to as* Nth-order (cross-cumulant) cycle frequencies *and* cyclic temporal cross-cumulant functions, *are the frequencies and coefficients, respectively, of the generalized Fourier series expansion of* $\mathcal{C}_{\boldsymbol{x}}(t, \boldsymbol{\tau})$ *that is,*

$$C_{\boldsymbol{x}}(\beta, \boldsymbol{\tau}) \triangleq \lim_{T\to\infty} \frac{1}{T} \int_{-T/2}^{T/2} \mathcal{C}_{\boldsymbol{x}}(t, \boldsymbol{\tau})\, e^{-j2\pi\beta t}\, \mathrm{d}t. \tag{2.84}$$

Furthermore, in (2.83a) and (2.83b),

$$B_{\boldsymbol{\tau}} \triangleq \{\beta \in \mathbb{R} : C_{\boldsymbol{x}}(\beta, \boldsymbol{\tau}) \neq 0\} \tag{2.85a}$$

$$= \bigcup_{k \in \mathbb{I}_c} \{\beta \in \mathbb{R} : \beta = \beta_k(\boldsymbol{\tau})\} \tag{2.85b}$$

is a countable set, $\mathbb{I}_c$ *is also countable, the real-valued functions* $\beta_k(\boldsymbol{\tau})$ *are referred to as* (cross-cumulant) lag-dependent cycle frequencies *and the complex-valued functions* $C_{\boldsymbol{x}}^{(k)}(\boldsymbol{\tau})$, *referred to as* generalized cyclic cross-cumulant functions, *are defined as*

$$C_{\boldsymbol{x}}^{(k)}(\boldsymbol{\tau}) \triangleq C_{\boldsymbol{x}}(\beta, \boldsymbol{\tau})|_{\beta=\beta_k(\boldsymbol{\tau})} \tag{2.86}$$

for all values of $\boldsymbol{\tau}$ such that two different N-dimensional varieties described by two different lag-dependent cycle frequencies do not intersect (Izzo and Napolitano 2002a), (Napolitano 2007a). It can be shown that (Izzo and Napolitano 1998a, 2002a)

$$C_{\boldsymbol{x}}(\beta, \boldsymbol{\tau}) = \sum_{k \in \mathbb{I}_c} C_{\boldsymbol{x}}^{(k)}(\boldsymbol{\tau})\, \delta_{\beta - \beta_k(\boldsymbol{\tau})} \tag{2.87}$$

That is, the N-dimensional varieties $\beta = \beta_k(\boldsymbol{\tau})$ with $\tau_N = 0$, $k \in \mathbb{I}_c$, describe the support of the cyclic cross-cumulant function $C_{\boldsymbol{x}}(\beta, \boldsymbol{\tau})$ in the N-dimensional $(\beta, \tau_1, \ldots, \tau_{N-1})$-space. □

The Nth-order cyclic temporal cross-cumulant function $C_{\boldsymbol{x}}(\beta, \boldsymbol{\tau})$ can be expressed in terms of Nth- and lower-order cyclic temporal cross-moment functions as (Izzo and Napolitano 1998a)

$$C_{\boldsymbol{x}}(\beta, \boldsymbol{\tau}) = \sum_{\mathrm{P}} \left[(-1)^{p-1}(p-1)! \sum_{\boldsymbol{\alpha}^{\mathrm{T}}\mathbf{1} = \beta} \prod_{i=1}^{p} R_{\boldsymbol{x}_{\mu_i}}(\alpha_{\mu_i}, \boldsymbol{\tau}_{\mu_i}) \right] \tag{2.88}$$

where α_{μ_i} are the cycle frequencies of the temporal cross-moment function of x_n, $n \in \mu_i$, and the second sum is extended over all vectors $\boldsymbol{\alpha} \triangleq [\alpha_{\mu_1}, \ldots, \alpha_{\mu_p}]$ such that $\boldsymbol{\alpha}^{\mathrm{T}}\mathbf{1} = \beta$. Equation (2.88) extends to GACS processes a known result for ACS processes (Gardner and Spooner 1994).

In (Izzo and Napolitano 1998a, 2005), it is shown that the set of the Nth-order cumulant cycle frequencies is contained in the set of the Nth-order moment cycle frequencies. That is

$$B_{\boldsymbol{\tau}} \subseteq A_{\boldsymbol{\tau}}. \tag{2.89}$$

More generally, for processes exhibiting Nth-order generalized almost-cyclostationarity the temporal cross-cumulant function is constituted by an almost-periodic component plus a residual term not containing any additive finite-strength sinewave component (Napolitano and Tesauro 2011):

$$\mathcal{C}_{\boldsymbol{x}}(t, \boldsymbol{\tau}) = \sum_{\beta \in B_{\boldsymbol{\tau}}} C_{\boldsymbol{x}}(\beta, \boldsymbol{\tau})\, e^{j2\pi\beta t} + c_{\boldsymbol{x}}(t, \boldsymbol{\tau}) \tag{2.90a}$$

$$= \sum_{k \in \mathbb{I}_c} C_{\boldsymbol{x}}^{(k)}(\boldsymbol{\tau})\, e^{j2\pi\beta_k(\boldsymbol{\tau})t} + c_{\boldsymbol{x}}(t, \boldsymbol{\tau}) \tag{2.90b}$$

with

$$\lim_{T \to \infty} \frac{1}{T} \int_{-T/2}^{T/2} c_{\boldsymbol{x}}(t, \boldsymbol{\tau})\, e^{-j2\pi\beta t}\, \mathrm{d}t = 0 \qquad \forall \beta \in \mathbb{R}. \tag{2.91}$$

That is, $\mathcal{C}_{\boldsymbol{x}}(t, \boldsymbol{\tau})$ is an almost-periodic function in the sense of Hartman and Ryll-Nardzewski (Section 1.2.5).

2.2.5 *Processes with Almost-Periodic Covariance*

In this section, processes with almost-periodic covariance function are briefly reviewed and their link with GACS processes is enlightened.

A continuous-time complex-valued process $x(t)$ is said to be a *process with almost-periodic covariance* (Lii and Rosenblatt 2006) if

$$\mathrm{E}\left\{x(t+s_1)\,x^*(t+s_2)\right\} = \sum_n R_x^{(n)}(s_1,s_2)\,e^{j2\pi\alpha_n(s_1,s_2)t}. \tag{2.92}$$

By making the variable change $t+s_2 = t'$ (hence $t+s_1 = t'+s_1-s_2$) in the left-hand side of (2.92), we get

$$\mathrm{E}\left\{x(t'+s_1-s_2)\,x^*(t')\right\} = \sum_n R_x^{(n)}(s_1-s_2,0)\,e^{j2\pi\alpha_n(s_1-s_2,0)t'} \tag{2.93}$$

where the right-hand side is obtained by using (2.92) with the replacements $s_1-s_2 \curvearrowright s_1$, $0 \curvearrowright s_2$, and $t' \curvearrowright t$. Thus, every process with almost-periodic covariance can be reduced to a second-order wide-sense GACS process by a time-variable change.

A continuous-time complex-valued process $x(t)$ is said to be with *covariance almost-periodic of two variables* (Swift 1996) if

$$\mathrm{E}\left\{x(t_1)\,x^*(t_2)\right\} = \sum_h \sum_k a(\lambda_h,\lambda'_k)\,e^{j(\lambda_h t_1-\lambda'_k t_2)} \tag{2.94}$$

An almost-periodic stochastic process

$$x(t) = \sum_k x_k e^{j2\pi\lambda_k t} \tag{2.95}$$

with x_k correlated random variables such that $\mathrm{E}\left\{x_k\,x_h^*\right\} = a(\lambda_h,\lambda'_k)$ has a covariance almost-periodic of two variables.

2.2.6 *Motivations and Examples*

In this section, motivations for adopting the GACS model and examples of GACS processes are presented. Applications to communications, radar, and sonar are treated in Chapter 7.

2.2.6.1 Doppler Channel due to Radial Acceleration

The GACS model turns out to be appropriate in mobile communications systems when the channel cannot be modeled as almost-periodically time-variant (Izzo and Napolitano 2002a,b). For example, the output $y(t)$ of the Doppler channel existing between transmitter and receiver with nonzero relative radial acceleration is GACS when the input signal $x(t)$ is ACS (Izzo and Napolitano 2002b) (Sections 7.4, 7.5.2, 7.8).

Let

$$z_x(t) = \mathrm{Re}\left\{x(t)\,e^{j2\pi f_c t}\right\} \tag{2.96}$$

be the transmitted signal. In the case of relative motion between transmitter and receiver, under mild assumptions the received signal can be written as (Section 7.1)

$$\begin{aligned} z_y(t) &= a\,z_x(t-D(t)) \\ &= a\,\mathrm{Re}\left\{x(t-D(t))\,e^{j2\pi f_c(t-D(t))}\right\}. \\ &\triangleq \mathrm{Re}\left\{y(t)\,e^{j2\pi f_c t}\right\} \end{aligned} \tag{2.97}$$

where a is attenuation and $D(t)$ is time-varying delay. In the case of transmitter and receiver with constant relative radial acceleration, the delay is quadratically time-varying (Section 7.4), (Kelly and Wishner 1965), (Rihaczek 1967):

$$D(t) \triangleq d_0 + d_1 t + d_2 t^2 \qquad d_2 \neq 0. \tag{2.98}$$

Under the "narrow-band" approximation, the time-varying component of the delay in the complex envelope $x(t - D(t))$ can be ignored (Section 7.5.2) (Kelly and Wishner 1965) obtaining the chirp-modulated signal:

$$y(t) = b\, x(t - d_0)\, e^{j2\pi\nu t}\, e^{j\pi\gamma t^2} \tag{2.99}$$

where $b = a e^{-j2\pi f_c d_0}$ is the complex gain, $\nu = -f_c d_1$ the frequency shift, and $\gamma = -2 f_c d_2$ the chirp rate. The autocorrelation function of $y(t)$ is (Section 7.8)

$$\mathrm{E}\Big\{ y(t+\tau)\, y^*(t) \Big\} = |b|^2\, \mathrm{E}\Big\{ x(t+\tau-d_0)\, x^*(t-d_0) \Big\}\; e^{j2\pi\nu\tau}\, e^{j\pi\gamma\tau^2}\, e^{j2\pi\gamma\tau t}. \tag{2.100}$$

If $x(t)$ is ACS, that is,

$$\mathrm{E}\Big\{ x(t+\tau)\, x^*(t) \Big\} = \sum_{n\in\mathbb{I}} R^{\alpha_n}_{xx^*}(\tau)\, e^{j2\pi\alpha_n t} \tag{2.101}$$

where $\mathbb{I}$ is a countable set and $\{\alpha_n\}_{n\in\mathbb{I}} = A_{xx^*}$ is the set of cycle frequencies, (2.100) specializes into

$$\mathrm{E}\Big\{ y(t+\tau)\, y^*(t) \Big\} = |b|^2 \sum_{n\in\mathbb{I}} R^{\alpha_n}_{xx^*}(\tau)\, e^{j2\pi\nu\tau}\, e^{j\pi\gamma\tau^2}\, e^{-j2\pi\alpha_n d_0}\, e^{j2\pi[\alpha_n+\gamma\tau]t} \tag{2.102}$$

That is, $y(t)$ is GACS with autocorrelation function

$$\mathrm{E}\Big\{ y(t+\tau)\, y^*(t) \Big\} = \sum_{n\in\mathbb{I}} R^{(n)}_{yy^*}(\tau)\, e^{j2\pi\eta_n(\tau)t} \tag{2.103}$$

lag dependent cycle frequencies

$$\eta_n(\tau) = \alpha_n + \gamma\tau\,, \qquad n \in \mathbb{I} \tag{2.104}$$

generalized cyclic autocorrelation functions

$$R^{(n)}_{yy^*}(\tau) = |b|^2\, R^{\alpha_n}_{xx^*}(\tau)\, e^{j2\pi\nu\tau}\, e^{j\pi\gamma\tau^2}\, e^{-j2\pi\alpha_n d_0}\,, \qquad n \in \mathbb{I} \tag{2.105}$$

and cyclic autocorrelation function

$$R_{yy^*}(\alpha,\tau) = \sum_{n\in\mathbb{I}} R^{(n)}_{yy^*}(\tau)\, \delta_{\alpha-\eta_n(\tau)}. \tag{2.106}$$

In the figures, supports are drawn as "checkerboard" plots where gray levels represent magnitude of generalized cyclic autocorrelation functions.

In Figures 2.2 and 2.3, the magnitude and the support of the cyclic autocorrelation function, as a function of α and τ, for both the input $x(t)$ and the output $y(t)$ are reported for $x(t)$ PAM signal with raised cosine pulse and stationary-white binary modulating sequence. The signal $x(t)$ is cyclostationary with cycle frequencies $\alpha_n = n/T_p$, $n \in \{0, \pm 1\}$ (Gardner 1985), where

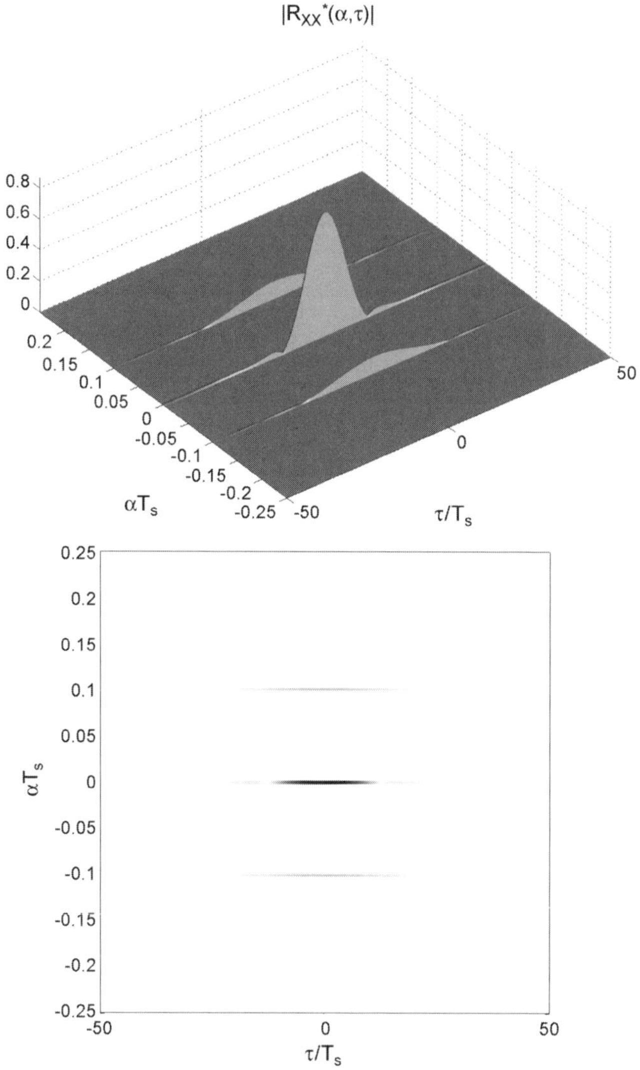

Figure 2.2 ACS input and GACS output of the Doppler channel existing between transmitter and receiver with constant relative radial acceleration. (Top) Magnitude and (bottom) support of the cyclic autocorrelation function $R_{xx^*}(\alpha, \tau)$, as a function of αT_s and τ / T_s, of the input ACS signal $x(t)$. *Source:* (Napolitano 2007a) © IEEE

T_p is the symbol period, and the support of the cyclic autocorrelation function is constituted by lines parallel to the τ axis (Figure 2.2 (bottom)). In contrast, the signal $y(t)$ is GACS with cyclic autocorrelation function whose support is described by the lag-dependent cycle frequencies $\eta_n(\tau)$ which are parallel lines with slope γ in the (α, τ) plane (Figure 2.3 (bottom)). Thus, they do not intersect each other. In Figures 2.2 and 2.3, $T_p = 10T_s$, with T_s the sampling period, and a raised cosine pulse with excess bandwidth $\eta = 0.85$ is considered.

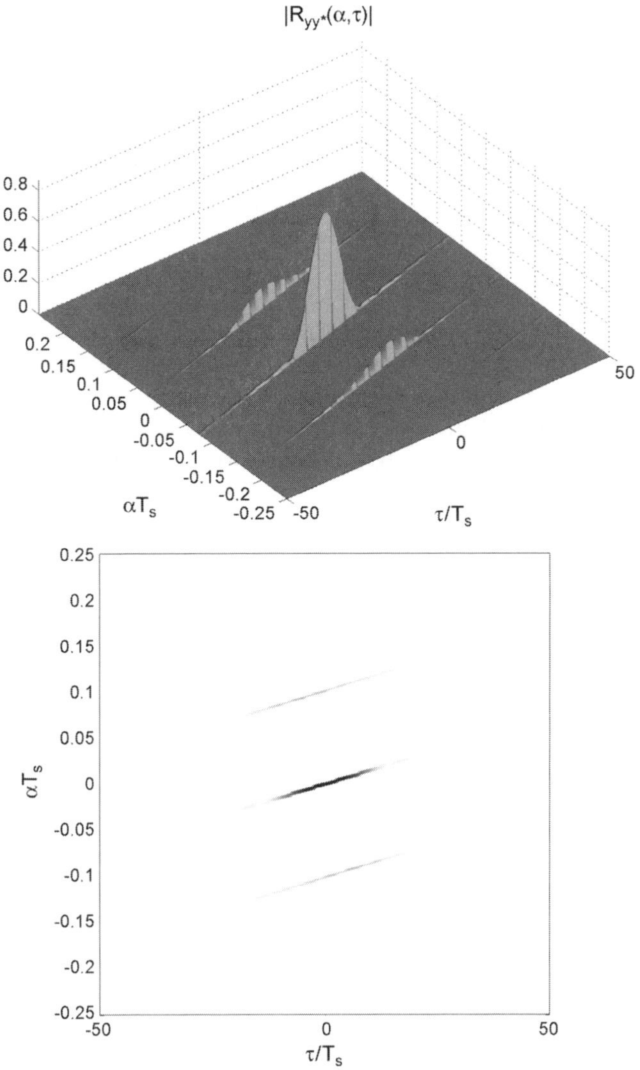

Figure 2.3 ACS input and GACS output of the Doppler channel existing between transmitter and receiver with constant relative radial acceleration. (Top) magnitude and (bottom) support of the cyclic autocorrelation function $R_{yy^*}(\alpha, \tau)$, as a function of αT_s and τ/T_s, of the output GACS signal $y(t)$. *Source:* (Napolitano 2007a) © IEEE

2.2.6.2 Communications Signals with Time-Varying Parameters

The GACS model is appropriate to describe communications signals with parameters, such as carrier frequency or baud rate, that are slowly varying functions of time. Specifically, in communications problems, signals can be modeled as ACS or GACS depending on the data-record length. In fact, if the data-record length is increased too much (e.g., in order to gain a better immunity against the effects of noise and interference), it can happen that the ACS model for the input signal is no longer appropriate but, rather, a GACS model needs to be considered since possible time variations of timing parameters of the signals (not evidenced with smaller data-record length) must be taken into account (Izzo and Napolitano 1998a, 2005). Consequently, increasing too much the data-record length without changing the statistical model for the signal, does not have, for example, the beneficial effect of improving the reliability of the signal cyclic statistic estimates but, rather, gives rise to biased cyclic statistics (Izzo and Napolitano 1998a, 2002b). Therefore, there exists an upper limit to the maximum usable data-record length and, consequently, there exists a limit to the minimum acceptable signal-to-noise ratio for cyclostationarity-based algorithms which are, in principle, intrinsically immune to the effects of noise and interference, provided that the data-record length approaches infinity. This fact puts some limitations on the performance obtainable with some signal processing algorithms adopted in communication applications where communications signals are modeled as ACS and constitutes a motivation to consider the class of the GACS signals.

Examples of communications signals with time-varying parameters that can be modeled as GACS are nonuniformly sampled signals (Izzo and Napolitano 1998a, 2002b, 2005) and modulated signals with sinusoidally-varying carrier frequency (Section 7.9.3). The former can be expressed as

$$x(t) \triangleq w(t) \sum_{k \in \mathbb{Z}} p(t - kT_p(t)) \tag{2.107}$$

where $p(t)$ is a finite-energy pulse and $T_p(t)$ is a slowly time-varying sampling period, and the latter can be written as

$$x(t) = w(t)\, \cos(2\pi(f_0 + \Delta \cos(2\pi f_m t))t) \tag{2.108}$$

where $\Delta \ll f_0$ and, in both examples, $w(t)$ is a stationary or ACS signal.

2.3 Linear Time-Variant Filtering of GACS Processes

Let $x_1(t)$ and $x_2(t)$ be jointly GACS processes, that is (Definition 2.2.10)

$$\mathrm{E}\left\{x_1(t+\tau_1)\; x_2^{(*)}(t+\tau_2)\right\} = \sum_{n \in \mathbb{I}_{\boldsymbol{x}}} R_{\boldsymbol{x}}^{(n)}(\tau_1 - \tau_2)\, e^{j2\pi\alpha_{\boldsymbol{x}}^{(n)}(\tau_1-\tau_2)\,(t+\tau_2)} \tag{2.109}$$

where $\boldsymbol{x} \triangleq [x_1 x_2^{(*)}]$, and let $y_1(t)$ and $y_2(t)$ be LTV filtered versions of $x_1(t)$ and $x_2(t)$, respectively, obtained by systems with impulse response functions $h_1(t,u)$ and $h_2(t,u)$.

Under assumption (1.47), substituting (2.109) into (1.46) leads to

$$\begin{aligned}
&\mathrm{E}\left\{y_1(t+\tau_1)\, y_2^{(*)}(t+\tau_2)\right\} \\
&= \int_{\mathbb{R}} \int_{\mathbb{R}} h_1(t+\tau_1, t+s_1)\, h_2^{(*)}(t+\tau_2, t+s_2) \\
&\qquad \mathrm{E}\left\{x_1(t+s_1)\; x_2^{(*)}(t+s_2)\right\} \,\mathrm{d}s_1\, \mathrm{d}s_2 \\
&= \sum_{n \in \mathbb{I}_{\boldsymbol{x}}} \int_{\mathbb{R}} \int_{\mathbb{R}} h_1(t+\tau_1, t+s_1)\, h_2^{(*)}(t+\tau_2, t+s_2) \\
&\qquad R_{\boldsymbol{x}}^{(n)}(s_1 - s_2)\, e^{j2\pi\alpha_{\boldsymbol{x}}^{(n)}(s_1-s_2)\,(t+s_2)} \,\mathrm{d}s_1\, \mathrm{d}s_2 .
\end{aligned} \tag{2.110}$$

Let $h_1(t, u)$ and $h_2(t, u)$ be LAPTV systems, that is, according to (1.107), with impulse-response functions

$$h_i(t + \tau_i, t + s_i) = \sum_{\sigma_i \in J_i} h_{\sigma_i}(\tau_i - s_i)\, e^{j2\pi\sigma_i(t+s_i)} \qquad i = 1, 2. \tag{2.111}$$

Substituting (2.111) into (2.110) leads to (Section 3.3)

$$\begin{aligned} &\mathrm{E}\left\{y_1(t+\tau_1)\, y_2^{(*)}(t+\tau_2)\right\} \\ &= \sum_{\sigma_1 \in J_1} \sum_{\sigma_2 \in J_2} \sum_{n \in \mathbb{I}_x} \int_{\mathbb{R}} R_{\boldsymbol{x}}^{(n)}(s)\, e^{j2\pi[\sigma_1 s + \alpha_{\boldsymbol{x}}^{(n)}(s) t]} \\ &\quad r_{\sigma_1\sigma_2(*)}^{\sigma_1 + (-)\sigma_2 + \alpha_{\boldsymbol{x}}^{(n)}(s)}(\tau_1 - \tau_2 - s)\, e^{j2\pi[\sigma_1 + (-)\sigma_2 + \alpha_{\boldsymbol{x}}^{(n)}(s)]\tau_2}\, \mathrm{d}s\, e^{j2\pi[\sigma_1 + (-)\sigma_2]t} \end{aligned} \tag{2.112}$$

where $r^{\gamma}_{\sigma_1\sigma_2(*)}(\tau)$ is defined in (1.115)

$$r^{\gamma}_{\sigma_1\sigma_2(*)}(\tau) \triangleq \int_{\mathbb{R}} h_{\sigma_1}(s+\tau)\, h_{\sigma_2}^{(*)}(s)\, e^{-j2\pi\gamma s}\, \mathrm{d}s. \tag{2.113}$$

Under assumptions

$$\sum_{n \in \mathbb{I}_x} \left| R_{\boldsymbol{x}}^{(n)}(\tau) \right| \in L^1(\mathbb{R}) \tag{2.114}$$

$$\sum_{\sigma_1 \in J_1} \|h_{\sigma_1}\|_{\infty} < \infty \tag{2.115}$$

$$\sum_{\sigma_2 \in J_2} \left| h_{\sigma_2}(\tau) \right| \in L^1(\mathbb{R}) \tag{2.116}$$

(where conditions (2.115) and (2.116) on $h_1(t, u)$ and $h_2(t, u)$ can be possibly interchanged) one obtains (Section 3.3)

$$\begin{aligned} R_{\boldsymbol{y}}(\beta, \tau) &\triangleq \left\langle \mathrm{E}\left\{y_1(t+\tau_1)\, y_2^{(*)}(t+\tau_2)\right\} e^{-j2\pi\beta t} \right\rangle_t \\ &= \sum_{\sigma_1 \in J_1} \sum_{\sigma_2 \in J_2} \sum_{n \in \mathbb{I}_x} \int_{\mathbb{R}} R_{\boldsymbol{x}}^{(n)}(s)\, r_{\sigma_1\sigma_2(*)}^{\sigma_1 + (-)\sigma_2 + \alpha_{\boldsymbol{x}}^{(n)}(s)}(\tau - s)\, e^{j2\pi\sigma_1 s} \\ &\quad \delta_{[\sigma_1 + (-)\sigma_2 + \alpha_{\boldsymbol{x}}^{(n)}(s) - \beta]}\, \mathrm{d}s. \end{aligned} \tag{2.117}$$

where $\boldsymbol{y} \triangleq [y_1 y_2^{(*)}]$.

Due to the presence of the Kronecker delta in the integrand function in (2.117), $R_{\boldsymbol{y}}(\beta, \tau)$ can be nonzero only if for some n, σ_1, and σ_2 the function $\sigma_1 + {}_{(-)}\sigma_2 + \alpha_{\boldsymbol{x}}^{(n)}(s) - \beta$ is nonzero in a set of values of s with positive Lebesgue measure. That is, only if $x_1(t)$ and $x_2(t)$ exhibit a joint ACS component in the cross-correlation function. In particular, from (2.117) with $x_1 \equiv x_2$ and $h_1 \equiv h_2$ LTI filters, accordingly with the results of Section 2.2.3, it follows that low-pass or band-pass filtering of a purely GACS signal (Section 2.2.2) leads to a zero-power signal (see Section 2.7.7 for a numerical example) (Izzo and Napolitano 2002a,b, 2005).

2.4 Estimation of the Cyclic Cross-Correlation Function

Let $x(t)$ and $y(t)$ be jointly GACS stochastic processes with cross-correlation function (2.31a)–(2.31c). In Sections 2.4–2.8 and 3.4–3.13, when it does not create ambiguity, for notation simplification we will put

$$\mathbb{I} \equiv \mathbb{I}_{yx^{(*)}} \quad \text{and} \quad \alpha_n(\tau) \equiv \alpha^{(n)}_{yx^{(*)}}(\tau).$$

From (2.31b), it follows that the knowledge of the cyclic cross-correlation function $R_{yx^{(*)}}(\alpha, \tau)$, as a function of the two variables α and τ, completely characterizes the second-order cross-moments of jointly GACS processes. For each τ, $R_{yx^{(*)}}(\alpha, \tau)$ is nonzero only for those values of α such that $\alpha = \alpha_n(\tau)$ for some $n \in \mathbb{I}_{yx^{(*)}}$ (see (2.39)). Moreover, for (α, τ) such that $\alpha = \alpha_n(\tau)$ for some $n \in \mathbb{I}_{yx^{(*)}}$, the magnitude and phase of $R_{yx^{(*)}}(\alpha, \tau)$ are the amplitude and phase of the finite-strength additive complex sinewave component at frequency α contained in the cross-moment $\mathrm{E}\{y(t+\tau)\, x^{(*)}(t)\}$ (see (2.31c) and (2.39)). Therefore, the problem of estimating second-order cross-moments of jointly GACS processes reduces to estimating the cyclic cross-correlation function as a function of the two variables (α, τ).

2.4.1 *The Cyclic Cross-Correlogram*

In this section, for jointly GACS processes, the cyclic cross-correlogram, the cyclic correlogram, and the conjugate cyclic correlogram are proposed as estimators of the cyclic cross-correlation function (2.33), the cyclic autocorrelation function (2.12), and the conjugate cyclic autocorrelation function (2.26), respectively. Moreover, their expected value and covariance are determined for finite data-record length (Napolitano 2007a).

Definition 2.4.1 *Let $\{x(t),\ t \in \mathbb{R}\}$ and $\{y(t),\ t \in \mathbb{R}\}$ be continuous-time stochastic processes. Their* cyclic cross-correlogram *at cycle frequency α is defined as*

$$R_{yx^{(*)}}(\alpha, \tau; t_0, T) \triangleq \int_{\mathbb{R}} w_T(t - t_0)\, y(t+\tau)\, x^{(*)}(t)\, e^{-j2\pi\alpha t}\, \mathrm{d}t \tag{2.118}$$

where $w_T(t)$ is a unit-area data-tapering window nonzero in $(-T/2, T/2)$. □

Note that since $w_T(t)$ has finite width, the integral in (2.118) is extended to $[t_0 - T/2, t_0 + T/2]$. Consequently, the cycle-frequency resolution is of the order of $1/T$.

By specializing (2.118) for $y(t) \equiv x(t)$ and $(*)$ present, one obtains the cyclic correlogram. For $y(t) \equiv x(t)$ and $(*)$ absent, one obtains the conjugate cyclic correlogram.

Assumption 2.4.2 Uniformly Almost-Periodic Statistics.

(a) The stochastic processes $y(t)$ and $x(t)$ are singularly and jointly (second-order) GACS in the wide sense, that is, according to (2.18) and (2.31c), for any choice of z_1 and z_2 in $\{x, x^*, y, y^*\}$

$$\mathrm{E}\{z_1(t+\tau_1)\, z_2(t+\tau_2)\} = \sum_{n \in \mathbb{I}_{z_1 z_2}} R^{(n)}_{z_1 z_2}(\tau_1 - \tau_2)\, e^{j2\pi\alpha^{(n)}_{z_1 z_2}(\tau_1-\tau_2)(t+\tau_2)} \tag{2.119}$$

(uniformly almost-periodic in t in the sense of Besicovitch (Besicovitch 1932)).

(b) *For any choice of* z_1 *in* $\{y, y^*\}$ *and* z_2 *in* $\{x, x^*\}$, *the fourth-order cumulant* $\text{cum}\{y(t+\tau_1), x^{(*)}(t+\tau_2), z_1(t+\tau_3), z_2(t)\}$ *can be expressed as*

$$\text{cum}\{y(t+\tau_1), x^{(*)}(t+\tau_2), z_1(t+\tau_3), z_2(t)\}$$
$$= \sum_{n \in \mathbb{I}_4} C^{(n)}_{yx^{(*)}z_1z_2}(\tau_1, \tau_2, \tau_3)\, e^{j2\pi\beta_n(\tau_1,\tau_2,\tau_3)t} \tag{2.120}$$

(uniformly almost-periodic in t *in the sense of Besicovitch), where* β_n *depends on* z_1, z_2, *and* $(*)$, *and cumulants of complex processes are defined according to (Spooner and Gardner 1994, App. A) (see also Section 1.4.2).* □

Assumption 2.4.3 Fourier Series Regularity.

(a) *For any choice of* z_1 *and* z_2 *in* $\{x, x^*, y, y^*\}$ *it results that*

$$\sum_{n \in \mathbb{I}_{z_1,z_2}} \left\| R^{(n)}_{z_1z_2} \right\|_\infty < \infty. \tag{2.121}$$

where $\|R\|_\infty \triangleq \text{ess}\sup_{\tau\in\mathbb{R}} |R(\tau)|$ *is the essential supremum of* $R(\tau)$ *(Champeney 1990).*

(b) *For any choice of* z_1 *in* $\{y, y^*\}$ *and* z_2 *in* $\{x, x^*\}$, *it results that*

$$\sum_{n \in \mathbb{I}_4} \left\| C^{(n)}_{yx^{(*)}z_1z_2} \right\|_\infty < \infty. \tag{2.122}$$

□

A necessary condition such that (2.121) holds is that there exists a positive number M such that

$$\sum_{n \in \mathbb{I}_{z_1,z_2}} \left| R^{(n)}_{z_1z_2}(\tau) \right| \leqslant M < \infty \tag{2.123}$$

uniformly with respect to τ.

Assumption 2.4.4 Fourth-Order Moment Boundedness. *The stochastic processes* $x(t)$ *and* $y(t)$ *have uniformly bounded fourth-order absolute moments. That is, for any* $z \in \{x, y\}$ *there exists a positive number* M_4 *such that*

$$\mathrm{E}\left\{|z(t)|^4\right\} \leqslant M_4 < \infty \quad \forall t \in \mathbb{R}. \tag{2.124}$$

□

Assumptions 2.4.2–2.4.4 regard time behavior and regularity of second- and fourth-order (joint) statistical functions of $x(t)$ and $y(t)$. Specifically, under such assumptions the second-order (cross-) moments of $x(t)$ and $y(t)$ and their fourth-order joint cumulants are limits of uniformly convergent sequences of trigonometric polynomials in t. Moreover, for each Fourier series in (2.119) and (2.120), the nth coefficient has amplitude approaching zero, as $n \to \infty$, sufficiently fast to assure that the infinite sums in (2.121) and (2.122) are convergent. A sufficient condition such that Assumption 2.4.4 holds is that the fourth-order moment functions of $x(t)$ and $y(t)$ are almost-periodic in t in the sense of Besicovitch.

Assumption 2.4.5 Data-Tapering Window Regularity. *$w_T(t)$ is a T-duration data-tapering window that can be expressed as*

$$w_T(t) = \frac{1}{T} a\left(\frac{t}{T}\right) \tag{2.125}$$

with $a(t) \in L^1(\mathbb{R}) \cap L^\infty(\mathbb{R})$, continuous almost everywhere (a.e.),

$$\int_{\mathbb{R}} a(t)\,\mathrm{d}t = 1 \tag{2.126}$$

$$\lim_{T\to\infty} a\left(\frac{t}{T}\right) = 1 \quad \forall t \in \mathbb{R}. \tag{2.127}$$

[Note that (2.127) is used only to prove (3.66)].

Let $A(f)$ be the Fourier transform of $a(t)$. In Section 3.5 (Lemma 3.5.1), it is shown that $A(f) \to 0$ as $|f| \to \infty$. Let us assume that there exists $\gamma > 0$ such that $A(f) = \mathcal{O}(|f|^{-\gamma})$ as $|f| \to \infty$.

In the proofs of the asymptotic properties of the discrete-time cyclic cross-correlogram (Section 2.6) we also need the assumptions that $a(t)$ is bounded and continuous in $(-1/2, 1/2)$ except, possibly, at $t = \pm 1/2$, $a(t)$ is Riemann integrable, and is differentiable almost everywhere (a.e.) with bounded first-order derivative $\dot{a}(t)$ (see Lemmas 3.10.1 and 3.10.2). □

Assumption 2.4.5 is easily verified by taking $a(t)$ with finite support $[-1/2, 1/2]$ and bounded (e.g., $a(t) = \mathrm{rect}(t)$). If $a(t)$ is continuous at $t = 0$, then from (2.127) it follows $a(0) = 1$.

The data-tapering window $w_T(t)$, strictly speaking, is a lag-product-tapering-window depending on the lag parameter τ. Its link with the signal-tapering window is discussed in Section 3.5, Fact 3.5.3.

By taking the expected value of the cyclic cross-correlogram (2.118) and using (2.31c), the following result is obtained, where the assumptions allow to interchange the order of expectation, integral, and sum operations.

Theorem 2.4.6 Expected Value of the Cyclic Cross-Correlogram (Napolitano 2007a, Theorem 3.1). *Let $y(t)$ and $x(t)$ be wide-sense jointly GACS stochastic processes with cross-correlation function (2.31c). Under Assumptions 2.4.2a (uniformly almost-periodic statistics), 2.4.3a (Fourier series regularity), and 2.4.5 (data-tapering window regularity), the expected value of the cyclic cross-correlogram $R_{yx^{(*)}}(\alpha, \tau; t_0, T)$ is given by*

$$\mathrm{E}\left\{R_{yx^{(*)}}(\alpha, \tau; t_0, T)\right\} = \sum_{n\in\mathbb{I}} R^{(n)}_{yx^{(*)}}(\tau)\, W_{\frac{1}{T}}(\alpha - \alpha_n(\tau))\, e^{-j2\pi[\alpha-\alpha_n(\tau)]t_0} \tag{2.128}$$

where $W_{\frac{1}{T}}(f)$ is the Fourier transform of $w_T(t)$.

Proof: See Section 3.4. □

The function $w_T(t)$ has duration T and, hence, $W_{\frac{1}{T}}(f)$ has a bandwidth of the order of $1/T$. Consequently, from (2.128), it follows that in the (α, τ)-plane the expected value of the cyclic cross-correlogram can be significantly different form zero within strips of width $1/T$ around the support curves $\alpha = \alpha_n(\tau)$, $n \in \mathbb{I}$, of the cyclic cross-correlation function. Moreover, the expected value along a given lag-dependent cycle-frequency curve $\alpha = \alpha_k(\tau)$ is influenced not only by $R^{(k)}_{yx^{(*)}}(\tau)$, but also by the values of the generalized cyclic cross-correlation functions $R^{(m)}_{yx^{(*)}}(\tau)$ relative to all the other lag-dependent cycle-frequency curves $\alpha = \alpha_m(\tau)$, $m \neq k$, the influence being stronger from curves $\alpha = \alpha_m(\tau)$ closer to $\alpha = \alpha_k(\tau)$ and with larger $|R^{(m)}_{yx^{(*)}}(\tau)|$. The effect becomes negligible as $T \to \infty$. Such a phenomenon, in the case of cyclic statistic estimates of ACS processes is referred to as *cyclic leakage* (Gardner 1987d).

By expressing the covariance of the second-order lag-product $y(t+\tau)\, x^{(*)}(t)$ in terms of second-order cross-moments and a fourth-order cumulant, the following result is obtained, where the assumptions allow to interchange the order of expectation, integral, and sum operations.

Theorem 2.4.7 Covariance of the Cyclic Cross-Correlogram (Napolitano 2007a, Theorem 3.2). *Let y(t) and x(t) be zero-mean wide-sense jointly GACS stochastic processes with cross-correlation function (2.31c). Under Assumptions 2.4.2 (uniformly almost-periodic statistics), 2.4.3 (Fourier series regularity), 2.4.4 (fourth-order moment boundedness), and 2.4.5 (data-tapering window regularity), the covariance of the cyclic cross-correlogram* $R_{yx^{(*)}}(\alpha, \tau; t_0, T)$ *is given by*

$$\operatorname{cov}\left\{R_{yx^{(*)}}(\alpha_1, \tau_1; t_1, T), R_{yx^{(*)}}(\alpha_2, \tau_2; t_2, T)\right\} = \mathcal{T}_1 + \mathcal{T}_2 + \mathcal{T}_3 \tag{2.129}$$

where

$$\mathcal{T}_1 \triangleq \sum_{n'} \sum_{n''} \int_{\mathbb{R}} R^{(n')}_{yy^*}(\tau_1 - \tau_2 + s)\, R^{(n'')}_{x^{(*)}x^{(*)*}}(s)\, e^{j2\pi\alpha^{(n')}_{yy^*}(\tau_1-\tau_2+s)\tau_2}\, e^{-j2\pi\alpha_1 s}$$
$$e^{-j2\pi\Phi_1(s)t_2}\, \frac{1}{T}\, r_{aa^*}\left(\Phi_1(s)\, T; \frac{t_2 - t_1 + s}{T}\right) \mathrm{d}s \tag{2.130}$$

$$\mathcal{T}_2 \triangleq \sum_{n'''} \sum_{n^{iv}} \int_{\mathbb{R}} R^{(n''')}_{yx^{(*)*}}(\tau_1 + s)\, R^{(n^{iv})}_{x^{(*)}y^*}(s - \tau_2)\, e^{j2\pi\alpha^{(n^{iv})}_{x^{(*)}y^*}(s-\tau_2)\tau_2}\, e^{-j2\pi\alpha_1 s}$$
$$e^{-j2\pi\Phi_2(s)t_2}\, \frac{1}{T}\, r_{aa^*}\left(\Phi_2(s)\, T; \frac{t_2 - t_1 + s}{T}\right) \mathrm{d}s \tag{2.131}$$

$$\mathcal{T}_3 \triangleq \sum_{n} \int_{\mathbb{R}} C^{(n)}_{yx^{(*)}y^*x^{(*)*}}(\tau_1 + s, s, \tau_2)\, e^{-j2\pi\alpha_1 s}$$
$$e^{-j2\pi\Phi_3(s)t_2}\, \frac{1}{T}\, r_{aa^*}\left(\Phi_3(s)\, T; \frac{t_2 - t_1 + s}{T}\right) \mathrm{d}s \tag{2.132}$$

with

$$r_{aa^*}(\beta; s) \triangleq \int_{\mathbb{R}} a(t+s)\, a^*(t)\, e^{-j2\pi\beta t}\, \mathrm{d}t \tag{2.133}$$

$$\Phi_1(s) \triangleq -[\alpha^{(n')}_{yy^*}(\tau_1 - \tau_2 + s) + \alpha^{(n'')}_{x^{(*)}x^{(*)*}}(s) - \alpha_1 + \alpha_2] \tag{2.134}$$

$$\Phi_2(s) \triangleq -[\alpha^{(n''')}_{yx^{(*)*}}(\tau_1 + s) + \alpha^{(n^{iv})}_{x^{(*)}y^*}(s - \tau_2) - \alpha_1 + \alpha_2] \tag{2.135}$$

$$\Phi_3(s) \triangleq -[\beta_n(\tau_1 + s, s, \tau_2) - \alpha_1 + \alpha_2]. \tag{2.136}$$

Proof: See Section 3.4. □

For complex-valued processes, both covariance and conjugate covariance are needed for a complete second-order wide-sense characterization (Picinbono 1996), (Picinbono and Bondon 1997), (Schreier and Scharf 2003a,b). The expression of the conjugate covariance of the cyclic cross-correlogram can be obtained by reasoning as for Theorem 2.4.7 and is reported in Section 3.7.

Finally, note that in the special case of ACS processes, Theorems 2.4.6 and 2.4.7 reduce to well-known results of (Hurd 1989a, 1991), (Hurd and Leśkow 1992b), (Dehay and Hurd 1994), (Genossar *et al.* 1994), (Dandawaté and Giannakis 1995).

2.4.2 Mean-Square Consistency of the Cyclic Cross-Correlogram

In this section, the cyclic cross-correlogram is shown to be a mean-square consistent estimator of the cyclic cross-correlation function (Napolitano 2007a).

Assumption 2.4.8 Mixing Conditions (I).

(a) *For any choice of z_1 and z_2 in $\{x, x^*, y, y^*\}$ it results that*

$$\sum_{n \in \mathbb{I}_{z_1, z_2}} \int_{\mathbb{R}} \left| R^{(n)}_{z_1 z_2}(s) \right| \mathrm{d}s < \infty. \tag{2.137}$$

(b) *For any choice of z_1 in $\{y, y^*\}$ and z_2 in $\{x, x^*\}$ and $\forall \tau_1, \tau_2 \in \mathbb{R}$ it results that*

$$\sum_{n \in \mathbb{I}_4} \int_{\mathbb{R}} \left| C^{(n)}_{yx^{(*)}z_1 z_2}(s + \tau_1, s, \tau_2) \right| \mathrm{d}s < \infty. \tag{2.138}$$

In the proofs of the asymptotic properties of the discrete-time cyclic cross-correlogram (Section 2.6) we also need the assumption that the functions $R^{(n)}_{z_1 z_2}(s)$ and $C^{(n)}_{yx^{(*)}z_1 z_2}$ $(s + \tau_1, s, \tau_2)$ are Riemann integrable. □

Assumptions 2.4.8a and 2.4.8b are referred to as mixing conditions and are generally satisfied if the involved stochastic processes have finite or practically finite memory, i.e., if $z_1(t)$ and $z_2(t+s)$ are asymptotically ($|s| \to \infty$) independent (Section 1.4.1), (Brillinger and Rosenblatt 1967). For example, under assumption (2.121), the function series $\sum_n |R^{(n)}_{z_1 z_2}(s)|$ is uniformly convergent due to the Weierstrass criterium (Smirnov 1964). In addition, in order to satisfy

(2.137), the function $\sum_n |R^{(n)}_{z_1 z_2}(s)|$ should be vanishing sufficiently fast as $|s| \to \infty$ in order to be summable. A sufficient condition is that there exists $\epsilon > 0$ such that $\sum_n |R^{(n)}_{z_1 z_2}(s)| = \mathcal{O}(|s|^{-1-\epsilon})$, where $\mathcal{O}$ is the "big oh" Landau symbol.

Assumption 2.4.8b is used to prove the asymptotic vanishing of the covariance of the cyclic cross-correlogram in Theorem 2.4.13. This condition, however, is not sufficient to obtain a bound for the covariance uniform with respect to α_1, α_2, τ_1, τ_2. Uniformity is obtained in Corollary 2.4.14 with the following further assumption.

Assumption 2.4.9 Mixing Conditions (II). *For any choice of z_1 in $\{y, y^*\}$ and z_2 in $\{x, x^*\}$ there exist functions $\phi^{(n)}(s)$ such that*

$$\sup_{\tau_1,\tau_2 \in \mathbb{R}} \left| C^{(n)}_{yx^{(*)} z_1 z_2}(s + \tau_1, s, \tau_2) \right| \leqslant \phi^{(n)}(s) \in L^1(\mathbb{R}) \tag{2.139}$$

with

$$\sum_{n \in \mathbb{I}_4} \int_{\mathbb{R}} \phi^{(n)}(s) \, \mathrm{d}s < \infty. \tag{2.140}$$

□

Assumption 2.4.9 implies Assumption 2.4.8b with the left-hand side of (2.138) uniformly bounded with respect to τ_1 and τ_2.

Assumption 2.4.10 Lack of Support Curve Clusters (I). *There is no cluster of support curves. That is, let*

$$\mathbb{J}_{\alpha,\tau} \triangleq \left\{ n \in \mathbb{I} : \alpha_n(\tau) = \alpha, \quad R^{(n)}_{yx^{(*)}}(\tau) \neq 0 \right\} \tag{2.141}$$

then for every α_0 and τ_0, for any $n \notin \mathbb{J}_{\alpha_0,\tau_0}$ no curve $\alpha_n(\tau)$ can be arbitrarily close to the value α_0 for $\tau = \tau_0$. Thus, for any α and τ it results in

$$h_{\alpha,\tau} \triangleq \inf_{n \notin \mathbb{J}_{\alpha,\tau}} |\alpha - \alpha_n(\tau)| > 0. \tag{2.142}$$

□

The set $\mathbb{J}_{\alpha_0,\tau_0}$ contains the only element $\bar{n}$ if $\alpha = \alpha_{\bar{n}}(\tau)$ is the only support curve such that $\alpha_0 = \alpha_{\bar{n}}(\tau_0)$. The set $\mathbb{J}_{\alpha_0,\tau_0}$ contains more elements if (α_0, τ_0) is a point where more support curves intercept each other (see (2.20) and (2.21)). The set $\mathbb{J}_{\alpha_0,\tau_0}$ is empty if $\alpha_0 \neq \alpha_n(\tau_0)$ $\forall n \in \mathbb{I}$.

Assumption 2.4.10 is used to establish the rate of convergence to zero of the bias and the asymptotic Normality of the cyclic cross-correlogram. In the special case of ACS signals, Assumption 2.4.10 means that there is no cycle frequency cluster point (see (Hurd 1991), (Dehay and Hurd 1994)).

By taking the limit of the expected value of the cyclic cross-correlogram (2.128) as the data-record length T approaches infinite, the following result is obtained, where the assumptions allow to interchange the order of limit and sum operations.

Theorem 2.4.11 Asymptotic Expected Value of the Cyclic Cross-Correlogram (Napolitano 2007a, Theorem 4.1). *Let $y(t)$ and $x(t)$ be wide-sense jointly GACS stochastic processes with cross-correlation function (2.31c). Under Assumptions 2.4.2a (uniformly almost-periodic statistics), 2.4.3a (Fourier series regularity), and 2.4.5 (data-tapering window regularity), the asymptotic expected value of the cyclic cross-correlogram $R_{yx^{(*)}}(\alpha, \tau; t_0, T)$ is given by*

$$\lim_{T\to\infty} \mathrm{E}\left\{R_{yx^{(*)}}(\alpha, \tau; t_0, T)\right\} = R_{yx^{(*)}}(\alpha, \tau). \tag{2.143}$$

Proof: See Section 3.5. □

Let $A(f)$ be the Fourier transform of $a(t)$ with rate of decay to zero $|f|^{-\gamma}$ for $|f| \to \infty$ as specified in Assumption 2.4.5. For the bias of the cyclic cross-correlogram

$$\text{bias}\left\{R_{yx^{(*)}}(\alpha, \tau; t_0, T)\right\} \triangleq \mathrm{E}\left\{R_{yx^{(*)}}(\alpha, \tau; t_0, T)\right\} - R_{yx^{(*)}}(\alpha, \tau). \tag{2.144}$$

the following result holds.

Theorem 2.4.12 Rate of Convergence of the Bias of the Cyclic Cross-Correlogram (Napolitano 2007a, Theorem 4.2). *Let $y(t)$ and $x(t)$ be wide-sense jointly GACS stochastic processes with cross-correlation function (2.31c). Under Assumptions 2.4.2a (uniformly almost-periodic statistics), 2.4.3a (Fourier series regularity), 2.4.5 (data-tapering window regularity), and 2.4.10 (lack of support curve clusters (I)), one obtains*

$$\lim_{T\to\infty} T^{\gamma} \left|\text{bias}\left\{R_{yx^{(*)}}(\alpha, \tau; t_0, T)\right\}\right| = \mathcal{O}(1). \tag{2.145}$$

Proof: See Section 3.5. □

We are interested in finding the maximum of the values of γ such that (2.145) holds, provided that such a maximum exists. This can be achieved, for example, by observing that if $a(t)$ is p times differentiable and the p-th order derivative $a^{(p)}(t)$ is summable, then $|A(f)| \leqslant \|a^{(p)}\|_{L^1}/((2\pi)^p|f|^p)$, where $\|\cdot\|_{L^1}$ denotes the L^1-norm. As further examples, $a(t) = \text{rect}(t) \Rightarrow A(f) = \text{sinc}(f) \Rightarrow \gamma = 1$; $a(t) = (1 - |t|)\,\text{rect}(t/2) \Rightarrow A(f) = \text{sinc}^2(f) \Rightarrow \gamma = 2$. Furthermore, the result of Theorem 2.4.12 can be extended with minor changes to the case $A(f) = \mathcal{O}(g(f))$ as $|f| \to \infty$, with $g(f)$ strictly increasing function with no order of infinity (e.g., $g(|f|) = |f| \ln |f|$).

In Fact 3.5.3, the link between the signal-tapering window and the lag-product-tapering window is established. Consequences of this link on the rate of convergence of bias are discussed in Section 3.5.

Starting from Theorem 2.4.7, the following result is obtained, where the assumptions allow to interchange the order of limit and sum operations.

Theorem 2.4.13 Asymptotic Covariance of the Cyclic Cross-Correlogram (Napolitano 2007a, Theorem 4.3). *Let $y(t)$ and $x(t)$ be zero-mean wide-sense jointly GACS stochastic processes with cross-correlation function (2.31c). Under Assumptions 2.4.2 (uniformly almost-periodic statistics), 2.4.3 (Fourier series regularity), 2.4.4 (fourth-order moment boundedness),*

2.4.5 (data-tapering window regularity), and 2.4.8 (mixing conditions (I)), the asymptotic covariance of the cyclic cross-correlogram $R_{yx^{(*)}}(\alpha, \tau; t_0, T)$ is given by

$$\lim_{T\to\infty} T\mathrm{cov}\left\{R_{yx^{(*)}}(\alpha_1, \tau_1; t_1, T), R_{yx^{(*)}}(\alpha_2, \tau_2; t_2, T)\right\} = \mathcal{T}_1' + \mathcal{T}_2' + \mathcal{T}_3' \tag{2.146}$$

where

$$\mathcal{T}_1' \triangleq \mathcal{E}_a \sum_{n'} \sum_{n''} \int_{\mathbb{R}} R^{(n')}_{yy^*}(\tau_1 - \tau_2 + s)\, R^{(n'')}_{x^{(*)}x^{(*)*}}(s) \; e^{j2\pi\alpha'_{n'}(\tau_1-\tau_2+s)\tau_2}\, e^{-j2\pi\alpha_1 s}\, \delta_{[\alpha'_{n'}(\tau_1-\tau_2+s)+\alpha''_{n''}(s)-\alpha_1+\alpha_2]}\, \mathrm{d}s \tag{2.147}$$

$$\mathcal{T}_2' \triangleq \mathcal{E}_a \sum_{n'''} \sum_{n^{iv}} \int_{\mathbb{R}} R^{(n''')}_{yx^{(*)*}}(\tau_1 + s)\, R^{(n^{iv})}_{x^{(*)}y^*}(s - \tau_2) \; e^{j2\pi\alpha^{iv}_{n^{iv}}(s-\tau_2)\tau_2}\, e^{-j2\pi\alpha_1 s}\, \delta_{[\alpha'''_{n'''}(\tau_1+s)+\alpha^{iv}_{n^{iv}}(s-\tau_2)-\alpha_1+\alpha_2]}\, \mathrm{d}s \tag{2.148}$$

$$\mathcal{T}_3' \triangleq \mathcal{E}_a \sum_{n} \int_{\mathbb{R}} C^{(n)}_{yx^{(*)}y^*x^{(*)*}}(\tau_1 + s, s, \tau_2)\, e^{-j2\pi\alpha_1 s} \delta_{[\beta_n(\tau_1+s,s,\tau_2)-\alpha_1+\alpha_2]}\, \mathrm{d}s \tag{2.149}$$

with $\mathcal{E}_a \triangleq \int_{\mathbb{R}} |a(u)|^2\, \mathrm{d}u$. In (2.147)–(2.149), for notation simplicity, $\alpha'_{n'}(\cdot) \equiv \alpha^{(n')}_{yy^*}(\cdot)$, $\alpha''_{n''}(\cdot) \equiv \alpha^{(n'')}_{x^{(*)}x^{(*)*}}(\cdot)$, $\alpha'''_{n'''}(\cdot) \equiv \alpha^{(n''')}_{yx^{(*)*}}(\cdot)$, and $\alpha^{iv}_{n^{iv}}(\cdot) \equiv \alpha^{(n^{iv})}_{x^{(*)}y^*}(\cdot)$.

Proof: See Section 3.5. □

Theorem 2.4.13 can also be proved by substituting the mixing condition in Assumption 2.4.8a with the mixing condition in Assumption 3.5.4 as explained in Section 3.5.

In Theorem 2.4.13, the terms in (2.147), (2.148), and (2.149) can give nonzero contribution only if the argument of the Kronecker deltas is zero in a set of values of s with a positive Lebesgue measure. For example, if the signals $x(t)$ and $y(t)$ are singularly and jointly purely GACS (Section 2.2.2), then in (2.147) with $\alpha_1 = \alpha_2$ and $\tau_1 = \tau_2$, the terms with $\alpha'_{n'}(s) = -\alpha''_{n''}(s)$ can give nonzero contribution.

From Theorem 2.4.11 it follows that the cyclic cross-correlogram (2.118), as a function of $(\alpha, \tau) \in \mathbb{R} \times \mathbb{R}$, is an asymptotically unbiased estimator of the cyclic cross-correlation function (2.33). Moreover, from Theorem 2.4.13 it follows that its variance

$$\mathrm{var}\{R_{yx^{(*)}}(\alpha, \tau; t_0, T)\} \triangleq \mathrm{cov}\{R_{yx^{(*)}}(\alpha, \tau; t_0, T), R_{yx^{(*)}}(\alpha, \tau; t_0, T)\} \tag{2.150}$$

is $\mathcal{O}(T^{-1})$ as $T \to \infty$. Therefore, since

$$\mathrm{E}\left\{\left|R_{yx^{(*)}}(\alpha, \tau; t_0, T) - R_{yx^{(*)}}(\alpha, \tau)\right|^2\right\} = \left|\mathrm{bias}\{R_{yx^{(*)}}(\alpha, \tau; t_0, T)\}\right|^2 + \mathrm{var}\{R_{yx^{(*)}}(\alpha, \tau; t_0, T)\} \tag{2.151}$$

the cyclic cross-correlogram is a mean-square consistent estimator of the cyclic cross-correlation function.

The expression of the asymptotic conjugate covariance of the cyclic cross-correlogram is reported in Section 3.7.

By specializing Theorems 2.4.11–2.4.13 for $y = x$ and $(*)$ present, we have that the cyclic correlogram is a mean-square consistent estimator of the cyclic autocorrelation function. For $y = x$ and $(*)$ absent, we have that the conjugate cyclic correlogram is a mean-square consistent estimator of the conjugate cyclic autocorrelation function.

For ACS processes, we obtain as a special case results of (Hurd 1989a, 1991), (Hurd and Leśkow 1992b), (Dehay and Hurd 1994), (Genossar *et al.* 1994), (Dandawaté and Giannakis 1995), (Schell 1995). In particular, in the asymptotic covariance (2.146), the terms $\mathcal{T}_1'$, $\mathcal{T}_2'$, and $\mathcal{T}_3'$ defined in (2.147), (2.148), and (2.149) specialize into

$$\mathcal{T}_1' \triangleq \mathcal{E}_a \sum_{n'} \int_{\mathbb{R}} R_{yy^*}^{\alpha'_{n'}}(\tau_1 - \tau_2 + s)\, R_{x^{(*)}x^{(*)*}}^{\alpha_1 - \alpha_2 - \alpha'_{n'}}(s)\, e^{j2\pi\alpha'_{n'}\tau_2}\, e^{-j2\pi\alpha_1 s}\, \mathrm{d}s \tag{2.152}$$

$$\mathcal{T}_2' \triangleq \mathcal{E}_a \sum_{n'''} \int_{\mathbb{R}} R_{yx^{(*)*}}^{\alpha'''_{n'''}}(\tau_1 + s)\, R_{x^{(*)}y^*}^{\alpha_1 - \alpha_2 - \alpha'''_{n'''}}(s - \tau_2)\, e^{j2\pi\alpha^{\prime v}_{n^{\prime v}}\tau_2}\, e^{-j2\pi\alpha_1 s}\, \mathrm{d}s \tag{2.153}$$

$$\mathcal{T}_3' \triangleq \mathcal{E}_a \int_{\mathbb{R}} C_{yx^{(*)}y^*x^{(*)*}}^{\alpha_1 - \alpha_2}(\tau_1 + s, s, \tau_2)\, e^{-j2\pi\alpha_1 s}\, \mathrm{d}s \tag{2.154}$$

where the sum over n' is extended to all cycle frequencies $\alpha'_{n'}$ of $R_{yy^*}^{\alpha'_{n'}}(s)$ such that $\alpha_1 - \alpha_2 - \alpha'_{n'}$ is a cycle frequency of $R_{x^{(*)}x^{(*)*}}^{\alpha''_{n''}}(s)$, the sum over n''' is extended to all cycle frequencies $\alpha'''_{n'''}$ of $R_{yx^{(*)*}}^{\alpha'''_{n'''}}(s)$ such that $\alpha_1 - \alpha_2 - \alpha'''_{n'''}$ is a cycle frequency of $R_{x^{(*)}y^*}^{\alpha^{\prime v}_{n^{\prime v}}}(s)$, and the last term is nonzero only if $\alpha_1 - \alpha_2$ is a (fourth-order) cycle frequency of $C_{yx^{(*)}y^*x^{(*)*}}^{\beta}(s_1, s_2, s_3)$.

The asymptotic bias (Theorem 2.4.12), in general, is not uniform with respect to α and τ. In fact, from (2.128) and the continuity of $W_{\frac{1}{T}}(\cdot)$ (Lemma 3.5.1a), $\mathrm{E}\left\{R_{yx^{(*)}}(\alpha, \tau; t_0, T)\right\}$ is a continuous function of α for each fixed τ at least in the case of $\mathbb{I}$ finite or for $|R_{yx^{(*)}}^{(n)}(\tau)|$ approaching zero sufficiently fast as $n \to \infty$. In contrast, since for each fixed τ the set A_τ defined in (2.32a) is countable, the function $R_{yx^{(*)}}(\alpha, \tau)$ is discontinuous in α (see also (2.39)). Thus, we have the convergence of a family of continuous functions to a discontinuous function and, hence, the convergence cannot be uniform. In particular, the closer the point (α, τ) is to a support curve, the slower the convergence of the estimator bias. An analogous result was found in (Genossar *et al.* 1994) for discrete-time asymptotically-mean cyclostationary processes. Furthermore, the asymptotic variance (Theorem 2.4.13) is not uniform with respect to α_1, α_2, τ_1, and τ_2. However, a bound for the covariance, uniform with respect to cycle frequencies and lag shifts can be obtained starting from Theorem 2.4.7 with the aid of Assumptions 2.4.8a and 2.4.9. It is the analogous for continuous-time GACS processes of a similar result proved in (Genossar *et al.* 1994) for discrete-time asymptotically-mean cyclostationary processes.

Corollary 2.4.14 (Napolitano 2007a, Corollary 4.1). *Let $y(t)$ and $x(t)$ be zero-mean wide-sense jointly GACS stochastic processes with cross-correlation function (2.31c). Under Assumptions 2.4.2 (uniformly almost-periodic statistics), 2.4.3 (Fourier series regularity), 2.4.4 (fourth-order moment boundedness), 2.4.5 (data-tapering window regularity), 2.4.8 (mixing conditions (I)), and 2.4.9 (mixing conditions (II)), there exists K such that*

$$\left|\mathrm{cov}\left\{R_{yx^{(*)}}(\alpha_1, \tau_1; t_1, T),\, R_{yx^{(*)}}(\alpha_2, \tau_2; t_2, T)\right\}\right| \leqslant \frac{K}{T} \tag{2.155}$$

uniformly with respect to α_1, α_2, τ_1, and τ_2.

Proof: See Section 3.5. □

2.4.3 Asymptotic Normality of the Cyclic Cross-Correlogram

In this section, the asymptotic complex Normality (Picinbono 1996), (van den Bos 1995) of the cyclic cross-correlogram is proved (Napolitano 2007a).

Let

$$z_i(t) \triangleq \left[y(t+\tau_i)\, x^{(*)}(t)\right]^{[*]_i}, \quad i = 1, \ldots, k \tag{2.156}$$

be second-order lag-product waveforms, with optional complex conjugations $[*]_i, i = 1, \ldots, k$, having kth-order cumulant

$$C_{z_1 \ldots z_k}(t; s_1, \ldots, s_{k-1}) \triangleq \text{cum}\,\{z_k(t), z_i(t+s_i),\; i = 1, \ldots, k-1\} \tag{2.157}$$

where the cumulant of complex processes is defined according to (Spooner and Gardner 1994, App. A) (see also Section 1.4.2).

Assumption 2.4.15 Mixing Conditions (III). *For every integer $k \geqslant 2$, $\tau_i \in \mathbb{R}$, $i = 1, \ldots, k$ and every conjugation configuration $[*]_1, \ldots, [*]_k$, there exists a positive summable function $\phi(s_1, \ldots, s_{k-1})$ (depending on τ_i, $i = 1, \ldots, k$ and the conjugation configuration) such that*

$$\left|C_{z_1 \ldots z_k}(t; s_1, \ldots, s_{k-1})\right| \leqslant \phi(s_1, \ldots, s_{k-1}) \in L^1(\mathbb{R}^{k-1}) \tag{2.158}$$

uniformly with respect to t. □

The cumulant of the lag products $z_i(t)$ can be expressed in terms of cumulants of $y(t+\tau_j)$ and $x(t)$. For this purpose, let us consider the $k \times 2$ table

$$\begin{array}{ll} y(t+\tau_1+s_1) & x^{(*)}(t+s_1) \\ y(t+\tau_2+s_2) & x^{(*)}(t+s_2) \\ \vdots & \vdots \\ y(t+\tau_{k-1}+s_{k-1}) & x^{(*)}(t+s_{k-1}) \\ y(t+\tau_k) & x^{(*)}(t) \end{array} \tag{2.159}$$

and a partition of its elements into disjoint sets $\{\nu_1, \ldots, \nu_p\}$, that is, $\nu_i \cap \nu_j = \emptyset$ for $i \neq j$ and $\cup_{i=1}^{p} \nu_i =$ the whole table. Then, let us denote each element of the table (2.159) by its row and column indices (ℓ, m). We say that the two sets ν_{i_1} and ν_{i_2} hook if there exist $(\ell_1, m_1) \in \nu_{i_1}$ and $(\ell_2, m_2) \in \nu_{i_2}$ such that $\ell_1 = \ell_2$, that is, the two sets contain (al least) two elements with the same row index. We say that the sets $\nu_{i'}$ and $\nu_{i''}$ communicate if there exists a sequence of sets $\nu_{i'} \equiv \nu_{i_1}, \nu_{i_2}, \ldots, \nu_{i_r} \equiv \nu_{i''}$ such that ν_{i_h} and $\nu_{i_{h+1}}$ hook for each h. A partition is said to be *indecomposable* if all its sets communicate (Leonov and Shiryaev 1959), (Brillinger 1965), (Brillinger and Rosenblatt 1967). It can be seen that if $r_1, \ldots, r_k$ denote the rows of the table (2.159), then the partition $\{\nu_1, \ldots, \nu_p\}$ is indecomposable if and only if there exist no sets $\nu_{i_1}, \ldots, \nu_{i_n}, (n < p)$, and rows $r_{j_1}, \ldots, r_{j_h}, (h < k)$, such that $r_{j_1} \cup \cdots \cup r_{j_h} = \nu_{i_1} \cup \cdots \cup \nu_{i_n}$; that is, if and only if we cannot build a rectangle of size $h \times 2$, with $h < k$, using some of the sets of the partition.

The cumulant $C_{z_1 \ldots z_k}(t; s_1, \ldots, s_{k-1})$ can be expressed as (Leonov and Shiryaev 1959), (Brillinger 1965), (Brillinger and Rosenblatt 1967)

$$C_{z_1 \ldots z_k} = \sum_{\nu} C_{\nu_1} \cdots C_{\nu_p} \tag{2.160}$$

where ν_i ($i = 1, \ldots, p$) are subsets of elements of the $k \times 2$ table (2.159), C_{ν_i} is the cumulant of the elements in ν_i, and the (finite) sum in (2.160) is extended over all indecomposable partitions of table (2.159), including the partition with only one element (see (3.45) for the case $k = 2$ and zero-mean processes). Therefore, Assumption 2.4.15 is satisfied provided that the cumulants up to order $2k$ of the elements taken from table in (2.159) are summable with respect to the variables $s_1, \ldots, s_{k-1}$ and a bound uniform with respect to t exists. That is

$$|C_{z_1 \ldots z_k}| \leqslant \sum_{\nu} |C_{\nu_1}| \cdots |C_{\nu_p}| \leqslant \phi. \tag{2.161}$$

Assumption 2.4.15 is referred to as a mixing condition and is fulfilled if the processes $x(t)$ and $y(t)$ have finite-memory or a memory decaying sufficiently fast. It generalizes to the nonstationary case, assumption made in (Brillinger 1965, eq. (5.5)) in the stationary case. Examples of processes satisfying Assumption 2.4.15 are communications signals with independent and identically distributed (i.i.d.) symbols or block coded symbols, processes with exponentially decaying joint probability density function, and finite-memory (time-variant) transformations of such processes.

For a single process $x(t)$ ($y \equiv x$), according to the results of Section 1.4.1, the asymptotic statistical independence implies that, for every ℓ,

$$\operatorname{cum}\{x(t),\ x(t+\tau_i),\ i = 1, \ldots, \ell - 1\} \to 0 \quad \text{as } \|\boldsymbol{\tau}\| \to \infty \tag{2.162}$$

where $\|\boldsymbol{\tau}\|^2 \triangleq \tau_1^2 + \cdots + \tau_{\ell-1}^2$. Thus, the summability condition (2.161) is satisfied provided that

$$\operatorname{cum}\{x(t),\ x(t+\tau_i),\ i = 1, \ldots, \ell - 1\} = \mathcal{O}\left(\|\boldsymbol{\tau}\|^{-(\ell-1+\epsilon)}\right) \tag{2.163}$$

for some $\epsilon > 0$, uniformly with respect to t, for every $\ell \leqslant 2k$.

Assumption 2.4.15 turns out to be verified if the stochastic processes $z_i(t)$, $i = 1, \ldots, k$ are jointly kth-order GACS (Section 2.2.4), (Izzo and Napolitano 1998a), that is

$$C_{z_1 \ldots z_k}(t; s_1, \ldots, s_{k-1}) = \sum_{n} C^{(n)}_{z_1 \cdots z_k}(s_1, \ldots, s_{k-1})\, e^{j2\pi\beta^{(n)}_{z_1 \cdots z_k}(s_1, \ldots, s_{k-1})t} \tag{2.164}$$

(uniformly almost-periodic in t in the sense of Besicovitch (Besicovitch 1932)) and, moreover,

$$\phi(s_1, \ldots, s_{k-1}) \triangleq \sum_{n} \left| C^{(n)}_{z_1 \cdots z_k}(s_1, \ldots, s_{k-1}) \right| \in L^1(\mathbb{R}^{k-1}). \tag{2.165}$$

Assumption 2.4.16 Cross-Moment Boundedness. *For every* $k \in \mathbb{N}$ *and every* $\{\ell_1, \ldots, \ell_n\} \subseteq \{1, \ldots, k\}$, *there exists a positive number* $M_{\ell_1 \cdots \ell_n}$ *such that*

$$\mathrm{E}\left\{|z_{\ell_1}(t_1) \cdots z_{\ell_n}(t_n)|\right\} \leqslant M_{\ell_1 \cdots \ell_n} < \infty \qquad \forall t_1, \ldots, t_n \in \mathbb{R}. \tag{2.166}$$

□

Assumption 2.4.16 means that the processes $y(t)$ and $x(t)$ have uniformly bounded $2k$th-order cross-moments for every k.

The proof of the zero-mean joint complex asymptotic Normality of the random variables

$$V_i^{(T)} \triangleq \sqrt{T}\ \left[R_{yx^{(*)}}(\alpha_i, \tau_i; t_i, T) - R_{yx^{(*)}}(\alpha_i, \tau_i)\right] \tag{2.167}$$

is given in Theorem 2.4.18 showing that asymptotically ($T \to \infty$)

1. $\text{cum}\{V_i^{(T)}\} \equiv \text{E}\{V_i^{(T)}\} = 0$
2a. the covariance $\text{cov}\{V_i^{(T)}, V_j^{(T)}\} \equiv \text{cum}\{V_i^{(T)}, V_j^{(T)*}\}$ is finite;
2b. the conjugate covariance $\text{cov}\{V_i^{(T)}, V_j^{(T)*}\} \equiv \text{cum}\{V_i^{(T)}, V_j^{(T)}\}$ is finite;
3. $\text{cum}\{V_{i_1}^{(T)[*]_1}, \ldots, V_{i_k}^{(T)[*]_k}\} = 0$ for $k \geqslant 3$;

with superscript $[*]_h$ denoting optional complex conjugation. In fact, in Section 1.4.2 it is shown that these are necessary and sufficient conditions for the joint asymptotic Normality of *complex* random variables, where cumulants of complex random variables are defined according to (Spooner and Gardner 1994, App. A) (see also (1.209) and Section 1.4.2 for a discussion on the usefulness of this definition). Condition 1 follows from Theorem 2.4.12 on the rate of decay to zero of the bias of the cyclic cross-correlogram. Condition 2a is a consequence of Theorem 2.4.13 on the asymptotic covariance of the cyclic cross-correlogram. Condition 2b follows form Theorem 3.7.2. Finally, Condition 3 follows from Lemma 2.4.17 on the rate of decay to zero of the joint cumulant of cyclic cross-correlograms.

The main assumptions used to prove these theorems and lemmas are regularity of the (generalized) Fourier series of the almost-periodic second- and fourth-order cumulants of the processes (Assumptions 2.4.2–2.4.3), short-range statistical dependence of the processes expressed in terms of summability of joint cumulants (Assumptions 2.4.8 and 2.4.15), lack of clusters of support curves (Assumption 2.4.10), and regularity of the data-tapering window (Assumption 2.4.5). In addition, Assumptions 2.4.4 and 2.4.16 on boundedness of moments are used for technicalities (application of the Fubini and Tonelli theorem).

Lemma 2.4.17 (Napolitano 2007a, Lemma 5.1). *Under Assumptions 2.4.5 (data-tapering window regularity), 2.4.15 (mixing conditions (III)), and 2.4.16 (cross-moment boundedness), for any $k \geqslant 2$ and $\epsilon > 0$ one obtains*

$$\lim_{T\to\infty} T^{k-1-\epsilon}\text{cum}\left\{R^{[*]_1}_{yx^{(*)}}(\alpha_1, \tau_1; t_1, T), \ldots, R^{[*]_k}_{yx^{(*)}}(\alpha_k, \tau_k; t_k, T)\right\} = 0. \tag{2.168}$$

Proof: See Section 3.6. □

Theorem 2.4.18 Asymptotic Joint Normality of the Cyclic Cross-Correlograms (Napolitano 2007a, Theorem 5.1). *Let $x(t)$ and $y(t)$ be zero mean. Under Assumptions 2.4.2 (uniformly almost-periodic statistics), 2.4.3 (Fourier series regularity), 2.4.4 (fourth-order moment boundedness), 2.4.5 (data-tapering window regularity), 2.4.8 (mixing conditions (I)), 2.4.10 (lack of support curve clusters (I)), 2.4.15 (mixing conditions (III)), 2.4.16 (cross-moment* *boundedness), and if* $\gamma > \frac{1}{2}$ *in Theorem 2.4.12, one obtains that, for every fixed* α_i, τ_i, t_i, $i = 1, \ldots, k$, *the random variables* $V_i^{(T)}$ *defined in (2.167)*

$$V_i^{(T)} \triangleq \sqrt{T}\,\left[R_{yx^{(*)}}(\alpha_i, \tau_i; t_i, T) - R_{yx^{(*)}}(\alpha_i, \tau_i)\right]$$

are asymptotically ($T \to \infty$) *zero-mean jointly complex Normal with asymptotic covariance matrix* $\boldsymbol{\Sigma}$ *with entries*

$$\Sigma_{ij} = \lim_{T\to\infty} \operatorname{cov}\left\{V_i^{(T)}, V_j^{(T)}\right\} \tag{2.169}$$

given by (2.146) and asymptotic conjugate covariance matrix $\boldsymbol{\Sigma}^{(c)}$ *with entries*

$$\Sigma_{ij}^{(c)} = \lim_{T\to\infty} \operatorname{cov}\left\{V_i^{(T)}, V_j^{(T)*}\right\} \tag{2.170}$$

given by (3.137).

Proof: See Section 3.6. □

Corollary 2.4.19 *Under the Assumptions for Theorem 2.4.18, one obtains that, for every fixed* α, τ, t_0, *the random variable* $\sqrt{T}\,\left[R_{yx^{(*)}}(\alpha, \tau; t_0, T) - R_{yx^{(*)}}(\alpha, \tau)\right]$ *is asymptotically zero-mean complex Normal:*

$$\sqrt{T}\,\left[R_{yx^{(*)}}(\alpha, \tau; t_0, T) - R_{yx^{(*)}}(\alpha, \tau)\right] \xrightarrow{\mathrm{d}} \mathcal{N}(0, \Sigma_{11}, \Sigma_{11}^{(c)}) \text{ as } T \to \infty \tag{2.171}$$

□

By specializing Theorem 2.4.18 and Corollary 2.4.19 for $y = x$ and $(*)$ present, we find that the cyclic correlogram is an asymptotically Normal estimator of the cyclic autocorrelation function. For $y = x$ and $(*)$ absent, we find that the conjugate cyclic correlogram is an asymptotically Normal estimator of the conjugate cyclic autocorrelation function. For ACS processes, we obtain as a special case the results of (Hurd and Leśkow 1992b), (Dehay and Hurd 1994), (Dandawaté and Giannakis 1995).

As a final remark, note that here it was preferred to describe the finite or practically finite memory of the involved processes by assuming summability of (joint) cumulants of the processes. These assumptions turn out to be more easily verifiable than φ-mixing assumptions (Billingsley 1968, p. 166), (Rosenblatt 1974, pp. 213–214), (Hurd and Leśkow 1992a,b) which are expressed in terms of a function $\varphi(s)$ which describes or controls the dependence between events separated by s time unites. In contrast, summability of cumulants can be derived from properties of the stochastic processes for which cumulants can be calculated (see e.g., (Gardner and Spooner 1994), (Spooner and Gardner 1994), (Napolitano 1995)). A discussion of the finite or practically finite memory of processes by using cumulants is given in Section 1.4.1 and (Brillinger 1965).

2.5 Sampling of GACS Processes

In this section, the problem of uniformly sampling GACS processes is addressed. Aliasing formulas for second-order cyclic cross-moments are derived. It is shown that uniformly sampling a continuous-time GACS process leads to a discrete-time ACS process. Moreover, it is shown that continuous-time GACS processes do not have a discrete-time counterpart, that is, discrete-time GACS processes do not exist.

Let

$$x_{\mathrm{d}}(n) \triangleq x(t)|_{t=nT_s} \qquad y_{\mathrm{d}}(n) \triangleq y(t)|_{t=nT_s} \tag{2.172}$$

be the discrete-time processes obtained by uniformly sampling with period $T_s = 1/f_s$ the continuous-time (jointly) GACS processes $x(t)$ and $y(t)$.

Definition 2.5.1 *The* cyclic cross-correlation function *of the discrete-time sequences $y_{\mathrm{d}}(n)$ and $x_{\mathrm{d}}(n)$ at cycle frequency $\widetilde{\alpha} \in [-1/2, 1/2)$ is defined as*

$$\widetilde{R}_{y_{\mathrm{d}}x_{\mathrm{d}}^{(*)}}(\widetilde{\alpha}, m) \triangleq \lim_{N\to\infty} \frac{1}{2N+1} \sum_{n=-N}^{N} \mathrm{E}\left\{y_{\mathrm{d}}(n+m)\, x_{\mathrm{d}}^{(*)}(n)\right\} e^{-j2\pi\widetilde{\alpha}n}. \tag{2.173}$$

□

The magnitude and phase of $\widetilde{R}_{y_{\mathrm{d}}x_{\mathrm{d}}^{(*)}}(\widetilde{\alpha}, m)$ are the amplitude and phase of the finite-strength additive complex sinewave component at frequency $\widetilde{\alpha}$ contained in the discrete-time cross-correlation $\mathrm{E}\{y_{\mathrm{d}}(n+m)\, x_{\mathrm{d}}^{(*)}(n)\}$.

Theorem 2.5.2 (Izzo and Napolitano 2003), (Napolitano 2009, Theorem 4.1). *Uniformly sampling continuous-time jointly GACS processes leads to discrete-time jointly ACS processes.*

Proof: Let be $A_{mT_s} = A_\tau|_{\tau=mT_s}$, with A_τ defined in (2.32a). Using (2.31b) and (2.31c) we have

$$\mathrm{E}\left\{y_{\mathrm{d}}(n+m)\, x_{\mathrm{d}}^{(*)}(n)\right\} \triangleq \mathrm{E}\left\{y(t+\tau)\, x^{(*)}(t)\right\}\Big|_{t=nT_s,\tau=mT_s} \tag{2.174a}$$

$$= \sum_{\alpha\in A_{mT_s}} R_{yx^{(*)}}(\alpha, mT_s)\, e^{j2\pi\alpha nT_s} \tag{2.174b}$$

$$= \sum_{k\in\mathbb{I}_{yx^{(*)}}} R_{yx^{(*)}}^{(k)}(mT_s)\, e^{j2\pi\alpha_k(mT_s)nT_s} \tag{2.174c}$$

$$= \sum_{\widetilde{\alpha}\in\widetilde{A}} \widetilde{R}_{y_{\mathrm{d}}x_{\mathrm{d}}^{(*)}}(\widetilde{\alpha}, m)\, e^{j2\pi\widetilde{\alpha}n} \tag{2.174d}$$

where, (2.174d) is obtained observing that for every m the discrete-time almost-periodic function of n in (2.174c) can be expressed by a (generalized) Fourier series expansion with coefficients $\widetilde{R}_{y_{\mathrm{d}}x_{\mathrm{d}}^{(*)}}(\widetilde{\alpha}, m)$, defined in (2.173) and, moreover, let

$$\widetilde{A}_m \triangleq \left\{\widetilde{\alpha} \in [-1/2, 1/2) \; : \; \widetilde{R}_{y_{\mathrm{d}}x_{\mathrm{d}}^{(*)}}(\widetilde{\alpha}, m) \neq 0\right\} \tag{2.175}$$

the set

$$\widetilde{A} \triangleq \bigcup_{m\in\mathbb{Z}} \widetilde{A}_m \tag{2.176}$$

is countable. Then, the sum in (2.174d) can be extended either to $\widetilde{A}_m$ or to $\widetilde{A}$. By substituting (2.174c) into (2.173) results in

$$\widetilde{R}_{y_d x_d^{(*)}}(\widetilde{\alpha}, m) = \sum_{k\in\mathbb{I}_{yx^{(*)}}} R^{(k)}_{yx^{(*)}}(mT_s)\, \delta_{[\widetilde{\alpha}-\alpha_k(mT_s)T_s] \bmod 1}. \tag{2.177}$$

In (2.177), mod b is the modulo b operation with values in $[-b/2, b/2)$, that is,

$$a \bmod b \triangleq \begin{cases} a \text{ Mod } b & \text{if } a \text{ Mod } b \in [0, b/2) \\ (a \text{ Mod } b) - b & \text{if } a \text{ Mod } b \in [b/2, b) \end{cases} \tag{2.178}$$

with Mod b being the usual modulo b operation with values in $[0, b)$. Thus, $\delta_{\widetilde{\alpha} \bmod 1} = 1$ if $\widetilde{\alpha} \in \mathbb{Z}$ and $\delta_{\widetilde{\alpha} \bmod 1} = 0$ if $\widetilde{\alpha} \notin \mathbb{Z}$. □

From (2.174d) it follows that the cross-correlation function of the discrete-time processes $y_d(n)$ and $x_d(n)$ obtained by uniformly sampling two continuous-time jointly GACS processes $y(t)$ and $x(t)$ is a (discrete-time) almost-periodic function of n that can be expressed by a (generalized) Fourier series with cycle frequencies not depending on the lag parameter m. That is, $y_d(n)$ and $x_d(n)$ are jointly ACS.

The function $\widetilde{R}_{y_d x_d^{(*)}}(\widetilde{\alpha}, m)$ is periodic in $\widetilde{\alpha}$ with period 1. It is linked to the cyclic cross-correlation function (2.33) and the generalized cyclic cross-correlation function (2.34)–(2.35) of the continuous-time signals $y(t)$ and $x(t)$ by the following result.

Theorem 2.5.3 Aliasing Formula for the Cyclic Cross-Correlation Function (Izzo and Napolitano 2003), (Napolitano 2009, Theorem 4.2). *Let $y_d(n)$ and $x_d(n)$ be the discrete-time processes defined in (2.172). It results that*

$$\widetilde{R}_{y_d x_d^{(*)}}(\widetilde{\alpha}, m) = \sum_{p=-\infty}^{+\infty} R_{yx^{(*)}}\Big((\widetilde{\alpha}+p)f_s, mT_s\Big) \tag{2.179a}$$

$$= \sum_{k\in\mathbb{I}_{yx^{(*)}}} R^{(k)}_{yx^{(*)}}(mT_s)\, \delta_{[\widetilde{\alpha}-\alpha_k(mT_s)T_s] \bmod 1}. \tag{2.179b}$$

Proof: See Section 3.8. □

Note that (2.179a) is the generalization to GACS signals of the analogous formula for ACS signals ((1.174) when $y \equiv x$).

In the following, consequences of Theorems 2.5.2 and 2.5.3 are analyzed. Furthermore, difficulties arising in cyclic spectral analysis of a continuous-time GACS process starting from the discrete-time process of its samples are discussed.

The set of cycle frequencies $\widetilde{\alpha}$ at lag m, accounting for (2.179a) can be expressed as

$$\widetilde{A}_m \triangleq \left\{ \widetilde{\alpha} \in [-1/2, 1/2) \; : \; \widetilde{R}_{y_d x_d^{(*)}}(\widetilde{\alpha}, m) \neq 0 \right\}$$
$$= \left\{ \widetilde{\alpha} \in [-1/2, 1/2) \; : \; \widetilde{\alpha} = (\alpha/f_s) \bmod 1, \alpha \in A_{mT_s} \right\}. \qquad (2.180)$$

The set $\widetilde{A}_m$ is countable and can contain cluster points (accumulation points) or be dense in $[-1/2, 1/2)$ if f_s is incommensurate with a countable infinity of cycle frequencies $\alpha \in A_{mT_s} = \{\alpha \in \mathbb{R} \, : \, R_{yx^{(*)}}(\alpha, mT_s) \neq 0\}$ (e.g., if $\alpha = k/T_0, k \in \mathbb{Z}$, and $T_0/T_s \notin \mathbb{Q}$).

From (2.179a) it follows that, in general, $\widetilde{R}_{y_d x_d^{(*)}}(\widetilde{\alpha}, m) \neq R_{yx^{(*)}}(\widetilde{\alpha} f_s, mT_s)$ due to the presence of aliasing in the cycle frequency domain. However, if the jointly GACS continuous-time signals $y(t)$ and $x(t)$ are such that at lag $\tau = mT_s$ there is no lag-dependent cycle frequency $\alpha_k(\tau)$, $k \in \mathbb{I}_{yx^{(*)}}$, such that $(\widetilde{\alpha} + p) f_s = \alpha_k(mT_s)$ for $p \neq 0$ then, by comparing (2.179b) with (2.39), it follows that $\widetilde{R}_{y_d x_d^{(*)}}(\widetilde{\alpha}, m) = R_{yx^{(*)}}(\widetilde{\alpha} f_s, mT_s)$. Such an equality could be difficult to be realized in the whole domain $(\widetilde{\alpha}, m) \in [-1/2, 1/2) \times \mathbb{Z}$ (Izzo and Napolitano 2003), as a consequence of the fact that GACS signals have the power spread over an infinite bandwidth (Sections 2.2.2 and 2.2.3), (Izzo and Napolitano 1998a, 2005). Furthermore, in Theorem 2.5.2, it is shown that the discrete-time signal constituted by the samples of a continuous-time GACS signal is a discrete-time ACS signal. Thus, discrete-time ACS signals can arise from sampling ACS and non-ACS continuous-time GACS signals. Moreover, in Section 2.7.6 it is shown that, starting from the sampled signal, the possible ACS or non-ACS nature of the continuous-time GACS signal can only be conjectured, provided that analysis parameters such as sampling period, padding factor, and data-record length are properly chosen. Thus, the results for discrete-time processes cannot be obtained straightforwardly from those of the continuous-time case as it can be made in the stationary case e.g., in (Brillinger and Rosenblatt 1967). In fact, unlike the case of WSS and ACS processes, continuous-time GACS processes do not have a discrete-time counterpart, that is, discrete-time GACS processes do not exist. This is a consequence of the fact that even if the set $A = \cup_{\tau \in \mathbb{R}} A_\tau$ can be uncountable, the set $\widetilde{A} = \cup_{m \in \mathbb{Z}} \widetilde{A}_m$ is always countable.

From the above facts it follows that the sampling frequency f_s cannot be easily chosen in order to avoid or limit aliasing, as it happens for bandlimited WSS and ACS signals (Napolitano 1995), (Izzo and Napolitano 1996), even if, in some cases, as for the chirp-modulated signal (2.99), analytical results can be obtained (Section 2.7.1). However, in the following it will be shown that the discrete-time cyclic cross-correlogram of the sampled process can be made as close as desired to samples of the continuous-time cyclic cross-correlation function by taking the sampling period sufficiently small and the data-record length sufficiently large.

2.6 Discrete-Time Estimator of the Cyclic Cross-Correlation Function

In this section, the discrete-time cyclic cross-correlogram of the discrete-time process obtained by uniformly sampling a continuous-time GACS process is shown to be a mean-square consistent and asymptotically complex Normal estimator of the continuous-time cyclic cross-correlation function as the data-record length approaches infinity and the sampling period approaches zero (Napolitano 2009).

Note that the results of Section 2.6 for discrete-time processes cannot be obtained straightforwardly by substituting integrals with sums into the results of Sections 2.4.1–2.4.3 which have been obtained for continuous-time processes as is made, in the stationary case, in (Brillinger and Rosenblatt 1967).

2.6.1 Discrete-Time Cyclic Cross-Correlogram

In this section, the discrete-time cyclic cross-correlogram is defined and its mean and covariance are evaluated for finite number of samples and sampling period.

Definition 2.6.1 *Let $y_{\mathrm{d}}(n)$ and $x_{\mathrm{d}}(n)$ be the discrete-time processes defined in (2.172). Their* discrete-time cyclic cross-correlogram (DT-CCC) *at cycle frequency $\widetilde{\alpha}$ is defined as*

$$\widetilde{R}_{y_{\mathrm{d}}x_{\mathrm{d}}^{(*)}}(\widetilde{\alpha}, m; n_0, N) \triangleq \sum_{n=-\infty}^{+\infty} v_N(n-n_0)\, y_{\mathrm{d}}(n+m)\, x_{\mathrm{d}}^{(*)}(n)\, e^{-j2\pi\widetilde{\alpha}n} \tag{2.181}$$

where

$$v_N(n) \triangleq \frac{1}{2N+1}\, a\left(\frac{n}{2N+1}\right) \tag{2.182}$$

is a data-tapering window nonzero in $\{-N, \ldots, N\}$, with $a(t)$ as in Assumption 2.4.5. □

Note that since $a(t)$ has finite width, the infinite sum in (2.181) is in fact a finite sum from $n_0 - N$ to $n_0 + N$.

By taking the expected value of the DT-CCC (2.181) and using (2.174c), the following result is obtained, where the assumptions allow to interchange the order of sum and expectation operations and the order of double-index sums.

Theorem 2.6.2 Expected Value of the Discrete-Time Cyclic Cross-Correlogram (Napolitano 2009, Theorem 5.1). *Let $x_{\mathrm{d}}(n)$ and $y_{\mathrm{d}}(n)$ be the discrete-time processes defined in (2.172). Under Assumptions 2.4.2a (uniformly almost-periodic statistics), 2.4.3a (Fourier series regularity), and 2.4.5 (data-tapering window regularity) on the continuous-time processes $x(t)$ and $y(t)$, the expected value of the DT-CCC (2.181) is given by*

$$\begin{aligned} &\mathrm{E}\left\{\widetilde{R}_{y_{\mathrm{d}}x_{\mathrm{d}}^{(*)}}(\widetilde{\alpha}, m; n_0, N)\right\} \\ &\quad = \sum_{k\in\mathbb{I}} R_{yx^{(*)}}^{(k)}(mT_s)\, V_{\frac{1}{N}}\left(\widetilde{\alpha} - \alpha_k(mT_s)T_s\right) e^{-j2\pi[\widetilde{\alpha}-\alpha_k(mT_s)T_s]n_0} \end{aligned} \tag{2.183}$$

where

$$\begin{aligned} V_{\frac{1}{N}}(\nu) &\triangleq \sum_{n=-\infty}^{+\infty} v_N(n)\, e^{-j2\pi\nu n} \\ &= \sum_{\ell=-\infty}^{+\infty} A\big((\nu-\ell)(2N+1)\big) \end{aligned} \tag{2.184}$$

with $A(f)$ Fourier transform of $a(t)$.

Proof: See Section 3.9. □

The function $V_{1/N}(\nu)$ is periodic with period 1 and has bandwidth of the order of $1/N$. Consequently, the cycle-frequency resolution is of the order of $1/N$ and aliasing in the cycle-frequency $\widetilde{\alpha}$ domain occurs. Specifically, from (2.183) it follows that in the $(\widetilde{\alpha}, m)$-domain the expected value of the DT-CCC can be significantly different from zero within intervals of width $1/N$ parallel to the $\widetilde{\alpha}$ axis and centered in the points $(\widetilde{\alpha}, m)$ with $\widetilde{\alpha} = \alpha_k(mT_s)T_s \bmod 1$, $k \in \mathbb{I}$. Such points correspond to scaled and aliased samples of the lag-dependent cycle frequency curves $\alpha = \alpha_k(\tau)$, that is, the support curves of the continuous-time cyclic cross-correlation function (see Figure 2.4a, where the magnitude of the cyclic autocorrelation function of the continuous-time chirp modulated PAM signal of Section 2.2.6 is represented). Moreover, the

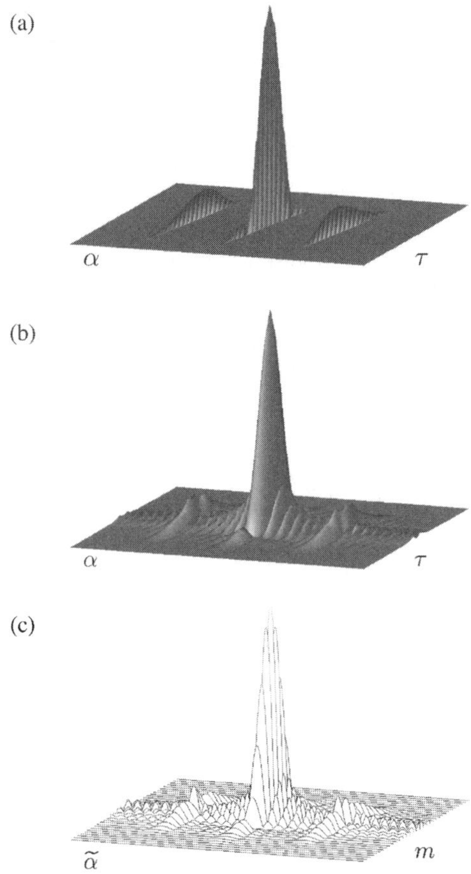

Figure 2.4 Chirp-modulated PAM signal: (a) Magnitude of the cyclic autocorrelation function as function of α and τ (see (2.104)–(2.106)); (b) magnitude of the expected value of the continuous-time cyclic correlogram as function of α and τ (cyclic leakage, see (2.128)); (c) magnitude of the expected value of the discrete-time cyclic correlogram as function of $\widetilde{\alpha} = \alpha T_s$ and $m = \tau/T_s$ (aliasing and cyclic leakage, see (2.183)). *Source:* (Napolitano 2009) © IEEE

expected value at $(\widetilde{\alpha}, m)$ is due essentially to all lag-dependent cycle frequencies $\alpha_k(\tau)$, such that

$$\left|(\widetilde{\alpha} + p) - \alpha_k(mT_s)T_s\right| < \frac{1}{N} \tag{2.185}$$

for some integer p and some $k \in \mathbb{I}$. That is, a leakage phenomenon among lag-dependent cycle-frequency curves which are sufficiently close (modulo f_s) to each other occurs (Figure 2.4c) (see also the discussion following Theorem 2.4.6 for the continuous-time estimator). Therefore, both aliasing and cyclic leakage effects are present. The effect of cyclic leakage becomes negligible as $N \to \infty$ and the effect of aliasing becomes negligible as $T_s \to 0$. In contrast, in the expected value of the continuous-time cyclic cross-correlogram, only cyclic leakage, without aliasing in the cycle frequency domain, is present (Figure 2.4b).

By expressing the covariance of the lag-product $y_{\mathrm{d}}(n+m)\, x_{\mathrm{d}}^{(*)}(n)$ in terms of second-order moments and a fourth-order cumulant, the following result can be proved, where the assumptions allow to interchange the order of sum and expectation operators and the order of multiple-index sum operations.

Theorem 2.6.3 Covariance of the Discrete-Time Cyclic Cross-Correlogram (Napolitano 2009, Theorem 5.2). *Let $x_{\mathrm{d}}(n)$ and $y_{\mathrm{d}}(n)$ be discrete-time zero-mean processes defined in (2.172). Under Assumptions 2.4.2 (uniformly almost-periodic statistics), 2.4.3 (Fourier series regularity), and 2.4.5 (data-tapering window regularity) on the continuous-time processes $x(t)$ and $y(t)$, the covariance of the DT-CCC (2.181) is given by*

$$\operatorname{cov}\left\{\widetilde{R}_{y_{\mathrm{d}}x_{\mathrm{d}}^{(*)}}(\widetilde{\alpha}_1, m_1; n_{01}, N),\, \widetilde{R}_{y_{\mathrm{d}}x_{\mathrm{d}}^{(*)}}(\widetilde{\alpha}_2, m_2; n_{02}, N)\right\} = \widetilde{\mathcal{T}}_1 + \widetilde{\mathcal{T}}_2 + \widetilde{\mathcal{T}}_3 \tag{2.186}$$

where

$$\begin{aligned}\widetilde{\mathcal{T}}_1 \triangleq \sum_{k'}\sum_{k''}\sum_{\ell=-\infty}^{+\infty} & R_{yy^*}^{(k')}((m_1 - m_2 + \ell)T_s)\, R_{x^{(*)}x^{(*)*}}^{(k'')}(\ell T_s) \\ & e^{j2\pi\alpha_{yy^*}^{(k')}((m_1-m_2+\ell)T_s)m_2T_s}\, e^{-j2\pi\widetilde{\alpha}_1\ell} \\ & e^{-j2\pi\widetilde{\Phi}_1 n_{02}}\, \frac{1}{2N+1}\, \widetilde{r}_{aa^*}\left(\widetilde{\Phi}_1, \ell - n_{01} + n_{02}; N\right)\end{aligned} \tag{2.187}$$

$$\begin{aligned}\widetilde{\mathcal{T}}_2 \triangleq \sum_{k'''}\sum_{k^{iv}}\sum_{\ell=-\infty}^{+\infty} & R_{yx^{(*)*}}^{(k''')}((m_1 + \ell)T_s)\, R_{x^{(*)}y^*}^{(k^{iv})}((\ell - m_2)T_s) \\ & e^{j2\pi\alpha_{x^{(*)}y^*}^{(k^{iv})}((\ell-m_2)T_s)m_2T_s}\, e^{-j2\pi\widetilde{\alpha}_1\ell} \\ & e^{-j2\pi\widetilde{\Phi}_2 n_{02}}\, \frac{1}{2N+1}\, \widetilde{r}_{aa^*}\left(\widetilde{\Phi}_2, \ell - n_{01} + n_{02}; N\right)\end{aligned} \tag{2.188}$$

$$\begin{aligned}\widetilde{\mathcal{T}}_3 \triangleq \sum_{k}\sum_{\ell=-\infty}^{+\infty} & C_{yx^{(*)}y^*x^{(*)*}}^{(k)}((m_1 + \ell)T_s, \ell T_s, m_2T_s)\, e^{-j2\pi\widetilde{\alpha}_1\ell} \\ & e^{-j2\pi\widetilde{\Phi}_3 n_{02}}\, \frac{1}{2N+1}\, \widetilde{r}_{aa^*}\left(\widetilde{\Phi}_3, \ell - n_{01} + n_{02}; N\right)\end{aligned} \tag{2.189}$$

with

$$\widetilde{r}_{aa^*}(\widetilde{\beta}, m; N) \triangleq \frac{1}{2N+1} \sum_{n=-\infty}^{+\infty} a\left(\frac{n+m}{2N+1}\right) a^*\left(\frac{n}{2N+1}\right) e^{-j2\pi\widetilde{\beta}n} \tag{2.190}$$

$$\widetilde{\Phi}_1 \triangleq [\widetilde{\alpha}_1 - \widetilde{\alpha}_2 - \alpha^{(k')}_{yy^*}((m_1 - m_2 + \ell)T_s)T_s - \alpha^{(k'')}_{x^{(*)}x^{(*)*}}(\ell T_s)T_s] \tag{2.191}$$

$$\widetilde{\Phi}_2 \triangleq [\widetilde{\alpha}_1 - \widetilde{\alpha}_2 - \alpha^{(k''')}_{yx^{(*)*}}((m_1 + \ell)T_s)T_s - \alpha^{(k^{iv})}_{x^{(*)}y^*}((\ell - m_2)T_s)T_s] \tag{2.192}$$

$$\widetilde{\Phi}_3 \triangleq [\widetilde{\alpha}_1 - \widetilde{\alpha}_2 - \beta_k((m_1 + \ell)T_s, \ell T_s, m_2 T_s)T_s]. \tag{2.193}$$

Proof: See Section 3.9. □

2.6.2 *Asymptotic Results as* $N \to \infty$

In this section, the asymptotic ($N \to \infty$) expected value and covariance of the DT-CCC are determined and its asymptotic complex Normality is proved.

Theorem 2.6.4 Asymptotic Expected Value of the Discrete-Time Cyclic Cross-Correlogram (Napolitano 2009, Theorem 6.1). *Let* $x_d(n)$ *and* $y_d(n)$ *be the discrete-time processes defined in (2.172). Under Assumptions 2.4.2a (uniformly almost-periodic statistics), 2.4.3a (Fourier series regularity), and 2.4.5 (data-tapering window regularity) on the continuous-time processes* $x(t)$ *and* $y(t)$*, the asymptotic (*$N \to \infty$*) expected value of the DT-CCC (2.181) is given by*

$$\lim_{N\to\infty} \mathrm{E}\left\{\widetilde{R}_{y_d x_d^{(*)}}(\widetilde{\alpha}, m; n_0, N)\right\} = \widetilde{R}_{y_d x_d^{(*)}}(\widetilde{\alpha}, m). \tag{2.194}$$

Proof: See Section 3.10. □

Theorem 2.6.5 Asymptotic Covariance of the Discrete-Time Cyclic Cross-Correlogram (Napolitano 2009, Theorem 6.2). *Let* $x_d(n)$ *and* $y_d(n)$ *be discrete-time zero-mean processes defined in (2.172). Under Assumptions 2.4.2 (uniformly almost-periodic statistics), 2.4.3 (Fourier series regularity), 2.4.5 (data-tapering window regularity), and 2.4.8 (mixing conditions (I)) on the continuous-time processes* $x(t)$ *and* $y(t)$*, the asymptotic (*$N \to \infty$*) covariance of the DT-CCC (2.181) is given by*

$$\lim_{N\to\infty} (2N+1) \operatorname{cov}\left\{\widetilde{R}_{y_d x_d^{(*)}}(\widetilde{\alpha}_1, m_1; n_{01}, N), \widetilde{R}_{y_d x_d^{(*)}}(\widetilde{\alpha}_2, m_2; n_{02}, N)\right\}$$
$$= \widetilde{T}'_1 + \widetilde{T}'_2 + \widetilde{T}'_3 \tag{2.195}$$

where

$$\widetilde{T}'_1 \triangleq \mathcal{E}_a \sum_{k'} \sum_{k''} \sum_{\ell=-\infty}^{+\infty} R^{(k')}_{yy^*}((m_1 - m_2 + \ell)T_s)\, R^{(k'')}_{x^{(*)}x^{(*)*}}(\ell T_s)$$
$$e^{j2\pi\alpha'_{k'}((m_1-m_2+\ell)T_s)m_2 T_s}\, e^{-j2\pi\widetilde{\alpha}_1\ell}$$
$$\delta_{[\widetilde{\alpha}_1-\widetilde{\alpha}_2-\alpha'_{k'}((m_1-m_2+\ell)T_s)T_s-\alpha''_{k''}(\ell T_s)T_s] \bmod 1} \tag{2.196}$$

$$\widetilde{T}'_2 \triangleq \mathcal{E}_a \sum_{k'''} \sum_{k^{i\nu}} \sum_{\ell=-\infty}^{+\infty} R^{(k''')}_{yx^{(*)}*}((m_1+\ell)T_s)\, R^{(k^{i\nu})}_{x^{(*)}y^*}((\ell-m_2)T_s)$$
$$e^{j2\pi\alpha^{i\nu}_{k^{i\nu}}((\ell-m_2)T_s)m_2T_s}\, e^{-j2\pi\widetilde{\alpha}_1\ell}$$
$$\delta_{[\widetilde{\alpha}_1-\widetilde{\alpha}_2-\alpha'''_{k'''}((m_1+\ell)T_s)T_s-\alpha^{i\nu}_{k^{i\nu}}((\ell-m_2)T_s)T_s]\text{mod }1} \tag{2.197}$$

$$\widetilde{T}'_3 \triangleq \mathcal{E}_a \sum_{k} \sum_{\ell=-\infty}^{+\infty} C^{(k)}_{yx^{(*)}y^*x^{(*)}*}((m_1+\ell)T_s, \ell T_s, m_2T_s)\, e^{-j2\pi\widetilde{\alpha}_1\ell}$$
$$\delta_{[\widetilde{\alpha}_1-\widetilde{\alpha}_2-\beta_k((m_1+\ell)T_s,\ell T_s,m_2T_s)T_s]\text{mod }1} \tag{2.198}$$

with $\delta_{\widetilde{\alpha}\text{ mod }1} = 1$ *if* $\widetilde{\alpha} \in \mathbb{Z}$ *and* $\delta_{\widetilde{\alpha}\text{ mod }1} = 0$ *otherwise, and*

$$\mathcal{E}_a \triangleq \int_{-1/2}^{1/2} |a(t)|^2\,\mathrm{d}t = \int_{\mathbb{R}} |a(t)|^2\,\mathrm{d}t. \tag{2.199}$$

In (2.196)–(2.198), for notation simplicity, $\alpha'_{k'}(\cdot) \equiv \alpha^{(k')}_{yy^*}(\cdot)$, $\alpha''_{k''}(\cdot) \equiv \alpha^{(k'')}_{x^{(*)}x^{(*)}*}(\cdot)$, $\alpha'''_{k'''}(\cdot) \equiv \alpha^{(k''')}_{yx^{(*)}*}(\cdot)$, *and* $\alpha^{i\nu}_{k^{i\nu}}(\cdot) \equiv \alpha^{(k^{i\nu})}_{x^{(*)}y^*}(\cdot)$.

Proof: See Section 3.10. □

From Theorem 2.6.4 it follows that the DT-CCC (2.181) is an asymptotically ($N \to \infty$), unbiased estimator of the discrete-time cyclic cross-correlation function (2.173). Moreover, from Theorem 2.6.5 it follows that the variance of the DT-CCC asymptotically vanishes. Consequently, the DT-CCC is a mean-square consistent estimator of the discrete-time cyclic cross-correlation function, that is, of an aliased version of the continuous-time cyclic cross-correlation function (see (2.179a)).

In order to establish the rate of convergence to zero of the bias of the DT-CCC and its asymptotic Normality, the following assumption of lack of clusters (accumulation points) of cycle frequencies is needed.

Assumption 2.6.6 Lack of Support Curve Clusters (II). *For every m, the set* $\widetilde{A}_m$ *defined in (2.180) does not contain any cluster of cycle frequencies. That is, let*

$$\mathbb{J}_{\widetilde{\alpha},m,T_s} \triangleq \left\{k \in \mathbb{I} \,:\, \alpha_k(mT_s)T_s = \widetilde{\alpha} \text{ mod } 1,\; R^{(k)}_{yx^{(*)}}(mT_s) \neq 0\right\} \tag{2.200}$$

then, for every $k \notin \mathbb{J}_{\widetilde{\alpha},m,T_s}$, *no curve* $\alpha_k(\tau)$ *is such that the value* $[\alpha_k(mT_s)T_s]$ mod 1 *can be arbitrarily close to the cycle frequency* $\widetilde{\alpha} \in [-1/2, 1/2)$. *Thus, for every* $\widetilde{\alpha}$ *and m it results that*

$$h_{\widetilde{\alpha},m,T_s} \triangleq \inf_{k \notin \mathbb{J}_{\widetilde{\alpha},m,T_s}} \left|\left[\widetilde{\alpha} - \alpha_k(mT_s)T_s\right] \text{ mod } 1\right| > 0. \tag{2.201}$$

□

The set $\mathbb{J}_{\widetilde{\alpha},m,T_s}$ is finite or at most countable and is the set of indices k such that, for the fixed $\widetilde{\alpha}$, m, and T_s we have $\widetilde{\alpha} + p = \alpha_k(mT_s)T_s$ for some $p \in \mathbb{Z}$. Assumption 2.6.6 means that the greatest lower bound (G.L.B.) or infimum of the distances $|\widetilde{\alpha} + p - \alpha_k(mT_s)T_s|$ is nonzero for every $p \in \mathbb{Z}$ and $k \notin \mathbb{J}_{\widetilde{\alpha},m,T_s}$. That is, the point $(\widetilde{\alpha}, m) \in [-1/2, 1/2) \times \mathbb{Z}$ is not a cluster of cycle frequencies. A sufficient condition assuring this is that the number of lag-dependent cycle frequencies is finite, that is, $\mathbb{I}$ is finite (see, e.g., the chirp-modulated PAM signal with raised cosine pulse in Section 2.2.6).

In the following, where necessary, the very mild assumption that for every fixed k the lag-dependent cycle frequency $\alpha_k(\tau)$ is bounded for finite τ will be made.

Let γ be the rate of decay to zero of $A(f)$ as specified in Assumption 2.4.5. We have the following result.

Theorem 2.6.7 Rate of Convergence of the Bias of the Discrete-Time Cyclic Cross-Correlogram (Napolitano 2009, Theorem 6.3). *Let $x_{\mathrm{d}}(n)$ and $y_{\mathrm{d}}(n)$ be the discrete-time processes defined in (2.172). Under Assumptions 2.4.2a (uniformly almost-periodic statistics), 2.4.3a (Fourier series regularity), 2.4.5 (data-tapering window regularity), and 2.6.6 (lack of support curve clusters (II)), on the continuous-time processes $x(t)$ and $y(t)$, and provided that $\gamma > 1$ (or $\gamma = 1$ in the special case of $a(t) = \mathrm{rect}(t)$), one obtains*

$$\lim_{N\to\infty} (2N+1)^{\gamma} \left| \mathrm{E}\left\{ \widetilde{R}_{y_{\mathrm{d}}x_{\mathrm{d}}^{(*)}}(\widetilde{\alpha}, m; n_0, N) \right\} - \widetilde{R}_{y_{\mathrm{d}}x_{\mathrm{d}}^{(*)}}(\widetilde{\alpha}, m) \right| = \mathcal{O}(1). \tag{2.202}$$

Proof: See Section 3.10. □

In the following, the asymptotic ($N \to \infty$) complex Normality of the DT-CCC will be proved.

Let

$$\begin{aligned} z_i(n) &\triangleq \left[y_{\mathrm{d}}(n+m_i)\, x_{\mathrm{d}}^{(*)}(n) \right]^{[*]_i} \\ &= \left[y(t+\tau_i)\, x^{(*)}(t) \right]^{[*]_i} \Big|_{t=nT_s, \tau_i=m_iT_s}, \qquad i = 1, \ldots, k \end{aligned} \tag{2.203}$$

be second-order discrete-time lag-product waveforms, with optional complex conjugations $[*]_i$, $i = 1, \ldots, k$, having kth-order cumulant $\mathrm{cum}\{z_k(n), z_i(n+m_i),\ i = 1, \ldots, k-1\}$ where the cumulant of complex processes is defined according to (Spooner and Gardner 1994, App. A) (see also Section 1.4.2).

Assumption 2.6.8 Mixing Conditions (IV). *For every integer $k \geqslant 2$, $m_i \in \mathbb{Z}$, $i = 1, \ldots, k$ and every conjugation configuration $[*]_1, \ldots, [*]_k$, there exists a positive bounded Riemann integrable (and hence also Lebesgue summable) function $\phi(s_1, \ldots, s_{k-1})$, $(s_1, \ldots, s_{k-1}) \in \mathbb{R}^{k-1}$, (depending on m_i, $i = 1, \ldots, k$ and the conjugation configuration) such that, for T_s fixed,*

$$|\mathrm{cum}\{z_k(n), z_i(n+r_i),\ i = 1, \ldots, k-1\}| \leqslant \phi(r_1T_s, \ldots, r_{k-1}T_s) \tag{2.204}$$

uniformly with respect to n. □

The function ϕ can be assumed Lebesgue integrable if it is defined in a compact. In fact, there exists a continuous function on the compact (hence Riemann integrable) close as desired to ϕ (Rudin 1987, p. 55) (Lusin Theorem).

Assumption 2.6.8 is the discrete-time counterpart of Assumption 2.4.15. Analogously to the continuous-time case, the cumulant cum $\{z_k(n), z_i(n+m_i),\ i=1,\dots,k-1\}$ can be expressed as in (2.160). Therefore, Assumption 2.6.8 is satisfied provided that the cumulants up to order $2k$ of the elements taken from table (2.159) are summable. In addition, as already observed in the continuous-time case, Assumption 2.6.8 is fulfilled if the processes $x(t)$ and $y(t)$ have finite-memory or memory decaying sufficiently fast. Processes of this kind include communications signals with independent and identically distributed (i.i.d.) symbols or block coded symbols, processes with exponentially decaying joint probability density function, and finite-memory (time-variant) transformations of such processes. Moreover, Assumption 2.6.8 turns out to be verified if the stochastic processes $z_i(t), i=1,\dots,k$ are jointly kth-order GACS satisfying condition (2.165).

Analogously to the continuous-time case, the proof of the zero-mean joint complex asymptotic Normality of the random variables

$$V_i^{(N)} \triangleq \sqrt{(2N+1)}\left[\widetilde{R}_{y_\mathrm{d}x_\mathrm{d}^{(*)}}(\widetilde{\alpha}_i, m_i; n_0, N) - \widetilde{R}_{y_\mathrm{d}x_\mathrm{d}^{(*)}}(\widetilde{\alpha}_i, m_i)\right] \tag{2.205}$$

is made in Theorem 2.6.10 showing that asymptotically ($N \to \infty$)

1. $\mathrm{cum}\{V_i^{(N)}\} \equiv \mathrm{E}\{V_i^{(N)}\} = 0$

2a. the covariance $\mathrm{cov}\{V_i^{(N)}, V_j^{(N)}\} \equiv \mathrm{cum}\{V_i^{(N)}, V_j^{(N)*}\}$ is finite;

2b. the conjugate covariance $\mathrm{cov}\{V_i^{(N)}, V_j^{(N)*}\} \equiv \mathrm{cum}\{V_i^{(N)}, V_j^{(N)}\}$ is finite;

3. $\mathrm{cum}\{V_{i_1}^{(N)[*]_1}, \dots, V_{i_k}^{(N)[*]_k}\} = 0$ for $k \geqslant 3$;

with superscript $[*]_h$ denoting optional complex conjugation. In fact, in Section 1.4.2 it is shown that these are necessary and sufficient conditions for the joint asymptotic Normality of *complex* random variables, where cumulants of complex random variables are defined according to (Spooner and Gardner 1994, App. A) (see also (1.209) and Section 1.4.2 for a discussion on the usefulness of this definition). Condition 1 follows from Theorem 2.6.7 on the rate of decay to zero of the bias of the DT-CCC. Condition 2a is a consequence of Theorem 2.6.5 on the asymptotic covariance of the DT-CCC. Condition 2b follows from Theorem 3.13.2. Finally, Condition 3 follows from Lemma 2.6.9 on the rate of decay to zero of the joint cumulant of DT-CCCs.

The main assumptions used to prove these theorems and lemmas are regularity of the (generalized) Fourier series of the almost-periodic second- and fourth-order cumulants of the underlying continuous-time GACS processes (Assumptions 2.4.2–2.4.3), short-range statistical dependence of the continuous-time GACS processes expressed in terms of summability of joint cumulants (Assumptions 2.4.8 and 2.6.8), and regularity of the data-tapering window (Assumption 2.4.5). A critical assumption used in the proof of Theorem 2.6.7 is lack of clusters of cycle frequencies of the discrete-time ACS processes (Assumptions 2.6.6). This can be obtained, for example, if the number of lag-dependent cycle frequencies $\alpha_k(\tau)$ is finite, that is, if the set $\mathbb{I}$ is finite.

Several differences exist between the assumptions used to prove the asymptotic properties of the continuous-time estimator (2.118) and those used for the discrete-time estimator (2.181). Specifically, Assumption 2.6.6 is stronger than the analogous Assumption 2.4.10 used to establish the rate of decay to zero of the bias of the continuous-time cyclic cross-correlogram and its asymptotic Normality. In fact, in (2.200) and (2.201) cycle frequencies have to be considered modulo f_s and (2.201) could not be verified if the set $\widetilde{A}_m$ defined in (2.180) contains cluster points. In addition, the needed rate of decay to zero of $A(f)$ as $|f| \to \infty$ is $\mathcal{O}(|f|^{-\gamma})$ with $\gamma > 1$ (or $\gamma = 1$ in the special case of $a(t) = \text{rect}(t)$). In contrast, in the continuous-time case, $\gamma > 1/2$ is sufficient (see Theorem 2.4.18). Finally, unlike the continuous-time case, Assumptions 2.4.4 and 2.4.16 on boundedness of moments (used for application of the Fubini and Tonelli theorem) are not needed in the discrete-time case.

Lemma 2.6.9 Rate of Convergence to Zero of Cumulants of Discrete-Time Cyclic Cross-Correlograms (Napolitano 2009, Lemma 6.1). *Under Assumptions 2.4.5 (data-tapering window regularity) and 2.6.8 (mixing conditions (IV)) for every $k \geqslant 2$ and $\epsilon > 0$ one obtains*

$$\lim_{N\to\infty} (2N+1)^{k-1-\epsilon} \text{cum}\left\{ \widetilde{R}^{[*]_i}_{y_d x_d^{(*)}}(\widetilde{\alpha}_i, m_i; n_{0i}, N),\ i = 1, \ldots, k \right\} = 0. \tag{2.206}$$

Proof: See Section 3.10. □

Theorem 2.6.10 Asymptotic Joint Normality of the Discrete-Time Cyclic Cross-Correlograms (Napolitano 2009, Theorem 6.4). *Let $x_d(n)$ and $y_d(n)$ be discrete-time zero-mean processes defined in (2.172). Under Assumptions 2.4.2 (uniformly almost-periodic statistics), 2.4.3 (Fourier series regularity), 2.4.5 (data-tapering window regularity), and 2.4.8 (mixing conditions (I)) on the continuous-time processes $x(t)$ and $y(t)$, under Assumptions 2.6.6 (lack of support curve clusters (II)) and 2.6.8 (mixing conditions (IV)) and provided that $\gamma > 1$ (or $\gamma = 1$ in the special case of $a(t) = \text{rect}(t)$), the random variables $V_i^{(N)}$ defined in (2.205)*

$$V_i^{(N)} \triangleq \sqrt{(2N+1)} \left[\widetilde{R}_{y_d x_d^{(*)}}(\widetilde{\alpha}_i, m_i; n_0, N) - \widetilde{R}_{y_d x_d^{(*)}}(\widetilde{\alpha}_i, m_i) \right]$$

are asymptotically (as $N \to \infty$) zero-mean jointly complex Normal with asymptotic covariance matrix $\boldsymbol{\Sigma}$ with entries

$$\Sigma_{ij} = \lim_{N\to\infty} \text{cov}\left\{ V_i^{(N)}, V_j^{(N)} \right\} \tag{2.207}$$

given by (2.195) and asymptotic conjugate covariance matrix $\boldsymbol{\Sigma}^{(c)}$ with entries

$$\Sigma_{ij}^{(c)} = \lim_{N\to\infty} \text{cov}\left\{ V_i^{(N)}, V_j^{(N)*} \right\} \tag{2.208}$$

given by (3.256).

Proof: See Section 3.10. □

2.6.3 Asymptotic Results as $N \to \infty$ and $T_s \to 0$

In order to obtain a reliable discrete-time estimate of samples of the continuous-time cyclic autocorrelation function, a further assumption is needed in order to control the amount of aliasing in (2.179a) and (2.179b) when the sampling period T_s approaches zero.

Assumption 2.6.11 Aliasing Series Summability. *For every $\widetilde{\alpha}$ and m there exists a sequence $\{M_p\}_{p\in\mathbb{Z}}$ of positive numbers independent of T_s such that*

1. $\left|R_{yx^{(*)}}\left((\widetilde{\alpha}+p)f_s, mT_s\right)\right| \leqslant M_p$
2. $\sum_{p=-\infty}^{+\infty} M_p < \infty$

A sufficient condition such that Assumption 2.6.11 is verified is that

$$R_{yx^{(*)}}(\alpha, \tau) = \mathcal{O}(|\alpha|^{-r}) \quad \text{as } |\alpha| \to \infty \tag{2.209}$$

with $r > 1$ uniformly w.r.t. τ. Every GACS process $x(t)$ with a finite number of lag-dependent cycle frequencies and bounded (conjugate) generalized cyclic autocorrelation functions satisfies (2.209) and consequently Assumption 2.6.11 with $y \equiv x$ since only a finite number of nonzero terms is present in the sum over p (see, e.g., the chirp-modulated PAM signal with raised cosine pulse of Section 2.2.6). The case $r = 1$ in (2.209) (which occurs for example for a PAM signal with rectangular pulse) needs to be treated separately.

Lemma 2.6.12 (Napolitano 2009, Lemma 7.1). *Under assumption (2.209) (from which Assumption 2.6.11 follows) one obtains*

$$\lim_{T_s\to 0} \sum_{\substack{p=-\infty \\ p\neq 0}}^{+\infty} R_{yx^{(*)}}\left((\widetilde{\alpha}+p)f_s, mT_s\right) = 0 \tag{2.210}$$

pointwise.

Proof: See Section 3.11. □

From Lemma 2.6.12 it follows that the aliasing terms ($p \neq 0$) in (2.179a) and (2.179b) can be made arbitrarily small by making the sampling period T_s sufficiently small. Furthermore, from Theorems 2.6.4 and 2.6.5 it follows that, for any fixed T_s, the DT-CCC approaches in the mean-square sense the aliased cyclic cross-correlation function (2.179a)–(2.179b) as the data-record length $T = (2N+1)T_s$ approaches infinity. Consequently, for T_s sufficiently small and N sufficiently large such that $T = (2N+1)T_s \gg 1$, the DT-CCC $\widetilde{R}_{y_d x_d^{(*)}}(\widetilde{\alpha}, m; n_0, N)$, at normalized cycle frequency $\widetilde{\alpha} = \alpha/f_s$ and normalized lag $m = \tau/T_s$, can be made arbitrarily close to samples of the continuous-time cyclic cross-correlation function $R_{yx^{(*)}}(\alpha, \tau)|_{\alpha=\widetilde{\alpha}f_s, \tau=mT_s}$ in the mean-square sense, provided that τ and α are multiples of T_s and $1/T_s$, respectively. Specifically, we have the following result.

Theorem 2.6.13 Mean-Square Consistency of the Discrete-Time Cyclic Cross-Correlogram (Napolitano 2009, Theorem 7.1). *Under Assumptions 2.4.2 (uniformly almost-periodic statistics), 2.4.3 (Fourier series regularity), 2.4.5 (data-tapering window regularity), 2.4.8 (mixing conditions (I)), and 2.6.11 (aliasing series summability) (i.e., the assumptions for*

Theorems 2.6.4, and 2.6.5, and Lemma 2.6.12) one obtains

$$\lim_{T_s \to 0} \lim_{N \to \infty} \mathrm{E}\left\{ \left| \widetilde{R}_{y_\mathrm{d} x_\mathrm{d}^{(*)}}(\widetilde{\alpha}, m; n_0, N) - R_{yx^{(*)}}(\alpha, \tau)|_{\alpha = \widetilde{\alpha} f_s, \tau = m T_s} \right|^2 \right\} = 0 \tag{2.211}$$

where the order of the two limits cannot be interchanged and the limit is not necessarily uniform with respect to $(\widetilde{\alpha}, m)$.

Proof: See Section 3.11. □

From Theorem 2.6.13 it follows that the mean-square error between the DT-CCC and samples of the continuous-time cyclic cross-correlation function can be made arbitrarily small, provided that the data-record length is sufficiently large and the sampling period is sufficiently small. However, the asymptotic result of Theorem 2.6.13 has the drawback that, for fixed $\widetilde{\alpha}$ and m, when $T_s \to 0$ it follows $\alpha = \widetilde{\alpha} f_s \to \infty$ and $\tau = m T_s \to 0$. Thus, this analysis turns out to be unhelpful if the asymptotic (as $N \to \infty$ and $T_s \to 0$) bias and covariance are needed and asymptotic Normality needs to be proved. An asymptotic analysis not suffering from such a drawback can be made starting from the hybrid cyclic cross-correlogram (Napolitano 2009).

Definition 2.6.14 *Let $y(t)$ and $x(t)$ be continuous-time processes. The* hybrid cyclic cross-correlogram (H-CCC) *at cycle frequency $\alpha \in \mathbb{R}$ and lag $\tau \in \mathbb{R}$ of $x_\mathrm{d}(n)$ defined as in (2.172) and*

$$y_{\mathrm{d}\tau}(n) \triangleq y(t+\tau)|_{t = n T_s} \qquad \tau \in \mathbb{R} \tag{2.212}$$

is defined as

$$\varrho_{y_\mathrm{d} x_\mathrm{d}^{(*)}}(\alpha, \tau; n_0, N, T_s) \triangleq \sum_{n=-\infty}^{+\infty} v_N(n - n_0)\, y_{\mathrm{d}\tau}(n)\, x_\mathrm{d}^{(*)}(n)\, e^{-j2\pi\alpha n T_s} \tag{2.213}$$

where $v_N(n)$ is the data-tapering window nonzero in $\{-N, \ldots, N\}$ defined in (2.182) (with $a(t)$ as in Assumption 2.4.5). □

Note that, unlike the DT-CCC of Definition 2.6.1, the lag parameter τ is not necessarily an integer multiple of T_s, is not assumed to be proportional to T_s, and α is a cycle frequency not normalized to f_s. Consequently, τ can be retained constant as $T_s \to 0$ avoiding the drawback $\tau \to 0$ as $T_s \to 0$ as in Theorem 2.6.13. Analogously, the cycle frequency of the continuous-time process does not need to be proportional to the sampling frequency f_s and can be kept constant as $f_s \to \infty$. It is worthwhile emphasizing that the hybrid cyclic cross-correlogram turns out to be useful just to analytically carry out the asymptotic analysis. In practice, the DT-CCC should be implemented since the available discrete-time samples are $y((n+m)T_s)$. Thus, for a fixed value of τ, when $T_s \to 0$, m should accordingly increase in order to keep constant the product $m T_s$. Analogously, $\widetilde{\alpha}$ should accordingly decrease in order to keep constant $\widetilde{\alpha} f_s$. Note that, however, the H-CCC can actually be computed by using Definition 3.11.1 of time-shifted discrete-time signal with noninteger time shift and its implementation described in Fact 3.11.2.

The expected value and covariance of the H-CCC can be obtained with minor changes from those of the DT-CCC.

For finite N and T_s, from Theorem 2.6.2 with the replacements $\tau \curvearrowright mT_s$ and $\alpha \curvearrowright \widetilde{\alpha} f_s$ we immediately get the *expected value of the H-CCC*:

$$\begin{aligned}
&\mathrm{E}\left\{\varrho_{y_\mathrm{d} x_\mathrm{d}^{(*)}}(\alpha, \tau; n_0, N, T_s)\right\} \\
&\quad = \sum_{k \in \mathbb{I}} R^{(k)}_{yx^{(*)}}(\tau)\, V_{\frac{1}{N}}\left((\alpha - \alpha_k(\tau))T_s\right) e^{-j2\pi[\alpha-\alpha_k(\tau)]n_0 T_s}
\end{aligned} \tag{2.214}$$

where $V_{\frac{1}{N}}(\nu)$ is defined in (2.184). The function $V_{1/N}(\nu)$ is periodic with period 1 and has a bandwidth of the order of $1/N$. Consequently, the cycle-frequency resolution is of the order of $1/(NT_s)$ and aliasing in the cycle-frequency α domain occurs. Specifically, from (2.214) it follows that the expected value of the H-CCC can be significantly different from zero within strips of width $1/(NT_s)$ around the curves in the (α, τ)-plane defined by $\alpha = \alpha_k(\tau) \bmod f_s$, $k \in \mathbb{I}$. Moreover, the expected value at (α, τ) is due essentially to all the lag-dependent cycle frequencies $\alpha_k(\tau)$ such that

$$|\alpha + pf_s - \alpha_k(\tau)| < \frac{1}{NT_s} \tag{2.215}$$

for some integer p and some $k \in \mathbb{I}$. That is, a leakage phenomenon among lag-dependent cycle-frequency curves which are sufficiently close (modulo f_s) to each other occurs (see also the discussion following Theorem 2.4.6 for the continuous-time estimator and Theorem 2.6.2 for the DT-CCC).

For finite N and T_s, from Theorem 2.6.3 we obtain the *covariance of the H-CCC*

$$\mathrm{cov}\left\{\varrho_{y_\mathrm{d} x_\mathrm{d}^{(*)}}(\alpha_1, \tau_1; n_{01}, N, T_s), \varrho_{y_\mathrm{d} x_\mathrm{d}^{(*)}}(\alpha_2, \tau_2; n_{02}, N, T_s)\right\} = \widetilde{\mathcal{T}}_1 + \widetilde{\mathcal{T}}_2 + \widetilde{\mathcal{T}}_3 \tag{2.216}$$

where $\widetilde{\mathcal{T}}_1$, $\widetilde{\mathcal{T}}_2$, and $\widetilde{\mathcal{T}}_3$ are defined in (2.187), (2.188), and (2.189), respectively, with the replacements $\tau_1 \curvearrowright m_1 T_s$, $\alpha_1 \curvearrowright \widetilde{\alpha}_1 f_s$, $\tau_2 \curvearrowright m_2 T_s$, $\alpha_2 \curvearrowright \widetilde{\alpha}_2 f_s$, and $\widetilde{r}_{aa^*}(\beta, m; N)$ (defined in (2.190)) substituted by $\widetilde{r}_{aa^*}(\beta T_s, \tau/T_s; N)$.

The asymptotic results as $N \to \infty$ for the H-CCC are obtained, with minor changes, as those for the DT-CCC. In particular, result (3.175) of Lemma 3.10.2 should be used.

By reasoning as in Theorem 2.6.4, we have that the *asymptotic ($N \to \infty$) expected value of the H-CCC* (2.213) is given by

$$\lim_{N\to\infty} \mathrm{E}\left\{\varrho_{y_\mathrm{d} x_\mathrm{d}^{(*)}}(\alpha, \tau; n_0, N, T_s)\right\} = \sum_{p=-\infty}^{+\infty} R_{yx^{(*)}}(\alpha + pf_s, \tau) \tag{2.217a}$$

$$= \sum_{k \in \mathbb{I}} R^{(k)}_{yx^{(*)}}(\tau)\, \delta_{[\alpha-\alpha_k(\tau)] \bmod f_s}. \tag{2.217b}$$

Furthermore, by reasoning as in Theorem 2.6.5, we have that the *asymptotic ($N \to \infty$) covariance of the H-CCC* (2.213) is given by

$$\begin{aligned}
&\lim_{N\to\infty} (2N+1)\,\mathrm{cov}\left\{\varrho_{y_\mathrm{d} x_\mathrm{d}^{(*)}}(\alpha_1, \tau_1; n_{01}, N, T_s), \varrho_{y_\mathrm{d} x_\mathrm{d}^{(*)}}(\alpha_2, \tau_2; n_{02}, N, T_s)\right\} \\
&\quad = \widetilde{\mathcal{T}}'_1 + \widetilde{\mathcal{T}}'_2 + \widetilde{\mathcal{T}}'_3
\end{aligned} \tag{2.218}$$

where $\widetilde{T}'_1$, $\widetilde{T}'_2$, and $\widetilde{T}'_3$ are defined in (2.196), (2.197), and (2.198), respectively, with the replacements $\tau_1 \curvearrowright m_1 T_s, \alpha_1 \curvearrowright \widetilde{\alpha}_1 f_s, \tau_2 \curvearrowright m_2 T_s, \alpha_2 \curvearrowright \widetilde{\alpha}_2 f_s$. Finally, the *rate of convergence of the bias of the H-CCC* as $N \to \infty$ is obtained by following the guidelines of Theorem 2.6.7:

$$\lim_{N\to\infty} (2N+1)^{\gamma} \left| \mathrm{E}\left\{ \varrho_{y_{\mathrm{d}} x_{\mathrm{d}}^{(*)}}(\alpha, \tau; n_0, N, T_s) \right\} - \sum_{p=-\infty}^{+\infty} R_{yx^{(*)}}(\alpha + p f_s, \tau) \right| = \mathcal{O}(1) \quad (2.219)$$

where γ is the rate of decay to zero of $A(f)$, the Fourier transform of the data-tapering window $a(t)$.

In order to establish the rate of convergence to zero of the bias of the H-CCC and its asymptotic Normality as $N \to \infty$ and $T_s \to 0$, an assumption on the lack of clusters of cycle frequencies different from Assumption 2.6.6 is needed. For this purpose, let us define the sets

$$A_\tau \triangleq \left\{ \alpha \in \mathbb{R} \;:\; R_{yx^{(*)}}(\alpha, \tau) \neq 0 \right\} \quad (2.220)$$

$$\bar{A}_\tau \triangleq \left\{ \beta \in [-f_{s/2}, f_{s/2}) \;:\; \sum_{p=-\infty}^{+\infty} R_{yx^{(*)}}(\beta + p f_s, \tau) \neq 0 \right\} \quad (2.221a)$$

$$= \left\{ \beta \in \mathbb{R} \;:\; \beta = \alpha \bmod f_s, \quad \alpha \in A_\tau \right\} \quad (2.221b)$$

$$= A_\tau \bmod f_s \quad (2.221c)$$

(strictly speaking, the second equality should be substituted by $\subseteq$).

Assumption 2.6.15 Lack of Support Curve Clusters (III). *For every τ, the cycle-frequency set $\bar{A}_\tau$ in (2.221c) does not contain any cluster of cycle frequencies. That is, let*

$$\mathbb{J}_{\alpha,\tau,T_s} \triangleq \left\{ k \in \mathbb{I} \;:\; \alpha_k(\tau) = \alpha \bmod f_s, \; R^{(k)}_{yx^{(*)}}(\tau) \neq 0 \right\} \quad (2.222)$$

then, for every $k \notin \mathbb{J}_{\alpha,\tau,T_s}$, no curve $\alpha_k(\tau)$ is such that the value $\alpha_k(\tau) \bmod f_s$ can be arbitrarily close to the cycle frequency α. Thus, for every α and τ it results that

$$h_{\alpha,\tau,T_s} \triangleq \inf_{k \notin \mathbb{J}_{\alpha,\tau,T_s}} |[\alpha - \alpha_k(\tau)]\, T_s \bmod 1| > 0. \quad (2.223)$$

□

Assumption 2.6.15 means that there is no cluster of lag-dependent cycle-frequency curves, where cycle frequencies are considered modulo f_s. Thus, it is stronger than Assumption 2.4.10. A sufficient condition assuring that Assumption 2.6.15 is satisfied is that the number of lag-dependent cycle-frequency curves is finite. That is, the set $\mathbb{I}$ is finite. Note that, Assumption 2.6.15 differs from Assumption 2.6.6 since the definitions of the sets $\mathbb{J}_{\alpha,\tau,T_s}$ and $\mathbb{J}_{\widetilde{\alpha},m,T_s}$ are different. Specifically, in the former the argument of the functions $\alpha_k(\cdot)$ does not depend on T_s whereas in the latter it does. Such a difference is fundamental in the study of the asymptotic properties as $T_s \to 0$.

In the following, asymptotic results as $N \to \infty$ and $T_s \to 0$ are provided. Condition $N \to \infty$ needs to find consistency for the discrete-time estimator, whereas condition $T_s \to 0$ assures lack of aliasing. Note that, in order to have asymptotically an infinitely long data-record length, condition $T = (2N+1)T_s \to \infty$ needs to be verified. Consequently, in the following asymptotic results, we have that first $N \to \infty$ and then $T_s \to 0$, that is, the order of the two limits as $N \to \infty$ and $T_s \to 0$ cannot be interchanged.

In the sequel, when necessary, the very mild assumption that the lag-dependent cycle frequencies $\alpha_k(\tau)$ are bounded for finite τ will be made.

Theorem 2.6.16 Asymptotic Expected Value of the Hybrid Cyclic Cross-Correlogram (Napolitano 2009, Theorem 7.2). *Let $y(t)$ and $x(t)$ be wide-sense jointly GACS stochastic processes with cross-correlation function (2.31c). Under Assumptions 2.4.2a (uniformly almost-periodic statistics), 2.4.3a (Fourier series regularity), 2.4.5 (data-tapering window regularity), and 2.6.11 (aliasing series summability), one obtains*

$$\lim_{T_s \to 0} \lim_{N \to \infty} \mathrm{E}\left\{ \varrho_{y_{\mathrm{d}} x_{\mathrm{d}}^{(*)}}(\alpha, \tau; n_0, N, T_s) \right\} = R_{yx^{(*)}}(\alpha, \tau) \tag{2.224}$$

where the order of the two limits cannot be interchanged.

Proof: See Section 3.11. □

Let $A(f)$ be the Fourier transform of $a(t)$. Under Assumption 2.4.5, there exists $\gamma > 0$ such that $A(f) = \mathcal{O}(|f|^{-\gamma})$ as $|f| \to \infty$.

For the bias of the H-CCC

$$\mathrm{bias}\left\{ \varrho_{y_{\mathrm{d}} x_{\mathrm{d}}^{(*)}}(\alpha, \tau; n_0, N, T_s) \right\} \triangleq \mathrm{E}\left\{ \varrho_{y_{\mathrm{d}} x_{\mathrm{d}}^{(*)}}(\alpha, \tau; n_0, N, T_s) \right\} - R_{yx^{(*)}}(\alpha, \tau) \tag{2.225}$$

the following result holds.

Theorem 2.6.17 Rate of Convergence of the Bias of the Hybrid Cyclic Cross-Correlogram (Napolitano 2009, Theorem 7.3). *Let $y(t)$ and $x(t)$ be wide-sense jointly GACS stochastic processes with cross-correlation function (2.31c). Under Assumptions 2.4.2a (uniformly almost-periodic statistics), 2.4.3a (Fourier series regularity), and 2.4.5 (data-tapering window regularity) assuming that the number of lag-dependent cycle frequencies is finite (so that also Assumptions 2.6.11 (aliasing series summability) and 2.6.15 (lack of support curve clusters (III)) are verified), and provided that $\gamma > 1$ (or $\gamma = 1$ in the special case of $a(t) = \mathrm{rect}(t)$), one obtains*

$$\lim_{T_s \to 0} \lim_{N \to \infty} \left[(2N+1)T_s\right]^{\gamma} \left| \mathrm{bias}\left\{ \varrho_{y_{\mathrm{d}} x_{\mathrm{d}}^{(*)}}(\alpha, \tau; n_0, N, T_s) \right\} \right| = \mathcal{O}(1) \tag{2.226}$$

where the order of the two limits cannot be interchanged.

Proof: See Section 3.11. □

Theorem 2.6.18 Asymptotic Covariance of the Hybrid Cyclic Cross-Correlogram (Napolitano 2009, Theorem 7.4). *Let $y(t)$ and $x(t)$ be zero-mean wide-sense jointly GACS stochastic* *processes with cross-correlation function (2.31c). Under Assumptions 2.4.2 (uniformly almost-periodic statistics), 2.4.3 (Fourier series regularity), 2.4.5 (data-tapering window regularity), and 2.4.8 (mixing conditions (I)) (with the sets $\mathbb{I}_{z_1 z_2}$ and $\mathbb{I}_4$ finite), the asymptotic ($N \to \infty$ and $T_s \to 0$ with $NT_s \to \infty$) covariance of the H-CCC (2.213) is given by*

$$\lim_{T_s \to 0} \lim_{N \to \infty} (2N+1) T_s \operatorname{cov}\left\{ \varrho_{y_\mathrm{d} x_\mathrm{d}^{(*)}}(\alpha_1, \tau_1; n_{01}, N, T_s), \varrho_{y_\mathrm{d} x_\mathrm{d}^{(*)}}(\alpha_2, \tau_2; n_{02}, N, T_s) \right\}$$
$$= \mathcal{T}_1' + \mathcal{T}_2' + \mathcal{T}_3' \tag{2.227}$$

with $\mathcal{T}_1'$, $\mathcal{T}_2'$, and $\mathcal{T}_3'$ given by (2.147), (2.148), and (2.149) (in Theorem 2.4.13), respectively, where the order of the two limits cannot be interchanged.

Proof: See Section 3.11. □

By comparing the results of Theorems 2.4.13 and 2.6.18 it follows that the asymptotic ($T \to \infty$) covariance of the continuous-time cyclic cross-correlogram $R_{yx^{(*)}}(\alpha, \tau; t_0, T)$ and the asymptotic ($N \to \infty$ and $T_s \to 0$ with $NT_s \to \infty$) covariance of the H-CCC $\varrho_{y_\mathrm{d} x_\mathrm{d}^{(*)}}(\alpha, \tau; n_0, N, T_s)$ are coincident. That is,

$$\lim_{T_s \to 0} \lim_{N \to \infty} (2N+1) T_s \operatorname{cov}\left\{ \varrho_{y_\mathrm{d} x_\mathrm{d}^{(*)}}(\alpha_1, \tau_1; n_{01}, N, T_s), \varrho_{y_\mathrm{d} x_\mathrm{d}^{(*)}}(\alpha_2, \tau_2; n_{02}, N, T_s) \right\}$$
$$= \lim_{T \to \infty} T \operatorname{cov}\left\{ R_{yx^{(*)}}(\alpha_1, \tau_1; t_1, T), R_{yx^{(*)}}(\alpha_2, \tau_2; t_2, T) \right\}. \tag{2.228}$$

From Theorem 2.6.16 it follows that the H-CCC is an asymptotically ($N \to \infty$ and $T_s \to 0$ with $NT_s \to \infty$) unbiased estimator of the continuous-time cyclic cross-correlation function. Moreover, from Theorem 2.6.18 it follows that the H-CCC has asymptotically vanishing variance. Therefore, the H-CCC is a mean-square consistent estimator of the continuous-time cyclic cross-correlation function.

The proof of the zero-mean joint complex asymptotic Normality of the random variables

$$V_i^{(N,T_s)} \triangleq \sqrt{(2N+1)T_s}\left[\varrho_{y_\mathrm{d} x_\mathrm{d}^{(*)}}(\alpha_i, \tau_i; n_0, N, T_s) - R_{yx^{(*)}}(\alpha_i, \tau_i) \right] \tag{2.229}$$

is given in Theorem 2.6.20 showing that asymptotically ($N \to \infty$ and $T_s \to 0$ with $NT_s \to \infty$)

1. $\operatorname{cum}\{V_i^{(N,T_s)}\} \equiv \mathrm{E}\{V_i^{(N,T_s)}\} = 0$

2a. the covariance $\operatorname{cov}\{V_i^{(N,T_s)}, V_j^{(N,T_s)}\} \equiv \operatorname{cum}\{V_i^{(N,T_s)}, V_j^{(N,T_s)*}\}$ is finite;

2b. the conjugate covariance $\operatorname{cov}\{V_i^{(N,T_s)}, V_j^{(N,T_s)*}\} \equiv \operatorname{cum}\{V_i^{(N,T_s)}, V_j^{(N,T_s)}\}$ is finite;

3. $\operatorname{cum}\{V_{i_1}^{(N,T_s)[*]_1}, \ldots, V_{i_k}^{(N,T_s)[*]_k}\} = 0$ for $k \geqslant 3$

with superscript $[*]_h$ denoting optional complex conjugation. Condition 1 follows from Theorem 2.6.17 on the rate of decay to zero of the bias of the H-CCC. Condition 2a is a consequence of Theorem 2.6.18 on the asymptotic covariance of the H-CCC. Condition 2b follows from Theorem 3.13.3. Finally, Condition 3 follows from Lemma 2.6.19 on the rate of decay to zero of the joint cumulant of H-CCCs.

Lemma 2.6.19 Rate of Convergence to Zero of Cumulants of Hybrid Cyclic Cross-Correlograms (Napolitano 2009, Lemma 7.2). *Under Assumptions 2.4.5 (data-tapering window regularity) and 2.6.8 (mixing conditions (IV)), for every $k \geqslant 2$ and $\epsilon > 0$ one obtains*

$$\lim_{T_s \to 0} \lim_{N \to \infty} [(2N+1)T_s]^{k-1-\epsilon} \text{cum}\left\{ \varrho^{[*]_i}_{y_d x_d^{(*)}}(\alpha_i, \tau_i; n_{0i}, N, T_s),\ i = 1, \ldots, k \right\} = 0 \quad (2.230)$$

where the order of the two limits cannot be interchanged.

Proof: See Section 3.11. □

Theorem 2.6.20 Asymptotic Joint Normality of the Hybrid Cyclic Cross-Correlograms (Napolitano 2009, Theorem 7.5). *Let $y(t)$ and $x(t)$ be zero-mean wide-sense jointly GACS stochastic processes with cross-correlation function (2.31c). Under Assumptions 2.4.2 (uniformly almost-periodic statistics), 2.4.3 (Fourier series regularity), 2.4.5 (data-tapering window regularity), and 2.4.8 (mixing conditions (I)) (with the sets $\mathbb{I}_{z_1 z_2}$ and $\mathbb{I}_4$ finite) and under Assumptions 2.6.8 (mixing conditions (IV)), 2.6.11 (aliasing series summability), and 2.6.15 (lack of support curve clusters (III)), and provided that $\gamma > 1$ (or $\gamma = 1$ in the special case of $a(t) = \text{rect}(t)$), the random variables $V_i^{(N,T_s)}$ defined in (2.229)*

$$V_i^{(N,T_s)} \triangleq \sqrt{(2N+1)T_s} \left[\varrho_{y_d x_d^{(*)}}(\alpha_i, \tau_i; n_0, N, T_s) - R_{yx^{(*)}}(\alpha_i, \tau_i) \right]$$

are asymptotically (as $N \to \infty$ and $T_s \to 0$ with $NT_s \to \infty$) zero-mean jointly complex Normal with asymptotic covariance matrix $\boldsymbol{\Sigma}$ with entries

$$\Sigma_{ij} = \lim_{T_s \to 0} \lim_{N \to \infty} \text{cov}\left\{ V_i^{(N,T_s)}, V_j^{(N,T_s)} \right\} \quad (2.231)$$

given by (2.227) and asymptotic conjugate covariance matrix $\boldsymbol{\Sigma}^{(c)}$ with entries

$$\Sigma_{ij}^{(c)} = \lim_{T_s \to 0} \lim_{N \to \infty} \text{cov}\left\{ V_i^{(N,T_s)}, V_j^{(N,T_s)*} \right\} \quad (2.232)$$

given by (3.260).

Proof: See Section 3.11. □

Theorems 2.6.17, 2.6.18, and 2.6.20 show that the H-CCC has the same asymptotic bias, covariance and distribution of the continuous-time cyclic cross-correlogram, provided that the aliasing in the cycle-frequency domain is controlled so that there is no cluster of cycle frequencies for the ACS discrete-time process. Thus, there is no loss in asymptotic performance by carrying out discrete-time estimation instead of continuous-time estimation of the cyclic cross-correlation function.

2.6.4 Concluding Remarks

Let us assume, without loss of generality, that $T \triangleq (2N+1)T_s$ and $t_0 \triangleq n_0 T_s$, with $N \in \mathbb{N}$ and $n_0 \in \mathbb{Z}$. The DT-CCC approaches in the mean-square sense the continuous-time cyclic cross-correlogram as the sampling period approaches zero. Specifically, we have the following result.

Theorem 2.6.21 Asymptotic Discrete-Time Cyclic Cross-Correlogram. *Let $x_{\mathrm{d}}(n)$ and $y_{\mathrm{d}}(n)$ be the discrete-time processes given by (2.172). Under Assumptions 2.4.2 (uniformly almost-periodic statistics), 2.4.3 (Fourier series regularity), 2.4.4 (fourth-order moment boundedness), and 2.4.5 (data-tapering window regularity) (assumptions of Theorem 2.4.7) on the continuous-time processes $x(t)$ and $y(t)$, one obtains*

$$\lim_{T_s \to 0} \mathrm{E}\Big\{\Big| \widetilde{R}_{y_{\mathrm{d}} x_{\mathrm{d}}^{(*)}}(\widetilde{\alpha}, m; n_0, N)$$
$$- R_{yx^{(*)}}(\alpha, \tau; t_0, T)\big|_{\alpha=\widetilde{\alpha} f_s, \tau=mT_s; t_0=n_0T_s, T=(2N+1)T_s}\Big|^2\Big\} = 0. \tag{2.233}$$

Proof: See Section 3.12. □

Remark 2.6.22 *In Theorems 2.6.4–2.6.21 the cycle frequencies $\widetilde{\alpha}$, $\widetilde{\alpha}_1$, and $\widetilde{\alpha}_2$ of the discrete-time ACS process are assumed to be fixed and the cycle-frequency resolution is $1/N$. If a bidimensional analysis with variable cycle frequency $\widetilde{\alpha}$ and discrete-lag m is carried out, then $\widetilde{\alpha}$ should vary continuously in $[-1/2, 1/2)$. However, in practical implementations of bidimensional analysis, $\widetilde{\alpha}$ assumes values $\widetilde{\alpha}_k^{(N)}$ uniformly distributed in a finite set $\mathcal{A}_N$ of points of $[-1/2, 1/2)$. Then, in order to prove mean-square consistency and asymptotic Normality, the assumption*

$$\Delta\widetilde{\alpha}^{(N)} \triangleq \widetilde{\alpha}_k^{(N)} - \widetilde{\alpha}_{k-1}^{(N)} = \mathrm{o}\Big(\frac{1}{N}\Big) \quad \text{as } N \to \infty \tag{2.234}$$

should be made, where o *denotes "small oh" Landau symbol, that is, $\Delta\widetilde{\alpha}^{(N)} = \mathrm{o}(1/N)$ as $N \to \infty$ means $\Delta\widetilde{\alpha}^{(N)} N \to 0$ as $N \to \infty$.*

For example, we can assume

$$\widetilde{\alpha}_k^{(N)} = \frac{k}{N^{\gamma}}, \quad k = -\left\lfloor \frac{N^{\gamma}}{2} \right\rfloor, -\left\lfloor \frac{N^{\gamma}}{2} \right\rfloor + 1, \dots, \left\lfloor \frac{N^{\gamma}}{2} \right\rfloor - 1 \tag{2.235}$$

with $\gamma > 1$ and $\lfloor \cdot \rfloor$ denoting the floor operation. Analogously, mean-square consistency and asymptotic Normality can be proved in the case of nonuniformly sampled cycle frequencies $\widetilde{\alpha}_k^{(N)}$ such that

$$\Delta\widetilde{\alpha}^{(N)} \triangleq \sup_k \left|\widetilde{\alpha}_k^{(N)} - \widetilde{\alpha}_{k-1}^{(N)}\right| = \mathrm{o}\Big(\frac{1}{N}\Big) \quad \text{as } N \to \infty. \tag{2.236}$$

2.6.4.1 Subsampling-Based Estimates

In the previous sections, under mild assumptions, it is shown that the continuous-time cyclic cross-correlogram of a GACS process is asymptotically complex Normal as $T \to \infty$. Analogous results are found for the DT-CCC as $N \to \infty$ and for the H-CCC as $N \to \infty$ and $T_s \to 0$ with $NT_s \to \infty$.

In order to perform a statistical test for presence of generalized almost cyclostationarity, covariance and conjugate covariance need to be known or estimated. If a complete knowledge of lag-dependent cycle frequencies and generalized cyclic second- and fourth-order statistics is available, then covariance and conjugate covariance can be computed, even if their expression generally involves infinite sums, and hence their value can only be approximated. If such *a priori* knowledge is only partial or not available, then subsampling techniques can be used in order to obtain estimates starting from a single realization of the process. For example, the subsampling technique (Politis 1998) provides the estimate

$$\mathrm{E}^{\star}\left\{\left[V^{(b)} - \mathrm{E}^{\star}\{V^{(b)}\}\right]^2\right\} \tag{2.237}$$

for the variance

$$\mathrm{E}\left\{\left[V^{(T)} - \mathrm{E}\{V^{(T)}\}\right]^2\right\} \tag{2.238}$$

where $\mathrm{E}^{\star}\{\cdot\}$ denotes a subsampling-based estimate of the expectation operator, b is the block size, and $T = kb$ for some k. Convergence is obtained when $b \to \infty$, $T \to \infty$, and $b/T \to 0$ assuming finite or practically finite memory for the processes (see (Lenart *et al.* 2008) for the cyclostationary case).

2.6.4.2 Estimation of Cyclic Higher-Order Statistics

In (Napolitano and Tesauro 2011), stochastic processes with higher-order statistical functions decomposable into an almost-periodic function plus a residual term not containing finite-strength additive sinewave components are considered (see (2.80a), (2.80b), (2.90a), and (2.90b)). These processes arise in mobile communications when ACS processes pass through time-varying channels. They include as special case the GACS processes.

For this class of processes, in (Napolitano and Tesauro 2011), the problem of estimating the Fourier coefficients of the (generalized) Fourier series expansion of the almost-periodic component of higher-order statistical functions of K possible distinct processes is addressed. The Kth-order cyclic cross-correlogram is proposed as an estimator of the Kth-order cyclic temporal cross-moment function. Expected value, cross-cumulants, and covariance of the cyclic cross-correlogram are derived for finite data-record length T, under mild regularity assumptions on the data-tapering window and the Fourier series convergence and under boundedness assumptions of moments and residual terms. Asymptotic ($T \to \infty$) expressions for expected value, cross-cumulants, and covariance, and the rate of convergence of bias of the cyclic cross-correlogram are derived under mixing conditions, residual term integrability assumption, and lack of clusters of lag-dependent cycle frequencies. Furthermore, the asymptotic ($T \to \infty$) joint complex Normality of cyclic correlograms at different lag vectors and cycle frequencies is proved. Thus, the Kth-order cyclic cross-correlogram is shown to be a mean-square consistent, asymptotically complex Normal estimator of the Kth-order cyclic temporal cross-moment function.

A Kth-order cyclic temporal cross-cumulant estimator is proposed as a combination of a finite number Q of products of Kth- and lower-order cyclic cross-correlograms. Asymptotic expected value and cross-cumulants of the cyclic cumulant estimator are derived and the asymptotic covariance is obtained as a special case. Moreover, the rate of convergence of bias is derived and the asymptotic joint complex Normality of estimators at different lag vectors and cycle frequencies is proved. In the limit as Q and the data-record length T approach infinity, the cyclic cross-cumulant estimator is proved to be mean-square consistent. Interestingly, it is found that the rate of decay to zero with the data-record length of the bias of the Kth-order cyclic cross-correlogram is $\mathcal{O}(T^{-\gamma})$, where $|f|^{-\gamma}$ is the rate of decay to zero as $|f| \to \infty$ of the Fourier transform of the data-tapering window. In contrast, the rate of decay to zero of the bias of the cyclic cross-cumulant estimator is always $\mathcal{O}(1/T)$. In addition, as in the almost-cyclostationary case, the cyclic cross-cumulant estimator requires the knowledge of the lower-order (moment) cycle-frequencies. Such *a priori* knowledge, however, is not necessary when K is the lowest-order of cyclostationarity exhibited by the process.

The proposed estimators have the same computational complexity of previously proposed estimators for ACS processes.

2.7 Numerical Results

In this section, simulation experiments are carried out, aimed at corroborating the theoretical results of the previous sections.

2.7.1 *Aliasing in Cycle-Frequency Domain*

In the simulation experiments, a GACS signal is obtained as the output of the Doppler channel existing between a transmitter and a receiver with constant relative radial acceleration, when the input signal is a cyclostationary PAM signal (Section 2.2.6.1). In this case, an analytical condition on the sampling frequency can be deduced, which limits aliasing. Starting from (2.179a) specialized for $x \equiv y$ and ACS signals, it follows that aliasing in the cycle-frequency domain can be avoided, provided that $f_s \geqslant 4B$, where B is the monolateral bandwidth of $x(t)$ (Theorem 1.3.7), (Izzo and Napolitano 1996). Since $2B$ is the maximum allowable cycle frequency for an ACS signal with bandwidth B, condition $f_s \geqslant 4B$ prevents replicas in (2.179a) with $p \neq 0$ from contributing to the discrete-time cyclic autocorrelation function in the "principal domain" $\widetilde{\alpha} \in [-1/2, 1/2)$. If $x(t)$ is not strictly bandlimited, with approximate monolateral bandwidth B, then $2B$ is the maximum cycle frequency at which $x(t)$ exhibits a significant degree of cyclostationarity. For the chirp-modulated signal $y(t)$ in (2.99) with $x(t)$ ACS, the effect of the chirp modulation is to rotate by an angle θ, where $\tan\theta = \gamma$, the support lines of the cyclic autocorrelation function of $x(t)$ (see (2.104) and Figures 2.2 (bottom) and 2.3 (bottom)). Consequently, denoted by τ_{corr} the maximum value of $|\tau|$ such that $R_{xx^*}(0, \tau)$ (and, hence, $R_{xx^*}(\alpha, \tau)\ \forall\alpha$) is significantly different from zero, we have that the maximum cycle frequency exhibited by the GACS signal $y(t)$ is $2B + \tau_{\text{corr}} \sin\theta$. Thus, the condition

$$f_s \geqslant 2(2B + \tau_{\text{corr}} \sin\theta) \tag{2.239}$$

is sufficient to prevent aliasing in the discrete-time cyclic autocorrelation function of the samples of $y(t)$.

2.7.2 *Simulation Setup*

In the experiments, unless otherwise specified, the input signal is a PAM signal with a raised cosine pulse with excess bandwidth $\eta = 0.85$, a stationary white binary modulating sequence, and symbol period $T_p = 10T_s$, where T_s is the sampling period. Such a signal passes through a Doppler channel which produces a delay $d_0 = 20T_s$, a frequency shift $\nu = 0.02/T_s$, and a chirp rate $\zeta = 1.5 \cdot 10^{-3}/T_s^2$. The resulting output GACS signal is the chirp-modulated signal (2.99). Since the PAM signal has a white modulating sequence and the raised cosine pulse decay sufficiently fast as $|t| \to \infty$, its cyclic autocorrelation functions are summable. Moreover, since this PAM signal is strictly bandlimited, then the number of cycle frequencies is finite and

Assumptions 2.4.3a and 2.4.8a turn out to be verified. Furthermore, similar considerations hold for the fourth-order cumulants. For the adopted values, condition (2.239) turns out to be verified.

2.7.3 Cyclic Correlogram Analysis with Varying N

The sample mean and the sample standard deviation, computed by 400 Monte Carlo runs, of the cyclic correlogram of the samples of $y(t)$ are evaluated for different data-record lengths $T = NT_s$ and using a rectangular data-tapering window. For $N = 2^9$, in Figure 2.5 (top) graph and

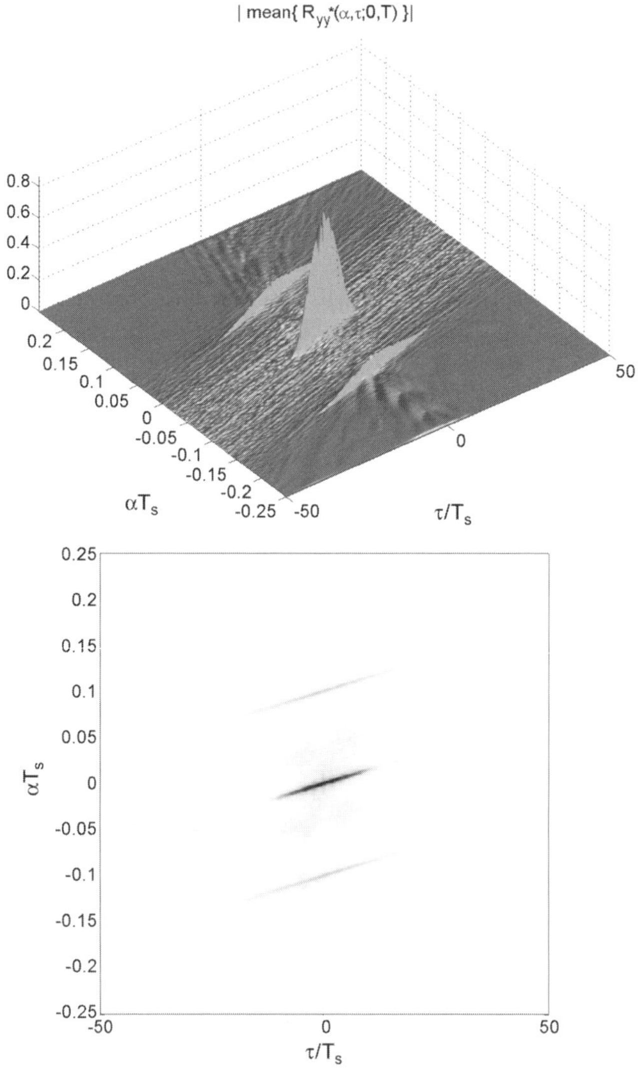

Figure 2.5 Cyclic correlogram of the chirp-modulated signal (2.99) computed by a data-record length $T = NT_s$ with $N = 2^9$: (Top) graph and (bottom) "checkerboard" plot of the magnitude of the sample mean of the cyclic correlogram as a function of αT_s and τ/T_s. *Source:* (Napolitano 2007a) © IEEE

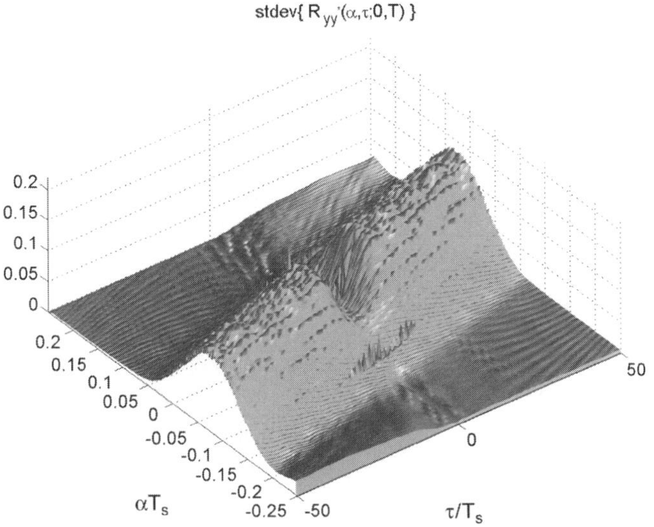

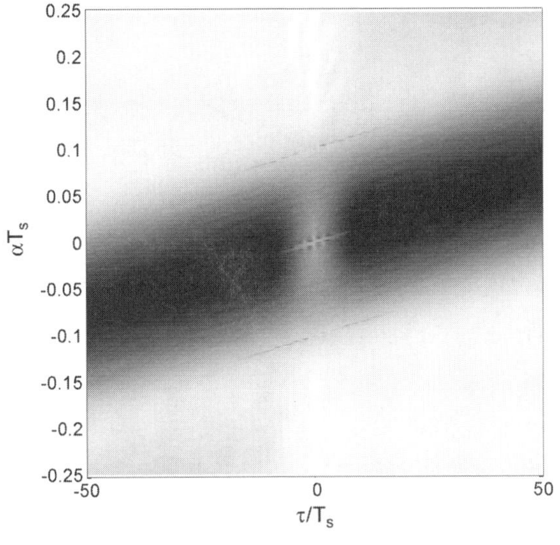

Figure 2.6 Cyclic correlogram of the chirp-modulated signal (2.99) computed by a data-record length $T = NT_s$ with $N = 2^9$: (Top) graph and (bottom) "checkerboard" plot of the sample standard deviation of the cyclic correlogram as a function of αT_s and τ / T_s. *Source:* (Napolitano 2007a) © IEEE

(bottom) "checkerboard" plot of the magnitude of the sample mean of the cyclic correlogram, and in Figure 2.6 (top) graph and (bottom) "checkerboard" plot of the sample standard deviation of the cyclic correlogram are reported as functions of αT_s and τ / T_s. The same quantities are reported in Figures 2.7 and 2.8 for $N = 2^{12}$. The adopted data-record lengths $N = 2^9$ and $N = 2^{12}$, correspond to about 52 and 411 symbols of the PAM signal. A discrete set $\mathcal{A}$ of values of $\widetilde{\alpha}$ is considered by taking 400 cycle-frequency values in the cycle-frequency interval $[-1/4, 1/4)$.

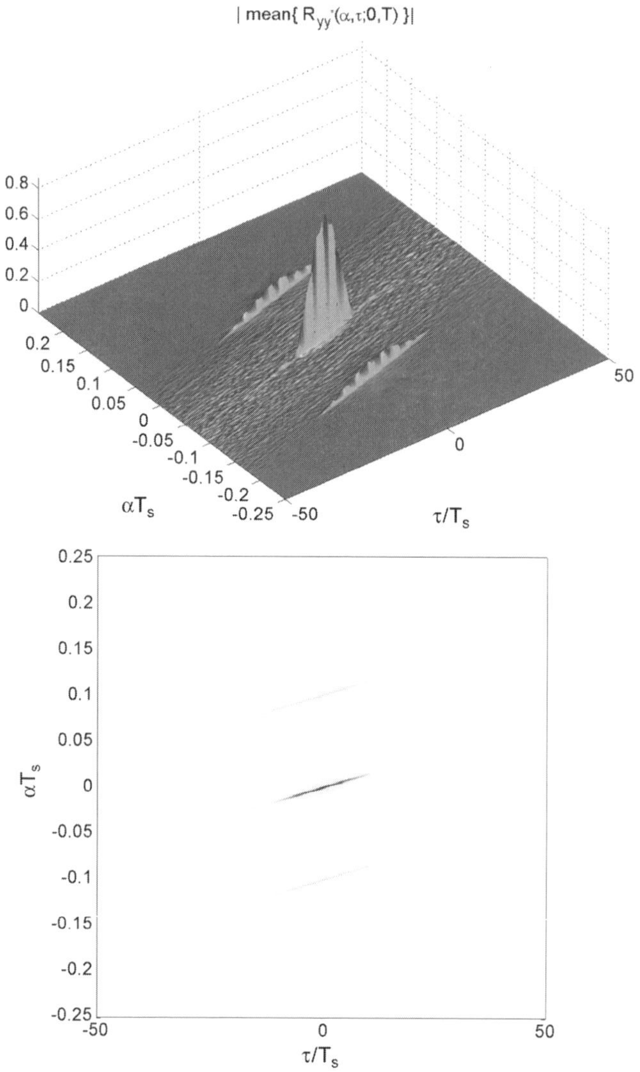

Figure 2.7 Cyclic correlogram of the chirp-modulated signal (2.99) computed by a data-record length $T = NT_s$ with $N = 2^{12}$: (Top) graph and (bottom) "checkerboard" plot of the magnitude of the sample mean of the cyclic correlogram as a function of αT_s and τ / T_s. *Source:* (Napolitano 2007a) © IEEE

The numerical results corroborate the theoretical results of Section 2.4.2. In fact, as the data-record is increased from $N = 2^9$ to $N = 2^{12}$, both bias and standard deviation decrease. Specifically, as regards the sample mean of the cyclic correlogram, the blurred region outside the support of the cyclic autocorrelation function exhibits reduced oscillations as N is increased (Figures 2.5 (top) and 2.7 (top)) and the sample mean approaches the true cyclic autocorrelation

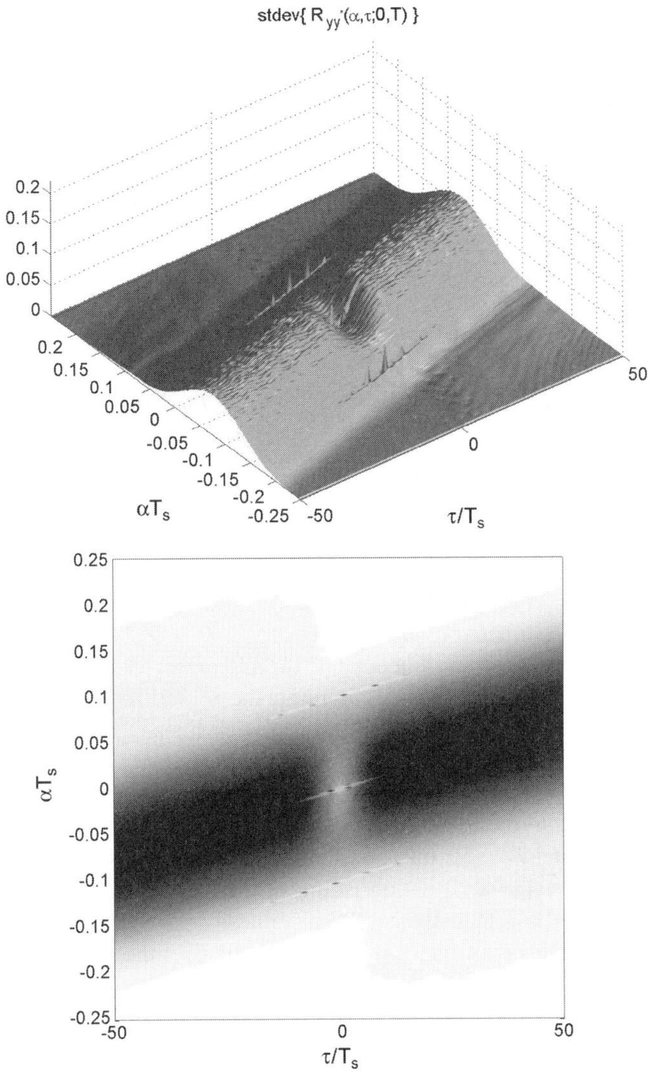

Figure 2.8 Cyclic correlogram of the chirp-modulated signal (2.99) computed by a data-record length $T = NT_s$ with $N = 2^{12}$: (Top) graph and (bottom) "checkerboard" plot of the sample standard deviation of the cyclic correlogram as a function of αT_s and τ/T_s. *Source:* (Napolitano 2007a) © IEEE

function (Figure 2.3 (top)). Moreover, the sample mean is significantly different from zero within thin strips around the true lag-dependent cycle frequencies and the strip width becomes narrower as the data-record is increased (see Figures 2.5 (bottom) and 2.7 (bottom) compared with Figure 2.3 (bottom)). Furthermore, the sample standard deviation decreases as N increases (Figures 2.6 (top) and 2.8 (top)).

2.7.4 Cyclic Correlogram Analysis with Varying N and T_s

In this section, results of experiments are reported to corroborate the results of Section 2.6.

The sample mean and the sample standard deviation, computed by 10^4 Monte Carlo runs, of the discrete-time cyclic correlogram of the ACS discrete-time signal obtained by uniformly sampling the continuous-time GACS signal are evaluated for different numbers of data samples N and sampling periods T_s. The overall data-record length is $T = NT_s$. For a rectangular signal-tapering window, the continuous-time lag-product tapering window and its Fourier transform are given by (Fact 3.5.3)

$$w_{T,\tau}(t) = \frac{1}{T} \operatorname{rect}\left(\frac{t-(T-\tau)/2}{T-|\tau|}\right) \operatorname{rect}\left(\frac{\tau}{2T}\right) \tag{2.240}$$

$$W_{\frac{1}{T},\tau}(f) = \left(1-\frac{|\tau|}{T}\right) \operatorname{rect}\left(\frac{\tau}{2T}\right) \operatorname{sinc}(f(T-|\tau|))\, e^{-j\pi f T}\, e^{j\pi f \tau} \tag{2.241}$$

respectively. Hence, according to the results of Section 2.6, the rate of decay of bias is $T^{-\gamma}$ with $\gamma = 1$ (Theorems 2.6.7 and 2.6.17). Moreover, the rate of decay to zero of standard deviation is $T^{-1/2}$ (Theorems 2.6.5 and 2.6.18).

In Figure 2.9, the results of the first experiment are reported. Specifically, the normalized (by T) real and imaginary parts of the sample bias of the discrete-time cyclic correlogram for $\widetilde{\alpha} = \bar{\zeta}m$, with $\bar{\zeta} \triangleq \zeta T_s^2$, estimated by a sampling period $T_s = T_p/6$ and (a) $N = 2^{11}$ and (b) $N = 2^{13}$ data samples are shown as functions of τ/T_s. Samples for non-integer values of τ/T_s are obtained by interpolation. The thin lines represent the normalized (by T) real and imaginary parts of the analytical bias evaluated by (2.183), the tick lines represent the normalized (by T) real and imaginary parts of the sample bias, and the shaded areas represent the normalized (by $T^{1/2}$) sample standard deviation of real and imaginary parts of the discrete-time cyclic correlogram. According to the theoretical results of Section 2.6, due to the presence of the normalizing factors T and $T^{1/2}$ for bias and standard deviation, respectively, the behavior of the estimated bias and standard deviation is almost the same for different values of N and different for few parts of $1/T$ from the analytical values.

In order to test the dependence on the sampling period T_s of the sample statistics, a second experiment is carried out (see Figure 2.10), where the sampling period $T_s = T_p/12$ is half that of the previous experiment and (a) $N = 2^{12}$ and (b) $N = 2^{14}$ data samples are taken in order to obtain the same overall data-record lengths $T = NT_s$ as in cases (a) and (b), respectively, of the first experiment. Similar considerations as for the first experiment hold about the behavior of the normalized sample bias and standard deviation. In particular, according to the theoretical results, the behavior of quantities reported in Figure 2.10 is almost the same as that of those reported in Figure 2.9.

Finally, by comparing the results with the analytic expression of the bias of the continuous-time estimator evaluated by (2.128), the result is that the effect of aliasing is negligible.

2.7.5 Discussion

From Theorems 2.4.11–2.4.13 it follows that the cyclic cross-correlogram is a mean-square consistent estimator of the cyclic cross-correlation function. Such a consistency result does not depend on the knowledge, or not, of the lag-dependent cycle-frequencies $\alpha_n(\tau)$. This could appear to contrast with a well-known result valid for (jointly) ACS processes. That is, for ACS processes, if the estimation of the cyclic cross-correlation function has to be performed at a

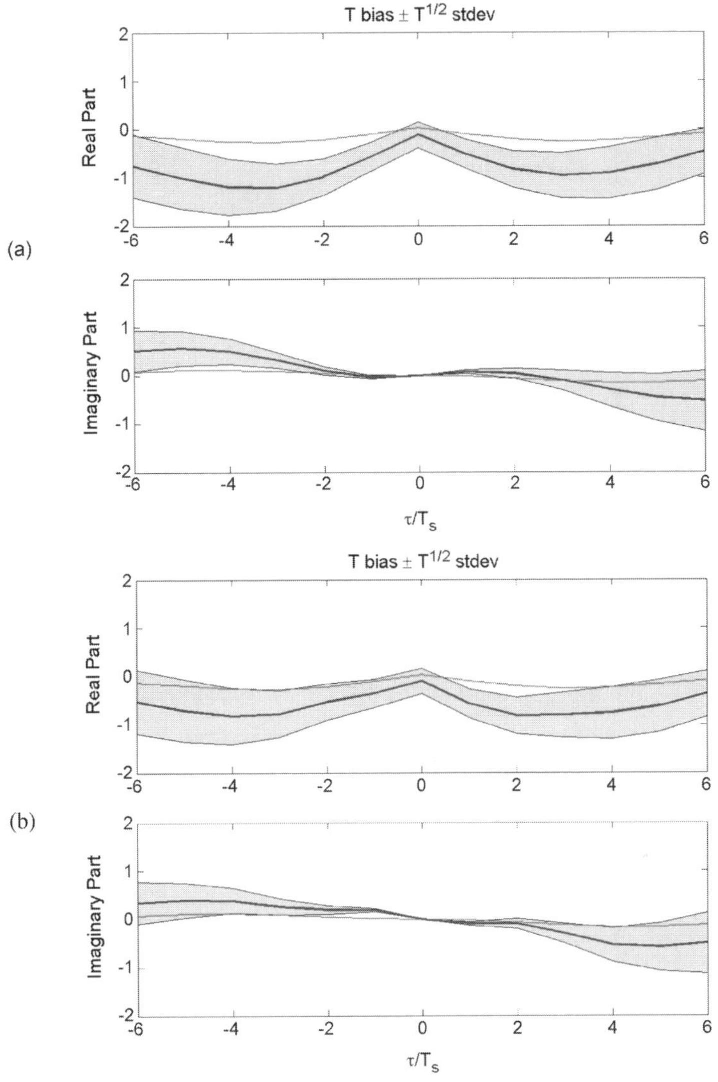

Figure 2.9 Normalized bias of the discrete-time cyclic correlogram as a function of τ/T_s, for $\widetilde{\alpha} = \bar{\zeta}m$, estimated by sampling period $T_s = T_p/6$ and (a) $N = 2^{11}$ and (b) $N = 2^{13}$ data samples. Thin line: Normalized (by T) actual value; Thick line: Normalized (by T) estimated value; Shaded area: Normalized (by $T^{1/2}$) sample standard deviation. *Source:* (Napolitano 2009) © IEEE

fixed cycle frequency, say α_0, then the not exact knowledge of the value of α_0 leads to a biased estimate. The results established in the chapter do not require the knowledge of the functions $\alpha_n(\tau)$ since the estimation of the cyclic cross-correlogram $R_{yx^{(*)}}(\alpha, \tau; t_0, T)$ as a function of the two variables (α, τ) is performed. From Theorem 2.4.6, it follows that the expected value of the cyclic cross-correlogram is significantly different from zero within strips of width $1/T$ around the support curves $\alpha = \alpha_n(\tau)$, $n \in \mathbb{I}$, of the cyclic cross-correlation function. Thus, in the limit as $T \to \infty$, the regions of the (α, τ) plane where $R_{yx^{(*)}}(\alpha, \tau; t_0, T)$ is significantly

(a)

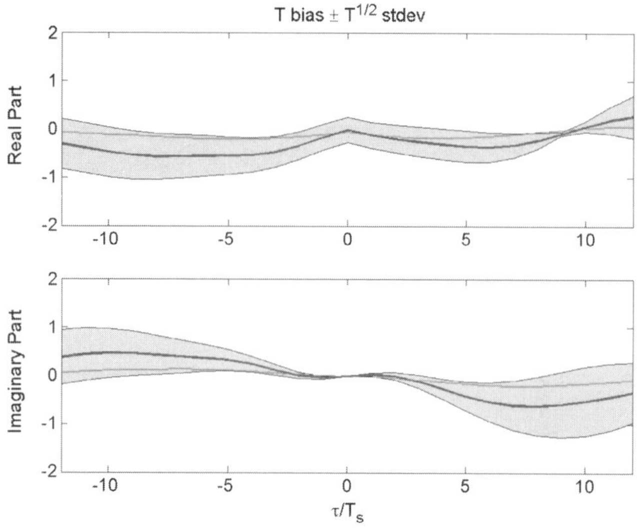

(b)

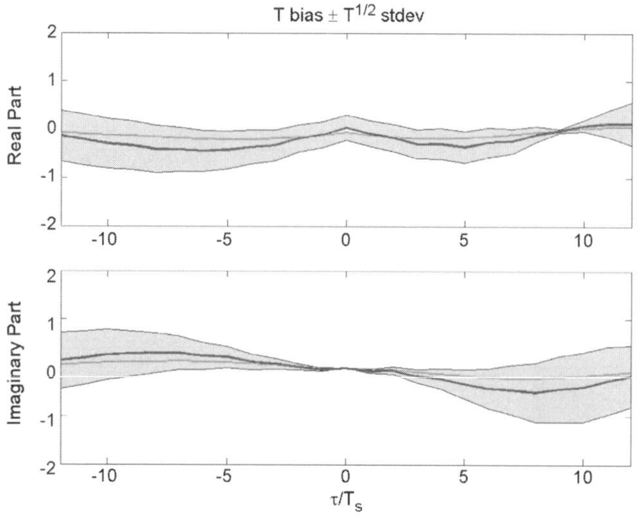

Figure 2.10 Normalized bias of the discrete-time cyclic correlogram as a function of τ/T_s, for $\widetilde{\alpha} = \bar{\zeta}m$, estimated by sampling period $T_s = T_p/12$ and (a) $N = 2^{12}$ and (b) $N = 2^{14}$ data samples. Thin line: Normalized (by T) actual value; Thick line: Normalized (by T) estimated value; Shaded area: Normalized (by $T^{1/2}$) sample standard deviation. *Source:* (Napolitano 2009) © IEEE

different from zero tend to the support curves of the cyclic cross-correlation function, that is, the curves $\alpha = \alpha_n(\tau)$, $n \in \mathbb{I}$ (see (2.39)). If the estimation of the cyclic cross-correlation function has to be performed along a fixed lag-dependent cycle frequency curve $\alpha = \alpha_n(\tau)$, that is, a generalized cyclic cross-correlation function has to be estimated, then the not exact knowledge of the function $\alpha_n(\tau)$ leads to an asymptotically biased estimate, in accordance with the well-known result for ACS processes. Moreover, in the discrete-time implementation of the estimator, the spacing between adjacent values of $\widetilde{\alpha}$ in the set $\widetilde{A}$ should decrease faster than $1/N$ as N increases.

If the lag-dependent cycle frequencies are unknown, a statistical test for the presence of generalized almost-cyclostationarity can be performed to estimate the unknown functions $\alpha_n(\tau)$ by exploiting the asymptotic complex Normality of the cyclic cross-correlogram. A point in the (α, τ) plane belongs to the estimated support if the magnitude of the cyclic cross-correlogram exceeds a threshold whose value has to be fixed in order to find assigned probabilities of false alarm or missed detection.

Note that a different behavior of statistical-function estimators is found for the class of the spectrally correlated stochastic processes (Chapter 4), (Napolitano 2003) that also extend the class of the ACS processes. Spectrally correlated processes have the Loève bifrequency spectrum with spectral masses concentrated on a countable set of curves in the bifrequency plane. The support curves of the Loève bifrequency spectrum play, for spectrally correlated processes, in the frequency domain, a role analogous to that played for GACS processes, in the time domain, by the lag-dependent cycle frequencies. The ACS processes are obtained as a special case of spectrally correlated processes when the separation between correlated spectral components can assume values only in a countable set (which is the set of the cycle frequencies). In such a case, the support curves of the Loève bifrequency spectrum are lines with unit slope. In Chapter 4 it is shown that, for spectrally correlated processes, when the location of the spectral masses is unknown, time- or frequency-smoothed versions of the periodogram do not provide estimates of the bifrequency spectral correlation density function that are asymptotically unbiased and with zero asymptotic variance. Moreover, there exists a tradeoff between the departure of the spectral-correlation-type nonstationarity from the almost cyclostationarity and the reliability of spectral correlation measurements obtainable by a single sample path.

Finally, note that the results of this chapter can be extended to the case in which the second-order cross-moments (2.119) and the fourth-order cumulant (2.120) in Assumption 2.4.2 and the kth-order cumulant (2.164) can be expressed as the sum of an almost-periodic component and a residual term not containing any additive sinewave component (see for example, (2.28b), (2.80a), (2.80b), (2.90a), and (2.90b)), provided that the residual terms are bounded and summable (or at least with finite principal-value integral) with respect to all the variables. For this class of processes, in (Napolitano and Tesauro 2011) continuous-time estimators are proposed for cyclic higher-order statistics.

2.7.6 *Conjecturing the Nonstationarity Type of the Continuous-Time Signal*

In Section 2.5, it is shown that both ACS and GACS that are not ACS continuous-time signals, by sampling, give rise to ACS discrete-time signals. Thus, the nonstationarity type of the subsumed continuous-time signal cannot be derived from that of the discrete-time signal of its samples. In this section, it is shown how the possible ACS nature of a continuous-time GACS signal can be conjectured from the behavior of the support of the estimated cyclic correlogram of the sampled ACS signal (Izzo and Napolitano 2003, 2005).

Let $x_{\mathrm{d}}(n) \triangleq x(t)|_{t=nT_s}$ be the discrete-time process obtained by uniformly sampling with period T_s a continuous-time GACS signal $x(t)$. According to (2.179b) we have

$$\widetilde{R}_{x_{\mathrm{d}}x_{\mathrm{d}}^{(*)}}(\widetilde{\alpha}, m) = \sum_{k \in \mathbb{I}_{xx^{(*)}}} R^{(k)}_{xx^{(*)}}(mT_s)\, \delta_{[\widetilde{\alpha} - \alpha^{(k)}_{xx^{(*)}}(mT_s)T_s] \bmod 1}. \tag{2.242}$$

Thus, the support of $\widetilde{R}_{x_{\mathrm{d}}x_{\mathrm{d}}^{(*)}}(\widetilde{\alpha}, m)$ (in the main cycle frequency domain) is

$$\operatorname{supp}\left\{\widetilde{R}_{x_{\mathrm{d}}x_{\mathrm{d}}^{(*)}}(\widetilde{\alpha}, m)\right\} = \bigcup_{k\in\mathbb{I}_{xx^{(*)}}} \Big\{(\widetilde{\alpha}, m) \in [-1/2, 1/2) \times \mathbb{Z} : \\ \widetilde{\alpha} - \alpha^{(k)}_{xx^{(*)}}(mT_s)T_s \bmod 1 = 0, \quad R^{(k)}_{xx^{(*)}}(mT_s) \neq 0\Big\} \tag{2.243}$$

Let the cyclic correlogram be evaluated by a N-point DFT-based algorithm with zero padding factor P. Thus, the cycle frequency step for the estimator is $\Delta\widetilde{\alpha} = 1/NP$. Therefore, if the parameters T_s, N, and P are such that the cycle frequency variations within a sampling period are smaller than twice the (normalized) cycle frequency step, that is, $\forall k \in \mathbb{I}_{xx^{(*)}}$

$$\left|\alpha^{(k)}_{xx^{(*)}}((m+1)T_s) - \alpha^{(k)}_{xx^{(*)}}(mT_s)\right| \leqslant 2\Delta\widetilde{\alpha}/T_s \\ \forall m \in \mathbb{Z} \;:\; R^{(k)}_{xx^{(*)}}(mT_s) \neq 0 \text{ and } R^{(k)}_{xx^{(*)}}((m+1)T_s) \neq 0 \tag{2.244}$$

then, due to the interpolation made by the graphic software, the support (2.243) will appear to be continuous in correspondence of the continuous tracts of the functions $\alpha^{(k)}_{xx^{(*)}}(\tau)$. Thus, in the case of piecewise constant functions $\alpha^{(k)}_{xx^{(*)}}(mT_s)$, according to (2.23), the continuous-time signal should be conjectured to be ACS. In contrast, when the $\alpha^{(k)}_{xx^{(*)}}(mT_s)$ are not piecewise constant, the continuous-time GACS signal should be conjectured to be not ACS.

The above-described conjecturing procedure can be found to be inadequate when the relationship involving analysis and signal parameters is not appropriate. In fact, if (2.244) is not satisfied, then the support (2.243) appears discontinuous also in correspondence of the continuous tracts of the functions $\alpha^{(k)}_{xx^{(*)}}(\tau)$. Specifically, it is constituted by small constant tracts and, hence, the continuous-time signal should be conjectured to be anyway ACS.

To illustrate how different relationships among analysis and signal parameters can lead to different conjectures on the nonstationarity type of the continuous-time signal, two experiments were conducted. A continuous-time chirp-modulated PAM signal $x(t)$ (see (2.99)) is sampled with period T_s. The PAM signal has full-duty-cycle rectangular pulse, stationary white binary modulating sequence, and symbol period $T_p = 8T_s$. According to (7.356), the lag-dependent cycle frequencies of $x(t)$ are linear functions with slope γ

$$\alpha^{(k)}_{xx^*}(\tau) = \gamma\tau + k/T_p \qquad k \in \mathbb{Z} \tag{2.245}$$

where $\{k/T_p\}_{k\in\mathbb{Z}}$ are the cycle frequencies of the PAM signal with rectangular pulse (Gardner 1985), (Gardner *et al.* 2006). Moreover, according to (7.357), the generalized cyclic autocorrelation functions of $x(t)$ are proportional to the cyclic autocorrelation functions of the PAM signal. From (2.245) it follows

$$\left|\alpha^{(k)}_{xx^*}((m+1)T_s) - \alpha^{(k)}_{xx^*}(mT_s)\right| = |\gamma|T_s \tag{2.246}$$

and (2.244) specializes into

$$|\gamma|T_s^2 \leqslant 2\Delta\widetilde{\alpha}. \tag{2.247}$$

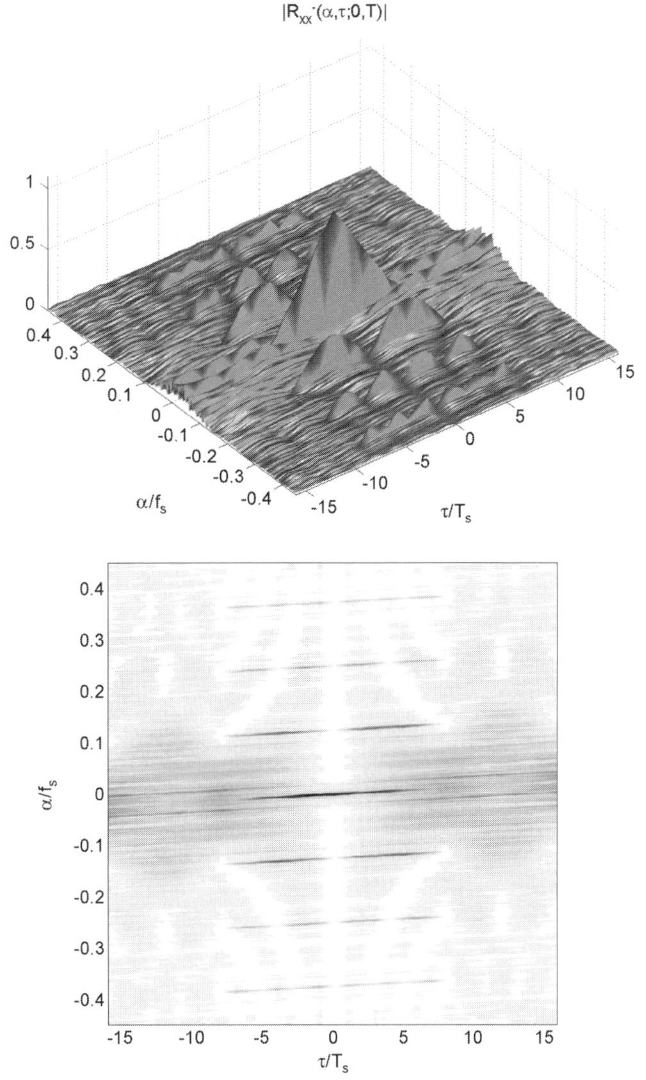

Figure 2.11 Chirp-modulated PAM signal. (Top) graph and (bottom) "checkerboard" plot of the magnitude of the cyclic correlogram as a function of αT_s and τ/T_s. The signal is conjectured to be GACS

The data-record length is $T = NT_s$ with $N = 2^{10}$ and the padding factor is $P = 1$. Therefore, $\Delta\widetilde{\alpha} = 1/NP = 2^{-10}$.

In the first experiment, the chirp modulation is with frequency shift $\nu = 0.02/T_s$ and chirp rate $\gamma = 1.5 \cdot 10^{-3}/T_s^2$. Consequently, (2.247) is satisfied and due to the behavior of the support (2.243), $x(t)$ is conjectured to be GACS (Figure 2.11). In the second experiment, $\nu = 0.02/T_s$ and $\gamma = 7.5 \cdot 10^{-3}/T_s^2$. Thus, (2.247) is not satisfied and $x(t)$ is conjectured to be ACS since the support appears as constituted by small constant tracts (Figure 2.12).

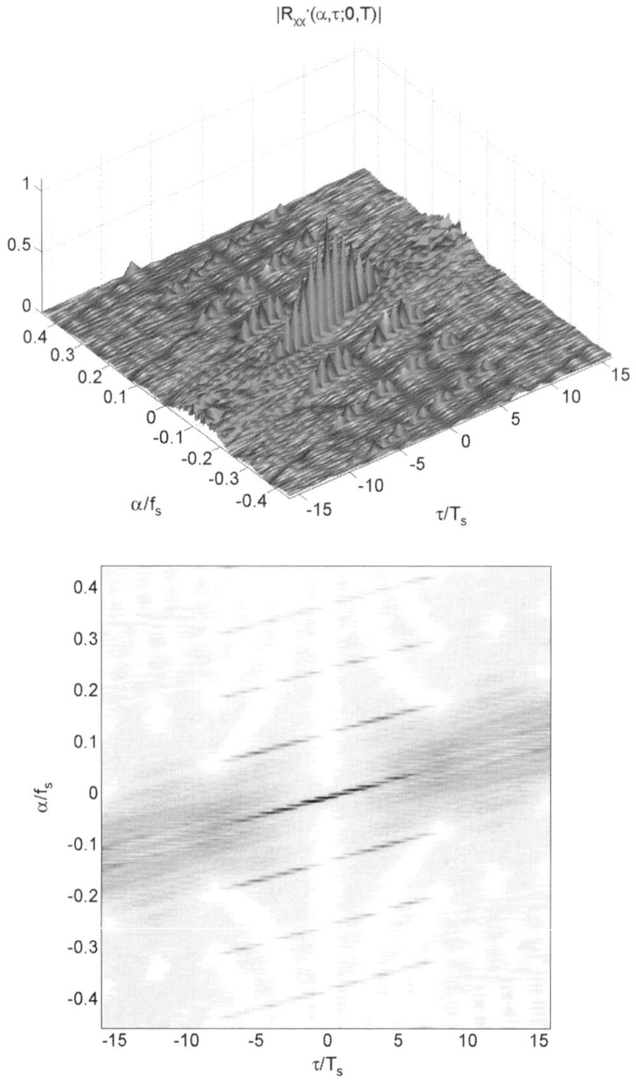

Figure 2.12 Chirp-modulated PAM signal. (Top) graph and (bottom) "checkerboard" plot of the magnitude of the cyclic correlogram as a function of αT_s and τ/T_s. The signal is conjectured to be ACS

Finally, it is worth emphasizing that variations of the (continuous-time) lag-dependent cycle frequencies can be evidenced by finite data-record analysis, provided that the cycle-frequency resolution $\Delta\alpha = 1/T$ is sufficiently fine, that is, the data-record length T is sufficiently large. For example, let us assume that the lag-dependent cycle frequency $\alpha^{(k)}_{xx^{(*)}}(\tau)$ is monotone within the interval $[\tau_a, \tau_b]$. The cycle-frequency dependence on the lag parameter τ when τ ranges in $[\tau_a, \tau_b]$ can be appreciated by an observation interval of length T provided that

$$\left|\alpha^{(k)}_{xx^{(*)}}(\tau_b) - \alpha^{(k)}_{xx^{(*)}}(\tau_a)\right| > \frac{1}{T} = \Delta\alpha. \qquad (2.248)$$

For the discrete-time estimator, based on N samples, where $T = NT_s$, condition (2.248) is equivalent to

$$|\widetilde{\alpha}_b - \widetilde{\alpha}_a| > \frac{T_s}{T} = \frac{1}{N} \tag{2.249}$$

where $\widetilde{\alpha}_a \triangleq \alpha^{(k)}_{xx^{(*)}}(m_a T_s) T_s$ and $\widetilde{\alpha}_b \triangleq \alpha^{(k)}_{xx^{(*)}}(m_b T_s) T_s$, with $m_a = \lfloor \tau_a / T_s \rfloor$, $m_b = \lfloor \tau_b / T_s \rfloor$, and $\lfloor \cdot \rfloor$ denoting the floor operation.

In the special case of continuous-time chirp-modulated PAM signal $x(t)$ (see (2.99)), condition (2.248) specializes into

$$|\alpha^{(k)}_{xx^*}(\tau_{\text{corr}}) - \alpha^{(k)}_{xx^*}(0)| = |\gamma| \tau_{\text{corr}} > \frac{1}{T} \tag{2.250}$$

and condition (2.249) into

$$|\widetilde{\gamma}| m_{\text{corr}} > \frac{1}{N} \tag{2.251}$$

where $\widetilde{\gamma} = \gamma T_s^2$, τ_{corr} is the maximum value of $|\tau|$ such that $R_{xx^*}(0, \tau)$ is significantly different from zero, and $m_{\text{corr}} = \lfloor \tau_{\text{corr}} / T_s \rfloor$.

If condition (2.248) is not satisfied $\forall k \in \mathbb{I}$, then the lag-dependent cycle frequencies are modeled as constant and the process $x(t)$, on the basis of the data-record length T, is modeled as ACS.

2.7.7 LTI Filtering of GACS Signals

In Section 2.3, it is shown that as a consequence of the input/output relationship (2.117), low-pass or band-pass filtering a purely GACS process leads to a zero power process. In this section, a simulation experiment is conducted aimed at illustrating this result.

The chirp-modulated PAM signal considered in Section 2.7.6 with chirp rate $\gamma = 1.5 \cdot 10^{-3}/T_s^2$ passes through a low-pass filter with harmonic response $H(f) = (1 + jf/B_h)^{-4}$, with $B_h = 2 \cdot 10^{-3}/T_s$. The cyclic correlogram of the output signal $y(t)$ is estimated on observation intervals $T = 2^9 T_s$ (Figure 2.13 (top)) and $T = 2^{12} T_s$ (Figure 2.13 (bottom)). The approach of the estimates to the identically zero value as the data-record length increases is evident.

2.8 Summary

In this chapter, GACS processes are characterized in the time domain in the strict and wide senses, at second- and higher-orders. Moreover, a heuristic characterization in the frequency domain is provided. The class of such nonstationary processes includes, as a special case, the ACS processes. Moreover, ACS processes filtered by Doppler channels and communications signals with time-varying parameters are further examples. The problem of estimating second-order statistical functions of (continuous-time) GACS processes is addressed. The cyclic cross-correlogram is proposed as an estimator of the cyclic cross-correlation function of jointly GACS processes and its expected value and covariance are determined for finite data-record length (Theorems 2.4.6 and 2.4.7). It is shown that, for GACS processes satisfying some mixing assumptions expressed in terms of summability of cumulants, the cyclic cross-correlogram is a mean-square consistent (Theorems 2.4.11 and 2.4.13) and asymptotically complex Normal

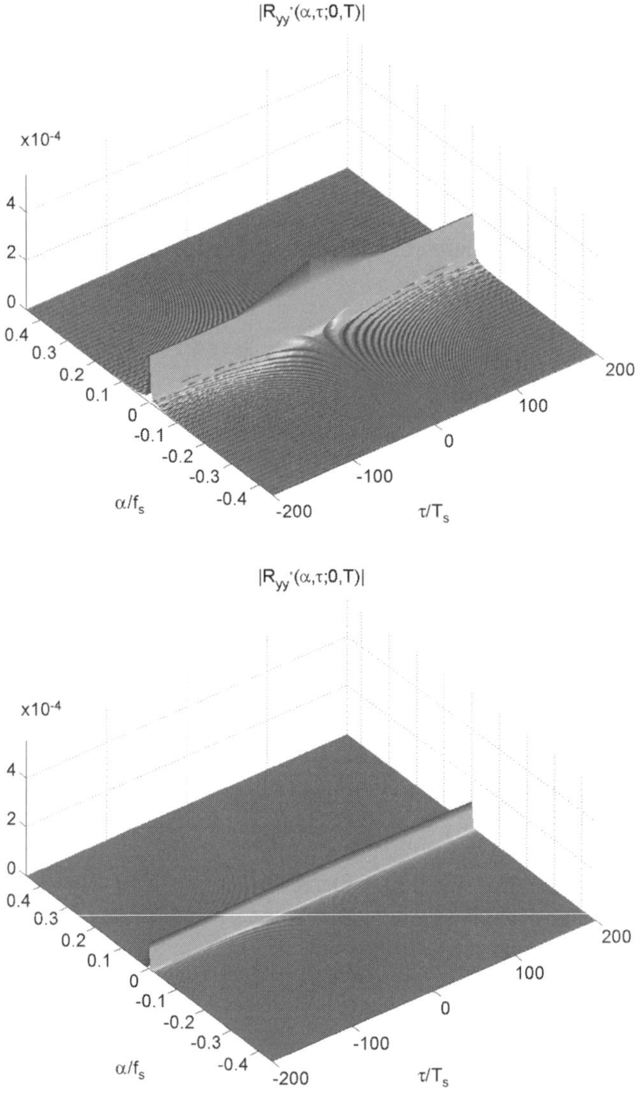

Figure 2.13 LTI filtered chirp-modulated PAM signal. Graph of the magnitude of the cyclic correlogram as a function of αT_s and τ / T_s. (Top) 2^9 samples and (bottom) 2^{12} samples

(Theorem 2.4.18) estimator of the cyclic cross-correlation function. Specifically, the covariance and conjugate covariance of the cyclic-cross correlogram are shown to approach zero as the reciprocal of the data-record length, when the data-record length approaches infinity. Moreover, the rate of convergence to zero of the bias is shown to depend on the rate of decay to zero of the Fourier transform of the data-tapering window (Theorem 2.4.12). An asymptotic bound for the covariance uniform with respect to cycle frequencies and lag shifts is also provided (Corollary 2.4.14). The mixing assumptions made are mild and are generally satisfied by processes with finite or practically finite memory. Then, well-known consistency and asymptotic-Normality results for ACS processes are shown to be obtained by specializing the results of this chapter.

It is shown that uniformly sampling a continuous-time GACS process leads to a discrete-time ACS process. Moreover, the class of the continuous-time GACS processes does not have a discrete-time counterpart. That is, discrete-time GACS processes do not exist.

The discrete-time jointly ACS processes obtained by uniformly sampling two continuous-time jointly GACS processes have been considered. It is shown that the discrete-time cyclic cross-correlation function of the jointly ACS processes is an aliased version of the continuous-time cyclic cross-correlation function of the jointly GACS processes (Theorem 2.5.3).

The DT-CCC of the jointly ACS sequences obtained by uniformly sampling two continuous-time jointly GACS process is proposed as an estimator of samples of the continuous-time cyclic cross-correlation function. Its mean value and covariance for finite number of data samples and finite sampling period are derived (Theorems 2.6.2 and 2.6.3). Then, asymptotic expected value and covariance as the data-sample number approaches infinity are derived (Theorems 2.6.4 and 2.6.5) and the DT-CCC is proved to be a mean-square consistent estimator of an aliased version of the cyclic cross-correlation function. Moreover, it is shown to be asymptotically complex Normal (Theorem 2.6.10). The H-CCC is defined and is adopted to analyze the asymptotic behavior of the DT-CCC as the number of data samples approaches infinity and the sampling period approaches zero, provided that the overall data-record length approaches infinity. The H-CCC is proved to be a mean-square consistent and asymptotically complex Normal estimator of the continuous-time cyclic cross-correlation function (Theorems 2.6.16, 2.6.18, and 2.6.20). In addition, it is proved that the rate of decay to zero of bias is equal to the rate of decay to zero of the Fourier transform of the data-tapering window (Theorem 2.6.17). Moreover, it is shown that the asymptotic performance of the H-CCC is the same as that of the continuous-time cyclic cross-correlogram. Thus, no loss in performance is obtained by making the cyclic spectral analysis of (jointly) GACS processes in discrete-time rather than continuous-time. The proved asymptotic Normality can be applied to find confidence intervals and establish statistical test for the presence of generalized almost cyclostationarity.

The example of a GACS signal received when a transmitted ACS signal passes through the Doppler channel existing between a transmitter and a receiver with constant relative radial acceleration is treated both analytically and by a simulation experiment. Numerical results are reported to validate the analytically derived rates of convergence to zero of bias and standard deviation of the cyclic cross-correlogram.

3

Complements and Proofs on Generalized Almost-Cyclostationary Processes

In this chapter, complements and proofs for the results of Chapter 2 are reported.

3.1 Proofs for Section 2.2.2 "Second-Order Wide-Sense Statistical Characterization"

3.1.1 Proof of Theorem 2.2.17 Mean-Square Integrability of GACS Processes

By substituting $t + \tau = t_1, t = t_2$ into (2.18) we have

$$\mathrm{E}\{x(t_1)\, x^*(t_2)\} = \sum_{n \in \mathbb{I}} R_{xx^*}^{(n)}(t_1 - t_2)\, e^{j2\pi\alpha_n(t_1 - t_2)t_2}. \tag{3.1}$$

Thus

$$\begin{aligned}
&\int_{-T/2}^{T/2} \int_{-T/2}^{T/2} \left|\mathrm{E}\{x(t_1)\, x^*(t_2)\}\right| \,\mathrm{d}t_1\, \mathrm{d}t_2 \\
&\qquad \leqslant \sum_{n \in \mathbb{I}} \int_{-T/2}^{T/2} \int_{-T/2}^{T/2} \left|R_{xx^*}^{(n)}(t_1 - t_2)\right| \,\mathrm{d}t_1\, \mathrm{d}t_2 \\
&\qquad = \sum_{n \in \mathbb{I}} \int_{-T/2}^{T/2} \int_{-T/2-t}^{T/2-t} \left|R_{xx^*}^{(n)}(\tau)\right| \,\mathrm{d}\tau\, \mathrm{d}t \\
&\qquad = \sum_{n \in \mathbb{I}} \int_{\mathbb{R}} \mathrm{rect}\left(\frac{t}{T}\right) \int_{\mathbb{R}} \mathrm{rect}\left(\frac{\tau + t}{T}\right) \left|R_{xx^*}^{(n)}(\tau)\right| \,\mathrm{d}\tau\, \mathrm{d}t
\end{aligned}$$

$$= \sum_{n \in \mathbb{I}} \int_{\mathbb{R}} \left| R_{xx^*}^{(n)}(\tau) \right| \int_{\mathbb{R}} \operatorname{rect}\left(\frac{\tau + t}{T}\right) \operatorname{rect}\left(\frac{t}{T}\right) \mathrm{d}t \, \mathrm{d}\tau$$

$$= T \sum_{n \in \mathbb{I}} \int_{-T}^{T} \left| R_{xx^*}^{(n)}(\tau) \right| \left(1 - \frac{|\tau|}{T}\right) \mathrm{d}\tau \tag{3.2}$$

where, in the first equality the variable changes $t_1 = t + \tau$ and $t_2 = t$ are made, and in the last equality the result

$$\int_{\mathbb{R}} \operatorname{rect}\left(\frac{\tau + t}{T}\right) \operatorname{rect}\left(\frac{t}{T}\right) \mathrm{d}t = T \left(1 - \frac{|\tau|}{T}\right) \operatorname{rect}\left(\frac{\tau}{2T}\right) \tag{3.3}$$

is used. Thus, accounting for Assumption 2.2.16, the sufficient condition of Theorem 2.2.15 turns out to be satisfied. □

3.1.2 *Proof of Fact 2.2.18*

$$\begin{aligned} R_{xx^*}(-\alpha, -\tau) &= \lim_{T \to \infty} \frac{1}{T} \int_{t_0 - T/2}^{t_0 + T/2} \mathrm{E}\left\{x(t - \tau)\, x^*(t)\right\} e^{j2\pi\alpha t} \, \mathrm{d}t \\ &= \lim_{T \to \infty} \frac{1}{T} \int_{t_0 - T/2 - \tau}^{t_0 + T/2 - \tau} \mathrm{E}\left\{x(t')\, x^*(t' + \tau)\right\} e^{j2\pi\alpha(t' + \tau)} \, \mathrm{d}t' \\ &= \left[\lim_{T \to \infty} \frac{1}{T} \int_{t_0 - T/2 - \tau}^{t_0 + T/2 - \tau} \mathrm{E}\left\{x(t' + \tau)\, x^*(t')\right\} e^{-j2\pi\alpha t'} \, \mathrm{d}t' \right]^* e^{j2\pi\alpha\tau} \end{aligned} \tag{3.4}$$

where, in the second equality, the variable change $t' = t - \tau$ is made. (3.4) is equivalent to (2.52) from which (2.54) follows.

From (2.54) we have

$$\alpha \in A_\tau \Leftrightarrow \exists n \in \mathbb{I}_{xx^*} \ : \ \alpha_{xx^*}^{(n)}(\tau) = \alpha \tag{3.5}$$

$$-\alpha \in -A_\tau \Leftrightarrow \exists n' \in \mathbb{I}_{xx^*} \ : \ \alpha_{xx^*}^{(n')}(-\tau) = -\alpha \tag{3.6}$$

Therefore, (2.55) follows. In addition, one has

$$\begin{aligned} R_{xx^*}^{(n)}(\tau) &= R_{xx^*}(\alpha, \tau)\big|_{\alpha = \alpha_{xx^*}^{(n)}(\tau)} \\ &= R_{xx^*}^*(-\alpha, -\tau)\, e^{j2\pi\alpha\tau}\Big|_{\alpha = -\alpha_{xx^*}^{(n')}(-\tau)} \end{aligned} \tag{3.7}$$

which is equivalent to (2.56).

3.1.3 *Proof of Fact 2.2.19*

$$\begin{aligned} R_{x^*x}(\alpha, \tau) &= \lim_{T \to \infty} \frac{1}{T} \int_{t_0 - T/2}^{t_0 + T/2} \mathrm{E}\left\{x^*(t + \tau)\, x(t)\right\} e^{-j2\pi\alpha t} \, \mathrm{d}t \\ &= \left[\lim_{T \to \infty} \frac{1}{T} \int_{t_0 - T/2}^{t_0 + T/2} \mathrm{E}\left\{x(t + \tau)\, x^*(t)\right\} e^{j2\pi\alpha t} \, \mathrm{d}t \right]^* \end{aligned} \tag{3.8}$$

from which we obtain (2.57). From (2.57) it follows that

$$\alpha \in A_{x^*x,\tau} \Leftrightarrow -\alpha \in A_{xx^*,\tau}. \tag{3.9}$$

Therefore

$$\exists n' \in \mathbb{I}_{x^*x},\ \exists n'' \in \mathbb{I}_{xx^*} \text{ such that } \alpha_{x^*x}^{(n')}(\tau) = \alpha,\ \alpha_{xx^*}^{(n'')}(\tau) = -\alpha \tag{3.10}$$

from which we have (2.58). In addition, one has

$$\begin{aligned} R_{x^*x}^{(n')}(\tau) &= R_{x^*x}(\alpha, \tau)\big|_{\alpha=\alpha_{x^*x}^{(n')}(\tau)} \\ &= R_{xx^*}^*(-\alpha, \tau)\big|_{\alpha=-\alpha_{xx^*}^{(n'')}(\tau)} \end{aligned} \tag{3.11}$$

which leads to (2.59).

3.1.4 Proof of Fact 2.2.20

$$\begin{aligned} R_{xx}(-\alpha, \tau) &= \lim_{T\to\infty} \frac{1}{T} \int_{t_0-T/2}^{t_0+T/2} \mathrm{E}\left\{x(t+\tau)\, x(t)\right\} e^{j2\pi\alpha t}\, \mathrm{d}t \\ &= \left[\lim_{T\to\infty} \frac{1}{T} \int_{t_0-T/2}^{t_0+T/2} \mathrm{E}\left\{x^*(t+\tau)\, x^*(t)\right\} e^{-j2\pi\alpha t}\, \mathrm{d}t\right]^* \end{aligned} \tag{3.12}$$

from which we obtain (2.60). From (2.60) it follows that

$$-\alpha \in A_{xx,\tau} \Leftrightarrow \alpha \in A_{x^*x^*,\tau}. \tag{3.13}$$

Therefore

$$\exists n' \in \mathbb{I}_{xx},\ \exists n'' \in \mathbb{I}_{x^*x^*} \text{ such that } \alpha_{xx}^{(n')}(\tau) = -\alpha,\ \alpha_{x^*x^*}^{(n'')}(\tau) = \alpha \tag{3.14}$$

from which we have (2.61). In addition, one has

$$\begin{aligned} R_{xx}^{(n')}(\tau) &= R_{xx}(\alpha, \tau)\big|_{\alpha=\alpha_{xx}^{(n')}(\tau)} \\ &= R_{x^*x^*}^*(-\alpha, \tau)\big|_{\alpha=-\alpha_{x^*x^*}^{(n'')}(\tau)} \end{aligned} \tag{3.15}$$

which leads to (2.62).

3.2 Proofs for Section 2.2.3 "Second-Order Spectral Characterization"

3.2.1 The μ Functional

In this section, the μ functional is introduced. It is defined as the functional whose value is the infinite-time average of the test function. The μ functional is formally characterized as the limit of approximating functions similar to the characterization of the Dirac delta as the limit of delta-approximating functions (Zemanian 1987, Section 1.3).

Let μ be the functional that associates to a test function φ its infinite-time average value. That is,

$$\mu : \varphi \to \bar{x} \triangleq \lim_{T\to\infty} \frac{1}{T} \int_{t-T/2}^{t+T/2} \varphi(\tau)\,\mathrm{d}\tau \tag{3.16}$$

provided that the limit exists (and, hence, is independent of t).

In the following, the μ functional is heuristically characterized through formal manipulations. Let it be

$$\mu_T(t) \triangleq \frac{1}{T}\mathrm{rect}\left(\frac{t}{T}\right) \tag{3.17}$$

where $\mathrm{rect}(t) = 1$ for $|t| \leqslant 1/2$ and $\mathrm{rect}(t) = 0$ otherwise. For any finite T one has

$$\frac{1}{T} \int_{t-T/2}^{t+T/2} \varphi(\tau)\,\mathrm{d}\tau = \int_{\mathbb{R}} \varphi(\tau)\mu_T(\tau - t)\,\mathrm{d}\tau = \int_{\mathbb{R}} \varphi(\tau)\mu_T(t-\tau)\,\mathrm{d}\tau. \tag{3.18}$$

Thus, in the limit as $T \to \infty$, we (rigorously) have

$$\bar{x} = \lim_{T\to\infty} \frac{1}{T} \int_{t-T/2}^{t+T/2} \varphi(\tau)\,\mathrm{d}\tau = \lim_{T\to\infty} \int_{\mathbb{R}} \varphi(\tau)\mu_T(t-\tau)\,\mathrm{d}\tau \tag{3.19}$$

and we can formally write

$$\bar{x} = \int_{\mathbb{R}} \varphi(\tau)\mu(t-\tau)\,\mathrm{d}\tau = \varphi(t) \otimes \mu(t) \tag{3.20}$$

with rhs independent of t, where $\mu(t)$ is formally defined as (see also (Silverman 1957))

$$\mu(t) \triangleq \lim_{T\to\infty} \mu_T(t). \tag{3.21}$$

In the sense of the ordinary functions, the limit in the rhs of (3.21) is the identically zero function. However, observing that for any finite T it results

$$\int_{\mathbb{R}} \mu_T(\tau - t)\,\mathrm{d}\tau = \frac{1}{T} \int_{t-T/2}^{t+T/2} \mathrm{d}t = 1 \tag{3.22}$$

then, in the limit as $T \to \infty$, we (rigorously) have

$$\lim_{T\to\infty} \int_{\mathbb{R}} \mu_T(\tau - t)\,\mathrm{d}\tau = 1 \tag{3.23}$$

and we can formally write

$$\int_{\mathbb{R}} \mu(\tau)\,\mathrm{d}\tau = \int_{\mathbb{R}} \lim_{T\to\infty} \mu_T(\tau - t)\,\mathrm{d}\tau = 1. \tag{3.24}$$

That is, $\mu(t)$ can be interpreted as the limit of a very tiny and large rectangular window with unit area. This limit, of course, is the identically zero function in spaces of ordinary functions. However $\mu(t)$ can be formally managed as an ordinary function satisfying (3.20) and (3.24) similar to the formal manipulations of the Dirac delta.

The Fourier transform of $\mu(t)$ can be formally derived by the following "limit passage." Accounting for the Fourier transform pair

$$\mu_T(t) = \frac{1}{T}\text{rect}\left(\frac{t}{T}\right) \overset{\mathcal{F}}{\longleftrightarrow} \text{sinc}(fT) = \frac{\sin(\pi f T)}{\pi f T} \tag{3.25}$$

we can formally write (Figure 3.1)

$$\mu(t) \triangleq \lim_{T\to\infty} \mu_T(t) \overset{\mathcal{F}}{\longleftrightarrow} \lim_{T\to\infty} \text{sinc}(fT) = \delta_f \tag{3.26}$$

where δ_f is the Kronecker delta, that is, $\delta_f = 1$ for $f = 0$ and $\delta_f = 0$ for $f \neq 0$. Thus, $\mu(t)$ cannot be expressed as ordinary inverse Fourier transform (Lebesgue integral) since the Kronecker delta δ_f is zero a.e.

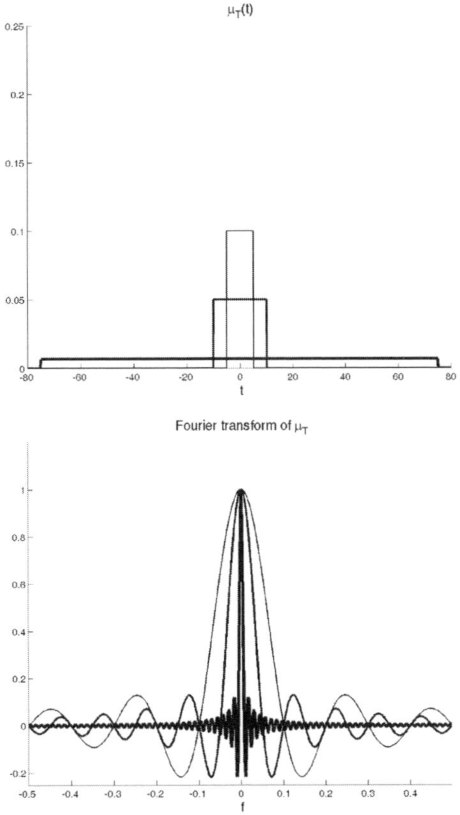

Figure 3.1 (Top) Function $\mu_T(t)$ and (bottom) its Fourier transform for increasing values of T (form thin line to thick line)

The following properties of the μ functional can be formally proved.

1. For every $t_0 \geqslant 0$, from (3.20) it follows that

$$\begin{aligned}
\varphi(t) \otimes \mu(t+t_0) &= \lim_{T\to\infty} \frac{1}{T} \int_{t+t_0-T/2}^{t+t_0+T/2} \varphi(\tau)\,\mathrm{d}\tau \\
&= \lim_{T\to\infty} \frac{1}{T} \left\{ -\int_{t-T/2}^{t+t_0-T/2} + \int_{t-T/2}^{t+T/2} + \int_{t+T/2}^{t+t_0+T/2} \right\} \varphi(\tau)\,\mathrm{d}\tau \\
&= \lim_{T\to\infty} \frac{1}{T} \int_{t-T/2}^{t+T/2} \varphi(\tau)\,\mathrm{d}\tau \\
&= \varphi(t) \otimes \mu(t)
\end{aligned} \tag{3.27}$$

where the fact that t_0 is finite has been accounted for. The case $t_0 < 0$ is similar.

2. If $x(t)$ contains the finite-strength additive sinewave component $x_\alpha\, e^{j2\pi\alpha t}$ (Section 6.3.1), then

$$\begin{aligned}
x(t) \otimes \left[\mu(t)\, e^{j2\pi\alpha t}\right] &= \int_{\mathbb{R}} x(\tau)\, \mu(t-\tau)\, e^{j2\pi\alpha(t-\tau)}\,\mathrm{d}\tau \\
&= \lim_{T\to\infty} \frac{1}{T} \int_{t-T/2}^{t+T/2} x(\tau)\, e^{-j2\pi\alpha\tau}\,\mathrm{d}\tau\, e^{j2\pi\alpha t} \\
&\triangleq x_\alpha\, e^{j2\pi\alpha t}
\end{aligned} \tag{3.28}$$

3. From (3.26) and using the modulation theorem for Fourier transform we have

$$\mu(t)\, e^{j2\pi\beta t} \overset{\mathcal{F}}{\longleftrightarrow} \delta_{f-\beta} \tag{3.29}$$

4. From (3.29) and using the duality theorem for Fourier transform we have

$$\delta_{t-t_0} \overset{\mathcal{F}}{\longleftrightarrow} \mu(f)\, e^{-j2\pi f t_0} \tag{3.30}$$

3.2.2 *Proof of Theorem 2.2.22 Loève Bifrequency Spectrum of GACS Processes*

From Theorem 2.2.7 ($t+\tau = t_1$ and $t = t_2$ in (2.18)) we have

$$\begin{aligned}
&\mathrm{E}\{x(t_1)\, x^{(*)}(t_2)\} \\
&\quad = \sum_{n\in\mathbb{I}} R^{(n)}_{xx^{(*)}}(t_1-t_2)\, e^{j2\pi\alpha_n(t_1-t_2)t_2} \\
&\quad = \sum_{n\in\mathbb{I}_d} R^{(n)}_{xx^{(*)}}(t_1-t_2)\, e^{j2\pi\beta_n t_2} + \sum_{n\in\mathbb{I}_c} R^{(n)}_{xx^{(*)}}(t_1-t_2)\, e^{j2\pi\alpha_n(t_1-t_2)t_2}
\end{aligned} \tag{3.31}$$

Thus, we formally have

$$
\begin{aligned}
&\mathrm{E}\left\{X(f_1)\, X^{(*)}(f_2)\right\} \\
&= \int_{\mathbb{R}} \int_{\mathbb{R}} \mathrm{E}\{x(t_1)\, x^{(*)}(t_2)\}\, e^{-j2\pi[f_1 t_1 + (-) f_2 t_2]}\, \mathrm{d}t_1\, \mathrm{d}t_2 \\
&= \sum_{n \in \mathbb{I}} \int_{\mathbb{R}} \int_{\mathbb{R}} R^{(n)}_{xx^{(*)}}(t_1 - t_2)\, e^{j2\pi\alpha_n(t_1-t_2)t_2}\, e^{-j2\pi[f_1 t_1 + (-) f_2 t_2]}\, \mathrm{d}t_1\, \mathrm{d}t_2 \\
&= \sum_{n \in \mathbb{I}} \int_{\mathbb{R}} R^{(n)}_{xx^{(*)}}(\tau)\, e^{-j2\pi f_1 \tau} \int_{\mathbb{R}} e^{j2\pi\alpha_n(\tau)t}\, e^{-j2\pi[f_1 + (-) f_2]t}\, \mathrm{d}t\, \mathrm{d}\tau \\
&= \sum_{n \in \mathbb{I}} \int_{\mathbb{R}} R^{(n)}_{xx^{(*)}}(\tau)\, e^{-j2\pi f_1 \tau}\, \delta(f_1 + (-) f_2 - \alpha_n(\tau))\, \mathrm{d}\tau \\
&= \sum_{n \in \mathbb{I}_d} \int_{\mathbb{R}} R^{(n)}_{xx^{(*)}}(\tau)\, e^{-j2\pi f_1 \tau}\, \mathrm{d}\tau\, \delta(f_1 + (-) f_2 - \beta_n) \\
&\quad + \sum_{n \in \mathbb{I}_c} \int_{\mathbb{R}} R^{(n)}_{xx^{(*)}}(\tau)\, e^{-j2\pi f_1 \tau} \\
&\quad \left|(\alpha_n^{-1})'(f_1 + (-) f_2)\right|\, \delta(\tau - \alpha_n^{-1}(f_1 + (-) f_2))\, \mathrm{d}\tau
\end{aligned}
\tag{3.32}
$$

where, in the third equality, the variable change $t_1 = t + \tau$ and $t_2 = t$ is made and in the fourth equality (2.71) and (2.72) are used. Then, (2.70) immediately follows.

3.3 Proofs for Section 2.3 "Linear Time-Variant Filtering of GACS Processes"

3.3.1 Proof of (2.112)

By substituting (2.111) into (2.110) we have

$$
\begin{aligned}
&\mathrm{E}\left\{y_1(t+\tau_1)\, y_2^{(*)}(t+\tau_2)\right\} \\
&= \int_{\mathbb{R}^2} \sum_{\sigma_1 \in J_1} h_{\sigma_1}(\tau_1 - s_1)\, e^{j2\pi\sigma_1(t+s_1)} \sum_{\sigma_2 \in J_2} h^{(*)}_{\sigma_2}(\tau_2 - s_2)\, e^{(-)j2\pi\sigma_2(t+s_2)} \\
&\quad \sum_{n \in \mathbb{I}_{\boldsymbol{x}}} R^{(n)}_{\boldsymbol{x}}(s_1 - s_2)\, e^{j2\pi\alpha^{(n)}_{\boldsymbol{x}}(s_1-s_2)\,(t+s_2)}\, \mathrm{d}s_1\, \mathrm{d}s_2 \\
&= \sum_{\sigma_1 \in J_1} \sum_{\sigma_2 \in J_2} \sum_{n \in \mathbb{I}_{\boldsymbol{x}}} \int_{\mathbb{R}} R^{(n)}_{\boldsymbol{x}}(s)\, e^{j2\pi[\sigma_1 s + \alpha^{(n)}_{\boldsymbol{x}}(s)t]} \\
&\quad \int_{\mathbb{R}} h_{\sigma_1}(\tau_1 - s - u)\, h^{(*)}_{\sigma_2}(\tau_2 - u)\, e^{j2\pi[\sigma_1 + (-)\sigma_2 + \alpha^{(n)}_{\boldsymbol{x}}(s)]u}\, \mathrm{d}u\, \mathrm{d}s\, e^{j2\pi[\sigma_1 + (-)\sigma_2]t}
\end{aligned}
\tag{3.33}
$$

where, in the second equality the variable changes $s_1 = s + u$, $s_2 = u$ are made. Furthermore, by making the variable change $v = \tau_2 - u$ into the inner integral in (3.33), we have

$$\int_{\mathbb{R}} h_{\sigma_1}(\tau_1 - s - u)\, h_{\sigma_2}^{(*)}(\tau_2 - u)\, e^{j2\pi[\sigma_1 + (-)\sigma_2 + \alpha_{\boldsymbol{x}}^{(n)}(s)]u}\, \mathrm{d}u$$
$$= \int_{\mathbb{R}} h_{\sigma_1}(\tau_1 - s - \tau_2 + v)\, h_{\sigma_2}^{(*)}(v)\, e^{-j2\pi[\sigma_1 + (-)\sigma_2 + \alpha_{\boldsymbol{x}}^{(n)}(s)]v}\, \mathrm{d}v$$
$$e^{j2\pi[\sigma_1 + (-)\sigma_2 + \alpha_{\boldsymbol{x}}^{(n)}(s)]\tau_2}$$
$$= r_{\sigma_1\sigma_2(*)}^{\sigma_1 + (-)\sigma_2 + \alpha_{\boldsymbol{x}}^{(n)}(s)}(\tau_1 - \tau_2 - s)\, e^{j2\pi[\sigma_1 + (-)\sigma_2 + \alpha_{\boldsymbol{x}}^{(n)}(s)]\tau_2} \tag{3.34}$$

where $r_{\sigma_1\sigma_2(*)}^{\gamma}(\tau)$ is defined in (2.113).

By substituting (3.34) into (3.33), we obtain (2.112).

Assumption (1.47) allows to interchange the order of integral and expectation operators (Fubini and Tonelli Theorem (Champeney 1990, Chapter 3)).

3.3.2 Proof of (2.117)

From (2.112) with $\tau_1 = \tau$ and $\tau_2 = 0$ it follows that

$$R_{\boldsymbol{y}}(\beta, \tau) \triangleq \left\langle \mathrm{E}\left\{ y_1(t + \tau)\, y_2^{(*)}(t) \right\} e^{-j2\pi\beta t} \right\rangle_t$$
$$= \sum_{\sigma_1 \in J_1} \sum_{\sigma_2 \in J_2} \sum_{n \in \mathbb{I}_{\boldsymbol{x}}} \int_{\mathbb{R}} R_{\boldsymbol{x}}^{(n)}(s)\, r_{\sigma_1\sigma_2(*)}^{\sigma_1 + (-)\sigma_2 + \alpha_{\boldsymbol{x}}^{(n)}(s)}(\tau - s)\, e^{j2\pi\sigma_1 s}$$
$$\left\langle e^{j2\pi[\sigma_1 + (-)\sigma_2 + \alpha_{\boldsymbol{x}}^{(n)}(s) - \beta]t} \right\rangle_t \mathrm{d}s \tag{3.35}$$

from which (2.117) immediately follows.

To obtain (3.35), under assumptions (2.114)–(2.116), the order of

$$\langle\cdot\rangle_t \triangleq \lim_{T \to \infty} \frac{1}{T} \int_{-T/2}^{T/2} (\cdot)\, \mathrm{d}t \quad \text{and} \quad \sum_{\sigma_1 \in J_1} \sum_{\sigma_2 \in J_2} \sum_{n \in \mathbb{I}_{\boldsymbol{x}}} \int_{\mathbb{R}} (\cdot)\, \mathrm{d}s$$

operations can be interchanged. In fact, the following inequalities hold

$$\left| \sum_{\sigma_1 \in J_1} \sum_{\sigma_2 \in J_2} \sum_{n \in \mathbb{I}_{\boldsymbol{x}}} R_{\boldsymbol{x}}^{(n)}(s)\, r_{\sigma_1\sigma_2(*)}^{\sigma_1 + (-)\sigma_2 + \alpha_{\boldsymbol{x}}^{(n)}(s)}(\tau - s)\, e^{j2\pi\sigma_1 s} \right|$$
$$\leqslant \sum_{n \in \mathbb{I}_{\boldsymbol{x}}} \left| R_{\boldsymbol{x}}^{(n)}(s) \right| \sum_{\sigma_1 \in J_1} \sum_{\sigma_2 \in J_2} \left| r_{\sigma_1\sigma_2(*)}^{\sigma_1 + (-)\sigma_2 + \alpha_{\boldsymbol{x}}^{(n)}(s)}(\tau - s) \right|$$
$$= \sum_{n \in \mathbb{I}_{\boldsymbol{x}}} \left| R_{\boldsymbol{x}}^{(n)}(s) \right| \sum_{\sigma_1 \in J_1} \sum_{\sigma_2 \in J_2} \left| \int_{\mathbb{R}} h_{\sigma_1}(\tau - s + v)\, h_{\sigma_2}^{(*)}(v)\, e^{-j2\pi[\sigma_1 + (-)\sigma_2 + \alpha_{\boldsymbol{x}}^{(n)}(s)]v}\, \mathrm{d}v \right|$$
$$\leqslant \sum_{n \in \mathbb{I}_{\boldsymbol{x}}} \left| R_{\boldsymbol{x}}^{(n)}(s) \right| \sum_{\sigma_1 \in J_1} \sum_{\sigma_2 \in J_2} \int_{\mathbb{R}} \left| h_{\sigma_1}(\tau - s + v) \right| \left| h_{\sigma_2}^{(*)}(v) \right| \mathrm{d}v$$
$$\leqslant \sum_{n \in \mathbb{I}_{\boldsymbol{x}}} \left| R_{\boldsymbol{x}}^{(n)}(s) \right| \sum_{\sigma_1 \in J_1} \| h_{\sigma_1} \|_\infty \sum_{\sigma_2 \in J_2} \int_{\mathbb{R}} \left| h_{\sigma_2}^{(*)}(v) \right| \mathrm{d}v \tag{3.36}$$

and

$$\left| \frac{1}{T} \int_{-T/2}^{T/2} e^{j2\pi[\sigma_1 +(-)\sigma_2 + \alpha_{\boldsymbol{x}}^{(n)}(s) - \beta]t} \, \mathrm{d}t \right| \leqslant \frac{1}{T} \int_{-T/2}^{T/2} \left| e^{j2\pi[\sigma_1 +(-)\sigma_2 + \alpha_{\boldsymbol{x}}^{(n)}(s) - \beta]t} \right| \mathrm{d}t = 1 \tag{3.37}$$

independent of T.

Therefore, conditions (2.114)–(2.116) allow to use the Fubini and Tonelli theorem (Champeney 1990, Chapter 3) to interchange the order of integrals in $\mathrm{d}t$ and $\mathrm{d}s$ for T finite. Moreover, the same conditions are sufficient to apply the dominated convergence theorem (Champeney 1990, Chapter 4) to interchange the order of limit and integral operations.

3.4 Proofs for Section 2.4.1 "The Cyclic Cross-Correlogram"

In this section, proofs of lemmas and theorems presented in Section 2.4.1 on the bias and covariance of the cyclic cross-correlogram are reported.

In the following, all the functions are assumed to be Lebesgue measurable. Consequently, without recalling the measurability assumption, we use the fact that if the functions φ_1 and φ_2 are such that $|\varphi_1| \leqslant |\varphi_2|$, φ_1 is measurable and φ_2 is integrable (i.e., φ_2 is measurable and $|\varphi_2|$ is integrable), then φ_1 is integrable (Prohorov and Rozanov 1989, p. 82). Furthermore, if $\varphi_1 \in L^1(\mathbb{R})$ and $\varphi_2 \in L^\infty(\mathbb{R})$, then $|\varphi_1 \, \varphi_2| \leqslant |\varphi_1| \, \|\varphi_2\|_\infty$ almost everywhere and, hence, $\varphi_1 \, \varphi_2 \in L^1(\mathbb{R})$.

3.4.1 Proof of Theorem 2.4.6 Expected Value of the Cyclic Cross-Correlogram

By using (2.31c) and (2.118) one has

$$\begin{aligned}
&\mathrm{E}\left\{R_{yx^{(*)}}(\alpha, \tau; t_0, T)\right\} \\
&\quad = \mathrm{E}\left\{\int_{\mathbb{R}} w_T(t - t_0)\, y(t + \tau)\, x^{(*)}(t)\, e^{-j2\pi\alpha t}\, \mathrm{d}t\right\} && (3.38a) \\
&\quad = \int_{\mathbb{R}} w_T(t - t_0)\, \mathrm{E}\left\{y(t + \tau)\, x^{(*)}(t)\right\} e^{-j2\pi\alpha t}\, \mathrm{d}t && (3.38b) \\
&\quad = \int_{\mathbb{R}} w_T(t - t_0) \sum_{n \in \mathbb{I}} R_{yx^{(*)}}^{(n)}(\tau)\, e^{j2\pi\alpha_n(\tau)t}\, e^{-j2\pi\alpha t}\, \mathrm{d}t && (3.38c) \\
&\quad = \sum_{n \in \mathbb{I}} R_{yx^{(*)}}^{(n)}(\tau) \int_{\mathbb{R}} w_T(t - t_0)\, e^{-j2\pi[\alpha - \alpha_n(\tau)]t}\, \mathrm{d}t && (3.38d)
\end{aligned}$$

from which (2.128) immediately follows.

In (3.38b), the interchange of statistical expectation and integral operations is justified by the Fubini and Tonelli theorem (Champeney 1990, Chapter 3). In fact, the cross-correlation is uniformly (with respect to t and τ) bounded since

$$\left| \mathrm{E}\left\{ \left| y(t + \tau)\, x^{(*)}(t) \right| \right\} \right|^2 \leqslant \mathrm{E}\left\{ |y(t + \tau)|^2 \right\} \mathrm{E}\left\{ |x(t)|^2 \right\} \tag{3.39}$$

and, accounting for Assumption 2.4.3a, for any $z \in \{x, y\}$

$$\mathrm{E}\left\{|z(t)|^2\right\} \leqslant \sum_{n \in \mathbb{I}_{zz^*}} \left|R_{zz^*}^{(n)}(0)\right| \leqslant \sum_{n \in \mathbb{I}_{zz^*}} \left\|R_{zz^*}^{(n)}\right\|_{\infty} < \infty. \tag{3.40}$$

Therefore,

$$\begin{aligned}
&\int_{\mathbb{R}} |w_T(t-t_0)|\ \mathrm{E}\left\{\left|y(t+\tau)\, x^{(*)}(t)\right|\right\} \mathrm{d}t \\
&\quad \leqslant \left[\sum_{n' \in \mathbb{I}_{xx^*}} \left\|R_{xx^*}^{(n')}\right\|_{\infty} \sum_{n'' \in \mathbb{I}_{yy^*}} \left\|R_{yy^*}^{(n'')}\right\|_{\infty}\right]^{1/2} \int_{\mathbb{R}} |a(s)|\, \mathrm{d}s < \infty
\end{aligned} \tag{3.41}$$

where Assumption 2.4.5 and the variable change $s = (t - t_0)/T$ are used.

The interchange of sum and integral operations to obtain (3.38d) is justified even if $\mathbb{I}$ is not finite by using the dominated convergence theorem (Champeney 1990, Chapter 4). Specifically, by denoting with $\{\mathbb{I}_k\}$ an increasing sequence of finite subsets of $\mathbb{I}$ such that $\mathbb{I}_h \subseteq \mathbb{I}_k$ for $h < k$ and $\lim_{k\to\infty} \mathbb{I}_k \triangleq \cup_{k\in\mathbb{N}} = \mathbb{I}$, and defining

$$F_n(t) \triangleq w_T(t-t_0) R_{yx^{(*)}}^{(n)}(\tau)\, e^{j2\pi\alpha_n(\tau)t}\, e^{-j2\pi\alpha t} \tag{3.42}$$

it results that

$$\lim_{k\to\infty} \sum_{n\in\mathbb{I}_k} \int_{\mathbb{R}} F_n(t)\,\mathrm{d}t = \lim_{k\to\infty} \int_{\mathbb{R}} \sum_{n\in\mathbb{I}_k} F_n(t)\,\mathrm{d}t = \int_{\mathbb{R}} \lim_{k\to\infty} \sum_{n\in\mathbb{I}_k} F_n(t)\,\mathrm{d}t. \tag{3.43}$$

In fact, the integrand function in the second term of equality (3.43) is bounded by a summable function of t not depending on k:

$$\begin{aligned}
\left|\sum_{n\in\mathbb{I}_k} F_n(t)\right| &\leqslant \sum_{n\in\mathbb{I}_k} |F_n(t)| \leqslant \sum_{n\in\mathbb{I}} |F_n(t)| \\
&= |w_T(t-t_0)| \sum_{n\in\mathbb{I}} \left|R_{yx^{(*)}}^{(n)}(\tau)\right| \\
&\leqslant |w_T(t-t_0)| \sum_{n\in\mathbb{I}} \left\|R_{yx^{(*)}}^{(n)}\right\|_{\infty} \in L^1(\mathbb{R})
\end{aligned} \tag{3.44}$$

where, in the last inequality, Assumptions 2.4.3a and 2.4.5 are used. □

3.4.2 *Proof of Theorem 2.4.7 Covariance of the Cyclic Cross-Correlogram*

For zero-mean stochastic processes $x(t)$ and $y(t)$ one obtains (Gardner 1985; Spooner and Gardner 1994)

$$\begin{aligned}
&\mathrm{cum}\left\{y(u_1+\tau_{y1}),\, x^{(*)}(u_1+\tau_{x1}),\, y^*(u_2+\tau_{y2}),\, x^{(*)*}(u_2+\tau_{x2})\right\} \\
&\quad = \mathrm{cov}\left\{y(u_1+\tau_{y1})\, x^{(*)}(u_1+\tau_{x1}),\ y(u_2+\tau_{y2})\, x^{(*)}(u_2+\tau_{x2})\right\} \\
&\qquad - \mathrm{E}\left\{y(u_1+\tau_{y1})\, y^*(u_2+\tau_{y2})\right\} \mathrm{E}\left\{x^{(*)}(u_1+\tau_{x1})\, x^{(*)*}(u_2+\tau_{x2})\right\} \\
&\qquad - \mathrm{E}\left\{y(u_1+\tau_{y1})\, x^{(*)*}(u_2+\tau_{x2})\right\} \mathrm{E}\left\{x^{(*)}(u_1+\tau_{x1})\, y^*(u_2+\tau_{y2})\right\}
\end{aligned} \tag{3.45}$$

where

$$\begin{aligned}\operatorname{cov}\{z_1, z_2\} &\triangleq \mathrm{E}\{(z_1 - \mathrm{E}\{z_1\})(z_2 - \mathrm{E}\{z_2\})^*\} \\ &= \mathrm{E}\{z_1 z_2^*\} - \mathrm{E}\{z_1\}\mathrm{E}\{z_2\}^* \end{aligned} \tag{3.46}$$

is the covariance of the complex random variables z_1 and z_2 and the cumulant of complex processes is defined according to (Spooner and Gardner 1994, Appendix A) (see also Section 1.4.2). Thus, accounting for (2.118), one obtains

$$\begin{aligned}&\operatorname{cov}\left\{R_{yx^{(*)}}(\alpha_1, \tau_1; t_1, T),\ R_{yx^{(*)}}(\alpha_2, \tau_2; t_2, T)\right\} \\ &= \operatorname{cov}\left\{\int_{\mathbb{R}} w_T(u_1 - t_1)\, y(u_1 + \tau_1)\, x^{(*)}(u_1)\, e^{-j2\pi\alpha_1 u_1}\, \mathrm{d}u_1,\right. \\ &\qquad \left.\int_{\mathbb{R}} w_T(u_2 - t_2)\, y(u_2 + \tau_2)\, x^{(*)}(u_2)\, e^{-j2\pi\alpha_2 u_2}\, \mathrm{d}u_2\right\} &\text{(3.47a)} \\ &= \int_{\mathbb{R}}\int_{\mathbb{R}} \operatorname{cov}\left\{y(u_1 + \tau_1)\, x^{(*)}(u_1),\ y(u_2 + \tau_2)\, x^{(*)}(u_2)\right\} \\ &\qquad w_T(u_1 - t_1)\, w_T^*(u_2 - t_2)\, e^{-j2\pi[\alpha_1 u_1 - \alpha_2 u_2]}\, \mathrm{d}u_1\, \mathrm{d}u_2 &\text{(3.47b)} \\ &= \int_{\mathbb{R}}\int_{\mathbb{R}} \left[\mathrm{E}\left\{y(u_1 + \tau_1)\, y^*(u_2 + \tau_2)\right\}\ \mathrm{E}\left\{x^{(*)}(u_1)\, x^{(*)*}(u_2)\right\}\right. \\ &\qquad + \mathrm{E}\left\{y(u_1 + \tau_1)\, x^{(*)*}(u_2)\right\}\ \mathrm{E}\left\{x^{(*)}(u_1)\, y^*(u_2 + \tau_2)\right\} \\ &\qquad \left.+ \operatorname{cum}\left\{y(u_1 + \tau_1), x^{(*)}(u_1), y^*(u_2 + \tau_2), x^{(*)*}(u_2)\right\}\right] \\ &\qquad w_T(u_1 - t_1)\, w_T^*(u_2 - t_2)\, e^{-j2\pi[\alpha_1 u_1 - \alpha_2 u_2]}\, \mathrm{d}u_1\, \mathrm{d}u_2 &\text{(3.47c)}\end{aligned}$$

By substituting (2.119) and (2.120) (Assumption 2.4.2) into (3.47c) and making the variable changes $u_1 = u$ and $u_2 = u - s$, it results in

$$\operatorname{cov}\left\{R_{yx^{(*)}}(\alpha_1, \tau_1; t_1, T), R_{yx^{(*)}}(\alpha_2, \tau_2; t_2, T)\right\} = \mathcal{D}_1 + \mathcal{D}_2 + \mathcal{D}_3 \tag{3.48}$$

where

$$\begin{aligned}\mathcal{D}_1 \triangleq \sum_{n'}\sum_{n''} \int_{\mathbb{R}}\int_{\mathbb{R}} & R_{yy^*}^{(n')}(\tau_1 - \tau_2 + s)\, R_{x^{(*)}x^{(*)*}}^{(n'')}(s) \\ & e^{j2\pi\alpha'_{n'}(\tau_1 - \tau_2 + s)(u - s + \tau_2)}\, e^{j2\pi\alpha''_{n''}(s)(u - s)} \\ & w_T(u - t_1)\, w_T^*(u - s - t_2)\, e^{-j2\pi[\alpha_1 u - \alpha_2 \cdot (u - s)]}\, \mathrm{d}u\, \mathrm{d}s \end{aligned} \tag{3.49}$$

$$\begin{aligned}\mathcal{D}_2 \triangleq \sum_{n'''}\sum_{n^{iv}} \int_{\mathbb{R}}\int_{\mathbb{R}} & R_{yx^{(*)*}}^{(n''')}(\tau_1 + s)\, R_{x^{(*)}y^*}^{(n^{iv})}(s - \tau_2) \\ & e^{j2\pi\alpha'''_{n'''}(\tau_1 + s)(u - s)}\, e^{j2\pi\alpha^{iv}_{n^{iv}}(s - \tau_2)(u - s + \tau_2)} \\ & w_T(u - t_1)\, w_T^*(u - s - t_2)\, e^{-j2\pi[\alpha_1 u - \alpha_2 \cdot (u - s)]}\, \mathrm{d}u\, \mathrm{d}s \end{aligned} \tag{3.50}$$

$$\begin{aligned}\mathcal{D}_3 \triangleq \sum_{n} \int_{\mathbb{R}}\int_{\mathbb{R}} & C_{yx^{(*)}y^*x^{(*)*}}^{(n)}(\tau_1 + s, s, \tau_2)\, e^{j2\pi\beta_n(\tau_1 + s, s, \tau_2)(u - s)} \\ & w_T(u - t_1)\, w_T^*(u - s - t_2)\, e^{-j2\pi[\alpha_1 u - \alpha_2 \cdot (u - s)]}\, \mathrm{d}u\, \mathrm{d}s \end{aligned} \tag{3.51}$$

Finally, by making the variable change $u/T = t' + (t_2 + s)/T$ into (3.49)–(3.51) and using (2.125) and (2.133), (2.129)–(2.132) easily follow. In (3.49)–(3.51), for notation simplicity, $\alpha'_{n'}(\cdot) \equiv \alpha^{(n')}_{yy^*}(\cdot)$, $\alpha''_{n''}(\cdot) \equiv \alpha^{(n'')}_{x^{(*)}x^{(*)*}}(\cdot)$, $\alpha'''_{n'''}(\cdot) \equiv \alpha^{(n''')}_{yx^{(*)*}}(\cdot)$, and $\alpha^{iv}_{n^{iv}}(\cdot) \equiv \alpha^{(n^{iv})}_{x^{(*)}y^*}(\cdot)$.

The interchange of the orders of $\text{cov}\{\cdot\}$ and integral operations to obtain (3.47b) is justified by the Fubini and Tonelli theorem (Champeney 1990, Chapter 3). In fact,

$$
\begin{aligned}
\text{cov}\Bigg\{&\int_{\mathbb{R}} w_T(u_1 - t_1)\, y(u_1 + \tau_1)\, x^{(*)}(u_1)\, e^{-j2\pi\alpha_1 u_1}\, \mathrm{d}u_1, \\
&\int_{\mathbb{R}} w_T(u_2 - t_2)\, y(u_2 + \tau_2)\, x^{(*)}(u_2)\, e^{-j2\pi\alpha_2 u_2}\, \mathrm{d}u_2\Bigg\} \\
&= \mathrm{E}\Bigg\{\int_{\mathbb{R}}\int_{\mathbb{R}} w_T(u_1 - t_1)\, y(u_1 + \tau_1)\, x^{(*)}(u_1)\, e^{-j2\pi\alpha_1 u_1} \\
&\qquad w_T^*(u_2 - t_2)\, y^*(u_2 + \tau_2)\, x^{(*)*}(u_2)\, e^{j2\pi\alpha_2 u_2}\, \mathrm{d}u_1\, \mathrm{d}u_2\Bigg\} \\
&\quad - \mathrm{E}\Bigg\{\int_{\mathbb{R}} w_T(u_1 - t_1)\, y(u_1 + \tau_1)\, x^{(*)}(u_1)\, e^{-j2\pi\alpha_1 u_1}\, \mathrm{d}u_1\Bigg\} \\
&\quad\; \mathrm{E}\Bigg\{\int_{\mathbb{R}} w_T^*(u_2 - t_2)\, y^*(u_2 + \tau_2)\, x^{(*)*}(u_2)\, e^{j2\pi\alpha_2 u_2}\, \mathrm{d}u_2\Bigg\}
\end{aligned}
\tag{3.52}
$$

The interchange of statistical expectation and single integral can be justified as in the proof of Theorem 2.4.6 accounting for Assumptions 2.4.3a and 2.4.5. As regards the interchange of statistical expectation and double integral, we have

$$
\begin{aligned}
&\left|\mathrm{E}\left\{\left|y(u_1 + \tau_1)\, x^{(*)}(u_1)\, y^*(u_2 + \tau_2)\, x^{(*)*}(u_2)\right|\right\}\right|^4 \\
&\leqslant \left|\mathrm{E}\left\{\left|y(u_1 + \tau_1)\, x^{(*)}(u_1)\right|^2\right\}\right|^2 \left|\mathrm{E}\left\{\left|y^*(u_2 + \tau_2)\, x^{(*)*}(u_2)\right|^2\right\}\right|^2 \\
&\leqslant \mathrm{E}\left\{\left|y(u_1 + \tau_1)\right|^4\right\} \mathrm{E}\left\{\left|x^{(*)}(u_1)\right|^4\right\} \\
&\quad \mathrm{E}\left\{\left|y^*(u_2 + \tau_2)\right|^4\right\} \mathrm{E}\left\{\left|x^{(*)*}(u_2)\right|^4\right\}
\end{aligned}
\tag{3.53}
$$

Thus, accounting for the uniform boundedness of the absolute fourth-order moments of $x(t)$ and $y(t)$ (Assumption 2.4.4) and Assumption 2.4.5, one has

$$
\begin{aligned}
\int_{\mathbb{R}}\int_{\mathbb{R}} &\mathrm{E}\left\{\left|y(u_1 + \tau_1)\, x^{(*)}(u_1)\, y^*(u_2 + \tau_2)\, x^{(*)*}(u_2)\right|\right\} \\
&\left|w_T(u_1 - t_1)\, w_T^*(u_2 - t_2)\right| \mathrm{d}u_1\, \mathrm{d}u_2 < \infty.
\end{aligned}
\tag{3.54}
$$

Therefore, the Fubini and Tonelli theorem (Champeney 1990, Chapter 3) can be used to obtain (3.47b) and (3.47c).

The interchange of sum and integral operations to obtain (3.48) is justified even if the sets $\mathbb{I}_{z_i z_j}$ and $\mathbb{I}$ are not finite. In fact, let us consider the term $\mathcal{D}_1$ defined in (3.49). Denote with $\{\mathbb{I}'_k\}$

and $\{\mathbb{I}''_k\}$ two increasing sequences of finite subsets of $\mathbb{I}_{yy^*}$ and $\mathbb{I}_{x^{(*)}x^{(*)*}}$, respectively, such that $\lim_{k\to\infty} \mathbb{I}'_k = \mathbb{I}_{yy^*}$ and $\lim_{k\to\infty} \mathbb{I}''_k = \mathbb{I}_{x^{(*)}x^{(*)*}}$, and define

$$F_{n'n''}(s,u) \triangleq R^{(n')}_{yy^*}(\tau_1 - \tau_2 + s)\, R^{(n'')}_{x^{(*)}x^{(*)*}}(s)\, e^{j2\pi\alpha'_{n'}(\tau_1-\tau_2+s)(u-s+\tau_2)}\, e^{j2\pi\alpha''_{n''}(s)(u-s)} \\ w_T(u-t_1)\, w_T^*(u-s-t_2)\, e^{-j2\pi[\alpha_1 u - \alpha_2\cdot(u-s)]}. \tag{3.55}$$

The result is that

$$\begin{aligned}\left|\sum_{n'\in\mathbb{I}'_h}\sum_{n''\in\mathbb{I}''_k} F_{n'n''}(s,u)\right| &\leqslant \sum_{n'\in\mathbb{I}_{yy^*}}\sum_{n''\in\mathbb{I}_{x^{(*)}x^{(*)*}}} \left|F_{n'n''}(s,u)\right| \\ &= \sum_{n'\in\mathbb{I}_{yy^*}} \left|R^{(n')}_{yy^*}(\tau_1-\tau_2+s)\right| \sum_{n''\in\mathbb{I}_{x^{(*)}x^{(*)*}}} \left|R^{(n'')}_{x^{(*)}x^{(*)*}}(s)\right| \\ &\quad \left|w_T(u-t_1)\, w_T^*(u-s-t_2)\right| \\ &\leqslant \sum_{n'\in\mathbb{I}_{yy^*}} \left\|R^{(n')}_{yy^*}\right\|_\infty \sum_{n''\in\mathbb{I}_{x^{(*)}x^{(*)*}}} \left\|R^{(n'')}_{x^{(*)}x^{(*)*}}\right\|_\infty \\ &\quad \left|w_T(u-t_1)\, w_T^*(u-s-t_2)\right| \in L^1(\mathbb{R}^2)\end{aligned} \tag{3.56}$$

where Assumptions 2.4.3a and 2.4.5 have been accounted for. Thus, the left-hand side of (3.56) is bounded by a summable function of (s,u) not depending on h and k. Therefore, the dominated convergence theorem (Champeney 1990, Chapter 4) can be applied as follows:

$$\begin{aligned}&\lim_{\substack{h\to\infty\\k\to\infty}} \sum_{n'\in\mathbb{I}'_h}\sum_{n''\in\mathbb{I}''_k} \int_{\mathbb{R}}\int_{\mathbb{R}} F_{n'n''}(s,u)\,\mathrm{d}s\,\mathrm{d}u \\ &\quad= \lim_{\substack{h\to\infty\\k\to\infty}} \int_{\mathbb{R}}\int_{\mathbb{R}} \sum_{n'\in\mathbb{I}'_h}\sum_{n''\in\mathbb{I}''_k} F_{n'n''}(s,u)\,\mathrm{d}s\,\mathrm{d}u \\ &\quad= \int_{\mathbb{R}}\int_{\mathbb{R}} \lim_{\substack{h\to\infty\\k\to\infty}} \sum_{n'\in\mathbb{I}'_h}\sum_{n''\in\mathbb{I}''_k} F_{n'n''}(s,u)\,\mathrm{d}s\,\mathrm{d}u.\end{aligned} \tag{3.57}$$

An analogous result can be found for the term $\mathcal{D}_2$ defined in (3.50). As regards the term $\mathcal{D}_3$ defined in (3.51), denote with $\{\mathbb{I}_k\}$ an increasing sequence of finite subsets of $\mathbb{I}_4$ such that $\lim_{k\to\infty} \mathbb{I}_k = \mathbb{I}_4$, and define

$$F_n(s,u) \triangleq C^{(n)}_{yx^{(*)}y^*x^{(*)*}}(s+\tau_1, s, \tau_2)\, e^{j2\pi\beta_n(s+\tau_1,s,\tau_2)(u-s)} \\ w_T(u-t_1)\, w_T^*(u-s-t_2)\, e^{-j2\pi[\alpha_1 u - \alpha_2\cdot(u-s)]}. \tag{3.58}$$

The result is that

$$\left|\sum_{n\in\mathbb{I}_k} F_n(s,u)\right| \leqslant \sum_{n\in\mathbb{I}_4} \|C^{(n)}_{yx^{(*)}y^*x^{(*)*}}\|_\infty\, |w_T(u-t_1)\, w_T^*(u-s-t_2)| \in L^1(\mathbb{R}^2) \tag{3.59}$$

where Assumption 2.4.3b has been accounted for. Hence, the dominated convergence theorem can be applied similarly as in (3.57).

3.5 Proofs for Section 2.4.2 "Mean-Square Consistency of the Cyclic Cross-Correlogram"

In this section, proofs of results presented in Section 2.4.2 on the mean-square consistency of the cyclic cross-correlogram are reported.

Lemma 3.5.1 *Let $a(t)$ be such that Assumption 2.4.5 is satisfied. We have the following.*

(a) *The Fourier transform*

$$A(f) \triangleq \int_{\mathbb{R}} a(t)\, e^{-j2\pi ft}\, \mathrm{d}t \tag{3.60}$$

is bounded, continuous, and infinitesimal as $|f| \to \infty$.

(b) *We have the result*

$$\lim_{T\to\infty} W_{\frac{1}{T}}(f) = \delta_f \tag{3.61}$$

where δ_f denotes Kronecker delta, that is, $\delta_f = 1$ if $f = 0$ and $\delta_f = 0$ if $f \neq 0$.

Proof: Since $a(t) \in L^1(\mathbb{R})$, item *(a)* is a consequence of the properties of the Fourier transforms (Champeney 1990). In Assumption 2.4.5 it is also assumed that there exists $\gamma > 0$ such that

$$A(f) = \mathcal{O}\left(|f|^{-\gamma}\right) \quad \text{as} \quad |f| \to \infty. \tag{3.62}$$

In reference to item *(b)*, observe that from (2.125) we have

$$\begin{aligned} W_{\frac{1}{T}}(f) &\triangleq \int_{\mathbb{R}} w_T(t)\, e^{-j2\pi ft}\, \mathrm{d}t \\ &= \int_{\mathbb{R}} \frac{1}{T} a\left(\frac{t}{T}\right) e^{-j2\pi ft}\, \mathrm{d}t \\ &= \int_{\mathbb{R}} a(s)\, e^{-j2\pi fTs}\, \mathrm{d}s \\ &= A(fT) \end{aligned} \tag{3.63}$$

where, in the third equality, the variable change $s = t/T$ is made. Thus, for $f = 0$,

$$W_{\frac{1}{T}}(0) = A(0) = \int_{\mathbb{R}} a(s)\, \mathrm{d}s = 1 \quad \forall T \in \mathbb{R} \tag{3.64}$$

(see (2.126)). For $f \neq 0$, since $a(t) \in L^1(\mathbb{R})$, we have

$$\lim_{T\to\infty} W_{\frac{1}{T}}(f) = \lim_{T\to\infty} \int_{\mathbb{R}} a(s)\, e^{-j2\pi fTs}\, \mathrm{d}s = 0 \tag{3.65}$$

for the Riemann-Lebesgue theorem (Champeney 1990, Chapter 3).

Furthermore, in the sense of distributions (generalized functions) (Champeney 1990; Zemanian 1987) from (2.127), it follows that

$$\lim_{T\to\infty} \int_{\mathbb{R}} a\left(\frac{t}{T}\right) e^{-j2\pi ft}\, \mathrm{d}t = \lim_{T\to\infty} T\, A(fT) = \delta(f) \tag{3.66}$$

where $\delta(\cdot)$ denotes Dirac delta. □

Lemma 3.5.2 *Let us consider the function $r_{aa^*}(\beta; s)$ defined in (2.133). Under Assumption 2.4.5, we have*

$$\lim_{T\to\infty} r_{aa^*}\left(\beta T; \frac{s}{T}\right) = \int_{\mathbb{R}} |a(t)|^2 \,\mathrm{d}t\, \delta_\beta. \tag{3.67}$$

Proof: For $\beta = 0$, one obtains

$$\begin{aligned}\lim_{T\to\infty} r_{aa^*}\left(0; \frac{s}{T}\right) &= \lim_{T\to\infty} \int_{\mathbb{R}} a\big(t + \frac{s}{T}\big)\, a^*(t)\,\mathrm{d}t \\ &= \int_{\mathbb{R}} \lim_{T\to\infty} a\big(t + \frac{s}{T}\big)\, a^*(t)\,\mathrm{d}t \\ &= \int_{\mathbb{R}} |a(t)|^2\,\mathrm{d}t \end{aligned}\tag{3.68}$$

where the interchange of limit and integral operations is allowed by the dominated convergence theorem. In fact, accounting for Assumption 2.4.5, the magnitude of the integrand function $|a(t + s/T)\, a^*(t)|$ is bounded by the summable function $|a(t)|\; \|a\|_\infty$ not depending on T. The third equality is a consequence of the a.e. continuity of $a(t)$. The last integral in (3.68) exists since $|a(t)|^2 \leqslant \|a\|_\infty |a(t)| \in L^1(\mathbb{R})$.

Let us consider, now, the case $\beta \neq 0$. We have

$$\begin{aligned}&\int_{\mathbb{R}} a\left(t + \frac{s}{T}\right) a^*(t)\, e^{-j2\pi\beta T t}\,\mathrm{d}t \\ &\quad = \int_{\mathbb{R}} \left[a\left(t + \frac{s}{T}\right) - a(t)\right] a^*(t)\, e^{-j2\pi\beta T t}\,\mathrm{d}t + \int_{\mathbb{R}} |a(t)|^2\, e^{-j2\pi\beta T t}\,\mathrm{d}t.\end{aligned}\tag{3.69}$$

Thus,

$$\begin{aligned}&\left|\int_{\mathbb{R}} a\left(t + \frac{s}{T}\right) a^*(t)\, e^{-j2\pi\beta T t}\,\mathrm{d}t\right| \\ &\quad \leqslant \left|\int_{\mathbb{R}} \left[a\left(t + \frac{s}{T}\right) - a(t)\right] a^*(t)\, e^{-j2\pi\beta T t}\,\mathrm{d}t\right| + \left|\int_{\mathbb{R}} |a(t)|^2\, e^{-j2\pi\beta T t}\,\mathrm{d}t\right| \\ &\quad \leqslant \int_{\mathbb{R}} \left|a\left(t + \frac{s}{T}\right) - a(t)\right| |a(t)|\,\mathrm{d}t + \left|\int_{\mathbb{R}} |a(t)|^2\, e^{-j2\pi\beta T t}\,\mathrm{d}t\right|.\end{aligned}\tag{3.70}$$

In reference to the first integral in right-hand side of (3.70) it results that

$$\left|a\left(t + \frac{s}{T}\right) - a(t)\right| |a(t)| \leqslant 2\|a\|_\infty\, |a(t)| \in L^1(\mathbb{R}). \tag{3.71}$$

That is, the integrand function is bounded by a summable function independent of T. Thus, by the dominated convergence theorem (Champeney 1990, Chapter 4) we have

$$\lim_{T\to\infty} \int_{\mathbb{R}} \left|a\left(t + \frac{s}{T}\right) - a(t)\right| |a(t)|\,\mathrm{d}t = \int_{\mathbb{R}} \lim_{T\to\infty} \left|a\left(t + \frac{s}{T}\right) - a(t)\right| |a(t)|\,\mathrm{d}t = 0 \tag{3.72}$$

since $a(t)$ is continuous a.e. (Assumption 2.4.5). In regard to the second integral in (3.70), since $|a(t)|^2 \leqslant \|a\|_\infty |a(t)| \in L^1(\mathbb{R})$ and $\beta \neq 0$, the Riemann-Lebesgue theorem (Champeney 1990, Chapter 3) can be applied:

$$\lim_{T\to\infty} \int_{\mathbb{R}} |a(t)|^2 \, e^{-j2\pi\beta T t} \, \mathrm{d}t = 0. \tag{3.73}$$

Therefore, for $\beta \neq 0$ one obtains

$$\lim_{T\to\infty} \int_{\mathbb{R}} a\left(t + \frac{s}{T}\right) a^*(t) \, e^{-j2\pi\beta T t} \, \mathrm{d}t = 0. \tag{3.74}$$

Analogously, for the function $r_{aa}(\beta, s)$ defined in (3.136) it results

$$\lim_{T\to\infty} r_{aa}\left(\beta T; \frac{s}{T}\right) = \int_{\mathbb{R}} a^2(t) \, \mathrm{d}t \; \delta_\beta. \tag{3.75}$$

□

Fact 3.5.3 Signal-Tapering Window Versus Lag-Product-Tapering Window. *The data-tapering window $w_T(t)$ in Assumption 2.4.5, strictly speaking, is a lag-product-tapering window depending on the lag parameter τ. Let $h_T(t)$ be the signal-tapering window. We have*

$$y(t+\tau)\, x^{(*)}(t)\, w_{T,\tau}(t) = y(t+\tau)\, h_T(t+\tau)\, x^{(*)}(t)\, h_T^{(*)}(t) \tag{3.76}$$

That is, the lag-product-tapering window $w_{T,\tau}(t)$ can be expressed in terms of the signal-tapering window $h_T(t)$ as follows:

$$w_{T,\tau}(t) = h_T(t+\tau)\, h_T^{(*)}(t) \tag{3.77}$$

Its Fourier transform is

$$\begin{aligned} W_{\frac{1}{T},\tau}(f) &= \left[H_{\frac{1}{T}}(f)\, e^{j2\pi f\tau} \right] \underset{f}{\otimes} H_{\frac{1}{T}}^{(*)}((-)f) \\ &= \int_{\mathbb{R}} H_{\frac{1}{T}}(\lambda)\, e^{j2\pi\lambda\tau}\, H_{\frac{1}{T}}^{(*)}((-)(f-\lambda))\, \mathrm{d}\lambda \end{aligned} \tag{3.78}$$

where $H_{\frac{1}{T}}(f)$ is the Fourier transform of $h_T(t)$ and $(-)$ is an optional minus sign linked to the optional complex conjugation $(*)$.

An analogous relation can be found for the normalized window $a(t)$ in (2.125). Let

$$w_{T,\tau}(t) \triangleq \frac{1}{T} a_\tau\left(\frac{t}{T}\right) \triangleq \frac{1}{T} \bar{h}\left(\frac{t+\tau}{T}\right) \bar{h}^{(*)}\left(\frac{t}{T}\right) \tag{3.79}$$

By Fourier transforming both sides we have

$$\begin{aligned} W_{\frac{1}{T},\tau}(f) = A_\tau(fT) &= \frac{1}{T}\left[T\,\bar{H}(fT)\,e^{j2\pi f\tau}\right] \underset{f}{\otimes} T\,\bar{H}^{(*)}((-)fT) \\ &= T\int_{\mathbb{R}} \bar{H}(\lambda T)\,e^{j2\pi\lambda\tau}\,\bar{H}^{(*)}((-)(f-\lambda)T)\,\mathrm{d}\lambda \\ &= \int_{\mathbb{R}} \bar{H}(\nu)\,e^{j2\pi\nu\tau/T}\,\bar{H}^{(*)}((-)(fT-\nu))\,\mathrm{d}\nu \end{aligned} \tag{3.80}$$

where $A_\tau(f)$ and $\bar{H}(f)$ are the Fourier transforms of $a_\tau(t)$ and $\bar{h}(t)$, respectively, and in the last equality the variable change $\nu = \lambda T$ is made. Thus, the rate of convergence to zero of $A_\tau(f)$ as $|f| \to \infty$ can be expressed in terms of the rate of convergence of $\bar{H}(f)$.

The functions $r_{aa^*}(\beta, s)$ and $r_{aa}(\beta, s)$ defined in (2.133) and (3.136), respectively, can be expressed in terms of the normalized signal-tapering window.

$$\begin{aligned} r_{aa^{[*]}}(\beta, s) &\triangleq \int_{\mathbb{R}} a_\tau(t+s)\,a_\tau^{[*]}(t)\,e^{-j2\pi\beta t}\,\mathrm{d}t \\ &= \int_{\mathbb{R}} \bar{h}(t+\tau+s)\,\bar{h}^{(*)}(t+s)\,\bar{h}^{[*]}(t+\tau)\,\bar{h}^{[*](*)}(t)\,e^{-j2\pi\beta t}\,\mathrm{d}t \end{aligned} \tag{3.81}$$

where $[*]$ represents an optional complex conjugation (different from $(*)$).

For a rectangular signal-tapering window $\bar{h}(t) = \mathrm{rect}(t)$, the continuous-time lag-product tapering window and its Fourier transform are given by

$$\begin{aligned} w_{T,\tau}(t) &\triangleq \frac{1}{T}\,\mathrm{rect}\left(\frac{t+\tau}{T}\right)\,\mathrm{rect}\left(\frac{t}{T}\right) \\ &= \frac{1}{T}\,\mathrm{rect}\left(\frac{t-(T-\tau)/2}{T-|\tau|}\right)\,\mathrm{rect}\left(\frac{\tau}{2T}\right) \end{aligned} \tag{3.82}$$

$$W_{\frac{1}{T},\tau}(f) = \left(1 - \frac{|\tau|}{T}\right)\,\mathrm{rect}\left(\frac{\tau}{2T}\right)\,\mathrm{sinc}(f(T-|\tau|))\,e^{-j\pi fT}\,e^{j\pi f\tau} \tag{3.83}$$

respectively. Therefore for $|\tau| < T$

$$W_{\frac{1}{T},\tau}(0) - 1 = \left(1 - \frac{|\tau|}{T}\right)\,\mathrm{rect}\left(\frac{\tau}{2T}\right) - 1 = -\frac{|\tau|}{T}\,\mathrm{rect}\left(\frac{\tau}{2T}\right). \tag{3.84}$$

That is,

$$W_{\frac{1}{T},\tau}(0) - 1 = \mathcal{O}\left(\frac{1}{T}\right) \quad \text{as } T \to \infty \tag{3.85}$$

which should be accounted for in the computation of the bias in Theorem 2.4.12.

Analogous considerations can be made for the discrete-time lag-product-tapering window in Definition 2.6.1. Let

$$\mathcal{R}_N(n) \triangleq \begin{cases} 1 & n \in \{0, 1, \dots, N-1\} \\ 0 & \text{otherwise} \end{cases} \tag{3.86}$$

$$\mathbf{1}_{\{A\}} \triangleq \begin{cases} 1 & A \text{ is truc} \\ 0 & \text{otherwise} \end{cases} \tag{3.87}$$

be the discrete-time casual rectangular window and the indicator function, respectively. The discrete-time casual counterparts of (3.82) and (3.83) are

$$\begin{aligned} v_{N,m}(n) &\triangleq \frac{1}{N} \mathcal{R}_N(n+m)\, \mathcal{R}_N(n) \\ &= \frac{1}{N} \mathcal{R}_{N-|m|}(n)\, \mathbf{1}_{\{|m| \leqslant N-1\}} \end{aligned} \tag{3.88}$$

$$\overset{\mathcal{F}}{\longleftrightarrow} V_{\frac{1}{N},m}(\nu) = \frac{1}{N} \frac{\sin(\pi\nu(N-|m|))}{\sin(\pi\nu)}\, e^{-j\pi\nu(N-|m|-1)}\, \mathbf{1}_{\{|m| \leqslant N-1\}} \tag{3.89}$$

It results that

$$\lim_{\nu \to 0} V_{\frac{1}{N},m}(\nu) = \frac{1}{N}(N-|m|)\, \mathbf{1}_{\{|m| \leqslant N-1\}} = \left(1 - \frac{|m|}{N}\right) \mathbf{1}_{\{|m| \leqslant N-1\}}. \tag{3.90}$$

Thus, for $N \to \infty$,

$$V_{\frac{1}{N},m}(0) - 1 = \mathcal{O}(N^{-1}). \tag{3.91}$$

3.5.1 Proof of Theorem 2.4.11 Asymptotic Expected Value of the Cyclic Cross-Correlogram

From (2.128) it follows that

$$\begin{aligned} &\lim_{T \to \infty} \mathrm{E}\left\{ R_{yx^{(*)}}(\alpha, \tau; t_0, T) \right\} \\ &\quad = \lim_{T \to \infty} \sum_{n \in \mathbb{I}} R^{(n)}_{yx^{(*)}}(\tau)\, W_{\frac{1}{T}}(\alpha - \alpha_n(\tau))\, e^{-j2\pi[\alpha - \alpha_n(\tau)]t_0} \\ &\quad = \sum_{n \in \mathbb{I}} R^{(n)}_{yx^{(*)}}(\tau) \lim_{T \to \infty} W_{\frac{1}{T}}(\alpha - \alpha_n(\tau))\, e^{-j2\pi[\alpha - \alpha_n(\tau)]t_0} \\ &\quad = \sum_{n \in \mathbb{I}} R^{(n)}_{yx^{(*)}}(\tau)\, \delta_{\alpha - \alpha_n(\tau)} \end{aligned} \tag{3.92}$$

where, in the last equality, (3.61) is used. Thus, accounting for (2.39), (2.143) immediately follows.

In (3.92), the interchange of limit and sum operations is justified since the function series

$$\sum_{n \in \mathbb{I}} R^{(n)}_{yx^{(*)}}(\tau)\, W_{\frac{1}{T}}(\alpha - \alpha_n(\tau))\, e^{-j2\pi[\alpha - \alpha_n(\tau)]t_0}$$

is uniformly convergent. In fact, from Lemma 3.5.1 the Fourier transform $A(f)$ of $a(t)$ is continuous and bounded. Hence, accounting for Assumption 2.4.3a, one obtains

$$\begin{aligned}\sum_{n\in\mathbb{I}} \left|R^{(n)}_{yx^{(*)}}(\tau)\right| \left|W_{\frac{1}{T}}(\alpha-\alpha_n(\tau))\right| &\leqslant \|A\|_\infty \sum_{n\in\mathbb{I}} \left|R^{(n)}_{yx^{(*)}}(\tau)\right| \\ &\leqslant \|A\|_\infty \sum_{n\in\mathbb{I}} \left\|R^{(n)}_{yx^{(*)}}\right\|_\infty \end{aligned} \tag{3.93}$$

with the right-hand side bounded and not depending on T. Thus, the function series is uniformly convergent due to the Weierstrass criterium (Johnsonbaugh and Pfaffenberger 2002, Theorem 62.6; Smirnov 1964).

3.5.2 Proof of Theorem 2.4.12 Rate of Convergence of the Bias of the Cyclic Cross-Correlogram

Accounting for (2.141), the cyclic cross-correlation function (2.39) can be written as

$$R_{yx^{(*)}}(\alpha,\tau) = \sum_{n\in\mathbb{I}} R^{(n)}_{yx^{(*)}}(\tau)\,\delta_{\alpha-\alpha_n(\tau)} = \sum_{n\in\mathbb{J}_{\alpha,\tau}} R^{(n)}_{yx^{(*)}}(\tau). \tag{3.94}$$

Moreover, from expression (2.128) of the expected value of the cyclic cross-correlogram, we have

$$\begin{aligned}\mathrm{E}\left\{R_{yx^{(*)}}(\alpha,\tau;t_0,T)\right\} = &\sum_{n\in\mathbb{J}_{\alpha,\tau}} R^{(n)}_{yx^{(*)}}(\tau)\,W_{\frac{1}{T}}(0) \\ &+ \sum_{n\notin\mathbb{J}_{\alpha,\tau}} R^{(n)}_{yx^{(*)}}(\tau)\,W_{\frac{1}{T}}(\alpha-\alpha_n(\tau))\,e^{-j2\pi[\alpha-\alpha_n(\tau)]t_0}\end{aligned} \tag{3.95}$$

and, hence,

$$\begin{aligned}&\mathrm{E}\left\{R_{yx^{(*)}}(\alpha,\tau;t_0,T)\right\} - R_{yx^{(*)}}(\alpha,\tau) \\ &\quad= \sum_{n\notin\mathbb{J}_{\alpha,\tau}} R^{(n)}_{yx^{(*)}}(\tau)\,W_{\frac{1}{T}}(\alpha-\alpha_n(\tau))\,e^{-j2\pi[\alpha-\alpha_n(\tau)]t_0} \\ &\quad\quad+\left[\sum_{n\in\mathbb{J}_{\alpha,\tau}} R^{(n)}_{yx^{(*)}}(\tau)\left(W_{\frac{1}{T}}(0)-1\right)\right] \\ &\quad= \sum_{n\notin\mathbb{J}_{\alpha,\tau}} R^{(n)}_{yx^{(*)}}(\tau)\,W_{\frac{1}{T}}(\alpha-\alpha_n(\tau))\,e^{-j2\pi[\alpha-\alpha_n(\tau)]t_0}\end{aligned} \tag{3.96}$$

where the fact that $W_{\frac{1}{T}}(0) = A(0) = 1$ (see (2.126)) is used. Therefore,

$$\begin{aligned}&\lim_{T\to\infty} T^\gamma\left[\mathrm{E}\left\{R_{yx^{(*)}}(\alpha,\tau;t_0,T)\right\} - R_{yx^{(*)}}(\alpha,\tau)\right] \\ &\quad= \lim_{T\to\infty} T^\gamma\left[\sum_{n\notin\mathbb{J}_{\alpha,\tau}} R^{(n)}_{yx^{(*)}}(\tau)\,W_{\frac{1}{T}}(\alpha-\alpha_n(\tau))\,e^{-j2\pi[\alpha-\alpha_n(\tau)]t_0}\right] \\ &\quad= \sum_{n\notin\mathbb{J}_{\alpha,\tau}} R^{(n)}_{yx^{(*)}}(\tau)\left[\lim_{T\to\infty} T^\gamma W_{\frac{1}{T}}(\alpha-\alpha_n(\tau))\right]e^{-j2\pi[\alpha-\alpha_n(\tau)]t_0}.\end{aligned} \tag{3.97}$$

Due to Assumption 2.4.5, there exits $\gamma > 0$ such that (3.62) holds. Then, accounting for (3.62) and (3.63), it follows that for $n \notin \mathbb{J}_{\alpha,\tau}$, that is $\alpha_n(\tau) \neq \alpha$, one obtains

$$\begin{aligned}
&\lim_{T\to\infty} T^\gamma W_{\frac{1}{T}}(\alpha - \alpha_n(\tau)) \\
&\quad = \lim_{T\to\infty} T^\gamma A((\alpha - \alpha_n(\tau))T) \\
&\quad = \lim_{T\to\infty} T^\gamma \,\mathcal{O}\left(\frac{1}{T^\gamma}\right) \\
&\quad = \mathcal{O}(1).
\end{aligned} \tag{3.98}$$

From (3.97), (3.98), and Assumption 2.4.3a, (2.145) immediately follows.

In (3.97), the interchange of limit and sum operations can be justified by the Weierstrass criterium (Johnsonbaugh and Pfaffenberger 2002, Theorem 62.6; Smirnov 1964). In fact, accounting for (2.142), from (3.62) it follows that

$$A\left((\alpha - \alpha_n(\tau))T\right) = \mathcal{O}\left(\frac{1}{h_{\alpha,\tau}^\gamma\, T^\gamma}\right) \quad \text{as } T \to \infty \tag{3.99}$$

uniformly with respect to n, $\forall n \notin \mathbb{J}_{\alpha,\tau}$. That is, there exists $T_{\alpha,\tau}$ independent of n, such that for $T > T_{\alpha,\tau}$ one obtains

$$\left|W_{\frac{1}{T}}(\alpha - \alpha_n(\tau))\right| = \left|A\left((\alpha - \alpha_n(\tau))T\right)\right| \leqslant \frac{K_{\alpha,\tau}}{T^\gamma} \qquad \forall n \notin \mathbb{J}_{\alpha,\tau} \tag{3.100}$$

for some $K_{\alpha,\tau}$ independent of n. Hence,

$$\begin{aligned}
&\sum_{n \notin \mathbb{J}_{\alpha,\tau}} \left\| T^\gamma R_{yx^{(*)}}^{(n)}(\tau)\, W_{\frac{1}{T}}(\alpha - \alpha_n(\tau))\, e^{-j2\pi[\alpha-\alpha_n(\tau)]t_0} \right\|_\infty \\
&\quad \leqslant K_{\alpha,\tau} \sum_{n \notin \mathbb{J}_{\alpha,\tau}} \left| R_{yx^{(*)}}^{(n)}(\tau) \right| \\
&\quad \leqslant K_{\alpha,\tau} \sum_{n \in \mathbb{I}} \left\| R_{yx^{(*)}}^{(n)} \right\|_\infty < \infty
\end{aligned} \tag{3.101}$$

where the first L^∞-norm is for functions of T and the second for functions of τ, and, in the last inequality, Assumption 2.4.3a has been accounted for. Therefore, for the Weierstrass criterium, the series of functions of T

$$\sum_{n \notin \mathbb{J}_{\alpha,\tau}} T^\gamma\, R_{yx^{(*)}}^{(n)}(\tau)\, W_{\frac{1}{T}}(\alpha - \alpha_n(\tau))\, e^{-j2\pi[\alpha-\alpha_n(\tau)]t_0} \tag{3.102}$$

is uniformly convergent and the order of limit and sum operations in (3.97) can be reversed.

Note that, in general, $K_{\alpha,\tau}$ depends on both α and τ. That is, the convergence of the bias is not uniform with respect to α and τ.

As a final remark, results in Fact 3.5.3 should be used for a refinement of this proof in the case of rectangular signal-tapering window. In fact, from (3.84) it follows that also the term

$$\sum_{n \in \mathbb{J}_{\alpha,\tau}} R^{(n)}_{yx^{(*)}}(\tau) \left(W_{\frac{1}{T}}(0) - 1 \right) = -\frac{|\tau|}{T} \operatorname{rect}\left(\frac{\tau}{2T}\right) \sum_{n \in \mathbb{J}_{\alpha,\tau}} R^{(n)}_{yx^{(*)}}(\tau) = \mathcal{O}\left(\frac{1}{T}\right) \tag{3.103}$$

should be accounted for in the computation of the bias in (3.96)–(3.98). That is, from (3.96)–(3.98) it follows

$$T^{\gamma} |\text{bias}| = \mathcal{O}(1) + T^{\gamma} \mathcal{O}\left(\frac{1}{T}\right) = \mathcal{O}(1) \tag{3.104}$$

when $\gamma = 1$ as for the rectangular signal-tapering window.

3.5.3 *Proof of Theorem 2.4.13 Asymptotic Covariance of the Cyclic Cross-Correlogram*

Let us consider the covariance expression (2.129). As regards the term $\mathcal{T}_1$ in (2.130), defined

$$\begin{aligned} F_{n'n''}(s;T) \triangleq\ & R^{(n')}_{yy^*}(\tau_1 - \tau_2 + s)\, R^{(n'')}_{x^{(*)}x^{(*)*}}(s) \\ & e^{j2\pi\alpha'_{n'}(\tau_1-\tau_2+s)(t_2+\tau_2)}\, e^{j2\pi\alpha''_{n''}(s)t_2}\, e^{-j2\pi\alpha_1\cdot(t_2+s)}\, e^{j2\pi\alpha_2 t_2} \\ & r_{aa^*}\left(-[\alpha'_{n'}(\tau_1 - \tau_2 + s) + \alpha''_{n''}(s) - \alpha_1 + \alpha_2]T; \frac{t_2 - t_1 + s}{T} \right) \end{aligned} \tag{3.105}$$

one obtains

$$\begin{aligned} \lim_{T\to\infty} T\, \mathcal{T}_1 &= \lim_{T\to\infty} \sum_{n'} \sum_{n''} \int_{\mathbb{R}} F_{n'n''}(s;T)\, \mathrm{d}s \\ &= \sum_{n'} \sum_{n''} \lim_{T\to\infty} \int_{\mathbb{R}} F_{n'n''}(s;T)\, \mathrm{d}s \\ &= \sum_{n'} \sum_{n''} \int_{\mathbb{R}} \lim_{T\to\infty} F_{n'n''}(s;T)\, \mathrm{d}s \\ &= \mathcal{T}'_1 \end{aligned} \tag{3.106}$$

where $\mathcal{T}'_1$ is defined in (2.147) and Lemma 3.5.2 has been accounted for.

The interchange of the order of limit and sum operations in (3.106) is allowed since the double series over n' and n'' of functions of T is uniformly convergent. In fact, we have

$$\begin{aligned} & \left| \int_{\mathbb{R}} F_{n'n''}(s;T)\, \mathrm{d}s \right| \\ & \quad \leqslant \int_{\mathbb{R}} \left| R^{(n')}_{yy^*}(\tau_1 - \tau_2 + s)\, R^{(n'')}_{x^{(*)}x^{(*)*}}(s) \right| \\ & \qquad \left| r_{aa^*}\left(-[\alpha'_{n'}(\tau_1 - \tau_2 + s) + \alpha''_{n''}(s) - \alpha_1 + \alpha_2]T; \frac{t_2 - t_1 + s}{T} \right) \right| \mathrm{d}s \\ & \quad \leqslant \left\| R^{(n'')}_{x^{(*)}x^{(*)*}} \right\|_{\infty} \int_{\mathbb{R}} \left| R^{(n')}_{yy^*}(\tau_1 - \tau_2 + s) \right| \mathrm{d}s\, \|a\|_{\infty} \int_{\mathbb{R}} |a(u)|\, \mathrm{d}u \end{aligned} \tag{3.107}$$

with the right-hand side not depending on T. In the last inequality in (3.107), the following inequality, which is a consequence of Assumption 2.4.5, is used

$$\begin{aligned}\left|r_{aa^*}\left(\beta T; \frac{s}{T}\right)\right| &= \left|\int_{\mathbb{R}} a\left(t + \frac{s}{T}\right) a^*(t)\, e^{-j2\pi\beta T t}\, \mathrm{d}t\right| \\ &\leqslant \int_{\mathbb{R}} \left|a\left(t + \frac{s}{T}\right) a^*(t)\right| \mathrm{d}t \\ &\leqslant \|a\|_\infty \int_{\mathbb{R}} |a(t)|\, \mathrm{d}t. \end{aligned} \tag{3.108}$$

Consequently,

$$\begin{aligned}\sum_{n'}\sum_{n''} \left\| \int_{\mathbb{R}} F_{n'n''}(s; \cdot)\, \mathrm{d}s \right\|_\infty &\leqslant \sum_{n'} \int_{\mathbb{R}} \left| R^{(n')}_{yy^*}(\tau_1 - \tau_2 + s) \right| \mathrm{d}s \sum_{n''} \left\| R^{(n'')}_{x^{(*)}x^{(*)*}} \right\|_\infty \\ &\quad \|a\|_\infty \int_{\mathbb{R}} |a(u)|\, \mathrm{d}u < \infty \end{aligned} \tag{3.109}$$

where, in the inequality, Assumptions 2.4.3a and 2.4.8a have been used. Then, from (3.109), it follows that the series of function of T

$$\sum_{n'}\sum_{n''} \int_{\mathbb{R}} F_{n'n''}(s; T)\, \mathrm{d}s$$

is uniformly convergent for the Weierstrass criterium (Johnsonbaugh and Pfaffenberger 2002, Theorem 62.6; Smirnov 1964).

The interchange of limit and integral operations in (3.106) is allowed by the dominated convergence theorem (Champeney 1990, Chapter 4). In fact, from (3.105) and (3.107) it follows that

$$|F_{n'n''}(s; T)| \leqslant \left| R^{(n')}_{yy^*}(\tau_1 - \tau_2 + s) \right| \left\| R^{(n'')}_{x^{(*)}x^{(*)*}} \right\|_\infty \|a\|_\infty \int_{\mathbb{R}} |a(u)|\, \mathrm{d}u. \tag{3.110}$$

That is, accounting for Assumption 2.4.8a, the integrand function in the second line of (3.106) is bounded by a summable function of s not depending on T.

Analogously, it can be shown that

$$\lim_{T\to\infty} T\, \mathcal{T}_2 = \mathcal{T}_2' \tag{3.111}$$

where $\mathcal{T}_2'$ is defined in (2.148).

As regards the term $\mathcal{T}_3$ defined in (2.132), defined

$$\begin{aligned} F_n(s; T) \triangleq\; & C^{(n)}_{yx^{(*)}y^*x^{(*)*}}(\tau_1 + s, s, \tau_2)\, e^{j2\pi[\beta_n(\tau_1+s,s,\tau_2)-\alpha_1+\alpha_2]t_2}\, e^{-j2\pi\alpha_1 s} \\ & r_{aa^*}\left(-[\beta_n(\tau_1 + s, s, \tau_2) - \alpha_1 + \alpha_2]T; \frac{t_2 - t_1 + s}{T}\right) \end{aligned} \tag{3.112}$$

one obtains

$$\begin{aligned}
\lim_{T\to\infty} T\,\mathcal{T}_3 &= \lim_{T\to\infty} \sum_n \int_{\mathbb{R}} F_n(s;T)\,\mathrm{d}s \\
&= \sum_n \lim_{T\to\infty} \int_{\mathbb{R}} F_n(s;T)\,\mathrm{d}s \\
&= \sum_n \int_{\mathbb{R}} \lim_{T\to\infty} F_n(s;T)\,\mathrm{d}s \\
&= \mathcal{T}_3'
\end{aligned} \tag{3.113}$$

where $\mathcal{T}_3'$ is defined in (2.149), and Lemma 3.5.2 has been accounted for.

The interchange of the order of limit and sum operations in (3.113) is allowed since the series over n of functions of T is uniformly convergent. In fact, accounting for (3.108), we have

$$\begin{aligned}
\left|\int_{\mathbb{R}} F_n(s;T)\,\mathrm{d}s\right| &\leqslant \int_{\mathbb{R}} \left|C^{(n)}_{yx^{(*)}y^*x^{(*)*}}(\tau_1+s,s,\tau_2)\right| \\
&\qquad \left|r_{aa^*}\left(-[\beta_n(\tau_1+s,s,\tau_2)-\alpha_1+\alpha_2]T;\frac{t_2-t_1+s}{T}\right)\right|\,\mathrm{d}s \\
&\leqslant \int_{\mathbb{R}} \left|C^{(n)}_{yx^{(*)}y^*x^{(*)*}}(\tau_1+s,s,\tau_2)\right|\,\mathrm{d}s\,\|a\|_\infty \int_{\mathbb{R}} |a(u)|\,\mathrm{d}u
\end{aligned} \tag{3.114}$$

with the right-hand side not depending on T. Consequently,

$$\begin{aligned}
\sum_n \left\|\int_{\mathbb{R}} F_n(s;\cdot)\,\mathrm{d}s\right\|_\infty &\leqslant \sum_n \int_{\mathbb{R}} \left|C^{(n)}_{yx^{(*)}y^*x^{(*)*}}(\tau_1+s,s,\tau_2)\right|\,\mathrm{d}s \\
&\qquad \|a\|_\infty \int_{\mathbb{R}} |a(u)|\,\mathrm{d}u < \infty
\end{aligned} \tag{3.115}$$

where, in the last inequality, Assumption 2.4.8b has been used. Then, from (3.115), it follows that the series of functions of T

$$\sum_n \int_{\mathbb{R}} F_n(s;T)\,\mathrm{d}s$$

is uniformly convergent for the Weierstrass criterium (Johnsonbaugh and Pfaffenberger 2002, Theorem 62.6; Smirnov 1964).

The interchange of limit and integral operations in (3.113) is allowed by the dominated convergence theorem (Champeney 1990, Chapter 4). In fact, from (3.112) and (3.114) it follows that

$$|F_n(s;T)| \leqslant \left|C^{(n)}_{yx^{(*)}y^*x^{(*)*}}(\tau_1+s,s,\tau_2)\right|\,\|a\|_\infty \int_{\mathbb{R}} |a(u)|\,\mathrm{d}u. \tag{3.116}$$

That is, accounting for Assumption 2.4.8b, the integrand function in the second line of (3.113) is bounded by a summable function of s not depending on T.

The proof of Theorem 2.4.13 can be carried out with minor changes by substituting Assumption 2.4.8a with the following:

Assumption 3.5.4 *For any choice of* $z_1, \dots, z_4$ *in* $\{x, x^*, y, y^*\}$ *and* $\forall \tau_1, \tau_2 \in \mathbb{R}$ *it results*

$$\sum_{n' \in \mathbb{I}_{z_1, z_2}} \sum_{n'' \in \mathbb{I}_{z_3, z_4}} \int_{\mathbb{R}} \left| R^{(n')}_{z_1 z_2}(s + \tau_1)\, R^{(n'')}_{z_3 z_4}(s + \tau_2) \right| \mathrm{d}s < \infty \tag{3.117}$$

□

By reasoning as in the comments following Assumption 2.4.8a, we have that, accounting for (2.121), the function series $\sum_n |R^{(n)}_{z_i z_j}(s)|$, $i, j \in \{1, \dots, 4\}$, are uniformly convergent due to the Weierstrass criterium (Johnsonbaugh and Pfaffenberger 2002, Theorem 62.6; Smirnov 1964). In addition, in order to satisfy (3.117), the function $\sum_n |R^{(n)}_{z_i z_j}(s)|$ should be vanishing sufficiently fast as $|s| \to \infty$ so that the product $\sum_{n'} |R^{(n')}_{z_1 z_2}(s + \tau_1)| \sum_{n''} |R^{(n'')}_{z_3 z_4}(s + \tau_2)|$ is summable. A sufficient condition is that there exists $\epsilon > 0$ such that $\sum_n |R^{(n)}_{z_i z_j}(s)| = \mathcal{O}(|s|^{-1/2-\epsilon/2})$ and, hence, the result is that $\sum_{n'} |R^{(n')}_{z_1 z_2}(s + \tau_1)| \sum_{n''} |R^{(n'')}_{z_3 z_4}(s + \tau_2)| = \mathcal{O}(|s|^{-1-\epsilon})$. In such a case, the process memory can vanish more slowly than as required to satisfy Assumption 2.4.8a (see the comments following Assumption 2.4.8).

If Assumption 3.5.4 is made instead of Assumption 2.4.8a, then (3.107), (3.109), and (3.110) should be replaced by

$$\left| \int_{\mathbb{R}} F_{n'n''}(s; T)\, \mathrm{d}s \right| \leqslant \int_{\mathbb{R}} \left| R^{(n')}_{yy^*}(\tau_1 - \tau_2 + s)\, R^{(n'')}_{x^{(*)} x^{(*)*}}(s) \right| \mathrm{d}s\, \|a\|_\infty \int_{\mathbb{R}} |a(u)|\, \mathrm{d}u \tag{3.118}$$

$$\sum_{n'} \sum_{n''} \left\| \int_{\mathbb{R}} F_{n'n''}(s; \cdot)\, \mathrm{d}s \right\|_\infty \leqslant \sum_{n'} \sum_{n''} \int_{\mathbb{R}} \left| R^{(n')}_{yy^*}(\tau_1 - \tau_2 + s)\, R^{(n'')}_{x^{(*)} x^{(*)*}}(s) \right| \mathrm{d}s\, \|a\|_\infty \int_{\mathbb{R}} |a(u)|\, \mathrm{d}u < \infty \tag{3.119}$$

$$|F_{n'n''}(s; T)| \leqslant \left| R^{(n')}_{yy^*}(\tau_1 - \tau_2 + s)\, R^{(n'')}_{x^{(*)} x^{(*)*}}(s) \right| \|a\|_\infty \int_{\mathbb{R}} |a(u)|\, \mathrm{d}u \tag{3.120}$$

respectively.

3.5.4 Proof of Corollary 2.4.14

Let us consider the term $\mathcal{T}_1$ defined in (2.130). One has

$$\begin{aligned} T\, |\mathcal{T}_1| &\leqslant \sum_{n'} \sum_{n''} \int_{\mathbb{R}} \left| R^{(n')}_{yy^*}(\tau_1 - \tau_2 + s)\, R^{(n'')}_{x^{(*)} x^{(*)*}}(s) \right| \\ &\quad \left| r_{aa^*}\left(-[\alpha'_{n'}(\tau_1 - \tau_2 + s) + \alpha''_{n''}(s) - \alpha_1 + \alpha_2] T; \frac{t_2 - t_1 + s}{T} \right) \right| \mathrm{d}s \\ &\leqslant \sum_{n'} \left\| R^{(n')}_{yy^*} \right\|_\infty \sum_{n''} \int_{\mathbb{R}} \left| R^{(n'')}_{x^{(*)} x^{(*)*}}(s) \right| \mathrm{d}s\, \|a\|_\infty \int_{\mathbb{R}} |a(t)|\, \mathrm{d}t \end{aligned} \tag{3.121}$$

where, in the second inequality, accounting for (2.133), the inequality

$$|r_{aa^*}(\beta;s)| \leqslant \|a\|_\infty \int_{\mathbb{R}} |a(t)|\, \mathrm{d}t \tag{3.122}$$

is used. The right-hand side of (3.121) is bounded due to Assumptions 2.4.3a, 2.4.5, and 2.4.8a. Analogously, with reference to the term $\mathcal{T}_2$ defined in (2.131), we get

$$T\,|\mathcal{T}_2| \leqslant \sum_{n'''} \left\| R^{(n''')}_{yx^{(*)*}} \right\|_\infty \sum_{n^{iv}} \int_{\mathbb{R}} \left| R^{(n^{iv})}_{x^{(*)}y^*}(s-\tau_2) \right| \mathrm{d}s\, \|a\|_\infty \int_{\mathbb{R}} |a(t)|\, \mathrm{d}t \tag{3.123}$$

with the right-hand side bounded and independent of τ_2 since the integral over $\mathbb{R}$ does not depend on shifts of the integrand function. Finally, with reference to the term $\mathcal{T}_3$ defined in (2.132), one has

$$T\,|\mathcal{T}_3| \leqslant \sum_{n} \int_{\mathbb{R}} \left| C^{(n)}_{yx^{(*)}y^*x^{(*)*}}(\tau_1+s, s, \tau_2) \right| \mathrm{d}s\, \|a\|_\infty \int_{\mathbb{R}} |a(t)|\, \mathrm{d}t \tag{3.124}$$

with the right-hand side uniformly bounded with respect to τ_1 and τ_2 due to Assumption 2.4.9.

3.6 Proofs for Section 2.4.3 "Asymptotic Normality of the Cyclic Cross-Correlogram"

In this section, proofs of results presented in Section 2.4.3 on the asymptotic Normality of the cyclic cross-correlogram are reported.

3.6.1 *Proof of Lemma 2.4.17*

By using (2.118), (2.125), and the multilinearity property of cumulants we have

$$\begin{aligned}
&\operatorname{cum}\left\{ R^{[*]_1}_{yx^{(*)}}(\alpha_1, \tau_1; t_1, T), \ldots, R^{[*]_k}_{yx^{(*)}}(\alpha_k, \tau_k; t_k, T) \right\} \\
&\quad = \operatorname{cum}\left\{ \int_{\mathbb{R}} w_T^{[*]_1}(u_1 - t_1)\, z_1(u_1)\, e^{-[-]_1 j2\pi\alpha_1 u_1}\, \mathrm{d}u_1, \right. \\
&\qquad \left. \ldots, \int_{\mathbb{R}} w_T^{[*]_k}(u_k - t_k)\, z_k(u_k)\, e^{-[-]_k j2\pi\alpha_k u_k}\, \mathrm{d}u_k \right\} \\
&\quad = \int_{\mathbb{R}} \cdots \int_{\mathbb{R}} \operatorname{cum}\{z_1(u_1), \ldots, z_k(u_k)\} \\
&\qquad w_T^{[*]_1}(u_1 - t_1) \cdots w_T^{[*]_k}(u_k - t_k)\, e^{-[-]_1 j2\pi\alpha_1 u_1} \cdots e^{-[-]_k j2\pi\alpha_k u_k}\, \mathrm{d}u_1 \cdots \mathrm{d}u_k \\
&\quad = \int_{\mathbb{R}^k} \operatorname{cum}\{z_k(u), z_i(u+s_i),\ i = 1, \ldots, k-1\} \\
&\qquad \prod_{i=1}^{k-1} \frac{1}{T} a^{[*]_i}\left(\frac{u + s_i - t_i}{T} \right) e^{-[-]_i j2\pi\alpha_i \cdot (u+s_i)} \\
&\qquad \frac{1}{T} a^{[*]_k}\left(\frac{u - t_k}{T} \right) e^{-[-]_k j2\pi\alpha_k u}\, \mathrm{d}s_1 \cdots \mathrm{d}s_{k-1}\, \mathrm{d}u
\end{aligned} \tag{3.125}$$

where $[-]_i$ is an optional minus sign which is linked to the optional complex conjugation $[*]_i$ and in the third equality the variable changes $u_k = u$, $u_i = u + s_i$, $i = 1, \ldots, k-1$ are made.

Thus,

$$
\begin{aligned}
&\left|\operatorname{cum}\left\{R^{[*]_1}_{yx^{(*)}}(\alpha_1, \tau_1; t_1, T), \ldots, R^{[*]_k}_{yx^{(*)}}(\alpha_k, \tau_k; t_k, T)\right\}\right| \\
&\quad \leqslant \frac{\|a\|_\infty^{k-1}}{T^{k-1}} \int_{\mathbb{R}^k} \left|C_{z_1 \cdots z_k}(u, s_1, \ldots, s_{k-1})\right| \frac{1}{T}\left|a\left(\frac{u-t_k}{T}\right)\right| \mathrm{d}s_1 \cdots \mathrm{d}s_{k-1}\,\mathrm{d}u \\
&\quad \leqslant \frac{\|a\|_\infty^{k-1}}{T^{k-1}} \int_{\mathbb{R}^{k-1}} \phi(s_1, \ldots, s_{k-1})\,\mathrm{d}s_1 \cdots \mathrm{d}s_{k-1} \int_{\mathbb{R}} |a(s)|\,\mathrm{d}s
\end{aligned} \tag{3.126}
$$

where in the second inequality the variable change $s = (u - t_k)/T$ is made and Assumption 2.4.15 is used.

Therefore, from (3.126), accounting for Assumptions 2.4.5 and 2.4.15, it immediately follows that, for every $k \geqslant 2$ and every $\epsilon > 0$, (2.168) holds.

The interchange of $\operatorname{cum}\{\cdot\}$ and integral operators in the second equality in (3.125) is allowed by the Fubini and Tonelli theorem (Champeney 1990, Chapter 3). In fact, by using Assumption 2.4.5 and the expression of a cumulant in terms of moments (1.209), (2.82b), the integrand function in the third term of equality (3.125) can be written as

$$
\sum_{\mathrm{P}} (-1)^{p-1}(p-1)! \prod_{i=1}^{p} \mathrm{E}\Big\{\prod_{\ell \in \mu_i} z_\ell(u_\ell)\Big\} \prod_{\ell=1}^{k} \frac{1}{T} a^{[*]_\ell}\left(\frac{u_\ell - t_\ell}{T}\right) e^{-j[-]_\ell 2\pi\alpha_\ell u_\ell}.
$$

Furthermore,

$$
\begin{aligned}
&\int_{\mathbb{R}^k} \sum_{\mathrm{P}} (-1)^{p-1}(p-1)! \prod_{i=1}^{p} \mathrm{E}\Big\{\prod_{\ell \in \mu_i} |z_\ell(u_\ell)|\Big\} \prod_{\ell=1}^{k} \frac{1}{T}\left|a\left(\frac{u_\ell - t_\ell}{T}\right)\right| \mathrm{d}u_1 \cdots \mathrm{d}u_k \\
&\quad = \sum_{\mathrm{P}} (-1)^{p-1}(p-1)! \prod_{i=1}^{p} \int_{\mathbb{R}^{|\mu_i|}} \mathrm{E}\Big\{\prod_{\ell \in \mu_i} |z_\ell(u_\ell)|\Big\} \prod_{\ell \in \mu_i} \frac{1}{T}\left|a\left(\frac{u_\ell - t_\ell}{T}\right)\right| \prod_{\ell \in \mu_i} \mathrm{d}u_\ell \\
&\quad \leqslant \sum_{\mathrm{P}} (-1)^{p-1}(p-1)! \prod_{i=1}^{p} M_{\mu_i} \prod_{\ell \in \mu_i} \frac{1}{T} \int_{\mathbb{R}} \left|a\left(\frac{u_\ell - t_\ell}{T}\right)\right| \mathrm{d}u_\ell < \infty
\end{aligned} \tag{3.127}
$$

where $|\mu_i|$ is the number of elements of μ_i, the fact that $\mu_i \cap \mu_j = \emptyset$ for $i \neq j$ is used and, accounting for Assumption 2.4.16, $M_{\mu_i} \triangleq M_{\ell_1, \ldots, \ell_n}$ if $\mu_i \triangleq \{\ell_1, \ldots, \ell_n\}$.

3.6.2 *Proof of Theorem 2.4.18 Asymptotic Joint Normality of the Cyclic Cross-Correlograms*

From Theorem 2.4.12 holding for $\gamma > \frac{1}{2}$ we have

$$
\lim_{T \to \infty} \mathrm{E}\left\{\sqrt{T}\left[R_{yx^{(*)}}(\alpha, \tau; t_0, T) - R_{yx^{(*)}}(\alpha, \tau)\right]\right\} = 0. \tag{3.128}
$$

From Theorem 2.4.13 it follows that the asymptotic covariance

$$\begin{aligned}
&\lim_{T\to\infty} \operatorname{cov}\left\{\sqrt{T}\, R_{yx^{(*)}}(\alpha_1, \tau_1; t_1, T), \sqrt{T}\, R_{yx^{(*)}}(\alpha_2, \tau_2; t_2, T)\right\} \\
&\qquad = \lim_{T\to\infty} \operatorname{cov}\left\{\sqrt{T}\left[R_{yx^{(*)}}(\alpha_1, \tau_1; t_1, T) - R_{yx^{(*)}}(\alpha_1, \tau_1)\right],\right. \\
&\qquad\qquad \left.\sqrt{T}\left[R_{yx^{(*)}}(\alpha_2, \tau_2; t_2, T) - R_{yx^{(*)}}(\alpha_2, \tau_2)\right]\right\}
\end{aligned} \tag{3.129}$$

is finite. Analogously, from Theorem 3.7.2 it follows that the asymptotic conjugate covariance is finite. Moreover, from Lemma 2.4.17 with $\epsilon = \frac{k}{2} - 1$ and $k \geqslant 3$ in (2.168), we have

$$\begin{aligned}
&\lim_{T\to\infty} T^{\frac{k}{2}} \operatorname{cum}\left\{R^{[*]_1}_{yx^{(*)}}(\alpha_1, \tau_1; t_1, T), \ldots, R^{[*]_k}_{yx^{(*)}}(\alpha_k, \tau_k; t_k, T)\right\} \\
&\quad = \lim_{T\to\infty} \operatorname{cum}\left\{\sqrt{T}\, R^{[*]_1}_{yx^{(*)}}(\alpha_1, \tau_1; t_1, T), \ldots, \sqrt{T}\, R^{[*]_k}_{yx^{(*)}}(\alpha_k, \tau_k; t_k, T)\right\} = 0
\end{aligned} \tag{3.130}$$

Since the value of the cumulant does not change by adding a constant to each of the random variables (Brillinger 1981, Theorem 2.3.1), we also have

$$\begin{aligned}
&\lim_{T\to\infty} \operatorname{cum}\left\{\sqrt{T}\left[R_{yx^{(*)}}(\alpha_1, \tau_1; t_1, T) - R_{yx^{(*)}}(\alpha_1, \tau_1)\right]^{[*]_1}, \ldots,\right. \\
&\qquad\qquad \left.\sqrt{T}\left[R_{yx^{(*)}}(\alpha_k, \tau_k; t_k, T) - R_{yx^{(*)}}(\alpha_k, \tau_k)\right]^{[*]_k}\right\} = 0.
\end{aligned} \tag{3.131}$$

That is, according to the results of Section 1.4.2, for every fixed α_i, τ_i, t_i, the random variables

$$\sqrt{T}\;\left[R_{yx^{(*)}}(\alpha_i, \tau_i; t_i, T) - R_{yx^{(*)}}(\alpha_i, \tau_i)\right]$$

$i = 1, \ldots, k$ are asymptotically ($T \to \infty$) zero-mean jointly complex Normal (Picinbono 1996; van den Bos 1995).

3.7 Conjugate Covariance

For complex-valued processes, both covariance and conjugate covariance are needed for a complete second-order wide-sense characterization (Picinbono 1996; Picinbono and Bondon 1997; Schreier and Scharf 2003a,b).

In this section, results analogous to those stated in Theorems 2.4.7 and 2.4.13 are presented for the conjugate covariance of the cyclic cross-correlogram.

By reasoning as for Theorem 2.4.7, we get the following result.

Theorem 3.7.1 Conjugate Covariance of the Cyclic Cross-Correlogram. *Let $y(t)$ and $x(t)$ be zero-mean wide-sense jointly GACS stochastic processes with cross-correlation* *function (2.31c). Under Assumptions 2.4.2–2.4.5, the conjugate covariance of the cyclic cross-correlogram $R_{yx^{(*)}}(\alpha, \tau; t_0, T)$ is given by*

$$\operatorname{cov}\left\{R_{yx^{(*)}}(\alpha_1, \tau_1; t_1, T), R^*_{yx^{(*)}}(\alpha_2, \tau_2; t_2, T)\right\} = \bar{\mathcal{T}}_1 + \bar{\mathcal{T}}_2 + \bar{\mathcal{T}}_3 \tag{3.132}$$

where

$$\begin{aligned}\bar{\mathcal{T}}_1 \triangleq \sum_{n'}\sum_{n''}\int_{\mathbb{R}} & R^{(n')}_{yy}(\tau_1 - \tau_2 + s)\, R^{(n'')}_{x^{(*)}x^{(*)}}(s) \\ & e^{j2\pi\alpha'_{n'}(\tau_1-\tau_2+s)(t_2+\tau_2)}\, e^{j2\pi\alpha''_{n''}(s)t_2}\, e^{-j2\pi\alpha_1\cdot(t_2+s)}\, e^{-j2\pi\alpha_2 t_2} \\ & \frac{1}{T} r_{aa}\left(-[\alpha'_{n'}(\tau_1 - \tau_2 + s) + \alpha''_{n''}(s) - \alpha_1 - \alpha_2]T; \frac{t_2 - t_1 + s}{T}\right) \mathrm{d}s \end{aligned}\tag{3.133}$$

$$\begin{aligned}\bar{\mathcal{T}}_2 \triangleq \sum_{n'''}\sum_{n^{iv}}\int_{\mathbb{R}} & R^{(n''')}_{yx^{(*)}}(\tau_1 + s)\, R^{(n^{iv})}_{x^{(*)}y}(s - \tau_2) \\ & e^{j2\pi\alpha'''_{n'''}(\tau_1+s)t_2}\, e^{j2\pi\alpha^{iv}_{n^{iv}}(s-\tau_2)(t_2+\tau_2)}\, e^{-j2\pi\alpha_1\cdot(t_2+s)}\, e^{-j2\pi\alpha_2 t_2} \\ & \frac{1}{T} r_{aa}\left(-[\alpha'''_{n'''}(\tau_1 + s) + \alpha^{iv}_{n^{iv}}(s - \tau_2) - \alpha_1 - \alpha_2]T; \frac{t_2 - t_1 + s}{T}\right) \mathrm{d}s \end{aligned}\tag{3.134}$$

$$\begin{aligned}\bar{\mathcal{T}}_3 \triangleq \sum_{n}\int_{\mathbb{R}} & C^{(n)}_{yx^{(*)}yx^{(*)}}(\tau_1 + s, s, \tau_2)\, e^{j2\pi[\beta_n(\tau_1+s,s,\tau_2)-\alpha_1-\alpha_2]t_2}\, e^{-j2\pi\alpha_1 s} \\ & \frac{1}{T} r_{aa}\left(-[\beta_n(\tau_1 + s, s, \tau_2) - \alpha_1 - \alpha_2]T; \frac{t_2 - t_1 + s}{T}\right) \mathrm{d}s \end{aligned}\tag{3.135}$$

with

$$r_{aa}(\beta; s) \triangleq \int_{\mathbb{R}} a(t + s)\, a(t)\, e^{-j2\pi\beta t}\, \mathrm{d}t. \tag{3.136}$$

In (3.133)–(3.135), for notational simplicity, $\alpha'_{n'}(\cdot) \equiv \alpha^{(n')}_{yy}(\cdot)$, $\alpha''_{n''}(\cdot) \equiv \alpha^{(n'')}_{x^{(*)}x^{(*)}}(\cdot)$, $\alpha'''_{n'''}(\cdot) \equiv \alpha^{(n''')}_{yx^{(*)}}(\cdot)$, and $\alpha^{iv}_{n^{iv}}(\cdot) \equiv \alpha^{(n^{iv})}_{x^{(*)}y}(\cdot)$. □

By reasoning as for Theorem 2.4.13, we get the following result.

Theorem 3.7.2 Asymptotic Conjugate Covariance of the Cyclic Cross-Correlogram. *Let $y(t)$ and $x(t)$ be zero-mean wide-sense jointly GACS stochastic processes with cross-correlation function (2.31c). Under Assumptions 2.4.2–2.4.5, and 2.4.8, the asymptotic conjugate covariance of the cyclic cross-correlogram $R_{yx^{(*)}}(\alpha, \tau; t_0, T)$ is given by*

$$\lim_{T\to\infty} T\operatorname{cov}\left\{R_{yx^{(*)}}(\alpha_1, \tau_1; t_1, T), R^*_{yx^{(*)}}(\alpha_2, \tau_2; t_2, T)\right\} = \bar{\mathcal{T}}'_1 + \bar{\mathcal{T}}'_2 + \bar{\mathcal{T}}'_3 \tag{3.137}$$

where

$$\begin{aligned}\bar{\mathcal{T}}'_1 \triangleq \bar{\mathcal{E}}_a \sum_{n'}\sum_{n''}\int_{\mathbb{R}} & R^{(n')}_{yy}(\tau_1 - \tau_2 + s)\, R^{(n'')}_{x^{(*)}x^{(*)}}(s) \\ & e^{j2\pi\alpha'_{n'}(\tau_1-\tau_2+s)\tau_2}\, e^{-j2\pi\alpha_1 s}\, \delta_{[\alpha'_{n'}(\tau_1-\tau_2+s)+\alpha''_{n''}(s)-\alpha_1-\alpha_2]}\, \mathrm{d}s \end{aligned}\tag{3.138}$$

$$\bar{\mathcal{T}}'_2 \triangleq \bar{\mathcal{E}}_a \sum_{n'''} \sum_{n^{\text{iv}}} \int_{\mathbb{R}} R^{(n''')}_{yx^{(*)}}(\tau_1 + s)\, R^{(n^{\text{iv}})}_{x^{(*)}y}(s - \tau_2)$$
$$e^{j2\pi\alpha^{\text{iv}}_{n^{\text{iv}}}(s-\tau_2)\tau_2}\, e^{-j2\pi\alpha_1 s}\, \delta_{[\alpha'''_{n'''}(\tau_1+s)+\alpha^{\text{iv}}_{n^{\text{iv}}}(s-\tau_2)-\alpha_1-\alpha_2]}\, \mathrm{d}s \tag{3.139}$$

$$\bar{\mathcal{T}}'_3 \triangleq \bar{\mathcal{E}}_a \sum_{n} \int_{\mathbb{R}} C^{(n)}_{yx^{(*)}yx^{(*)}}(\tau_1 + s, s, \tau_2)\, e^{-j2\pi\alpha_1 s} \delta_{[\beta_n(\tau_1+s,s,\tau_2)-\alpha_1-\alpha_2]}\, \mathrm{d}s \tag{3.140}$$

with $\bar{\mathcal{E}}_a \triangleq \int_{\mathbb{R}} a^2(u)\, \mathrm{d}u$. □

3.8 Proofs for Section 2.5 "Sampling of GACS Processes"

3.8.1 *Proof of Theorem 2.5.3 Aliasing Formula for the Cyclic Cross-Correlation Function*

Equation (2.179b) is coincident with (2.177). Equation (2.179a) is obtained by specializing to second-order (Izzo and Napolitano 2003, eq. (49)) (or (Izzo and Napolitano 2005, eq. (4.30))). The proof is analogous to the proof of (1.174) for ACS processes.

Once (2.179a) is proved, an alternative proof of (2.179b) is obtained as follows. By substituting (2.39) into (2.179a) one has

$$\widetilde{R}_{y_{\mathrm{d}}x^{(*)}_{\mathrm{d}}}(\widetilde{\alpha}, m) = \sum_{p=-\infty}^{+\infty} \sum_{k \in \mathbb{I}_{yx^{(*)}}} R^{(k)}_{yx^{(*)}}(mT_s)\, \delta_{\widetilde{\alpha} f_s - \alpha_k(mT_s) + pf_s} \tag{3.141}$$

from which (2.179b) easily follows, provided that condition

$$\sum_{p=-\infty}^{+\infty} \sum_{k \in \mathbb{I}_{yx^{(*)}}} \left| R^{(k)}_{yx^{(*)}}(mT_s) \right| \delta_{\widetilde{\alpha} f_s - \alpha_k(mT_s) + pf_s} < \infty \tag{3.142}$$

is satisfied in order to allow the interchange of the order of the sum operations (Johnsonbaugh and Pfaffenberger 2002, Theorem 29.4).

3.9 Proofs for Section 2.6.1 "Discrete-Time Cyclic Cross-Correlogram"

3.9.1 *Proof of Theorem 2.6.2 Expected Value of the Discrete-Time Cyclic Cross-Correlogram*

By taking the expected value of both sides in (2.181) and using (2.174c) we get

$$\mathrm{E}\left\{ \widetilde{R}_{y_{\mathrm{d}}x^{(*)}_{\mathrm{d}}}(\widetilde{\alpha}, m; n_0, N) \right\}$$
$$= \sum_{n=-\infty}^{+\infty} v_N(n - n_0)\, \mathrm{E}\{y_{\mathrm{d}}(n + m)\, x^{(*)}_{\mathrm{d}}(n)\}\, e^{-j2\pi\widetilde{\alpha} n}$$

$$
\begin{aligned}
&= \sum_{n=-\infty}^{+\infty} v_N(n-n_0) \sum_{k\in\mathbb{I}} R^{(k)}_{yx^{(*)}}(mT_s)\, e^{j2\pi\alpha_k(mT_s)nT_s}\, e^{-j2\pi\widetilde{\alpha}n} \\
&= \sum_{k\in\mathbb{I}} R^{(k)}_{yx^{(*)}}(mT_s) \sum_{n=-\infty}^{+\infty} v_N(n-n_0)\, e^{-j2\pi[\widetilde{\alpha}-\alpha_k(mT_s)T_s]n}
\end{aligned}
\tag{3.143}
$$

from which (2.183) immediately follows.

The data-tapering window $v_N(n)$ is finite length (see (2.182)). Thus, in the first equality the sum over n is finite and can be freely interchanged with the expectation operator. In the second equality, Assumption 2.4.2a is used. In the third equality the order of the two sums can be interchanged since the double-index series over k and n is absolutely convergent (Johnsonbaugh and Pfaffenberger 2002, Theorem 29.4). In fact,

$$
\begin{aligned}
&\sum_n \sum_k \left| R^{(k)}_{yx^{(*)}}(mT_s)\, v_N(n-n_0)\, e^{-j2\pi[\widetilde{\alpha}-\alpha_k(mT_s)T_s]n} \right| \\
&\quad \leqslant \sum_n |v_N(n-n_0)| \sum_k \left| R^{(k)}_{yx^{(*)}}(mT_s) \right| \\
&\quad \leqslant \|a\|_\infty \sum_k \left\| R^{(k)}_{yx^{(*)}} \right\|_\infty < \infty
\end{aligned}
\tag{3.144}
$$

where Assumption 2.4.3a, (2.182), and the inequality (Assumption 2.4.5)

$$
\begin{aligned}
\left| V_{\frac{1}{N}}(\nu) \right| &\leqslant \sum_{n=-\infty}^{+\infty} |v_N(n)| \\
&= \frac{1}{2N+1} \sum_{n=-N}^{N} \left| a\left(\frac{n}{2N+1}\right) \right| \\
&\leqslant \frac{1}{2N+1} \sum_{n=-N}^{N} \|a\|_\infty \\
&= \|a\|_\infty
\end{aligned}
\tag{3.145}
$$

are used.

By substituting (2.182) into the first line of (2.184) we get

$$
\begin{aligned}
V_{\frac{1}{N}}(\nu) &= \sum_{n=-\infty}^{+\infty} \frac{1}{2N+1}\, a\left(\frac{n}{2N+1}\right) e^{-j2\pi\nu n} \\
&= \sum_{n=-\infty}^{+\infty} \frac{1}{2N+1} \int_{\mathbb{R}} A(f)\, e^{j2\pi f n/(2N+1)}\, \mathrm{d}f\, e^{-j2\pi\nu n} \\
&= \frac{1}{2N+1} \int_{\mathbb{R}} A(f) \sum_{n=-\infty}^{+\infty} e^{-j2\pi[\nu - f/(2N+1)]n}\, \mathrm{d}f
\end{aligned}
$$

$$= \frac{1}{2N+1} \int_{\mathbb{R}} A(f) \sum_{\ell=-\infty}^{+\infty} \delta\Big(\nu - \frac{f}{2N+1} - \ell\Big) \, \mathrm{d}f$$

$$= \int_{\mathbb{R}} A(f) \sum_{\ell=-\infty}^{+\infty} \delta\Big((\nu - \ell)(2N+1) - f\Big) \, \mathrm{d}f \tag{3.146}$$

from which the right-hand side of (2.184) immediately follows. In (3.146), in the fourth equality the Poisson's sum formula (Zemanian 1987, p. 189)

$$\sum_{n=-\infty}^{+\infty} e^{-j2\pi(f/f_s)n} = f_s \sum_{\ell=-\infty}^{+\infty} \delta(f - \ell f_s) \tag{3.147}$$

and in the fifth equality the scaling property of the Dirac delta $\delta(bt) = \delta(t)/|b|$ (Zemanian 1987, p. 27) are used. Furthermore, equalities should be intended in the sense of distributions (generalized functions) (Zemanian 1987).

3.9.2 Proof of Theorem 2.6.3 Covariance of the Discrete-Time Cyclic Cross-Correlogram

For zero-mean stochastic processes, the covariance of the lag product can be expressed in terms of second-order moments and a fourth-order cumulant (see (3.45)). Then, using the multilinearity property of covariance, one obtains

$$\begin{aligned}
&\operatorname{cov}\Big\{ \widetilde{R}_{y_\mathrm{d} x_\mathrm{d}^{(*)}}(\widetilde{\alpha}_1, m_1; n_{01}, N), \widetilde{R}_{y_\mathrm{d} x_\mathrm{d}^{(*)}}(\widetilde{\alpha}_2, m_2; n_{02}, N) \Big\} \\
&= \operatorname{cov}\Big\{ \sum_{n_1=n_{01}-N}^{n_{01}+N} v_N(n_1 - n_{01}) \, y_\mathrm{d}(n_1 + m_1) \, x_\mathrm{d}^{(*)}(n_1) \, e^{-j2\pi\widetilde{\alpha}_1 n_1}, \\
&\quad \sum_{n_2=n_{02}-N}^{n_{02}+N} v_N(n_2 - n_{02}) \, y_\mathrm{d}(n_2 + m_2) \, x_\mathrm{d}^{(*)}(n_2) \, e^{-j2\pi\widetilde{\alpha}_2 n_2} \Big\} \\
&= \sum_{n_1=n_{01}-N}^{n_{01}+N} \sum_{n_2=n_{02}-N}^{n_{02}+N} v_N(n_1 - n_{01}) \, v_N^*(n_2 - n_{02}) \, e^{-j2\pi[\widetilde{\alpha}_1 n_1 - \widetilde{\alpha}_2 n_2]} \\
&\quad \operatorname{cov}\Big\{ y_\mathrm{d}(n_1 + m_1) \, x_\mathrm{d}^{(*)}(n_1), y_\mathrm{d}(n_2 + m_2) \, x_\mathrm{d}^{(*)}(n_2) \Big\} \\
&= \sum_{n_1=n_{01}-N}^{n_{01}+N} \sum_{n_2=n_{02}-N}^{n_{02}+N} v_N(n_1 - n_{01}) \, v_N^*(n_2 - n_{02}) \, e^{-j2\pi[\widetilde{\alpha}_1 n_1 - \widetilde{\alpha}_2 n_2]} \\
&\quad \Big[\mathrm{E}\left\{ y_\mathrm{d}(n_1 + m_1) \, y_\mathrm{d}^*(n_2 + m_2) \right\} \mathrm{E}\Big\{ x_\mathrm{d}^{(*)}(n_1) \, x_\mathrm{d}^{(*)*}(n_2) \Big\} \\
&\quad + \mathrm{E}\Big\{ y_\mathrm{d}(n_1 + m_1) \, x_\mathrm{d}^{(*)*}(n_2) \Big\} \mathrm{E}\Big\{ x_\mathrm{d}^{(*)}(n_1) \, y_\mathrm{d}^*(n_2 + m_2) \Big\} \\
&\quad + \operatorname{cum}\Big\{ y_\mathrm{d}(n_1 + m_1), x_\mathrm{d}^{(*)}(n_1), y_\mathrm{d}^*(n_2 + m_2), x_\mathrm{d}^{(*)*}(n_2) \Big\} \Big].
\end{aligned} \tag{3.148}$$

By substituting (2.119) and (2.120) (Assumption 2.4.2) into (3.148), accounting for (2.182) and making the variable changes $n_1 = n$ and $n_2 = n - \ell$, it results in

$$\operatorname{cov}\left\{\widetilde{R}_{y_{\mathrm{d}}x_{\mathrm{d}}^{(*)}}(\widetilde{\alpha}_1, m_1; n_{01}, N), \widetilde{R}_{y_{\mathrm{d}}x_{\mathrm{d}}^{(*)}}(\widetilde{\alpha}_2, m_2; n_{02}, N)\right\} = \widetilde{\mathcal{D}}_1 + \widetilde{\mathcal{D}}_2 + \widetilde{\mathcal{D}}_3 \tag{3.149}$$

where

$$\begin{aligned}\widetilde{\mathcal{D}}_1 \triangleq & \sum_{n=n_{01}-N}^{n_{01}+N} \sum_{\ell=n-n_{02}-N}^{n-n_{02}+N} \frac{1}{2N+1}\, a\left(\frac{n-n_{01}}{2N+1}\right) \\ & \frac{1}{2N+1}\, a^*\left(\frac{n-\ell-n_{02}}{2N+1}\right) e^{-j2\pi[\widetilde{\alpha}_1 n - \widetilde{\alpha}_2(n-\ell)]} \\ & \sum_{k'}\sum_{k''} R_{yy^*}^{(k')}((m_1-m_2+\ell)T_s)\, R_{x^{(*)}x^{(*)*}}^{(k'')}(\ell T_s) \\ & e^{j2\pi\alpha'_{k'}((m_1-m_2+\ell)T_s)(n-\ell+m_2)T_s}\, e^{j2\pi\alpha''_{k''}(\ell T_s)(n-\ell)T_s}\end{aligned} \tag{3.150}$$

$$\begin{aligned}\widetilde{\mathcal{D}}_2 \triangleq & \sum_{n=n_{01}-N}^{n_{01}+N} \sum_{\ell=n-n_{02}-N}^{n-n_{02}+N} \frac{1}{2N+1}\, a\left(\frac{n-n_{01}}{2N+1}\right) \\ & \frac{1}{2N+1}\, a^*\left(\frac{n-\ell-n_{02}}{2N+1}\right) e^{-j2\pi[\widetilde{\alpha}_1 n - \widetilde{\alpha}_2(n-\ell)]} \\ & \sum_{k'''}\sum_{k^{iv}} R_{yx^{(*)*}}^{(k''')}((m_1+\ell)T_s)\, R_{x^{(*)}y^*}^{(k^{iv})}((\ell-m_2)T_s) \\ & e^{j2\pi\alpha'''_{k'''}((m_1+\ell)T_s)(n-\ell)T_s}\, e^{j2\pi\alpha^{iv}_{n^{iv}}((\ell-m_2)T_s)(n-\ell+m_2)T_s}\end{aligned} \tag{3.151}$$

$$\begin{aligned}\widetilde{\mathcal{D}}_3 \triangleq & \sum_{n=n_{01}-N}^{n_{01}+N} \sum_{\ell=n-n_{02}-N}^{n-n_{02}+N} \frac{1}{2N+1}\, a\left(\frac{n-n_{01}}{2N+1}\right) \\ & \frac{1}{2N+1}\, a^*\left(\frac{n-\ell-n_{02}}{2N+1}\right) e^{-j2\pi[\widetilde{\alpha}_1 n - \widetilde{\alpha}_2(n-\ell)]} \\ & \sum_{k} C_{yx^{(*)}y^*x^{(*)*}}^{(k)}((m_1+\ell)T_s, \ell T_s, m_2 T_s) \\ & e^{j2\pi\beta_k((m_1+\ell)T_s,\ell T_s,m_2 T_s)(n-\ell)T_s}\end{aligned} \tag{3.152}$$

In (3.150)–(3.152), for notation simplicity, $\alpha'_{k'}(\cdot) \equiv \alpha_{yy^*}^{(k')}(\cdot)$, $\alpha''_{k''}(\cdot) \equiv \alpha_{x^{(*)}x^{(*)*}}^{(k'')}(\cdot)$, $\alpha'''_{k'''}(\cdot) \equiv \alpha_{yx^{(*)*}}^{(k''')}(\cdot)$, and $\alpha^{iv}_{k^{iv}}(\cdot) \equiv \alpha_{x^{(*)}y^*}^{(k^{iv})}(\cdot)$. The sums over n and ℓ can be extended from $-\infty$ to $+\infty$, they remain the same finite sums due to the presence of the product of the two finite-length data-tapering windows $a(\cdot)$ and $a^*(\cdot)$. Then, by making the variable change $n = n' + n_{02} + \ell$ into (3.150)–(3.152) and using (2.190), (2.186)–(2.189) easily follow.

The interchange of covariance and sum operations in the second equality of (3.148) is allowed since the sums are finite (due to the finite support of $v_N(n)$). The interchange of the order of sums to get (3.149)–(3.152) and then (2.186)–(2.189) is justified since the four-index series over (k', k'', ℓ, n) in (3.150) and over (k''', k^{iv}, ℓ, n) in (3.151) and the three-index series over (k, ℓ, n) in (3.152) are absolutely convergent (Johnsonbaugh and Pfaffenberger 2002, Theorem 29.4).

In fact, as regards the term $\widetilde{\mathcal{D}}_1$ in (3.150), one has

$$\sum_n \sum_\ell \sum_{k'} \sum_{k''} \left| \frac{1}{2N+1} a\left(\frac{n-n_{01}}{2N+1}\right) \frac{1}{2N+1} a^*\left(\frac{n-\ell-n_{02}}{2N+1}\right) e^{-j2\pi[\widetilde{\alpha}_1 n - \widetilde{\alpha}_2(n-\ell)]} \right.$$
$$\cdot R_{yy^*}^{(k')}((m_1-m_2+\ell)T_s)\, R_{x^{(*)}x^{(*)*}}^{(k'')}(\ell T_s)$$
$$\left. \cdot e^{j2\pi\alpha'_{k'}((m_1-m_2+\ell)T_s)(n-\ell+m_2)T_s}\, e^{j2\pi\alpha''_{k''}(\ell T_s)(n-\ell)T_s} \right|$$
$$\leqslant \frac{1}{2N+1}\sum_n \left| a\left(\frac{n-n_{01}}{2N+1}\right)\right| \frac{1}{2N+1}\sum_\ell \left| a\left(\frac{n-\ell-n_{02}}{2N+1}\right)\right|$$
$$\cdot \sum_{k'} \left\| R_{yy^*}^{(k')} \right\|_\infty \sum_{k''} \left\| R_{x^{(*)}x^{(*)*}}^{(k'')} \right\|_\infty$$
$$\leqslant \|a\|_\infty^2 \sum_{k'} \left\| R_{yy^*}^{(k')} \right\|_\infty \sum_{k''} \left\| R_{x^{(*)}x^{(*)*}}^{(k'')} \right\|_\infty < \infty \tag{3.153}$$

where the sums over n and ℓ are at most over $2N+1$ terms, and Assumptions 2.4.3a and 2.4.5 are used. The term $\widetilde{\mathcal{D}}_2$ in (3.151) can be treated analogously. As regards the term $\widetilde{\mathcal{D}}_3$ in (3.152), one has

$$\sum_n \sum_\ell \sum_k \left| \frac{1}{2N+1} a\left(\frac{n-n_{01}}{2N+1}\right) \frac{1}{2N+1} a^*\left(\frac{n-\ell-n_{02}}{2N+1}\right) e^{-j2\pi[\widetilde{\alpha}_1 n - \widetilde{\alpha}_2(n-\ell)]} \right.$$
$$\left. \cdot C_{yx^{(*)}y^*x^{(*)*}}^{(k)}((m_1+\ell)T_s, \ell T_s, m_2 T_s)\, e^{j2\pi\beta_k((m_1+\ell)T_s,\ell T_s,m_2T_s)(n-\ell)T_s} \right|$$
$$\leqslant \frac{1}{2N+1}\sum_n \left| a\left(\frac{n-n_{01}}{2N+1}\right)\right| \frac{1}{2N+1}\sum_\ell \left| a\left(\frac{n-\ell-n_{02}}{2N+1}\right)\right| \sum_k \left\| C_{yx^{(*)}y^*x^{(*)*}}^{(k)} \right\|_\infty$$
$$\leqslant \|a\|_\infty^2 \sum_k \left\| C_{yx^{(*)}y^*x^{(*)*}}^{(k)} \right\|_\infty < \infty \tag{3.154}$$

where Assumptions 2.4.3b and 2.4.5 are used.

3.10 Proofs for Section 2.6.2 "Asymptotic Results as $N \to \infty$"

Lemma 3.10.1 *Let $V_{\frac{1}{N}}(\nu)$ be given by (2.184) with $a(t)$ as in Assumption 2.4.5. It results that*

$$\lim_{N\to\infty} V_{\frac{1}{N}}(\nu) = \sum_{p=-\infty}^{+\infty} \delta_{\nu-p} = \delta_{\nu \bmod 1}. \tag{3.155}$$

Proof: Since $a(t)$ has finite support $[-1/2, 1/2)$, to prove (3.155) is equivalent to prove that

$$\lim_{N\to\infty} \sum_{n=-N}^{N} \frac{1}{2N+1} a\left(\frac{n}{2N+1}\right) e^{-j2\pi\nu n} = \begin{cases} 1 & \nu \in \mathbb{Z} \\ 0 & \nu \notin \mathbb{Z} \end{cases} \tag{3.156}$$

Let us consider first the case $\nu \in \mathbb{Z}$.
Since $a(t)$ is Riemann integrable (Assumption 2.4.5), we have

$$\lim_{N\to\infty} \sum_{n=-N}^{N} a\left(\frac{n}{2N+1}\right) \frac{1}{2N+1} = \int_{-1/2}^{1/2} a(t)\,\mathrm{d}t = \int_{\mathbb{R}} a(t)\,\mathrm{d}t = 1. \tag{3.157}$$

Let us consider now the case $\nu \notin \mathbb{Z}$, $\nu > 0$.
For every $\bar{\nu} \in \mathbb{R}$ the function $a(t)e^{-j2\pi\bar{\nu}t}$ is Riemann integrable and we have

$$\lim_{N\to\infty} \sum_{n=-N}^{N} a\left(\frac{n}{2N+1}\right) e^{-j2\pi\bar{\nu}n/(2N+1)} \frac{1}{2N+1} = \int_{-1/2}^{1/2} a(t)\, e^{-j2\pi\bar{\nu}t}\,\mathrm{d}t \tag{3.158}$$

that is, $\forall \epsilon_1 > 0 \;\; \exists N_{\epsilon_1}$ such that for $N > N_{\epsilon_1}$

$$\left| \sum_{n=-N}^{N} a\left(\frac{n}{2N+1}\right) e^{-j2\pi\bar{\nu}n/(2N+1)} \frac{1}{2N+1} - \int_{-1/2}^{1/2} a(t)\, e^{-j2\pi\bar{\nu}t}\,\mathrm{d}t \right| < \epsilon_1 \tag{3.159}$$

(not necessarily uniformly with respect to $\bar{\nu}$).

Since $a(t) \in L^1(\mathbb{R})$, the Riemann-Lebesgue theorem (Champeney 1990, Chapter 3) can be applied. Thus,

$$\lim_{\bar{\nu}\to\infty} \int_{-1/2}^{1/2} a(t)\, e^{-j2\pi\bar{\nu}t}\,\mathrm{d}t = 0 \tag{3.160}$$

that is, $\forall \epsilon_2 > 0 \;\; \exists \bar{\nu}_{\epsilon_2}$ such that for $\bar{\nu} > \bar{\nu}_{\epsilon_2}$

$$\left| \int_{-1/2}^{1/2} a(t)\, e^{-j2\pi\bar{\nu}t}\,\mathrm{d}t \right| < \epsilon_2 \tag{3.161}$$

Let us consider, now, the following inequality:

$$\begin{aligned}
&\left| \sum_{n=-N}^{N} a\left(\frac{n}{2N+1}\right) e^{-j2\pi\nu n} \frac{1}{2N+1} \right| \\
&= \left| \sum_{n=-N}^{N} a\left(\frac{n}{2N+1}\right) e^{-j2\pi\nu n} \frac{1}{2N+1} - \int_{-1/2}^{1/2} a(t)\, e^{-j2\pi\nu(2N+1)t}\,\mathrm{d}t \right. \\
&\qquad \left. + \int_{-1/2}^{1/2} a(t)\, e^{-j2\pi\nu(2N+1)t}\,\mathrm{d}t \right| \\
&\leqslant \left| \sum_{n=-N}^{N} a\left(\frac{n}{2N+1}\right) e^{-j2\pi\nu n} \frac{1}{2N+1} - \int_{-1/2}^{1/2} a(t)\, e^{-j2\pi\nu(2N+1)t}\,\mathrm{d}t \right| \\
&\qquad + \left| \int_{-1/2}^{1/2} a(t)\, e^{-j2\pi\nu(2N+1)t}\,\mathrm{d}t \right|
\end{aligned} \tag{3.162}$$

Since $\nu > 0$, take ϵ_2 arbitrarily small in (3.161), and define N_{ϵ_2} such that $\bar{\nu}_{\epsilon_2} = \nu(2N_{\epsilon_2} + 1)$. Fix $\bar{\nu} > \bar{\nu}_{\epsilon_2}$ in (3.158) and take ϵ_1 arbitrarily small and $N > N_{\epsilon_1}$ in (3.159). Let $N > \max(N_{\epsilon_1}, N_{\epsilon_2})$ and $\bar{\nu} = \nu(2N + 1)$. It results in $\bar{\nu} > \bar{\nu}_{\epsilon_2}$ in (3.161) and $N > N_{\epsilon_1}$ in (3.159). Thus, the right-hand side of (3.162) is bounded by $\epsilon_1 + \epsilon_2$ with ϵ_1 and ϵ_2 arbitrarily small, which proves (3.156) for $\nu > 0$. Finally, note that the result holds for every non-integer ν since the function $V_{\frac{1}{N}}(\nu)$ is periodic in ν with period 1.

As a final remark, note that Lemma 3.10.1 can also be obtained as corollary of the following Lemma 3.10.2 with $b(t) = \text{rect}(t)$, $m = 0$. To prove Lemma 3.10.1, however, the assumption $|\dot{a}(t)|$ bounded is not used. □

Lemma 3.10.2 *Let $a(t)$ be bounded and continuous in $(-1/2, 1/2)$ except, possibly, at $t = \pm 1/2$, derivable almost-everywhere with bounded derivative, and let $b(t) \in L^1(\mathbb{R})$. Let both $a(t)$ and $b(t)$ be with finite support $[-1/2, 1/2]$, and such that for every $\tau \in [-1, 1]$ the function $a(t+\tau)\, b^*(t)$ is Riemann integrable in $[-1/2, 1/2]$. We have*

$$\lim_{N\to\infty} \frac{1}{2N+1} \sum_{n=-\infty}^{+\infty} a\left(\frac{n+m}{2N+1}\right) b^*\left(\frac{n}{2N+1}\right) e^{-j2\pi\widetilde{\alpha}n}$$
$$= \int_{-1/2}^{1/2} a(t)\, b^*(t)\, \mathrm{d}t\, \delta_{\widetilde{\alpha} \bmod 1}. \tag{3.163}$$

Proof: The function $b(t)$ has finite support $[-1/2, 1/2]$. Thus the sum in (3.163) can be extended from $-N$ to N.

Let $\widetilde{\alpha} \in \mathbb{Z}$. The following inequality holds

$$\left| \frac{1}{2N+1} \sum_{n=-N}^{N} a\left(\frac{n+m}{2N+1}\right) b^*\left(\frac{n}{2N+1}\right) - \int_{-1/2}^{1/2} a(t)\, b^*(t)\, \mathrm{d}t \right|$$
$$= \left| \frac{1}{2N+1} \sum_{n=-N}^{N} a\left(\frac{n+m}{2N+1}\right) b^*\left(\frac{n}{2N+1}\right) \right.$$
$$- \frac{1}{2N+1} \sum_{n=-N}^{N} a\left(\frac{n}{2N+1}\right) b^*\left(\frac{n}{2N+1}\right)$$
$$\left. + \frac{1}{2N+1} \sum_{n=-N}^{N} a\left(\frac{n}{2N+1}\right) b^*\left(\frac{n}{2N+1}\right) - \int_{-1/2}^{1/2} a(t)\, b^*(t)\, \mathrm{d}t \right|$$
$$\leqslant \left| \frac{1}{2N+1} \sum_{n=-N}^{N} \left[a\left(\frac{n+m}{2N+1}\right) - a\left(\frac{n}{2N+1}\right) \right] b^*\left(\frac{n}{2N+1}\right) \right|$$
$$+ \left| \frac{1}{2N+1} \sum_{n=-N}^{N} a\left(\frac{n}{2N+1}\right) b^*\left(\frac{n}{2N+1}\right) - \int_{-1/2}^{1/2} a(t)\, b^*(t)\, \mathrm{d}t \right| \tag{3.164}$$

As regards the first term in the right-hand side of (3.164), observe that the first-order derivative $\dot{a}(t)$ exists a.e. and is bounded (Assumption 2.4.5). Therefore, for almost all t_1 and t_2

there exists $\bar{t} \in [t_1, t_2]$ such that $a(t_2) - a(t_1) = \dot{a}(t)|_{t=\bar{t}}(t_2 - t_1)$ and, hence, $|a(t_2) - a(t_1)| \leqslant \| \dot{a} \|_\infty |t_2 - t_1|$. Consequently,

$$\left| \frac{1}{2N+1} \sum_{n=-N}^{N} \left[a\left(\frac{n+m}{2N+1}\right) - a\left(\frac{n}{2N+1}\right) \right] b^*\left(\frac{n}{2N+1}\right) \right|$$
$$\leqslant \frac{\|b\|_\infty}{2N+1} \sum_{n=-N}^{N} \left| a\left(\frac{n+m}{2N+1}\right) - a\left(\frac{n}{2N+1}\right) \right|$$
$$\leqslant \frac{\|b\|_\infty}{2N+1} (2N+1) \| \dot{a} \|_\infty \frac{|m|}{2N+1}. \tag{3.165}$$

As regards the second term in the right-hand side of (3.164), observe that the function $a(t+\tau)\, b^*(t)$ is Riemann integrable in $[-1/2, 1/2]$ so that, for $\tau = 0$,

$$\lim_{N\to\infty} \sum_{n=-N}^{N} a\left(\frac{n}{2N+1}\right) b^*\left(\frac{n}{2N+1}\right) \frac{1}{2N+1} = \int_{-1/2}^{1/2} a(t)\, b^*(t)\, \mathrm{d}t. \tag{3.166}$$

That is, $\forall \epsilon > 0\ \exists N_\epsilon$ such that for $N > N_\epsilon$ one has

$$\left| \sum_{n=-N}^{N} a\left(\frac{n}{2N+1}\right) b^*\left(\frac{n}{2N+1}\right) \frac{1}{2N+1} - \int_{-1/2}^{1/2} a(t)\, b^*(t)\, \mathrm{d}t \right| < \epsilon. \tag{3.167}$$

Therefore, the right-hand side of (3.164) is bounded by

$$\|b\|_\infty \| \dot{a} \|_\infty \frac{|m|}{2N+1} + \epsilon$$

which can be made arbitrarily small, provided that N is sufficiently large.

This proves (3.163) for $\widetilde{\alpha} \in \mathbb{Z}$.

Let $\widetilde{\alpha} \notin \mathbb{Z}$. For every $\bar{\nu} \in \mathbb{R}$, $\tau \in \mathbb{R}$, the function $a(t+\tau)\, b^*(t)\, e^{-j2\pi\bar{\nu}t}$ is Riemann integrable in $[-1/2, 1/2]$. Thus

$$\lim_{N\to\infty} \sum_{n=-N}^{N} a\left(\frac{n}{2N+1} + \tau\right) b^*\left(\frac{n}{2N+1}\right) e^{-j2\pi\bar{\nu}n/(2N+1)} \frac{1}{2N+1}$$
$$= \int_{-1/2}^{1/2} a(t+\tau)\, b^*(t)\, e^{-j2\pi\bar{\nu}t}\, \mathrm{d}t \tag{3.168}$$

that is, $\forall \epsilon_1 > 0\ \ \exists N_{\epsilon_1}(\bar{\nu}, \tau)$ such that for $N > N_{\epsilon_1}(\bar{\nu}, \tau)$

$$\left| \sum_{n=-N}^{N} a\left(\frac{n}{2N+1} + \tau\right) b^*\left(\frac{n}{2N+1}\right) e^{-j2\pi\bar{\nu}n/(2N+1)} \frac{1}{2N+1} \right.$$
$$\left. - \int_{-1/2}^{1/2} a(t+\tau)\, b^*(t)\, e^{-j2\pi\bar{\nu}t}\, \mathrm{d}t \right| < \epsilon_1 \tag{3.169}$$

not necessarily uniformly with respect to $\bar{\nu}$ and τ. Since $a(t) \in L^{\infty}(\mathbb{R})$ and $b(t) \in L^{1}(\mathbb{R})$, it results in $|a(t+\tau)\, b^*(t)| \leqslant \|a\|_{\infty} |b(t)| \in L^{1}(\mathbb{R})$. Thus, the Riemann-Lebesgue theorem (Champeney 1990, Section 3.7) can be applied:

$$\lim_{\bar{\nu}\to\infty} \int_{-1/2}^{1/2} a(t+\tau)\, b^*(t)\, e^{-j2\pi\bar{\nu}t}\, \mathrm{d}t = 0 \tag{3.170}$$

that is, $\forall \epsilon_2 > 0 \;\; \exists \bar{\nu}_{\epsilon_2}$ such that for $\bar{\nu} > \bar{\nu}_{\epsilon_2}$

$$\left| \int_{-1/2}^{1/2} a(t+\tau)\, b^*(t)\, e^{-j2\pi\bar{\nu}t}\, \mathrm{d}t \right| < \epsilon_2 \tag{3.171}$$

not necessarily uniformly with respect to τ.

For arbitrarily small ϵ_2, fix $\bar{\nu} > \bar{\nu}_{\epsilon_2}$ so that (3.171) holds. Once $\bar{\nu}$ is fixed, for arbitrarily small ϵ_1 fix $N > N_{\epsilon_1}(\bar{\nu}, \tau)$ such that (3.169) holds. Thus, by setting $\nu \triangleq \bar{\nu}/(2N+1)$, $\nu \neq 0$, in (3.169) and (3.171), we have the following inequality

$$\begin{aligned}
&\left| \frac{1}{2N+1} \sum_{n=-N}^{N} a\Big(\frac{n}{2N+1} + \tau\Big)\, b^*\Big(\frac{n}{2N+1}\Big)\, e^{-j2\pi\nu n} \right| \\
&\quad = \left| \frac{1}{2N+1} \sum_{n=-N}^{N} a\Big(\frac{n}{2N+1} + \tau\Big)\, b^*\Big(\frac{n}{2N+1}\Big)\, e^{-j2\pi\nu n} \right. \\
&\qquad - \int_{-1/2}^{1/2} a(t+\tau)\, b^*(t)\, e^{-j2\pi\nu(2N+1)t}\, \mathrm{d}t \\
&\qquad \left. + \int_{-1/2}^{1/2} a(t+\tau)\, b^*(t)\, e^{-j2\pi\nu(2N+1)t}\, \mathrm{d}t \right| \\
&\quad \leqslant \left| \frac{1}{2N+1} \sum_{n=-N}^{N} a\Big(\frac{n}{2N+1} + \tau\Big)\, b^*\Big(\frac{n}{2N+1}\Big)\, e^{-j2\pi\nu n} \right. \\
&\qquad \left. - \int_{-1/2}^{1/2} a(t+\tau)\, b^*(t)\, e^{-j2\pi\nu(2N+1)t}\, \mathrm{d}t \right| \\
&\qquad + \left| \int_{-1/2}^{1/2} a(t+\tau)\, b^*(t)\, e^{-j2\pi\nu(2N+1)t}\, \mathrm{d}t \right| .
\end{aligned} \tag{3.172}$$

Let N_{ϵ_2} such that $\nu(2N_{\epsilon_2}+1) \geqslant \bar{\nu}_{\epsilon_2}$. For $N > \max(N_{\epsilon_1}, N_{\epsilon_2})$, the right-hand side of (3.172) is bounded by $\epsilon_1 + \epsilon_2$ with ϵ_1 and ϵ_2 arbitrarily small. Thus, for $\nu \neq 0$ (and hence for ν non-integer due to the periodicity in ν) we have

$$\lim_{N\to\infty} \frac{1}{2N+1} \sum_{n=-N}^{N} a\Big(\frac{n}{2N+1} + \tau\Big)\, b^*\Big(\frac{n}{2N+1}\Big)\, e^{-j2\pi\nu n} = 0 \qquad \forall \tau \in \mathbb{R}. \tag{3.173}$$

Accounting for (3.165) and (3.172) (with $\tau = 0$), we have the following inequality.

$$
\begin{aligned}
&\left| \frac{1}{2N+1} \sum_{n=-N}^{N} a\Big(\frac{n+m}{2N+1}\Big)\, b^*\Big(\frac{n}{2N+1}\Big)\, e^{-j2\pi\widetilde{\alpha}n} \right| \\
&\leqslant \left| \frac{1}{2N+1} \sum_{n=-N}^{N} a\Big(\frac{n+m}{2N+1}\Big)\, b^*\Big(\frac{n}{2N+1}\Big)\, e^{-j2\pi\widetilde{\alpha}n} \right. \\
&\qquad \left. - \frac{1}{2N+1} \sum_{n=-N}^{N} a\Big(\frac{n}{2N+1}\Big)\, b^*\Big(\frac{n}{2N+1}\Big)\, e^{-j2\pi\widetilde{\alpha}n} \right| \\
&\qquad + \left| \frac{1}{2N+1} \sum_{n=-N}^{N} a\Big(\frac{n}{2N+1}\Big)\, b^*\Big(\frac{n}{2N+1}\Big)\, e^{-j2\pi\widetilde{\alpha}n} \right| \\
&\leqslant \frac{\|b\|_\infty}{2N+1} \sum_{n=-N}^{N} \left| a\Big(\frac{n+m}{2N+1}\Big) - a\Big(\frac{n}{2N+1}\Big) \right| + \epsilon_1 + \epsilon_2 \\
&\leqslant \frac{\|b\|_\infty}{2N+1} \, \| \dot{a} \|_\infty \, |m| + \epsilon_1 + \epsilon_2 .
\end{aligned}
\tag{3.174}
$$

The rhs of (3.174) can be made arbitrarily small. This proves (3.163) for $\widetilde{\alpha} \notin \mathbb{Z}$.

This proof is simplified when a slightly modified version of this Lemma is used in Section 3.11 for the proof of the asymptotic covariance of the hybrid cyclic cross-correlogram. Specifically, we have the following.

$$
\begin{aligned}
&\lim_{N\to\infty} \frac{1}{2N+1} \sum_{n=-\infty}^{+\infty} a\Big(\frac{n}{2N+1} + \frac{\tau}{(2N+1)T_s}\Big)\, b^*\Big(\frac{n}{2N+1}\Big)\, e^{-j2\pi\alpha n T_s} \\
&= \int_{-1/2}^{1/2} a(t)\, b^*(t)\, \mathrm{d}t \; \delta_{\alpha \bmod f_s} .
\end{aligned}
\tag{3.175}
$$

□

3.10.1 Proof of Theorem 2.6.4 Asymptotic Expected Value of the Discrete-Time Cyclic Cross-Correlogram

From (2.183) it follows

$$
\begin{aligned}
&\lim_{N\to\infty} \mathrm{E}\left\{ \widetilde{R}_{y_\mathrm{d} x_\mathrm{d}^{(*)}}(\widetilde{\alpha}, m; n_0, N) \right\} \\
&= \lim_{N\to\infty} \sum_{k\in\mathbb{I}} R^{(k)}_{yx^{(*)}}(mT_s)\, V_{\frac{1}{N}}\Big(\widetilde{\alpha} - \alpha_k(mT_s)T_s\Big)\, e^{-j2\pi[\widetilde{\alpha}-\alpha_k(mT_s)T_s]n_0} \\
&= \sum_{k\in\mathbb{I}} R^{(k)}_{yx^{(*)}}(mT_s) \lim_{N\to\infty} V_{\frac{1}{N}}\Big(\widetilde{\alpha} - \alpha_k(mT_s)T_s\Big)\, e^{-j2\pi[\widetilde{\alpha}-\alpha_k(mT_s)T_s]n_0} \\
&= \sum_{k\in\mathbb{I}} R^{(k)}_{yx^{(*)}}(mT_s) \sum_{p=-\infty}^{+\infty} \delta_{\widetilde{\alpha}-\alpha_k(mT_s)T_s+p}
\end{aligned}
\tag{3.176}
$$

where, in the third equality, Lemma 3.10.1 is used.

In the second equality, the interchange of the limit and sum operations is justified by the Weierstrass criterium (Johnsonbaugh and Pfaffenberger 2002, Theorem 62.6; Smirnov 1964) since the series of functions of N

$$\sum_{k \in \mathbb{I}} R^{(k)}_{yx^{(*)}}(mT_s)\, V_{\frac{1}{N}}\left(\widetilde{\alpha} - \alpha_k(mT_s)T_s\right) e^{-j2\pi[\widetilde{\alpha}-\alpha_k(mT_s)T_s]n_0}$$

is uniformly convergent. In fact, by using (3.145) and Assumptions 2.4.3a and 2.4.5, we have

$$\sum_{k \in \mathbb{I}} \left| R^{(k)}_{yx^{(*)}}(mT_s)\right| \left| V_{\frac{1}{N}}\left(\widetilde{\alpha} - \alpha_k(mT_s)T_s\right)\right| \leqslant \|a\|_\infty \sum_{k \in \mathbb{I}} \left\| R^{(k)}_{yx^{(*)}}(mT_s)\right\|_\infty \tag{3.177}$$

with the right-hand side bounded and independent of N.

3.10.2 *Proof of Theorem 2.6.5 Asymptotic Covariance of the Discrete-Time Cyclic Cross-Correlogram*

Let us consider the covariance expression (2.186). As regards the term $\widetilde{\mathcal{T}}_1$ defined in (2.187), let

$$\begin{aligned}
&F_{k'k''\ell}(N) \\
&\quad \triangleq R^{(k')}_{yy^*}((m_1 - m_2 + \ell)T_s)\, R^{(k'')}_{x^{(*)}x^{(*)*}}(\ell T_s) \\
&\qquad e^{j2\pi\alpha'_{k'}((m_1-m_2+\ell)T_s)(n_{02}+m_2)T_s}\, e^{j2\pi\alpha''_{k''}(\ell T_s)n_{02}T_s}\, e^{-j2\pi\widetilde{\alpha}_1\cdot(n_{02}+\ell)}\, e^{j2\pi\widetilde{\alpha}_2 n_{02}} \\
&\qquad \widetilde{r}_{aa^*}\left([\widetilde{\alpha}_1 - \widetilde{\alpha}_2 - \alpha'_{k'}((m_1 - m_2 + \ell)T_s)T_s - \alpha''_{k''}(\ell T_s)T_s], \ell - n_{01} + n_{02}; N\right)
\end{aligned} \tag{3.178}$$

It results that

$$\begin{aligned}
\lim_{N\to\infty} (2N+1)\widetilde{\mathcal{T}}_1 &= \lim_{N\to\infty} \sum_{k'}\sum_{k''}\sum_{\ell} F_{k'k''\ell}(N) \\
&= \sum_{k'}\sum_{k''}\sum_{\ell} \lim_{N\to\infty} F_{k'k''\ell}(N) \\
&= \widetilde{\mathcal{T}}'_1
\end{aligned} \tag{3.179}$$

where $\widetilde{\mathcal{T}}'_1$ is defined in (2.196) and Lemma 3.10.2 (with $a \equiv b$) has been accounted for.

The interchange of the order of limit and sum operations in (3.179) is allowed since the three-index series over (k', k'', ℓ) of functions of N is uniformly convergent for the Weierstrass criterium (Johnsonbaugh and Pfaffenberger 2002, Theorem 62.6), (Smirnov 1964). In fact,

$$\begin{aligned}
&\sum_{k'}\sum_{k''}\sum_{\ell}|F_{k'k''\ell}(N)| \\
&\quad = \sum_{k'}\sum_{k''}\sum_{\ell}\left|R^{(k')}_{yy^*}((m_1-m_2+\ell)T_s)\right| \left|R^{(k'')}_{x^{(*)}x^{(*)*}}(\ell T_s)\right| \\
&\quad \left|\widetilde{r}_{aa^*}([\widetilde{\alpha}_1-\widetilde{\alpha}_2-\alpha'_{k'}((m_1-m_2+\ell)T_s)T_s-\alpha''_{k''}(\ell T_s)T_s],\ell-n_{01}+n_{02};N)\right| \\
&\quad \leqslant \|a\|^2_\infty \sum_{k'}\left\|R^{(k')}_{yy^*}\right\|_\infty \sum_{k''}\sum_{\ell}\left|R^{(k'')}_{x^{(*)}x^{(*)*}}(\ell T_s)\right|
\end{aligned} \tag{3.180}$$

with the right-hand side independent of N, where the inequality

$$\left|\widetilde{r}_{aa^*}(\widetilde{\beta},m;N)\right| = \left|\frac{1}{2N+1}\sum_{n=-\infty}^{+\infty} a\left(\frac{n+m}{2N+1}\right)a^*\left(\frac{n}{2N+1}\right)e^{-j2\pi\widetilde{\beta}n}\right| \leqslant \|a\|^2_\infty \tag{3.181}$$

is used. The inequality in (3.181) follows from the fact that the sum over n is at most on $2N+1$ nonzero terms. The right-hand side of (3.180) is bounded due to Assumptions 2.4.3a and 2.4.8a. In particular, from Assumption 2.4.8a it follows that

$$\sum_k \int_{\mathbb{R}}\left|R^{(k)}_{z_1z_2}(s)\right|\,\mathrm{d}s<\infty \Rightarrow \sum_k\sum_\ell\left|R^{(k)}_{z_1z_2}(\ell T_s)\right|<\infty \tag{3.182}$$

In fact, $\sum_\ell |R^{(k)}_{z_1z_2}(\ell T_s)|T_s$ is the Riemann sum for the integral $\int_{\mathbb{R}}|R^{(k)}_{z_1z_2}(s)|\,\mathrm{d}s$, with the function $|R^{(k)}_{z_1z_2}(s)|$ assumed to be Riemann integrable.

Analogously, it can be shown that

$$\lim_{N\to\infty}(2N+1)\widetilde{\mathcal{T}}_2=\widetilde{\mathcal{T}}'_2 \tag{3.183}$$

where $\widetilde{\mathcal{T}}'_2$ is defined in (2.197).

As regards the term $\widetilde{\mathcal{T}}_3$ defined in (2.189), let

$$\begin{aligned}
F_{k\ell}(N) &\triangleq C^{(k)}_{yx^{(*)}y^*x^{(*)*}}((m_1+\ell)T_s,\ell T_s,m_2T_s) \\
&\quad e^{j2\pi[\beta_k((m_1+\ell)T_s,\ell T_s,m_2T_s)T_s-\widetilde{\alpha}_1+\widetilde{\alpha}_2]n_{02}}\,e^{-j2\pi\widetilde{\alpha}_1\ell} \\
&\quad \widetilde{r}_{aa^*}\left([\widetilde{\alpha}_1-\widetilde{\alpha}_2-\beta_k((m_1+\ell)T_s,\ell T_s,m_2T_s)T_s],\ell-n_{01}+n_{02};N\right)
\end{aligned} \tag{3.184}$$

It results that

$$\begin{aligned}
\lim_{N\to\infty}(2N+1)\widetilde{\mathcal{T}}_3 &= \lim_{N\to\infty}\sum_k\sum_\ell F_{k\ell}(N) \\
&= \sum_k\sum_\ell \lim_{N\to\infty}F_{k\ell}(N) \\
&= \widetilde{\mathcal{T}}'_3
\end{aligned} \tag{3.185}$$

where $\widetilde{\mathcal{T}}'_3$ is defined in (2.198) and Lemma 3.10.2 (with $a\equiv b$) has been accounted for.

The interchange of the order of limit and sum operations in (3.185) is allowed since the double-index series over k and ℓ of functions of N is uniformly convergent for the Weierstrass criterium (Johnsonbaugh and Pfaffenberger 2002, Theorem 62.6), (Smirnov 1964). In fact,

$$\begin{aligned}\sum_k \sum_\ell |F_{k\ell}(N)| &= \sum_k \sum_\ell \left| C^{(k)}_{yx^{(*)}y^*x^{(*)*}}((m_1+\ell)T_s, \ell T_s, m_2 T_s) \right| \\ &\quad \left| \widetilde{r}_{aa^*}\left([\widetilde{\alpha}_1 - \widetilde{\alpha}_2 - \beta_k((m_1+\ell)T_s, \ell T_s, m_2 T_s)T_s], \ell - n_{01} + n_{02}; N \right) \right| \\ &\leqslant \|a\|^2_\infty \sum_k \sum_\ell \left| C^{(k)}_{yx^{(*)}y^*x^{(*)*}}((m_1+\ell)T_s, \ell T_s, m_2 T_s) \right| \end{aligned} \tag{3.186}$$

with the right-hand side independent of N, where the inequality (3.181) is used. The right-hand side of (3.186) is bounded due to Assumption 2.4.8b from which it follows that

$$\begin{aligned} &\sum_k \int_{\mathbb{R}} \left| C^{(k)}_{yx^{(*)}y^*x^{(*)*}}(m_1 T_s + s, s, m_2 T_s) \right| \, \mathrm{d}s < \infty \\ &\quad \Rightarrow \sum_k \sum_\ell \left| C^{(k)}_{yx^{(*)}y^*x^{(*)*}}((m_1+\ell)T_s, \ell T_s, m_2 T_s) \right| < \infty \end{aligned} \tag{3.187}$$

under the assumption that the function $|C^{(k)}_{yx^{(*)}y^*x^{(*)*}}(m_1 T_s + s, s, m_2 T_s)|$ is Riemann integrable with respect to s.

As a final remark, observe that the sum over ℓ in (3.180) is finite. In fact, $a(t)$ has finite support $[-1/2, 1/2]$ and, hence, $\widetilde{r}_{aa^*}(\widetilde{\beta}, m; N) \neq 0$ only for $-2N \leqslant m \leqslant 2N$. Consequently, in (3.180), $-2N \leqslant \ell - n_{01} + n_{02} \leqslant 2N$. This fact, however, is not sufficient to make the right-hand side of (3.180) independent of N. Therefore, Assumption 2.4.8a needs to be exploited.

Lemma 3.10.3 *Let $\epsilon \in [-1/2, 1/2]$, $\epsilon \neq 0$, and $\gamma > 1$. Thus*

$$\sum_{\ell=-\infty}^{+\infty} \frac{1}{|\epsilon - \ell|^\gamma} \leqslant \frac{1}{|\epsilon|^\gamma} + C_\gamma \tag{3.188}$$

where $C_\gamma \triangleq \dfrac{2\gamma - 1}{\gamma - 1} 2^\gamma$.

Proof: For $\epsilon \in [-1/2, 1/2]$, $\epsilon \neq 0$ one has

$$\begin{aligned} \sum_{\ell=1}^{+\infty} \frac{1}{(\ell - \epsilon)^\gamma} &= \frac{1}{(1-\epsilon)^\gamma} + \sum_{\ell=2}^{+\infty} \frac{1}{(\ell - \epsilon)^\gamma} \\ &\leqslant \frac{1}{(1-\epsilon)^\gamma} + \int_1^{+\infty} \frac{1}{(\lambda - \epsilon)^\gamma} \, \mathrm{d}\lambda \\ &= \frac{1}{(1-\epsilon)^\gamma} + \frac{1}{-\gamma + 1} \left[\frac{1}{(\lambda - \epsilon)^{\gamma - 1}} \right]_1^{+\infty} \\ &= \frac{1}{(1-\epsilon)^\gamma} + \frac{1}{\gamma - 1} \frac{1}{(1-\epsilon)^{\gamma-1}} \end{aligned} \tag{3.189}$$

Thus,

$$
\begin{aligned}
\sum_{\ell=-\infty}^{+\infty} \frac{1}{|\epsilon-\ell|^{\gamma}} &= \frac{1}{|\epsilon|^{\gamma}} + \sum_{\ell=-\infty}^{-1} \frac{1}{|\epsilon-\ell|^{\gamma}} + \sum_{\ell=1}^{+\infty} \frac{1}{|\epsilon-\ell|^{\gamma}} \\
&= \frac{1}{|\epsilon|^{\gamma}} + \sum_{\ell=1}^{+\infty} \frac{1}{(\ell+\epsilon)^{\gamma}} + \sum_{\ell=1}^{+\infty} \frac{1}{(\ell-\epsilon)^{\gamma}} \\
&\leqslant \frac{1}{|\epsilon|^{\gamma}} + \frac{1}{(1+\epsilon)^{\gamma}} + \frac{1}{\gamma-1}\frac{1}{(1+\epsilon)^{\gamma-1}} + \frac{1}{(1-\epsilon)^{\gamma}} + \frac{1}{\gamma-1}\frac{1}{(1-\epsilon)^{\gamma-1}} \\
&\leqslant \frac{1}{|\epsilon|^{\gamma}} + \frac{1}{(1/2)^{\gamma}} + \frac{1}{\gamma-1}\frac{1}{(1/2)^{\gamma-1}} + \frac{1}{(1/2)^{\gamma}} + \frac{1}{\gamma-1}\frac{1}{(1/2)^{\gamma-1}} \\
&= \frac{1}{|\epsilon|^{\gamma}} + \frac{2\gamma-1}{\gamma-1} 2^{\gamma}
\end{aligned}
\tag{3.190}
$$

□

3.10.3 *Proof of Theorem 2.6.7 Rate of Convergence of the Bias of the Discrete-Time Cyclic Cross-Correlogram*

From (2.179b) and (2.200) it follows

$$
\begin{aligned}
\widetilde{R}_{y_{\mathrm{d}} x_{\mathrm{d}}^{(*)}}(\widetilde{\alpha}, m) &= \sum_{k\in\mathbb{I}} R^{(k)}_{yx^{(*)}}(mT_s)\, \delta_{[\widetilde{\alpha}-\alpha_k(mT_s)T_s] \bmod 1} \\
&= \sum_{k\in\mathbb{J}_{\widetilde{\alpha},m,T_s}} R^{(k)}_{yx^{(*)}}(mT_s)
\end{aligned}
\tag{3.191}
$$

and from (2.183) we have

$$
\mathrm{E}\left\{\widetilde{R}_{y_{\mathrm{d}} x_{\mathrm{d}}^{(*)}}(\widetilde{\alpha}, m; n_0, N)\right\} = \sum_{k\in\mathbb{I}} R^{(k)}_{yx^{(*)}}(mT_s)\, V_{\frac{1}{N}}\left(\widetilde{\alpha}-\alpha_k(mT_s)T_s\right) e^{-j2\pi[\widetilde{\alpha}-\alpha_k(mT_s)T_s]n_0}
$$

Therefore,

$$
\begin{aligned}
&\left|\mathrm{E}\left\{\widetilde{R}_{y_{\mathrm{d}} x_{\mathrm{d}}^{(*)}}(\widetilde{\alpha}, m; n_0, N)\right\} - \widetilde{R}_{y_{\mathrm{d}} x_{\mathrm{d}}^{(*)}}(\widetilde{\alpha}, m)\right| \\
&\quad = \Bigg| \sum_{k\in\mathbb{J}_{\widetilde{\alpha},m,T_s}} R^{(k)}_{yx^{(*)}}(mT_s)\, V_{\frac{1}{N}}(0) \\
&\qquad + \sum_{k\notin\mathbb{J}_{\widetilde{\alpha},m,T_s}} R^{(k)}_{yx^{(*)}}(mT_s)\, V_{\frac{1}{N}}\left(\widetilde{\alpha}-\alpha_k(mT_s)T_s\right) e^{-j2\pi[\widetilde{\alpha}-\alpha_k(mT_s)T_s]n_0} \\
&\qquad - \sum_{k\in\mathbb{J}_{\widetilde{\alpha},m,T_s}} R^{(k)}_{yx^{(*)}}(mT_s)\Bigg| \\
&\quad \leqslant \sum_{k\in\mathbb{J}_{\widetilde{\alpha},m,T_s}} \left|R^{(k)}_{yx^{(*)}}(mT_s)\right| \left|1 - V_{\frac{1}{N}}(0)\right|
\end{aligned}
$$

$$+ \sum_{k \notin \mathbb{J}_{\widetilde{\alpha},m,T_s}} \left| R^{(k)}_{yx^{(*)}}(mT_s) \right| \left| V_{\frac{1}{N}}\left(\widetilde{\alpha} - \alpha_k(mT_s)T_s\right) \right|$$

$$\leqslant \sum_{k \in \mathbb{I}} \left\| R^{(k)}_{yx^{(*)}} \right\|_{\infty} \left| 1 - V_{\frac{1}{N}}(0) \right|$$

$$+ \sum_{k \notin \mathbb{J}_{\widetilde{\alpha},m,T_s}} \left| R^{(k)}_{yx^{(*)}}(mT_s) \right| \left| V_{\frac{1}{N}}\left(\widetilde{\alpha} - \alpha_k(mT_s)T_s\right) \right| \tag{3.192}$$

Let us consider the first term in the rhs of (3.192). Accounting for (2.184) we have

$$V_{\frac{1}{N}}(0) = A(0) + \sum_{\substack{\ell=-\infty \\ \ell \neq 0}}^{+\infty} A\Big(\ell(2N+1)\Big) \tag{3.193}$$

where $A(f)$ is the Fourier transform of $a(t)$ and $A(0) = 1$ (see (3.64)). According to Assumption 2.4.5, we have $|A(f)| \leqslant K/|f|^{\gamma}$ for $|f|$ sufficiently large, say, $|f| > f^*$. Thus, for N sufficiently large such that $2N + 1 > f^*$ (and for $|\ell| \geqslant 1$), we have

$$\begin{aligned}
\left| 1 - V_{\frac{1}{N}}(0) \right| &= \Big| \sum_{\substack{\ell=-\infty \\ \ell \neq 0}}^{+\infty} A\Big(\ell(2N+1)\Big) \Big| \\
&\leqslant \sum_{\substack{\ell=-\infty \\ \ell \neq 0}}^{+\infty} \Big| A\Big(\ell(2N+1)\Big) \Big| \\
&\leqslant \frac{K}{(2N+1)^{\gamma}} \sum_{\substack{\ell=-\infty \\ \ell \neq 0}}^{+\infty} \frac{1}{|\ell|^{\gamma}} \\
&= \frac{K_{\gamma}}{(2N+1)^{\gamma}}
\end{aligned} \tag{3.194}$$

where the sum in the third line is convergent when $\gamma > 1$. Note that in the special case of a rectangular data-tapering window $a(t) = \text{rect}(t) \Rightarrow A(f) = \text{sinc}(f)$. Even if $\gamma = 1$, the result is that $A(\ell(2N + 1)) = 0$ for $\ell \neq 0$, N integer. Thus, $|1 - V_{\frac{1}{N}}(0)| = 0$.

Let us now consider the second term in the rhs of (3.192).

From (2.184) we have

$$V_{\frac{1}{N}}(\widetilde{\alpha} - \alpha_k(mT_s)T_s) = \sum_{\ell=-\infty}^{+\infty} A\Big((\widetilde{\alpha} - \alpha_k(mT_s)T_s - \ell)(2N+1)\Big) \tag{3.195}$$

with $(\widetilde{\alpha} - \alpha_k(mT_s)T_s - \ell)$ never equal to zero since $\widetilde{\alpha} - \alpha_k(mT_s)T_s \neq$ integer when $k \notin \mathbb{J}_{\widetilde{\alpha},m,T_s}$. Moreover, under Assumption 2.6.6,

$$h_{\widetilde{\alpha},m,T_s} \triangleq \inf_{k \notin \mathbb{J}_{\widetilde{\alpha},m,T_s}} \Big| [\widetilde{\alpha} - \alpha_k(mT_s)T_s] \bmod 1 \Big| > 0. \tag{3.196}$$

Let N be sufficiently large such that $h_{\widetilde{\alpha},m,T_s}(2N+1) > f^*$. Thus, for all $k \notin \mathbb{J}_{\widetilde{\alpha},m,T_s}$ and for all $\ell \in \mathbb{Z}$ it results in $|\widetilde{\alpha} - \alpha_k(mT_s)T_s - \ell|(2N+1) > f^*$ and, hence,

$$\left|A\Big((\widetilde{\alpha} - \alpha_k(mT_s)T_s - \ell)(2N+1)\Big)\right| \leqslant \frac{K}{|\widetilde{\alpha} - \alpha_k(mT_s)T_s - \ell|^{\gamma}(2N+1)^{\gamma}} \tag{3.197}$$

By taking $\epsilon_k = (\widetilde{\alpha} - \alpha_k(mT_s)T_s) \bmod 1$ in Lemma 3.10.3 ($\epsilon_k \in [-1/2, 1/2)$, $\epsilon_k \neq 0$ for $k \notin \mathbb{J}_{\widetilde{\alpha},m,T_s}$), for every fixed $\widetilde{\alpha}$, m, T_s, and k one obtains

$$\begin{aligned}\sum_{\ell=-\infty}^{+\infty} \frac{1}{|\widetilde{\alpha} - \alpha_k(mT_s)T_s - \ell|^{\gamma}} &= \sum_{\ell=-\infty}^{+\infty} \frac{1}{|(\widetilde{\alpha} - \alpha_k(mT_s)T_s) \bmod 1 - \ell|^{\gamma}} \\ &\leqslant \frac{1}{|(\widetilde{\alpha} - \alpha_k(mT_s)T_s) \bmod 1|^{\gamma}} + C_{\gamma}\end{aligned} \tag{3.198}$$

Thus, by using (3.194)–(3.198) into (3.192), we have

$$\begin{aligned}&\left|\mathrm{E}\left\{\widetilde{R}_{y_{\mathrm{d}}x_{\mathrm{d}}^{(*)}}(\widetilde{\alpha}, m; n_0, N)\right\} - \widetilde{R}_{y_{\mathrm{d}}x_{\mathrm{d}}^{(*)}}(\widetilde{\alpha}, m)\right| \\ &\quad\leqslant \sum_{k\in\mathbb{I}} \left\|R^{(k)}_{yx^{(*)}}\right\|_{\infty} \frac{K_{\gamma}}{(2N+1)^{\gamma}} \\ &\qquad + \sum_{k \notin \mathbb{J}_{\widetilde{\alpha},m,T_s}} \left|R^{(k)}_{yx^{(*)}}(mT_s)\right| \frac{K}{(2N+1)^{\gamma}} \left[\frac{1}{|(\widetilde{\alpha} - \alpha_k(mT_s)T_s) \bmod 1|^{\gamma}} + C_{\gamma}\right] \\ &\quad\leqslant \sum_{k\in\mathbb{I}} \left\|R^{(k)}_{yx^{(*)}}\right\|_{\infty} \frac{1}{(2N+1)^{\gamma}} \left[\frac{K}{(h_{\widetilde{\alpha},m,T_s})^{\gamma}} + K_{\gamma} + KC_{\gamma}\right]\end{aligned} \tag{3.199}$$

where, in the last inequality, Assumption 2.6.6 ($h_{\widetilde{\alpha},m,T_s} > 0$) is used. Thus, (2.202) immediately follows.

For a rectangular lag-product-tapering window, $V_{\frac{1}{N}}(0) = 1$, $K_{\gamma} = 0$ can be taken in (3.194), and (3.199) (with $\gamma = 1$) must be accordingly modified .

For a rectangular signal-tapering window, the corresponding lag-product-tapering window $v_{N,m}(n)$ has Fourier transform $V_{\frac{1}{N},m}(\nu)$ such that (Fact 3.5.3)

$$1 - V_{\frac{1}{N},m}(0) = \mathcal{O}\Big(\frac{1}{N}\Big). \tag{3.200}$$

Thus, (3.200) and (3.199) (with $\gamma = 1$) lead to

$$\left|\mathrm{E}\left\{\widetilde{R}_{y_{\mathrm{d}}x_{\mathrm{d}}^{(*)}}(\widetilde{\alpha}, m; n_0, N)\right\} - \widetilde{R}_{y_{\mathrm{d}}x_{\mathrm{d}}^{(*)}}(\widetilde{\alpha}, m)\right| = \mathcal{O}\Big(\frac{1}{N}\Big). \tag{3.201}$$

3.10.4 Proof of Lemma 2.6.9 Rate of Convergence to Zero of Cumulants of Discrete-Time Cyclic Cross-Correlograms

By using (2.181), (2.182), and the multilinearity property of cumulants we have

$$
\begin{aligned}
&\operatorname{cum}\Big\{ \widetilde{R}^{[*]_i}_{y_{\rm d} x^{(*)}_{\rm d}}(\widetilde{\alpha}_i, m_i; n_{0i}, N),\ \ i = 1, \dots, k \Big\} \\
&= \operatorname{cum}\Big\{ \sum_{n=n_{0i}-N}^{n_{0i}+N} v_N^{[*]_i}(n - n_{0i})\, z_i(n)\, e^{-[-]_i j2\pi\widetilde{\alpha}_i n},\ \ i = 1, \dots, k \Big\} \\
&= \sum_{n_1=n_{01}-N}^{n_{01}+N} \cdots \sum_{n_k=n_{0k}-N}^{n_{0k}+N} \left[\prod_{i=1}^{k} v_N^{[*]_i}(n_i - n_{0i})\, e^{-[-]_i j2\pi\widetilde{\alpha}_i n_i} \right] \\
&\quad \operatorname{cum}\Big\{ z_1(n_1), \dots, z_k(n_k) \Big\} \\
&= \sum_{n\in\mathbb{Z}} \sum_{\boldsymbol{r}\in\mathbb{Z}^{k-1}} \left(\prod_{i=1}^{k-1} \frac{1}{2N+1}\, a^{[*]_i}\left(\frac{n + r_i - n_{0i}}{2N+1}\right) e^{-[-]_i j2\pi\widetilde{\alpha}_i (n+r_i)} \right. \\
&\quad \left. \frac{1}{2N+1}\, a^{[*]_k}\left(\frac{n - n_{0k}}{2N+1}\right) e^{-[-]_k j2\pi\widetilde{\alpha}_k n} \right) \\
&\quad \operatorname{cum}\Big\{ z_k(n), z_i(n + r_i),\ i = 1, \dots, k-1 \Big\}
\end{aligned}
\tag{3.202}
$$

where, in the third equality, the variable changes $n_k = n$ and $n_i = n + r_i, i = 1, \dots, k-1$ are made, $\boldsymbol{r} \triangleq [r_1, \dots, r_{k-1}]$, and all the sums are finite since the function $a(\cdot)$ has finite support (Assumption 2.4.5). Note that the order of finite sums can be freely interchanged. From (3.202) it follows that

$$
\begin{aligned}
&\left| \operatorname{cum}\Big\{ \widetilde{R}^{[*]_i}_{y_{\rm d} x^{(*)}_{\rm d}}(\widetilde{\alpha}_i, m_i; n_{0i}, N),\ \ i = 1, \dots, k \Big\} \right| \\
&\leqslant \frac{\|a\|_\infty^{k-1}}{(2N+1)^{k-1}} \sum_{\boldsymbol{r}} \phi(r_1 T_s, \dots, r_{k-1} T_s) \sum_{n} \frac{1}{2N+1} \left| a\left(\frac{n - n_{0k}}{2N+1}\right) \right|
\end{aligned}
\tag{3.203}
$$

where Assumption 2.6.8 is used and all the sums converge. In fact, $\sum_{\boldsymbol{r}} \phi(r_1 T_s, \dots, r_{k-1} T_s) T_s$ is the Riemann sum for $\int_{\mathbb{R}^{k-1}} \phi(s_1, \dots, s_{k-1})\, \mathrm{d}s_1 \cdots \mathrm{d}s_{k-1}$ and $\sum_n |a((n - n_{0k})/(2N+1))|/(2N+1)$ is the Riemann sum for $\int_{\mathbb{R}} |a(t)|\, \mathrm{d}t$. Thus, (2.206) immediately follows.

This lemma is the discrete-time counterpart of Lemma 2.4.17 on the rate of convergence to zero of the cumulant of continuous-time cyclic cross-correlograms. Note that in the discrete-time case Assumption 2.4.16 is not used. In fact, such assumption is used in the continuous-time case in Lemma 2.4.17 to interchange the order of integrals. In the discrete-time case, as a consequence of the finite support of $a(\cdot)$, all the sums are finite and their order can be freely interchanged. □

3.10.5 Proof of Theorem 2.6.10 Asymptotic Joint Normality of the Discrete-Time Cyclic Cross-Correlograms

From Theorem 2.6.7 holding for $\gamma > 1$ (or $\gamma = 1$ if $a(t) = \text{rect}(t)$) we have

$$\lim_{N\to\infty} \mathrm{E}\left\{ \sqrt{2N+1} \left[\widetilde{R}_{y_\mathrm{d} x_\mathrm{d}^{(*)}}(\widetilde{\alpha}, m; n_0, N) - \widetilde{R}_{y_\mathrm{d} x_\mathrm{d}^{(*)}}(\widetilde{\alpha}, m) \right] \right\} = 0. \tag{3.204}$$

From Theorem 2.6.5 it follows that the asymptotic covariance

$$\begin{aligned} \lim_{N\to\infty} \mathrm{cov}\Big\{ & \sqrt{2N+1} \left[\widetilde{R}_{y_\mathrm{d} x_\mathrm{d}^{(*)}}(\widetilde{\alpha}_1, m_1; n_{01}, N) - \widetilde{R}_{y_\mathrm{d} x_\mathrm{d}^{(*)}}(\widetilde{\alpha}_1, m_1) \right], \\ & \sqrt{2N+1} \left[\widetilde{R}_{y_\mathrm{d} x_\mathrm{d}^{(*)}}(\widetilde{\alpha}_2, m_2; n_{02}, N) - \widetilde{R}_{y_\mathrm{d} x_\mathrm{d}^{(*)}}(\widetilde{\alpha}_2, m_2) \right] \Big\} \end{aligned} \tag{3.205}$$

is finite. From Theorem 3.13.2 it follows that the asymptotic conjugate covariance is finite. Moreover, from Lemma 2.6.9 with $\epsilon = \frac{k}{2} - 1$ and $k \geqslant 3$ we have

$$\begin{aligned} & \lim_{N\to\infty} (2N+1)^{k/2} \mathrm{cum}\left\{ \widetilde{R}^{[*]_i}_{y_\mathrm{d} x_\mathrm{d}^{(*)}}(\widetilde{\alpha}_i, m_i; n_{0i}, N),\ i = 1, \ldots, k \right\} \\ & = \lim_{N\to\infty} \mathrm{cum}\left\{ \sqrt{2N+1}\ \widetilde{R}^{[*]_i}_{y_\mathrm{d} x_\mathrm{d}^{(*)}}(\widetilde{\alpha}_i, m_i; n_{0i}, N),\ i = 1, \ldots, k \right\} \\ & = \lim_{N\to\infty} \mathrm{cum}\left\{ \sqrt{2N+1} \left[\widetilde{R}_{y_\mathrm{d} x_\mathrm{d}^{(*)}}(\widetilde{\alpha}_i, m_i; n_{0i}, N) - \widetilde{R}_{y_\mathrm{d} x_\mathrm{d}^{(*)}}(\widetilde{\alpha}_i, m_i) \right]^{[*]_i}, i = 1, \ldots, k \right\} \\ & = 0 \end{aligned} \tag{3.206}$$

where the second equality holds since the value of the cumulant does not change by adding a constant to each of the random variables (Brillinger 1981, Theorem 2.3.1).

Thus, according to the results of Section 1.4.2, for every fixed $\widetilde{\alpha}_i, m_i, n_{0i}$ the random variables $\sqrt{2N+1}[\widetilde{R}_{y_\mathrm{d} x_\mathrm{d}^{(*)}}(\widetilde{\alpha}_i, m_i; n_{0i}, N) - \widetilde{R}_{y_\mathrm{d} x_\mathrm{d}^{(*)}}(\widetilde{\alpha}_i, m_i)]$, $i = 1, \ldots, k$, are asymptotically ($N \to \infty$) zero-mean jointly complex Normal (Picinbono 1996; van den Bos 1995).

3.11 Proofs for Section 2.6.3 "Asymptotic Results as $N \to \infty$ and $T_s \to 0$"

3.11.1 Proof of Lemma 2.6.12

Condition (2.209) holding uniformly w.r.t. τ assures that Assumption 2.6.11 is verified. The numbers M_p, possibly depending on $\widetilde{\alpha}$, are independent of T_s. Under Assumption 2.6.11, the Weierstrass M-test (Johnsonbaugh and Pfaffenberger 2002, Theorem 62.6) assures the uniform convergence of the series of functions of T_s

$$\sum_{p=-\infty}^{+\infty} R_{yx^{(*)}}\Big((\widetilde{\alpha} + p) f_s, m T_s \Big).$$

Therefore, the limit operation can be interchanged with the infinite sum

$$\lim_{T_s \to 0} \sum_{\substack{p=-\infty \\ p \neq 0}}^{+\infty} R_{yx^{(*)}}\left((\widetilde{\alpha} + p) f_s, mT_s\right) = \sum_{\substack{p=-\infty \\ p \neq 0}}^{+\infty} \lim_{T_s \to 0} R_{yx^{(*)}}\left((\widetilde{\alpha} + p) f_s, mT_s\right) = 0 \quad (3.207)$$

where, in the second equality the sufficient condition (2.209) for Assumption 2.6.11 is used.

3.11.2 *Proof of Theorem 2.6.13 Mean-Square Consistency of the Discrete-Time Cyclic Cross-Correlogram*

From Lemma 2.6.12 we have

$$\forall \epsilon_1 > 0 \quad \exists T_{\epsilon_1} > 0 \,:\, 0 < T_s < T_{\epsilon_1} \Rightarrow \left| \sum_{\substack{p=-\infty \\ p \neq 0}}^{+\infty} R_{yx^{(*)}}((\widetilde{\alpha} + p) f_s, mT_s) \right| < \epsilon_1 \quad (3.208)$$

(not necessarily uniformly with respect to $\widetilde{\alpha}$ and m).

From Theorems 2.6.4 and 2.6.5 it follows that, for every fixed T_s, $\widetilde{\alpha}$, and m,

$$\lim_{N \to \infty} \mathrm{E}\left\{ \left| \widetilde{R}_{y_d x_d^{(*)}}(\widetilde{\alpha}, m; n_0, N) - \widetilde{R}_{y_d x_d^{(*)}}(\widetilde{\alpha}, m) \right|^2 \right\} = 0 \quad (3.209)$$

that is,

$$\begin{aligned} &\forall \epsilon_2 > 0 \quad \exists N_{\epsilon_2} > 0 \,: \\ &\quad N > N_{\epsilon_2} \Rightarrow \mathrm{E}\left\{ \left| \widetilde{R}_{y_d x_d^{(*)}}(\widetilde{\alpha}, m; n_0, N) - \widetilde{R}_{y_d x_d^{(*)}}(\widetilde{\alpha}, m) \right|^2 \right\} < \epsilon_2 \end{aligned} \quad (3.210)$$

Therefore, for $0 < T_s < T_{\epsilon_1}$ and $N > N_{\epsilon_2}$ we have

$$\begin{aligned} &\mathrm{E}\left\{ \left| \widetilde{R}_{y_d x_d^{(*)}}(\widetilde{\alpha}, m; n_0, N) - R_{yx^{(*)}}(\widetilde{\alpha} f_s, mT_s) \right|^2 \right\} \\ &\quad = \mathrm{E}\left\{ \left| \widetilde{R}_{y_d x_d^{(*)}}(\widetilde{\alpha}, m; n_0, N) - \widetilde{R}_{y_d x_d^{(*)}}(\widetilde{\alpha}, m) \right.\right. \\ &\qquad \left.\left. + \sum_{\substack{p=-\infty \\ p \neq 0}}^{+\infty} R_{yx^{(*)}}((\widetilde{\alpha} + p) f_s, mT_s) \right|^2 \right\} \\ &\quad \leqslant \mathrm{E}\left\{ \left| \widetilde{R}_{y_d x_d^{(*)}}(\widetilde{\alpha}, m; n_0, N) - \widetilde{R}_{y_d x_d^{(*)}}(\widetilde{\alpha}, m) \right|^2 \right\} \\ &\qquad + \left| \sum_{\substack{p=-\infty \\ p \neq 0}}^{+\infty} R_{yx^{(*)}}((\widetilde{\alpha} + p) f_s, mT_s) \right|^2 \\ &\quad \leqslant \epsilon_2 + \epsilon_1^2 \end{aligned} \quad (3.211)$$

with ϵ_1 and ϵ_2 arbitrarily small. Note that the limit is not necessarily uniform in $(\widetilde{\alpha}, m)$.

Equation (3.209) holds for fixed (and finite) $T_s > 0$. Therefore, in (2.211) we have that first $N \to \infty$ and then $T_s \to 0$. This order of the two limits is in agreement with the result of (Dehay 2007) where $T_s = N^{-\delta}$ with $0 < \delta < 1$.

3.11.3 Noninteger Time-Shift

Let us consider a discrete-time signal $x(n)$ with Fourier transform $X(\nu)$.

$$x(n) = \int_{-1/2}^{1/2} X(\nu)\, e^{j2\pi\nu n}\, \mathrm{d}\nu \overset{\mathcal{F}}{\longleftrightarrow} X(\nu) = \sum_{n=-\infty}^{+\infty} x(n)\, e^{-j2\pi\nu n} \tag{3.212}$$

Definition 3.11.1 *The time-shifted version of $x(n)$ with noninteger time-shift μ is defined as*

$$x_\mu(n) \triangleq x(n+\mu) \triangleq \int_{-1/2}^{1/2} X(\nu)\, e^{j2\pi\nu(n+\mu)}\, \mathrm{d}\nu \tag{3.213}$$

□

Fact 3.11.2 *If $x(n)$ is a finite N-length sequence defined for $n \in \{0, 1, \ldots, N-1\}$, the time-shifted version of $x(n)$ with noninteger time-shift μ, can be expressed by the* exact interpolation formula

$$\begin{aligned} x_\mu(n) &= \int_{-1/2}^{1/2} X(\nu)\, e^{j2\pi\nu(n+\mu)}\, \mathrm{d}\nu \\ &= \int_{-1/2}^{1/2} \left[\sum_{k=0}^{N-1} \frac{1}{N} X\left(\frac{k}{N}\right) D_{\frac{1}{N}}\left(\nu - \frac{k}{N}\right) \right] e^{j2\pi\nu(n+\mu)}\, \mathrm{d}\nu \\ &= \sum_{k=0}^{N-1} \frac{1}{N} X\left(\frac{k}{N}\right) \int_{-1/2}^{1/2} D_{\frac{1}{N}}\left(\nu - \frac{k}{N}\right) e^{j2\pi\nu(n+\mu)}\, \mathrm{d}\nu \\ &= \sum_{k=0}^{N-1} \frac{1}{N} X\left(\frac{k}{N}\right) e^{j2\pi(k/N)(n+\mu)}\, \mathcal{R}_N(n+\mu). \end{aligned} \tag{3.214}$$

In (3.214),

$$X\left(\frac{k}{N}\right) = \mathrm{DFT}[x(n)] \triangleq \sum_{n=0}^{N-1} x(n)\, e^{-j2\pi(k/N)n} \qquad k = 0, 1, \ldots, N-1 \tag{3.215}$$

is the discrete Fourier transform (DFT) of $x(n)$ and

$$\mathcal{R}_N(n) \triangleq \begin{cases} 1 & n \in \{0, 1, \ldots, N-1\} \\ 0 & \text{otherwise} \end{cases} \tag{3.216a}$$

$$\overset{\mathcal{F}}{\longleftrightarrow} \quad D_{\frac{1}{N}}(\nu) = e^{-j\pi\nu(N-1)} \frac{\sin(\pi\nu N)}{\sin(\pi\nu)} \tag{3.216b}$$

with $D_{\frac{1}{N}}(\nu)$ *referred to as Dirichlet kernel. According to Definition 3.11.1, we have*

$$\mathcal{R}_N(n+\mu) \triangleq \int_{-1/2}^{1/2} D_{\frac{1}{N}}(\nu)\, e^{j2\pi\nu(n+\mu)}\, \mathrm{d}\nu. \tag{3.217}$$

Alternatively, the time-shifted version of $x(n)$ *with noninteger time-shift* μ*, can be expressed by the* exact interpolation formula

$$\begin{aligned} x_\mu(n) &= \int_{-1/2}^{1/2} X(\nu)\, e^{j2\pi\nu(n+\mu)}\, \mathrm{d}\nu \\ &= \int_{-1/2}^{1/2} \sum_{m=0}^{N-1} x(m)\, e^{-j2\pi\nu m}\, e^{j2\pi\nu(n+\mu)}\, \mathrm{d}\nu \\ &= \sum_{m=0}^{N-1} x(m) \int_{-1/2}^{1/2} e^{j2\pi\nu(n-m+\mu)}\, \mathrm{d}\nu \\ &= \sum_{m=0}^{N-1} x(m)\, \mathrm{sinc}(n-m+\mu). \end{aligned} \tag{3.218}$$

□

3.11.4 *Proof of Theorem 2.6.16 Asymptotic Expected Value of the Hybrid Cyclic Cross-Correlogram*

Use a simple modification of Lemma 2.6.12 with mT_s replaced by τ and then use (2.217a).

3.11.5 *Proof of Theorem 2.6.17 Rate of Convergence of the Bias of the Hybrid Cyclic Cross-Correlogram*

Let us consider the set

$$\mathbb{J}_{\alpha,\tau,T_s} \triangleq \left\{k \in \mathbb{I} \,:\, \alpha_k(\tau) = \alpha \bmod f_s,\ R^{(k)}_{yx^{(*)}}(\tau) \neq 0\right\} \tag{3.219}$$

defined in (2.222). Since $\mathbb{I}$ is assumed to be finite and the functions $\alpha_k(\tau)$ are bounded for finite τ, the sampling period T_s can be chosen sufficiently small such that

$$\max_{k\in\mathbb{I}} |(\alpha - \alpha_k(\tau))T_s| < \frac{1}{2}. \tag{3.220}$$

that is, mod f_s is not necessary in the definition (3.219). Moreover, if $k \notin \mathbb{J}_{\alpha,\tau,T_s}$, then $\alpha_k(\tau) \neq \alpha$. Thus, for T_s sufficiently small (depending on α and τ) the result is that

$$0 < |(\alpha - \alpha_k(\tau))T_s| < \frac{1}{2} \qquad \forall k \notin \mathbb{J}_{\alpha,\tau,T_s} \tag{3.221}$$

In addition, since $\mathbb{I}$ is finite, for T_s sufficiently small $(\alpha - \alpha_k(\tau))T_s = 0$ only if $\alpha = \alpha_k(\tau)$. Therefore, for T_s sufficiently small (depending on α and τ) we have

$$\begin{aligned} \mathbb{J}_{\alpha,\tau,T_s} &\triangleq \left\{k \in \mathbb{I} \,:\, \alpha_k(\tau) = \alpha \bmod f_s,\ R^{(k)}_{yx^{(*)}}(\tau) \neq 0\right\} \\ &= \left\{k \in \mathbb{I} \,:\, \alpha_k(\tau) = \alpha,\ R^{(k)}_{yx^{(*)}}(\tau) \neq 0\right\} \\ &\triangleq \mathbb{J}_{\alpha,\tau} \end{aligned} \tag{3.222}$$

finite set not depending on T_s. Consequently,

$$\lim_{T_s \to 0} \mathbb{J}_{\alpha,\tau,T_s} = \mathbb{J}_{\alpha,\tau}. \tag{3.223}$$

Since for T_s small we have that $\mathbb{J}_{\alpha,\tau,T_s} = \mathbb{J}_{\alpha,\tau}$ finite set independent of T_s, with regard to the quantity h_{α,τ,T_s} defined in (2.223), for T_s small we have

$$\begin{aligned} h_{\alpha,\tau,T_s} &\triangleq \inf_{k \notin \mathbb{J}_{\alpha,\tau}} |[\alpha - \alpha_k(\tau)]\, T_s \bmod 1| \\ &= \min_{k \notin \mathbb{J}_{\alpha,\tau}} |[\alpha - \alpha_k(\tau)]\, T_s \bmod 1| \\ &= \min_{k \notin \mathbb{J}_{\alpha,\tau}} |[\alpha - \alpha_k(\tau)]\, T_s| \end{aligned} \tag{3.224}$$

where in the second equality inf is substituted by min since $\mathbb{I}$ (and also $\mathbb{J}_{\alpha,\tau} \subseteq \mathbb{I}$) is finite, and the third equality holds for T_s sufficiently small (depending on α and τ) such that (3.221) holds. Therefore,

$$\begin{aligned} \lim_{T_s \to 0} h_{\alpha,\tau,T_s} &= \lim_{T_s \to 0} \min_{k \notin \mathbb{J}_{\alpha,\tau}} |[\alpha - \alpha_k(\tau)]\, T_s| \\ &= \min_{k \notin \mathbb{J}_{\alpha,\tau}} \left| \lim_{T_s \to 0} [\alpha - \alpha_k(\tau)]\, T_s \right| \\ &= 0 \end{aligned} \tag{3.225}$$

provided that the functions $\alpha_k(\tau)$ are bounded for finite τ. In the second equality, lim and min operations can be interchanged since min is over a finite set not depending on T_s (in general, lim and inf operations cannot be inverted if inf is over an infinite set). In addition,

$$\begin{aligned} \lim_{T_s \to 0} \frac{h_{\alpha,\tau,T_s}}{T_s} &= \lim_{T_s \to 0} \frac{1}{T_s} \min_{k \notin \mathbb{J}_{\alpha,\tau}} |[\alpha - \alpha_k(\tau)]\, T_s| \\ &= \min_{k \notin \mathbb{J}_{\alpha,\tau}} |\alpha - \alpha_k(\tau)| \\ &\triangleq C_{\alpha,\tau} > 0 \end{aligned} \tag{3.226}$$

Let us define for notation simplicity

$$\begin{aligned} R &\triangleq R_{yx^{(*)}}(\alpha, \tau) \\ \widetilde{R} &\triangleq \sum_{p=-\infty}^{+\infty} R_{yx^{(*)}}(\alpha + p f_s, \tau) \\ \varrho_{N,T_s} &\triangleq \varrho_{y_{\mathrm{d}} x^{(*)}_{\mathrm{d}}}(\alpha, \tau; n_0, N, T_s) \end{aligned}$$

It results that

$$\begin{aligned}\left|\text{bias}\{\varrho_{N,T_s}\}\right| &\triangleq \left|\mathrm{E}\{\varrho_{N,T_s}\} - R\right| \\ &= \left|\mathrm{E}\{\varrho_{N,T_s}\} - \widetilde{R} + \widetilde{R} - R\right| \\ &\leqslant \left|\mathrm{E}\{\varrho_{N,T_s}\} - \widetilde{R}\right| + \left|\widetilde{R} - R\right| \end{aligned} \tag{3.227}$$

Under the assumption of finite $\mathbb{I}$, Assumption 2.6.11 is verified and the thesis of Lemma 2.6.12 is also verified since in the right-hand side of

$$\left|\widetilde{R} - R\right| = \sum_{\substack{p=-\infty \\ p\neq 0}}^{+\infty} R_{yx^{(*)}}(\alpha + pf_s, \tau) \tag{3.228}$$

the sum is identically zero for T_s sufficiently small (depending on τ), that is, for f_s sufficiently large so that, for fixed τ,

$$f_s > \max_k |\alpha_k(\tau)| \tag{3.229}$$

where the maximum exists since the number of lag-dependent cycle frequencies is finite and each function $\alpha_k(\tau)$ is assumed to be bounded for finite τ.

From the counterpart for the H-CCC of Theorem 2.6.7, it follows that (see (3.199))

$$\left|\mathrm{E}\{\varrho_{N,T_s}\} - \widetilde{R}\right| \leqslant \frac{\mathcal{R}_\infty}{(2N+1)^\gamma}\left[\frac{K}{(h_{\alpha,\tau,T_s})^\gamma} + C'_\gamma\right] \tag{3.230}$$

where $C'_\gamma \triangleq K_\gamma + KC_\gamma$ and

$$\mathcal{R}_\infty \triangleq \sum_{k\in\mathbb{I}} \left\| R^{(k)}_{yx^{(*)}} \right\|_\infty . \tag{3.231}$$

Consequently, from (3.226), (3.227), and (3.230), we have

$$\begin{aligned}&\lim_{T_s\to 0}\lim_{N\to\infty} [T_s(2N+1)]^\gamma \left|\mathrm{E}\{\varrho_{N,T_s}\} - R\right| \\ &\leqslant \lim_{T_s\to 0} \mathcal{R}_\infty \left[K\left(\frac{T_s}{h_{\alpha,\tau,T_s}}\right)^\gamma + C'_\gamma T_s^\gamma\right] \\ &= \frac{\mathcal{R}_\infty K}{C^\gamma_{\alpha,\tau}}\end{aligned} \tag{3.232}$$

where $C_{\alpha,\tau} \triangleq \min_{k\notin\mathbb{J}_{\alpha,\tau}} |\alpha - \alpha_k(\tau)|$ and the order of the two limits cannot be inverted.

3.11.6 Proof of Theorem 2.6.18 Asymptotic Covariance of the Hybrid Cyclic Cross-Correlogram

Let us consider the limit

$$\lim_{T_s\to 0}\lim_{N\to\infty}(2N+1)T_s \operatorname{cov}\left\{\varrho_{y_\mathrm{d}x_\mathrm{d}^{(*)}}(\alpha_1,\tau_1;n_{01},N,T_s),\varrho_{y_\mathrm{d}x_\mathrm{d}^{(*)}}(\alpha_2,\tau_2;n_{02},N,T_s)\right\}$$
$$=\lim_{T_s\to 0} T_s\left[\widetilde{\mathcal{T}}'_1+\widetilde{\mathcal{T}}'_2+\widetilde{\mathcal{T}}'_3\right] \tag{3.233}$$

where $\widetilde{\mathcal{T}}'_1$, $\widetilde{\mathcal{T}}'_2$, and $\widetilde{\mathcal{T}}'_3$ are defined in (2.196), (2.197), and (2.198), respectively, with the replacements $\tau_1 \curvearrowright m_1 T_s$, $\alpha_1 \curvearrowright \widetilde{\alpha}_1 f_s$, $\tau_2 \curvearrowright m_2 T_s$, and $\alpha_2 \curvearrowright \widetilde{\alpha}_2 f_s$.

As regards the term $\widetilde{\mathcal{T}}'_1$, we have

$$\lim_{T_s\to 0} T_s\,\widetilde{\mathcal{T}}'_1 = \lim_{T_s\to 0}\mathcal{E}_a\sum_{k'}\sum_{k''}\sum_{\ell=-\infty}^{+\infty} R^{(k')}_{yy^*}(\tau_1-\tau_2+\ell T_s)\,R^{(k'')}_{x^{(*)}x^{(*)*}}(\ell T_s)$$
$$e^{j2\pi\alpha'_{k'}(\tau_1-\tau_2+\ell T_s)\tau_2}\,e^{-j2\pi\alpha_1\ell T_s}\,T_s$$
$$\delta_{[\alpha_1-\alpha_2-\alpha'_{k'}(\tau_1-\tau_2+\ell T_s)-\alpha''_{k''}(\ell T_s)]T_s \bmod 1}$$
$$=\mathcal{E}_a\sum_{k'}\sum_{k''}\int_{\mathbb{R}} R^{(k')}_{yy^*}(\tau_1-\tau_2+s)\,R^{(k'')}_{x^{(*)}x^{(*)*}}(s)$$
$$e^{j2\pi\alpha'_{k'}(\tau_1-\tau_2+s)\tau_2}\,e^{-j2\pi\alpha_1 s}\,\delta_{[\alpha_1-\alpha_2-\alpha'_{k'}(\tau_1-\tau_2+s)-\alpha''_{k''}(s)]}\,\mathrm{d}s \tag{3.234}$$

with the right-hand side coincident with $\mathcal{T}'_1$ defined in (2.147). In (3.234), we used the fact that if $\alpha_k(\tau)$ are bounded and $\mathbb{I}_{yy^*}$ and $\mathbb{I}_{x^{(*)}x^{(*)*}}$ are finite sets, for T_s sufficiently small, one has

$$[\alpha_1-\alpha_2-\alpha'_{k'}(\tau_1-\tau_2+\ell T_s)-\alpha''_{k''}(\ell T_s)]T_s = 0 \bmod 1$$
$$\Leftrightarrow [\alpha_1-\alpha_2-\alpha'_{k'}(\tau_1-\tau_2+\ell T_s)-\alpha''_{k''}(\ell T_s)] = 0$$
$$\Rightarrow \delta_{[\alpha_1-\alpha_2-\alpha'_{k'}(\tau_1-\tau_2+\ell T_s)-\alpha''_{k''}(\ell T_s)]T_s \bmod 1}$$
$$=\delta_{[\alpha_1-\alpha_2-\alpha'_{k'}(\tau_1-\tau_2+\ell T_s)-\alpha''_{k''}(\ell T_s)]}. \tag{3.235}$$

In addition, the integrand function has been assumed to be Riemann integrable.

Note that the right-hand side of (3.234) can be nonzero only if $[\alpha_1-\alpha_2-\alpha'_{k'}(\tau_1-\tau_2+s)-\alpha''_{k''}(s)]=0$ in a set of values of s with positive Lebesgue measure.

Analogous results hold for terms $\widetilde{\mathcal{T}}'_2$, and $\widetilde{\mathcal{T}}'_3$.

3.11.7 Proof of Lemma 2.6.19 Rate of Convergence to Zero of Cumulants of Hybrid Cyclic Cross-Correlograms

From (3.203) it follows that

$$\begin{aligned}
&\lim_{N\to\infty} (2N+1)^{k-1} \left| \operatorname{cum}\left\{ \widetilde{R}^{[*]_i}_{y_\mathrm{d} x_\mathrm{d}^{(*)}}(\widetilde{\alpha}_i, m_i; n_{0i}, N),\ i=1,\ldots,k \right\} \right| \\
&\quad \leqslant \|a\|_\infty^{k-1} \lim_{N\to\infty} \sum_{r} \phi(r_1 T_s, \ldots, r_{k-1} T_s) \lim_{N\to\infty} \sum_{n} \frac{1}{2N+1} \left| a\left(\frac{n-n_{0k}}{2N+1}\right)\right| \\
&\quad = \|a\|_\infty^{k-1} \sum_{r\in\mathbb{Z}^{k-1}} \phi(r_1 T_s, \ldots, r_{k-1} T_s) \int_{\mathbb{R}} |a(s)| \,\mathrm{d}s
\end{aligned} \tag{3.236}$$

where the sums over $r_i, i = 1, \ldots, k-1$, and n in the second term range over sets $\{r_{i,\min}, \ldots, r_{i,\max}\}$ and $\{n_{\min}, \ldots, n_{\max}\}$ with extremes $r_{i,\min}$, $r_{i,\max}$, $n_{\min}$, $n_{\max}$ depending on N and such that, as $N \to \infty$, $r_{i,\min}$ and $n_{\min}$ approach $-\infty$ and $r_{i,\max}$ and $n_{\max}$ approach $+\infty$.

The rhs of (3.236) does not depend on $\widetilde{\alpha}_i$, m_i. Therefore, the same inequality holds by replacing in the lhs $\widetilde{R}^{[*]_i}_{y_\mathrm{d} x_\mathrm{d}^{(*)}}(\widetilde{\alpha}_i, m_i; n_{0i}, N)$ with $\varrho^{[*]_i}_{y_\mathrm{d} x_\mathrm{d}^{(*)}}(\alpha_i, \tau_i; n_{0i}, N, T_s)$ and we also have

$$\begin{aligned}
&\lim_{T_s\to 0} \lim_{N\to\infty} [(2N+1)T_s]^{k-1} \left| \operatorname{cum}\left\{ \varrho^{[*]_i}_{y_\mathrm{d} x_\mathrm{d}^{(*)}}(\alpha_i, \tau_i; n_{0i}, N, T_s),\ i=1,\ldots,k \right\} \right| \\
&\quad \leqslant \|a\|_\infty^{k-1} \lim_{T_s\to 0} \sum_{r\in\mathbb{Z}^{k-1}} \phi(r_1 T_s, \ldots, r_{k-1} T_s)\, T_s^{k-1} \int_{\mathbb{R}} |a(s)| \,\mathrm{d}s \\
&\quad = \|a\|_\infty^{k-1} \int_{\mathbb{R}^{k-1}} \phi(s_1, \ldots, s_{k-1}) \,\mathrm{d}s_1 \cdots \mathrm{d}s_{k-1} \int_{\mathbb{R}} |a(s)| \,\mathrm{d}s
\end{aligned} \tag{3.237}$$

Thus, for $k \geqslant 2$ and $\epsilon > 0$, we obtain (2.230), where the order of the two limits cannot be interchanged (that is, $NT_s \to \infty$).

Note that the $(k-1)$-dimensional Riemann sum in the second line converges to the $(k-1)$-dimensional Riemann integral in the third line if the Riemann-integrable function ϕ is sufficiently regular. In fact, the limit as $T_s \to 0$ of the infinite sum in the second line is not exactly the definition of a Riemann integral over an infinite $(k-1)$-dimensional interval. However, the function ϕ in Assumption 2.6.8 can always be chosen such that this Riemann integral exists. □

3.11.8 Proof of Theorem 2.6.20 Asymptotic Joint Normality of the Hybrid Cyclic Cross-Correlograms

By following the guidelines of the proof of Theorem 2.6.10, let

$$V_i^{(N,T_s)} \triangleq \sqrt{(2N+1)T_s}\,[\varrho_{y_\mathrm{d} x_\mathrm{d}^{(*)}}(\alpha_i, \tau_i; n_{0i}, N, T_s) - R_{yx^{(*)}}(\alpha_i, \tau_i)], \qquad i = 1, \ldots, k. \tag{3.238}$$

From Theorem 2.6.17 holding for $\gamma > 1$ (or $\gamma = 1$ if $a(t) = \text{rect}(t)$), we have

$$\lim_{T_s \to 0} \lim_{N \to \infty} \mathrm{E}\left\{ V_i^{(N,T_s)} \right\} = 0. \tag{3.239}$$

From Theorem 2.6.18, we have that

$$\lim_{T_s \to 0} \lim_{N \to \infty} \operatorname{cov}\left\{ V_i^{(N,T_s)}, V_j^{(N,T_s)} \right\}$$

is finite. From Theorem 3.13.3, we have that

$$\lim_{T_s \to 0} \lim_{N \to \infty} \operatorname{cov}\left\{ V_i^{(N,T_s)}, V_j^{(N,T_s)*} \right\}$$

is finite. From Lemma 2.6.19 with $\epsilon = \frac{k}{2} - 1$ and $k \geqslant 3$, we have

$$\lim_{T_s \to 0} \lim_{N \to \infty} \operatorname{cum}\left\{ \left[V_i^{(N,T_s)} \right]^{[*]_i}, \; i = 1, \dots, k \right\} = 0. \tag{3.240}$$

Thus, according to the results of Section 1.4.2, for every fixed α_i, τ_i, n_{0i} the random variables $V_i^{(N,T_s)}$, $i = 1, \dots, k$, are asymptotically ($N \to \infty$ and $T_s \to 0$ with $NT_s \to \infty$) zero-mean jointly complex Normal (Picinbono 1996; van den Bos 1995).

3.12 Proofs for Section 2.6.4 "Concluding Remarks"

3.12.1 Proof of Theorem 2.6.21 Asymptotic Discrete-Time Cyclic Cross-Correlogram

Let $t_0 = n_0 T_s$, $T = (2N+1)T_s$ fixed.

$$\begin{aligned}
&\mathrm{E}\left\{ \left| \widetilde{R}_{y_{\mathrm{d}} x_{\mathrm{d}}^{(*)}}(\widetilde{\alpha}, m; n_0, N) - R_{yx^{(*)}}(\widetilde{\alpha} f_s, mT_s; t_0, T) \right|^2 \right\} \\
&= \mathrm{E}\left\{ \left| \sum_{n=-\infty}^{+\infty} v_N(n - n_0)\, y_{\mathrm{d}}(n+m)\, x_{\mathrm{d}}^{(*)}(n)\, e^{-j2\pi\widetilde{\alpha} n} \right.\right. \\
&\qquad \left.\left. - \int_{\mathbb{R}} w_T(t - n_0 T_s)\, y(t + mT_s)\, x^{(*)}(t)\, e^{-j2\pi\widetilde{\alpha} f_s t}\, \mathrm{d}t \right|^2 \right\} \\
&= \mathrm{E}\left\{ \left| \sum_{n=n_0-N}^{n_0+N} \frac{1}{2N+1}\, \psi(nT_s) - \int_{t_0-T/2}^{t_0+T/2} \frac{1}{T}\, \psi(t)\, \mathrm{d}t \right|^2 \right\}
\end{aligned} \tag{3.241}$$

where

$$\psi(t) \triangleq a\left(\frac{t - n_0 T_s}{(2N+1)T_s} \right) y(t + mT_s)\, x^{(*)}(t)\, e^{-j2\pi\widetilde{\alpha} f_s t}. \tag{3.242}$$

The stochastic function $\psi(t)$ is mean-square Riemann-integrable in $(t_0 - T/2, t_0 + T/2)$. That is, in the limit as the sampling period T_s approaches zero (and, hence, $N \to \infty$ so that $(2N + 1)T_s = T$ is constant), in (3.241) we have

$$\lim_{T_s \to 0} \mathrm{E}\left\{\left| \sum_{n=n_0-N}^{n_0+N} \psi(nT_s)\, T_s - \int_{t_0-T/2}^{t_0+T/2} \psi(t)\, \mathrm{d}t \right|^2\right\} = 0. \tag{3.243}$$

In fact, a necessary and sufficient condition such that (3.243) holds is (Loève 1963, Chapter X) (see also Theorem 2.2.15)

$$\int_{t_0-T/2}^{t_0+T/2} \int_{t_0-T/2}^{t_0+T/2} |\mathrm{cov}\,\{\psi(t_1), \psi(t_2)\}|\; \mathrm{d}t_1\, \mathrm{d}t_2 < \infty \tag{3.244}$$

and the summability of $\mathrm{cov}\,\{\psi(t_1), \psi(t_2)\}$ can be proved under Assumptions 2.4.2–2.4.5 by following the proof of Theorem 2.4.7 (see (3.47b), (3.47c), and (3.52)–(3.54)).

Remark 3.12.1 *Let*

$$R \triangleq R_{yx^{(*)}}(\alpha, \tau)|_{\alpha=\widetilde{\alpha} f_s, \tau=mT_s} \tag{3.245}$$

$$R_T \triangleq R_{yx^{(*)}}(\alpha, \tau; t_0, T)|_{\alpha=\widetilde{\alpha} f_s, \tau=mT_s; t_0=n_0T_s, T=(2N+1)T_s} \tag{3.246}$$

$$\widetilde{R}_N \triangleq \widetilde{R}_{y_\mathrm{d}x_\mathrm{d}^{(*)}}(\widetilde{\alpha}, m; n_0, N) \tag{3.247}$$

Then,

$$\begin{aligned}\mathrm{E}\left\{|\widetilde{R}_N - R|^2\right\} &= \mathrm{E}\left\{|\widetilde{R}_N - R_T + R_T - R|^2\right\} \\ &\leqslant \mathrm{E}\left\{|\widetilde{R}_N - R_T|^2\right\} + \mathrm{E}\left\{|R_T - R|^2\right\}\end{aligned} \tag{3.248}$$

is not a useful bound to prove that $\mathrm{E}\{|\widetilde{R}_N - R|^2\} \to 0$ *as* $T_s \to 0$ *and* $T \to \infty$ *since the limits*

$$\lim_{T_s \to 0} \mathrm{E}\left\{|\widetilde{R}_N - R_T|^2\right\} = 0 \quad \textit{(Theorem 2.6.21)} \tag{3.249}$$

$$\lim_{T \to \infty} \mathrm{E}\left\{|R_T - R|^2\right\} = 0 \quad \textit{(Theorems 2.4.11 and 2.4.13 and eq. (2.151))} \tag{3.250}$$

are not uniform with respect to T *and* T_s*, respectively. This is in agreement with the fact that in Theorem 2.6.13 we have first* $N \to \infty$ *and then* $T_s \to 0$.

3.13 Discrete-Time and Hybrid Conjugate Covariance

A complete wide-sense second-order characterization of complex processes requires the knowledge of both covariance and conjugate covariance (Picinbono 1996; Picinbono and Bondon 1997), (Schreier and Scharf 2003a,b).

By reasoning as in Theorems 2.6.3, 2.6.5, and 2.6.18, respectively, the following theorems can be proved.

Theorem 3.13.1 Conjugate Covariance of the Discrete-Time Cyclic Cross-Correlogram. *Let* $x_\mathrm{d}(n)$ *and* $y_\mathrm{d}(n)$ *be zero-mean discrete-time processes defined in (2.172). Under Assumptions*

2.4.2, 2.4.3, and 2.4.5 on the continuous-time processes $x(t)$ and $y(t)$, the conjugate covariance of the DT-CCC (2.181) is given by

$$\text{cov}\left\{\widetilde{R}_{y_d x_d^{(*)}}(\widetilde{\alpha}_1, m_1; n_{01}, N), \widetilde{R}^*_{y_d x_d^{(*)}}(\widetilde{\alpha}_2, m_2; n_{02}, N)\right\} = \bar{\bar{\mathcal{T}}}_1 + \bar{\bar{\mathcal{T}}}_2 + \bar{\bar{\mathcal{T}}}_3 \tag{3.251}$$

where

$$\begin{aligned}
\bar{\bar{\mathcal{T}}}_1 &\triangleq \sum_{k'} \sum_{k''} \sum_{\ell=-\infty}^{+\infty} R_{yy}^{(k')}((m_1 - m_2 + \ell)T_s)\, R_{x^{(*)}x^{(*)}}^{(k'')}(\ell T_s) \\
&\quad e^{j2\pi\alpha_{yy}^{(k')}((m_1-m_2+\ell)T_s)m_2T_s}\, e^{-j2\pi\widetilde{\alpha}_1\ell} \\
&\quad e^{-j2\pi\bar{\Phi}_1 n_{02}} \frac{1}{2N+1}\, \widetilde{r}_{aa}\left(\bar{\Phi}_1, \ell - n_{01} + n_{02}; N\right) \qquad (3.252) \\
\bar{\bar{\mathcal{T}}}_2 &\triangleq \sum_{k'''} \sum_{k^{iv}} \sum_{\ell=-\infty}^{+\infty} R_{yx^{(*)}}^{(k''')}((m_1 + \ell)T_s)\, R_{x^{(*)}y}^{(k^{iv})}((\ell - m_2)T_s) \\
&\quad e^{j2\pi\alpha_{x^{(*)}y}^{(k^{iv})}((\ell-m_2)T_s)m_2T_s}\, e^{-j2\pi\widetilde{\alpha}_1\ell} \\
&\quad e^{-j2\pi\bar{\Phi}_2 n_{02}} \frac{1}{2N+1}\, \widetilde{r}_{aa}\left(\bar{\Phi}_2, \ell - n_{01} + n_{02}; N\right) \qquad (3.253) \\
\bar{\bar{\mathcal{T}}}_3 &\triangleq \sum_{k} \sum_{\ell=-\infty}^{+\infty} C_{yx^{(*)}yx^{(*)}}^{(k)}((m_1 + \ell)T_s, \ell T_s, m_2T_s)\, e^{-j2\pi\widetilde{\alpha}_1\ell} \\
&\quad e^{-j2\pi\bar{\Phi}_3 n_{02}} \frac{1}{2N+1}\, \widetilde{r}_{aa}\left(\bar{\Phi}_3, \ell - n_{01} + n_{02}; N\right) \qquad (3.254)
\end{aligned}$$

with

$$\widetilde{r}_{aa}(\widetilde{\beta}, m; N) \triangleq \frac{1}{2N+1} \sum_{n=-\infty}^{+\infty} a\left(\frac{n+m}{2N+1}\right) a\left(\frac{n}{2N+1}\right) e^{-j2\pi\widetilde{\beta}n} \tag{3.255}$$

and

$$\begin{aligned}
\bar{\Phi}_1 &\triangleq [\widetilde{\alpha}_1 + \widetilde{\alpha}_2 - \alpha_{yy}^{(k')}((m_1 - m_2 + \ell)T_s)T_s - \alpha_{x^{(*)}x^{(*)}}^{(k'')}(\ell T_s)T_s] \\
\bar{\Phi}_2 &\triangleq [\widetilde{\alpha}_1 + \widetilde{\alpha}_2 - \alpha_{yx^{(*)}}^{(k''')}((m_1 + \ell)T_s)T_s - \alpha_{x^{(*)}y}^{(k^{iv})}((\ell - m_2)T_s)T_s] \\
\bar{\Phi}_3 &\triangleq [\widetilde{\alpha}_1 + \widetilde{\alpha}_2 - \beta_k((m_1 + \ell)T_s, \ell T_s, m_2T_s)T_s].
\end{aligned}$$

□

Theorem 3.13.2 Asymptotic ($N \to \infty$) Conjugate Covariance of the Discrete-Time Cyclic Cross-Correlogram. *Let $x_d(n)$ and $y_d(n)$ be zero-mean discrete-time processes defined in (2.172). Under Assumptions 2.4.2–2.4.8 on the continuous-time processes $x(t)$ and $y(t)$, the* *asymptotic ($N \to \infty$) conjugate covariance of the DT-CCC (2.181) is given by*

$$\lim_{N\to\infty} (2N+1)\,\mathrm{cov}\left\{ \widetilde{R}_{y_{\mathrm{d}} x_{\mathrm{d}}^{(*)}}(\widetilde{\alpha}_1, m_1; n_{01}, N), \widetilde{R}^*_{y_{\mathrm{d}} x_{\mathrm{d}}^{(*)}}(\widetilde{\alpha}_2, m_2; n_{02}, N) \right\}$$
$$= \bar{\bar{\mathcal{T}}}'_1 + \bar{\bar{\mathcal{T}}}'_2 + \bar{\bar{\mathcal{T}}}'_3 \tag{3.256}$$

where

$$\bar{\bar{\mathcal{T}}}'_1 \triangleq \bar{\mathcal{E}}_a \sum_{k'} \sum_{k''} \sum_{\ell=-\infty}^{+\infty} R_{yy}^{(k')}((m_1 - m_2 + \ell)T_s)\, R_{x^{(*)}x^{(*)}}^{(k'')}(\ell T_s)$$
$$e^{j2\pi\alpha'_{k'}((m_1-m_2+\ell)T_s)m_2 T_s}\, e^{-j2\pi\widetilde{\alpha}_1 \ell}$$
$$\delta_{[\widetilde{\alpha}_1+\widetilde{\alpha}_2-\alpha'_{k'}((m_1-m_2+\ell)T_s)T_s-\alpha''_{k''}(\ell T_s)T_s] \bmod 1} \tag{3.257}$$

$$\bar{\bar{\mathcal{T}}}'_2 \triangleq \bar{\mathcal{E}}_a \sum_{k'''} \sum_{k^{iv}} \sum_{\ell=-\infty}^{+\infty} R_{yx^{(*)}}^{(k''')}((m_1 + \ell)T_s)\, R_{x^{(*)}y}^{(k^{iv})}((\ell - m_2)T_s)$$
$$e^{j2\pi\alpha^{iv}_{k^{iv}}((\ell-m_2)T_s)m_2 T_s}\, e^{-j2\pi\widetilde{\alpha}_1 \ell}$$
$$\delta_{[\widetilde{\alpha}_1+\widetilde{\alpha}_2-\alpha'''_{k'''}((m_1+\ell)T_s)T_s-\alpha^{iv}_{k^{iv}}((\ell-m_2)T_s)T_s] \bmod 1} \tag{3.258}$$

$$\bar{\bar{\mathcal{T}}}'_3 \triangleq \bar{\mathcal{E}}_a \sum_{k} \sum_{\ell=-\infty}^{+\infty} C_{yx^{(*)}yx^{(*)}}^{(k)}((m_1 + \ell)T_s, \ell T_s, m_2 T_s)\, e^{-j2\pi\widetilde{\alpha}_1 \ell}$$
$$\delta_{[\widetilde{\alpha}_1+\widetilde{\alpha}_2-\beta_k((m_1+\ell)T_s,\ell T_s,m_2 T_s)T_s] \bmod 1} \tag{3.259}$$

with $\bar{\mathcal{E}}_a \triangleq \int_{-1/2}^{1/2} a^2(t)\,\mathrm{d}t$. In (3.257)–(3.259), for notation simplicity, $\alpha'_{k'}(\cdot) \equiv \alpha_{yy}^{(k')}(\cdot)$, $\alpha''_{k''}(\cdot) \equiv \alpha_{x^{(*)}x^{(*)}}^{(k'')}(\cdot)$, $\alpha'''_{k'''}(\cdot) \equiv \alpha_{yx^{(*)}}^{(k''')}(\cdot)$, and $\alpha^{iv}_{k^{iv}}(\cdot) \equiv \alpha_{x^{(*)}y}^{(k^{iv})}(\cdot)$. □

Theorem 3.13.3 Asymptotic ($N \to \infty$ and $T_s \to 0$) Conjugate Covariance of the Hybrid Cyclic Cross-Correlogram. *Let $y(t)$ and $x(t)$ be zero-mean wide-sense jointly GACS stochastic processes with cross-correlation function (2.31c). Under Assumptions 2.4.2–2.4.8, the asymptotic ($N \to \infty$ and $T_s \to 0$ with $NT_s \to \infty$) conjugate covariance of the H-CCC (2.213) is given by*

$$\lim_{T_s\to 0} \lim_{N\to\infty} (2N+1)T_s\,\mathrm{cov}\left\{ \varrho_{y_{\mathrm{d}} x_{\mathrm{d}}^{(*)}}(\alpha_1, \tau_1; n_{01}, N, T_s), \varrho^*_{y_{\mathrm{d}} x_{\mathrm{d}}^{(*)}}(\alpha_2, \tau_2; n_{02}, N, T_s) \right\}$$
$$= \bar{\mathcal{T}}'_1 + \bar{\mathcal{T}}'_2 + \bar{\mathcal{T}}'_3 \tag{3.260}$$

with $\bar{\mathcal{T}}'_1$, $\bar{\mathcal{T}}'_2$, and $\bar{\mathcal{T}}'_3$ given by (3.138), (3.139), and (3.140) (in Theorem 3.7.2), respectively, where the order of the two limits cannot be interchanged. □

4

Spectrally Correlated Processes

In this chapter, the class of the spectrally correlated (SC) stochastic processes is characterized and the problem of its statistical function estimation addressed. Processes belonging to this class exhibit a Loève bifrequency spectrum with spectral masses concentrated on a countable set of support curves in the bifrequency plane. Thus, such processes have spectral components that are correlated. The introduced class generalizes that of the almost-cyclostationary (ACS) processes that are obtained as a special case when the separation between correlated spectral components assumes values only in a countable set. In such a case, the support curves are lines with unit slope. SC processes properly model the output of Doppler channels when the product of the signal bandwidth and the data-record length is not too smaller than the ratio of the medium propagation speed and the radial speed between transmitter and receiver. Thus they find application in wide-band or ultra-wideband mobile communications. For SC processes, the amount of spectral correlation existing between two separate spectral components is characterized by the bifrequency spectral correlation density function which is the density of the Loève bifrequency spectrum on its support curves. When the location of the support curves is unknown, the time-smoothed bifrequency cross-periodogram provides a reliable (low bias and variance) single-sample-path-based estimate of the bifrequency spectral correlation density function in those points of the bifrequency plane where the slope of the support curves is not too far from unity. Moreover, a trade-off exists between the departure of the nonstationarity from the almost-cyclostationarity and the reliability of spectral correlation measurements obtainable by a single sample-path. Furthermore, in general, the estimate accuracy cannot be improved as wished by increasing the data-record length and the spectral resolution. If the location of a support curve is known, the cross-periodogram frequency smoothed along the given support curve and properly normalized provides a mean-square consistent and asymptotically Normal estimator of the density of the Loève bifrequency spectrum along that curve. Furthermore, well-known consistency results for ACS processes can be obtained by specializing the results for SC processes. It is shown that uniformly sampling a continuous-time SC process leads to a discrete-time SC process. Conditions on the sampling frequency and the process bandwidth are provided to avoid aliasing in the Loève bifrequency spectrum, and its density, of the discrete-time sampled signal. Finally, multirate processing of SC processes is treated.

Generalizations of Cyclostationary Signal Processing: Spectral Analysis and Applications, First Edition.

4.1 Introduction

It is well known that wide-sense stationary (WSS) stochastic processes do not exhibit correlation between spectral components at distinct frequencies. That is, by passing a WSS process throughout two bandpass filters with nonoverlapping passbands, and then frequency shifting the two output processes to a common band, one obtains two uncorrelated stochastic processes (Gardner 1987d). Equivalently, for WSS processes, the Loève bifrequency spectrum (Loève 1963) (also called dual-frequency spectrum (Øigård *et al.* 2006) or cointensity spectrum (Middleton 1967)) has support contained in the main diagonal of the bifrequency plane. The presence of spectral correlation between spectral components at distinct frequencies is an indicator of the nonstationarity of the process (Loève 1963), (Gardner 1987d), (Hurd and Gerr 1991), (Genossar 1992), (Dehay and Hurd 1994), (Napolitano 2003), (Dmochowski *et al.* 2009). When correlation exists only between spectral components that are separated by quantities belonging to a countable set of values, the process is said almost-cyclostationary (ACS) or almost-periodically correlated (Gardner 1985, 1987d), (Gardner *et al.* 2006), (Hurd and Miamee 2007). The values of the separation between correlated spectral components are called cycle frequencies and are the frequencies of the (generalized) Fourier series expansion of the almost-periodically time-variant statistical autocorrelation function (Gardner 1987d), (Gardner 1991b), (Dehay and Hurd 1994). For ACS processes, the Loève bifrequency spectrum has the support contained in lines parallel to the main diagonal of the bifrequency plane and its values on such lines are described by the spectral correlation density functions (Hurd 1989a, 1991), (Hurd and Gerr 1991).

A new class of nonstationary stochastic processes, the spectrally correlated (SC) processes, has been introduced and characterized in (Napolitano 2003). SC processes exhibit Loève bifrequency spectrum with spectral masses concentrated on a countable set of support curves in the bifrequency plane. Thus, ACS processes are obtained as a special case of SC processes when the support curves are lines with unit slope. In communications, SC processes are obtained when ACS processes pass through Doppler channels that induce frequency warping on the input signal (Franaszek and Liu 1967). For example, let us consider the case of relative motion between transmitter and receiver in the presence of moving reflecting objects. If the involved relative radial speeds are constant within the observation interval and the product signal-bandwidth times data-record length is not much smaller than the ratios of the medium propagation speed and the radial speeds, then the resulting multipath Doppler channel introduces a different complex gain, time-delay, frequency shift, and nonunit time-scale factor for each path (Napolitano 2003). Since the time-scale factors are nonunit, when an ACS process passes through such a channel, the output signal is SC (Napolitano 2003). Therefore, the SC model is appropriate in high mobility scenarios in modern communication systems where wider and wider bandwidths are considered to get higher and higher bit rates and, moreover, large data-record lengths are used for blind channel identification or equalization algorithms or for detection techniques in highly noise- and interference-corrupted environments. Further situations where nonunit time-scale factors should be accounted for can be encountered in radar and sonar applications (Van Trees 1971, pp. 339–340), communications with wide-band and ultra wide-band (UWB) signals (Jin *et al.* 1995), (Schlotz 2002), space communications (Oberg 2004), and underwater acoustics (Munk *et al.* 1995). In all these cases, SC processes are appropriate models for the signals involved (Napolitano 2003). In this chapter, it is shown that SC processes with nonlinear support curves in the bifrequency plane occur in spectral analysis with nonuniform frequency spacing (Oppenheim *et al.* 1971), (Oppenheim and Johnson 1972), (Braccini and Oppenheim 1974), or signal processing algorithms exploiting frequency warping techniques as those considered in (Makur and Mitra 2001), (Franz *et al.* 2003). It is also shown that jointly SC processes occur in the presence of some linear time-variant systems as those described in (Franaszek 1967), (Franaszek and Liu 1967), (Liu and Franaszek 1969), (Claasen and Mecklenbräuker 1982). In (Øigård *et al.* 2006), it is shown that fractional Brownian motion (fBm) processes have Loève bifrequency spectrum with spectral masses concentrated on three lines of the bifrequency plane. Finally, in discrete-time, ACS processes and their multirate processed versions turn out to be jointly SC (Napolitano 2010a).

SC processes are characterized at second-order in the frequency domain in terms of Loève bifrequency spectrum. The amount of spectral correlation existing between two separate spectral components is characterized by the bifrequency spectral correlation density function, that is, the density of the Loève bifrequency spectrum on its support curves. Then the estimation problem for such a density is considered.

Spectral estimation techniques have been developed for stationary, ACS, and some classes of nonstationary processes. For WSS processes, the power spectrum can be consistently estimated by the frequency-smoothed periodogram, provided that some mixing assumptions regulating the memory of the process are satisfied (Brillinger 1981), (Thomson 1982). The spectral-correlation density estimation problem has been addressed in the case of ACS processes in (Gardner 1986a), (Gardner 1987d), (Hurd 1989a), (Hurd and Leśkow 1992a), (Dandawaté and Giannakis 1994), (Dehay and Hurd 1994), (Genossar *et al.* 1994), (Gerr and Allen 1994), (Dehay and Leśkow 1996), (Sadler and Dandawaté 1998). The proposed techniques are based on time- or frequency smoothing the cyclic periodogram and provide consistent estimators of the spectral correlation density function under very mild conditions (e.g., the finite or approximately finite memory of the stochastic process). These estimators are the extensions to ACS signals of the well known power-spectrum estimators for WSS signals (Brillinger 1981), (Thomson 1982).

The correlation existing between separated spectral components, that in the following will be referred to as the *spectral correlation* property, is adopted by several authors in detecting and characterizing nonstationary processes. In (Gardner 1988a), (Gardner 1991c), it is shown that spectral correlation measurements can be reliable (small bias and variance) only if the nonstationarity is of almost-periodic nature (almost-cyclostationarity) or of known form. The former case is considered in (Gardner 1987d), (Hurd 1991), (Hurd and Gerr 1991), (Gerr and Allen 1994), and (Varghese and Donohue 1994). To the latter case belong the techniques of the radial and Doppler smoothing presented in (Allen and Hobbs 1992) and the estimator proposed in (Lii and Rosenblatt 2002). They provide consistent estimators of the spectral correlation density function for processes whose Loève bifrequency spectrum has the support on known lines with not necessarily unit slopes. The case of transient processes is treated in (Hurd 1988).

The problem of estimating the bifrequency spectral correlation density function is addressed at first when the location of the spectral masses is unknown. The bifrequency cross-periodogram is shown to be an estimator of the bifrequency spectral correlation density function asymptotically biased with nonzero asymptotic variance and the bias can be compensated only if the location of the spectral masses in the bifrequency plane is known. The time-smoothed cross-periodogram is considered as an estimator of the bifrequency spectral correlation density function since it represents the finite-time-averaged cross-correlation between finite-bandwidth spectral components of the process. It is shown that time-smoothing the cross-periodogram is the only practicable way of smoothing when the location of the spectral masses is unknown. In contrast, the frequency-smoothing technique can be adopted only if such a location is known (Allen and Hobbs 1992; Lii and Rosenblatt 2002).

Bias and variance of the time-smoothed bifrequency cross-periodogram are derived in the case of finite bandwidth of the spectral components and finite observation interval. Moreover asymptotic results as the bandwidth approaches zero and the observation interval approaches infinity are derived. It is shown that the bifrequency spectral correlation density function of SC processes that are not ACS can be estimated with some degree of reliability only if the departure of the nonstationarity from the almost-cyclostationarity is not too large. In particular, if in the neighborhood of a given point of the bifrequency plane the slope of the support curve of the Loève bifrequency spectrum is not too far from unity, then the smoothing product can be large enough to obtain a small variance and maintaining, at the same time, small bias. Then, a trade-off exists between the departure of the nonstationarity from the almost-cyclostationarity and the reliability of spectral correlation measurements obtainable by a single sample path. Moreover, in general, the estimate accuracy cannot be improved as wished by increasing the data-record length and the spectral resolution.

In (Gardner 1988a, 1991b), it is shown that spectral correlation measurements can be reliable only if the nonstationarity is almost cyclostationarity or of known form. Consequently, accordingly with the results in (Napolitano 2003), for SC processes that are not ACS, reliable estimates of spectral correlation density cannot be obtained if the location of the support curves is unknown.

For ACS processes, frequency smoothing the cross-periodogram along the unit-slope support lines of the bifrequency Loève spectrum leads to the frequency-smoothed cyclic periodogram which is a mean-square consistent and asymptotically Normal estimator of the cyclic spectrum (Gardner 1986a, 1987d), (Alekseev 1988), (Hurd 1989a), (Hurd 1991), (Hurd and Gerr 1991), (Hurd and Leśkow 1992a), (Dandawaté and Giannakis 1994), (Dehay and Hurd 1994), (Gerr and Allen 1994), (Genossar *et al.* 1994), (Dehay and Leśkow 1996), (Sadler and Dandawaté 1998).

Spectral analysis of nonstationary processes is addressed in (Hurd 1988), (Gardner 1991b), (Genossar 1992) with reference to the spectral-coherence estimation problem. For nonstationary processes having the Loève bifrequency spectrum not concentrated on sets of zero Lebesgue measure in the plane, the only possibility to get consistent estimates of the spectral density is to resort to more realizations or sample paths (Lii and Rosenblatt 2002), (Soedjack 2002). For SC processes, by following the main idea of smoothing the periodogram as for WSS and ACS processes, in the special case of known support curves constituted by lines, in (Chiu 1986), (Allen and Hobbs 1992), (Lii and Rosenblatt 1998), and (Lii and Rosenblatt 2002), the cross-periodogram smoothed along a given known support line has been proposed as estimator of the spectral correlation density on this support line. In (Lii and Rosenblatt 2002), the case of finite number of support lines is considered.

The problem of estimation of the density of Loève bifrequency spectrum (or spectral correlation density) of SC processes is addressed in this chapter in the case of known support curves. Specifically, the cross-periodogram frequency smoothed along a known given support curve is proposed as estimator of the spectral correlation density on this support curve. It is shown that the frequency-smoothed cross-periodogram asymptotically (as the data-record length approaches infinity and the spectral resolution approaches zero) approaches the product of the spectral correlation density function and a function of frequency and slope of the support curve.

Since the support curve is assumed to be known, such a multiplicative bias term can be compensated. Moreover, it is shown that the asymptotic covariance of the frequency smoothed cross-periodogram approaches zero. Therefore, the properly normalized frequency-smoothed cross-periodogram is a mean-square consistent estimator of the spectral correlation density function. Moreover, it is shown to be asymptotically complex Normal. Thus, such a result can be exploited to design statistical tests for presence of spectral correlation. Finally, it is shown that the well-known result for ACS processes that the frequency-smoothed cyclic periodogram is a mean-square consistent and asymptotically Normal estimator of the cyclic spectrum (Gardner 1986a, 1987d), (Alekseev 1988), (Hurd 1989a, 1991), (Hurd and Gerr 1991), (Hurd and Leśkow 1992a), (Dandawaté and Giannakis 1994), (Dehay and Hurd 1994), (Gerr and Allen 1994), (Genossar *et al.* 1994), (Dehay and Leśkow 1996), (Sadler and Dandawaté 1998) can be obtained as a special case of the results for SC processes.

A definition of band limitedness for nonstationary processes is given and, according to such definition, band-limited (continuous-time) SC processes are characterized. Then, discrete-time SC processes are introduced and characterized. It is shown that a discrete-time SC process can be obtained by uniformly sampling a continuous-time SC process and its Loève bifrequency spectrum is constituted by the superposition of replicas of the Loève bifrequency spectrum of the continuous-time SC process. It is also shown that for strictly band-limited SC processes a sufficient condition to avoid overlapping replicas in the Loève bifrequency spectrum is that the sampling frequency f_s is at least two times the process bandwidth which is the classical Shannon condition. However, more strict conditions on the sampling frequency need to be considered if a mapping between frequencies of the density of Loève bifrequency spectrum of the continuous-time process and frequencies of the density of Loève bifrequency spectrum of the discrete-time process must hold in the whole principal frequency domain (Napolitano 2011).

The chapter is organized as follows. In Section 4.2, SC stochastic processes are defined and characterized and motivating examples are provided. Bias and covariance are determined in Section 4.4 for the bifrequency cross-periodogram and in Section 4.5 for the time-smoothed bifrequency cross-periodogram in the case of unknown support curves. Proofs of lemmas and theorems of Sections 4.4 and 4.5 are reported in Sections 5.2 and 5.3, respectively. Bias and covariance are determined in Section 4.6 for the frequency-smoothed cross-periodogram in the case of known support curves. In Section 4.7, the properly normalized frequency-smoothed cross-periodogram is shown to be mean-square consistent and its asymptotic bias and variance are determined. Moreover, it is shown to be asymptotically complex Normal. Proofs of the results presented in Sections 4.6, 4.7.1, and 4.7.2 are reported in Sections 5.4, 5.5, and 5.6, respectively. Results obtained in the case of assumptions alternative to those made in this chapter are briefly presented in Section 5.7. Some issues concerning complex processes are addressed in Section 5.8. A discussion on the established results is made in Section 4.7.3. In Section 4.8, discrete-time SC processes are defined and characterized. Sampling theorems for SC processes are provided in Section 4.9. Proofs of theorems are reported in Section 5.10. Illustrative examples are presented in Section 4.9.3. Multirate processing of SC processes is treated in Section 4.10 and proofs of the results are reported in Section 5.11. In Section 4.11, discrete-time estimators for the spectral cross-correlation density are considered. In Section 4.12 simulation results on the spectral correlation density estimation are reported. Finally, in Section 4.13, spectral analysis with nonuniform frequency spacing is addressed. A Summary is given in Section 4.14.

4.2 Characterization of SC Stochastic Processes

In this section, spectrally correlated stochastic processes are introduced (Definitions 4.2.4–4.2.8) and characterized (Theorems 4.2.7 and 4.2.9). Moreover, examples of applications where such processes occur are presented.

4.2.1 Second-Order Characterization

Definition 4.2.1 *A covariance function* $\mathcal{R}(t_1, t_2)$ *is said to be* harmonizable *if there exists a spectral covariance function* $\gamma(f_1, f_2)$ *of bounded variation on* $\mathbb{R} \times \mathbb{R}$

$$\int_{\mathbb{R}^2} |\,\mathrm{d}\gamma(f_1, f_2)| < \infty \tag{4.1}$$

such that

$$\mathcal{R}(t_1, t_2) = \int_{\mathbb{R}^2} e^{j2\pi(f_1 t_1 - f_2 t_2)}\,\mathrm{d}\gamma(f_1, f_2) \tag{4.2}$$

where the integral is a Fourier-Stieltjes transform (Loève 1963). □

Definition 4.2.2 *A second-order stochastic process* $\{x(t),\ t \in \mathbb{R}\}$ *is said to be* (strongly) harmonizable *if there exists a second-order stochastic process* $\chi(f)$ *with increments* $\mathrm{d}\chi(f)$ *having covariance function* $\mathrm{E}\{\mathrm{d}\chi(f_1)\,\mathrm{d}\chi^*(f_2)\} = \mathrm{d}\gamma(f_1, f_2)$ *with* $\gamma(f_1, f_2)$ *of bounded variation on* $\mathbb{R} \times \mathbb{R}$ *such that*

$$x(t) = \int_{\mathbb{R}} e^{j2\pi f t}\,\mathrm{d}\chi(f) \tag{4.3}$$

with probability one (Loève 1963). □

The function $\chi(f)$ is named *integrated spectrum* (Gardner 1985, Section 10.1.2).

In (Loève 1963), it is shown that a necessary condition for a stochastic process to be harmonizable is that it is second-order continuous (or mean-square continuous) (Definition 2.2.11, Theorem 2.2.12). Moreover, it is shown that a stochastic process is harmonizable if and only if its covariance function is harmonizable. In fact, convergence of integrals in (4.2) and (4.3) is in the mean-square sense. In (Hurd 1973), the harmonizability of processes obtained by some processing of other harmonizable processes is studied.

In the following, the spectral characterization of stochastic processes will be provided in the sense of distributions (generalized functions) (Gelfand and Vilenkin 1964, Chapter 3), (Henniger 1970). Instead of considering Fourier-Stieltjes transforms as in (4.2) and (4.3), integrals have to be intended as Lebesgue integrals if the argument is a summable function and as formal representations for functional operators if the argument is a distribution (e.g., a Dirac delta) (Gardner 1985, Section 10.1.2), (Papoulis 1991, Section 12-4). Therefore, $\chi(f)$ and $\gamma(f_1, f_2)$ are assumed to have zero singular component (Cramér 1940).

Furthermore, all the ordinary functions are assumed to be Lebesgue measurable and considerations made at the beginning of Section 3.4 hold.

Definition 4.2.3 *Let* $\{x(t),\ t \in \mathbb{R}\}$ *be a second-order complex-valued harmonizable stochastic process. Its* bifrequency spectral correlation function *or* Loève bifrequency spectrum *(Loève 1963), (Thomson 1982), also called* generalized spectrum *in (Gerr and Allen 1994),* cointensity spectrum *in (Middleton 1967), or* dual frequency spectral correlation *in (Hanssen and Scharf 2003), is defined as*

$$\mathcal{S}_{xx^{(*)}}(f_1, f_2) \triangleq \mathrm{E}\left\{X(f_1)\, X^{(*)}(f_2)\right\} . \tag{4.4}$$

In (4.4),

$$X(f) \triangleq \int_{\mathbb{R}} x(t)\, e^{-j2\pi ft}\, \mathrm{d}t \tag{4.5}$$

is the Fourier transform of $x(t)$ *and is assumed to exist (at least) in the sense of distributions (Gelfand and Vilenkin 1964, Chapter 3), (Henniger 1970) w.p.1.. Superscript* $(*)$ *denotes an optional complex conjugation.* □

The functions defined in (4.4) and (4.5) can be linked to those in (4.2) and (4.3). In fact, even if $\gamma(f_1, f_2)$ and $\chi(f)$ contain first kind discontinuities (jumps), in the sense of distributions we can formally write (Gardner 1985, Section 10.1.2), (Papoulis 1991, Section 12-4)

$$\mathrm{d}\chi(f) = X(f)\,\mathrm{d}f \tag{4.6}$$

$$\begin{aligned} \mathrm{d}\gamma(f_1, f_2) &= \mathrm{E}\{\mathrm{d}\chi(f_1)\,\mathrm{d}\chi^*(f_2)\} \\ &= \mathrm{E}\{X(f_1)X^*(f_2)\}\,\mathrm{d}f_1\,\mathrm{d}f_2 . \end{aligned} \tag{4.7}$$

Such representation does not hold in the presence of singular components (Hurd and Miamee 2007, p. 197).

Note that, two functions are defined in (4.4) depending on the choice of conjugating or not the second term even if, in general, only the function with the conjugation is named bifrequency spectrum and is considered in the literature. Both possibilities on $(*)$ need to be considered, however, in dealing with nonstationary complex signals (Thomson 1982), (Picinbono and Bondon 1997), (Schreier and Scharf 2003a), (Schreier and Scharf 2003b).

The Loève bifrequency spectrum (4.4) is the statistical correlation between the spectral components of $x(t)$ at frequencies f_1 and f_2. It is well known that for second-order WSS stochastic processes the bifrequency spectrum (with $(*)$ present) is zero unless $f_1 = f_2$. Then, the presence of spectral correlation for $f_1 \neq f_2$ can be adopted to detect the presence of nonstationarity in the data (Thomson 1982), (Hurd 1988, 1991), (Hurd and Gerr 1991), (Allen and Hobbs 1992), (Genossar 1992), (Dehay and Hurd 1994), (Gerr and Allen 1994), (Varghese and Donohue 1994).

Definition 4.2.4 *A complex-valued second-order harmonizable stochastic process* $\{x(t),\ t \in \mathbb{R}\}$*, is said to be* spectrally correlated *if its Loève bifrequency spectrum can be expressed as*

$$\mathcal{S}_{xx^{(*)}}(f_1, f_2) = \sum_{n \in \mathbb{I}_{xx^{(*)}}} S^{(n)}_{xx^{(*)}}(f_1)\, \delta(f_2 - \Psi^{(n)}_{xx^{(*)}}(f_1)) , \tag{4.8}$$

where $\delta(\cdot)$ *denotes Dirac delta,* $\mathbb{I}_{xx^{(*)}}$ *is a countable set, the curves* $f_2 = \Psi^{(n)}_{xx^{(*)}}(f_1)$, $n \in \mathbb{I}_{xx^{(*)}}$, *describe the support of* $\mathcal{S}_{xx^{(*)}}(f_1, f_2)$, *and the functions* $S^{(n)}_{xx^{(*)}}(f_1)$, *referred to as* spectral correlation density functions, *describe the density of the Loève spectrum on its support curves.* □

According to (Lii and Rosenblatt 2002) where the special case of support lines is considered, the curve $f_2 = \Psi^{(n)}_{xx^{(*)}}(f_1)$ is said a support curve if $S^{(n)}_{xx^{(*)}}(f_1) \neq 0$ in a set of positive one-dimensional Lebesgue measure.

Note that the functions $\Psi^{(n)}_{xx^{(*)}}(\cdot)$ depend on the choice made for $(*)$, can always be chosen invertible if no correlation exists along lines with $f_1 =$ constant, but cannot be arbitrary since the symmetry condition $\mathcal{S}_{xx^{(*)}}(f_1, f_2) = \mathcal{S}_{xx^{(*)}}(f_2, f_1)^{(*)}$ needs to be satisfied (Picinbono and Bondon 1997). Furthermore, the functions $\Psi^{(n)}_{xx^{(*)}}(\cdot)$ and $S^{(n)}_{xx^{(*)}}(\cdot)$ can always be chosen such that the set

$$\mathbb{I}(\eta) \triangleq \left\{ f_1 \in \mathbb{R} \,:\, \eta = \Psi^{(n')}_{xx^{(*)}}(f_1) = \Psi^{(n'')}_{xx^{(*)}}(f_1) \text{ for some } n' \neq n'' \right\} \tag{4.9}$$

is at most countable $\forall \eta \in \mathbb{R}$. In such a case, the condition of bounded variation of the covariance $\gamma(f_1, f_2)$ can be written as

$$\int_{\mathbb{R}} \int_{\mathbb{R}} \left| \mathrm{E}\left\{ X(f_1)\, X^{(*)}(f_2) \right\} \right| \mathrm{d}f_1 \, \mathrm{d}f_2 = \sum_{n \in \mathbb{I}_{xx^{(*)}}} \int_{\mathbb{R}} \left| S^{(n)}_{xx^{(*)}}(f_1) \right| \mathrm{d}f_1 < \infty \tag{4.10}$$

which is the generalization to SC processes of eq. (4.4) in (Hurd 1991) derived for ACS processes (see also (10) in (Hurd 1989a) and (13) in (Ogura 1971) for the case of cyclostationary processes).

From Definition 4.2.4, it follows that the process $x(t)$ does not contain a deterministic or random almost-periodic component (Section 1.2.7). In fact, if

$$x(t) = \xi(t) + \sum_k x_k \, e^{j2\pi\lambda_k t} \tag{4.11}$$

with $\xi(t)$ zero-mean and SC according to Definition 4.2.4 and x_k deterministic coefficients or random variables, then, denoting by $\Xi(f)$ the Fourier transform in the sense of distributions of $\xi(t)$, we have the following expression for the Loève bifrequency spectrum of $x(t)$:

$$\begin{aligned}
\mathcal{S}_{xx^{(*)}}(f_1, f_2) &= \mathrm{E}\left\{ \left[\Xi(f_1) + \sum_{k_1} x_{k_1} \, \delta(f_1 - \lambda_{k_1}) \right] \left[\Xi(f_2) + \sum_{k_2} x_{k_2} \, \delta(f_2 - \lambda_{k_2}) \right]^{(*)} \right\} \\
&= \sum_{n \in \mathbb{I}} S^{(n)}_{\xi\xi^{(*)}}(f_1) \, \delta\left(f_2 - \Psi^{(n)}_{\xi\xi^{(*)}}(f_1) \right) \\
&\quad + \sum_{k_1} \mathrm{E}\left\{ x_{k_1} \, \Xi^{(*)}(f_2) \right\} \delta(f_1 - \lambda_{k_1})
\end{aligned}$$

$$+\sum_{k_2} \mathrm{E}\left\{\Xi(f_1)\, x_{k_2}^{(*)}\right\} \delta(f_2 - \lambda_{k_2})$$
$$+\sum_{k_1}\sum_{k_2} \mathrm{E}\left\{x_{k_1}\, x_{k_2}^{(*)}\right\} \delta(f_1 - \lambda_{k_1})\, \delta(f_2 - \lambda_{k_2})\,. \tag{4.12}$$

where

$$\mathrm{E}\left\{\Xi(f_1)\, \Xi^{(*)}(f_2)\right\} = \sum_{n \in \mathbb{I}} S_{\xi\xi^{(*)}}^{(n)}(f_1)\, \delta\left(f_2 - \Psi_{\xi\xi^{(*)}}^{(n)}(f_1)\right) \tag{4.13}$$

is the Loève bifrequency spectrum of the SC process $\xi(t)$.

From Definition 4.2.4, it follows that SC processes are those nonstationary processes for which the Loève bifrequency spectrum is constituted by spectral masses located on a countable set of support curves in the bifrequency plane (f_1, f_2). If the process $x(t)$ contains a deterministic or random almost-periodic component, then also isolated points and lines parallel to the frequency axis are present in the support of the Loève bifrequency spectrum. Strictly speaking, terms with $\delta(f_1 - \lambda_{k_1})$ can be accommodated by (4.8). This case, however, will be excluded in the following by assuming the functions $\Psi_{\xi\xi^{(*)}}^{(n)}(\cdot)$ invertible.

More generally, the Loève bifrequency spectrum can be constituted by both an impulsive and a continuous part.

Definition 4.2.5 *A complex-valued second-order harmonizable stochastic process* $\{x(t),\ t \in \mathbb{R}\}$ *with no finite-strength additive sinewave components, is said to* exhibit spectral correlation *if its Loève bifrequency spectrum can be expressed as*

$$\mathcal{S}_{xx^{(*)}}(f_1, f_2) = \sum_{n \in \mathbb{I}_{xx^{(*)}}} S_{xx^{(*)}}^{(n)}(f_1)\, \delta(f_2 - \Psi_{xx^{(*)}}^{(n)}(f_1)) + \mathcal{S}_{xx^{(*)}}^{(c)}(f_1, f_2) \tag{4.14}$$

where $\mathcal{S}_{xx^{(*)}}^{(c)}(f_1, f_2)$ *is a continuous function.* □

Nonstationary processes with $\mathcal{S}_{xx^*}(f_1, f_2) = \mathcal{S}_{xx^*}^{(c)}(f_1, f_2)$ are treated in (Soedjack 2002), where a consistent estimator of $\mathcal{S}_{xx^*}^{(c)}(f_1, f_2)$ is obtained by using several realizations of the stochastic process. In (Allen and Hobbs 1997) a consistent estimator based on a single realization is proposed for the case of nonstationary white noise.

A single SC stochastic process is also said singularly SC if a distinction is necessary with the joint spectral correlation property of two different processes.

Definition 4.2.6 *Two complex-valued second-order harmonizable stochastic processes* $\{y(t),\ t \in \mathbb{R}\}$ *and* $\{x(t),\ t \in \mathbb{R}\}$ *are said* jointly SC *if their bifrequency cross-spectrum* $\mathcal{S}_{yx^{(*)}}(f_1, f_2)$ *is constituted by spectral masses located on a countable set of support curves in the bifrequency plane* (f_1, f_2)*, that is,*

$$\begin{aligned}\mathcal{S}_{yx^{(*)}}(f_1, f_2) &\triangleq \mathrm{E}\left\{Y(f_1)\, X^{(*)}(f_2)\right\} \\ &= \sum_{n \in \mathbb{I}_{yx^{(*)}}} S_{yx^{(*)}}^{(n)}(f_1)\, \delta\left(f_2 - \Psi_{yx^{(*)}}^{(n)}(f_1)\right)\end{aligned} \tag{4.15}$$

where $\mathbb{I}_{yx^{(*)}}$ *is a countable set, the functions* $\Psi^{(n)}_{yx^{(*)}}(\cdot)$ *can always be chosen invertible, and the functions* $S^{(n)}_{yx^{(*)}}(f_1)$ *are referred to as the spectral cross-correlation density functions.* □

In the following, for notation simplicity, when this does not generate ambiguity, we will put

$$\Psi^{(n)}(f_1) \equiv \Psi^{(n)}_{yx^{(*)}}(f_1) \qquad S^{(n)}(f_1) \equiv S^{(n)}_{yx^{(*)}}(f_1) \qquad \mathbb{I} \equiv \mathbb{I}_{yx^{(*)}} .$$

Moreover, with reference to definitions given in the following Theorem 4.2.7, we will put

$$\Phi^{(n)}(f_2) \equiv \Phi^{(n)}_{yx^{(*)}}(f_2) \qquad G^{(n)}(f_2) \equiv G^{(n)}_{yx^{(*)}}(f_2) .$$

Accounting for the fact that $\delta(f_2 - \Psi^{(n)}(f_1)) = |\Phi^{(n)\prime}(f_2)|\, \delta(f_1 - \Phi^{(n)}(f_2))$ and $\delta(f_1 - \Phi^{(n)}(f_2)) = |\Psi^{(n)\prime}(f_1)|\, \delta(f_2 - \Psi^{(n)}(f_1))$ (Zemanian 1987, Section 1.7), with $\Phi^{(n)}(\cdot)$ denoting the inverse function of $\Psi^{(n)}(\cdot)$, and $\Phi^{(n)\prime}(\cdot)$ and $\Psi^{(n)\prime}(\cdot)$ the derivatives of $\Phi^{(n)}(\cdot)$ and $\Psi^{(n)}(\cdot)$, respectively, and by using the sampling property of the Dirac delta , the following result can be proved.

Theorem 4.2.7 *Let* $\{y(t),\ t \in \mathbb{R}\}$ *and* $\{x(t),\ t \in \mathbb{R}\}$ *be continuous-time complex-valued second-order jointly harmonizable stochastic processes not containing any additive finite-strength sinewave component. If the processes are jointly spectrally correlated, then their Loève bifrequency cross-spectrum can be expressed in one of the two equivalent forms*

$$\mathcal{S}_{yx^{(*)}}(f_1, f_2) = \sum_{n\in\mathbb{I}} S^{(n)}(f_1)\, \delta\left(f_2 - \Psi^{(n)}(f_1)\right) \tag{4.16a}$$

$$= \sum_{n\in\mathbb{I}} G^{(n)}(f_2)\, \delta\left(f_1 - \Phi^{(n)}(f_2)\right) . \tag{4.16b}$$

In (4.16a) and (4.16b), $\delta(\cdot)$ *is Dirac delta,* $\mathbb{I}$ *is a countable set, the curves* $f_2 = \Psi^{(n)}(f_1)$ *or, equivalently,* $f_1 = \Phi^{(n)}(f_2)$*, describe the support of* $\mathcal{S}_{yx^{(*)}}(f_1, f_2)$*, and the complex-valued functions* $S^{(n)}(f_1)$ *and* $G^{(n)}(f_2)$*, called* spectral correlation density functions, *represent the density of the Loève bifrequency spectrum on its support curves. The functions* $S^{(n)}(f_1)$ *and* $G^{(n)}(f_2)$ *are not unambiguously determined since each of them can be expressed as sum of more functions with non overlapping supports. Assuming that the functions* $\Psi^{(n)}(f_1)$ *and* $\Phi^{(n)}(f_2)$ *are not constant within intervals of positive Lebesgue measure, the functions* $S^{(n)}(f_1)$ *and* $G^{(n)}(f_2)$ *can be chosen with supports such that* $\Psi^{(n)}(\cdot)$ *and* $\Phi^{(n)}(\cdot)$ *are invertible functions and the curves* $f_2 = \Psi^{(n)}(f_1)$ *and* $f_2 = \Psi^{(n')}(f_1)$ *for* $n \neq n'$ *intersect at most in a countable set of points. If the real-valued functions* $\Psi^{(n)}(\cdot)$ *are assumed to be differentiable,* $\Phi^{(n)}(\cdot)$ *are their inverse and are also differentiable, then*

$$S^{(n)}(f_1) = \left|\Psi^{(n)\prime}(f_1)\right| G^{(n)}\left(\Psi^{(n)}(f_1)\right) \tag{4.17a}$$

$$G^{(n)}(f_2) = \left|\Phi^{(n)\prime}(f_2)\right| S^{(n)}\left(\Phi^{(n)}(f_2)\right) \tag{4.17b}$$

where superscript $'$ *denotes derivative.* □

The differentiability assumption for both $\Psi^{(n)}(\cdot)$ and $\Phi^{(n)}(\cdot)$ prevents that their derivatives are zero or infinity. This assumption is fulfilled in all practical cases considered in Section 4.2.4 except the fBm. Such an assumption can be relaxed by considering differentiability almost everywhere and the possibility of derivatives equal to zero or infinity, even in frequency intervals with nonzero Lebesgue measure, as it happens in (4.12) or for the fBm discussed in Section 4.2.4. In such a case, results similar to those subsequently derived hold, but proofs are more cumbersome.

The assumption that $y(t)$ and $x(t)$ do not contain additive finite-strength sinewave components, say $y_k \, e^{j2\pi \bar{f}_{k,y}t}$ and $x_k \, e^{j2\pi \bar{f}_{k,x}t}$ prevents that the Loève bifrequency cross-spectrum (4.15) contains spectral masses concentrated in isolated points $(\bar{f}_{k,x}, \bar{f}_{k,y})$ of the bifrequency plane or on lines parallel to the frequency axes. That is, the Loève bifrequency cross-spectrum does not contain additive terms as $\mathrm{E}\{y_k \, x_k^{(*)}\} \, \delta(f_1 - \bar{f}_{k,x}) \, \delta(f_2 - \bar{f}_{k,y})$, $g_2(f_2) \, \delta(f_1 - \bar{f}_{k,x})$, and $g_1(f_1) \, \delta(f_2 - \bar{f}_{k,y})$, with $g_1(f_1)$ and $g_2(f_2)$ functions depending on $x(t)$ and $y(t)$ (see (4.12) for the case $y \equiv x$).

The bounded variation condition (4.1), which is consequence of the harmonizability assumption, in the case of jointly SC processes can be written as

$$\begin{aligned} \int_{\mathbb{R}^2} \left| \mathrm{d}\gamma_{yx^{(*)}}(f_1, f_2) \right| &= \sum_{n \in \mathbb{I}} \int_{\mathbb{R}} \left| S^{(n)}(f_1) \right| \, \mathrm{d}f_1 \\ &= \sum_{n \in \mathbb{I}} \int_{\mathbb{R}} \left| G^{(n)}(f_2) \right| \, \mathrm{d}f_2 < \infty. \end{aligned} \tag{4.18}$$

Definition 4.2.8 *Given two complex-valued second-order harmonizable jointly SC stochastic processes* $\{y(t), \; t \in \mathbb{R}\}$ *and* $\{x(t), \; t \in \mathbb{R}\}$, *their* bifrequency spectral cross-correlation density function *is defined as*

$$\bar{\mathbb{S}}_{yx^{(*)}}(f_1, f_2) \triangleq \sum_{n \in \mathbb{I}} S^{(n)}(f_1) \, \delta_{f_2 - \Psi^{(n)}(f_1)} \tag{4.19}$$

where δ_f *denotes Kronecker delta (i.e.,* $\delta_f = 1$ *for* $f = 0$ *and* $\delta_f = 0$ *otherwise).* □

The function $\bar{\mathbb{S}}_{yx^{(*)}}(f_1, f_2)$ is named *bifrequency spectral cross-correlation density function* since it is the generalization to SC processes (and two frequency arguments) of the spectral cross-correlation density function defined for ACS processes in (Gardner 1986a), (Gardner 1987d), and (Gardner 1991b). For any pair (f_1, f_2), $\bar{\mathbb{S}}_{yx^{(*)}}(f_1, f_2)$ is given by the sum of all $S^{(n)}(f_1)$ such that $f_2 = \Psi^{(n)}(f_1)$. Thus, it represents the density of the spectral cross-correlation $\mathrm{E}\{Y(f_1)X^{(*)}(f_2)\}$ between the spectral components at frequencies f_1 and f_2 of $y(t)$ and $x(t)$, respectively.

The arguments of the Dirac deltas in (4.16a) and (4.16b) contain (possibly nonlinear) functions $\Psi^{(n)}(f_1)$ and $\Phi^{(n)}(f_2)$, respectively. Thus, two kinds of bifrequency spectral-correlation densities should be considered, depending on which one of the two frequency variables, f_1 or f_2, is considered as independent in the argument of $\delta(\cdot)$. Specifically, under the assumption that there is no cluster of support curves, by integrating both sides of (4.16a) w.r.t. f_2 and both sides of (4.16b) w.r.t. f_1, the following two *bifrequency spectral cross-correlation density (SCD) functions* can be defined:

$$\bar{\mathcal{S}}^{(1)}_{yx^{(*)}}(f_1, f_2) \triangleq \lim_{\Delta f \to 0+} \int_{f_2-\Delta f/2}^{f_2+\Delta f/2} \mathcal{S}_{yx^{(*)}}(f_1, \lambda_2)\, \mathrm{d}\lambda_2$$

$$= \sum_{n\in\mathbb{I}} S^{(n)}(f_1)\, \delta_{f_2-\Psi^{(n)}(f_1)} \tag{4.20a}$$

$$\bar{\mathcal{S}}^{(2)}_{yx^{(*)}}(f_1, f_2) \triangleq \lim_{\Delta f \to 0+} \int_{f_1-\Delta f/2}^{f_1+\Delta f/2} \mathcal{S}_{yx^{(*)}}(\lambda_1, f_2)\, \mathrm{d}\lambda_1$$

$$= \sum_{n\in\mathbb{I}} G^{(n)}(f_2)\, \delta_{f_1-\Phi^{(n)}(f_2)} \tag{4.20b}$$

In the sequel, unlike otherwise specified, the term bifrequency spectral correlation density will be used for the function $\bar{\mathcal{S}}^{(1)}_{yx^{(*)}}(f_1, f_2)$ that will be denoted by $\bar{\mathcal{S}}_{yx^{(*)}}(f_1, f_2)$ if it does not create ambiguity.

Representation (4.16a) for the bifrequency cross-spectrum must be used if spectral correlation exists along lines with $f_2 =$ constant, whereas representation (4.16b) must be used if spectral correlation exists along lines with $f_1 =$ constant. Let be

$$x(t) = \sum_k x_k\, e^{j2\pi\lambda_k t} + \xi(t) \tag{4.21}$$

$$y(t) = \sum_k y_k\, e^{j2\pi\mu_k t} + \eta(t) \tag{4.22}$$

with $\xi(t)$ and $\eta(t)$ zero-mean and SC according to Definition 4.2.4 and x_k and y_k deterministic coefficients or random variables statistically independent of $\xi(t)$ and $\eta(t)$ $\forall t$. The most general expression for the bifrequency cross-spectrum of $x(t)$ and $y(t)$ which accommodates also possible spectral correlation along lines parallel to the f_1 and f_2 axes (when some of the $\Psi^{(n_1)}_{\xi\eta^{(*)}}(\cdot)$ and/or $\Phi^{(n_2)}_{\xi\eta^{(*)}}(\cdot)$ functions are constant) and a continuous term, can be obtained by splitting the support curves into two disjoint sets:

$$\begin{aligned}\mathcal{S}_{yx^{(*)}}(f_1, f_2) = &\sum_{n_1\in\mathbb{I}_1} S^{(n_1)}_{\xi\eta^{(*)}}(f_1)\, \delta\left(f_2 - \Psi^{(n)}_{\xi\eta^{(*)}}(f_1)\right) \\ &+ \sum_{n_2\in\mathbb{I}_2} G^{(n_2)}_{\xi\eta^{(*)}}(f_2)\, \delta\left(f_1 - \Phi^{(n_2)}_{\xi\eta^{(*)}}(f_2)\right) \\ &+ \sum_{k_1}\sum_{k_2} \mathrm{E}\left\{y_{k_1}\, x^{(*)}_{k_2}\right\}\, \delta(f_1 - \mu_{k_1})\, \delta(f_2 - \lambda_{k_2}) \\ &+ \mathcal{S}^{(c)}_{yx^{(*)}}(f_1, f_2)\end{aligned} \tag{4.23}$$

where the decomposition is not unique and $\mathcal{S}^{(c)}_{yx^{(*)}}(f_1, f_2)$ is a continuous function.

Accounting for (4.15) and the sampling property of the Dirac delta, the following result can be proved.

Theorem 4.2.9 Second-Order Temporal Cross-Moment of Jointly SC Processes. *Let* $\{y(t),\ t \in \mathbb{R}\}$ *and* $\{x(t),\ t \in \mathbb{R}\}$ *be complex-valued second-order harmonizable jointly SC* *stochastic processes with Loève bifrequency cross-spectrum (4.16a). The second-order temporal cross-moment of* $y(t)$ *and* $x(t)$ *is given by*

$$\mathcal{R}_{yx^{(*)}}(t_1, t_2) \triangleq \mathrm{E}\{y(t_1)\, x^{(*)}(t_2)\} \tag{4.24a}$$

$$= \int_{\mathbb{R}^2} \mathcal{S}_{yx^{(*)}}(f_1, f_2)\, e^{j2\pi(f_1 t_1 + (-) f_2 t_2)}\, \mathrm{d}f_1\, \mathrm{d}f_2 \tag{4.24b}$$

$$= \sum_{n \in \mathbb{I}_{yx^{(*)}}} \int_{\mathbb{R}} S^{(n)}_{yx^{(*)}}(f_1)\, e^{j2\pi(f_1 t_1 + (-)\Psi^{(n)}_{yx^{(*)}}(f_1) t_2)}\, \mathrm{d}f_1 \tag{4.24c}$$

$$= \sum_{n \in \mathbb{I}_{yx^{(*)}}} R^{(n)}_{yx^{(*)}}(t_1) \underset{t_1}{\otimes} \Xi^{(n)}_{yx^{(*)}}(t_1, t_2) \tag{4.24d}$$

where $(-)$ *is an optional minus sign to be considered only if* $(*)$ *is present,* $\otimes_{t_1}$ *denotes convolution with respect to* t_1,

$$R^{(n)}_{yx^{(*)}}(t_1) \triangleq \int_{\mathbb{R}} S^{(n)}_{yx^{(*)}}(f_1)\, e^{j2\pi f_1 t_1}\, \mathrm{d}f_1 \tag{4.25}$$

$$\Xi^{(n)}_{yx^{(*)}}(t_1, t_2) \triangleq \int_{\mathbb{R}} e^{(-)j2\pi\Psi^{(n)}_{yx^{(*)}}(f_1) t_2}\, e^{j2\pi f_1 t_1}\, \mathrm{d}f_1 \tag{4.26}$$

and the Fourier transforms are in the sense of distributions (Gelfand and Vilenkin 1964), (Henniger 1970), (Zemanian 1987).

Proof: *See Section 5.1.* □

Theorem 4.2.10 Time-Variant Cross-Spectrum of Jointly SC Processes. *Let* $\{y(t),\ t \in \mathbb{R}\}$ *and* $\{x(t),\ t \in \mathbb{R}\}$ *be complex-valued second-order harmonizable jointly SC stochastic processes with Loève bifrequency cross-spectrum (4.16a). The time-variant cross-spectrum of* $y(t)$ *and* $x(t)$ *is given by*

$$\mathcal{S}_{yx^{(*)}}(t, f) \triangleq \int_{\mathbb{R}} \mathrm{E}\left\{y(t+\tau)\, x^{(*)}(t)\right\} e^{-j2\pi f\tau}\, \mathrm{d}\tau \tag{4.27a}$$

$$= \sum_{n \in \mathbb{I}} S^{(n)}_{yx^{(*)}}(f)\, e^{j2\pi[f + (-)\Psi^{(n)}_{yx^{(*)}}(f)]t} \tag{4.27b}$$

Proof: *See Section 5.1.* □

From (4.27b) it follows that the time-variant cross-spectrum of jointly SC processes is an almost periodic function of t with both coefficient and frequencies depending on f.

The class of the SC stochastic processes includes, as a special case, the class of the second-order wide-sense ACS processes which, in turn, includes the class of the second-order WSS processes. For ACS processes, the support of the bifrequency spectrum in the bifrequency plane is constituted by lines with unit slope (Hurd 1989a), (Dehay and Hurd 1994). Consequently, the separation between correlated spectral components assumes values belonging to a countable set. Such values, called cycle frequencies, are the frequencies of the Fourier series expansion of the almost-periodically time-variant statistical autocorrelation function (Gardner 1987d), (Dehay and Hurd 1994). Thus, for ACS processes, a one-to-one correspondence exists between elements $n \in \mathbb{I}_{xx^{(*)}}$ and cycle frequencies α, and

$$\Psi^{(n)}_{xx^{(*)}}(f_1) = (-)(\alpha - f_1). \tag{4.28}$$

Furthermore, the spectral correlation density functions $S^{(n)}_{xx^{(*)}}(f)$ are coincident with the (conjugate if $(*)$ is absent) cyclic spectra $S^{\alpha}_{xx^{(*)}}(f)$. Therefore,

$$\mathcal{S}_{xx^{(*)}}(f_1, f_2) = \sum_{\alpha \in A_{xx^{(*)}}} S^{\alpha}_{xx^{(*)}}(f_1)\, \delta(f_2 - (-)(\alpha - f_1)) \tag{4.29}$$

$$\mathcal{R}_{xx^{(*)}}(t_1, t_2) = \sum_{\alpha \in A_{xx^{(*)}}} R^{\alpha}_{xx^{(*)}}(t_1 - t_2)\, e^{j2\pi\alpha t_2} \tag{4.30}$$

where $A_{xx^{(*)}}$ denotes the countable set of the (conjugate) cycle frequencies and $R^{\alpha}_{xx^{(*)}}(\tau)$ are the (conjugate) cyclic autocorrelation functions, that is, the inverse Fourier transforms of the (conjugate) cyclic spectra $S^{\alpha}_{xx^{(*)}}(f_1)$. Moreover, for ACS processes the time-variant spectrum (4.27a) is an almost-periodic function of t with coefficients depending on f and equal to the cyclic spectra and frequencies not depending on f and coincident with the cycle frequencies (see (1.90)). Note that the terms SC and ACS are used as synonyms in (Gardner 1987d), (Gardner 1991b), where it is considered only the case in which the separation between correlated spectral components can assume values in a countable set. WSS processes can be obtained as a special case of ACS processes when the set A_{xx^*} contains the only element $\alpha = 0$. In such a case, by specializing (4.29) and (4.30), one has

$$\mathcal{S}_{xx^*}(f_1, f_2) = S^{0}_{xx^*}(f_1)\, \delta(f_2 - f_1) \tag{4.31}$$

$$\mathcal{R}_{xx^*}(t_1, t_2) = R^{0}_{xx^*}(t_1 - t_2) \tag{4.32}$$

where $S^{0}_{xx^*}(f_1)$ is the power spectral density function of $x(t)$. From (4.31) and (4.32) it is evident that a zero-mean stochastic process $x(t)$ is WSS (i.e., $\mathcal{R}_{xx^*}(t_1, t_2)$ is a function of $t_1 - t_2$) if and only if no correlation exists between spectral components that are separated (i.e., $\mathcal{S}_{xx^*}(f_1, f_2)$ is concentrated on the diagonal $f_2 = f_1$ of the bifrequency plane). If the process is also circular (i.e., $\mathcal{R}_{xx}(t_1, t_2) \equiv 0$ (Picinbono and Bondon 1997)), then it also results $\mathcal{S}_{xx}(f_1, f_2) \equiv 0$.

In Figures 4.1(a) and 4.1(b), two examples of supports of Loève bifrequency spectra in the bifrequency plane are represented. The case of an ACS process is considered in Figure 4.1(a) and that of an SC not ACS process in Figure 4.1(b).

4.2.2 *Relationship among ACS, GACS, and SC Processes*

In Chapter 2, the class of the generalized almost-cyclostationary (GACS) processes (Izzo and Napolitano 1998b) is characterized. GACS processes exhibit an almost-periodically time-variant statistical autocorrelation function whose (generalized) Fourier series expansion has coefficients and frequencies (cycle frequencies) depending on the lag parameter. ACS processes are obtained as special case when the frequencies are constant with respect to the lag parameter.

SC processes have a Loève bifrequency spectrum with spectral masses concentrated on a countable set of curves in the bifrequency plane. ACS processes are obtained as special case of

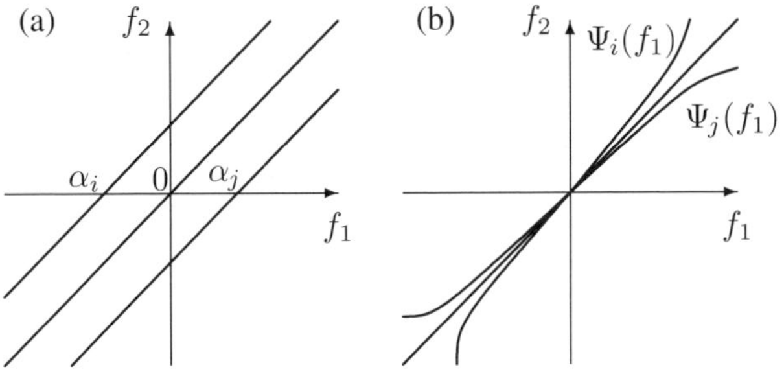

Figure 4.1 Support in the bifrequency plane of the Loève bifrequency spectrum of (a) an ACS process and (b) a SC process

SC processes when the support curves are lines with unit slope (Hurd 1989a), (Hurd and Gerr 1991), (Dehay and Hurd 1994), (see (4.29)). For ACS processes, correlation exists only between spectral components that are separated by quantities belonging to a countable set of values, the cycle frequencies. In such a case, the (conjugate) autocorrelation function $\mathrm{E}\{x(t_1)\, x^{(*)}(t_2)\}$ obtained as double inverse Fourier transform of the Loève bifrequency spectrum (4.4), as a function of $t = t_2$ and $\tau = t_1 - t_2$, is almost periodic in t with frequencies equal to the cycle frequencies and independent of τ (see (4.30)).

Therefore, ACS processes are obtained by the intersection of the class of the GACS processes and the class of the SC processes (Figure 4.2). In particular, ACS processes are the subclass of the GACS processes that exhibit the spectral correlation property.

4.2.3 Higher-Order Statistics

In this section, the higher-order characterization of SC processes is carried out in the frequency domain. Specifically, higher-order spectral moments and cumulants are defined and their relationships established.

Let $x_i(t)$, $(i = 1, \ldots, N)$, be (possibly distinct) complex-valued harmonizable processes with Cramèr representation

$$x_i(t) = \int_{\mathbb{R}} e^{j2\pi f t}\, \mathrm{d}\chi_i(f). \tag{4.33}$$

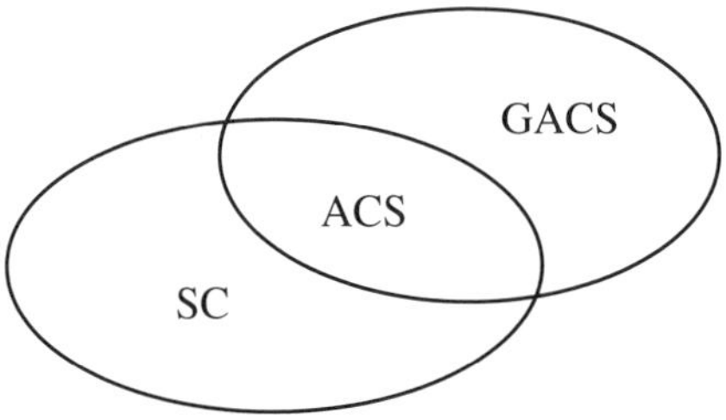

Figure 4.2 Relationship among GACS, SC, and ACS process classes

Since the processes are denoted as distinct, without lack of generality no optional complex conjugation is necessary in the notation of the following definitions. In fact, if $x_i(t) = x_0^*(t)$, then $\mathrm{d}\chi_i(f) = \mathrm{d}\chi_0^*(-f)$.

Accordingly with the (second-order) harmonizability definition (Definition 4.2.1) and the definitions in (Brillinger and Rosenblatt 1967), (Brillinger 1974), (Hanssen and Scharf 2003), for higher-order moment and cumulants, we have the following, where cumulants of complex variables and processes are defined according to (Spooner and Gardner 1994, App. A), (Napolitano 2007a), and Section 1.4.2.

Definition 4.2.11 *The processes* $x_i(t)$, $(i = 1, \ldots, N)$ *are said to be* jointly harmonizable for the moment *if there exists a spectral moment function* $\mathrm{E}\left\{\prod_{n=1}^{N} \mathrm{d}\chi_n(f_n)\right\}$ *such that*

$$\mathrm{E}\left\{\prod_{n=1}^{N} x_n(t_n)\right\} = \int_{\mathbb{R}} e^{j2\pi[f_1 t_1 + \cdots + f_N t_N]}\, \mathrm{E}\left\{\prod_{n=1}^{N} \mathrm{d}\chi_n(f_n)\right\} \tag{4.34}$$

with

$$\int_{\mathbb{R}} \left| \mathrm{E}\left\{\prod_{n=1}^{N} \mathrm{d}\chi_n(f_n)\right\}\right| < \infty. \tag{4.35}$$

Definition 4.2.12 *The processes* $x_i(t)$, $(i = 1, \ldots, N)$ *are said to be* jointly harmonizable for the cumulant *if there exists a spectral cumulant function* $\mathrm{cum}\{\mathrm{d}\chi_1(f_1), \ldots, \mathrm{d}\chi_N(f_N)\}$ *such that*

$$\mathrm{cum}\{x_1(t_1), \ldots, x_N(t_N)\} = \int_{\mathbb{R}} e^{j2\pi[f_1 t_1 + \cdots + f_N t_N]}\, \mathrm{cum}\{\mathrm{d}\chi_1(f_1), \ldots, \mathrm{d}\chi_N(f_N)\} \tag{4.36}$$

with

$$\int_{\mathbb{R}} |\mathrm{cum}\{\mathrm{d}\chi_1(f_1), \ldots, \mathrm{d}\chi_N(f_N)\}| < \infty. \tag{4.37}$$

Accounting for the relationships existing between moments and cumulants of real-valued random variables and processes (Leonov and Shiryaev 1959), (Brillinger 1965), (Brillinger and Rosenblatt 1967), and their extension to the complex case (Spooner and Gardner 1994, App. A), (Napolitano 2007a), (Section 1.4.2), one obtains

$$\begin{aligned}&\mathrm{cum}\{\mathrm{d}\chi_1(f_1), \ldots, \mathrm{d}\chi_N(f_N)\}\\ &\quad = \sum_{\mathrm{P}} \left[(-1)^{p-1}(p-1)! \prod_{i=1}^{p} \mathrm{E}\left\{\prod_{n \in \mu_i} \mathrm{d}\chi_n(f_n)\right\}\right]\end{aligned} \tag{4.38}$$

where P is the set of distinct partitions of $\{1, \ldots, N\}$, each constituted by the subsets $\{\mu_i,\ i = 1, \ldots, p\}$. Moreover, it results that

$$\mathrm{E}\left\{\prod_{n=1}^{N} \mathrm{d}\chi_n(f_n)\right\} = \sum_{\mathrm{P}} \left[\prod_{i=1}^{p} \mathrm{cum}\left\{\mathrm{d}\chi_n(f_n),\ n \in \mu_i\right\}\right]. \tag{4.39}$$

Definition 4.2.13 *Let $x_i(t)$, $(i = 1, \ldots, N)$ be processes jointly harmonizable for the cumulant. The processes are said to be* (Nth-order) jointly SC *if*

$$\operatorname{cum}\Big\{ \mathrm{d}\chi_1(f_1), \ldots, \mathrm{d}\chi_N(f_N)\Big\} = \sum_{k\in\mathbb{I}} \mathrm{d}\mathcal{P}_{\boldsymbol{x}}^{(k)}(f_1, \ldots, f_{N-1})\, \delta\Big(f_N - \Psi_{\boldsymbol{x}}^{(k)}(f_1, \ldots, f_{N-1})\Big)\, \mathrm{d}f_N \tag{4.40}$$

where $\boldsymbol{x} \triangleq [x_1, \ldots, x_N]$ and $\mathcal{P}_{\boldsymbol{x}}^{(k)}(f_1, \ldots, f_{N-1})$ does not contain jumps. If, in addition, $\mathcal{P}_{\boldsymbol{x}}^{(k)}(f_1, \ldots, f_{N-1})$ does not have singular component, $\mathcal{P}_{\boldsymbol{x}}^{(k)}(f_1, \ldots, f_{N-1})$ is absolutely continuous and

$$\mathrm{d}\mathcal{P}_{\boldsymbol{x}}^{(k)}(f_1, \ldots, f_{N-1}) = P_{\boldsymbol{x}}^{(k)}(f_1, \ldots, f_{N-1})\, \mathrm{d}f_1 \cdots \mathrm{d}f_{N-1}. \tag{4.41}$$

It results that

$$P_{\boldsymbol{x}}^{(k)}(f_1, \ldots, f_{N-1}) = \lim_{T\to\infty} \frac{1}{T} \operatorname{cum}\Big\{ X_1^{(T)}(t, f_1), \ldots, X_{N-1}^{(T)}(t, f_{N-1}), X_N^{(T)}\Big(t, \Psi_{\boldsymbol{x}}^{(k)}(f_1, \ldots, f_{N-1})\Big)\Big\} \tag{4.42}$$

where

$$\begin{aligned} X_n^{(T)}(t, f_n) &\triangleq \int_{t-T/2}^{t+T/2} x_n(u)\, e^{-j2\pi f_n u}\, \mathrm{d}u \\ &= \int_{\mathbb{R}} x_n(u) \operatorname{rect}\Big(\frac{u-t}{T}\Big)\, e^{-j2\pi f_n u}\, \mathrm{d}u \\ &= \Big[x_n(t)\, e^{-j2\pi f_n t}\Big] \otimes \operatorname{rect}\Big(\frac{t}{T}\Big) \end{aligned} \tag{4.43}$$

By letting $h_B(t) \triangleq B \operatorname{rect}(Bt)$ with $B = 1/T$, (4.42) becomes

$$P_{\boldsymbol{x}}^{(k)}(f_1, \ldots, f_{N-1}) = \lim_{B\to 0} \frac{1}{B^{N-1}} \operatorname{cum}\Big\{ \Big[x_1(t)\, e^{-j2\pi f_1 t}\Big] \otimes h_B(t), \ldots, \Big[x_{N-1}(t)\, e^{-j2\pi f_{N-1} t}\Big] \otimes h_B(t), \Big[x_N(t)\, e^{-j2\pi \Psi_{\boldsymbol{x}}^{(k)}(f_1, \ldots, f_{N-1}) t}\Big] \otimes h_B(t)\Big\} \tag{4.44}$$

which is the Nth-order cross-cumulant of low-pass filtered versions of the frequency-shifted signals $x_n(t)\, e^{-j2\pi f_n t}$, $n = 1, \ldots, N$, when the frequency shifts are such that

$$f_N = \Psi_{\boldsymbol{x}}^{(k)}(f_1, \ldots, f_{N-1}) \tag{4.45}$$

and the bandwidth B of the low-pass filters $h_B(t)$ approaches zero.

In the special case of jointly ACS processes, (4.42) and (4.44) reduce to (13.19) and (13.20) of (Gardner *et al.* 2006), respectively.

Cumulants are invariant with respect to permutation of the random variables. Consequently, let $\{i_1, \ldots, i_N\}$ be any permutation of $\{1, \ldots, N\}$, one has

$$
\begin{aligned}
&\operatorname{cum}\Big\{ \mathrm{d}\chi_1(f_1), \ldots, \mathrm{d}\chi_N(f_N) \Big\} \\
&\quad = \operatorname{cum}\Big\{ \mathrm{d}\chi_{i_1}(f_{i_1}), \ldots, \mathrm{d}\chi_{i_N}(f_{i_N}) \Big\} \\
&\quad = \sum_{k\in\mathbb{I}} \mathrm{d}\mathcal{P}^{(k)}_{\boldsymbol{x}_i}(f_{i_1}, \ldots, f_{i_{N-1}})\, \delta\left(f_{i_N} - \Psi^{(k)}_{\boldsymbol{x}_i}(f_{i_1}, \ldots, f_{i_{N-1}})\right) \mathrm{d}f_{i_N}
\end{aligned}
\tag{4.46}
$$

where $\boldsymbol{x}_i \triangleq [x_{i_1}, \ldots, x_{i_N}]$.

Theorem 4.2.14 *For every partition $\{\mu_1, \ldots, \mu_p\}$ of $\{1, \ldots, N\}$, let $x_n(t)$, $n \in \mu_i$, be jointly SC. That is,*

$$
\operatorname{cum}\Big\{ \mathrm{d}\chi_n(f_n),\ n \in \mu_i \Big\} = \sum_{k_i\in\mathbb{I}_{\mu_i}} \mathrm{d}\mathcal{P}^{(k_i)}_{\boldsymbol{x}_{\mu_i}}(\boldsymbol{f}'_{\mu_i})\, \delta\Big(f_{r_i} - \Psi^{(k_i)}_{\boldsymbol{x}_{\mu_i}}(\boldsymbol{f}'_{\mu_i})\Big)\, \mathrm{d}f_{r_i}
\tag{4.47}
$$

where the functions $\mathcal{P}^{(k_i)}_{\boldsymbol{x}_{\mu_i}}(\boldsymbol{f}'_{\mu_i})$ do not contain jumps. In (4.47), $\boldsymbol{x}_{\mu_i}$ is the $|\mu_i|$-dimensional vector whose components are those of $\boldsymbol{x}$ having indices in μ_i, with $|\mu_i|$ the number of elements in μ_i, r_i represents the last element in μ_i, and each partition is ordered so that μ_p always contains N as last element (i.e., $r_p = N$). Moreover, $\boldsymbol{f}_{\mu_i} \triangleq [\boldsymbol{f}'^{\mathrm{T}}_{\mu_i}, f_{r_i}]^{\mathrm{T}}$ and $\mathbb{I}_{\mu_i}$ is a countable set. If, in addition, the functions $\mathcal{P}^{(k_i)}_{\boldsymbol{x}_{\mu_i}}(\boldsymbol{f}'_{\mu_i})$ do not have singular components, they are are absolutely continuous and

$$
\mathrm{d}\mathcal{P}^{(k_i)}_{\boldsymbol{x}_{\mu_i}}(\boldsymbol{f}'_{\mu_i}) = P^{(k_i)}_{\boldsymbol{x}_{\mu_i}}(\boldsymbol{f}'_{\mu_i})\, \mathrm{d}\boldsymbol{f}'_{\mu_i}
\tag{4.48}
$$

where $\mathrm{d}\boldsymbol{f}'_{\mu_i} \triangleq \prod_{n\in\mu'_i} \mathrm{d}f_n$, with $\mu'_i \triangleq \mu_i - \{r_i\}$.

By substituting (4.47) into (4.39) the result is that

$$
\begin{aligned}
&\mathrm{E}\Big\{ \prod_{n=1}^{N} \mathrm{d}\chi_n(f_n) \Big\} \\
&\quad = \sum_{\mathrm{P}} \Bigg[\sum_{k_1\in\mathbb{I}_{\mu_1}} \cdots \sum_{k_p\in\mathbb{I}_{\mu_p}} \mathrm{d}\mathcal{P}^{(k_1)}_{\boldsymbol{x}_{\mu_1}}(\boldsymbol{f}'_{\mu_1}) \cdots \mathrm{d}\mathcal{P}^{(k_p)}_{\boldsymbol{x}_{\mu_p}}(\boldsymbol{f}'_{\mu_p}) \\
&\qquad \delta\Big(f_{r_1} - \Psi^{(k_1)}_{\boldsymbol{x}_{\mu_1}}(\boldsymbol{f}'_{\mu_1})\Big) \cdots \delta\Big(f_{r_p} - \Psi^{(k_p)}_{\boldsymbol{x}_{\mu_p}}(\boldsymbol{f}'_{\mu_p})\Big)\, \mathrm{d}f_{r_1} \cdots \mathrm{d}f_{r_p} \Bigg]
\end{aligned}
\tag{4.49a}
$$

$$
\begin{aligned}
&\quad = \sum_{\mathrm{P}} \Bigg[\sum_{k_1\in\mathbb{I}_{\mu_1}} \cdots \sum_{k_p\in\mathbb{I}_{\mu_p}} P^{(k_1)}_{\boldsymbol{x}_{\mu_1}}(\boldsymbol{f}'_{\mu_1}) \cdots P^{(k_p)}_{\boldsymbol{x}_{\mu_p}}(\boldsymbol{f}'_{\mu_p}) \\
&\qquad \delta\Big(f_{r_1} - \Psi^{(k_1)}_{\boldsymbol{x}_{\mu_1}}(\boldsymbol{f}'_{\mu_1})\Big) \cdots \delta\Big(f_{r_p} - \Psi^{(k_p)}_{\boldsymbol{x}_{\mu_p}}(\boldsymbol{f}'_{\mu_p})\Big) \Bigg] \mathrm{d}f_1 \cdots \mathrm{d}f_N
\end{aligned}
\tag{4.49b}
$$

where the second equality holds if all $\mathcal{P}^{(k_i)}_{\boldsymbol{x}_{\mu_i}}(\boldsymbol{f}'_{\mu_i})$ are absolutely continuous. □

In the absence of singular components, the following notation can be used (Section 1.1.2 and Definition 4.2.3)

$$\mathrm{E}\left\{\prod_{n=1}^{N} \mathrm{d}\chi_n(f_n)\right\} = \mathrm{E}\left\{\prod_{n=1}^{N} X_n(f_n)\right\} \mathrm{d}f_1 \cdots \mathrm{d}f_N \tag{4.50}$$

$$\mathrm{cum}\left\{\mathrm{d}\chi_n(f_n),\ n = 1, \ldots, N\right\} = \mathrm{cum}\left\{X_n(f_n),\ n = 1, \ldots, N\right\} \mathrm{d}f_1 \cdots \mathrm{d}f_N. \tag{4.51}$$

The above definitions and results specialize to first- and second-order as follows.

A stochastic process $x(t)$ is said to be *first-order SC* if its *spectral mean* is given by

$$\mathrm{E}\left\{\mathrm{d}\chi(f)\right\} = \sum_k \overline{x}^{(k)}\, \delta(f - \bar{f}_k)\, \mathrm{d}f \tag{4.52}$$

That is, $x(t)$ is an almost periodic function

$$x(t) = \sum_k x_k\, e^{j2\pi \bar{f}_k t} \tag{4.53}$$

with possibly random coefficients x_k with $\mathrm{E}\{x_k\} = \overline{x}^{(k)}$ and deterministic frequencies $\bar{f}_k$.

Two jointly harmonizable processes $x_1(t)$ and $x_2(t)$ are said to be *jointly SC* if their *spectral covariance* function is given by

$$\begin{aligned} &\mathrm{E}\left\{\mathrm{d}\chi_1(f_1)\, \mathrm{d}\chi_2(f_2)\right\} - \mathrm{E}\left\{\mathrm{d}\chi_1(f_1)\right\} \mathrm{E}\left\{\mathrm{d}\chi_2(f_2)\right\} \\ &\quad = \sum_k \mathrm{d}\mathcal{P}^{(k)}_{x_1x_2}(f_1)\, \delta\left(f_2 - \Psi^{(k)}_{x_1x_2}(f_1)\right) \mathrm{d}f_2 \end{aligned} \tag{4.54}$$

where $\mathcal{P}^{(k)}_{x_1x_2}(f_1)$ does not contain jumps. If, in addition, $\mathcal{P}^{(k)}_{x_1x_2}(f_1)$ has zero singular component, then $\mathcal{P}^{(k)}_{x_1x_2}(f_1)$ is absolutely continuous and we have

$$\mathrm{d}\mathcal{P}^{(k)}_{x_1x_2}(f_1) = P^{(k)}_{x_1x_2}(f_1)\, \mathrm{d}f_1. \tag{4.55}$$

Note that in the above definition, unlike the definition in Theorem 4.2.7, it is not assumed that the processes have no finite-strength additive sinewave components. In fact, since a covariance function is considered, the contributions of the Fourier transform of possible almost-periodic components present in the processes $x_1(t)$ and $x_2(t)$ are subtracted in the covariance definition.

From (4.49a) it follows that the Nth-order spectral moment function $\mathrm{E}\left\{\prod_{n=1}^{N} \mathrm{d}\chi_n(f_n)\right\}$ is given by the sum of the term (4.40) corresponding to the (only) partition with $p = 1$ in (4.49a) and the term corresponding to all other partitions with $p \geqslant 2$. The former, containing only one Dirac delta, is refereed to as the *pure Nth-order spectrally correlated component* of the processes $x_1, \cdots x_N$. The latter is constituted by the sum of terms each containing the product of at least two Dirac deltas with different frequency variables. For this reason, the Nth-order spectral moment function $\mathrm{E}\left\{\prod_{n=1}^{N} \mathrm{d}\chi_n(f_n)\right\}$ is referred to as the *impure Nth-order spectrally* *correlated component* of the processes $x_1, \cdots x_N$. In fact, a nonzero $\mathrm{E}\left\{\prod_{n=1}^{N} \mathrm{d}\chi_n(f_n)\right\}$ is also due to terms with $p \geqslant 2$ in (4.40), which arise from products of lower-than-Nth-order spectrally correlated components.

In contrast, the contribution of term (4.40) is not generated by products of lower-order spectrally correlated components. Moreover, according to (4.38), it can be obtained by depurating the spectral moment function $\mathrm{E}\left\{\prod_{n=1}^{N} \mathrm{d}\chi_n(f_n)\right\}$ (the term with $p = 1$ in (4.38)) from all possible products of lower-order spectrally correlated components.

Note that, in the special case of ACS processes, these considerations are the frequency-domain counterparts of the interpretation of the Nth-order temporal cumulant of an ACS signal as the pure additive finite-strength sinewave components that can be generated by the Nth-order lag-product of the signal (Gardner and Spooner 1994), (Spooner and Gardner 1994). However, the interpretation of pure and impure spectrally correlated components applies to a class of stochastic processes wider than that of the ACS process, namely the class of the SC processes.

4.2.4 *Motivating Examples*

Nonstationary (possibly jointly) SC processes that are not (possibly jointly) ACS can arise from linear not-almost-periodically time-variant transformations of ACS processes. For example, an ACS signal transmitted by a moving source and received by two stationary sensors, in the case of constant radial speed, gives rise to two signals that are jointly SC but are not jointly ACS (Allen and Hobbs 1992). Moreover, reverberation mechanisms generate coherency relationships ensemblewise between spectral components (Middleton 1967). In the following, examples where the SC model is appropriate to describe stochastic processes are discussed. Applications to communications, radar, and sonar are treated in Chapter 7.

4.2.4.1 Multipath Doppler Channel

Let us consider an ACS process passing through a multipath Doppler channel, that is, a linear time-variant system such that for the input complex-envelope signal $x(t)$ the output complex-envelope $y(t)$ is given by (Section 7.6.1)

$$y(t) = \sum_{k=1}^{K} a_k \, x(s_k t - d_k) \, e^{j2\pi\nu_k t} \tag{4.56}$$

where, for each path of the channel, a_k is the (possibly complex) scaling amplitude, d_k the delay, s_k the time-scale factor, and ν_k the frequency shift. Such a model is appropriate for describing the multipath channel when, for each path, the relative radial speeds between transmitter, receiver, and reflecting moving objects can be considered constant within the observation interval (Van Trees 1971) (Section 7.3). For example, in the case of moving transmitter and stationary receiver (e.g., transmitter is a mobile device and receiver is a base station), it results in $s_k = c/(c + v_k)$ and $\nu_k = -f_c v_k/(c + v_k)$, where c is the medium propagation speed, v_k is the relative radial speed for the kth-path, and f_c the carrier frequency of the real signals.

Accounting for the Fourier transform of (4.56), it can be shown that the Loève bifrequency spectrum of the output $y(t)$ is given by (Section 7.7.2)

$$\begin{aligned} &\mathrm{E}\left\{Y(f_1)\, Y^*(f_2)\right\} \\ &= \sum_{k_1=1}^{K} \sum_{k_2=1}^{K} \frac{a_{k_1} a_{k_2}^*}{|s_{k_1}|\,|s_{k_2}|}\, e^{-j2\pi(f_1-\nu_{k_1})d_{k_1}/s_{k_1}}\, e^{j2\pi(f_2-\nu_{k_2})d_{k_2}/s_{k_2}} \\ &\quad \mathrm{E}\left\{X\left(\frac{f_1-\nu_{k_1}}{s_{k_1}}\right) X^*\left(\frac{f_2-\nu_{k_2}}{s_{k_2}}\right)\right\}. \end{aligned} \tag{4.57}$$

Furthermore, by substituting into (4.57) the expression (4.29) of the Loève bifrequency spectrum of an ACS process, it results in (Section 7.7.2)

$$\begin{aligned} &\mathrm{E}\left\{Y(f_1)\, Y^*(f_2)\right\} \\ &= \sum_{k_1=1}^{K} \sum_{k_2=1}^{K} \sum_{\alpha\in A_{xx^*}} \frac{a_{k_1} a_{k_2}^*}{|s_{k_1}|} \\ &\quad e^{-j2\pi(f_1-\nu_{k_1})d_{k_1}/s_{k_1}}\, e^{j2\pi(f_1/s_{k_1}-\alpha-\nu_{k_1}/s_{k_1})d_{k_2}} \\ &\quad S^{\alpha}_{xx^*}\left(\frac{f_1-\nu_{k_1}}{s_{k_1}}\right) \delta\left(f_2 - \frac{s_{k_2}}{s_{k_1}} f_1 - \nu_{k_2} + s_{k_2}\alpha + \frac{s_{k_2}}{s_{k_1}}\nu_{k_1}\right) \end{aligned} \tag{4.58}$$

That is, when an ACS process passes throughout a multipath Doppler channel, the output process is SC and the support in the bifrequency plane of its Loève bifrequency spectrum is constituted by lines with slopes s_{k_2}/s_{k_1} (Figure 4.3).

In Section 7.5.1 it is shown that time-scale factors s_k can be considered unity in the argument of the complex envelope $x(\cdot)$ in (4.56) if the condition

$$BT \ll c/v_k + 1 \quad \forall k \tag{4.59}$$

is fulfilled, where B is input-signal bandwidth and T is the data-record length (see also (Van Trees 1971)). In such a case, the channel can be modelled as linear almost-periodically time variant and $y(t)$ is ACS. However, in modern communication systems, wider and wider bandwidths are required to get higher and higher bit rates. Moreover, large data-record lengths are necessary for blind channel identification techniques or detection algorithms in highly noise- and interference-corrupted environments. That is, there are practical situations where the BT product does not satisfy (4.59) and, hence, the time-scale factors cannot be considered unity in the argument of the complex envelope $x(\cdot)$ in (4.56) so that the output process $y(t)$ must be modelled as SC. For example, let us consider the direct-sequence spread-spectrum (DSSS) transmitted signal in a code-division multiple access (CDMA) system

$$x(t) = \sum_{k=-\infty}^{+\infty} b_k \sum_{n=0}^{N_c-1} c_n\, q(t - nT_c - kT_b) \tag{4.60}$$

where $\{b_k\}_{k\in\mathbb{Z}}$, with $b_k \in \{\pm 1\}$, is the information-bit sequence, $\{c_n\}_{n=0,\dots,N_c-1}$, with $c_n \in \{\pm 1\}$, the spreading-code sequence, T_c the chip period, N_c the number of chip per bit, $T_b = N_c T_c$ the bit period, and $q(t)$ is a T_c-duration rectangular pulse. By assuming an approximate

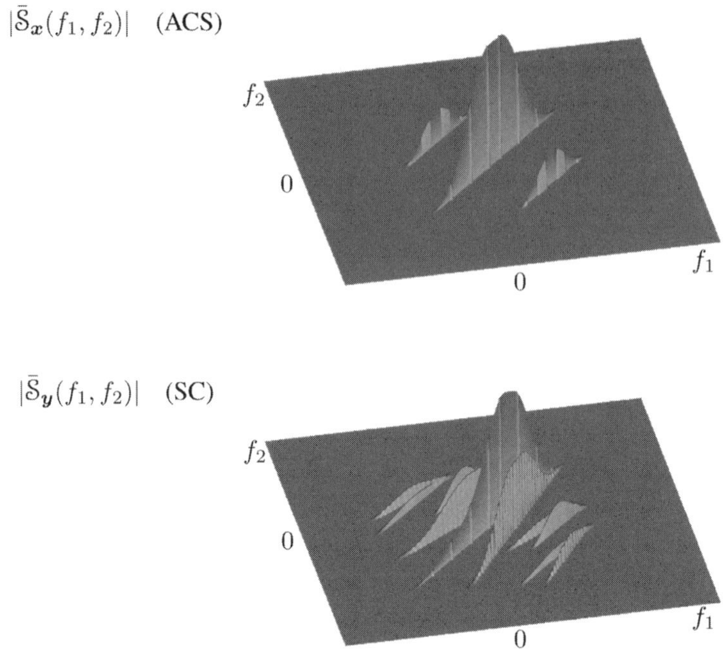

Figure 4.3 ACS input and SC output of the multipath Doppler channel existing between a stationary transmitter and a moving receiver with constant relative radial speeds for each path. Magnitude of the bifrequency spectral correlation density (top) of the input ACS signal $x(t)$ and (bottom) of the output SC signal $y(t)$ as functions of f_1 and f_2

bandwidth $B \simeq 1/T_c$ for $x(t)$ in (4.60), condition (4.59) becomes

$$BT \simeq N_b N_c \ll c/v_k + 1, \qquad \forall k \tag{4.61}$$

where $N_b = T/T_b = T/(N_c T_c)$ is the number of processed bits in the data-record of length T. Therefore, if $c \simeq 3 \cdot 10^8$ m s^{-1}, v_k=100 km h^{-1}, and $N_c = 512$, then (4.61) leads to $N_b \ll 2 \cdot 10^4$. That is, if the maximum number of processed bits exceeds few hundreds, then the received signal must be modelled as SC instead of as ACS.

Further situations where condition (4.59) is not satisfied and, hence, nonunit time-scale factors should be accounted for, can be encountered in sonar applications (Van Trees 1971, pp. 339–340), acoustic detection of aircrafts (Ferguson 1999), time-delay and Doppler estimation of wide-band signals (Jin *et al.* 1995), and ocean acoustic tomography (Munk *et al.* 1995).

4.2.4.2 MIMO Multipath Doppler Channel

Let us consider a multi-input multi-output (MIMO) system with sources $x_\ell(t)$, $\ell = 1, \ldots, L$, and received signals $y_m(t)$, $m = 1, \ldots, M$, on a sensor array, where each link $\ell \to m$ is constituted by a multipath Doppler channel:

$$y_m(t) = \sum_{\ell=1}^{L} \sum_{k=1}^{K} a_{k\ell m}\, x_\ell(s_{k\ell m} t - d_{k\ell m})\, e^{j2\pi \nu_{k\ell m} t}. \tag{4.62}$$

In (4.62), for each path k of the $\ell \to m$ link, $a_{k\ell m}$ is the (possibly complex) scaling amplitude, $d_{k\ell m}$ the delay, $s_{k\ell m}$ the time-scale factor, and $\nu_{k\ell m}$ the frequency shift. K is the maximum number of paths over all links $\ell \to m$ (some of the $a_{k\ell m}$ are possibly zero).

If the sources are jointly ACS, that is

$$\mathrm{E}\left\{X_{\ell_1}(f_1)\, X^*_{\ell_2}(f_2)\right\} = \sum_{\alpha \in A_{\ell_1 \ell_2}} S^{\alpha}_{x_{\ell_1} x^*_{\ell_2}}(f_1)\, \delta(f_2 - f_1 + \alpha) \tag{4.63}$$

then the sensor signals are jointly SC, that is,

$$\begin{aligned}
&\mathrm{E}\left\{Y_{m_1}(f_1)\, Y^*_{m_2}(f_2)\right\} \\
&= \sum_{\ell_1=1}^{L} \sum_{\ell_2=1}^{L} \sum_{k_1=1}^{K} \sum_{k_2=1}^{K} \sum_{\alpha \in A_{\ell_1 \ell_2}} \frac{a_{[1]} a^*_{[2]}}{|s_{[1]}|} \\
&\quad e^{-j2\pi(f_1 - \nu_{[1]}) d_{[1]}/s_{[1]}}\, e^{j2\pi(f_1/s_{[1]} - \alpha - \nu_{[1]}/s_{[1]}) d_{[2]}} \\
&\quad S^{\alpha}_{x_{\ell_1} x^*_{\ell_2}}\left(\frac{f_1 - \nu_{[1]}}{s_{[1]}}\right) \delta\left(f_2 - \frac{s_{[2]}}{s_{[1]}} f_1 - \nu_{[2]} + s_{[2]}\alpha + \frac{s_{[2]}}{s_{[1]}} \nu_{[1]}\right)
\end{aligned} \tag{4.64}$$

where, for notation simplicity, $a_{[i]} \equiv a_{k_i \ell_i m_i}$, $d_{[i]} \equiv d_{k_i \ell_i m_i}$, $s_{[i]} \equiv s_{k_i \ell_i m_i}$, and $\nu_{[i]} \equiv \nu_{k_i \ell_i m_i}$.

For $K = 1$, $L = 1$, and $M = 2$, we obtain as a special case the model considered in (Allen and Hobbs 1992).

4.2.4.3 SC Processes with Non-Linear Support Curves

Spectrally correlated stochastic processes with nonlinear functions $\Psi^{(n)}_{xx^{(*)}}(\cdot)$ can be obtained by feeding with ACS processes the "stationary linear time-varying systems" considered in (Claasen and Mecklenbräuker 1982) or considering frequency-warped wide-sense stationary or ACS processes. Furthermore, jointly SC processes with nonlinear functions $\Psi^{(n)}_{yx^{(*)}}(\cdot)$ are the input and output signals of the "stationary linear time-varying systems" or the linear time-variant systems described in (Franaszek 1967), (Franaszek and Liu 1967), (Liu and Franaszek 1969). Spectral analysis with nonuniform frequency spacing (Oppenheim *et al.* 1971), (Oppenheim and Johnson 1972), (Braccini and Oppenheim 1974), or spectral analysis by frequency warping techniques (von Schroeter 1999), (Makur and Mitra 2001), (Franz *et al.* 2003), can be shown to give rise to (jointly) SC processes with non linear support curves when the analyzed process is WSS or ACS (Section 4.13).

4.2.4.4 Fractional Brownian Motion

Definition 4.2.15 *A real-valued stochastic process $x(t)$ is said to be* self-similar *(Mandelbrot and Van Ness 1968) if*

$$x(st) \stackrel{\mathrm{d}}{=} s^H\, x(t) \qquad t \in \mathbb{R}, \quad \forall s > 0 \tag{4.65}$$

where $\stackrel{\mathrm{d}}{=}$ denotes equality in distribution and $H > 0$ is the similarity exponent. □

Fact 4.2.16 *Equation (4.65) implies that*

$$x(0) = 0 \quad \text{almost surely} \tag{4.66}$$

$$\mathrm{E}\{x(t)\} = s^{-H}\mathrm{E}\{x(st)\} \tag{4.67}$$

$$\mathrm{E}\left\{|x(t)|^p\right\} = \mathrm{E}\left\{|x(1)|^p\right\}\,|t|^{pH} \qquad p \in \mathbb{N} \tag{4.68}$$

$$\mathrm{E}\{x(t_1)\,x(t_2)\} = s^{-2H}\mathrm{E}\{x(st_1)\,x(st_2)\} \tag{4.69}$$

□

Definition 4.2.17 Fractional Brownian Motion (fBm). *The fractional Brownian motion process $B_H(t)$ is defined by (Mandelbrot and Van Ness 1968)*

$$B_H(t) - B_H(0) = \frac{1}{\Gamma(H+1/2)}\left[\int_{-\infty}^{0} [|t-s|^{H-1/2} - |s|^{H-1/2}]\,\mathrm{d}B(s) + \int_0^t |t-s|^{H-1/2}\,\mathrm{d}B(s)\right] \tag{4.70}$$

where $B(t)$ is a zero-mean standard Brownian motion or Wiener process. □

In (4.70), $\Gamma(\cdot)$ is the Gamma function (NIST 2010, eq. 5.2.1) and the parameter $H \in (0, 1)$ is the Hurst parameter. Standard Brownian motion is obtained for $H = 1/2$. The process $B_H(t)$ is a zero-mean nonstationary Gaussian process with stationary and self-similar increments, that is,

$$B_H(t+s\tau) - B_H(t) \stackrel{\mathrm{d}}{=} s^H\, B_H(\tau) \tag{4.71}$$

where the sense of equality is in distribution. Since $B_H(t)$ is a zero-mean nonstationary Gaussian process, it is completely characterized by its autocorrelation function or bifrequency Loève spectrum.

Theorem 4.2.18 Autocorrelation of fBm (Flandrin 1989; Wornell 1993). *Let $x(t) = B_H(t)$. It results that*

$$\mathrm{E}\{x(t_1)\,x(t_2)\} = \frac{V_H}{2}\left[|t_1|^{2H} + |t_2|^{2H} - |t_1 - t_2|^{2H}\right] \tag{4.72}$$

where

$$V_H = \Gamma(1-2H)\,\frac{\cos(\pi H)}{\pi H}. \tag{4.73}$$

□

By substituting (4.72) into (1.16) (with (∗) present), using the Fourier transform (Champeney 1990, pp. 137–138) (in the sense of distributions)

$$\int_{\mathbb{R}} |t|^{\lambda-1}\, e^{-j2\pi f t}\, \mathrm{d}t = \frac{2\Gamma(\lambda)\,\cos(\lambda\pi/2)}{(2\pi)^{\lambda}}\,|f|^{-\lambda} \qquad \lambda > 0, \quad \lambda \notin \mathbb{N} \tag{4.74}$$

and accounting for the relationships $\Gamma(z+1) = z\Gamma(z)$ (NIST 2010, eq. 5.5.1) and $\Gamma(z)\Gamma(1-z) = \pi/\sin(\pi z)$, $z \notin \mathbb{Z}$ (NIST 2010, eq. 5.5.3), the following result can be proved by taking $\lambda = 2H+1$.

Theorem 4.2.19 Bifrequency Loève Spectrum of fBm (Øigård *et al.* 2006). *Let* $x(t) = B_H(t)$. *It results that*

$$\begin{aligned} \mathrm{E}\left\{X(f_1)\,X^*(f_2)\right\} &= |2\pi f_1|^{-(2H+1)}\,\delta(f_1 - f_2) \\ &\quad -|2\pi f_2|^{-(2H+1)}\,\delta(f_1) - |2\pi f_1|^{-(2H+1)}\,\delta(f_2). \end{aligned} \tag{4.75}$$

□

From Theorem 4.2.19 it follows that $x(t)$ is a SC process with spectral masses concentrated on the three lines $f_1 = f_2$, $f_1 = 0$, $f_2 = 0$, and (4.75) turns out to be a special case of (4.23). Moreover, the power spectrum, defined as the Fourier transform of the time-averaged autocorrelation function (1.17) is given by the following result.

Corollary 4.2.20 *Let* $x(t) = B_H(t)$. *Its power spectrum is given by (Flandrin 1989)*

$$S_x(f) \triangleq \int_{\mathbb{R}} \langle \mathrm{E}\{x(t+\tau)\,x(t)\}\rangle_t\, e^{-j2\pi f\tau}\,\mathrm{d}\tau = \frac{S_x(1)}{|f|^{\gamma}} \qquad \gamma = 2H+1. \tag{4.76}$$

□

4.3 Linear Time-Variant Filtering of SC Processes

4.3.1 FOT-Deterministic Linear Systems

In this section, the problem of LTV filtering of SC processes is addressed. The class of LTV systems considered here is that of the FOT-deterministic linear systems characterized in Section 6.3.8. FOT-deterministic linear systems are defined as those linear systems that transform input almost-periodic functions into output almost-periodic functions.

In Section 6.3.8 it is shown that the system transmission function of a FOT-deterministic linear system can be written as

$$H(f, \lambda) = \sum_{\sigma \in \Omega} G_\sigma(\lambda)\, \delta(f - \varphi_\sigma(\lambda)) \tag{4.77a}$$

$$= \sum_{\sigma \in \Omega} H_\sigma(f)\, \delta(\lambda - \psi_\sigma(f)) \tag{4.77b}$$

where Ω is a countable set. The functions $\varphi_\sigma(\cdot)$ are assumed to be invertible and differentiable, with inverse functions $\psi_\sigma(\cdot)$ also differentiable and referred to as *frequency mapping functions*. The functions $G_\sigma(\cdot)$ and $H_\sigma(\cdot)$ are linked by the relationships

$$H_\sigma(f) = |\psi'_\sigma(f)|\, G_\sigma(\psi_\sigma(f)) \tag{4.78}$$

$$G_\sigma(\lambda) = |\varphi'_\sigma(\lambda)|\, H_\sigma(\varphi_\sigma(\lambda)) \tag{4.79}$$

with $\psi'_\sigma(\cdot)$ and $\varphi'_\sigma(\cdot)$ denoting the derivative of $\psi_\sigma(\cdot)$ and $\varphi_\sigma(\cdot)$, respectively.

In Section 6.3.8 it is shown that the impulse-response function of FOT deterministic linear systems can be expressed as

$$h(t, u) = \sum_{\sigma \in \Omega} \int_{\mathbb{R}} G_\sigma(\lambda)\, e^{j2\pi\varphi_\sigma(\lambda)t} e^{-j2\pi\lambda u}\, \mathrm{d}\lambda \tag{4.80a}$$

$$= \sum_{\sigma \in \Omega} \int_{\mathbb{R}} H_\sigma(f)\, e^{-j2\pi\psi_\sigma(f)u} e^{j2\pi ft}\, \mathrm{d}f. \tag{4.80b}$$

By substituting (4.77a) and (4.77b) into the input/output relationship in the frequency domain (1.43) one has (Section 6.3.8)

$$Y(f) = \sum_{\sigma \in \Omega} \int_{\mathbb{R}} G_\sigma(\lambda)\, \delta(f - \varphi_\sigma(\lambda))\, X(\lambda)\, \mathrm{d}\lambda \tag{4.81a}$$

$$= \sum_{\sigma \in \Omega} H_\sigma(f)\, X(\psi_\sigma(f)). \tag{4.81b}$$

Analogously, by substituting (4.80a) and (4.80b) into the input/output relationship in the time domain (1.41) one has (Section 6.3.8)

$$y(t) = \sum_{\sigma \in \Omega} \int_{\mathbb{R}} G_\sigma(\lambda)\, X(\lambda)\, e^{j2\pi\varphi_\sigma(\lambda)t}\, \mathrm{d}\lambda \tag{4.82a}$$

$$= \sum_{\sigma \in \Omega} h_\sigma(t) \otimes x_{\psi_\sigma}(t) \tag{4.82b}$$

where $\otimes$ denotes convolution and

$$x_{\psi_\sigma}(t) \triangleq \int_{\mathbb{R}} X(\psi_\sigma(f))\, e^{j2\pi ft}\, \mathrm{d}f. \tag{4.83}$$

In other words, the output of FOT-deterministic LTV systems is constituted by frequency warped and then LTI filtered versions of the input (Figure 4.4).

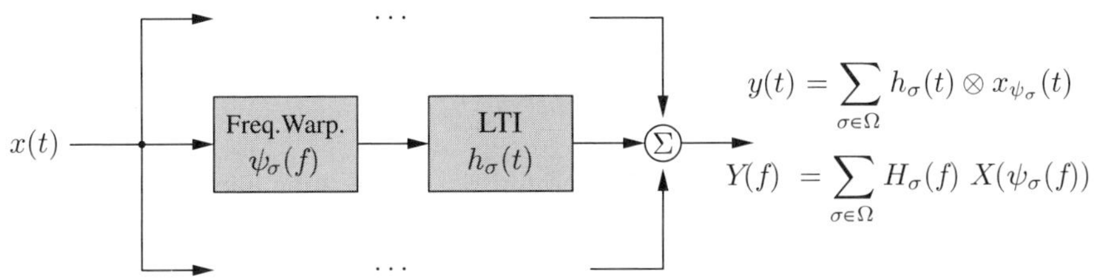

Figure 4.4 FOT-deterministic LTV systems: Realization and corresponding input/output relations in time and frequency domains. The first system is LTV operating frequency warping, the second system is a LTI filter.

It can be shown that the parallel and cascade concatenation of FOT-deterministic LTV systems is still a FOT-deterministic LTV system.

The subclass of FOT-deterministic LTV systems obtained by considering Ω containing only one element was studied, in the stochastic process framework, in (Franaszek 1967) and (Franaszek and Liu 1967) with reference to the continuous-time case and in (Liu and Franaszek 1969) with reference to the discrete-time case. The most important property of these systems, as evidenced in (Franaszek 1967), (Franaszek and Liu 1967), and (Liu and Franaszek 1969), is that they preserve in the output the wide-sense stationarity of the input random process (Section 4.13).

4.3.1.1 LAPTV Systems

The class of FOT deterministic LTV systems includes that of the linear almost-periodically time-variant (LAPTV) systems (Section 1.3.3) which, in turn, includes, as special cases, linear periodically time-variant (LPTV) and linear time-invariant (LTI) systems. For LAPTV systems, the frequency mapping functions $\psi_\sigma(f)$ are linear with unit slope, that is,

$$\psi_\sigma(f) = f - \sigma, \qquad \sigma \in \Omega \tag{4.84a}$$

$$\varphi_\sigma(\lambda) = \lambda + \sigma, \qquad \sigma \in \Omega \tag{4.84b}$$

and then the impulse-response function can be expressed as (see (1.107))

$$h(t, u) = \sum_{\sigma\in\Omega} h_\sigma(t - u) e^{j2\pi\sigma u}. \tag{4.85}$$

4.3.1.2 Time-Scale Changing

Systems performing time-scale changing are FOT deterministic. The impulse-response function is given by

$$h(t, u) = \delta(u - st) \tag{4.86}$$

where $s \neq 0$ is the time-scale factor, the set Ω contains just one element,

$$\psi_\sigma(f) = \frac{f}{s} \tag{4.87}$$

and

$$H_\sigma(f) = \frac{1}{|s|}. \tag{4.88}$$

Decimators and interpolators are FOT deterministic discrete-time linear systems (Izzo and Napolitano 1998b) (Section 4.10).

4.3.2 SC Signals and FOT-Deterministic Systems

Let $x_1(t)$ and $x_2(t)$ be jointly SC signals, that is

$$\mathrm{E}\left\{X_1(f_1)\, X_2^{(*)}(f_2)\right\} = \sum_{n\in\mathbb{I}} S_{\boldsymbol{x}}^{(n)}(f_1)\,\delta\left(f_2 - \Psi_{\boldsymbol{x}}^{(n)}(f_1)\right) \tag{4.89}$$

where $\boldsymbol{x} \triangleq [x_1 x_2^{(*)}]$, and let $h_1(t,u)$ and $h_2(t,u)$ be deterministic FOT systems, that is, accordingly with (4.77a) and (4.77b), with transmission functions

$$H_i(f,\lambda) = \sum_{\ell_i\in\mathbb{J}_i} G_i^{(\ell_i)}(\lambda)\,\delta\left(f - \varphi_i^{(\ell_i)}(\lambda)\right) \tag{4.90a}$$

$$= \sum_{\ell_i\in\mathbb{J}_i} H_i^{(\ell_i)}(f)\,\delta\left(\lambda - \psi_i^{(\ell_i)}(f)\right) \qquad i=1,2 \tag{4.90b}$$

where $\psi_i^{(\ell_i)}(\cdot)$ is invertible, $\varphi_i^{(\ell_i)}(\cdot)$ is its inverse, and both functions are assumed to be differentiable. According to (4.81b), we have

$$Y_i(f) = \sum_{\ell_i\in\mathbb{J}_i} H_i^{(\ell_i)}(f)\,X_i\left(\psi_i^{(\ell_i)}(f)\right) \qquad i=1,2 \tag{4.91}$$

In Section 5.1 it is proved that

$$\begin{aligned}
&\mathrm{E}\left\{Y_1(f_1)\, Y_2^{(*)}(f_2)\right\} \\
&= \sum_{\ell_1\in\mathbb{J}_1}\sum_{\ell_2\in\mathbb{J}_2} H_1^{(\ell_1)}(f_1)\, H_2^{(\ell_2)}(f_2)^{(*)}\,\mathrm{E}\left\{X_1\left(\psi_1^{(\ell_1)}(f_1)\right)\, X_2^{(*)}\left(\psi_2^{(\ell_2)}(f_2)\right)\right\} \\
&= \sum_{\ell_1\in\mathbb{J}_1}\sum_{\ell_2\in\mathbb{J}_2}\sum_{n\in\mathbb{I}} H_1^{(\ell_1)}(f_1)\, H_2^{(\ell_2)}\left(\varphi_2^{(\ell_2)}\left(\Psi_{\boldsymbol{x}}^{(n)}\left(\psi_1^{(\ell_1)}(f_1)\right)\right)\right)^{(*)} \\
&\quad S_{\boldsymbol{x}}^{(n)}\left(\psi_1^{(\ell_1)}(f_1)\right)\left|\varphi_2^{(\ell_2)\prime}\left(\Psi_{\boldsymbol{x}}^{(n)}\left(\psi_1^{(\ell_1)}(f_1)\right)\right)\right| \\
&\quad \delta\left(f_2 - \varphi_2^{(\ell_2)}\left(\Psi_{\boldsymbol{x}}^{(n)}\left(\psi_1^{(\ell_1)}(f_1)\right)\right)\right)
\end{aligned} \tag{4.92}$$

From (4.92) it follows that $y_1(t)$ and $y_2(t)$ are jointly SC. In particular, if $x_1 \equiv x_2$ and $h_1 \equiv h_2$ (and, hence, $y_1 \equiv y_2$), we obtain the notable result that FOT-deterministic linear systems transform SC signals into SC signals. That is, the class of the SC signals is closed under FOT-deterministic linear transformations.

By specializing (4.92) to jointly ACS signals and LAPTV systems one obtains the results of Section 1.3.3.

4.4 The Bifrequency Cross-Periodogram

In this section, for jointly SC processes, the bifrequency cross-periodogram is defined and its expected value (Lemma 3.1) and covariance (Lemma 3.2) are derived. Moreover, the bifrequency cross-periodogram is shown to be an estimator of the bifrequency spectral cross-correlation density function $\bar{\mathbb{S}}_{yx^{(*)}}(f_1, f_2)$ asymptotically biased (Theorem 3.1) and with nonzero asymptotic variance (Theorem 3.2).

Definition 4.4.1 *Given two stochastic processes $\{y(t),\ t \in \mathbb{R}\}$ and $\{x(t),\ t \in \mathbb{R}\}$, their bifrequency cross-periodogram is defined as*

$$I_{yx^{(*)}}(t; f_1, f_2)_{\frac{1}{\Delta f}} \triangleq \Delta f\, Y_{\frac{1}{\Delta f}}(t, f_1)\, X^{(*)}_{\frac{1}{\Delta f}}(t, f_2) \tag{4.93}$$

where $Y_{\frac{1}{\Delta f}}(t, f_1)$ and $X_{\frac{1}{\Delta f}}(t, f_2)$ are the short-time Fourier transforms (STFTs) of $y(t)$ and $x(t)$, respectively, defined according to

$$Z_{\frac{1}{\Delta f}}(t, f) \triangleq \int_{\mathbb{R}} z(u)\, b_{\frac{1}{\Delta f}}(u - t)\, e^{-j2\pi f u}\, \mathrm{d}u \tag{4.94a}$$

$$= \left[B_{\Delta f}(f)\, e^{-j2\pi f t}\right] \otimes Z(f)$$

$$= \int_{\mathbb{R}} Z(\nu)\, B_{\Delta f}(f - \nu)\, e^{-j2\pi(f-\nu)t}\, \mathrm{d}\nu \tag{4.94b}$$

with $Z(f)$ denoting the (generalized) Fourier transform of $z(t)$ and $b_{\frac{1}{\Delta f}}(t)$ a $(1/\Delta f)$-duration data-tapering window whose Fourier transform $B_{\Delta f}(f)$ has bandwidth Δf. □

Assuming $b_{\frac{1}{\Delta f}}(t)$ be an even function, the STFT (4.94a) is a low-pass filtered version of a frequency-shifted version of $z(t)$ with the bandwidth Δf of the low-pass filter approximately equal to the reciprocal of the width of the data-tapering window. Thus, the function of t, $Z_{\frac{1}{\Delta f}}(t, f)$ is the superposition of the spectral components of $z(t)$ within the spectral interval $[f - \Delta f/2, f + \Delta f/2]$ (see (4.94b)).

In (Hurd 1973), it is shown that if $z(t)$ is harmonizable, its spectral covariance is of bounded variation (see (4.1)), and $b_{\frac{1}{\Delta f}}(t)$ is a Fourier-Stiltjes transform, then the STFT $Z_{\frac{1}{\Delta f}}(t, f)$ is harmonizable.

In the following, some assumptions are made that allow to interchange the order of expectations, sum, and integral operations in the derivations of the expressions for expected value and covariance of the bifrequency cross-periodogram, the time-smoothed bifrequency cross-periodogram, the frequency-smoothed periodogram, and their asymptotic expressions as the data-record length approaches infinity and the spectral resolution approaches zero.

Assumption 4.4.2 SC Statistics.

(a) *The second-order harmonizable stochastic processes $\{y(t),\ t \in \mathbb{R}\}$ and $\{x(t),\ t \in \mathbb{R}\}$ are singularly and jointly (second-order) spectrally correlated, that is, for any choice of z_1*

and z_2 *in* $\{x, x^*, y, y^*\}$ *it results that*

$$\mathcal{S}_{z_1 z_2}(f_1, f_2) = \sum_{n \in \mathbb{I}_{z_1 z_2}} S^{(n)}_{z_1 z_2}(f_1)\, \delta\left(f_2 - \Psi^{(n)}_{z_1 z_2}(f_1)\right) \tag{4.95}$$

where $\mathbb{I}_{z_1 z_2}$ *is a countable set.*

(b) *The fourth-order spectral cumulant* $\operatorname{cum}\left\{Y(f_1),\ X^{(*)}(f_2),\ Y^*(f_3),\ X^{(*)*}(f_4)\right\}$ *can be expressed as (Section 4.2.3)*

$$\operatorname{cum}\left\{Y(f_1),\ X^{(*)}(f_2),\ Y^*(f_3),\ X^{(*)*}(f_4)\right\}$$
$$= \sum_{n \in \mathbb{I}_4} P^{(n)}_{yx^{(*)}y^*x^{(*)*}}(f_1, f_2, f_3)\, \delta\left(f_4 - \Psi^{(n)}_{yx^{(*)}y^*x^{(*)*}}(f_1, f_2, f_3)\right) \tag{4.96}$$

where $\mathbb{I}_4$ *is a countable set and the cumulant of complex random variables is defined according to (Spooner and Gardner 1994; Napolitano 2007a) (see also Section 1.4.2).*

□

The functions $S^{(n)}_{z_1 z_2}(f_1)$ and $\Psi^{(n)}_{z_1 z_2}(f_1)$ in (4.95) can always be chosen such that, for $m \neq n$, $\Psi^{(m)}_{z_1 z_2}(f_1) = \Psi^{(n)}_{z_1 z_2}(f_1)$ at most in a set of zero Lebesgue measure in $\mathbb{R}$. Analogously, the functions $P^{(n)}_{yx^{(*)}y^*x^{(*)*}}(f_1, f_2, f_3)$ and $\Psi^{(n)}_{yx^{(*)}y^*x^{(*)*}}(f_1, f_2, f_3)$ in (4.96) can always be chosen such that, for $m \neq n$, $\Psi^{(m)}_{yx^{(*)}y^*x^{(*)*}}(f_1, f_2, f_3) = \Psi^{(n)}_{yx^{(*)}y^*x^{(*)*}}(f_1, f_2, f_3)$ at most in a set of zero Lebesgue measure in $\mathbb{R}^3$.

Assumption 4.4.3 Series Regularity.

(a) *For any choice of* z_1 *and* z_2 *in* $\{x, x^*, y, y^*\}$, *the functions* $S^{(n)}_{z_1 z_2}(f)$ *in (4.95) are almost everywhere (a.e.) continuous, in* $L^\infty(\mathbb{R})$ *and, moreover, such that*

$$\sum_{n \in \mathbb{I}_{z_1 z_2}} \|S^{(n)}_{z_1 z_2}\|_\infty < \infty \tag{4.97}$$

where $\|S\|_\infty \triangleq \operatorname{ess\,sup}_{f \in \mathbb{R}} |S(f)|$ *is the essential supremum of* $S(f)$ *(Champeney 1990).*

(b) *The functions* $P^{(n)}_{yx^{(*)}y^*x^{(*)*}}(f_1, f_2, f_3)$ *in (4.96) are a.e. continuous, in* $L^\infty(\mathbb{R}^3)$ *and, moreover, such that*

$$\sum_{n \in \mathbb{I}_4} \|P^{(n)}_{yx^{(*)}y^*x^{(*)*}}\|_\infty < \infty \tag{4.98}$$

where $\|P\|_\infty \triangleq \operatorname{ess\,sup}_{\boldsymbol{f} \in \mathbb{R}^3} |P(\boldsymbol{f})|$.

□

Note that, since the Fourier transform of a summable function is continuous, bounded, and infinitesimal at infinity (Champeney 1990), a sufficient condition assuring Assumption 4.4.3a holds, is

$$\sum_{n \in \mathbb{I}_{z_1 z_2}} R^{(n)}_{z_1 z_2}(\tau) \in L^1(\mathbb{R}) \tag{4.99}$$

where $R^{(n)}_{z_1 z_2}(\tau)$ is the inverse Fourier transform of $S^{(n)}_{z_1 z_2}(f)$. Furthermore, a sufficient condition assuring Assumption 4.4.3b holds, is

$$\sum_{n\in\mathbb{I}_4} C^{(n)}_{yx^{(*)}y^*x^{(*)*}}(\tau_1, \tau_2, \tau_3) \in L^1(\mathbb{R}^3) \tag{4.100}$$

where $C^{(n)}_{yx^{(*)}y^*x^{(*)*}}(\tau_1, \tau_2, \tau_3)$ is the inverse Fourier transform of $P^{(n)}_{yx^{(*)}y^*x^{(*)*}}(f_1, f_2, f_3)$.

In the special case of ACS processes, sufficient conditions assuring Assumption 4.4.3a holds are derived in (Alekseev 1988). Moreover, (4.99) and (4.100) reduce to the well-known summability conditions on the second- and fourth-order cumulants as in (Hurd 1989a), (Hurd 1991), (Dandawaté and Giannakis 1994), (Dehay and Hurd 1994), (Sadler and Dandawaté 1998).

Assumption 4.4.4 Support-Curve Regularity (I). *For any choice of z_1 and z_2 in $\{x, x^*, y, y^*\}$, the functions $\Psi^{(n)}_{z_1 z_2}(f)$ in (4.95) are a.e. derivable with a.e. continuous derivatives.* □

Assumption 4.4.5 Data-Tapering Window Regularity. *$b_{\frac{1}{\Delta f}}(t)$ is a $(1/\Delta f)$-duration data-tapering window with Fourier transform $B_{\Delta f}(f)$ such that*

$$b_{\frac{1}{\Delta f}}(t) = w_b(\Delta f\, t) \tag{4.101}$$

$$\lim_{\Delta f\to 0} w_b(\Delta f\, t) = 1 \qquad \forall t \in \mathbb{R} \tag{4.102}$$

$$B_{\Delta f}(f) = \frac{1}{\Delta f}\, W_B\left(\frac{f}{\Delta f}\right) \tag{4.103}$$

where $W_B(f)$, the Fourier transform of $w_b(t)$, is a.e. continuous and regular as $|f| \to \infty$, $W_B(f) \in L^1(\mathbb{R}) \cap L^\infty(\mathbb{R})$, and $\int_{\mathbb{R}} W_B(f)\,\mathrm{d}f = w_b(0) = 1$. From (4.102), it follows that

$$\lim_{\Delta f\to 0} \frac{1}{\Delta f}\, W_B\left(\frac{f}{\Delta f}\right) = \delta(f) \tag{4.104}$$

$$\lim_{\Delta f\to 0} W_B\left(\frac{f}{\Delta f}\right) = W_B(0)\,\delta_f \tag{4.105}$$

where $\delta(f)$ is Dirac delta and δ_f is Kronecker delta, (that is, $\delta_f = 1$ if $f = 0$ and $\delta_f = 0$ if $f \neq 0$), and the first limit should be intended in the sense of distributions (generalized functions) (Zemanian 1987). □

The general case of complex-valued data-tapering windows is considered. In fact, in (Politis 2005) it is shown that, by the adoption of appropriate complex-valued tapers, the bias of spectral estimators can be reduced by orders of magnitude. Moreover, in (Lahiri 2003) it is shown that different asymptotic independence properties of the Discrete Fourier Transform (DFT), and hence of the periodogram, can be obtained with different data-tapers.

By taking the statistical expectation of the bifrequency cross-periodogram (4.93) with the expressions of the STFTs (4.94b) of $x(t)$ and $y(t)$ substituted into, one obtains the following result.

Lemma 4.4.6 Expected Value of the Bifrequency Cross-Periodogram (Napolitano 2003, Lemma 3.1). *Let $\{y(t),\ t \in \mathbb{R}\}$ and $\{x(t),\ t \in \mathbb{R}\}$ be second-order harmonizable jointly SC stochastic processes with bifrequency cross-spectrum (4.15). Under Assumptions 4.4.3a (series regularity) and 4.4.5 (data-tapering window regularity), the expected value of the bifrequency cross-periodogram (4.93) is given by*

$$\begin{aligned} & \mathrm{E}\left\{I_{yx^{(*)}}(t; f_1, f_2)_{\frac{1}{\Delta f}}\right\} \\ & \quad = \Delta f \sum_{n \in \mathbb{I}_{yx^{(*)}}} \int_{\mathbb{R}} S^{(n)}_{yx^{(*)}}(\nu)\, \mathcal{B}_{\Delta f}(f_1 - \nu; t)\, \mathcal{B}^{(*)}_{\Delta f}\left(f_2 - \Psi^{(n)}_{yx^{(*)}}(\nu); t\right) \mathrm{d}\nu \end{aligned} \tag{4.106}$$

where

$$\mathcal{B}_{\Delta f}(f; t) \triangleq B_{\Delta f}(f)\, e^{-j2\pi f t}. \tag{4.107}$$

Proof: See Section 5.2. □

By expressing the covariance of the bifrequency cross-periodogram in terms of second-order moments and a fourth-order cumulant of the STFTs of $x(t)$ and $y(t)$, the following result is proved.

Lemma 4.4.7 Covariance of the Bifrequency Cross-Periodogram (Napolitano 2003, Lemma 3.2). *Let $\{y(t),\ t \in \mathbb{R}\}$ and $\{x(t),\ t \in \mathbb{R}\}$ be second-order harmonizable zero-mean singularly and jointly SC stochastic processes with bifrequency spectra and cross-spectra (4.95). Under Assumptions 4.4.2 (SC statistics), 4.4.3 (series regularity), and 4.4.5 (data-tapering window regularity), the covariance of the bifrequency cross-periodogram (4.93) is given by*

$$\mathrm{cov}\left\{I_{yx^{(*)}}(t_1; f_{y1}, f_{x1})_{\frac{1}{\Delta f}},\ I_{yx^{(*)}}(t_2; f_{y2}, f_{x2})_{\frac{1}{\Delta f}}\right\} = \mathcal{T}_1 + \mathcal{T}_2 + \mathcal{T}_3 \tag{4.108}$$

with

$$\begin{aligned} \mathcal{T}_1 \triangleq (\Delta f)^2 & \sum_{n' \in \mathbb{I}'} \int_{\mathbb{R}} S^{(n')}_{yy^*}(\nu_{y1})\, \mathcal{B}_{\Delta f}(f_{y1} - \nu_{y1}; t_1) \\ & \mathcal{B}^{*}_{\Delta f}\left(f_{y2} - \Psi^{(n')}_{yy^*}(\nu_{y1}); t_2\right) \mathrm{d}\nu_{y1} \\ & \sum_{n'' \in \mathbb{I}''} \int_{\mathbb{R}} S^{(n'')}_{x^{(*)}x^{(*)*}}(\nu_{x1})\, \mathcal{B}^{(*)}_{\Delta f}(f_{x1} - \nu_{x1}; t_1) \\ & \mathcal{B}^{(*)*}_{\Delta f}\left(f_{x2} - \Psi^{(n'')}_{x^{(*)}x^{(*)*}}(\nu_{x1}); t_2\right) \mathrm{d}\nu_{x1} \end{aligned} \tag{4.109}$$

$$\mathcal{T}_2 \triangleq (\Delta f)^2 \sum_{n''' \in \mathbb{I}'''} \int_{\mathbb{R}} S^{(n''')}_{yx^{(*)*}}(\nu_{y1})\, \mathcal{B}_{\Delta f}(f_{y1} - \nu_{y1}; t_1)$$

$$\mathcal{B}^{(*)*}_{\Delta f}\left(f_{x2} - \Psi^{(n''')}_{yx^{(*)*}}(\nu_{y1}); t_2\right)\, \mathrm{d}\nu_{y1}$$

$$\sum_{n^{iv} \in \mathbb{I}^{iv}} \int_{\mathbb{R}} S^{(n^{iv})}_{x^{(*)}y^*}(\nu_{x1})\, \mathcal{B}^{(*)}_{\Delta f}(f_{x1} - \nu_{x1}; t_1)$$

$$\mathcal{B}^{*}_{\Delta f}\left(f_{y2} - \Psi^{(n^{iv})}_{x^{(*)}y^*}(\nu_{x1}); t_2\right)\, \mathrm{d}\nu_{x1} \tag{4.110}$$

$$\mathcal{T}_3 \triangleq (\Delta f)^2 \sum_{n \in \mathbb{I}_4} \int_{\mathbb{R}} \int_{\mathbb{R}} \int_{\mathbb{R}} P^{(n)}_{yx^{(*)}y^*x^{(*)*}}(\nu_{y1}, \nu_{x1}, \nu_{y2})$$

$$\mathcal{B}_{\Delta f}(f_{y1} - \nu_{y1}; t_1)\, \mathcal{B}^{(*)}_{\Delta f}(f_{x1} - \nu_{x1}; t_1)\, \mathcal{B}^{*}_{\Delta f}(f_{y2} - \nu_{y2}; t_2)$$

$$\mathcal{B}^{(*)*}_{\Delta f}\left(f_{x2} - \Psi^{(n)}_{yx^{(*)}y^*x^{(*)*}}(\nu_{y1}, \nu_{x1}, \nu_{y2}); t_2\right)\, \mathrm{d}\nu_{y1}\, \mathrm{d}\nu_{x1}\, \mathrm{d}\nu_{y2} \tag{4.111}$$

where, for notation simplicity, $\mathbb{I}' = \mathbb{I}_{yy^*}$, $\mathbb{I}'' = \mathbb{I}_{x^{(*)}x^{(*)*}}$, $\mathbb{I}''' = \mathbb{I}_{yx^{(*)*}}$, *and* $\mathbb{I}^{iv} = \mathbb{I}_{x^{(*)}y^*}$.

Proof: See Section 5.2. □

In the following, Theorems 4.4.8 and 4.4.9 provide the asymptotic expected value and covariance, respectively, of the cross-periodogram (4.93) when the spectral resolution $\Delta f \to 0$, that is, when the data-record length $1/\Delta f \to \infty$.

Theorem 4.4.8 Asymptotic Expected Value of the Bifrequency Cross-Periodogram (Napolitano 2003, Theorem 3.1). *Let* $\{y(t),\ t \in \mathbb{R}\}$ *and* $\{x(t),\ t \in \mathbb{R}\}$ *be second-order harmonizable jointly SC stochastic processes with bifrequency cross-spectrum (4.15). Under Assumptions 4.4.2a (SC statistics), 4.4.3a (series regularity), 4.4.4 (support-curve regularity (I)), and 4.4.5 (data-tapering window regularity), the asymptotic* ($\Delta f \to 0$) *expected value (4.106) of the bifrequency cross-periodogram (4.93) is given by*

$$\lim_{\Delta f \to 0} \mathrm{E}\left\{ I_{yx^{(*)}}(t; f_1, f_2)_{\frac{1}{\Delta f}} \right\} = \sum_{n \in \mathbb{I}_{yx^{(*)}}} S^{(n)}_{yx^{(*)}}(f_1)\, \mathcal{E}^{(n)}(f_1)\, \delta_{f_2 - \Psi^{(n)}_{yx^{(*)}}(f_1)} \tag{4.112}$$

where

$$\mathcal{E}^{(n)}(f_1) \triangleq \int_{\mathbb{R}} W_B(\lambda)\, W_B^{(*)}\left(\lambda \Psi^{(n)\,\prime}_{yx^{(*)}}(f_1)\right) \mathrm{d}\lambda. \tag{4.113}$$

Proof: See Section 5.2. □

The presence of the multiplicative term $\mathcal{E}^{(n)}(f_1)$ in (4.112) implies that, in general, the cross-periodogram of jointly SC processes is an asymptotically biased estimator of the bifrequency spectral cross-correlation density function $\bar{\mathcal{S}}_{yx^{(*)}}(f_1, f_2)$ (see (4.19)). However, if the functions $\Psi^{(n)}_{yx^{(*)}}(f_1)$ are known, the bias is known.

In the special case of jointly ACS processes, there is a one-to-one correspondence between indices $n \in \mathbb{I}_{yx^{(*)}}$ and cycle frequencies $\alpha \in A_{yx^{(*)}}$, $\Psi^{(n)}_{yx^{(*)}}(f_1) = (-)(\alpha - f_1)$, $S^{(n)}_{yx^{(*)}}(f_1) = S^{\alpha}_{yx^{(*)}}(f_1)$, and (4.106) reduces to the well known result (Gardner 1987d), (Hurd 1989a), (Hurd and Leśkow 1992a), (Dandawaté and Giannakis 1994), (Dehay and Hurd 1994), (Dehay and

Leśkow 1996), (Sadler and Dandawaté 1998)

$$\lim_{\Delta f \to 0} \mathrm{E}\left\{ I_{yx^{(*)}}(t; f_1, f_2)_{\frac{1}{\Delta f}} \right\} = \begin{cases} S^{\alpha}_{yx^{(*)}}(f_1) \displaystyle\int_{\mathbb{R}} W_B(\lambda)\, W_B^{(*)}(-_{(-)}\lambda)\, \mathrm{d}\lambda\,, \\ f_2 = _{(-)}(\alpha - f_1), \quad \alpha \in A_{yx^{(*)}} \\ \\ 0, \qquad\qquad \text{otherwise,} \end{cases} \tag{4.114}$$

where $A_{yx^{(*)}}$ is the set of the (conjugate) cycle frequencies of the (conjugate) statistical cross-correlation function of $y(t)$ and $x(t)$. Therefore, the cyclic cross-periodogram $I_{yx^{(*)}}(t; f_1, _{(-)}(\alpha - f_1))_{\frac{1}{\Delta f}}$ is an asymptotically unbiased (but for a known scaling factor) estimator of the (conjugate) cross cyclic spectrum $S^{\alpha}_{yx^{(*)}}(f_1)$, provided that the cycle frequency α is perfectly known.

Theorem 4.4.9 Asymptotic Covariance of the Bifrequency Cross-Periodogram (Napolitano 2003, Theorem 3.2). *Let $\{y(t),\ t \in \mathbb{R}\}$ and $\{x(t),\ t \in \mathbb{R}\}$ be second-order harmonizable zero-mean singularly and jointly SC stochastic processes with bifrequency cross-spectrum (4.15). Under Assumptions 4.4.2 (SC statistics), 4.4.3 (series regularity), 4.4.4 (support-curve regularity (I)), 4.4.5 (data-tapering window regularity), the asymptotic ($\Delta f \to 0$) covariance (4.108) of the cross-periodogram (4.93) is given by*

$$\lim_{\Delta f \to 0} \operatorname{cov}\left\{ I_{yx^{(*)}}(t_1; f_{y1}, f_{x1})_{\frac{1}{\Delta f}}\,,\; I_{yx^{(*)}}(t_2; f_{y2}, f_{x2})_{\frac{1}{\Delta f}} \right\} = \mathcal{T}_1{}' + \mathcal{T}_2{}' \tag{4.115}$$

where

$$\mathcal{T}_1{}' \triangleq \sum_{n' \in \mathbb{I}'} S^{(n')}_{yy^*}(f_{y1})\, \mathcal{E}_1^{(n')}(f_{y1})\, \delta_{f_{y2} - \Psi^{(n')}_{yy^*}(f_{y1})}$$
$$\sum_{n'' \in \mathbb{I}''} S^{(n'')}_{x^{(*)}x^{(*)*}}(f_{x1})\, \mathcal{E}_2^{(n'')}(f_{x1})\, \delta_{f_{x2} - \Psi^{(n'')}_{x^{(*)}x^{(*)*}}(f_{x1})} \tag{4.116}$$

$$\mathcal{T}_2{}' \triangleq \sum_{n''' \in \mathbb{I}'''} S^{(n''')}_{yx^{(*)*}}(f_{y1})\, \mathcal{E}_3^{(n''')}(f_{y1})\, \delta_{f_{x2} - \Psi^{(n''')}_{yx^{(*)*}}(f_{y1})}$$
$$\sum_{n^{iv} \in \mathbb{I}^{iv}} S^{(n^{iv})}_{x^{(*)}y^*}(f_{x1})\, \mathcal{E}_4^{(n^{iv})}(f_{x1})\, \delta_{f_{y2} - \Psi^{(n^{iv})}_{x^{(*)}y^*}(f_{x1})}. \tag{4.117}$$

In (4.116) and (4.117),

$$\mathcal{E}_1^{(n')}(f) = \int_{\mathbb{R}} W_B(\lambda) W_B^*\left(\lambda \Psi^{(n')\prime}_{yy^*}(f)\right) \mathrm{d}\lambda$$
$$\mathcal{E}_2^{(n'')}(f) = \int_{\mathbb{R}} W_B^{(*)}(\lambda) W_B^{(*)*}\left(\lambda \Psi^{(n'')}_{x^{(*)}x^{(*)*}}{}'(f)\right) \mathrm{d}\lambda$$
$$\mathcal{E}_3^{(n''')}(f) = \int_{\mathbb{R}} W_B(\lambda) W_B^{(*)*}\left(\lambda \Psi^{(n''')}_{yx^{(*)*}}{}'(f)\right) \mathrm{d}\lambda$$
$$\mathcal{E}_4^{(n^{iv})}(f) = \int_{\mathbb{R}} W_B^{(*)}(\lambda) W_B^*\left(\lambda \Psi^{(n^{iv})}_{x^{(*)}y^*}{}'(f)\right) \mathrm{d}\lambda.$$

Proof: See Section 5.2. □

From Theorems 4.4.8 and 4.4.9 it follows that the bifrequency cross-periodogram of SC processes is an inconsistent estimator of the bifrequency spectral cross-correlation density function $\bar{\mathcal{S}}_{yx^{(*)}}(f_1, f_2)$.

In the special case of jointly ACS processes, accounting for (1.101)) we have

$$\Psi^{(n')}_{yy^*}(f_{y1}) = -(\alpha' - f_{y1}), \quad \alpha' \in A_{yy^*} \tag{4.118}$$

$$\Psi^{(n'')}_{x^{(*)}x^{(*)*}}(f_{x1}) = -_{(-)}(\alpha'' - _{(-)}f_{x1}), \quad \alpha'' \in A_{x^{(*)}x^{(*)*}} \tag{4.119}$$

$$\Psi^{(n''')}_{yx^{(*)*}}(f_{y1}) = -_{(-)}(\alpha''' - f_{y1}), \quad \alpha''' \in A_{yx^{(*)*}} \tag{4.120}$$

$$\Psi^{(n^{\text{iv}})}_{x^{(*)}y^*}(f_{x1}) = -(\alpha^{\text{iv}} - _{(-)}f_{x1}), \quad \alpha^{\text{iv}} \in A_{x^{(*)}y^*}. \tag{4.121}$$

Then, by substituting (4.118)–(4.121) into (4.116) and (4.117) and the result into (4.115), it follows that (corrected version of (Napolitano 2003, eq. (57)))

$$\begin{aligned}
\lim_{\Delta f \to 0} \operatorname{cov}\left\{ I_{yx^{(*)}}(t_1; f_{y1}, _{(-)}(\alpha_1 - f_{y1}))_{\frac{1}{\Delta f}}, \; I_{yx^{(*)}}(t_2; f_{y2}, _{(-)}(\alpha_2 - f_{y2})_{\frac{1}{\Delta f}} \right\} \\
= \mathcal{E}_1^2 \, S_{yy^*}^{f_{y1}-f_{y2}}(f_{y1}) \, S_{x^{(*)}x^{(*)*}}^{\alpha_1 - f_{y1} - \alpha_2 + f_{y2}}((-)(\alpha_1 - f_{y1})) \\
+ \mathcal{E}_3 \mathcal{E}_3^{(*)} \, S_{yx^{(*)*}}^{f_{y1} - \alpha_2 + f_{y2}}(f_{y1}) \, S_{x^{(*)}y^*}^{-f_{y2} + \alpha_1 - f_{y1}}((-)(\alpha_1 - f_{y1}))
\end{aligned} \tag{4.122}$$

in accordance with the results of (Hurd 1989a), (Dandawaté and Giannakis 1994), (Dehay and Hurd 1994), (Dehay and Leśkow 1996), (Sadler and Dandawaté 1998, eq. (13)). In (4.122), α_1 and $\alpha_2 \in A_{yx^{(*)}}$, and $\mathcal{E}_1 = \mathcal{E}_2$ and $\mathcal{E}_3 = \mathcal{E}_4^{(*)}$ are defined by substituting (4.118)–(4.121) into the expressions of $\mathcal{E}_1^{(n')}(f), \ldots, \mathcal{E}_4^{(n^{\text{iv}})}(f)$.

4.5 Measurement of Spectral Correlation – Unknown Support Curves

In this section, for jointly SC processes, the time-smoothed bifrequency cross-periodogram is considered as an estimator of the bifrequency spectral cross-correlation density function $\bar{\mathcal{S}}_{yx^{(*)}}(f_1, f_2)$ when the location of the support curves is unknown. For such an estimator, bias (Lemma 4.5.5) and covariance (Lemma 4.5.6) are determined. Moreover, its asymptotic biasedness and consistency are discussed (Theorems 4.5.7 and 4.5.9 and Corollary 4.5.8) (Napolitano 2001, 2003).

Definition 4.5.1 *Given two stochastic processes $\{y(t), \; t \in \mathbb{R}\}$ and $\{x(t), \; t \in \mathbb{R}\}$, their time-smoothed bifrequency cross-periodogram is defined as*

$$S_{yx^{(*)}}(t; f_1, f_2)_{\frac{1}{\Delta f}, T} \triangleq \Delta f \left[Y_{\frac{1}{\Delta f}}(t, f_1) \, X^{(*)}_{\frac{1}{\Delta f}}(t, f_2) \right] \underset{t}{\otimes} a_T(t) \tag{4.123}$$

where $Y_{1/\Delta f}(t, f_1)$ and $X_{1/\Delta f}(t, f_2)$ are the STFTs of $y(t)$ and $x(t)$, respectively, defined according to (4.94a), and $a_T(t)$ is a T-duration time-smoothing window. □

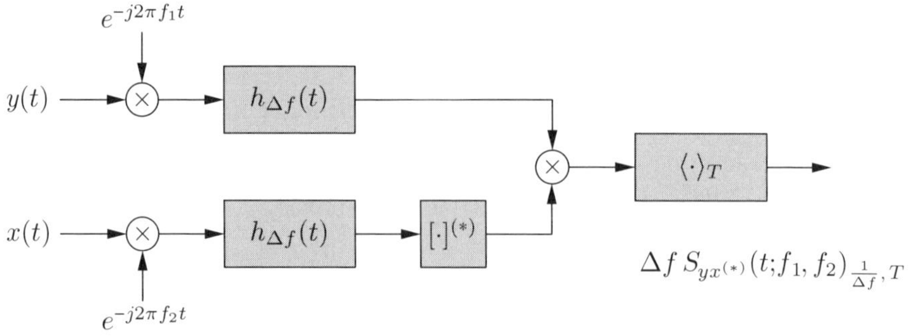

Figure 4.5 Spectral cross-correlation analyzer

Accordingly with the definition of STFT (4.94a), $Y_{1/\Delta f}(t, f_1)$ represents the output of a low-pass filter with impulse-response function $b_{1/\Delta f}(t)$ (assumed to be an even function) with bandwidth Δf when the input is the frequency-shifted version $y(t)e^{-j2\pi f_1 t}$ of $y(t)$. An analogous interpretation holds for $X_{1/\Delta f}(t, f_2)$. Therefore, the time-smoothed bifrequency cross-periodogram represents the finite-time-averaged cross-correlation (with zero lag) between the spectral components of $y(t)$ and $x(t)$ in the bands $(f_1 - \Delta f/2, f_1 + \Delta f/2)$ and $(f_2 - \Delta f/2, f_2 + \Delta f/2)$, respectively, normalized to $1/\Delta f$. The spectral cross-correlation analyzer can be realized by frequency shifting $y(t)$ by f_1 and $x(t)$ by f_2, passing such frequency-shifted versions through two low-pass filters $h_{\Delta f}(t)$ with bandwidth Δf and unity band-pass height, and then correlating the output signals (Figure 4.5) ($h_{\Delta f}(t) = b_{1/\Delta f}(t)$ in (4.94a)).

From Theorems 4.4.8 and 4.4.9 it follows that, for SC processes, the bifrequency cross-periodogram is an inconsistent estimator (with known bias only if the functions $\Psi^{(n)}$ are known) of the bifrequency spectral cross-correlation density function. For ACS processes it is well known that a frequency-smoothed version of the cross-periodogram provides a consistent estimator of the spectral cross-correlation density function, provided that the cycle frequencies are perfectly known (Gardner 1986a), (Gardner 1987d), (Hurd 1989a), (Hurd and Leśkow 1992a), (Dandawaté and Giannakis 1994), (Dehay and Hurd 1994), (Gerr and Allen 1994), (Dehay and Leśkow 1996), (Sadler and Dandawaté 1998). Such a result cannot be extended, in general, to SC processes that are not ACS. In (Allen and Hobbs 1992) and (Lii and Rosenblatt 2002), it is shown that smoothing the cross-periodogram in the frequency domain provides a consistent estimator of the spectral cross-correlation density function if the support curves $f_2 = \Psi^{(n)}(f_1)$ are lines (with not necessarily unit slopes). Furthermore, in (Allen and Hobbs 1992) and (Lii and Rosenblatt 2002) it is shown that the proposed frequency-smoothing techniques are effective only if the location of the support lines is known. That is, in general, frequency-smoothing techniques to reduce the variance of the cross-periodogram cannot be adopted if the location of the support curves of the Loève bifrequency spectrum is unknown. This result is in accordance with the more general result of (Gardner 1988a), (Gardner 1991b), where it is shown that reliable (low bias and variance) estimates of statistical functions of nonstationary processes can be obtained only if the nonstationarity is of almost-periodic nature (almost-cyclostationarity) or of known form.

Motivated by the interpretation of the time-smoothed bifrequency cross-periodogram $S_{yx^{(*)}}(t; f_1, f_2)_{1/\Delta f, T}$ as time-averaged spectral cross-correlation and by the asymptotic equivalence of time- and frequency-smoothing the cross-periodogram in the case of ACS processes (Gardner 1987d), the function $S_{yx^{(*)}}(t; f_1, f_2)_{1/\Delta f, T}$ is proposed as an estimator of the bifrequency spectral cross-correlation density function for SC processes when the location of the support curves is unknown. Moreover, its asymptotic bias and variance are discussed and a constraint on the reliability of the estimate is found.

Assumption 4.5.2 Time-Smoothing Window Regularity. *$a_T(t)$ is a T-duration time-smoothing window that can be expressed as*

$$a_T(t) = \frac{1}{T} w_a\left(\frac{t}{T}\right) \tag{4.124}$$

with $w_a(t) \in L^1(\mathbb{R})$ and $\lim_{T\to\infty} w_a(t/T) = 1$. □

Assumption 4.5.3 Lack of Support-Curve Clusters (I). *There is no cluster of support curves. That is, let*

$$\mathbb{J}_{\eta, f_1} \triangleq \left\{ n \in \mathbb{I} \,:\, \eta = \Psi^{(n)}_{yx^{(*)}}(f_1),\; S^{(n)}_{yx^{(*)}}(f_1) \neq 0 \right\} \tag{4.125}$$

then for every η_0 and f_{10}, the set $\mathbb{J}_{\eta_0, f_{10}}$ is finite (or empty) and for any $n \notin \mathbb{J}_{\eta_0, f_{10}}$ no curve $\Psi^{(n)}_{yx^{(*)}}(f_1)$ can be arbitrarily close to the value η_0 for $f_1 = f_{10}$. That is, for any η and f_1 it results in

$$h_{\eta, f_1} \triangleq \inf_{n \notin \mathbb{J}_{\eta, f_1}} \left| \eta - \Psi^{(n)}_{yx^{(*)}}(f_1) \right| > 0. \tag{4.126}$$

□

In the special case of ACS processes, Assumption 4.5.3 means that there is no cluster point of second-order cycle frequencies (Dehay and Hurd 1994).

Assumption 4.5.4 Support-Curve Regularity (II). *The second-order derivatives $\Psi^{(n)\prime\prime}$ of the functions $\Psi^{(n)}$, $n \in \mathbb{I}$, exist a.e. and are uniformly bounded.* □

Lemma 4.5.5 Expected Value of the Time-Smoothed Bifrequency Cross-Periodogram (Napolitano 2003, Lemma 4.1). *Let $\{y(t),\ t \in \mathbb{R}\}$ and $\{x(t),\ t \in \mathbb{R}\}$ be second-order harmonizable jointly SC stochastic processes with bifrequency cross-spectrum (4.15). Under Assumptions 4.4.3a (series regularity), 4.4.5 (data-tapering window regularity), and 4.5.2 (time-smoothing window regularity), the expected value of the time-smoothed cross-periodogram (4.123) is given by*

$$\mathrm{E}\left\{ S_{yx^{(*)}}(t; f_1, f_2)_{\frac{1}{\Delta f}, T} \right\} = \Delta f \sum_{n \in \mathbb{I}} \int_{\mathbb{R}} S^{(n)}(\nu_1)\, \mathcal{B}_{\Delta f}(f_1 - \nu_1; t)\, \mathcal{B}^{(*)}_{\Delta f}(f_2 - \Psi^{(n)}(\nu_1); t)$$
$$A_{\frac{1}{T}}(\nu_1 - f_1 +_{(-)} (\Psi^{(n)}(\nu_1) - f_2))\, \mathrm{d}\nu_1 \tag{4.127}$$

where $A_{\frac{1}{T}}(f)$ is the Fourier transform of $a_T(t)$.

Proof: See Section 5.3. □

Lemma 4.5.6 Covariance of the Time-Smoothed Bifrequency Cross-Periodogram (Napolitano 2003, Lemma 4.2). *Let $\{y(t),\ t \in \mathbb{R}\}$ and $\{x(t),\ t \in \mathbb{R}\}$ be second-order harmonizable zero-mean singularly and jointly SC stochastic processes with bifrequency cross-spectrum (4.15). Under Assumptions 4.4.2 (SC statistics), 4.4.3 (series regularity), 4.4.5 (data-tapering window regularity), and 4.5.2 (time-smoothing window regularity), the covariance of the time-smoothed cross-periodogram (4.123) is given by*

$$\operatorname{cov}\left\{S_{yx^{(*)}}(t_1; f_{y1}, f_{x1})_{\frac{1}{\Delta f},T},\ S_{yx^{(*)}}(t_2; f_{y2}, f_{x2})_{\frac{1}{\Delta f},T}\right\} = \mathcal{T}_1'' + \mathcal{T}_2'' + \mathcal{T}_3'' \tag{4.128}$$

where

$$\begin{aligned}
\mathcal{T}_1'' \triangleq (\Delta f)^2 \sum_{n' \in \mathbb{I}'} \sum_{n'' \in \mathbb{I}''} \int_{\mathbb{R}} \int_{\mathbb{R}} & S_{yy^*}^{(n')}(\nu_{y1})\, S_{x^{(*)}x^{(*)*}}^{(n'')}(\nu_{x1}) \\
& \mathcal{B}_{\Delta f}(f_{y1} - \nu_{y1}; t_1)\, \mathcal{B}_{\Delta f}^{*}(f_{y2} - \Psi_{yy^*}^{(n')}(\nu_{y1}); t_2) \\
& \mathcal{B}_{\Delta f}^{(*)}(f_{x1} - \nu_{x1}; t_1)\, \mathcal{B}_{\Delta f}^{(*)*}(f_{x2} - \Psi_{x^{(*)}x^{(*)*}}^{(n'')}(\nu_{x1}); t_2) \\
& A_{\frac{1}{T}}(\nu_{y1} - f_{y1} +_{(-)} (\nu_{x1} - f_{x1})) \\
& A_{\frac{1}{T}}^{*}(\Psi_{yy^*}^{(n')}(\nu_{y1}) - f_{y2} +_{(-)} (\Psi_{x^{(*)}x^{(*)*}}^{(n'')}(\nu_{x1}) - f_{x2}))\, \mathrm{d}\nu_{y1}\, \mathrm{d}\nu_{x1}
\end{aligned} \tag{4.129}$$

$$\begin{aligned}
\mathcal{T}_2'' \triangleq (\Delta f)^2 \sum_{n''' \in \mathbb{I}'''} \sum_{n^{iv} \in \mathbb{I}^{iv}} \int_{\mathbb{R}} \int_{\mathbb{R}} & S_{yx^{(*)*}}^{(n''')}(\nu_{y1})\, S_{x^{(*)}y^*}^{(n^{iv})}(\nu_{x1}) \\
& \mathcal{B}_{\Delta f}(f_{y1} - \nu_{y1}; t_1)\, \mathcal{B}_{\Delta f}^{(*)*}(f_{x2} - \Psi_{yx^{(*)*}}^{(n''')}(\nu_{y1}); t_2) \\
& \mathcal{B}_{\Delta f}^{(*)}(f_{x1} - \nu_{x1}; t_1)\, \mathcal{B}_{\Delta f}^{*}(f_{y2} - \Psi_{x^{(*)}y^*}^{(n^{iv})}(\nu_{x1}); t_2) \\
& A_{\frac{1}{T}}(\nu_{y1} - f_{y1} +_{(-)} (\nu_{x1} - f_{x1})) \\
& A_{\frac{1}{T}}^{*}(_{(-)}(\Psi_{yx^{(*)*}}^{(n''')}(\nu_{y1}) - f_{x2}) + \Psi_{x^{(*)}y^*}^{(n^{iv})}(\nu_{x1}) - f_{y2})\, \mathrm{d}\nu_{y1}\, \mathrm{d}\nu_{x1}
\end{aligned} \tag{4.130}$$

$$\begin{aligned}
\mathcal{T}_3'' \triangleq (\Delta f)^2 \sum_{n \in \mathbb{I}_4} \int_{\mathbb{R}} \int_{\mathbb{R}} \int_{\mathbb{R}} & P_{yx^{(*)}y^*x^{(*)*}}^{(n)}(\nu_{y1}, \nu_{x1}, \nu_{y2}) \\
& \mathcal{B}_{\Delta f}(f_{y1} - \nu_{y1}; t_1)\, \mathcal{B}_{\Delta f}^{(*)}(f_{x1} - \nu_{x1}; t_1) \\
& \mathcal{B}_{\Delta f}^{*}(f_{y2} - \nu_{y2}; t_2)\, \mathcal{B}_{\Delta f}^{(*)*}(f_{x2} - \Psi_4^{(n)}(\nu_{y1}, \nu_{x1}, \nu_{y2}); t_2) \\
& A_{\frac{1}{T}}(\nu_{y1} - f_{y1} +_{(-)} (\nu_{x1} - f_{x1})) \\
& A_{\frac{1}{T}}^{*}(\nu_{y2} - f_{y2} +_{(-)} (\Psi_4^{(n)}(\nu_{y1}, \nu_{x1}, \nu_{y2}) - f_{x2}))\, \mathrm{d}\nu_{y1}\, \mathrm{d}\nu_{x1}\, \mathrm{d}\nu_{y2}
\end{aligned} \tag{4.131}$$

where $\Psi_4^{(n)} \equiv \Psi_{yx^{(*)}y^*x^{(*)*}}^{(n)}$ *for notation simplicity.*

Proof: See Section 5.3. □

Theorem 4.5.7 Asymptotic Expected Value of the Time-Smoothed Bifrequency Cross-Periodogram (Napolitano 2003, Theorem 4.1). *Let $\{y(t),\ t \in \mathbb{R}\}$ and $\{x(t),\ t \in \mathbb{R}\}$ be second-order harmonizable jointly SC stochastic processes with bifrequency cross-spectrum (4.15).*

Under Assumptions 4.4.3a (series regularity), 4.4.4 (support-curve regularity (I)), 4.4.5 (data-tapering window regularity), and 4.5.2 (time-smoothing window regularity), the asymptotic ($T \to \infty$, $\Delta f \to 0$, with $T\Delta f \to \infty$) expected value (4.127) of the time-smoothed cross-periodogram (4.123) is given by

$$\lim_{\Delta f \to 0} \lim_{T \to \infty} \mathrm{E}\left\{ S_{yx^{(*)}}(t; f_1, f_2)_{\frac{1}{\Delta f}, T} \right\}$$
$$= W_A(0) \sum_{n \in \mathbb{I}} S^{(n)}(f_1)\, \mathcal{E}^{(n)}(f_1)\, \delta_{f_2 - \Psi^{(n)}(f_1)}\, \bar{\delta}_{1+(-)\Psi^{(n)\prime}(f_1)} \tag{4.132}$$

where $\bar{\delta}_{g(f)} = 1$ if $g(f) = 0$ in a neighborhood of f and $\bar{\delta}_{g(f)} = 0$ otherwise ($\bar{\delta}_{g(f)} = 0$ also if $g(\nu) = 0$ for $\nu = f$ but $g(\nu) \neq 0$ for $\nu \neq f$). [In (Napolitano 2003, Theorem 4.1), $\bar{\delta}$ is erroneously indicated by δ].

Proof: See Section 5.3. □

In (4.132) the order of the two limits cannot be interchanged. In fact, only for $T \to \infty$ and $\Delta f \to 0$ with $T\Delta f \to \infty$ the asymptotic variance vanishes (Theorem 4.5.9).

From Theorem 4.5.7 it follows that the asymptotic expectation of the time-smoothed cross-periodogram can be nonzero only for those (f_1, f_2) such that $f_2 = \Psi^{(n)}(f_1)$ with $\Psi^{(n)}$ having unit slope in a neighborhood of f_1 for some $n \in \mathbb{I}$. Thus, in the case of SC processes that are not ACS, the time-smoothed cross-periodogram is an asymptotically biased estimator of the bifrequency spectral cross-correlation density function (4.19). Such a result is in accordance with that derived in (Lii and Rosenblatt 2002) for the Daniell-like estimate in the special case of spectral support constituted by lines.

For jointly ACS processes, $\Psi^{(n)\prime}(f_1) = 1$ if $(*)$ is present and $\Psi^{(n)\prime}(f_1) = -1$ otherwise. Thus, the second Kronecker delta in (4.132) is always unity and the time-smoothed cross-periodogram is an asymptotically unbiased estimator of $\bar{\mathcal{S}}_{yx^{(*)}}(f_1, f_2)$ (but for a known multiplicative factor depending on the data-tapering and time-smoothing windows). More specifically, for $\Psi^{(n)}(f_1) = {}_{(-)}(\alpha - f_1)$ and $S^{(n)}(f_1) = S^{\alpha}_{yx^{(*)}}(f_1)$, (4.132) reduces to the well-known result (Gardner 1987d), (Hurd 1989a), (Hurd and Leśkow 1992a), (Dandawaté and Giannakis 1994), (Dehay and Hurd 1994), (Sadler and Dandawaté 1998)

$$\lim_{\Delta f \to 0} \lim_{T \to \infty} \mathrm{E}\left\{ S_{yx^{(*)}}(t; f_1, f_2)_{\frac{1}{\Delta f}, T} \right\} = \begin{cases} S^{\alpha}_{yx^{(*)}}(f_1)\, W_A(0) \displaystyle\int_{\mathbb{R}} W_B(\lambda)\, W_B^{(*)}(-{}_{(-)}\lambda)\, \mathrm{d}\lambda\,, \\ f_2 = {}_{(-)}(\alpha - f_1), \quad \alpha \in A_{yx^{(*)}} \\ 0, \quad \text{otherwise.} \end{cases} \tag{4.133}$$

In the following, the bias of the time-smoothed bifrequency cross-periodogram is analyzed in points belonging to the neighborhoods of the support curves where the slope of the curve is not too far from unity and, hence, accounting for Theorem 4.5.7, a small bias could be expected for finite T and Δf, provided that, however, T is sufficiently large and Δf sufficiently small.

Corollary 4.5.8 (Napolitano 2003, Corollary 4.1). *Let $\{y(t),\ t \in \mathbb{R}\}$ and $\{x(t),\ t \in \mathbb{R}\}$ be second-order harmonizable jointly SC stochastic processes with bifrequency cross-spectrum*

(4.15). Under Assumptions 4.4.3a (series regularity), 4.4.4 (support-curve regularity (I)), 4.4.5 (data-tapering window regularity), 4.5.2 (time-smoothing window regularity), 4.5.3 (lack of support-curve clusters (I)), and 4.5.4 (support-curve regularity (II)), if at the point (f_1, f_2) of the bifrequency plane it results that

$$\left| f_2 - \Psi^{(n)}(f_1) \right| \ll \frac{1}{T} \tag{4.134}$$

and

$$\left| 1 + (-)\Psi^{(n)\prime}(f_1) \right| \ll \frac{1}{T\Delta f} \tag{4.135}$$

for $n \in \mathbb{I}_0 \subseteq \mathbb{I}$, then

$$\mathrm{E}\left\{ S_{yx^{(*)}}(t; f_1, f_2)_{\frac{1}{\Delta f}, T} \right\} \simeq W_A(0) \sum_{n \in \mathbb{I}_0} S^{(n)}(f_1)\, \mathcal{E}^{(n)}(f_1) \tag{4.136}$$

Proof: See Section 5.3. □

Corollary 4.5.8 means that nonzero contribution to the expected value of the time-smoothed bifrequency cross-periodogram in the point (f_1, f_2) is given only by spectral densities $S^{(n)}(f_1)$ corresponding to support curves such that the value of $\Psi^{(n)}(f_1)$ is close to f_2 (in the sense of (4.134)) and the slope in (f_1, f_2) is close to ± 1 (in the sense of (4.135)).

Theorem 4.5.9 Asymptotic Covariance of the Time-Smoothed Bifrequency Cross-Periodogram (Napolitano 2003, Theorem 4.2). *Let $\{y(t),\ t \in \mathbb{R}\}$ and $\{x(t),\ t \in \mathbb{R}\}$ be two second-order harmonizable zero-mean singularly and jointly SC stochastic processes with bifrequency cross-spectrum (4.15). Under Assumptions 4.4.2 (SC statistics), 4.4.3 (series regularity), 4.4.4 (support-curve regularity (I)), 4.4.5 (data-tapering window regularity), and 4.5.2 (time-smoothing window regularity), the asymptotic ($T \to \infty$, $\Delta f \to 0$, with $T\Delta f \to \infty$) covariance (4.128) of the time-smoothed cross-periodogram (4.123) is such that*

$$\lim_{\Delta f \to 0} \lim_{T \to \infty} T\Delta f \left| \operatorname{cov}\left\{ S_{yx^{(*)}}(t_1; f_{y1}, f_{x1})_{\frac{1}{\Delta f}, T},\ S_{yx^{(*)}}(t_2; f_{y2}, f_{x2})_{\frac{1}{\Delta f}, T} \right\} \right| \leqslant \mathcal{T}_1''' + \mathcal{T}_2''' \tag{4.137}$$

where

$$\mathcal{T}_1''' \triangleq \|W_A\|_\infty \sum_{n' \in \mathbb{I}'} \sum_{n'' \in \mathbb{I}''} \|S_{yy^*}^{(n')}\|_\infty \, \|S_{x^{(*)}x^{(*)*}}^{(n'')}\|_\infty \, \mathcal{E}^{(n',n'')}(f_{y1}, f_{x1})\, \delta_{f_{y2} - \Psi_{yy^*}^{(n')}(f_{y1})}\, \delta_{f_{x2} - \Psi_{x^{(*)}x^{(*)*}}^{(n'')}(f_{x1})} \tag{4.138}$$

$$\mathcal{T}_2''' \triangleq \|W_A\|_\infty \sum_{n''' \in \mathbb{I}'''} \sum_{n^{iv} \in \mathbb{I}^{iv}} \|S_{yx^{(*)*}}^{(n''')}\|_\infty \, \|S_{x^{(*)}y^*}^{(n^{iv})}\|_\infty \, \mathcal{E}^{(n''',n^{iv})}(f_{y1}, f_{x1})\, \delta_{f_{x2} - \Psi_{yx^{(*)*}}^{(n''')}(f_{y1})}\, \delta_{f_{y2} - \Psi_{x^{(*)}y^*}^{(n^{iv})}(f_{x1})} \tag{4.139}$$

with

$$\mathcal{E}^{(n_1,n_2)}(f_{y1}, f_{x1}) \triangleq \int_{\mathbb{R}} \left| \mathcal{W}^{(n_1,n_2)}(\lambda; f_{y1}, f_{x1}) \right| \,\mathrm{d}\lambda$$

$$(n_1, n_2) \in \{(n', n''), (n''', n^{iv})\} \tag{4.140}$$

$$\mathcal{W}^{(n',n'')}(\lambda; f_{y1}, f_{x1}) \triangleq W_B(-_{(-)}\lambda)\, W_B^*\left(-_{(-)}\lambda \Psi^{(n')}_{yy^*}{}'(f_{y1})\right) W_B^{(*)}(\lambda)\, W_B^{(*)*}\left(\lambda \Psi^{(n'')}_{x^{(*)}x^{(*)*}}{}'(f_{x1})\right) \tag{4.141}$$

$$\mathcal{W}^{(n''',n^{iv})}(\lambda; f_{y1}, f_{x1}) \triangleq W_B(-_{(-)}\lambda)\, W_B^{(*)*}\left(-_{(-)}\lambda \Psi^{(n''')}_{yx^{(*)*}}{}'(f_{y1})\right) W_B^{(*)}(\lambda)\, W_B^*\left(\lambda \Psi^{(n^{iv})}_{x^{(*)}y^*}{}'(f_{x1})\right) \tag{4.142}$$

Proof: See Section 5.3. □

In the special case of singularly and jointly ACS processes, by substituting (4.118)–(4.121) into (4.129)–(4.131), one obtains the known result (Dandawaté and Giannakis 1994), (Dehay and Hurd 1994), (Sadler and Dandawaté 1998, eq. (23)) (corrected version of (Napolitano 2003, eq. (76)))

$$\begin{aligned}
&\lim_{\Delta f \to 0} \lim_{T\to\infty} T\Delta f \operatorname{cov}\Big\{ S_{yx^{(*)}}(t_1; f_{y1}, {}_{(-)}(\alpha_1 - f_{y1}))_{\frac{1}{\Delta f}, T} ,\\
&\qquad S_{yx^{(*)}}(t_2; f_{y2}, {}_{(-)}(\alpha_2 - f_{y2}))_{\frac{1}{\Delta f}, T} \Big\} \\
&= W_A(0)\, \mathcal{E}_1^{(4)}\, S_{yy^*}^{f_{y1}-f_{y2}}(f_{y1})\, S_{x^{(*)}x^{(*)*}}^{\alpha_1 - f_{y1} - \alpha_2 + f_{y2}}({}_{(-)}(\alpha_1 - f_{y1})) \\
&\quad + W_A(0)\, \mathcal{E}_2^{(4)}\, S_{yx^{(*)*}}^{f_{y1}-\alpha_2+f_{y2}}(f_{y1})\, S_{x^{(*)}y^*}^{-f_{y2}+\alpha_1-f_{y1}}({}_{(-)}(\alpha_1 - f_{y1}))
\end{aligned} \tag{4.143}$$

where α_1 and $\alpha_2 \in A_{yx^{(*)}}$,

$$\mathcal{E}_1^{(4)} \triangleq \int_{\mathbb{R}} \mathcal{W}^{(n',n'')}(\lambda; f_{y1}, f_{x1})\,\mathrm{d}\lambda$$

$$\mathcal{E}_2^{(4)} \triangleq \int_{\mathbb{R}} \mathcal{W}^{(n''',n^{iv})}(\lambda; f_{y1}, f_{x1})\,\mathrm{d}\lambda$$

with $\mathcal{W}^{(n',n'')}$ and $\mathcal{W}^{(n''',n^{iv})}$ given by (4.141) and (4.142), respectively, with (4.118)–(4.121) substituted into.

From Theorem 4.5.9 it follows that, even if

$$T\Delta f \operatorname{cov}\{S_{yx^{(*)}}(t_1; f_{y1}, f_{x1})_{1/\Delta f, T} ,\ S_{yx^{(*)}}(t_2; f_{y2}, f_{x2})_{1/\Delta f, T}\}$$

is not regular as $T \to \infty$ and $\Delta f \to 0$ with $T\Delta f \to \infty$, the result is that

$$\operatorname{cov}\{S_{yx^{(*)}}(t_1; f_{y1}, f_{x1})_{1/\Delta f, T}, S_{yx^{(*)}}(t_2; f_{y2}, f_{x2})_{1/\Delta f, T}\} = \mathcal{O}\left(\frac{1}{T\Delta f}\right) \tag{4.144}$$

where $\mathcal{O}$ denotes the "big oh" Landau symbol. Therefore, for $T\Delta f \gg 1$ the variance of the estimate (4.123) is $\mathcal{O}(1/T\Delta f)$. Moreover, from Corollary 4.5.8 one has that the bias at the point (f_1, f_2) with $f_2 = \Psi^{(n)}(f_1)$ is negligible, provided that the condition

$$\sup_{n\in\mathbb{I}} \left| {}^{(-)}\Psi^{(n)\prime}(f_1) + 1 \right| \ll \frac{1}{T\Delta f} \tag{4.145}$$

is satisfied. Thus, in the case of departure of the nonstationarity from the almost-cyclostationarity, that is, ${}^{(-)}\Psi^{(n)\prime}(f_1)$ different from 1 for some $n \in \mathbb{I}$ and some f_1, the variance cannot be made arbitrarily small (e.g., augmenting the collect time T) without obtaining a significant bias. On the contrary, when the stochastic processes $x(t)$ and $y(t)$ are jointly ACS, it results in ${}^{(-)}\Psi^{(n)\prime}(f_1) = 1\ \forall n$ and $\forall f_1$ and, hence, (4.145) is always satisfied allowing one to obtain an estimator of the bifrequency spectral cross-correlation density function which is mean-square consistent (that is, asymptotically unbiased and with vanishing asymptotic variance). Thus, there exists a trade-off between the departure of the nonstationarity from the almost-cyclostationarity and the estimate accuracy obtainable by single sample-path measurements. The larger is such a departure, that is, the larger is the left-hand side of (4.145) for some f_1, the smaller is the allowed value of $T\Delta f$ and, hence, the larger is the estimate variance. Note that, such a trade-off is not the usual bias-variance trade-off as for (stationary and) ACS processes where, for a fixed data-record length, the variance of the cyclic spectrum estimate can be reduced by augmenting the bandwidth of the frequency-smoothing window paying the price of a bias increase. Moreover, unlike the case of ACS processes, for SC processes the estimator performance cannot be improved as wanted by increasing the data-record length and the spectral resolution.

It is worthwhile emphasizing that, in the special case where the $\Psi^{(n)}(f_1)$ are known linear functions, the frequency-smoothing based techniques presented in (Allen and Hobbs 1992) and (Lii and Rosenblatt 2002) provide consistent estimators of the spectral cross-correlation density functions. However, note that the proposed estimator (4.123), unlike the estimators in (Allen and Hobbs 1992) and (Lii and Rosenblatt 2002), can be adopted when the functions $\Psi^{(n)}(f_1)$ are not necessarily linear. Moreover, the estimator (4.123) does not require the a priori knowledge of the shape and location of the support curves $f_2 = \Psi^{(n)}(f_1)$. The paid price for such a lack of knowledge is the absence of asymptotic unbiasedness and consistency of the estimator. This is in accordance with the result stated in (Gardner 1988a), (Gardner 1991b), where it is shown that spectral correlation measurements can be reliable only when the nonstationarity is of almost-periodic nature (almost-cyclostationarity) or known form. The results of this section establish quantitatively how accurate can be spectral correlation measurements made by a single sample-path of a spectrally correlated stochastic process.

In the case of ACS processes, the lack of knowledge of the location of the support lines is equivalent to the fact that cycle frequencies are unknown. In such a case, asymptotically unbiased estimates of the cyclic spectra cannot be obtained. For ACS processes, however, from Theorems 4.5.7 and 4.5.9 it follows that the estimate of the whole bifrequency spectral cross-correlation density function performed by the time-smoothed cross-periodogram is mean-square consistent. In such a case, in the limit as $T \to \infty$ and $\Delta f \to 0$ with $T\Delta f \to \infty$, the regions of the bifrequency plane where the time-smoothed cross-periodogram is significantly different from zero tend to the support lines and, hence, the unknown cycle frequencies can potentially be estimated. From Theorems 4.5.7 and 4.5.9 it follows that an analogous result does not hold for SC processes that are not ACS.

Finally, note that the theorems on the behavior of bias and variance of the estimators (4.93) and (4.123) can be extended to estimate the density of the impulsive term in the Loève bifrequency spectrum of processes exhibiting joint spectral correlation (Definition 4.2.5), provided that the continuous terms $\mathbb{S}^{(c)}_{z_1 z_2}(f_1, f_2)$ are bounded in $\mathbb{R}^2$ for any choice of z_1 and z_2 in $\{x, x^*, y, y^*\}$ and the continuous term in $\text{cum}\{Y(f_1), X^{(*)}(f_2), Y^*(f_3), X^{(*)*}(f_4)\}$ is bounded in $\mathbb{R}^4$.

4.6 The Frequency-Smoothed Cross-Periodogram

It is well known that the power spectrum of WSS processes can be consistently estimated by the frequency-smoothed periodogram, provided that some mixing assumptions regulating the memory of the process are satisfied (Parzen 1957), (Brillinger 1981), (Thomson 1982). Moreover, for ACS processes, the frequency-smoothed cyclic periodogram at a given cycle frequency is a mean-square consistent and asymptotically Normal estimator of the cyclic spectrum at that cycle frequency (Alekseev 1988), (Hurd 1989a), (Hurd 1991), (Hurd and Gerr 1991), (Hurd and Leśkow 1992a), (Dandawaté and Giannakis 1994), (Dehay and Hurd 1994), (Gerr and Allen 1994), (Dehay and Leśkow 1996), (Sadler and Dandawaté 1998). In both cases of WSS and ACS processes, the frequency-smoothing procedure consists in considering frequency averages of the cross-periodogram along the support lines (with unity slope) of the Loève bifrequency spectrum. This technique was extended in (Allen and Hobbs 1992), (Lii and Rosenblatt 1998), and (Lii and Rosenblatt 2002) to the case of SC processes with known support lines with not necessarily unit slope.

By following the above idea, in this section the cross-periodogram frequency-smoothed along a known given support curve is proposed as estimator of the spectral correlation density on this support curve (Napolitano 2007c).

According to Definition 4.4.1, the *bifrequency cross-periodogram* of two stochastic processes $\{y(t),\ t \in \mathbb{R}\}$ and $\{x(t),\ t \in \mathbb{R}\}$ observed for $t \in (-T/2, T/2)$ is defined as

$$I_{yx^{(*)}}(t; f_1, f_2)_T \triangleq \frac{1}{T} Y_T(t, f_1)\, X_T^{(*)}(t, f_2). \tag{4.146}$$

Under Assumptions 4.4.3a and 4.4.5 (with $\Delta f = 1/T$), the expected value of the bifrequency cross-periodogram (4.146) is given in Lemma 4.4.6 (eq. (4.106) with $\Delta f = 1/T$). Under Assumptions 4.4.2, 4.4.3, and 4.4.5 (with $\Delta f = 1/T$), the covariance of the bifrequency cross-periodogram (4.146) is given in Lemma 4.4.7 (eq. (4.108) with $\Delta f = 1/T$). Under Assumptions 4.4.2a, 4.4.3a, 4.4.4, and 4.4.5 (with $\Delta f = 1/T$), the asymptotic ($T \to \infty$) expected value of the bifrequency cross-periodogram of two jointly SC processes is given in Theorem 4.4.8 (eq. (4.112) with $\Delta f = 1/T$). Finally, under Assumptions 4.4.2, 4.4.3, 4.4.4, and 4.4.5 (with $\Delta f = 1/T$), the asymptotic ($T \to \infty$) covariance of the bifrequency cross-periodogram of two zero-mean jointly SC processes is given in Theorem 4.4.9 (eq. (4.115) with $\Delta f = 1/T$). These results show that, as for WSS and ACS processes, the properly normalized cross-periodogram of SC processes is an asymptotically unbiased but not consistent estimator of the spectral correlation density function.

Definition 4.6.1 *Given two jointly SC stochastic processes* $\{y(t),\ t \in \mathbb{R}\}$ *and* $\{x(t),\ t \in \mathbb{R}\}$, *their frequency-smoothed cross-periodogram along the support curve* $f_2 = \Psi^{(n)}_{yx^{(*)}}(f_1)$ *is defined as*

$$S^{(n)}_{yx^{(*)}}(t; f_1)_{T,\Delta f} \triangleq \left[I_{yx^{(*)}}(t; f_1, f_2)_T \Big|_{f_2=\Psi^{(n)}_{yx^{(*)}}(f_1)} \right] \underset{f_1}{\otimes} A_{\Delta f}(f_1) \tag{4.147}$$

where $A_{\Delta f}(f_1)$ *is a* Δf*-bandwidth frequency-smoothing window and* $\underset{f_1}{\otimes}$ *denotes convolution with respect to* f_1. □

Assumption 4.6.2 Frequency-Smoothing Window Regularity. $A_{\Delta f}(f)$ *is a* Δf*-bandwidth frequency-smoothing window such that*

$$A_{\Delta f}(f) = \frac{1}{\Delta f} W_A\left(\frac{f}{\Delta f}\right) \tag{4.148}$$

with $W_A(f)$ *a.e. continuous,* $W_A(f) \in L^1(\mathbb{R}) \cap L^\infty(\mathbb{R})$, $\int_{\mathbb{R}} W_A(f)\,\mathrm{d}f = 1$, *and*

$$\lim_{\Delta f \to 0} \frac{1}{\Delta f} W_A\left(\frac{f}{\Delta f}\right) = \delta(f). \tag{4.149}$$

□

By using the expressions of the expected value (Lemma 4.4.6) and covariance (Lemma 4.4.7) of the bifrequency cross-periodogram, the following results are obtained, where the made assumptions allow the interchange of the order of expectation, integral, and sum operations.

Theorem 4.6.3 Expected Value of the Frequency-Smoothed Cross-Periodogram. *Let* $\{y(t),\ t \in \mathbb{R}\}$ *and* $\{x(t),\ t \in \mathbb{R}\}$ *be second-order harmonizable jointly SC stochastic processes with bifrequency cross-spectrum (4.15). Under Assumptions 4.4.3a (series regularity), 4.4.5 (data-tapering window regularity) (with* $\Delta f = 1/T$*), and 4.6.2 (frequency-smoothing window regularity), the expected value of the frequency-smoothed cross-periodogram (4.147) is given by*

$$\mathrm{E}\left\{ S^{(n)}_{yx^{(*)}}(t; f)_{T,\Delta f} \right\} = \sum_{m \in \mathbb{I}} \int_{\mathbb{R}} S^{(m)}_{yx^{(*)}}(\nu)\, K_{T,\Delta f}(f, \nu; n, m; t)\,\mathrm{d}\nu \tag{4.150}$$

where

$$K_{T,\Delta f}(f, \nu; n, m; t) \triangleq \frac{1}{T} \int_{\mathbb{R}} \mathcal{B}_{\frac{1}{T}}(\lambda - \nu; t)\, \mathcal{B}^{(*)}_{\frac{1}{T}}\left(\Psi^{(n)}_{yx^{(*)}}(\lambda) - \Psi^{(m)}_{yx^{(*)}}(\nu); t\right)$$

$$A_{\Delta f}(f - \lambda)\,\mathrm{d}\lambda. \tag{4.151}$$

Proof: See Section 5.4. □

The expected value of the frequency-smoothed cross-periodogram along a given support curve $f_2 = \Psi^{(n)}_{yx^{(*)}}(f_1)$ is influenced not only by $S^{(n)}_{yx^{(*)}}(f_1)$, but also by the spectral cross-correlation densities $S^{(m)}_{yx^{(*)}}(f_1)$ relative to all the others support curves $f_2 = \Psi^{(m)}_{yx^{(*)}}(f_1), m \neq n$, the influence being stronger from support curves $f_2 = \Psi^{(m)}_{yx^{(*)}}(f_1)$ closer to $f_2 = \Psi^{(n)}_{yx^{(*)}}(f_1)$ and with larger $|S^{(m)}_{yx^{(*)}}(f_1)|$. Such a phenomenon is similar to the cyclic leakage phenomenon occurring in cyclic spectra estimates of ACS processes (Gardner 1986a), (Gardner 1987d) and the leakage phenomenon occurring in the estimation of cyclic statistics of GACS processes (Theorem 2.4.6), (Napolitano 2007a). Moreover, accounting for (4.151), in the sum over m in (4.150), all terms such that $\left|\Psi^{(m)}_{yx^{(*)}}(f) - \Psi^{(n)}_{yx^{(*)}}(f)\right| \leqslant 1/T$, give a significantly nonzero contribution. Thus, in correspondence of the at most countable set of values of f in which $\Psi^{(m)}_{yx^{(*)}}(f) = \Psi^{(n)}_{yx^{(*)}}(f)$ for $n \neq m$, we have spikes in the spectral density estimate. In the special case of linear support curves, this behavior was observed in (Lii and Rosenblatt 2002).

Theorem 4.6.4 Covariance of the Frequency-Smoothed Cross-Periodogram. *Let $\{y(t),\ t \in \mathbb{R}\}$ and $\{x(t),\ t \in \mathbb{R}\}$ be second-order harmonizable zero-mean singularly and jointly SC stochastic processes with bifrequency spectra and cross-spectra (4.95). Under Assumptions 4.4.2 (SC statistics), 4.4.3 (series regularity), 4.4.5 (data-tapering window regularity) (with $\Delta f = 1/T$), and 4.6.2 (frequency-smoothing window regularity) the covariance of the frequency-smoothed cross-periodogram (4.147) is given by*

$$\operatorname{cov}\left\{S^{(n_1)}_{yx^{(*)}}(t_1; f_1)_{T,\Delta f},\ S^{(n_2)}_{yx^{(*)}}(t_2; f_2)_{T,\Delta f}\right\} = \mathcal{T}_1'' + \mathcal{T}_2'' + \mathcal{T}_3'' \tag{4.152}$$

where

$$\begin{aligned}
\mathcal{T}_1'' \triangleq \frac{1}{T^2} \int_{\mathbb{R}} \int_{\mathbb{R}} \sum_{n' \in \mathbb{I}'} \int_{\mathbb{R}} & S^{(n')}_{yy^*}(\nu_{y1})\, \mathcal{B}_{\frac{1}{T}}(\lambda_1 - \nu_{y1}; t_1) \\
& \mathcal{B}^*_{\frac{1}{T}}\left(\lambda_2 - \Psi^{(n')}_{yy^*}(\nu_{y1}); t_2\right)\, \mathrm{d}\nu_{y1} \\
& \sum_{n'' \in \mathbb{I}''} \int_{\mathbb{R}} S^{(n'')}_{x^{(*)}x^{(*)*}}(\nu_{x1})\, \mathcal{B}^{(*)}_{\frac{1}{T}}\left(\Psi^{(n_1)}_{yx^{(*)}}(\lambda_1) - \nu_{x1}; t_1\right) \\
& \mathcal{B}^{(*)*}_{\frac{1}{T}}\left(\Psi^{(n_2)}_{yx^{(*)}}(\lambda_2) - \Psi^{(n'')}_{x^{(*)}x^{(*)*}}(\nu_{x1}); t_2\right)\, \mathrm{d}\nu_{x1} \\
& A_{\Delta f}(f_1 - \lambda_1)\, A^*_{\Delta f}(f_2 - \lambda_2)\, \mathrm{d}\lambda_1\, \mathrm{d}\lambda_2
\end{aligned} \tag{4.153}$$

$$\begin{aligned}
\mathcal{T}_2'' \triangleq \frac{1}{T^2} \int_{\mathbb{R}} \int_{\mathbb{R}} \sum_{n''' \in \mathbb{I}'''} \int_{\mathbb{R}} & S^{(n''')}_{yx^{(*)*}}(\nu_{y1})\, \mathcal{B}_{\frac{1}{T}}(\lambda_1 - \nu_{y1}; t_1) \\
& \mathcal{B}^{(*)*}_{\frac{1}{T}}\left(\Psi^{(n_2)}_{yx^{(*)}}(\lambda_2) - \Psi^{(n''')}_{yx^{(*)*}}(\nu_{y1}); t_2\right)\, \mathrm{d}\nu_{y1} \\
& \sum_{n^{iv} \in \mathbb{I}^{iv}} \int_{\mathbb{R}} S^{(n^{iv})}_{x^{(*)}y^*}(\nu_{x1})\, \mathcal{B}^{(*)}_{\frac{1}{T}}\left(\Psi^{(n_1)}_{yx^{(*)}}(\lambda_1) - \nu_{x1}; t_1\right) \\
& \mathcal{B}^*_{\frac{1}{T}}\left(\lambda_2 - \Psi^{(n^{iv})}_{x^{(*)}y^*}(\nu_{x1}); t_2\right)\, \mathrm{d}\nu_{x1}
\end{aligned}$$

$$A_{\Delta f}(f_1 - \lambda_1)\, A^*_{\Delta f}(f_2 - \lambda_2)\, d\lambda_1\, d\lambda_2 \tag{4.154}$$

$$\begin{aligned} \mathcal{T}_3'' \triangleq \frac{1}{T^2} \int_{\mathbb{R}} \int_{\mathbb{R}} \sum_{n \in \mathbb{I}_4} \int_{\mathbb{R}} \int_{\mathbb{R}} \int_{\mathbb{R}} & P^{(n)}_{yx^{(*)}y^*x^{(*)*}}(\nu_{y1}, \nu_{x1}, \nu_{y2}) \\ & \mathcal{B}_{\frac{1}{T}}(\lambda_1 - \nu_{y1}; t_1)\, \mathcal{B}^{(*)}_{\frac{1}{T}}\left(\Psi^{(n_1)}_{yx^{(*)}}(\lambda_1) - \nu_{x1}; t_1\right) \mathcal{B}^*_{\frac{1}{T}}(\lambda_2 - \nu_{y2}; t_2) \\ & \mathcal{B}^{(*)*}_{\frac{1}{T}}\left(\Psi^{(n_2)}_{yx^{(*)}}(\lambda_2) - \Psi^{(n)}_{yx^{(*)}y^*x^{(*)*}}(\nu_{y1}, \nu_{x1}, \nu_{y2}); t_2\right) d\nu_{y1}\, d\nu_{x1}\, d\nu_{y2} \\ & A_{\Delta f}(f_1 - \lambda_1)\, A^*_{\Delta f}(f_2 - \lambda_2)\, d\lambda_1\, d\lambda_2 \end{aligned} \tag{4.155}$$

Proof: See Section 5.4. □

4.7 Measurement of Spectral Correlation – Known Support Curves

In this section, for jointly SC processes the properly normalized frequency-smoothed cross-periodogram along a given support curve is considered as an estimator of the spectral cross-correlation density function on this support curve. Bias and covariance are determined. Moreover asymptotic unbiasedness, consistency, and asymptotic complex Normality are proved.

4.7.1 Mean-Square Consistency of the Frequency-Smoothed Cross-Periodogram

Assumption 4.7.1 Lack of Support-Curve Clusters (II). *Let*

$$\begin{aligned} \mathbb{J}_{n,f} \triangleq \Big\{ m \in \mathbb{I} - \{n\} \;:\; & \Psi^{(m)}_{yx^{(*)}}(f) = \Psi^{(n)}_{yx^{(*)}}(f), \\ & S^{(m)}_{yx^{(*)}}(f) \neq 0,\; S^{(n)}_{yx^{(*)}}(f) \neq 0 \Big\} \end{aligned} \tag{4.156}$$

be the set of indices of the curves $\Psi^{(m)}_{yx^{(*)}}(\cdot)$, $m \neq n$, that intercept $\Psi^{(n)}_{yx^{(*)}}(\cdot)$ in f.

There is no cluster of support curves. That is, for every $n \in \mathbb{I}$ and $f \in \mathbb{R}$, the set $\mathbb{J}_{n,f}$ is finite (or empty) and no curve $\Psi^{(m)}_{yx^{(*)}}(\cdot)$ with $m \notin \mathbb{J}_{n,f}$, $m \neq n$, can be arbitrarily close to the curve $\Psi^{(n)}_{yx^{(*)}}(\cdot)$ in f. That is, for every $n \in \mathbb{I}$ and $f \in \mathbb{R}$

$$h_{n,f} \triangleq \inf_{\substack{m \notin \mathbb{J}_{n,f} \\ m \neq n}} \left| \Psi^{(m)}_{yx^{(*)}}(f) - \Psi^{(n)}_{yx^{(*)}}(f) \right| > 0. \tag{4.157}$$

□

Assumption 4.7.1 is a slightly different formulation of the equivalent Assumption 4.5.3.

Assumption 4.7.2 Data-Tapering Window Regularity. *The data-tapering window satisfies the regularity conditions of Assumption 4.4.5. In addition, the first-order derivative* W_B *exists a.e. and, where it exists, is uniformly bounded,* $f^p\, W_B(f) \in L^1(\mathbb{R})$, $p = 1, 2$, *and there exists* $\gamma \geqslant 1$ *such that* $W_B(f) = \mathcal{O}(|f|^{-\gamma})$ *as* $|f| \to \infty$. □

Assumption 4.7.3 Frequency-Smoothing Window Regularity. *The frequency-smoothing window satisfies the regularity conditions of Assumption 4.6.2. In addition,* $f^p\, W_A(f) \in L^1(\mathbb{R})$, $p = 1, 2$. □

Assumption 4.7.4 Spectral Cross-Correlation Density Regularity. *The spectral cross-correlation density functions* $S^{(n)}_{yx^{(*)}}(f)$, $n \in \mathbb{I}$, *are a.e. derivable with uniformly bounded derivative.* □

Starting from the expressions of the expected value (Theorem 4.6.3) and covariance (Theorem 4.6.4) of the frequency-smoothed cross-periodogram, the following results are obtained, where the made assumptions allow the interchange of the order of sum, integral, and limit operations.

Theorem 4.7.5 Asymptotic Expected Value of the Frequency-Smoothed Cross-Periodogram. *Let* $\{y(t),\ t \in \mathbb{R}\}$ *and* $\{x(t),\ t \in \mathbb{R}\}$ *be second-order harmonizable jointly SC stochastic processes with bifrequency cross-spectrum (4.15). Under Assumptions 4.4.3a (series regularity), 4.4.4 (support-curve regularity (I)), 4.4.5 (data-tapering window regularity) (with* $\Delta f = 1/T$*), and 4.6.2 (frequency-smoothing window regularity), the asymptotic (*$T \to \infty$, $\Delta f \to 0$, *with* $T\Delta f \to \infty$*) expected value of the frequency-smoothed cross-periodogram (4.147) is given by*

$$\lim_{\Delta f \to 0} \lim_{T \to \infty} \mathrm{E}\left\{ S^{(n)}_{yx^{(*)}}(t; f)_{T,\Delta f} \right\} = S^{(n)}_{yx^{(*)}}(f)\, \mathcal{E}^{(n)}(f) \tag{4.158}$$

where

$$\mathcal{E}^{(n)}(f) \triangleq \int_{\mathbb{R}} W_B(\lambda)\, W_B^{(*)}\left(\lambda\, \Psi^{(n)}_{yx^{(*)}}{}'(f)\right) \mathrm{d}\lambda \tag{4.159}$$

with $\Psi^{(n)}_{yx^{(*)}}{}'(f)$ *denoting the first-order derivative of* $\Psi^{(n)}_{yx^{(*)}}(f)$.

Proof: See Section 5.5. □

From Theorem 4.7.5 it follows that the asymptotic expected value of the cross-periodogram frequency-smoothed along a known support curve $f_2 = \Psi^{(n)}_{yx^{(*)}}(f_1)$ is equal to the product of the spectral correlation density $S^{(n)}_{yx^{(*)}}(f)$ along the same curve and a multiplicative known bias term $\mathcal{E}^{(n)}(f)$. Such a term does not depend on f if the support curve $f_2 = \Psi^{(n)}_{yx^{(*)}}(f_1)$ is a line (with any slope). In particular, it does not depend on f in the ACS case.

The contribution to the expected value of $S^{(n)}_{yx^{(*)}}(t; f)_{T,\Delta f}$ of spectral cross-correlation densities of support curves such that $\Psi^{(m)}_{yx^{(*)}}(f_1) = \Psi^{(n)}_{yx^{(*)}}(f_1)$, $m \neq n$, is present for finite T and Δf. It disappears in the limit as $T \to \infty$ and $\Delta f \to 0$, provided that the order of these limits is not interchanged ($T\Delta f \to \infty$). In fact, for $T \to \infty$ and Δf fixed, the kernel $K_{T,\Delta f}(f, \nu; n, m; t)$ in (4.151) becomes smaller and smaller for $m \neq n$, even if $f = \nu$ (see also (Lii and Rosenblatt 2002, Figure 3) for the case of support lines).

Theorem 4.7.6 Rate of Convergence of the Bias of the Frequency-Smoothed Cross-Periodogram. *Let $\{y(t),\ t \in \mathbb{R}\}$ and $\{x(t),\ t \in \mathbb{R}\}$ be second-order harmonizable jointly SC stochastic processes with bifrequency cross-spectrum (4.15). Under Assumptions 4.4.3a (series regularity), 4.4.4 (support-curve regularity (I)), 4.4.5 (data-tapering window regularity) (with $\Delta f = 1/T$), 4.5.4 (support-curve regularity (II)), 4.6.2 (frequency-smoothing window regularity), 4.7.1 (lack of support-curve clusters (II)), under the further regularity conditions on the data-tapering and frequency-smoothing windows Assumptions 4.7.2 and 4.7.3, and the spectral cross-correlation density regularity Assumption 4.7.4, asymptotically ($T \to \infty$, $\Delta f \to 0$, with $T\Delta f \to \infty$) for every $f \notin \mathbb{J}_{n,f}$, that is, for all points f where two or more different curves do not intercept, one obtains*

$$\begin{aligned} \text{bias} &\triangleq \mathrm{E}\left\{ S^{(n)}_{yx^{(*)}}(t; f)_{T,\Delta f} \right\} - S^{(n)}_{yx^{(*)}}(f)\, \mathcal{E}^{(n)}(f) \\ &= \mathcal{O}\left(\frac{1}{T}\right) + \mathcal{O}(\Delta f) + \mathcal{O}\left(T(\Delta f)^2\right) \end{aligned} \tag{4.160}$$

provided that $T(\Delta f)^2 \to 0$.

Proof: See Section 5.5. □

Theorem 4.7.7 Asymptotic Covariance of the Frequency-Smoothed Cross-Periodogram. *Let $\{y(t),\ t \in \mathbb{R}\}$ and $\{x(t),\ t \in \mathbb{R}\}$ be second-order harmonizable zero-mean singularly and jointly SC stochastic processes with bifrequency spectra and cross-spectra (4.95). Under Assumptions 4.4.2 (SC statistics), 4.4.3 (series regularity), 4.4.4 (support-curve regularity (I)), 4.4.5 (data-tapering window regularity) (with $\Delta f = 1/T$), and 4.6.2 (frequency-smoothing window regularity), the asymptotic ($T \to \infty$, $\Delta f \to 0$, with $T\Delta f \to \infty$) covariance of the frequency-smoothed cross-periodogram (4.147) is such that*

$$\lim_{\Delta f \to 0} \lim_{T \to \infty} (T\Delta f) \operatorname{cov}\left\{ S^{(n_1)}_{yx^{(*)}}(t_1; f_1)_{T,\Delta f},\ S^{(n_2)}_{yx^{(*)}}(t_2; f_2)_{T,\Delta f} \right\} = \mathcal{T}_1''' + \mathcal{T}_2''' \tag{4.161}$$

where

$$\begin{aligned} \mathcal{T}_1''' \triangleq \sum_{n' \in \mathbb{I}'} \sum_{n'' \in \mathbb{I}''} & S^{(n')}_{yy^*}(f_1)\, S^{(n'')}_{x^{(*)}x^{(*)*}}\left(\Psi^{(n_1)}_{yx^{(*)}}(f_1)\right) \\ & \mathcal{J}^{(n',n'')}_1(f_1)\, \bar{\delta}_{\Psi^{(n_2)}_{yx^{(*)}}\left(\Psi^{(n')}_{yy^*}(f_1)\right) - \Psi^{(n'')}_{x^{(*)}x^{(*)*}}\left(\Psi^{(n_1)}_{yx^{(*)}}(f_1)\right)} \\ & \mathcal{J}^{(n')}_1(f_1)\, \delta_{f_2 - \Psi^{(n')}_{yy^*}(f_1)} \end{aligned} \tag{4.162}$$

$$\mathcal{T}_2''' \triangleq \sum_{n''' \in \mathbb{I}'''} \sum_{n^{iv} \in \mathbb{I}^{iv}} S^{(n''')}_{yx^{(*)*}}(f_1)\, S^{(n^{iv})}_{x^{(*)}y^*}\left(\Psi^{(n_1)}_{yx^{(*)}}(f_1)\right)$$
$$\mathcal{J}_2^{(n''',n^{iv})}(f_1)\, \bar{\delta}_{\Psi^{(n_2)}_{yx^{(*)}}\left(\Psi^{(n^{iv})}_{x^{(*)}y^*}\left(\Psi^{(n_1)}_{yx^{(*)}}(f_1)\right)\right)-\Psi^{(n''')}_{yx^{(*)*}}(f_1)}$$
$$\mathcal{J}_2^{(n^{iv})}(f_1)\, \delta_{f_2-\Psi^{(n^{iv})}_{x^{(*)}y^*}\left(\Psi^{(n_1)}_{yx^{(*)}}(f_1)\right)} \tag{4.163}$$

with

$$\mathcal{J}_1^{(n',n'')}(f_1) \triangleq \int_{\mathbb{R}}\int_{\mathbb{R}}\int_{\mathbb{R}} W_B^{(*)*}\left(\Psi^{(n_2)}_{yx^{(*)}}{}'\left(\Psi^{(n')}_{yy^*}(f_1)\right)\left[-\Psi^{(n')}_{yy^*}{}'(f_1)\,\nu_{y1}+\lambda_2\right]\right.$$
$$\left. +\,\Psi^{(n'')}_{x^{(*)}x^{(*)*}}{}'\left(\Psi^{(n_1)}_{yx^{(*)}}(f_1)\right)\,\nu_{x1}\right)$$
$$W_B(\nu_{y1})\,W_B^{(*)}(\nu_{x1})\,W_B^*(\lambda_2)\,\mathrm{d}\nu_{y1}\,\mathrm{d}\nu_{x1}\,\mathrm{d}\lambda_2 \tag{4.164}$$

$$\mathcal{J}_1^{(n')}(f_1) \triangleq \int_{\mathbb{R}} W_A(\lambda_1)\,W_A^*\left(\lambda_1\Psi^{(n')}_{yy^*}{}'(f_1)\right)\mathrm{d}\lambda_1 \tag{4.165}$$

$$\mathcal{J}_2^{(n''',n^{iv})}(f_1) \triangleq \int_{\mathbb{R}}\int_{\mathbb{R}}\int_{\mathbb{R}} W_B^{(*)*}\left(\Psi^{(n_2)}_{yx^{(*)}}{}'\left(\Psi^{(n^{iv})}_{x^{(*)}y^*}\left(\Psi^{(n_1)}_{yx^{(*)}}(f_1)\right)\right)\right.$$
$$\left.\left[-\Psi^{(n^{iv})}_{x^{(*)}y^*}{}'\left(\Psi^{(n_1)}_{yx^{(*)}}(f_1)\right)\,\nu_{x1}+\lambda_2\right]+\Psi^{(n''')}_{yx^{(*)*}}{}'(f_1)\,\nu_{y1}\right)$$
$$W_B(\nu_{y1})\,W_B^{(*)}(\nu_{x1})\,W_B^*(\lambda_2)\,\mathrm{d}\nu_{y1}\,\mathrm{d}\nu_{x1}\,\mathrm{d}\lambda_2 \tag{4.166}$$

$$\mathcal{J}_2^{(n^{iv})}(f_1) \triangleq \int_{\mathbb{R}} W_A(\lambda_1)\,W_A^*\left(\lambda_1\left[\Psi^{(n^{iv})}_{x^{(*)}y^*}\left(\Psi^{(n_1)}_{yx^{(*)}}(f_1)\right)\right]'\right)\mathrm{d}\lambda_1 \tag{4.167}$$

and $\bar{\delta}_{g(f)} = 1$ *if* $g(f) = 0$ *in a neighborhood of* f *and* $\bar{\delta}_{g(f)} = 0$ *otherwise (then also if* $g(\nu) = 0$ *for* $\nu = f$ *but* $g(\nu) \neq 0$ *for* $\nu \neq f$*).*

Proof: See Section 5.5. □

From Theorem 4.7.5 it follows that the frequency-smoothed cross-periodogram $S^{(n)}_{yx^{(*)}}(t, f)_{T,\Delta f}$ (smoothed along the known support curve $\Psi^{(n)}_{yx^{(*)}}(f)$), normalized by $\mathcal{E}^{(n)}(f)$, is an asymptotically unbiased estimator of the spectral cross-correlation density function along the same curve. Moreover, from Theorem 4.7.7 with $n_1 = n_2$, $f_1 = f_2$, and $t_1 = t_2$, it follows that the frequency-smoothed cross-periodogram has asymptotically vanishing variance of the order $\mathcal{O}((T\Delta f)^{-1})$. Therefore, the properly normalized frequency-smoothed cross-periodogram is a mean-square consistent estimator of the spectral correlation density. That is,

$$\lim_{\Delta f \to 0}\lim_{T\to\infty} \mathrm{E}\left\{\left|S^{(n)}_{yx^{(*)}}(t; f)_{T,\Delta f} - S^{(n)}_{yx^{(*)}}(f)\,\mathcal{E}^{(n)}(f)\right|^2\right\} = 0. \tag{4.168}$$

In the special case of jointly ACS processes, the support curves $\Psi^{(n')}_{yy^*}(f_1)$, $\Psi^{(n'')}_{x^{(*)}x^{(*)*}}(f_1)$, $\Psi^{(n''')}_{yx^{(*)*}}(f_1)$, and $\Psi^{(n^{iv})}_{x^{(*)}y^*}(f_1)$ are given by (4.118), (4.119), (4.120), and (4.118), respectively.

By substituting these expressions and

$$\Psi^{(n_i)}_{yx^{(*)}}(f_i) = (-)(\alpha_i - f_i) \qquad \alpha_i \in A_{yx^{(*)}},\ i = 1, 2$$

into the asymptotic covariance expression (4.161), an expression equivalent, but for a multiplicative constant not depending on cycle and spectral frequencies, to the asymptotic covariance of the time-smoothed cyclic cross-periodogram (4.143) is obtained. In fact, in $\mathcal{T}_1'''$ in (4.162) one has

$$\begin{aligned} &\delta_{f_2 - \Psi^{(n')}_{yy^*}(f_1)} = \delta_{[f_2 + \alpha' - f_1]} \\ \Rightarrow\quad & f_2 = \Psi^{(n')}_{yy^*}(f_1) \quad \text{and} \quad \alpha' = f_1 - f_2 \qquad (4.169) \\ &\bar{\delta}_{\Psi^{(n_2)}_{yx^{(*)}}(f_2) - \Psi^{(n'')}_{x^{(*)}x^{(*)*}}\left(\Psi^{(n_1)}_{yx^{(*)}}(f_1)\right)} = \bar{\delta}_{[(-)(\alpha_2 - f_2) + (-)\alpha'' - (-)(\alpha_1 + f_1)]} \\ \Rightarrow\quad & \alpha'' = \alpha_1 - f_1 - \alpha_2 + f_2 \qquad (4.170) \end{aligned}$$

and in $\mathcal{T}_2'''$ in (4.163) one has

$$\begin{aligned} &\delta_{f_2 - \Psi^{(n^{iv})}_{x^{(*)}y^*}\left(\Psi^{(n_1)}_{yx^{(*)}}(f_1)\right)} = \delta_{[f_2 + \alpha^{iv} - (-)(-)(\alpha_1 - f_1)]} \\ \Rightarrow\quad & f_2 = \Psi^{(n^{iv})}_{x^{(*)}y^*}\left(\Psi^{(n_1)}_{yx^{(*)}}(f_1)\right) \quad \text{and} \quad \alpha^{iv} = \alpha_1 - f_1 - f_2 \qquad (4.171) \\ &\bar{\delta}_{\Psi^{(n_2)}_{yx^{(*)}}(f_2) - \Psi^{(n''')}_{yx^{(*)*}}(f_1)} = \bar{\delta}_{[(-)(\alpha_2 - f_2) + (-)(\alpha''' - f_1)]} \\ \Rightarrow\quad & \alpha''' = f_1 - \alpha_2 + f_2 \qquad (4.172) \end{aligned}$$

This result is in agreement with the asymptotic equivalence between time- and frequency-smoothed cyclic periodogram proved in (Gardner 1987d) for ACS signals.

4.7.2 *Asymptotic Normality of the Frequency-Smoothed Cross-Periodogram*

Assumption 4.7.8 Spectral Cumulants. *For any choice of V_i, $i = 1, \ldots, k$ in $\{X, X^*, Y, Y^*\}$, $V_i(f)$ are the (generalized) Fourier transforms of k processes kth-order jointly spectrally correlated (Section 4.2.3). That is,*

$$\begin{aligned} &\operatorname{cum}\{V_1(f_1), \ldots, V_k(f_k)\} \\ &\quad = \sum_{\ell} P^{(\ell)}_{V_1 \cdots V_k}(f_1, \ldots, f_{k-1})\, \delta\left(f_k - \Psi^{(\ell)}_{V_1 \cdots V_k}(f_1, \ldots, f_{k-1})\right) \end{aligned} \qquad (4.173)$$

with

$$\sum_{\ell} \| P^{(\ell)}_{V_1 \cdots V_k} \|_\infty < \infty \qquad (4.174)$$

where the cumulant of complex random variables is defined according to (Spooner and Gardner 1994) (see also (Napolitano 2007a) and Section 1.4.2). □

Assumption 4.7.9 Derivatives of Support Functions. *For every $n \in \mathbb{I}$ the support function $\Psi^{(n)}_{yx^{(*)}}(\cdot)$ is invertible. Both $\Psi^{(n)}_{yx^{(*)}}(\cdot)$ and its inverse $\Phi^{(n)}_{yx^{(*)}}(\cdot)$ are differentiable with uniformly bounded derivative.* □

In the stationary case the periodogram is distributed as χ_1^2 and two periodograms are jointly χ_2^2 (that is, are exponentially distributed) (Brillinger 1981, Theorem 5.2.6 p. 126) and the periodogram is asymptotically ($T \to \infty$) exponentially distributed.

In Section 1.4.2, it is shown that the Nth-order cumulant for complex random variables as defined in (Spooner and Gardner 1994, App. A) is zero for $N \geqslant 3$ when the random variables are jointly complex Normal (Napolitano 2007a). This results is exploited in the following to prove the asymptotic complex Normality of the frequency-smoothed cross-periodogram.

Lemma 4.7.10 Cumulants of Frequency-Smoothed Cross-Periodograms. *Let $\{y(t),\ t \in \mathbb{R}\}$ and $\{x(t),\ t \in \mathbb{R}\}$ be second-order harmonizable zero-mean singularly and jointly SC stochastic processes with bifrequency spectra and cross-spectra (4.95). Under Assumptions 4.4.5 (data-tapering window regularity), 4.6.2 (frequency-smoothing window regularity), 4.7.8 (spectral cumulants), 4.7.9 (derivatives of support functions), and assuming the spectral cross-correlation densities $S^{(n)}_{yx^{(*)}}(f)$ continuous a.e. (a weaker condition w.r.t. that in Assumption 4.7.4 (spectral cross-correlation density regularity)), for any $k \geqslant 2$, one obtains*

$$\lim_{\Delta f \to 0} \lim_{T \to \infty} (T\Delta f)^{k/2} \operatorname{cum}\left\{ S^{(n_1)}_{yx^{(*)}}(t_1, f_1)^{[*]_1}_{T,\Delta f}, \ldots, S^{(n_k)}_{yx^{(*)}}(t_k, f_k)^{[*]_k}_{T,\Delta f} \right\}$$
$$= \begin{cases} \mathcal{O}(1) & k = 2 \\ 0 & k \geqslant 3 \end{cases} \tag{4.175}$$

where $[*]_i$ denotes ith optional complex conjugation and the order of the two limits cannot be interchanged ($T \to \infty$ and $\Delta f \to 0$ with $T\Delta f \to \infty$).

Proof: See Section 5.6. □

Theorem 4.7.11 Asymptotic Joint Complex Normality of the Frequency-Smoothed Cross-Periodograms. *Under the assumptions for Theorem 4.7.6 (rate of convergence of the bias of the frequency-smoothed cross-periodogram) and Lemma 4.7.10 (cumulants of frequency-smoothed cross-periodograms), if $\Delta f \equiv \Delta f_T = T^{-a}$, with $3/5 < a < 1$, it follows that for every fixed n_i, f_i, t_i the random variables*

$$V_i^{(T,\Delta f)} \triangleq \sqrt{T\Delta f}\left[S^{(n_i)}_{yx^{(*)}}(t_i, f_i)_{T,\Delta f} - \mathcal{E}^{(n_i)}(f)\, S^{(n_i)}_{yx^{(*)}}(f_i) \right] \tag{4.176}$$

are asymptotically ($T \to \infty$ and $\Delta f \to 0$ with $T\Delta f \to \infty$) zero-mean jointly complex Normal with asymptotic covariance matrix $\boldsymbol{\Sigma}$ with entries

$$\Sigma_{ij} = \lim_{\Delta f \to 0} \lim_{T \to \infty} \operatorname{cov}\left\{ V_i^{(T,\Delta f)}, V_j^{(T,\Delta f)} \right\} \tag{4.177}$$

given by (4.161) and asymptotic conjugate covariance matrix $\boldsymbol{\Sigma}^{(c)}$ *with entries*

$$\Sigma_{ij}^{(c)} = \lim_{\Delta f \to 0} \lim_{T \to \infty} \operatorname{cov}\left\{ V_i^{(T,\Delta f)}, V_j^{(T,\Delta f)*} \right\} \tag{4.178}$$

given by (5.165).

Proof: See Section 5.6. □

Corollary 4.7.12 Asymptotic Complex Normality of the Frequency-Smoothed Cross-Periodogram. *Under the assumptions of Theorem 4.7.11, it follows that for every fixed* n, f, t, *the random variable* $\sqrt{T\Delta f}[S_{yx^{(*)}}^{(n)}(t, f)_{T,\Delta f} - S_{yx^{(*)}}^{(n)}(f)\, \mathcal{E}^{(n)}(f)]$ *is asymptotically zero-mean complex Normal:*

$$\sqrt{T\Delta f}\left[S_{yx^{(*)}}^{(n)}(t, f)_{T,\Delta f} - S_{yx^{(*)}}^{(n)}(f)\, \mathcal{E}^{(n)}(f) \right] \overset{\mathrm{d}}{\longrightarrow} \mathcal{N}(0, \Sigma_{11}, \Sigma_{11}^{(c)}) \tag{4.179}$$

as $T \to \infty$ *and* $\Delta f \to 0$ *with* $T\Delta f \to \infty$. □

From the asymptotic Normality and the expression of the asymptotic covariance of the frequency-smoothed cross-periodogram, it follows the asymptotic independence of the frequency-smoothed cross-periodograms for frequencies separated of at least Δf.

4.7.3 Final Remarks

In this section, some remarks are made on the results of Sections 4.5, 4.7.1, and 4.7.2. Specifically, it is shown how some assumptions made to obtain results of Sections 4.5, 4.7.1, and 4.7.2 can be relaxed or modified.

- In the asymptotic results of Theorems 4.7.5 and 4.7.7, we have that $T \to \infty$ and $\Delta f \to 0$, in this order, so that $T\Delta f \to \infty$. Therefore, we can take Δf finite and fixed, make $T \to \infty$, and then make $\Delta f \to 0$. Analogously, we could consider $\Delta f = \Delta f_T$ such that $\Delta f_T \to 0$ and $T\Delta f_T \to \infty$ as $T \to \infty$. This approach is adopted in Theorem 4.7.11 and in (Dandawaté and Giannakis 1994) and (Sadler and Dandawaté 1998) for ACS processes.
- In general, a rectangular data-tapering window cannot be used in the previous results since it does not satisfy Assumption 4.4.5. In fact,

$$w_b(t) = \operatorname{rect}(t) \overset{\mathcal{F}}{\longleftrightarrow} W_B(f) = \operatorname{sinc}(f) \notin L^1(\mathbb{R})$$

  However, a rectangular window can be adopted if more assumptions are made on the stochastic processes x and y. For example, if the number of support curves is finite, as it happens for WSS processes and strictly band-limited cyclostationary processes. The former, have only one support line (the main diagonal if $(*)$ is present). The latter have a finite number of cycle frequencies $\alpha_n = n/T_0$ and, hence, a finite number of support lines.

The triangular data-tapering window satisfies Assumption 4.4.5. In fact,

$$w_b(t) = (1 - 2|t|)\,\mathrm{rect}(t) \overset{\mathcal{F}}{\longleftrightarrow} W_B(f) = \frac{1}{2}\,\mathrm{sinc}^2(f/2) \in L^1(\mathbb{R})$$

In general, any data-tapering window continuous and with possibly discontinuous first order derivative has Fourier transform with rate of decay to zero which is at least $\mathcal{O}(|f|^{-2})$ and hence is summable.

- The proofs of the Theorems of Sections 4.4–4.7.1 can be carried out with minor changes by substituting Assumption 4.4.3 with the following one.

Assumption 4.7.13 Spectral-Density Summability.

(a) For any choice of z_1 and z_2 in $\{x, x^*, y, y^*\}$, the functions $S^{(n)}_{z_1 z_2}$ in (4.95) are almost everywhere (a.e.) continuous and such that

$$\sum_{n \in \mathbb{I}_{z_1 z_2}} \int_{\mathbb{R}} \left| S^{(n)}_{z_1 z_2}(f) \right| \mathrm{d}f < \infty. \tag{4.180}$$

(b) The functions $P^{(n)}_{yx^{(*)}y^*x^{(*)*}}(f_1, f_2, f_3)$ in (4.96) are a.e. continuous and such that

$$\sum_{n \in \mathbb{I}_4} \int_{\mathbb{R}^3} \left| P^{(n)}_{yx^{(*)}y^*x^{(*)*}}(f_1, f_2, f_3) \right| \mathrm{d}f_1\, \mathrm{d}f_2\, \mathrm{d}f_3 < \infty. \tag{4.181}$$

□

Assumption 4.7.13a is similar to Assumption A1 in (Lii and Rosenblatt 2002), where, however, only a finite set of support curves (lines) are considered.

If Assumption 4.7.13 is made instead of Assumption 4.4.3, then several bounds derived in the proofs of Theorems of Sections 4.4–4.7.1 (see Sections 5.2–5.5) should be modified as shown in Section 5.7. In this case, $W_B(f)$ does not need to be summable and we can take $w_b(t) = \mathrm{rect}(t)$ (i.e., $W_B(f) = \mathrm{sinc}(f) \notin L^1(\mathbb{R})$).

- If there is no cluster of support curves (Assumption 4.5.3) and $W_B(f)$ and $W_A(f)$ are both rigorously band-limited, then all the sums in (4.150), (4.153), (4.154), and (4.155) reduce to finite sums.

In fact, let $\Lambda_1 = \lambda_1$ or $\Psi^{(n_1)}_{yx^{(*)}}(\lambda_1)$, and $\Lambda_2 = \lambda_2$ or $\Psi^{(n_2)}_{yx^{(*)}}(\lambda_2)$, and let $\Psi^{(n)} = \Psi^{(n')}_{yy^*}$ or $\Psi^{(n'')}_{x^{(*)}x^{(*)*}}$ or $\Psi^{(n''')}_{yx^{(*)*}}$ or $\Psi^{(n^{iv})}_{x^{(*)}y^*}$. In the case of W_B rigorously bandlimited, in the arguments of the integrals in (4.150), (4.153), (4.154), and (4.155) there are products of the kind

$$\mathrm{rect}\left(\frac{\Lambda_2 - \Psi^{(n)}(\nu)}{1/T}\right)\, \mathrm{rect}\left(\frac{\Lambda_1 - \nu}{1/T}\right)$$

with ν integration variable.

For fixed Λ_1 and Λ_2, each product is nonzero only in correspondence of a finite number of values of n since $\nu \in (\Lambda_1 - 1/(2T), \Lambda_1 + 1/(2T))$ due to the presence of the second rect. This result is still valid if Λ_1 and Λ_2 range in a finite interval of values, that is, if also W_A is rigorously bandlimited.

Thus, Assumption 4.4.3 can be relaxed. Specifically, (4.97) and (4.98) are not necessary.

This situation occurs in the simulation experiment in Section 4.12.2 and in (Napolitano 2003, Section V), where the Fourier transform $B_{\Delta f}$ of the data-tapering window is rectangular and, for the PAM signal with full duty-cycle rectangular pulse, (4.97) (or (Napolitano 2003, eq. (36))) is not satisfied.

4.8 Discrete-Time SC Processes

In this section, discrete-time spectrally correlated processes are defined and characterized (Napolitano 2011). For the sake of generality, a joint characterization of two processes $x_1(n)$ and $x_2(n)$ in terms of cross-statistics is provided. The characterization of a single process can be obtained as a special case by taking $x_1 \equiv x_2$.

The discrete-time processes $x_1(n)$ and $x_2(n)$ are said to be second-order jointly harmonizable if (Loève 1963)

$$\mathrm{E}\left\{x_1(n_1)\, x_2^{(*)}(n_2)\right\} = \int_{[-1/2,1/2]^2} e^{j2\pi[\nu_1 n_1 + (-)\nu_2 n_2]}\, \mathrm{d}\widetilde{\gamma}_{\boldsymbol{x}}(\nu_1, \nu_2) \tag{4.182}$$

with $\widetilde{\gamma}_{\boldsymbol{x}}(\nu_1, \nu_2)$ spectral covariance of bounded variation:

$$\int_{[-1/2,1/2]^2} \left| \mathrm{d}\widetilde{\gamma}_{\boldsymbol{x}}(\nu_1, \nu_2)\right| < \infty. \tag{4.183}$$

Under the harmonizability assumption, we have

$$x_i(n) = \int_{-1/2}^{1/2} e^{j2\pi\nu n}\, \mathrm{d}\chi_i(\nu) \tag{4.184}$$

where $\chi_i(\nu)$ is the integrated spectrum of $x_i(n)$. We can formally write $\mathrm{d}\chi_i(\nu) = X_i(\nu)\,\mathrm{d}\nu$ (Gardner 1985, Chapter 10.1.2), (Papoulis 1991, Chapter 12-4), where

$$X_i(\nu) = \sum_{n\in\mathbb{Z}} x_i(n)\, e^{-j2\pi\nu n} \tag{4.185}$$

is the Fourier transform of $x_i(n)$ to be considered in the sense of distributions (Gelfand and Vilenkin 1964, Chapter 3), (Henniger 1970), provided that $\chi_i(\nu)$ does not contain singular component (Hurd and Miamee 2007).

Definition 4.8.1 *Let $x_1(n)$ and $x_2(n)$ be discrete-time complex-valued second-order jointly harmonizable stochastic processes. Their Loève bifrequency cross-spectrum is defined as*

$$\widetilde{\mathbb{S}}_{\boldsymbol{x}}(\nu_1, \nu_2) \triangleq \mathrm{E}\left\{X_1(\nu_1)\, X_2^{(*)}(\nu_2)\right\} \tag{4.186}$$

where subscript $\boldsymbol{x}$ denotes $[x_1, x_2^{(*)}]$ and, in the sense of distributions (Gelfand and Vilenkin 1964, Chapter 3), (Henniger 1970),

$$\mathrm{d}\widetilde{\gamma}_{\boldsymbol{x}}(\nu_1, \nu_2) = \mathrm{E}\left\{X_1(\nu_1)\, X_2^{(*)}(\nu_2)\right\} \mathrm{d}\nu_1\, \mathrm{d}\nu_2 \tag{4.187}$$

provided that $\chi_i(\nu)$ and $\widetilde{\gamma}_{\boldsymbol{x}}(\nu_1, \nu_2)$ do not contain singular components. □

Definition 4.8.2 *Let $x_1(n)$ and $x_2(n)$ be discrete-time complex-valued second-order jointly harmonizable stochastic processes not containing any additive finite-strength sinewave component. The processes are said to be jointly spectrally correlated if their Loève bifrequency cross-spectrum can be expressed as*

$$\mathrm{E}\left\{X_1(\nu_1)\, X_2^{(*)}(\nu_2)\right\} = \sum_{k\in\mathbb{I}} \widetilde{S}_x^{(k)}(\nu_1)\, \widetilde{\delta}\left(\nu_2 - \widetilde{\Psi}_x^{(k)}(\nu_1)\right) \tag{4.188}$$

where $\mathbb{I}$ is a countable set, $\widetilde{\delta}(\nu) \triangleq \sum_{p\in\mathbb{Z}} \delta(\nu - p)$, and $\widetilde{S}_x^{(k)}(\nu)$ and $\widetilde{\Psi}_x^{(k)}(\nu)$ are complex- and real-valued, respectively, periodic functions of ν with period 1. □

From (4.188) it follows that discrete-time jointly SC processes have spectral masses concentrated on the countable set of support curves

$$\nu_2 = \widetilde{\Psi}_x^{(k)}(\nu_1) \bmod 1 \qquad k \in \mathbb{I} \tag{4.189}$$

where mod 1 is the modulo 1 operation with values in $[-1/2, 1/2)$. Moreover, the spectral mass distribution is periodic with period 1 in both frequency variables ν_1 and ν_2.

Two equivalent representations for the Loève bifrequency cross-spectrum hold, as stated by the following result.

Theorem 4.8.3 Characterization of Discrete-Time Jointly Spectrally Correlated Processes (Napolitano 2011, Theorem 3.1). *Let $x_1(n)$ and $x_2(n)$ be discrete-time jointly SC processes. Their Loève bifrequency cross-spectrum (4.188) can be expressed in the two equivalent forms*

$$\widetilde{\mathcal{S}}_x(\nu_1, \nu_2) = \sum_{k\in\mathbb{I}} \widetilde{S}_x^{(k)}(\nu_1)\, \widetilde{\delta}\left(\nu_2 - \widetilde{\Psi}_x^{(k)}(\nu_1)\right) \tag{4.190a}$$

$$= \sum_{k\in\mathbb{I}} \widetilde{G}_x^{(k)}(\nu_2)\, \widetilde{\delta}\left(\nu_1 - \widetilde{\Phi}_x^{(k)}(\nu_2)\right) \tag{4.190b}$$

with

$$\widetilde{S}_x^{(k)}(\nu_1) = \left|\widetilde{\Psi}_x^{(k)\prime}(\nu_1)\right| \widetilde{G}_x^{(k)}\left(\widetilde{\Psi}_x^{(k)}(\nu_1)\right) \tag{4.191a}$$

$$\widetilde{G}_x^{(k)}(\nu_2) = \left|\widetilde{\Phi}_x^{(k)\prime}(\nu_2)\right| \widetilde{S}_x^{(k)}\left(\widetilde{\Phi}_x^{(k)}(\nu_2)\right) \tag{4.191b}$$

provided that $\widetilde{\Psi}_x^{(k)}(\cdot)$ is locally invertible in every interval of width 1, $\widetilde{\Phi}_x^{(k)}(\cdot)$ is the periodic replication with period 1 of one of the local inverses, and both $\widetilde{\Psi}_x^{(k)}(\cdot)$ and $\widetilde{\Phi}_x^{(k)}(\cdot)$ are differentiable.

Proof: See Section 5.9. □

For discrete-time jointly SC processes with Loève bifrequency cross-spectrum (4.190a), (4.190b), the bounded variation condition (4.183) reduces to

$$\int_{[-1/2,1/2]^2} \left| \mathrm{d}\widetilde{\gamma}_x(\nu_1, \nu_2) \right| = \sum_{k \in \mathbb{I}} \int_{-1/2}^{1/2} \left| \widetilde{S}_x^{(k)}(\nu_1) \right| \mathrm{d}\nu_1 = \sum_{k \in \mathbb{I}} \int_{-1/2}^{1/2} \left| \widetilde{G}_x^{(k)}(\nu_2) \right| \mathrm{d}\nu_2 < \infty. \tag{4.192}$$

From (4.190a) and (4.190b), according to notation in (Lii and Rosenblatt 2002), the following alternative representation for the Loève bifrequency cross-spectrum of discrete-time jointly SC processes can be easily proved.

$$\widetilde{\mathbb{S}}_x(\nu_1, \nu_2) = \sum_{k \in \mathbb{I}} S_x^{(k)}(\nu_1 \bmod 1)\, \delta\left([\nu_2 - \Psi_x^{(k)}(\nu_1 \bmod 1)] \bmod 1 \right) \tag{4.193a}$$

$$= \sum_{k \in \mathbb{I}} G_x^{(k)}(\nu_2 \bmod 1)\, \delta\left([\nu_1 - \Phi_x^{(k)}(\nu_2 \bmod 1)] \bmod 1 \right) \tag{4.193b}$$

where the functions $\Psi_x^{(k)}(\cdot)$, $S_x^{(k)}(\cdot)$, $\Phi_x^{(k)}(\cdot)$, and $G_x^{(k)}(\cdot)$ are such that

$$\widetilde{\Psi}_x^{(k)}(\nu_1) = \Psi_x^{(k)}(\nu_1 \bmod 1) \tag{4.194}$$

$$\widetilde{S}_x^{(k)}(\nu_1) = S_x^{(k)}(\nu_1 \bmod 1) \tag{4.195}$$

$$\widetilde{\Phi}_x^{(k)}(\nu_2) = \Phi_x^{(k)}(\nu_2 \bmod 1) \tag{4.196}$$

$$\widetilde{G}_x^{(k)}(\nu_2) = G_x^{(k)}(\nu_2 \bmod 1). \tag{4.197}$$

Without lack of generality, it can be assumed that two support curves $\nu_2 = \widetilde{\Psi}_x^{(k)}(\nu_1) \bmod 1$ and $\nu_2 = \widetilde{\Psi}_x^{(k')}(\nu_1) \bmod 1$, with $k \neq k'$, intersect at most in a finite or countable set of points (ν_1, ν_2).

In the general case of support functions $\widetilde{\Psi}_x^{(k)}(\cdot)$ with nonunit slope for every k, analogously to the continuous-time case (see (4.20a) and (4.20b)), two spectral cross-correlation density functions (4.191a) and (4.191b) should be considered, depending on which one of ν_1 and ν_2 is taken as independent variable in the argument of the Dirac deltas in (4.190a) and (4.190b)

From (4.182), (4.190a), and (4.190b), it follows that the second-order cross-moment of jointly SC processes can be expressed as

$$\mathrm{E}\left\{ x_1(n_1)\, x_2^{(*)}(n_2) \right\} = \sum_{k \in \mathbb{I}} \int_{-1/2}^{1/2} \widetilde{S}_x^{(k)}(\nu_1)\, e^{j2\pi[\nu_1 n_1 + (-)\widetilde{\Psi}_x^{(k)}(\nu_1) n_2]}\, \mathrm{d}\nu_1 \tag{4.198a}$$

$$= \sum_{k \in \mathbb{I}} \int_{-1/2}^{1/2} \widetilde{G}_x^{(k)}(\nu_2)\, e^{j2\pi[\widetilde{\Phi}_x^{(k)}(\nu_2) n_1 + (-)\nu_2 n_2]}\, \mathrm{d}\nu_2. \tag{4.198b}$$

Second-order jointly ACS signals in the wide-sense are characterized by an almost-periodic cross-correlation function. That is (Section 1.3.8)

$$\mathrm{E}\left\{x_1(n+m)\, x_2^{(*)}(n)\right\} = \sum_{\widetilde{\alpha} \in \widetilde{\mathcal{A}}} \widetilde{\widetilde{R}}_{\boldsymbol{x}}^{\alpha}(m)\, e^{j2\pi\widetilde{\alpha}n} \tag{4.199}$$

where the Fourier coefficients

$$\begin{aligned}\widetilde{\widetilde{R}}_{\boldsymbol{x}}^{\alpha}(m) &\triangleq \left\langle \mathrm{E}\left\{x_1(n+m)\, x_2^{(*)}(n)\right\}\, e^{-j2\pi\widetilde{\alpha}n}\right\rangle_n \\ &\triangleq \lim_{N\to\infty} \frac{1}{2N+1} \sum_{n=-N}^{N} \mathrm{E}\left\{x_1(n+m)\, x_2^{(*)}(n)\right\}\, e^{-j2\pi\widetilde{\alpha}n}\end{aligned} \tag{4.200}$$

are referred to as cyclic cross-correlation functions and

$$\widetilde{\mathcal{A}} \triangleq \left\{\widetilde{\alpha} \in [-1/2, 1/2) \,:\, \widetilde{\widetilde{R}}_{\boldsymbol{x}}^{\alpha}(m) \not\equiv 0\right\} \tag{4.201}$$

is the countable set of cycle frequencies $\widetilde{\alpha}$ in the principal domain $[-1/2, 1/2)$. By double Fourier transforming both sides of (4.199) with $n_1 = n + m$ and $n_2 = n$, the following expression for the Loève bifrequency cross-spectrum is obtained

$$\mathrm{E}\left\{X_1(\nu_1)\, X_2^{(*)}(\nu_2)\right\} = \sum_{\widetilde{\alpha} \in \widetilde{\mathcal{A}}} \widetilde{\widetilde{S}}_{\boldsymbol{x}}^{\alpha}(\nu_1)\, \widetilde{\delta}(\nu_2 - {\scriptstyle (-)}(\widetilde{\alpha} - \nu_1)) \tag{4.202}$$

where

$$\widetilde{\widetilde{S}}_{\boldsymbol{x}}^{\alpha}(\nu) = \sum_{m \in \mathbb{Z}} \widetilde{\widetilde{R}}_{\boldsymbol{x}}^{\alpha}(m)\, e^{-j2\pi\nu m} \tag{4.203}$$

are the cyclic spectra. From (4.202) it follows that discrete-time jointly ACS processes are obtained as a special case of jointly SC processes when the spectral support curves are lines with slope ± 1 in the principal frequency domain $(\nu_1, \nu_2) \in [-1/2, 1/2]^2$. If the set $\widetilde{\mathcal{A}}$ contains the only element $\widetilde{\alpha} = 0$, then the cross-correlation function (4.199) does not depend on n (for the considered choice of $(*)$) and the processes $x_1(n)$ and $x_2(n)$ are jointly WSS. In such a case, the Loève bifrequency cross-spectrum has support contained in a diagonal of the principal frequency domain.

More generally, the processes $x_1(n)$ and $x_2^{(*)}(n)$ are said to exhibit joint almost-cyclostationarity at cycle frequency $\widetilde{\alpha}_0$ if the (conjugate) cross-correlation function is not necessarily an almost-periodic function of n but contains a finite-strength additive sinewave component at frequency $\widetilde{\alpha}_0$. In such a case, the cyclic cross-correlation function (4.200) is nonzero for $\widetilde{\alpha} = \widetilde{\alpha}_0$.

In (Akkarakaran and Vaidyanathan 2000), (Izzo and Napolitano 1998b), it is shown that multirate transformations, such as expansion and decimation, of ACS processes lead to ACS processes with different cyclostationarity properties. In contrast, in Section 4.10 it is shown that a discrete-time ACS process and its expanded or decimated version are jointly SC.

4.9 Sampling of SC Processes

In this section, a sampling theorem for SC processes is stated to link the Loève bifrequency spectrum of the sampled discrete-time process with that of the continuous-time process. An analogous result is established in terms of densities. Furthermore, in the case of strictly band-limited continuous-time SC processes, sufficient conditions on the sampling frequency to avoid aliasing effects are provided (Napolitano 2011).

For a WSS process, two equivalent conditions describe the absence of aliasing in the Loève bifrequency spectrum of the sampled signal. The former is the non overlap of replicas in the power spectrum of the sampled signal (Lloyd 1959). The latter is that the mappings $\nu_i = f_i/f_s$, $i = 1, 2$, between frequencies $f_i \in [-f_s/2, f_s/2]$ of the density of Loève bifrequency spectrum of the continuous-time process and frequencies ν_i of the density of Loève bifrequency spectrum of the discrete-time process hold $\forall \nu_1 \in [-1/2, 1/2]$, on the main diagonal $\nu_2 = \nu_1$. In fact, on the main diagonal the density of Loève bifrequency spectrum is coincident with the power spectrum. It is well known that both conditions are satisfied provided that the sampling frequency is equal to or greater than twice the process bandwidth. In this section it is shown that for SC processes, unlike the case of WSS processes, a sampling frequency equal to twice the signal bandwidth is not sufficient to assure that the mappings $\nu_i = f_i/f_s$ hold for the density on *every support curve* $\forall \nu_1 \in [-1/2, 1/2]$ (or $\forall \nu_2 \in [-1/2, 1/2]$). A sufficient condition is derived on the sampling frequency to assure that these mappings hold and known results for ACS processes (Napolitano 1995), (Izzo and Napolitano 1996), (Gardner *et al.* 2006) are obtained as special cases.

It is worth underlining that the problem of avoiding aliasing in the spectral correlation density of the sampled process should not be confused with the problem of signal reconstruction from its samples. This latter problem is addressed, for example, in (Lloyd 1959) for stationary processes and in (Gardner 1972), (Lee 1978) for nonstationary processes. In (Lloyd 1959), for stationary processes it is shown that the lack of overlap of replicas in the bifrequency Loève spectrum of the sampled process is a necessary and sufficient condition to allow the continuous-time process reconstruction with appropriate sense of convergence. In the case of base-band signals, this condition is the classical Shannon condition on the sampling frequency. In (Lee 1978), it is shown that for nonstationary processes such a condition is only sufficient. Finally, note that the class of processes considered here, namely the SC and ACS processes, have sample paths that are finite-power functions. Thus, known results for $L^2(\mathbb{R})$ functions (Garcia *et al.* 2001), (Song *et al.* 2007) or $L^p(\mathbb{R})$ functions cannot be exploited.

4.9.1 Band-Limitedness Property

It is well known that for WSS processes the strictly band-limitedness condition allows one to avoid aliasing after uniform sampling, provided that the sampling frequency exceeds twice the process bandwidth. Nonstationary processes should be carefully handled. In particular, not every nonstationary structure is compatible with the band-limitedness property. ACS processes can be strictly band limited (Gardner *et al.* 2006). In contrast, in Sections 2.2.2 and 2.2.3 (see also (Napolitano 2007a) and (Napolitano 2009)) it is shown that GACS processes cannot be strictly band limited. In the sequel it is proved that SC processes can be strictly band limited (Napolitano 2011). For this purpose, a definition of strict band limitedness is given for the general case of nonstationary processes. Specifically, a nonstationary process is said to be strictly band limited when it passes unaltered through an ideal strictly band-limited filter.

Let $h^{(b)}(t)$ be the impulse response function of the ideal low-pass filter with harmonic response $H^{(b)}(f) = \text{rect}(f/2b)$.

Definition 4.9.1 *A continuous-time nonstationary stochastic process $x_a(t)$ is said to be* (almost-surely) strictly band limited *with bandwidth B if B is the infimum of the values of b such that*

$$x_a(t) \otimes h^{(b)}(t) = x_a(t) \tag{4.204}$$

almost surely (a.s.) $\forall t$, where $\otimes$ denotes convolution. □

If $x_a(t)$ is a second-order process, then (4.204) holding a.s. $\forall t$ is equivalent to

$$\mathrm{E}\{|x_a(t) \otimes h^{(b)}(t) - x_a(t)|^2\} = 0 \qquad \forall t. \tag{4.205}$$

The condition of Definition 4.9.1 can be relaxed by requiring that (4.204) holds for almost all t. If (4.204) is satisfied, it follows that (see (5.179) in Section 5.10)

$$\mathrm{E}\left\{X_a(f_1)\, X_a^*(f_2)\right\} = 0 \qquad \text{for } |f_1| > B,\ |f_2| > B \tag{4.206}$$

which is the band-limitedness condition for nonstationary processes considered in (Gardner 1972). Such a condition can also be obtained by following (Lii and Rosenblatt 2002) in terms of the spectral measure $F_{x_a}(\mathrm{d}f_1, \mathrm{d}f_2) = \mathrm{d}\gamma_{x_a}(f_1, f_2)$ which is a set function (Hurd and Miamee 2007). Let A be a Borel subset of $\mathbb{R}^2$. The process $x_a(t)$ is strictly band limited in $[-B, B]$ if

$$B = \inf_b \left\{|F_{x_a}(A)| = 0 \text{ whenever } A \cap [-b, b]^2 = \emptyset\right\}. \tag{4.207}$$

If (4.206) holds, in (Gardner 1972) it is shown that for a sampling period $T_s < 1/(2B)$ the process $x_a(t)$ admits the mean-square equivalent "sample representation"

$$\mathrm{E}\left\{\left|x_a(t) - \sum_{n\in\mathbb{Z}} x_a(nT_s)\mathrm{sinc}(t/T_s - n)\right|^2\right\} = 0 \qquad \forall t \in \mathbb{R} \tag{4.208}$$

where $\mathrm{sinc}(t) \triangleq \sin(\pi t)/(\pi t)$. Further band-limitedness definitions for nonstationary processes are considered in (Zakai 1965), (Cambanis and Masry 1976).

In Sections 2.2.2 and 2.2.3 it is shown that GACS processes cannot be strictly band limited (see also (Izzo and Napolitano 2002a) and (Izzo and Napolitano 2002b)). In contrast, for the SC processes the following results hold.

Let $x_{a,i}(t)$, $(i = 1, 2)$, be continuous-time jointly SC processes with Loève bifrequency cross-spectrum

$$\mathcal{S}_{\boldsymbol{x}_a}(f_1, f_2) = \sum_{k \in \mathbb{I}_a} S^{(k)}_{\boldsymbol{x}_a}(f_1)\, \delta\left(f_2 - \Psi^{(k)}_{\boldsymbol{x}_a}(f_1)\right) \tag{4.209a}$$

$$= \sum_{k \in \mathbb{I}_a} G^{(k)}_{\boldsymbol{x}_a}(f_2)\, \delta\left(f_1 - \Phi^{(k)}_{\boldsymbol{x}_a}(f_2)\right) \tag{4.209b}$$

where $\boldsymbol{x}_a \triangleq [x_{a,1}, x^{(*)}_{a,2}]$.

Theorem 4.9.2 Strictly Band-Limited Spectrally Correlated Processes (Napolitano 2011, Theorem 2.1). *Let $x_{a,i}(t)$, $(i = 1, 2)$, be jointly SC processes with Loève bifrequency cross-spectrum (4.209a), (4.209b). If the processes are strictly band-limited with bandwidths B_i and if the points of the intersections of the support curves are non-dense (in the meaning assumed in the proof), the result is that $\forall k \in \mathbb{I}_a$*

$$S^{(k)}_{\boldsymbol{x}_a}(f_1) = 0 \quad \text{if } |f_1| > B_1 \text{ or } \left|\Psi^{(k)}_{\boldsymbol{x}_a}(f_1)\right| > B_2 \quad \text{a.e.} \tag{4.210a}$$

$$G^{(k)}_{\boldsymbol{x}_a}(f_2) = 0 \quad \text{if } |f_2| > B_2 \text{ or } \left|\Phi^{(k)}_{\boldsymbol{x}_a}(f_2)\right| > B_1 \quad \text{a.e.} \tag{4.210b}$$

Moreover, the functions $S^{(k)}_{\boldsymbol{x}_a}(f_1)$ and $G^{(k)}_{\boldsymbol{x}_a}(f_2)$ can be suitably modified so that (4.210a) and (4.210b) hold everywhere since with this modification the cross-correlation (4.24b) (with $[y, x^{(*)}] = [x_{a,1}, x^{(*)}_{a,2}]$ and (4.209a) or (4.209b) substituted into) remains unaltered. In addition, $\Psi^{(k)}_{\boldsymbol{x}_a}(f_1)$ and $\Phi^{(k)}_{\boldsymbol{x}_a}(f_2)$ can be assumed to be zero for $f_1 \notin [-B_1, B_1]$ and $f_2 \notin [-B_2, B_2]$, respectively. In this case, $\Psi^{(k)}_{\boldsymbol{x}_a}(f_1)$ is considered to be invertible from the support of $S^{(k)}_{\boldsymbol{x}_a}(f_1)$ to the support of $G^{(k)}_{\boldsymbol{x}_a}(f_2)$ for each $k \in \mathbb{I}_a$.

Conversely, if at least one of (4.210a) or (4.210b) holds, then also the other one holds and the processes $x_{a,i}(t)$ are strictly band-limited with bandwidth B_i.

Proof: See Section 5.10. □

Corollary 4.9.3 Support of the Spectral Cross-Correlation Density Functions (Napolitano 2011, Corollary 2.1). *Let $x_{a,i}(t)$, $(i = 1, 2)$, be strictly band-limited jointly SC processes with bandwidths B_i. In the expressions of the Loève bifrequency cross-spectrum (4.209a), (4.209b) $\forall k \in \mathbb{I}_a$ it results that*

$$\begin{aligned} &\operatorname{supp}\left\{S^{(k)}_{\boldsymbol{x}_a}(f_1)\, \delta\left(f_2 - \Psi^{(k)}_{\boldsymbol{x}_a}(f_1)\right)\right\} \\ &\triangleq \left\{(f_1, f_2) \in \mathbb{R} \times \mathbb{R} \,:\, S^{(k)}_{\boldsymbol{x}_a}(f_1) \neq 0,\ f_2 = \Psi^{(k)}_{\boldsymbol{x}_a}(f_1)\right\} \\ &\subseteq \left\{(f_1, f_2) \in \mathbb{R} \times \mathbb{R} \,:\, |f_1| \leqslant B_1, \right. \\ &\qquad \left. |f_2| \leqslant B_2,\ f_2 = \Psi^{(k)}_{\boldsymbol{x}_a}(f_1)\right\} \end{aligned} \tag{4.211a}$$

$$
\begin{aligned}
&\operatorname{supp}\Big\{ G^{(k)}_{\boldsymbol{x}_a}(f_2)\, \delta\Big(f_1 - \Phi^{(k)}_{\boldsymbol{x}_a}(f_2)\Big) \Big\} \\
&\triangleq \Big\{ (f_1, f_2) \in \mathbb{R}\times\mathbb{R} \,:\, G^{(k)}_{\boldsymbol{x}_a}(f_2) \neq 0,\ f_1 = \Phi^{(k)}_{\boldsymbol{x}_a}(f_2) \Big\} \\
&\subseteq \Big\{ (f_1, f_2) \in \mathbb{R}\times\mathbb{R} \,:\, |f_1| \leqslant B_1, \\
&\quad |f_2| \leqslant B_2,\ f_1 = \Phi^{(k)}_{\boldsymbol{x}_a}(f_2) \Big\}
\end{aligned} \tag{4.211b}
$$

where supp means support. □

4.9.2 Sampling Theorems

Theorem 4.9.4 Loève Bifrequency Cross-Spectrum of Sampled Jointly Spectrally Correlated Processes (Napolitano 2011, Theorem 4.1). *Let $x_{a,i}(t)$, $(i = 1, 2)$, be continuous-time jointly SC processes with Loève bifrequency cross-spectrum (4.209a), (4.209b) and let*

$$
x_i(n) \triangleq x_{a,i}(t)|_{t=nT_{si}}, \qquad n \in \mathbb{Z}, \qquad (i = 1, 2) \tag{4.212}
$$

be the discrete-time processes obtained by uniformly sampling $x_{a,i}(t)$ with sampling periods $T_{si} \triangleq 1/f_{si}$. The processes $x_i(n)$ are discrete-time jointly SC with Loève bifrequency cross-spectrum given by

$$
\begin{aligned}
\mathrm{E}\Big\{ X_1(\nu_1)\, X_2^{(*)}(\nu_2) \Big\} = \sum_{k\in\mathbb{I}_a} \frac{1}{T_{s1}} \sum_{p_1\in\mathbb{Z}} S^{(k)}_{\boldsymbol{x}_a}((\nu_1 - p_1) f_{s1}) \\
\sum_{p_2\in\mathbb{Z}} \delta\Big(\nu_2 - p_2 - \Psi^{(k)}_{\boldsymbol{x}_a}((\nu_1 - p_1) f_{s1})\, T_{s2} \Big)
\end{aligned} \tag{4.213}
$$

and the bounded variation condition (4.183) specializes to

$$
\frac{1}{T_{s1}} \sum_{k\in\mathbb{I}_a} \sum_{p_1\in\mathbb{Z}} \int_{-1/2}^{1/2} \Big| S^{(k)}_{\boldsymbol{x}_a}((\nu_1 - p_1) f_{s1}) \Big|\, \mathrm{d}\nu_1 < \infty. \tag{4.214}
$$

Proof: See Section 5.10. □

By comparing (4.213) with (4.209a) it follows that the Loève bifrequency cross-spectrum of the discrete-time signals is constituted by replicas of that of the continuous-time signals. In addition, according to Theorem 4.9.4, it follows that the discrete-time processes $x_i(n)$ are jointly SC. In fact, their Loève bifrequency cross-spectrum exhibits spectral masses concentrated on a countable set of support curves and is periodic with period 1 in both variables ν_1 and ν_2, even if in general the functions $\widetilde{S}^{(k)}_{\boldsymbol{x}}(\nu_1)$ and $\widetilde{\Psi}^{(k)}_{\boldsymbol{x}}(\nu_1)$ of Theorem 4.8.3 cannot be easily expressed in terms of $S^{(k)}_{\boldsymbol{x}_a}(f_1)$ and $\Psi^{(k)}_{\boldsymbol{x}_a}(f_1)$. Furthermore, at every point $(\nu_1, \nu_2) \in [-1/2, 1/2)^2$, nonzero contribution to the Loève bifrequency cross-spectrum is given by the terms in (4.213) such that

$$
(\nu_2 - p_2) f_{s2} = \Psi^{(k)}_{\boldsymbol{x}_a}((\nu_1 - p_1) f_{s1}) \tag{4.215}
$$

and

$$S_{x_a}^{(k)}((\nu_1 - p_1)f_{s1}) \neq 0 \tag{4.216}$$

for some $(p_1, p_2, k) \in \mathbb{Z} \times \mathbb{Z} \times \mathbb{I}_a$. Specifically, the support curves in the bifrequency principal domain $(\nu_1, \nu_2) \in [-1/2, 1/2)^2$ are described by the set

$$\mathbb{S} \triangleq \bigcup_{k \in \mathbb{I}_a} \bigcup_{p_1 \in \mathbb{Z}} \Big\{ (\nu_1, \nu_2) \in [-1/2, 1/2)^2 : \\ \nu_2 = \left[\Psi_{x_a}^{(k)}((\nu_1 - p_1)f_{s1})T_{s2} \right] \bmod 1,\ S_{x_a}^{(k)}((\nu_1 - p_1)f_{s1}) \neq 0 \Big\}. \tag{4.217}$$

In general, aliasing effects are present. The set $\mathbb{S}$ can contain clusters of curves or can be dense in the square $[-1/2, 1/2)^2$ if, for some $\nu_1 \in [-1/2, 1/2)$, f_{s2} is incommensurate with a countable infinity of values of $\Psi_{x_a}^{(k)}((\nu_1 - p_1)\, f_{s1})$, $k \in \mathbb{I}_a$, $p_1 \in \mathbb{Z}$, and in correspondence of these values the spectral densities $S_{x_a}^{(k)}((\nu_1 - p_1)f_{s1})$ are nonzero. In such a case, $S_{x_a}^{(k)}(f_1)$ is non identically zero for $|f_1| > B$ for some B.

In the special case of support lines $\Psi_{x_a}^{(k)}(f) = a^{(k)} f + b^{(k)}$, single sampling frequency $f_{s1} = f_{s2} = f_s$, and if frequencies ν_1 are evaluated on the bins of a N-point discrete Fourier transform (DFT), that is, $\nu_1 = n_1/N$, $n_1 = -N/2, \ldots, (N-1)/2$, ($N$ even), then

$$\Psi_{x_a}^{(k)}((\nu_1 - p_1)\, f_s)\Big|_{\nu_1 = n_1/N} = a^{(k)} \frac{n_1}{N} f_s - a^{(k)} p_1 f_s + b^{(k)} \qquad k \in \mathbb{I}_a,\ p_1 \in \mathbb{Z}. \tag{4.218}$$

Thus, if $\mathbb{I}_a$ is infinite and $a^{(k)}$ is irrational and/or $b^{(k)}/f_s$ is irrational for some k, then the restriction of $\mathbb{S}$ to $\nu_1 = n_1/N$ is dense (in ν_2) due to the presence of clusters of lines in $\mathbb{S}$, provided that the spectral density $S_{x_a}^{(k)}(f_1)$ is not identically zero for $|f_1| > B$, for some B. In (Lii and Rosenblatt 2002), only condition $a^{(k)}$ irrational is considered since the special case of $\mathbb{I}_a$ finite is treated, so that only the term $a^{(k)} p_1 f_s$ can make the set $\mathbb{S}$ dense when $a^{(k)}$ is irrational and $p_1 \in \mathbb{Z}$.

In (Lii and Rosenblatt 2002), it is explained that the aliasing problem (in the sense of spectral mass folding onto $[-1/2, 1/2]^2$) is a consequence of the relationship (see also (5.185))

$$\mathrm{d}\widetilde{\gamma}_x(\nu_1, \nu_2) = \frac{1}{T_{s1} T_{s2}} \sum_{p_1 \in \mathbb{Z}} \sum_{p_2 \in \mathbb{Z}} \mathrm{d}\gamma_{x_a}((\nu_1 - p_1)f_{s1}, (\nu_2 - p_2)f_{s2}) \tag{4.219}$$

This aliasing effect is illustrated in the numerical experiments in (Lii and Rosenblatt 2002), (Napolitano 2010b) and Section 4.9.3.

Theorem 4.9.5 Loève Bifrequency Cross-Spectrum of Sampled Strictly Band-Limited Jointly SC Processes (Napolitano 2011, Theorem 4.2). *Let $x_{a,i}(t)$, $i = 1, 2$, be continuous-time jointly SC processes with Loève bifrequency cross-spectrum (4.209a), (4.209b) strictly band-limited with bandwidths B_i and functions $\Psi_{x_a}^{(k)}(\cdot)$ invertible. If $f_{si} \geqslant 2B_i$, then the processes $x_i(n)$ defined in (4.212) are discrete-time jointly SC having Loève bifrequency cross-spectrum*

(4.190a) and (4.190b) with

$$\widetilde{S}_{\boldsymbol{x}}^{(k)}(\nu_1) \triangleq \frac{1}{T_{s1}} \sum_{p \in \mathbb{Z}} S_{\boldsymbol{x}_a}^{(k)}((\nu_1 - p) f_{s1}) \tag{4.220a}$$

$$\widetilde{G}_{\boldsymbol{x}}^{(k)}(\nu_2) \triangleq \frac{1}{T_{s2}} \sum_{p \in \mathbb{Z}} G_{\boldsymbol{x}_a}^{(k)}((\nu_2 - p) f_{s2}) \tag{4.220b}$$

$$\widetilde{\Psi}_{\boldsymbol{x}}^{(k)}(\nu_1) \triangleq T_{s2} \sum_{p \in \mathbb{Z}} \Psi_{\boldsymbol{x}_a}^{(k)}((\nu_1 - p) f_{s1}) \tag{4.221a}$$

$$\widetilde{\Phi}_{\boldsymbol{x}}^{(k)}(\nu_2) \triangleq T_{s1} \sum_{p \in \mathbb{Z}} \Phi_{\boldsymbol{x}_a}^{(k)}((\nu_2 - p) f_{s2}) \tag{4.221b}$$

and all periodic functions (4.220a)–(4.221b) are sums of nonoverlapping replicas. *In such a case, the bounded variation condition (4.214) is equivalent to the bounded variation condition for the continuous-time processes (4.18). Moreover, if $\Psi_{\boldsymbol{x}_a}^{(k)}(\cdot)$ and $\Phi_{\boldsymbol{x}_a}^{(k)}(\cdot)$ are also differentiable, then the assumptions for Theorem 4.8.3 are verified and (4.191a) and (4.191b) hold.*

Proof: See Section 5.10. □

The support function $\Psi_{\boldsymbol{x}_a}^{(k)}(f_1)$ and the corresponding spectral correlation density $S_{\boldsymbol{x}_a}^{(k)}(f_1)$ are said to be exactly identifiable starting from the samples (4.212) if and only if the equalities

$$\widetilde{S}_{\boldsymbol{x}}^{(k)}(\nu_1) = \frac{1}{T_{s1}} S_{\boldsymbol{x}_a}^{(k)}(f_1)|_{f_1 = \nu_1 f_{s1}} \tag{4.222}$$

$$\widetilde{\Psi}_{\boldsymbol{x}}^{(k)}(\nu_1) = T_{s2} \Psi_{\boldsymbol{x}_a}^{(k)}(f_1)|_{f_1 = \nu_1 f_{s1}} \tag{4.223}$$

hold $\forall \nu_1 \in [-1/2, 1/2]$. Under the assumption $f_{si} \geqslant 2B_i$ of Theorem 4.9.5, the relationships (4.222) and (4.223) hold only for $\nu_1 \in \text{suppc}\left\{S_{\boldsymbol{x}_a}^{(k)}(\nu_1 f_{s1})\right\}$ which, in general, is a proper subset of $[-1/2, 1/2]$, where suppc means "support curve of". Analogously,

$$\widetilde{G}_{\boldsymbol{x}}^{(k)}(\nu_2) = \frac{1}{T_{s2}} G_{\boldsymbol{x}_a}^{(k)}(f_2)|_{f_2 = \nu_2 f_{s2}} \tag{4.224}$$

$$\widetilde{\Phi}_{\boldsymbol{x}}^{(k)}(\nu_2) = T_{s1} \Phi_{\boldsymbol{x}_a}^{(k)}(f_2)|_{f_2 = \nu_2 f_{s2}} \tag{4.225}$$

hold only for $\nu_2 \in \text{suppc}\left\{G_{\boldsymbol{x}_a}^{(k)}(\nu_2 f_{s2})\right\}$ which, in general, is a proper subset of $[-1/2, 1/2]$. Thus, for (jointly) SC processes, (4.222), (4.223) defined by the mapping $\nu_1 = f_1/f_{s1}$ do not hold for every support curve $\forall \nu_1 \in [-1/2, 1/2]$ and (4.224), (4.225) defined by the mapping $\nu_2 = f_2/f_{s2}$ do not hold for every support curve $\forall \nu_2 \in [-1/2, 1/2]$. In contrast, for (jointly) WSS processes, spectral masses are present only on the main diagonal of the bifrequency plane and condition $f_{si} \geqslant 2B_i$ assures that (4.222)–(4.225) defined by the mappings $\nu_i = f_i/f_{si}$ hold for the unique support line and the corresponding density $\forall \nu_i \in [-1/2, 1/2]$, $i = 1, 2$.

By considering a more stringent condition on the sampling frequencies, (4.222)–(4.225) determined by the $\nu_i \leftrightarrow f_i$ mappings can be made valid for every support curve $\forall \nu_i \in [-1/2, 1/2]$, $i = 1, 2$. We have the following result.

Theorem 4.9.6 Density of Loève Bifrequency Cross-Spectrum of Sampled Strictly Band-Limited Jointly SC Processes (Napolitano 2011, Theorem 4.3). *Let $x_{a,i}(t)$, $i = 1, 2$, be continuous-time jointly SC processes with Loève bifrequency cross-spectrum (4.209a), (4.209b) and let the processes be strictly band limited with bandwidths B_i. If*

$$\begin{aligned} &f_{s1} \geqslant 2B_1 \quad \text{and} \\ &f_{s2} \geqslant B_2 + \max\left\{B_2, \sup_k \sup_{\nu_1 \in [-1/2, 1/2]} |\Psi^{(k)}_{x_a}(\nu_1 f_{s1})|\right\} \end{aligned} \tag{4.226}$$

then conditions (4.222) and (4.223) hold $\forall \nu_1 \in [-1/2, 1/2]$. Analogously, if

$$\begin{aligned} &f_{s2} \geqslant 2B_2 \quad \text{and} \\ &f_{s1} \geqslant B_1 + \max\left\{B_1, \sup_k \sup_{\nu_2 \in [-1/2, 1/2]} |\Phi^{(k)}_{x_a}(\nu_2 f_{s2})|\right\} \end{aligned} \tag{4.227}$$

then conditions (4.224) and (4.225) hold $\forall \nu_2 \in [-1/2, 1/2]$.

Proof: See Section 5.10. □

For $\Psi^{(k)}_{x_a}(\cdot)$ and $\Phi^{(k)}_{x_a}(\cdot)$ strictly increasing or decreasing functions, the sup over ν_1 and ν_2 in (4.226) and (4.227) are achieved for ν_i equal to $1/2$ or $-1/2$.

If the sufficient conditions (4.226) and (4.227) are satisfied then, accounting for (4.222)–(4.225), support curves and cross-spectral densities of the continuous-time signals can be reconstructed from those of the discrete-time signals by the frequency rescalings $\nu_1 = f_1/f_{s1}$ and $\nu_2 = f_2/f_{s2}$ as follows.

$$S^{(k)}_{x_a}(f_1) = \begin{cases} T_{s1}\, \widetilde{S}^{(k)}_{x}(f_1/f_{s1}) = T_{s1}\, \overset{\approx(1)}{S_x}(f_1/f_{s1}, \widetilde{\Psi}^{(k)}_{x}(f_1/f_{s1})), \\ \quad f_1 \in [-f_{s1}/2, f_{s1}/2] \\ 0, \qquad \text{otherwise} \end{cases} \tag{4.228}$$

$$\Psi^{(k)}_{x_a}(f_1) = \begin{cases} f_{s2}\, \widetilde{\Psi}^{(k)}_{x}(f_1/f_{s1}) \;\; f_1 \in [-f_{s1}/2, f_{s1}/2] \\ \quad \text{arbitrary}, \qquad \text{otherwise} \end{cases} \tag{4.229}$$

$$G^{(k)}_{x_a}(f_2) = \begin{cases} T_{s2}\, \widetilde{G}^{(k)}_{x}(f_2/f_{s2}) = T_{s2}\, \overset{\approx(2)}{S_x}(\widetilde{\Phi}^{(k)}_{x}(f_2/f_{s2}), f_2/f_{s2}), \\ \quad f_2 \in [-f_{s2}/2, f_{s2}/2] \\ 0, \qquad \text{otherwise} \end{cases} \tag{4.230}$$

$$\Phi^{(k)}_{x_a}(f_2) = \begin{cases} f_{s1}\, \widetilde{\Phi}^{(k)}_{x}(f_2/f_{s2}) \;\; f_2 \in [-f_{s2}/2, f_{s2}/2] \\ \quad \text{arbitrary}, \qquad \text{otherwise} \end{cases} \tag{4.231}$$

where the second equality in (4.228) and (4.230) holds in all points where support curves do not intersect and $\overset{\approx(1)}{S_x}(\nu_1, \nu_2)$ and $\overset{\approx(2)}{S_x}(\nu_1, \nu_2)$ are the bifrequency spectral cross-correlation densities of the Loève bifrequency cross-spectrum (4.190a), (4.190b)

$$\overset{\approx(1)}{S_x}(\nu_1, \nu_2) \triangleq \lim_{\Delta\nu \to 0+} \int_{\nu_2-\Delta\nu/2}^{\nu_2+\Delta\nu/2} \widetilde{\mathbb{S}}_x(\nu_1, \mu_2)\, d\mu_2$$
$$= \sum_{k \in \mathbb{I}} \widetilde{S}_x^{(k)}(\nu_1)\, \delta_{[\nu_2 - \widetilde{\Psi}_x^{(k)}(\nu_1)] \bmod 1} \tag{4.232a}$$

$$\overset{\approx(2)}{S_x}(\nu_1, \nu_2) \triangleq \lim_{\Delta\nu \to 0+} \int_{\nu_1-\Delta\nu/2}^{\nu_1+\Delta\nu/2} \widetilde{\mathbb{S}}_x(\mu_1, \nu_2)\, d\mu_1$$
$$= \sum_{k \in \mathbb{I}} \widetilde{G}_x^{(k)}(\nu_2)\, \delta_{[\nu_1 - \widetilde{\Phi}_x^{(k)}(\nu_2)] \bmod 1} \tag{4.232b}$$

defined analogously to their counterparts (4.20a) and (4.20b) for continuous-time processes.

Solutions of inequalities (4.226) and (4.227) can be easily found in the special case where $\Psi_{x_a}^{(k)}(\cdot)$ and $\Phi_{x_a}^{(k)}(\cdot)$ are linear functions with unit slope. Specifically, for a single ACS process $x_{a,i}(t) = x_a(t)$, $f_{si} = f_s$, $i = 1, 2$, and $(*)$ present, one has $\Psi_{x_a}^{(k)}(f_1) = f_1 - \alpha_k$. Thus, $\sup_k \sup_{\nu_1 \in [-1/2,1/2]} |\Psi_{x_a}^{(k)}(\nu_1 f_s)| = \sup_k |f_s/2 - \alpha_k| = f_s/2 + 2B$, and (4.226) becomes $f_s \geqslant B + f_s/2 + 2B$, that is, $f_s \geqslant 6B$. This condition, however, can be relaxed to $f_s \geqslant 4B$ (Section 1.3.9) by exploiting the fact that the support curves are parallel lines (with unit slope) (Gardner *et al.* 2006), (Izzo and Napolitano 1996) and by using bounds for supports of cyclic spectra.

As a concluding remark, let us consider GACS processes, which are the other class of processes that, as the class of the SC processes, extends that of the ACS processes. The discrete-time analysis of GACS processes differs from that of SC processes. In fact, in Section 2.5 it is shown that continuous-time GACS processes, unlike WSS, ACS, and SC processes, do not have a discrete-time counterpart and uniformly sampling a continuous-time GACS process leads to a discrete-time ACS process. In addition, in Sections 2.2.2 and 2.2.3 it is shown that a GACS process cannot be strictly band limited and is not harmonizable unless in the special case it reduces to an ACS process (Izzo and Napolitano 2002b; Napolitano 2007a).

4.9.3 *Illustrative Examples*

In this section, illustrative examples of the results of Section 4.9 are presented. These results which are expressed in terms of Loève bifrequency spectra, can be equivalently formulated in terms of spectral cross-correlation densities (4.20a), (4.20b), (4.232a), and (4.232b) by replacing Dirac with Kronecker deltas.

In the first experiment, a pulse-amplitude-modulated (PAM) signal $x_a(t)$ is considered with stationary white modulating sequence, raised cosine pulse with excess bandwidth η, and symbol period T_p. It is second-order cyclostationary with period T_p and strictly band limited with bandwidth $(1+\eta)/(2T_p)$. Thus, an upper bound for the bandwidth is $B = 1/T_p$ and the signal has three cycle frequencies $\alpha_h = h/T_p, h \in \{0, \pm 1\}$ (Gardner *et al.* 2006). A frequency-warped version of $x_a(t)$, namely

$$y_a(t) = \int_{\mathbb{R}} X_a(\psi(f))\, e^{j2\pi ft}\, \mathrm{d}f \tag{4.233}$$

is also considered, with frequency-warping function $\psi(f) = B_m \tan^{-1}(f/B_s)$. This kind of frequency warping is used in spectral analysis with nonuniform frequency spacing on the frequency axis (Oppenheim *et al.* 1971), (Oppenheim and Johnson 1972), (Braccini and Oppenheim 1974). Transformation (4.233) preserves in $y_a(t)$ the possible wide-sense stationarity of $x_a(t)$ (Franaszek 1967), (Franaszek and Liu 1967), (Liu and Franaszek 1969). In contrast, if $x_a(t)$ is ACS, then $y_a(t)$ is SC (Section 4.13). Moreover, the signals $y_a(t)$ and $x_a(t)$ turn out to be jointly SC with Loève bifrequency cross-spectrum having support curves (Section 4.13)

$$\Psi^{(k)}_{y_a x_a^*}(f_1) = \psi(f_1) - \alpha_k\,, \qquad k \in \{0, \pm 1\}. \tag{4.234}$$

Computing the bifrequency SCD (4.20a) between $y_a(t)$ and $x_a(t)$ leads to

$$\bar{S}^{(1)}_{y_a x_a^*}(f_1, f_2) = \sum_{k=-1}^{1} S^{\alpha_k}_{x_a x_a^*}\Big(\psi(f_1)\Big)\, \delta_{f_2 - \psi(f_1) + \alpha_k} \tag{4.235}$$

where $S^{\alpha_k}_{x_a x_a^*}(f_1)$ denote the cyclic spectra of the PAM signal $x_a(t)$.

In the figures, supports of impulsive functions are drawn as "checkerboard" plots with gray levels representing the magnitude of the density of the impulsive functions. In addition, for notation simplicity, the bifrequency SCD of signals a and b is denoted by $S_{ab^*}(\cdot,\cdot)$.

In Figure 4.6, (top) magnitude and (bottom) support of the bifrequency SCD of the continuous-time signal $x_a(t)$ are reported as functions of f_1/f_r and f_2/f_r, where f_r is a fixed reference frequency. In Figure 4.7, (top) magnitude and (bottom) support of the bifrequency SCD in (4.235) of the continuous-time signals $y_a(t)$ and $x_a(t)$ are reported as functions of f_1/f_r and f_2/f_r. In the experiment, $\eta = 0.85$, $T_p = 4/f_r$, $B_m = 0.21\ f_r$ and $B_s = 0.1\ f_r$.

A unique sampling frequency $f_s = 1/T_s = f_{s1} = f_{s2}$ is used for both signals. In Figures 4.8 and 4.9, the bifrequency SCD of the discrete-time signals $y(n) \triangleq y_a(t)|_{t=nT_s}$ and $x(n) = x_a(t)|_{t=nT_s}$, as a function of ν_1 and ν_2 is reported for two values of the sampling frequency f_s. Specifically, in Figure 4.8 (top) magnitude and (bottom) support of the bifrequency SCD for $f_s = 2B$ are reported and in Figure 4.9 (top) magnitude and (bottom) support of the bifrequency SCD for $f_s = 4B$ are reported. According to the results of Theorem 4.9.5, condition $f_s = 2B$ assures non-overlapping replicas (Figure 4.8). However, the SCD of the discrete-time signals along a support curve is not always a scaled version of the SCD of the continuous-time signals along the corresponding support curve $\forall \nu_1 \in [-1/2, 1/2]$. Specifically, in Figure 4.10, the slice of the magnitude of the SCD of the continuous-time signals $y_a(t)$ and $x_a(t)$ (thin solid line) along $f_2 = B_m \tan^{-1}(f_1/B_s) - \alpha_k \equiv \Psi^{(k)}_a(f_1)$ for $\alpha_k = 1/T_p$ and of the rescaled ($\nu_1 = f_1/f_s$) SCD of the discrete-time signals $y(n) \triangleq y_a(t)|_{t=nT_s}$ and $x(n) \triangleq x_a(t)|_{t=nT_s}$ (thick dashed line) along $\nu_2 = [B_m \tan^{-1}(\nu_1 f_s/B_s) - (1/T_p)]/f_s \equiv \Psi^{(k)}(\nu_1)$, as a function of f_1/f_r is reported for (top) $f_s = 2B$ and (bottom) $f_s = 4B$. In case $f_s = 2B$, the left thick-line replica R_1 is exactly overlapped to the thin-line curve but a further aliasing thick-line replica R_2 is present. The two replicas correspond to the supports with the same labels R_1 and R_2 in Figure 4.8 (bottom). In contrast, in case $f_s = 4B$ only one thick-line replica (labeled R_1 in Figure 4.9 (bottom))

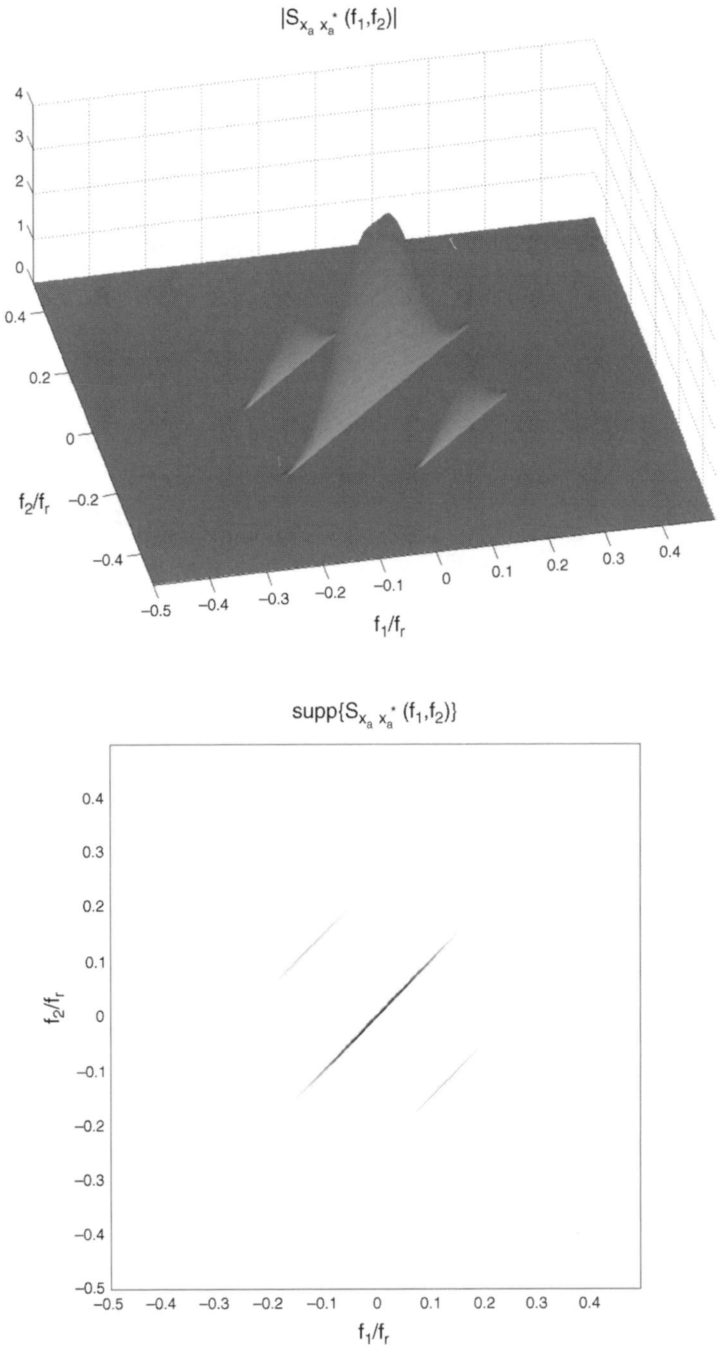

Figure 4.6 (Continuous-time) PAM signal $x_a(t)$ with raised cosine pulse. (Top) magnitude and (bottom) support of the bifrequency SCD as a function of f_1/f_r and f_2/f_r

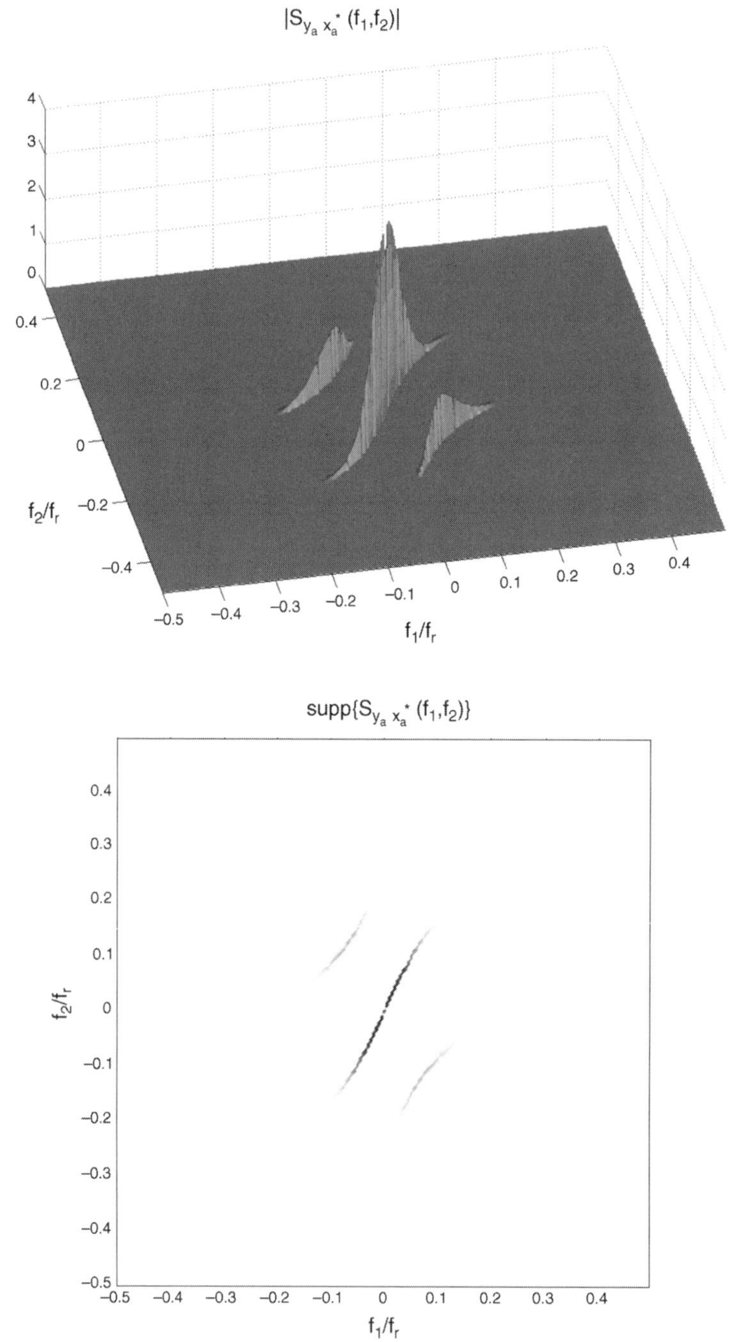

Figure 4.7 (Continuous-time) PAM signal $x_a(t)$ with raised cosine pulse and its frequency-warped version $y_a(t)$ with frequency-warping function $\psi(f) = B_m \tan^{-1}(f/B_s)$. (Top) magnitude and (bottom) support of the bifrequency SCD of $y_a(t)$ and $x_a(t)$ as a function of f_1/f_r and f_2/f_r. *Source:* (Napolitano 2011) © IEEE

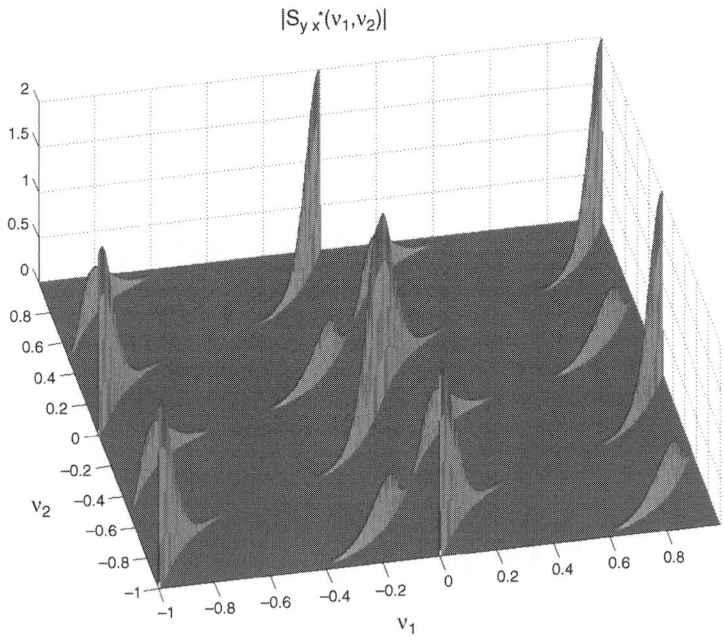

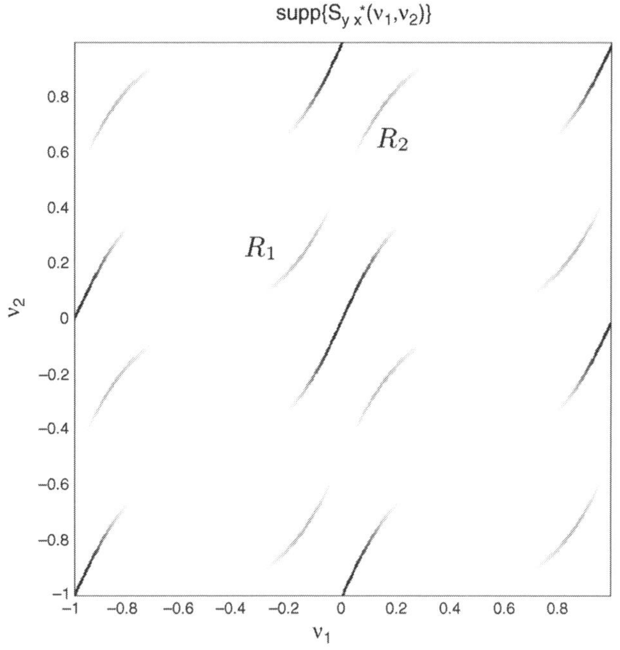

Figure 4.8 (Continuous-time) PAM signal $x_a(t)$ with raised cosine pulse and its frequency-warped version $y_a(t)$ with frequency-warping function $\psi(f) = B_m \tan^{-1}(f/B_s)$. Bifrequency SCD of the discrete-time signals $y(n) \triangleq y_a(t)|_{t=nT_s}$ and $x(n) = x_a(t)|_{t=nT_s}$, as a function of ν_1 and ν_2. (Top) magnitude and (bottom) support for $f_s = 2B$. *Source:* (Napolitano 2011) © IEEE

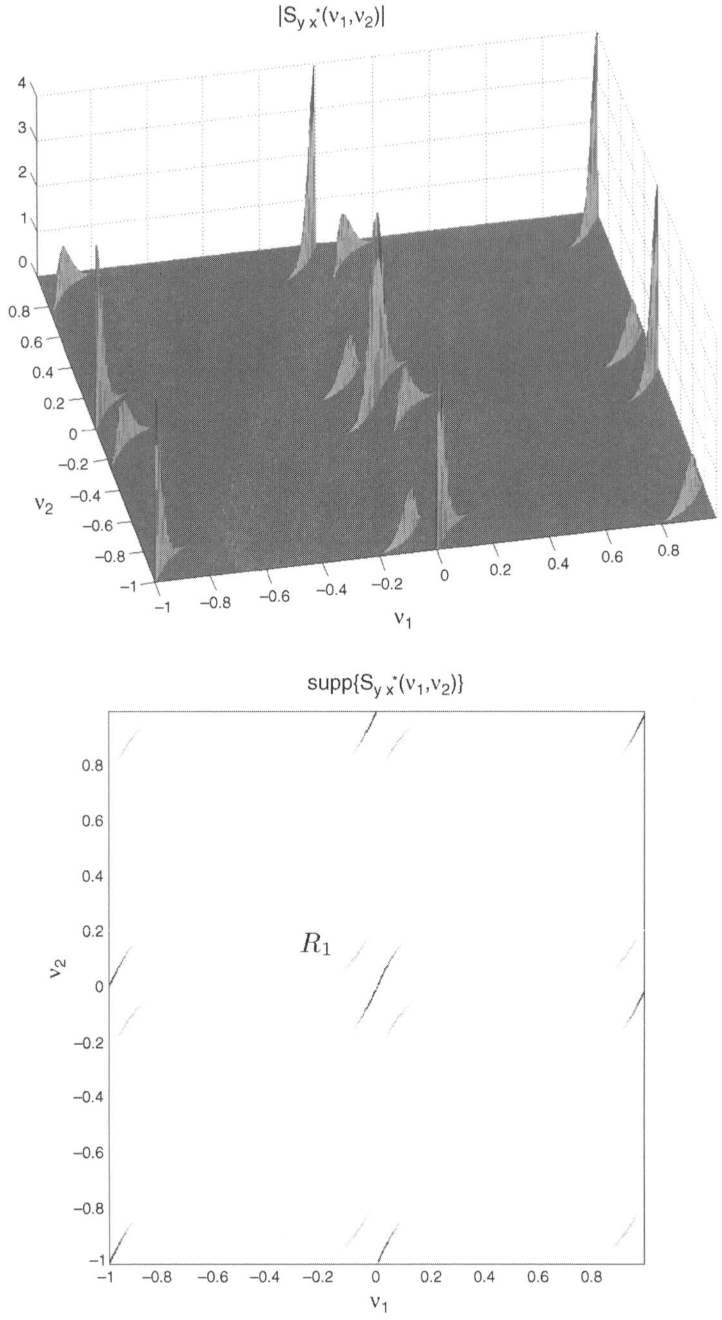

Figure 4.9 (Continuous-time) PAM signal $x_a(t)$ with raised cosine pulse and its frequency-warped version $y_a(t)$ with frequency-warping function $\psi(f) = B_m \tan^{-1}(f/B_s)$. Bifrequency SCD of the discrete-time signals $y(n) \triangleq y_a(t)|_{t=nT_s}$ and $x(n) = x_a(t)|_{t=nT_s}$, as a function of ν_1 and ν_2. (Top) magnitude and (bottom) support for $f_s = 4B$. *Source:* (Napolitano 2011) © IEEE

is present which exactly overlaps the thin-line curve. In fact, according to Theorem 4.9.6, condition $f_s = 4B = f_r$ is sufficient to prevent the presence of aliasing replicas so that (4.228) holds (with $f_{s1} = f_s$). Specifically, from (4.226) with $f_{s1} = f_{s2} = f_s$ and $B = 1/T_p$ we have

$$\begin{aligned}
&\left\{f_s \ : \ (4.226) \ \text{holds}\right\} \\
&= \left\{f_s \ : \ f_s \geqslant B + \max\left\{B, \max_k |B_m \ \tan^{-1}(f_s/2B_s) - \alpha_k|\right\}\right. \\
&= \left\{f_s \ : \ f_s \geqslant B + \max\left\{B, B_m \ \tan^{-1}(f_s/2B_s) + B\right\}\right. \\
&= \left\{f_s \ : \ f_s \geqslant 2B + B_m \ \tan^{-1}(f_s/2B_s)\right\} \\
&\supseteq \left\{f_s \ : \ f_s \geqslant 2B + B_m \pi/2\right\} \\
&= \left\{f_s \ : \ f_s \geqslant 0.83 \ f_r\right\}
\end{aligned} \tag{4.236}$$

where the last inclusion follows from inequality $0 \leqslant \tan^{-1}(x) < \pi/2$ for $x \geqslant 0$.

Note that for $f_s = 2B \simeq f_r/2$ (Figure 4.10 (top)) the mapping $\nu_1 = f_1/f_{s1}$ between the arguments of the SCDs of the continuous- and discrete-time signals holds only for $f_1 \in [-f_s/2, 0]$ and $\nu_1 \in [-1/2, 0]$ whereas for $f_s = 4B \simeq f_r$ (Figure 4.10 (bottom)) such mapping holds for $f_1 \in [-f_s/2, f_s/2]$ and $\nu_1 \in [-1/2, 1/2]$.

In the second experiment, a PAM signal $x_a(t)$ with full-duty-cycle rectangular pulse and baud-rate $1/T_p$ is considered. Such a signal is second-order cyclostationary with period T_p but is not strictly band limited. It has approximate bandwidth $B = 1/T_p$ and presents cycle frequencies $\alpha_h = h/T_p, h \in \mathbb{Z}$. When it passes through a multipath Doppler channel, the output signal $y_a(t)$ given in (4.56) is SC with Loève bifrequency spectrum given by (4.58) (see also Section 7.7.2). Since the number of cycle frequencies of the continuous-time signal is infinite and the cyclic spectra are not identically zero outside some frequency interval, the discrete-time signal is SC with infinite support curves in the principal spectral domain. However, in the experiment, a finite number N_c of support curves is considered by taking only those curves for which the SCD is significantly different from zero. This number N_c depends on the (finite) number N_r of replicas used to evaluate the aliased bifrequency SCD function. In the experiment, $N_r = 5$ and $N_c = 25$ are used and $T_p = 4/f_r$. The multipath Doppler channel is characterized by $K = 2$, $a_1 = 1$, $d_1 = 2.5T_s$, $\nu_1 = 0.05/Ts$, $s_1 = 0.85$, $a_2 = 0.7$, $d_2 = 6.8T_s$, $\nu_2 = -0.03/T_s$, and $s_2 = 1.25$.

In Figures 4.11 and 4.12, for the continuous-time input signal $x_a(t)$ and output signal $y_a(t)$, respectively, (top) magnitude and (bottom) support of the bifrequency SCD are reported as functions of f_1/f_r and f_2/f_r. In Figures 4.13 and 4.14, (top) magnitude and (bottom) support for $f_s = 2B$ and $f_s = 5B$, respectively, of the bifrequency SCD of the discrete-time signal $y(n) \triangleq y_a(t)|_{t=nT_s}$, as functions of ν_1 and ν_2 are reported. In Figure 4.15, the slice of the magnitude of the SCD of the continuous-time signal $y_a(t)$ along $f_2 = f_1/s_1 - (1/T_p) \equiv \Psi_a^{(k)}(f_1)$ (thin solid line) and of the rescaled ($\nu_1 = f_1/f_s$) SCD of the discrete-time signal $y(n) \triangleq y_a(t)|_{t=nT_s}$ along $\nu_2 = \nu_1/s_1 - 1/(T_p f_s) \equiv \Psi^{(k)}(\nu_1)$ (thick dashed line), as functions of f_1/f_r are reported for (a) $f_s = 2B \simeq 0.685 f_r$ and (b) $f_s = 5B \simeq 1.71 f_r$. In this experiment, the sufficient condition

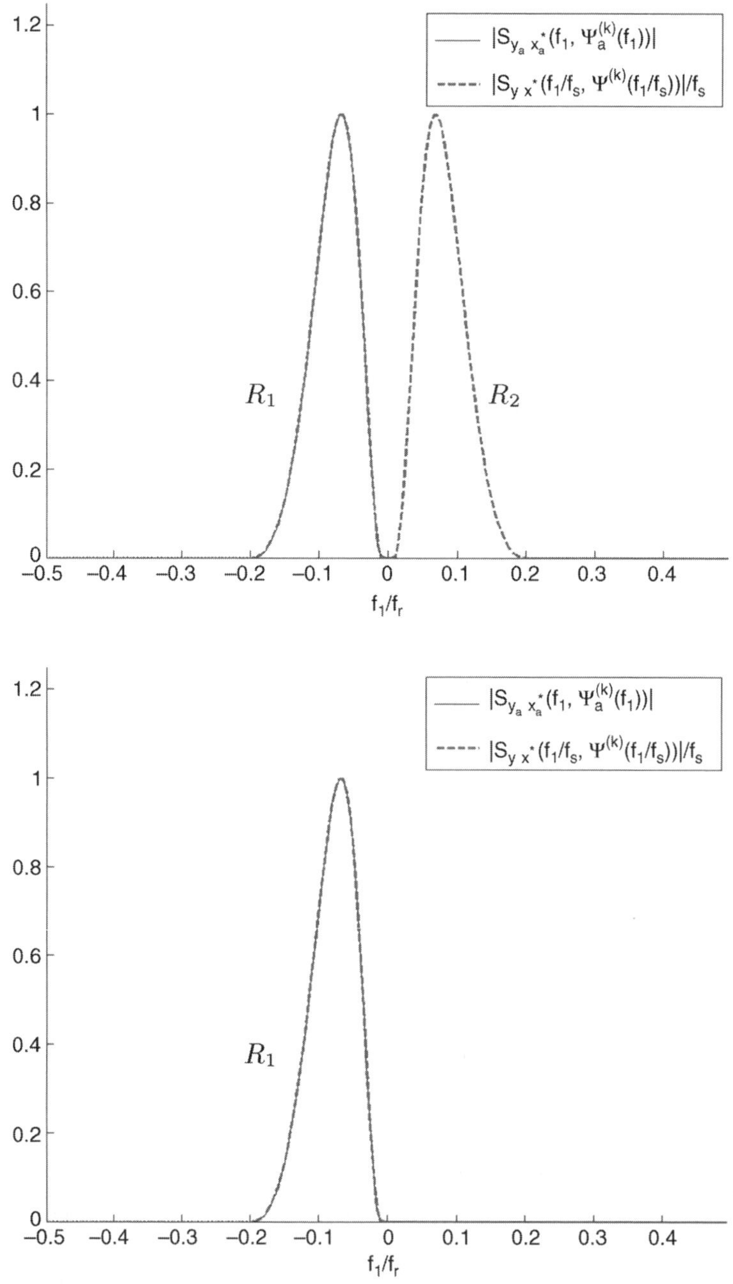

Figure 4.10 (Continuous-time) PAM signal $x_a(t)$ with raised cosine pulse and its frequency-warped version $y_a(t)$ with frequency-warping function $\psi(f) = B_m \tan^{-1}(f/B_s)$. Slice of the magnitude of the SCD of the continuous-time signals $y_a(t)$ and $x_a(t)$ along $f_2 = \psi(f_1) - (1/T_p) \equiv \Psi_a^{(k)}(f_1)$ (thin solid line) and of the rescaled ($\nu_1 = f_1/f_s$) SCD of the discrete-time signals $y(n) \triangleq y_a(t)|_{t=nT_s}$ and $x(n) \triangleq x_a(t)|_{t=nT_s}$ along $\nu_2 = [\psi(\nu_1 f_s) - 1/T_p]/f_s \equiv \Psi^{(k)}(\nu_1)$ (thick dashed line) as a function of f_1/f_r. (Top) $f_s = 2B$, $(f_s/2 \simeq f_r/4)$, and (bottom) $f_s = 4B$, $(f_s/2 \simeq f_r/2)$. *Source:* (Napolitano 2011) © IEEE

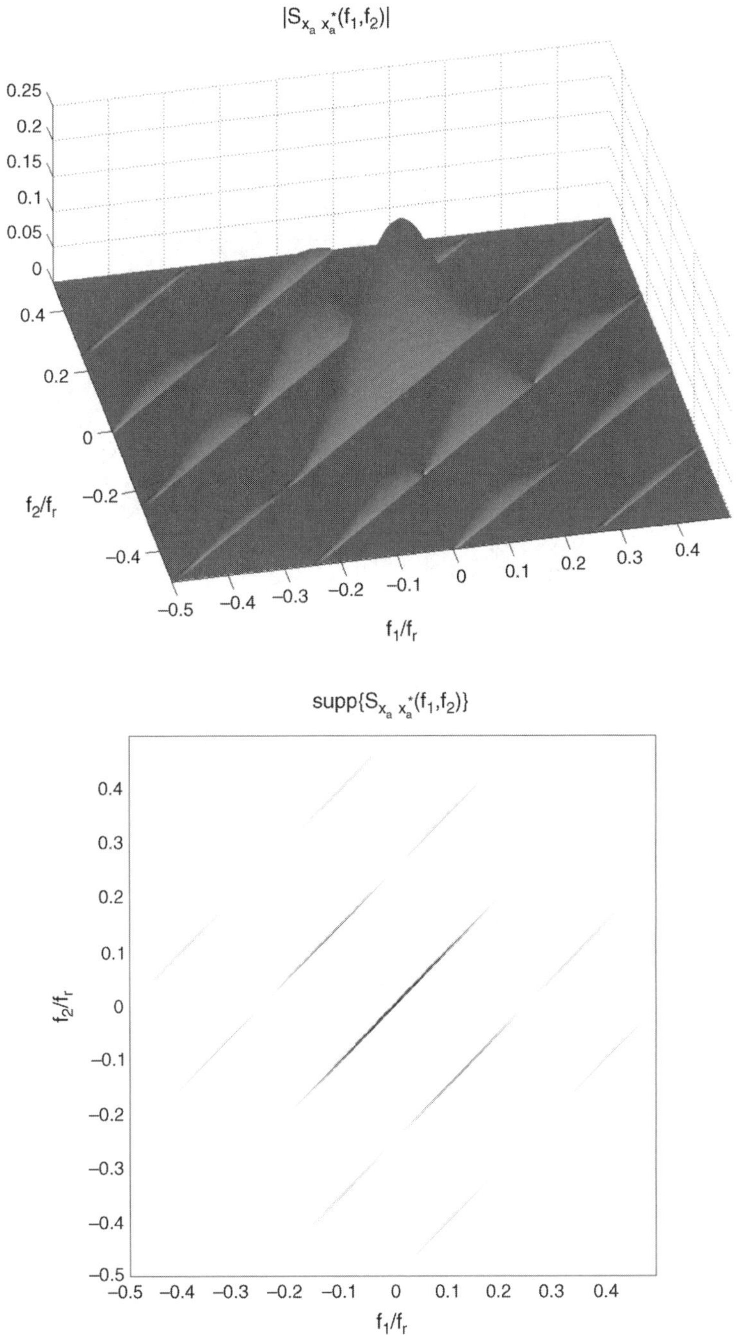

Figure 4.11 (Continuous-time) PAM signal $x_a(t)$ with rectangular pulse. (Top) magnitude and (bottom) support of the bifrequency SCD as a function of f_1/f_r and f_2/f_r

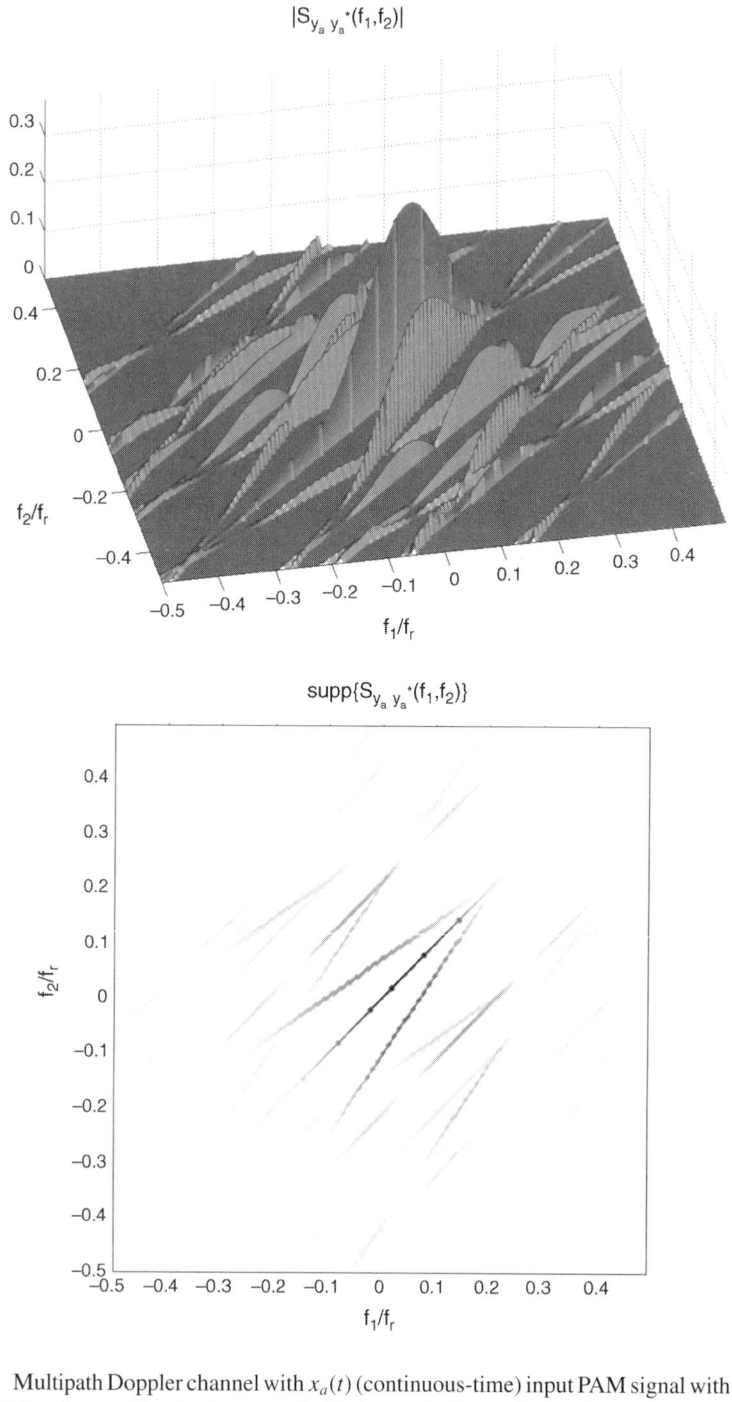

Figure 4.12 Multipath Doppler channel with $x_a(t)$ (continuous-time) input PAM signal with rectangular pulse and $y_a(t)$ (continuous-time) output signal. (Top) magnitude and (bottom) support of the bifrequency SCD of $y_a(t)$, as a function of f_1/f_r and f_2/f_r. *Source:* (Napolitano 2011) © IEEE

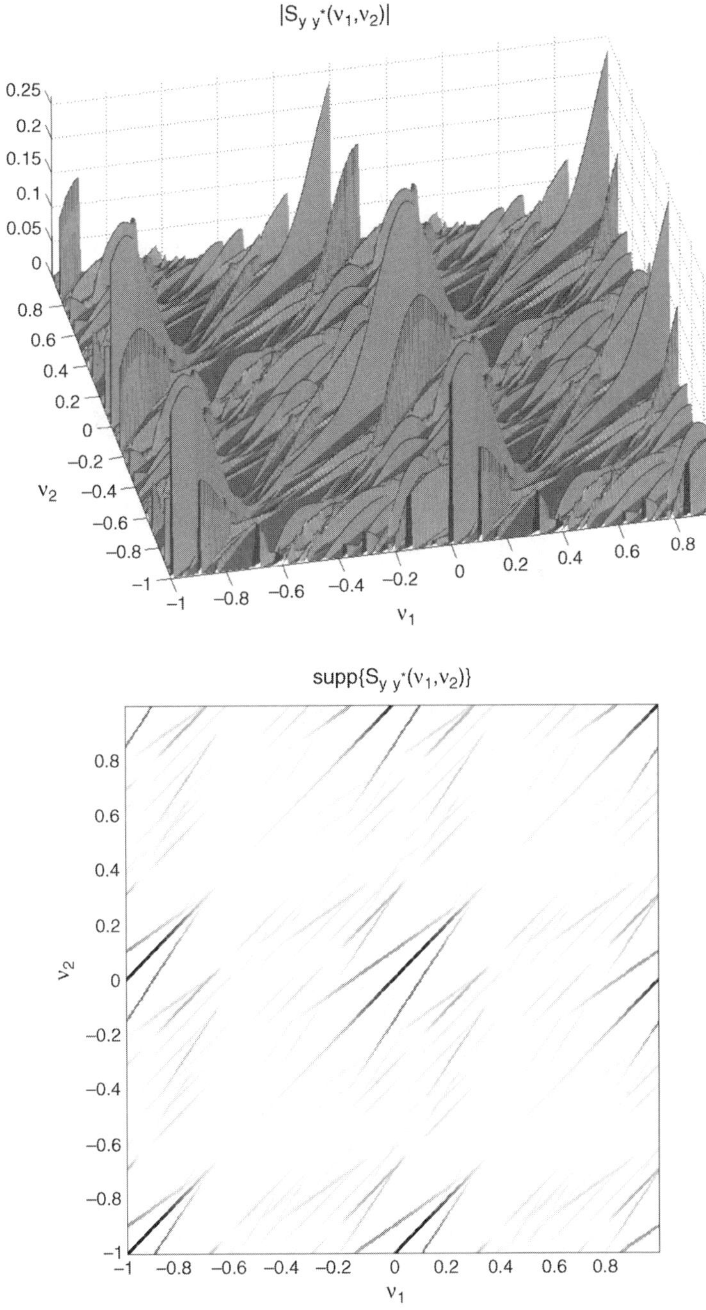

Figure 4.13 Multipath Doppler channel with $x_a(t)$ (continuous-time) input PAM signal with rectangular pulse and $y_a(t)$ (continuous-time) output signal. Bifrequency SCD of the discrete-time signal $y(n) \triangleq y_a(t)|_{t=nT_s}$, as a function of ν_1 and ν_2. (Top) magnitude and (bottom) support for $f_s = 2B$. *Source:* (Napolitano 2011) © IEEE

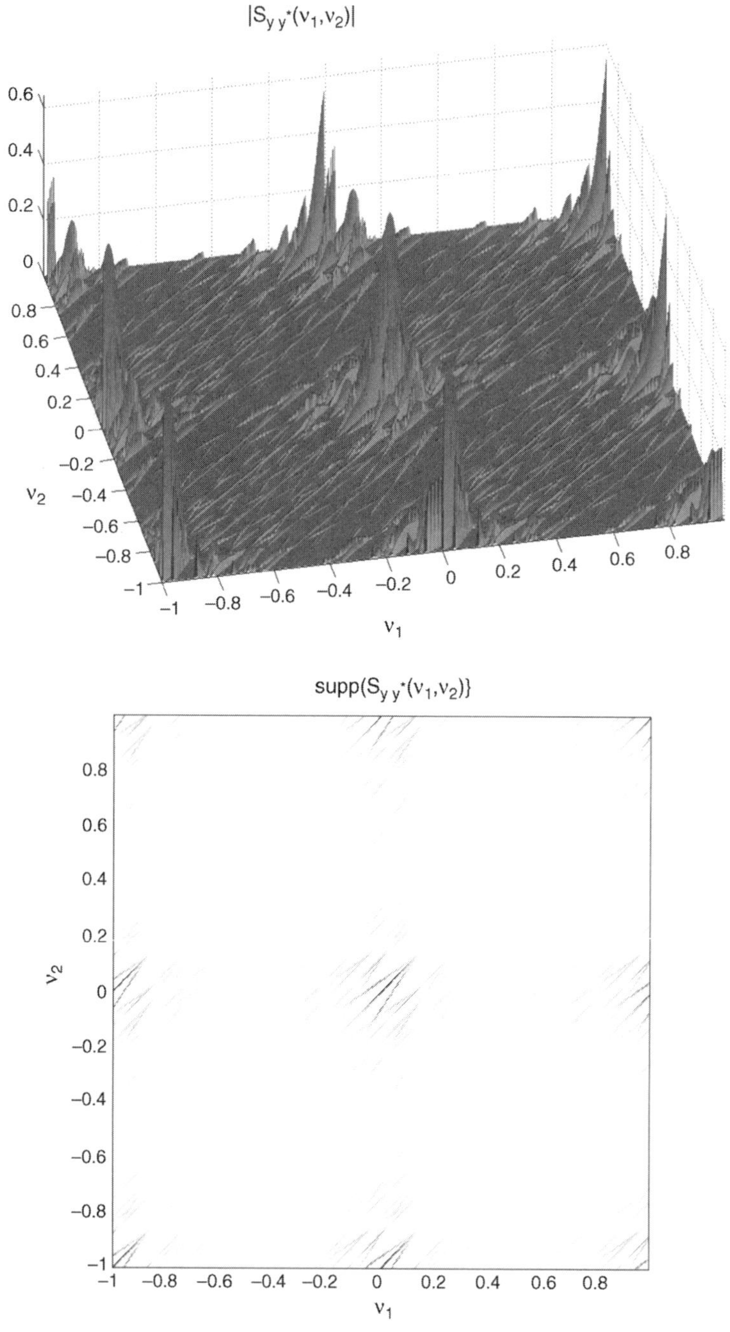

Figure 4.14 Multipath Doppler channel with $x_a(t)$ (continuous-time) input PAM signal with rectangular pulse and $y_a(t)$ (continuous-time) output signal. Bifrequency SCD of the discrete-time signal $y(n) \triangleq y_a(t)|_{t=nT_s}$, as a function of ν_1 and ν_2. (Top) magnitude and (bottom) support for $f_s = 5B$. *Source:* (Napolitano 2011) © IEEE

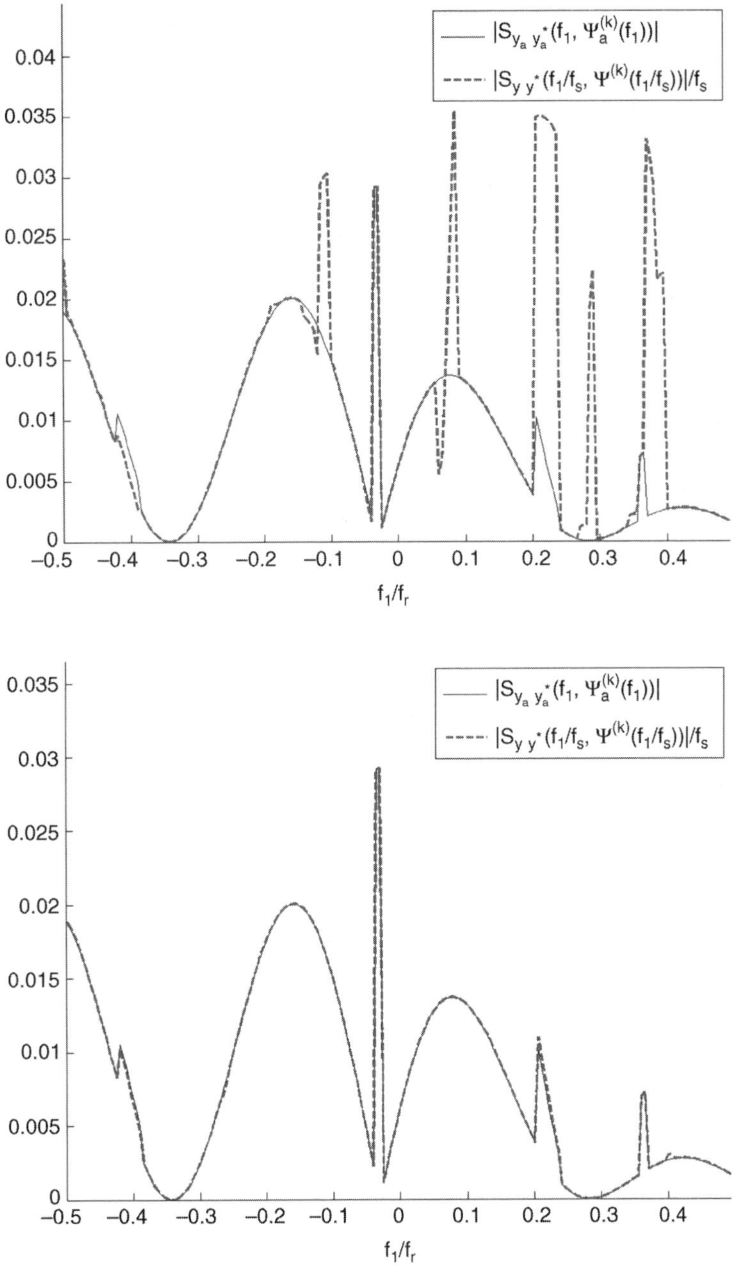

Figure 4.15 Multipath Doppler channel with $x_a(t)$ (continuous-time) input PAM signal with rectangular pulse and $y_a(t)$ (continuous-time) output signal. Slice of the magnitude of the SCD of the continuous-time signal $y_a(t)$ along $f_2 = f_1/s_1 - (1/T_p) \equiv \Psi_a^{(k)}(f_1)$ (thin solid line) and of the rescaled ($\nu_1 = f_1/f_s$) SCD of the discrete-time signal $y(n) \triangleq y_a(t)|_{t=nT_s}$ along $\nu_2 = \nu_1/s_1 - 1/(T_p f_s) \equiv \Psi^{(k)}(\nu_1)$ (thick dashed line) as functions of f_1/f_r. (Top) $f_s = 2B$, ($f_s/2 \simeq 0.34 f_r$), and (bottom) $f_s = 5B$, ($f_s/2 \simeq 0.85 f_r$). *Source:* (Napolitano 2011) © IEEE

$f_s = 2B$ of Theorem 4.9.5 does not assure a complete non overlap of replicas since the signal $x_a(t)$ is not strictly band limited. For $f_s = 5B$ replicas turn out to be sufficiently separate and also (4.228) is practically verified.

4.10 Multirate Processing of Discrete-Time Jointly SC Processes

In multirate digital signal processing, systems constituted by complicate interconnections of interpolators, decimators, and linear time-invariant (LTI) filters acts on input stochastic processes (Crochiere and Rabiner 1981), (Vaidyanathan 1990). Since interpolators and decimators are linear time-variant systems, the stationarity or nonstationarity properties of the input processes turn out to be modified at the output. For example, the expanded version of a WSS process is cyclostationary with period of cyclostationarity equal to the expansion factor (Sathe and Vaidyanathan 1993).

The effects of multirate systems on second-order WSS and cyclostationary processes are analyzed in (Sathe and Vaidyanathan 1993). In (Akkarakaran and Vaidyanathan 2000), conditions that preserve at the output of an interpolation filter the wide-sense stationarity of the input sequence are derived. Moreover, conditions for the joint wide-sense stationarity of the outputs of interpolation filters are also obtained. Higher-order statistics of ACS processes are considered in (Izzo and Napolitano 1998b).

In (Sathe and Vaidyanathan 1993), (Izzo and Napolitano 1998b), and (Akkarakaran and Vaidyanathan 2000), multirate operations on a single process are addressed and no cross-statistics of a process and its expanded or decimated version are considered. Moreover, no cross-statistics are considered between processes obtained by multirate elaborations of the same process with different rates.

Expansions with different rates transforms jointly WSS, jointly ACS, and jointly SC processes into jointly SC processes. An analogous result is obtained by decimation with different rates.

In this section, the effects of expansion and decimation operations on jointly SC processes are analyzed. Specifically, the way in which the Loève bifrequency cross-spectrum of input jointly SC processes transforms into the Loève bifrequency cross-spectrum of the output processes is derived for expansors and decimators which are the basic building blocks of multirate systems. The class of the jointly SC processes is shown to be closed under linear time variant transformations realized by interconnecting expansors and decimators. The case of input jointly ACS processes is analyzed in detail. It is proved that even if the input processes are jointly ACS, in general the output processes are jointly SC. Known results for ACS processes are obtained as special cases.

The problem of suppressing imaging and aliasing replicas in the Loève bifrequency cross-spectra of interpolated or decimated processes is addressed. Specifically, the reconstruction (but for two frequency scale factors) of an image-free Loève bifrequency cross-spectrum of two jointly SC processes when they passes through two interpolator filters is considered. Then, the reconstruction of a frequency-scaled version of the Loève bifrequency cross-spectrum of two jointly SC processes starting from the outputs of two decimation filters is addressed. Finally, the problem of reconstructing a frequency-scaled image- and alias-free version of the Loève bifrequency cross-spectrum of two jointly SC processes starting from their fractionally sampled versions is treated.

4.10.1 Expansion

4.10.1.1 LTV System Characterization

The L-fold expanded version of the discrete-time signal $x(n)$ is defined as

$$x_I(n) = x\left[\frac{n}{L}\right] \triangleq \begin{cases} x\left(\dfrac{n}{L}\right) & n = kL, k \in \mathbb{Z} \\ 0 & \text{otherwise} \end{cases} \tag{4.237}$$

Thus, it results that

$$x(n) = x_I(nL) \tag{4.238}$$

from which it follows that the Fourier transform $X_I(\nu)$ of $x_I(n)$ is given by

$$X_I(\nu) = X(\nu L). \tag{4.239}$$

In Section 5.11 it is shown that expansion is a linear time-variant transformation (Figure 4.16)

$$x_I(n) = \sum_{\ell=-\infty}^{+\infty} h_I(n, \ell)\, x(\ell) \tag{4.240}$$

with impulse-response function

$$h_I(n, \ell) = \delta_{\ell - n/L} \tag{4.241}$$

and transmission function

$$H_I(\nu, \lambda) \triangleq \sum_{n\in\mathbb{Z}} \sum_{\ell\in\mathbb{Z}} h_I(n, \ell)\, e^{-j2\pi(\nu n - \lambda\ell)} \tag{4.242a}$$

$$= \widetilde{\delta}_\lambda(\lambda - \nu L) \tag{4.242b}$$

$$= \frac{1}{L} \sum_{p=0}^{L-1} \widetilde{\delta}_\nu\left(\nu - \frac{\lambda - p}{L}\right) \tag{4.242c}$$

where

$$\widetilde{\delta}_\lambda(\lambda) \triangleq \sum_{k\in\mathbb{Z}} \delta(\lambda - k) \qquad \widetilde{\delta}_\nu(\nu) \triangleq \sum_{k\in\mathbb{Z}} \delta(\nu - k) \tag{4.243}$$

and the subscript in notation $\widetilde{\delta}_\lambda(\lambda)$ and $\widetilde{\delta}_\nu(\nu)$ indicates the variable with respect to which periodization is made and needs to avoid ambiguities when a function of two variables λ and ν is considered.

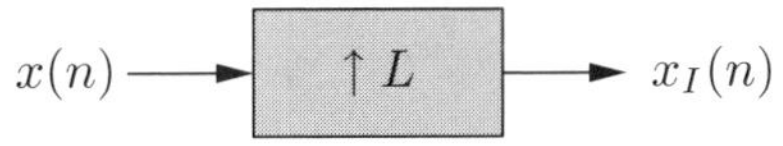

Figure 4.16 L-fold expansion

The function $H_I(\nu, \lambda)$ is periodic in ν with period $1/L$ and periodic in λ with period 1. Systems performing L-fold expansion are a special case of discrete-time FOT-deterministic LTV systems (Section 4.3.1) not LAPTV.

4.10.1.2 Input/Output Statistics

Let $x_{Ii}(n)$ be the L_i-fold expanded version of $x_i(n)$, $(i = 1, 2)$. That is

$$x_{Ii}(n) = x_i\left[\frac{n}{L_i}\right] \triangleq \begin{cases} x_i\left(\dfrac{n}{L_i}\right) & n = kL_i, k \in \mathbb{Z} \\ 0 & \text{otherwise} \end{cases} \tag{4.244}$$

whose Fourier transform is

$$X_{Ii}(\nu) = X_i(\nu L_i). \tag{4.245}$$

If $x_1(n)$ and $x_2(n)$ are jointly SC with Loève bifrequency cross-spectrum (4.190a), the Loève bifrequency cross-spectrum of the expanded processes $x_{I1}(n)$ and $x_{I2}(n)$ (Figure 4.17) is given by

$$\begin{aligned}
&\mathrm{E}\left\{X_{I1}(\nu_1)\, X_{I2}^{(*)}(\nu_2)\right\} \\
&\quad = \mathrm{E}\left\{X_1(\nu_1 L_1)\, X_2^{(*)}(\nu_2 L_2)\right\} \\
&\quad = \sum_{k\in\mathbb{I}} \widetilde{S}_{\boldsymbol{x}}^{(k)}(\nu_1 L_1)\, \widetilde{\delta}\left(\nu_2 L_2 - \widetilde{\Psi}_{\boldsymbol{x}}^{(k)}(\nu_1 L_1)\right) && (4.246\text{a}) \\
&\quad = \sum_{k\in\mathbb{I}} \widetilde{S}_{\boldsymbol{x}}^{(k)}(\nu_1 L_1)\, \frac{1}{L_2} \sum_{h\in\mathbb{Z}} \delta\left(\nu_2 - \frac{h}{L_2} - \frac{1}{L_2}\widetilde{\Psi}_{\boldsymbol{x}}^{(k)}(\nu_1 L_1)\right) && (4.246\text{b}) \\
&\quad = \sum_{k\in\mathbb{I}} \widetilde{S}_{\boldsymbol{x}}^{(k)}(\nu_1 L_1)\, \frac{1}{L_2} \sum_{p=0}^{L_2-1} \widetilde{\delta}_{\nu_2}\left(\nu_2 - \frac{p}{L_2} - \frac{1}{L_2}\widetilde{\Psi}_{\boldsymbol{x}}^{(k)}(\nu_1 L_1)\right) && (4.246\text{c})
\end{aligned}$$

where, in the third equality, the scaling property of the Dirac delta is used (Zemanian 1987, Section 1.7) and in the fourth equality (5.191) is accounted for. Equivalently, accounting for

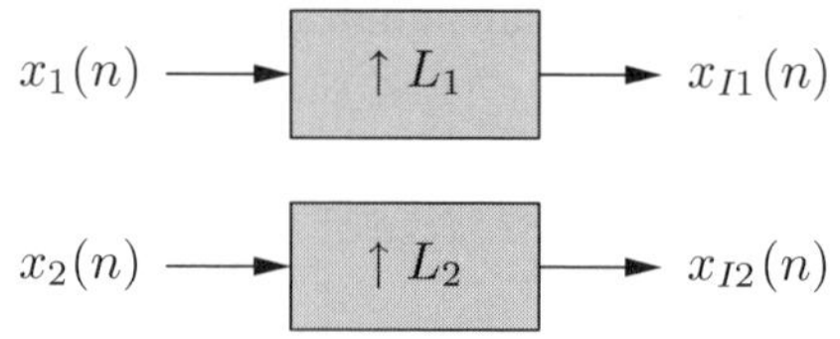

Figure 4.17 Expansion of two processes with different expansion factors

(4.190b), we also have

$$\mathrm{E}\left\{X_{I1}(\nu_1)\, X_{I2}^{(*)}(\nu_2)\right\}$$

$$= \sum_{k\in\mathbb{I}} \widetilde{G}_x^{(k)}(\nu_2 L_2)\, \widetilde{\delta}\Big(\nu_1 L_1 - \widetilde{\Phi}_x^{(k)}(\nu_2 L_2)\Big) \tag{4.247a}$$

$$= \sum_{k\in\mathbb{I}} \widetilde{G}_x^{(k)}(\nu_2 L_2)\, \frac{1}{L_1} \sum_{h\in\mathbb{Z}} \delta\Big(\nu_1 - \frac{h}{L_1} - \frac{1}{L_1}\widetilde{\Phi}_x^{(k)}(\nu_2 L_2)\Big) \tag{4.247b}$$

$$= \sum_{k\in\mathbb{I}} \widetilde{G}_x^{(k)}(\nu_2 L_2)\, \frac{1}{L_1} \sum_{p=0}^{L_1-1} \widetilde{\delta}_{\nu_1}\Big(\nu_1 - \frac{p}{L_1} - \frac{1}{L_1}\widetilde{\Phi}_x^{(k)}(\nu_2 L_2)\Big). \tag{4.247c}$$

From (4.246a)–(4.246c) and (4.247a)–(4.247c) it follows that the expanded processes are jointly SC. In addition, their Loève bifrequency cross-spectrum is periodic in ν_1 with period $1/L_1$ and in ν_2 with period $1/L_2$.

In the special case of $x_1(n)$ and $x_2(n)$ jointly ACS, accounting for (4.202), the Loève bifrequency cross-spectrum of the expanded processes $x_{I1}(n)$ and $x_{I2}(n)$ can be expressed as

$$\mathrm{E}\left\{X_{I1}(\nu_1)\, X_{I2}^{(*)}(\nu_2)\right\}$$

$$= \sum_{\widetilde{\alpha}\in\widetilde{A}} \widetilde{\widetilde{S}}_x^{\alpha}(\nu_1 L_1)\, \widetilde{\delta}(\nu_2 L_2 - {}_{(-)}(\widetilde{\alpha} - \nu_1 L_1)) \tag{4.248a}$$

$$= \sum_{\widetilde{\alpha}\in\widetilde{A}} \widetilde{\widetilde{S}}_x^{\alpha}(\nu_1 L_1)\, \frac{1}{L_2} \sum_{h\in\mathbb{Z}} \delta\Big(\nu_2 - \frac{h}{L_2} - {}_{(-)}\Big(\frac{\widetilde{\alpha}}{L_2} - \frac{L_1}{L_2}\nu_1\Big)\Big) \tag{4.248b}$$

$$= \sum_{\widetilde{\alpha}\in\widetilde{A}} \widetilde{\widetilde{S}}_x^{\alpha}(\nu_1 L_1)\, \frac{1}{L_2} \sum_{p=0}^{L_2-1} \widetilde{\delta}_{\nu_2}\Big(\nu_2 - \frac{p}{L_2} - {}_{(-)}\Big(\frac{\widetilde{\alpha}}{L_2} - \frac{L_1}{L_2}\nu_1\Big)\Big). \tag{4.248c}$$

In (4.248a)–(4.248c), the cyclic spectra $\widetilde{\widetilde{S}}_x^{\alpha}(\nu_1 L_1)$ are periodic in ν_1 with period $1/L_1$ and the delta train is periodic in ν_1 with period $1/L_1$ and is periodic in ν_2 with period $1/L_2$. Therefore, $\mathrm{E}\{X_{I1}(\nu_1)\, X_{I2}^{(*)}(\nu_2)\}$ is periodic in ν_1 with period $1/L_1$ and in ν_2 with period $1/L_2$. In addition, from (4.248c) it follows that the Loève bifrequency cross-spectrum of the expanded processes $x_{I1}(n)$ and $x_{I2}(n)$ in the principal domain $(\nu_1, \nu_2) \in [-1/2, 1/2)^2$ has support curves (for the case ${}_{(-)}$ present)

$$\nu_2 = (L_1/L_2)\nu_1 - \widetilde{\alpha}/L_2 + p/L_2 \bmod 1 \qquad \widetilde{\alpha} \in \widetilde{A},\ p \in \{0, 1, \ldots, L_2 - 1\} \tag{4.249}$$

that is, lines with slope L_1/L_2. Consequently, $x_{I1}(n)$ and $x_{I2}(n)$ are jointly SC. In particular, for a single process $x_1(n) \equiv x_2(n) \equiv x(n)$, by taking $L_1 = 1$, that is $x_{I1}(n) \equiv x(n)$, it follows that the ACS process $x(n)$ and its expanded version $x_{I2}(n)$ are jointly SC with support lines with slope $1/L_2$. In the special case of $x(n)$ WSS, $x(n)$ and $x_{I2}(n)$ are jointly SC with support constituted by a unique line.

In Section 5.11 it is shown that when $x_1(n)$ and $x_2(n)$ are jointly ACS the cross-correlation function of $x_{I1}(n)$ and $x_{I2}(n)$ is given by

$$\begin{aligned}
&\mathrm{E}\left\{x_{I1}(n+m)\,x_{I2}^{(*)}(n)\right\} \\
&= \int_{[-1/2,1/2]^2} \mathrm{E}\left\{X_{I1}(\nu_1)\,X_{I2}^{(*)}(\nu_2)\right\} e^{j2\pi[\nu_1(n+m)+(-)\nu_2 n]}\,\mathrm{d}\nu_1\,\mathrm{d}\nu_2 \\
&= \sum_{\widetilde{\alpha}\in\widetilde{\mathcal{A}}} e^{j2\pi(\widetilde{\alpha}/L_2)n}\,\widetilde{\widetilde{R}}_{\boldsymbol{x}}^{\alpha}\left[\frac{(1-L_1/L_2)n+m}{L_1}\right]\delta_{n \bmod L_2}
\end{aligned} \tag{4.250}$$

where the Kronecker delta $\delta_{n \bmod L_2}$ is defined in (5.190) and $\widetilde{\widetilde{R}}_{\boldsymbol{x}}^{\alpha}[\ell/L_2]$ is defined according to (4.244), that is,

$$\widetilde{\widetilde{R}}_{\boldsymbol{x}}^{\alpha}\left[\frac{\ell}{L}\right] \triangleq \begin{cases} \widetilde{\widetilde{R}}_{\boldsymbol{x}}^{\alpha}(\ell/L) & \ell = kL, k\in\mathbb{Z} \\ 0 & \text{otherwise} \end{cases} \tag{4.251}$$

When $L_1 \neq L_2$, the argument of the cyclic cross-correlation function $\widetilde{\widetilde{R}}_{\boldsymbol{x}}^{\alpha}(\cdot)$ contains the time variable n. Consequently, under the mild assumption that $\widetilde{\widetilde{R}}_{\boldsymbol{x}}^{\alpha}(\cdot) \in \ell_1(\mathbb{Z})$, it follows that, for every fixed m, the cross-correlation function in (4.250) as a function of n does not contain any finite-strength additive sinewave component. That is, $x_{I1}(n)$ and $x_{I2}^{(*)}(n)$ do not exhibit joint almost-cyclostationarity at any cycle frequency:

$$\widetilde{\widetilde{R}}^{\alpha}_{x_{I1}x_{I2}^{(*)}}(m) \triangleq \lim_{N\to\infty}\frac{1}{2N+1}\sum_{n=-N}^{N}\mathrm{E}\left\{x_{I1}(n+m)\,x_{I2}^{(*)}(n)\right\} e^{-j2\pi\widetilde{\alpha}n} = 0 \qquad \forall\widetilde{\alpha}\in\mathbb{R}. \tag{4.252}$$

In particular, their time-averaged cross-correlation $\widetilde{R}^{0}_{x_{I1}x_{I2}^{(*)}}(m)$ is identically zero. Moreover, from (4.250) it follows that $x_{I1}(n)$ and $x_{I2}(n)$ are jointly ACS if and only if $L_1 = L_2 = L$. In such a case, their cyclic cross-correlation functions are (Section 5.11)

$$\widetilde{\widetilde{R}}^{\alpha}_{x_{I1}x_{I2}^{(*)}}(m) = \frac{1}{L}\widetilde{\widetilde{R}}_{\boldsymbol{x}}^{\alpha L}\left[\frac{m}{L}\right] \qquad \widetilde{\alpha}L \bmod 1 \in \widetilde{\mathcal{A}}. \tag{4.253}$$

Furthermore, since from (4.253) it follows

$$\widetilde{\widetilde{R}}_{\boldsymbol{x}}^{\alpha}(m) = L\,\widetilde{\widetilde{R}}^{\alpha/L}_{x_{I1}x_{I2}^{(*)}}(mL) \tag{4.254}$$

accordingly with the Fourier-transform pair (4.244) and (4.245), the cyclic cross-spectra of $x_{I1}(n)$ and $x_{I2}(n)$ are

$$\widetilde{\widetilde{S}}^{\alpha}_{x_{I1}x_{I2}^{(*)}}(\nu) = \frac{1}{L}\widetilde{\widetilde{S}}_{\boldsymbol{x}}^{\alpha L}(\nu L) \qquad \widetilde{\alpha}L \bmod 1 \in \widetilde{\mathcal{A}}. \tag{4.255}$$

In the special case of $x_1 \equiv x_2$, (4.253) and (4.255) reduce to (Izzo and Napolitano 1998b, eqs. (32) and (34)) specialized to second order.

4.10.2 Sampling

Let $x_{\delta i}(n)$ be the sampled version of $x_i(n)$, with sampling period M_i $(i = 1, 2)$. That is

$$x_{\delta i}(n) \triangleq x_i(n) \sum_{\ell=-\infty}^{+\infty} \delta_{n-\ell M_i} = \sum_{\ell=-\infty}^{+\infty} x_i(\ell M_i)\, \delta_{n-\ell M_i}. \tag{4.256}$$

Accounting for (5.190), the Fourier transforms of the sampled processes are

$$X_{\delta i}(\nu) = X_i(\nu) \otimes \frac{1}{M_i} \sum_{p=0}^{M_i-1} \widetilde{\delta}\left(\nu - \frac{p}{M_i}\right) = \frac{1}{M_i} \sum_{p=0}^{M_i-1} X_i\left(\nu - \frac{p}{M_i}\right) \tag{4.257}$$

where $\otimes$ denotes periodic convolution with period 1. Consequently, if $x_1(n)$ and $x_2(n)$ are jointly SC with with Loève bifrequency cross-spectrum (4.190a), then the Loève bifrequency cross-spectrum of $x_{\delta 1}(n)$ and $x_{\delta 2}(n)$ is

$$\begin{aligned} \mathrm{E}\left\{X_{\delta 1}(\nu_1)\, X^{(*)}_{\delta 2}(\nu_2)\right\} &= \frac{1}{M_1 M_2} \sum_{p_1=0}^{M_1-1} \sum_{p_2=0}^{M_2-1} \mathrm{E}\left\{X_1\left(\nu_1 - \frac{p_1}{M_1}\right) X_2\left(\nu_2 - \frac{p_2}{M_2}\right)\right\} \\ &= \frac{1}{M_1 M_2} \sum_{p_1=0}^{M_1-1} \sum_{p_2=0}^{M_2-1} \sum_{k\in\mathbb{I}} \widetilde{S}^{(k)}_{x}\left(\nu_1 - \frac{p_1}{M_1}\right) \\ &\quad \widetilde{\delta}_{\nu_2}\left(\nu_2 - \frac{p_2}{M_2} - \widetilde{\Psi}^{(k)}_{x}\left(\nu_1 - \frac{p_1}{M_1}\right)\right). \end{aligned} \tag{4.258}$$

Equivalently, accounting for (4.190b) we also have

$$\begin{aligned} \mathrm{E}\left\{X_{\delta 1}(\nu_1)\, X^{(*)}_{\delta 2}(\nu_2)\right\} &= \frac{1}{M_1 M_2} \sum_{p_1=0}^{M_1-1} \sum_{p_2=0}^{M_2-1} \sum_{k\in\mathbb{I}} \widetilde{G}^{(k)}_{x}\left(\nu_2 - \frac{p_2}{M_2}\right) \\ &\quad \widetilde{\delta}_{\nu_1}\left(\nu_1 - \frac{p_1}{M_1} - \widetilde{\Phi}^{(k)}_{x}\left(\nu_2 - \frac{p_2}{M_2}\right)\right). \end{aligned} \tag{4.259}$$

Thus, the sampled processes $x_{\delta 1}(n)$ and $x_{\delta 2}(n)$ are jointly SC. In addition, the Loève bifrequency cross-spectrum (4.258) (or (4.259)) is periodic in ν_1 with period $1/M_1$ and in ν_2 with period $1/M_2$.

In the special case where $x_1(n)$ and $x_2(n)$ are jointly ACS, accounting for (4.202), it results in

$$\begin{aligned} \mathrm{E}\left\{X_{\delta 1}(\nu_1)\, X^{(*)}_{\delta 2}(\nu_2)\right\} &= \frac{1}{M_1 M_2} \sum_{p_1=0}^{M_1-1} \sum_{p_2=0}^{M_2-1} \sum_{\widetilde{\alpha}\in\widetilde{A}} \widetilde{S}^{\widetilde{\alpha}}_{x}\left(\nu_1 - \frac{p_1}{M_1}\right) \\ &\quad \widetilde{\delta}\left(\nu_2 - \frac{p_2}{M_2} - _{(-)}\left(\widetilde{\alpha} - \nu_1 + \frac{p_1}{M_1}\right)\right). \end{aligned} \tag{4.260}$$

That is, the sampled processes are in turn jointly ACS for every value of M_1 and M_2. Such a result is not surprising since sampling is a linear periodically time-variant transformation (Napolitano 1995). In addition, if $M_1 = M_2 = M$, in Section 5.11 it is shown that the cyclic cross-correlation function of $x_{\delta 1}(n)$ and $x_{\delta 2}(n)$ is given by

$$\widetilde{\widetilde{R}}^{\alpha}_{x_{\delta 1} x^{(*)}_{\delta 2}}(m) = \frac{1}{M} \sum_{q=0}^{M-1} \widetilde{\widetilde{R}}^{\alpha - q/M}_{\boldsymbol{x}}(m)\, \delta_{m \bmod M}. \tag{4.261}$$

By Fourier transforming both sides of (4.261) and using (5.190), we have the following expression for the cyclic cross-spectrum

$$\begin{aligned} \widetilde{\widetilde{S}}^{\alpha}_{x_{\delta 1} x^{(*)}_{\delta 2}}(\nu) &= \frac{1}{M} \sum_{q=0}^{M-1} \widetilde{\widetilde{S}}^{\alpha - q/M}_{\boldsymbol{x}}(\nu) \otimes \frac{1}{M} \sum_{p=0}^{M-1} \widetilde{\delta}\left(\nu - \frac{p}{M}\right) \\ &= \frac{1}{M^2} \sum_{p=0}^{M-1} \sum_{q=0}^{M-1} \widetilde{\widetilde{S}}^{\alpha - q/M}_{\boldsymbol{x}}\left(\nu - \frac{p}{M}\right). \end{aligned} \tag{4.262}$$

where $\otimes$ denotes periodic convolution with period 1, and the fact that $\widetilde{\delta}(\nu)$ is the unit element for the periodic convolution with period 1 is used.

4.10.3 Decimation

4.10.3.1 LTV System Characterization

The M-fold decimated version $x_D(n)$ of the discrete-time signal $x(n)$, its link with the sampled signal $x_\delta(n)$, and the frequency-domain counterpart of these relationships are

$$x_D(n) = x(nM) = x_\delta(nM) \tag{4.263}$$

$$\Leftrightarrow \quad x_\delta(n) = x_D\left[\frac{n}{M}\right] \tag{4.264}$$

$$\overset{\mathcal{F}}{\longleftrightarrow} \quad X_\delta(\nu) = X_D(\nu M) \tag{4.265}$$

$$\Leftrightarrow \quad X_D(\nu) = X_\delta\left(\frac{\nu}{M}\right) = \frac{1}{M} \sum_{p=0}^{M-1} X\left(\frac{\nu - p}{M}\right) \tag{4.266}$$

where (4.265) is obtained by (4.264) accounting for (4.239).

In Section 5.11 it is shown that decimation is a linear time-variant transformation (Figure 4.18)

$$x_D(n) = \sum_{\ell=-\infty}^{+\infty} h_D(n, \ell)\, x(\ell) \tag{4.267}$$

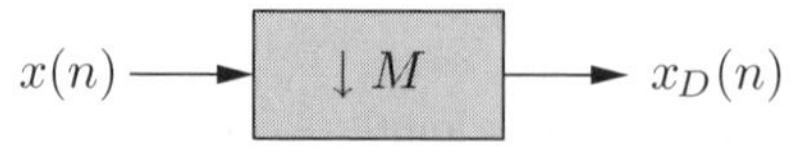

Figure 4.18 M-fold decimation

with impulse-response function

$$h_D(n, \ell) = \delta_{\ell - nM} \tag{4.268}$$

and transmission function

$$H_D(\nu, \lambda) \triangleq \sum_{n \in \mathbb{Z}} \sum_{\ell \in \mathbb{Z}} h_D(n, \ell)\, e^{-j2\pi(\nu n - \lambda \ell)} \tag{4.269a}$$

$$= \widetilde{\delta}_\nu(\nu - \lambda M) \tag{4.269b}$$

$$= \frac{1}{M} \sum_{p=0}^{M-1} \widetilde{\delta}_\lambda \left(\lambda - \frac{\nu - p}{M}\right) \tag{4.269c}$$

where $\widetilde{\delta}_\lambda(\lambda)$ and $\widetilde{\delta}_\nu(\nu)$ are defined in (4.243).

The function $H_D(\nu, \lambda)$ is periodic in ν with period 1 and periodic in λ with period $1/M$. Systems performing M-fold decimation are a special case of discrete-time FOT-deterministic LTV systems (Section 4.3.1) not LAPTV.

4.10.3.2 Input/Output Statistics

Let $x_{Di}(n)$ be the decimated version of $x_i(n)$, with decimation factor M_i $(i = 1, 2)$ (Figure 4.19). That is,

$$x_{Di}(n) = x_i(nM_i) = x_{\delta i}(nM_i). \tag{4.270}$$

The cross-correlation functions of decimated, sampled, and original signals are linked by the relationship

$$\mathrm{E}\left\{x_{D1}(n+m)\, x_{D2}^{(*)}(n)\right\} = \mathrm{E}\left\{x_1((n+m)M_1)\, x_2^{(*)}(nM_2)\right\}$$
$$= \mathrm{E}\left\{x_{\delta 1}((n+m)M_1)\, x_{\delta 2}^{(*)}(nM_2)\right\}. \tag{4.271}$$

Since the sampled process is an expanded version of the decimated process, that is $x_{\delta i}(n) = x_{Di}[n/M_i]$, accounting for (4.245), the Fourier transform of the sampled and decimated processes are linked by $X_{\delta i}(\nu) = X_{Di}(\nu M_i)$. Thus, using (4.257), the Fourier transforms of the decimated processes are

$$X_{Di}(\nu) = X_{\delta i}\left(\frac{\nu}{M_i}\right) = \frac{1}{M_i} \sum_{p=0}^{M_i - 1} X_i\left(\frac{\nu - p}{M_i}\right). \tag{4.272}$$

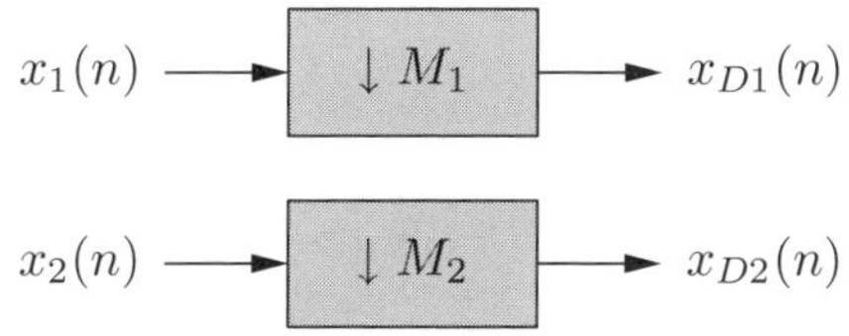

Figure 4.19 Decimation of two processes with different decimation factors

Therefore, if $x_1(n)$ and $x_2(n)$ are jointly SC with Loève bifrequency cross-spectrum (4.190a), by using (4.258) and (4.272), the Loève bifrequency cross-spectrum of $x_{D1}(n)$ and $x_{D2}(n)$ can be expressed as

$$\mathrm{E}\left\{X_{D1}(\nu_1)\,X_{D2}^{(*)}(\nu_2)\right\}$$
$$= \mathrm{E}\left\{X_{\delta 1}\left(\frac{\nu_1}{M_1}\right)X_{\delta 2}^{(*)}\left(\frac{\nu_2}{M_2}\right)\right\}$$
$$= \frac{1}{M_1 M_2}\sum_{p_1=0}^{M_1-1}\sum_{p_2=0}^{M_2-1}\sum_{k\in\mathbb{I}}\widetilde{S}_x^{(k)}\left(\frac{\nu_1-p_1}{M_1}\right)\widetilde{\delta}\left(\frac{\nu_2-p_2}{M_2}-\widetilde{\Psi}_x^{(k)}\left(\frac{\nu_1-p_1}{M_1}\right)\right) \tag{4.273a}$$
$$= \frac{1}{M_1}\sum_{p_1=0}^{M_1-1}\sum_{k\in\mathbb{I}}\widetilde{S}_x^{(k)}\left(\frac{\nu_1-p_1}{M_1}\right)\sum_{h\in\mathbb{Z}}\delta\left(\nu_2-h-M_2\widetilde{\Psi}_x^{(k)}\left(\frac{\nu_1-p_1}{M_1}\right)\right) \tag{4.273b}$$
$$= \frac{1}{M_1}\sum_{p_1=0}^{M_1-1}\sum_{k\in\mathbb{I}}\widetilde{S}_x^{(k)}\left(\frac{\nu_1-p_1}{M_1}\right)\widetilde{\delta}_{\nu_2}\left(\nu_2-M_2\widetilde{\Psi}_x^{(k)}\left(\frac{\nu_1-p_1}{M_1}\right)\right) \tag{4.273c}$$

where, in the third equality, the scaling property of the Dirac delta (Zemanian 1987, Section 1.7) and identity (5.192) are used. Analogously, accounting for (4.259) one obtains the equivalent expressions

$$\mathrm{E}\left\{X_{D1}(\nu_1)\,X_{D2}^{(*)}(\nu_2)\right\}$$
$$= \frac{1}{M_1 M_2}\sum_{p_1=0}^{M_1-1}\sum_{p_2=0}^{M_2-1}\sum_{k\in\mathbb{I}}\widetilde{G}_x^{(k)}\left(\frac{\nu_2-p_2}{M_2}\right)\widetilde{\delta}\left(\frac{\nu_1-p_1}{M_1}-\widetilde{\Phi}_x^{(k)}\left(\frac{\nu_2-p_2}{M_2}\right)\right) \tag{4.274a}$$
$$= \frac{1}{M_2}\sum_{p_2=0}^{M_2-1}\sum_{k\in\mathbb{I}}\widetilde{G}_x^{(k)}\left(\frac{\nu_2-p_2}{M_2}\right)\sum_{h\in\mathbb{Z}}\delta\left(\nu_1-h-M_1\widetilde{\Phi}_x^{(k)}\left(\frac{\nu_2-p_2}{M_2}\right)\right) \tag{4.274b}$$
$$= \frac{1}{M_2}\sum_{p_2=0}^{M_2-1}\sum_{k\in\mathbb{I}}\widetilde{G}_x^{(k)}\left(\frac{\nu_2-p_2}{M_2}\right)\widetilde{\delta}_{\nu_1}\left(\nu_1-M_1\widetilde{\Phi}_x^{(k)}\left(\frac{\nu_2-p_2}{M_2}\right)\right). \tag{4.274c}$$

From (4.273a)–(4.273c) and (4.274a)–(4.274c) it follows that the decimated versions of jointly SC processes are jointly SC. In addition, unlike the case of sampled processes, their Loève bifrequency cross-spectrum is periodic with period 1 in both variables ν_1 and ν_2.

In the special case where $x_1(n)$ and $x_2(n)$ are jointly ACS, then accounting for (4.202) it results in

$$\mathrm{E}\left\{X_{D1}(\nu_1)\,X_{D2}^{(*)}(\nu_2)\right\}$$
$$= \frac{1}{M_1}\sum_{p_1=0}^{M_1-1}\sum_{\widetilde{\alpha}\in\widetilde{A}}\widetilde{S_x^{\widetilde{\alpha}}}\left(\frac{\nu_1-p_1}{M_1}\right)\widetilde{\delta}_{\nu_2}\left(\nu_2-{}_{(-)}\left(\widetilde{\alpha}M_2-(\nu_1-p_1)\frac{M_2}{M_1}\right)\right). \tag{4.275}$$

The cyclic spectra $\widetilde{S_x^{\widetilde{\alpha}}}((\nu_1-p_1)/M_1)$ are periodic in ν_1 with period M_1 and the delta train is periodic in both ν_1 and ν_2 with period 1. In addition, from (4.275) it follows that the Loève bifrequency cross-spectrum of the decimated processes $x_{D1}(n)$ and $x_{D2}(n)$ in the principal domain $(\nu_1, \nu_2) \in [-1/2, 1/2)^2$ has support curves (in the case $(-)$ present)

$$\begin{aligned} \nu_2 = (M_2/M_1)\nu_1 - \widetilde{\alpha} M_2 - (M_2/M_1)p_1 \bmod 1 \\ \widetilde{\alpha} \in \widetilde{\mathcal{A}}, \;\; p_1 \in \{0, 1, \ldots, M_1 - 1\} \end{aligned} \tag{4.276}$$

that is, lines with slope M_2/M_1. Consequently, unlike the case of sampled processes, the decimated processes $x_{D1}(n)$ and $x_{D2}(n)$ are jointly SC. In particular, for a single process $x_1(n) \equiv x_2(n) = x(n)$, by taking $M_1 = 1$, that is $x_{D1}(n) \equiv x(n)$, it follows that the ACS process $x(n)$ and its decimated version $x_{D2}(n)$ are jointly SC with support lines with slope M_2. In the special case of $x(n)$ WSS, the support line is unique.

In Section 5.11 it is shown that, when $x_1(n)$ and $x_2(n)$ are jointly ACS, the cross-correlation function of $x_{D1}(n)$ and $x_{D2}(n)$ is given by

$$\begin{aligned} & \mathrm{E}\left\{x_{D1}(n+m)\, x_{D2}^{(*)}(n)\right\} \\ &= \int_{[-1/2,1/2]^2} \mathrm{E}\left\{X_{D1}(\nu_1)\, X_{D2}^{(*)}(\nu_2)\right\} e^{j2\pi[\nu_1(n+m)+(-)\nu_2 n]}\, \mathrm{d}\nu_1\, \mathrm{d}\nu_2 \\ &= \sum_{\widetilde{\alpha} \in \widetilde{\mathcal{A}}} e^{j2\pi\widetilde{\alpha} M_2 n}\, \widetilde{R}_{\boldsymbol{x}}^{\widetilde{\alpha}}\Big((M_1 - M_2)n + M_1 m\Big). \end{aligned} \tag{4.277}$$

By reasoning as for (4.250), we have that if $\widetilde{R}_{\boldsymbol{x}}^{\widetilde{\alpha}}(\cdot) \in \ell^1(\mathbb{Z})$ then for $M_1 \neq M_2$ the right-hand side of (4.277), as a function of n, does not contain any finite-strength additive sinewave component. That is, for $M_1 \neq M_2$ the processes $x_{D1}(n)$ and $x_{D2}^{(*)}(n)$ do not exhibit joint almost-cyclostationarity at any cycle frequency. From (4.275) and (4.277) it follows that $x_{D1}(n)$ and $x_{D2}(n)$ are jointly ACS if and only if $M_1 = M_2 = M$. In such a case, their cyclic cross-correlation function is (Section 5.11)

$$\widetilde{R}^{\widetilde{\alpha}}_{x_{D1} x_{D2}^{(*)}}(m) = \sum_{q=0}^{M-1} \widetilde{R}_{\boldsymbol{x}}^{(\widetilde{\alpha}-q)/M}(mM) \tag{4.278}$$

and, accounting for the Fourier-transform pair (4.270) and (4.272), their cyclic cross-spectrum can be expressed as

$$\widetilde{S}^{\widetilde{\alpha}}_{x_{D1} x_{D2}^{(*)}}(\nu) = \frac{1}{M} \sum_{p=0}^{M-1} \sum_{q=0}^{M-1} \widetilde{S}_{\boldsymbol{x}}^{(\widetilde{\alpha}-q)/M}\left(\frac{\nu - p}{M}\right). \tag{4.279}$$

In the special case $x_1 \equiv x_2$, (4.278) and (4.279) reduce to (Izzo and Napolitano 1998b, eqs. (22) and (24)) specialized to second order. In addition, by comparing (4.261) with (4.278) the following relationship between cyclic cross-correlations of decimated and sampled signals can be established

$$\widetilde{R}^{\widetilde{\alpha}}_{x_{D1} x_{D2}^{(*)}}(m) = M\, \widetilde{R}^{\widetilde{\alpha}/M}_{x_{\delta 1} x_{\delta 2}^{(*)}}(mM) \tag{4.280}$$

or, equivalently

$$\widetilde{\widetilde{R}}^{\alpha}_{x_{\delta 1} x^{(*)}_{\delta 2}}(m) = \frac{1}{M} \widetilde{\widetilde{R}}^{\alpha M}_{x_{D1} x^{(*)}_{D2}} \left[\frac{m}{M} \right]. \tag{4.281}$$

Moreover, by comparing (4.262) with (4.279) it results in

$$\widetilde{\widetilde{S}}^{\alpha}_{x_{D1} x^{(*)}_{D2}}(\nu) = M \, \widetilde{\widetilde{S}}^{\alpha/M}_{x_{\delta 1} x^{(*)}_{\delta 2}} \left(\frac{\nu}{M} \right). \tag{4.282}$$

4.10.4 Expansion and Decimation

Let $x_{I1}(n)$ and $x_{D2}(n)$ be the L_1-fold expanded and M_2-fold decimated versions of $x_1(n)$ and $x_2(n)$, respectively (Figure 4.20).

If $x_1(n)$ and $x_2(n)$ are jointly SC processes with Loève bifrequency cross-spectrum (4.190a), accounting for (4.245) and (4.272) and using and (5.192), the Loève bifrequency cross-spectrum of $x_{I1}(n)$ and $x_{D2}(n)$ can be expressed as

$$\begin{aligned}
& \mathrm{E}\left\{ X_{I1}(\nu_1) \, X^{(*)}_{D2}(\nu_2) \right\} \\
& = \frac{1}{M_2} \sum_{p=0}^{M_2-1} \mathrm{E}\left\{ X_1(\nu_1 L_1) \, X^{(*)}_2\left(\frac{\nu_2 - p}{M_2} \right) \right\} \\
& = \frac{1}{M_2} \sum_{p=0}^{M_2-1} \sum_{k \in \mathbb{I}} \widetilde{S}^{(k)}_{\boldsymbol{x}}(\nu_1 L_1) \, \widetilde{\delta}\left(\frac{\nu_2 - p}{M_2} - \widetilde{\Psi}^{(k)}_{\boldsymbol{x}}(\nu_1 L_1) \right) && (4.283a) \\
& = \sum_{k \in \mathbb{I}} \widetilde{S}^{(k)}_{\boldsymbol{x}}(\nu_1 L_1) \sum_{h \in \mathbb{Z}} \delta\left(\nu_2 - h - M_2 \widetilde{\Psi}^{(k)}_{\boldsymbol{x}}(\nu_1 L_1) \right) && (4.283b) \\
& = \sum_{k \in \mathbb{I}} \widetilde{S}^{(k)}_{\boldsymbol{x}}(\nu_1 L_1) \, \widetilde{\delta}_{\nu_2}\left(\nu_2 - M_2 \widetilde{\Psi}^{(k)}_{\boldsymbol{x}}(\nu_1 L_1) \right) && (4.283c)
\end{aligned}$$

or equivalently, using (4.190b), as

$$\begin{aligned}
& \mathrm{E}\left\{ X_{I1}(\nu_1) \, X^{(*)}_{D2}(\nu_2) \right\} \\
& = \frac{1}{M_2} \sum_{p=0}^{M_2-1} \sum_{k \in \mathbb{I}} \widetilde{G}^{(k)}_{\boldsymbol{x}}\left(\frac{\nu_2 - p}{M_2} \right) \widetilde{\delta}\left(\nu_1 L_1 - \widetilde{\Phi}^{(k)}_{\boldsymbol{x}}\left(\frac{\nu_2 - p}{M_2} \right) \right) && (4.284a)
\end{aligned}$$

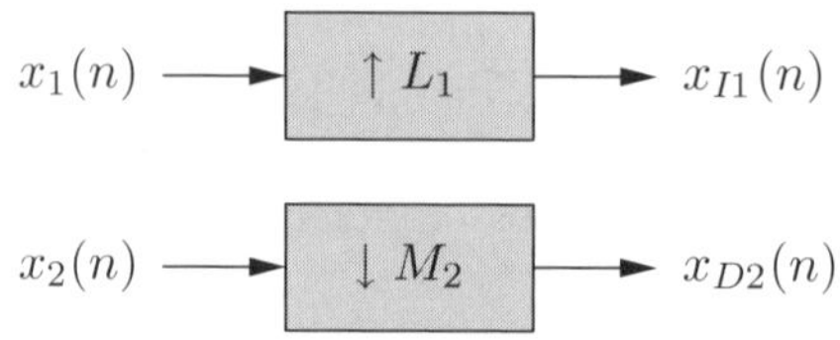

Figure 4.20 Expansion and decimation of two processes

$$= \frac{1}{M_2 L_1} \sum_{p=0}^{M_2-1} \sum_{k \in \mathbb{I}} \widetilde{G}_{\boldsymbol{x}}^{(k)}\left(\frac{\nu_2 - p}{M_2}\right) \sum_{h \in \mathbb{Z}} \delta\left(\nu_1 - \frac{h}{L_1} - \frac{1}{L_1}\widetilde{\Phi}_{\boldsymbol{x}}^{(k)}\left(\frac{\nu_2 - p}{M_2}\right)\right) \tag{4.284b}$$

$$= \frac{1}{M_2 L_1} \sum_{p_1=0}^{L_1-1} \sum_{p_2=0}^{M_2-1} \sum_{k \in \mathbb{I}} \widetilde{G}_{\boldsymbol{x}}^{(k)}\left(\frac{\nu_2 - p_2}{M_2}\right)$$

$$\widetilde{\delta}_{\nu_1}\left(\nu_1 - \frac{p_1}{L_1} - \frac{1}{L_1}\widetilde{\Phi}_{\boldsymbol{x}}^{(k)}\left(\frac{\nu_2 - p_2}{M_2}\right)\right) \tag{4.284c}$$

Thus $x_{I1}(n)$ and $x_{I2}(n)$ are jointly SC. In addition, from (4.283a)–(4.283c) and (4.284a)–(4.284c) it follows that their Loève bifrequency cross-spectrum is periodic in ν_1 with period $1/L_1$ and in ν_2 with period 1.

If $x_1(n)$ and $x_2(n)$ are jointly ACS processes, (4.283c) specializes into

$$\mathrm{E}\left\{X_{I1}(\nu_1)\, X_{D2}^{(*)}(\nu_2)\right\} = \sum_{\widetilde{\alpha} \in \widetilde{A}} \widetilde{\widetilde{S}}_{\boldsymbol{x}}^{\alpha}(\nu_1 L_1)\, \widetilde{\delta}_{\nu_2}\left(\nu_2 - {}_{(-)}(\widetilde{\alpha} M_2 - \nu_1 L_1 M_2)\right). \tag{4.285}$$

The cyclic spectra $\widetilde{\widetilde{S}}_{\boldsymbol{x}}^{\alpha}(\nu_1 L_1)$ are periodic in ν_1 with period $1/L_1$ and the delta train is periodic in ν_2 with period 1 and in ν_1 with period $1/(L_1 M_2)$. Therefore, $\mathrm{E}\{X_{I1}(\nu_1)\, X_{D2}^{(*)}(\nu_2)\}$ is periodic in ν_1 with period $1/L_1$ and in ν_2 with period 1. In addition, from (4.285) it follows that $x_{I1}(n)$ and $x_{D2}(n)$ are jointly SC for every value of L_1 and M_2 except in the trivial case $L_1 = M_2 = 1$.

When $x_1(n)$ and $x_2(n)$ are jointly ACS, the cross-correlation function of $x_{L1}(n)$ and $x_{D2}(n)$ can be expressed as (Section 5.11)

$$\mathrm{E}\left\{x_{I1}(n+m)\, x_{D2}^{(*)}(n)\right\}$$

$$= \int_{[-1/2,1/2]^2} \mathrm{E}\left\{X_{I1}(\nu_1)\, X_{D2}^{(*)}(\nu_2)\right\} e^{j2\pi[\nu_1(n+m)+(-)\nu_2 n]}\, d\nu_1\, d\nu_2$$

$$= \sum_{\widetilde{\alpha} \in \widetilde{A}} e^{j2\pi\widetilde{\alpha} M_2 n}\, \widetilde{\widetilde{R}}_{\boldsymbol{x}}^{\alpha}\left[\frac{(1 - L_1 M_2)n + m}{L_1}\right]. \tag{4.286}$$

Accordingly with the result in the frequency domain, the cross-correlation is an almost-periodic function of n for every fixed m if and only if $L_1 = M_2 = 1$. In such a case, $x_{L1}(n) \equiv x_1(n)$ and $x_{D2}(n) \equiv x_2(n)$ are jointly ACS. In contrast, if at least one of L_1 and M_2 is not unit, $x_{L1}(n)$ and $x_{D2}(n)$ are jointly SC. Under the mild assumption that $\widetilde{\widetilde{R}}_{\boldsymbol{x}}^{\alpha}(\cdot) \in \ell_1(\mathbb{Z})$, for every fixed m, the cross-correlation function in (4.286), as a function of n, does not contain any finite-strength additive sinewave component. That is, $x_{I1}(n)$ and $x_{D2}^{(*)}(n)$ do not exhibit joint almost-cyclostationarity at any cycle frequency.

4.10.5 Strictly Band-Limited SC Processes

Let $H_b(\nu) \triangleq \sum_{m \in \mathbb{Z}} \mathrm{rect}((\nu - m)/(2b))$, with $\mathrm{rect}(\nu) = 1$ if $|\nu| \leqslant 1/2$ and $\mathrm{rect}(\nu) = 0$ otherwise, be the harmonic response of an ideal low-pass filter with monolateral bandwidth $b < 1/2$. A discrete-time nonstationary process $x(n)$ is said to be strictly band-limited with bandwidth B if B is the smallest value of bandwidth b such that $x(n)$ passes unaltered through a filter with harmonic response $H_b(\nu)$. In Section 4.9.1, the a.s. band-limitedness definition is given with reference to continuous-time processes. This definition can be extended with obvious changes to discrete-time processes.

Let $x_1(n)$ and $x_2(n)$ be jointly SC processes strictly band-limited with bandwidths B_1 and B_2, respectively. In Section 5.11 it is shown that for every term of the sum over $k \in \mathbb{I}$ in the Loève bifrequency cross-spectrum (4.190a) one has

$$\begin{aligned}
\operatorname{supp}&\left\{\widetilde{S}_{\boldsymbol{x}}^{(k)}(\nu_1)\,\widetilde{\delta}\left(\nu_2-\widetilde{\Psi}_{\boldsymbol{x}}^{(k)}(\nu_1)\right)\right\}\\
&\subseteq\Big\{(\nu_1,\nu_2)\in\mathbb{R}\times\mathbb{R}\,:\,|\nu_1 \bmod 1|\leqslant B_1,\\
&\qquad \left|\widetilde{\Psi}_{\boldsymbol{x}}^{(k)}(\nu_1)\right|\leqslant B_2,\ \nu_2=\widetilde{\Psi}_{\boldsymbol{x}}^{(k)}(\nu_1) \bmod 1\Big\} &\text{(4.287a)}\\
&=\Big\{(\nu_1,\nu_2)\in\mathbb{R}\times\mathbb{R}\,:\,|\nu_1 \bmod 1|\leqslant B_1,\\
&\qquad |\nu_2 \bmod 1|\leqslant B_2,\ \nu_2=\widetilde{\Psi}_{\boldsymbol{x}}^{(k)}(\nu_1) \bmod 1\Big\} &\text{(4.287b)}
\end{aligned}$$

where mod 1 is the modulo 1 operation with values in $[-1/2, 1/2)$. Analogously, starting from (4.190b), for every $k \in \mathbb{I}$, one has

$$\begin{aligned}
\operatorname{supp}&\left\{\widetilde{G}_{\boldsymbol{x}}^{(k)}(\nu_2)\,\widetilde{\delta}\left(\nu_1-\widetilde{\Phi}_{\boldsymbol{x}}^{(k)}(\nu_2)\right)\right\}\\
&\subseteq\Big\{(\nu_1,\nu_2)\in\mathbb{R}\times\mathbb{R}\,:\,|\nu_2 \bmod 1|\leqslant B_2,\\
&\qquad \left|\widetilde{\Phi}_{\boldsymbol{x}}^{(k)}(\nu_2)\right|\leqslant B_1,\ \nu_1=\widetilde{\Phi}_{\boldsymbol{x}}^{(k)}(\nu_2) \bmod 1\Big\} &\text{(4.288a)}\\
&=\Big\{(\nu_1,\nu_2)\in\mathbb{R}\times\mathbb{R}\,:\,|\nu_2 \bmod 1|\leqslant B_2,\\
&\qquad |\nu_1 \bmod 1|\leqslant B_1,\ \nu_1=\widetilde{\Phi}_{\boldsymbol{x}}^{(k)}(\nu_2) \bmod 1\Big\}. &\text{(4.288b)}
\end{aligned}$$

4.10.6 *Interpolation Filters*

Figure 4.21 presents two interpolation filters. Each interpolation filter ($i = 1, 2$) is constituted by a L_i-fold interpolator followed by a LTI filter with bandwidth W_i whose purpose is to obtain image-free interpolation (Crochiere and Rabiner 1981; Vaidyanathan 1990). An additional LTI filter with bandwidth B_i precedes the L_i-fold interpolator to strictly bandlimit the input process $x_{\text{in},i}(n)$ in order to avoid overlapping among images in the Loève bifrequency cross-spectrum.

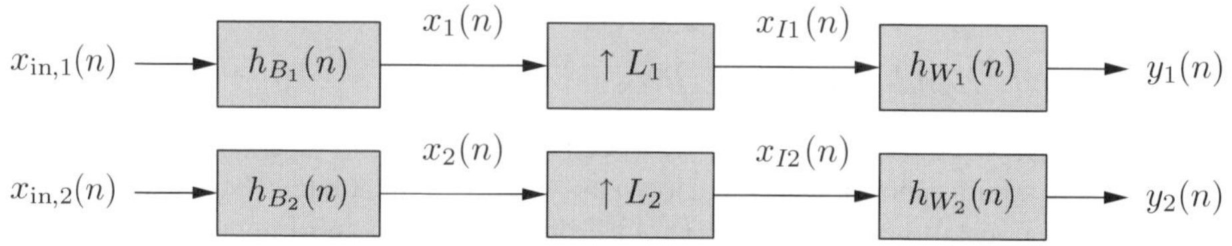

Figure 4.21 Interpolation filters

In this section, the effects of interpolation filters on jointly SC processes are analyzed and sufficient conditions are established to assure that the Loève bifrequency cross-spectrum of the interpolated processes $y_1(n)$ and $y_2(n)$ is a frequency-scaled image-free version of that of the band-limited processes $x_1(n)$ and $x_2(n)$.

If $x_{\text{in},1}(n)$ and $x_{\text{in},2}(n)$ are jointly SC processes, then $x_1(n)$ and $x_2(n)$ are jointly SC processes strictly bandlimited with bandwidths B_1 and B_2, respectively (Section 4.3.2). Accounting for (4.246c), the Loève bifrequency cross-spectrum of $y_1(n)$ and $y_2(n)$ is given by

$$\begin{aligned}\mathrm{E}\left\{Y_1(\nu_1)\, Y_2^{(*)}(\nu_2)\right\} &= H_{W_1}(\nu_1)\, H_{W_2}(\nu_2)\, \mathrm{E}\left\{X_{I1}(\nu_1)\, X_{I2}^{(*)}(\nu_2)\right\} \\ &= H_{W_1}(\nu_1)\, H_{W_2}(\nu_2) \sum_{k\in\mathbb{I}} \widetilde{S}_{\boldsymbol{x}}^{(k)}(\nu_1 L_1) \\ &\qquad \frac{1}{L_2}\sum_{p=0}^{L_2} \widetilde{\delta}_{\nu_2}\left(\nu_2 - \frac{p}{L_2} - \frac{1}{L_2}\widetilde{\Psi}_{\boldsymbol{x}}^{(k)}(\nu_1 L_1)\right)\end{aligned} \tag{4.289}$$

In Section 5.11 it is shown that

$$\frac{B_i}{L_i} \leqslant W_i \leqslant \frac{1}{2L_i} \qquad i = 1, 2 \tag{4.290}$$

is a sufficient condition to assure that the Loève bifrequency cross-spectrum $\mathrm{E}\{Y_1(\nu_1)\, Y_2^{(*)}(\nu_2)\}$ is a frequency-scaled image-free version of $\mathrm{E}\{X_1(\nu_1)\, X_2^{(*)}(\nu_2)\}$.

Condition (4.290) is independent of the shape of the support curves $\nu_2 = \widetilde{\Psi}_{\boldsymbol{x}}^{(k)}(\nu_1)$. In the special case of a single ACS process and $L_1 = L_2 = L$, is less restrictive than (Izzo and Napolitano 1998b, eq. (56)) specialized to second-order. In fact, (Izzo and Napolitano 1998b, eq. (56)) assures the lack of images in the densities of spectral correlation (the cyclic spectra) for $\nu_1 \in [-1/(2L), 1/(2L)]$ and cycle frequencies $\widetilde{\alpha} \in [-1/(2L), 1/(2L)]$. In Section 5.11, it is shown that

$$1 \geqslant B_2 + \max\left\{B_2, \sup_k \sup_{\nu_1\in[-1/2,1/2]} \left|\widetilde{\Psi}_{\boldsymbol{x}}^{(k)}(\nu_1)\right|\right\} \tag{4.291}$$

or, equivalently,

$$1 \geqslant B_1 + \max\left\{B_1, \sup_k \sup_{\nu_2\in[-1/2,1/2]} \left|\widetilde{\Phi}_{\boldsymbol{x}}^{(k)}(\nu_2)\right|\right\} \tag{4.292}$$

is a sufficient condition such that the density of Loève bifrequency cross-spectrum (4.289) along every support curve is a frequency-scaled image-free version of the corresponding density of Loève bifrequency cross-spectrum $\mathrm{E}\{X_1(\nu_1)\, X_2^{(*)}(\nu_2)\}$ for every $\nu_1 \in [-1/(2L_1), 1/(2L_1)]$. Conditions (4.291) and (4.292) are independent of L_1 and L_2 and correspond to sufficient conditions to avoid aliasing in spectral cross-correlation densities in the case of sampling continuous-time SC processes (see (4.226) and (4.227)).

4.10.7 Decimation Filters

In Figure 4.22, two decimation filters are represented. Each decimation filter ($i = 1, 2$) is constituted by a M_i-fold decimator preceded by a LTI filter to strictly bandlimit the input

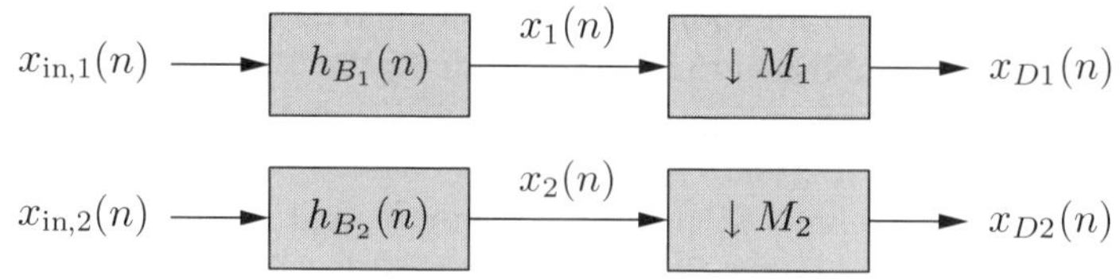

Figure 4.22 Decimation filters

process $x_{\text{in},i}(n)$ in order to avoid aliasing among replicas in the Loève bifrequency cross-spectrum. In this section, the effects of decimation filters on jointly SC processes are analyzed and sufficient conditions are established to assure that the Loève bifrequency cross-spectrum of the decimated processes $x_{D1}(n)$ and $x_{D2}(n)$ is a frequency-scaled alias-free version of that of the band-limited processes $x_1(n)$ and $x_2(n)$.

If $x_{\text{in},1}(n)$ and $x_{\text{in},2}(n)$ are jointly SC processes, then $x_1(n)$ and $x_2(n)$ are jointly SC processes strictly bandlimited with bandwidths B_1 and B_2, respectively (Section 4.3.2). The Loève bifrequency cross-spectrum of $x_{D1}(n)$ and $x_{D2}(n)$ is given by (4.273a)–(4.273c).

In Section 5.11 it is shown that

$$M_i B_i \leqslant \frac{1}{2} \quad i = 1, 2 \tag{4.293}$$

is a sufficient condition to assure that the Loève bifrequency cross-spectrum $\mathrm{E}\{X_{D1}(\nu_1)\, X_{D2}^{(*)}(\nu_2)\}$ in (4.273a)–(4.273c) is a frequency-scaled alias-free version of $\mathrm{E}\{X_1(\nu_1)\, X_2^{(*)}(\nu_2)\}$.

Condition (4.293) is the Shannon condition to avoid aliasing in the sampled discrete-time processes. It is independent of the shape of the support curves $\nu_2 = \widetilde{\Psi}_x^{(k)}(\nu_1)$. Moreover, in the special case of a single ACS process and $M_1 = M_2$, is less restrictive than (Izzo and Napolitano 1998b, eq. (40)) specialized to second-order. In fact, (Izzo and Napolitano 1998b, eq. (40)) assures the lack of aliasing replicas in the densities of spectral correlation (the cyclic spectra) for $\nu_1 \in [-1/2, 1/2]$ and cycle frequencies $\widetilde{\alpha} \in [-1/2, 1/2]$.

In Section 5.11, it is shown that

$$1 \geqslant M_2 B_2 + M_2 \max\Big\{ B_2, \sup_k \sup_{\nu_1 \in [-1/(2M_1), 1/(2M_1)]} \Big|\widetilde{\Psi}_x^{(k)}(\nu_1)\Big| \Big\} \tag{4.294}$$

or, equivalently,

$$1 \geqslant M_1 B_1 + M_1 \max\Big\{ B_1, \sup_k \sup_{\nu_2 \in [-1/(2M_2), 1/(2M_2)]} \Big|\widetilde{\Phi}_x^{(k)}(\nu_2)\Big| \Big\} \tag{4.295}$$

is a sufficient condition such that the density of Loève bifrequency cross-spectrum (4.273a)–(4.273c) along every support curve is a frequency-scaled aliasing-free version of the corresponding density of Loève bifrequency cross-spectrum $\mathrm{E}\{X_1(\nu_1)\, X_2^{(*)}(\nu_2)\}$ for every $\nu_1 \in [-1/2, 1/2]$. Conditions (4.294) and (4.295) are the discrete-time counterpart of sufficient conditions to avoid aliasing in spectral cross-correlation densities in the case of sampling continuous-time SC processes (see (4.226) and (4.227)).

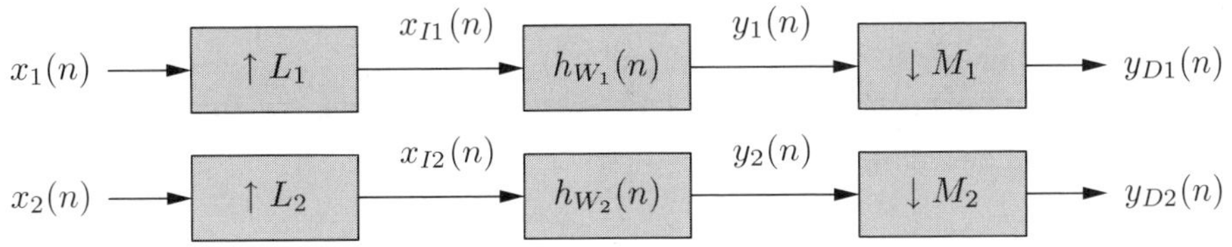

Figure 4.23 Fractional sampling rate converters

4.10.8 Fractional Sampling Rate Converters

In Figure 4.23, two systems are represented, that perform sampling rate conversions by noninteger (rational) numbers M_i/L_i, $(i = 1, 2)$. Each input process $x_i(n)$ is assumed to be strictly band limited with bandwidth B_i. An ideal low-pass filter with bandwidth W_i follows the L_i-fold interpolator in order to cancel spectral images.

Let $x_1(n)$ and $x_2(n)$ be jointly SC processes with Loève bifrequency cross-spectrum (4.190a). In Section 5.11 it is shown that

$$\begin{aligned} &\mathrm{E}\left\{Y_{D1}(\nu_1)\, Y_{D2}^{(*)}(\nu_2)\right\} \\ &= \frac{1}{M_1 M_2} \sum_{p_1=0}^{M_1-1} \sum_{p_2=0}^{M_2-1} H_{W_1}\left(\frac{\nu_1 - p_1}{M_1}\right) H_{W_2}\left(\frac{\nu_2 - p_2}{M_2}\right) \\ &\sum_{k \in \mathbb{I}} \widetilde{S}_{x}^{(k)}\left((\nu_1 - p_1)\frac{L_1}{M_1}\right) \widetilde{\delta}\left((\nu_2 - p_2)\frac{L_2}{M_2} - \widetilde{\Psi}_{x}^{(k)}\left((\nu_1 - p_1)\frac{L_1}{M_1}\right)\right). \end{aligned} \tag{4.296}$$

Let $x_1(n)$ and $x_2(n)$ be strictly band limited with bandwidths B_1 and B_2, respectively. According to the results of Section 4.10.6, $B_i/L_i \leqslant W_i \leqslant 1/(2L_i)$, $(i = 1, 2)$, is a sufficient condition to assure that the Loève bifrequency cross-spectrum $\mathrm{E}\{Y_1(\nu_1)\, Y_2^{(*)}(\nu_2)\}$ is a frequency-scaled image-free version of $\mathrm{E}\{X_1(\nu_1)\, X_2^{(*)}(\nu_2)\}$. In addition, since $y_1(n)$ and $y_2(n)$ are jointly SC with bandwidths W_1 and W_2, according to the results of Section 4.10.7, $M_i W_i \leqslant 1/2$, $(i = 1, 2)$, is a sufficient condition to assure that the Loève bifrequency cross-spectrum $\mathrm{E}\{Y_{D1}(\nu_1)\, Y_{D2}^{(*)}(\nu_2)\}$ is a frequency-scaled aliasing-free version of $\mathrm{E}\{Y_1(\nu_1)\, Y_2^{(*)}(\nu_2)\}$. Therefore,

$$\frac{B_i}{L_i} \leqslant W_i \leqslant \frac{1}{2} \min\left\{\frac{1}{L_i}, \frac{1}{M_i}\right\} \quad (i = 1, 2) \tag{4.297}$$

is a sufficient condition to assure that the Loève bifrequency cross-spectrum $\mathrm{E}\{Y_{D1}(\nu_1)\, Y_{D2}^{(*)}(\nu_2)\}$ is a frequency-scaled image- and aliasing-free version of $\mathrm{E}\{X_1(\nu_1)\, X_2^{(*)}(\nu_2)\}$.

4.11 Discrete-Time Estimators of the Spectral Cross-Correlation Density

Let

$$x_{\mathrm{d}}(n) \triangleq x(t)|_{t=nT_s} \qquad y_{\mathrm{d}}(n) \triangleq y(t)|_{t=nT_s} \tag{4.298}$$

be the discrete-time sequences obtained by uniformly sampling with period $T_s = 1/f_s$ the continuous-time (jointly) SC processes $x(t)$ and $y(t)$.

The *discrete-time frequency-smoothed cross-periodogram* along the support curve $\nu_2 = \widetilde{\Psi}^{(k)}_{y_d x_d^{(*)}}(\nu_1)_{\text{mod } 1}$ is defined as

$$\widetilde{S}^{(k)}_{y_d x_d^{(*)}}(n; \nu_1)_{N,\Delta\nu} \triangleq \left[\frac{1}{2N+1} \; Y_N(n, \nu_1) \, X_N^{(*)}(n, \nu_2) \Big|_{\nu_2 = \widetilde{\Psi}^{(k)}_{y_d x_d^{(*)}}(\nu_1)_{\text{mod } 1}} \right] \underset{\nu_1}{\otimes} \widetilde{A}_{\Delta\nu}(\nu_1) \quad (4.299)$$

where $\otimes_{\nu_1}$ denotes periodic convolution (with period 1) with respect to ν_1, $Y_N(n, \nu_1)$ and $X_N(n, \nu_2)$ are discrete-time STFTs defined according to

$$Z_N(n, \nu) \triangleq \sum_{\ell=n-N}^{n+N} z(\ell) \, e^{-j2\pi\nu\ell} \quad (4.300)$$

and

$$\widetilde{A}_{\Delta\nu}(\nu) \triangleq \frac{1}{\Delta\nu} \widetilde{W}\left(\frac{\nu}{\Delta\nu}\right) \quad (4.301)$$

is a *frequency-smoothing window* with $\widetilde{W}(\nu)$ summable in $[-1/2, 1/2]$ and such that

$$\lim_{\Delta\nu \to 0} \frac{1}{\Delta\nu} \widetilde{W}\left(\frac{\nu}{\Delta\nu}\right) = \widetilde{\delta}(\nu). \quad (4.302)$$

In the absence of aliasing, statistical functions of discrete-time sampled processes in the principal frequency domain are scaled versions of statistical functions of the continuous-time processes (Sections 4.9.2 and 7.7.5). In such a case, consistency results for the discrete-time frequency-smoothed cross-periodogram can be proved with obvious changes analogously to the case of continuous-time processes. Moreover, even in the case of non-strictly band-limited continuous-time processes, consistency results can be proved provided that the amount of aliasing is controlled by taking the sampling period sufficiently small. In such a case, under suitable conditions, the result is that

$$\lim_{T_s \to 0} \lim_{\Delta\nu \to 0} \lim_{N \to \infty} \mathrm{E}\left\{ \left| \widetilde{S}^{(k)}_{y_d x_d^{(*)}}(n; \nu)_{N,\Delta\nu} - S^{(k)}_{yx^{(*)}}(f) \, \mathcal{E}^{(k)}(f) \Big|_{f=\nu f_s} \right|^2 \right\} = 0 \quad (4.303)$$

where the order of the three limits cannot be interchanged.

4.12 Numerical Results

In this section, simulation results are reported to corroborate the theoretical results of the previous sections on the spectral cross-correlation density estimation.

4.12.1 Simulation Setup

The spectral cross-correlation density function between complex-envelope signals $x(t)$ and $y(t)$ is measured, where $x(t)$ and $y(t)$ are the input and output signals, respectively, of a linear time-variant system that models the channel between transmitter and receiver in relative motion with constant relative radial speed, that is (Section 7.3)

$$y(t) = a\, x(st - d)\, e^{j2\pi\nu t}. \tag{4.304}$$

In (4.304), a is the (possibly complex) scaling amplitude, d the delay, s the time-scale factor, and ν the frequency shift.

For an ACS input process $x(t)$ the Loève bifrequency cross-spectrum between $y(t)$ and $x(t)$ is given by (see (7.276))

$$\begin{aligned} &\mathrm{E}\left\{Y(f_1)\, X^*(f_2)\right\} \\ &= \frac{a}{|s|}\, e^{-j2\pi(f_1-\nu)d/s}\, \mathrm{E}\left\{X\left(\frac{f_1-\nu}{s}\right)\, X^*(f_2)\right\} \\ &= \frac{a}{|s|}\, e^{-j2\pi(f_1-\nu)d/s} \sum_{\beta\in A_{xx^*}} S_{xx^*}^{\beta}\left(\frac{f_1-\nu}{s}\right)\, \delta\left(f_2 - \frac{f_1}{s} + \beta + \frac{\nu}{s}\right) \end{aligned} \tag{4.305}$$

where $S_{xx^*}^{\beta}(f)$ and A_{xx^*} are the cyclic spectra and the set of cycle frequencies, respectively, of $x(t)$. Therefore, if $s \neq 1$, $x(t)$ and $y(t)$ are not jointly ACS (even if they are singularly ACS) but are jointly SC and exhibit joint spectral correlation on curves $f_2 = \Psi^{(n)}(f_1)$ which are lines with slope $1/s$ in the bifrequency plane.

In the simulation experiments, both time and frequency are discretized with sampling increments $T_s = T/N$ and $\Delta F = 1/T$, respectively, where T is the data-record length and N the number of samples. The channel parameters are fixed at $a = 1$, $d = 0$, $\nu = 0$, and $s = 0.99$. The input signal $x(t)$ is a binary PAM signal with full-duty-cycle rectangular pulse and bit rate $1/T_p = 1/8T_s$. Thus, $x(t)$ has an approximate bandwidth $B = 1/8T_s$.

4.12.2 Unknown Support Curves

In this section, the support curves of the Loève bifrequency cross-spectrum are assumed to be unknown. Hence, the time-smoothed cross periodogram is adopted as estimator of the bifrequency spectral cross-correlation density (4.5).

The time-smoothed cross-periodogram (4.123) (with $(*)$ present) is obtained by taking $B_{\Delta f}(f)$ and $a_T(t)$ both rectangular windows (see discussion in Section 4.7.3). Its performance is evaluated by the sample mean and the sample standard deviation computed on the basis of 100 Monte Carlo trials. Two experiments are conducted with increasing data-record length aimed at illustrating the asymptotic behavior of the estimator. In both the experiments, $\Delta f = 1/2^6 T_s$ and the time-smoothed cross-periodogram is computed for $(f_1, f_2) \in [-0.025/T_s, 0.025/T_s] \times [-0.025/T_s,\ 0.025/T_s]$. Moreover, to better highlight slopes of the support lines close to unity, all the graphs are reported as functions of (f_1, α), with $\alpha \triangleq f_1 - f_2$. A support line with slope $1/s$ in the (f_1, f_2) plane becomes a line with slope $1 - 1/s$ in the (f_1, α) plane.

In the first experiment, the number of samples is $N = 2^{11}$. Therefore, $BT \simeq 256 \not\ll 1/|1 - s| = 100$ and, hence, s cannot be approximated by 1 (Section 7.5.1). Moreover, $T\Delta f = 32$ and, hence, $|-\Psi^{(n)\prime}(f_1) + 1| = |1 - 1/s| \simeq 0.01 < 1/T\Delta f$, $\forall n \in \mathbb{I}$, that is, condition (4.135) of Corollary 4.5.8 is practically verified. In Figure 4.24, (a) graph and (b) "checkerboard" plot of the magnitude of the sample mean of the time-smoothed periodogram $S_{xx^*}(0; f_1, f_1 - \alpha)_{1/\Delta f,T}$ of the ACS input signal $x(t)$ are reported as functions of (f_1, α). A portion of the power

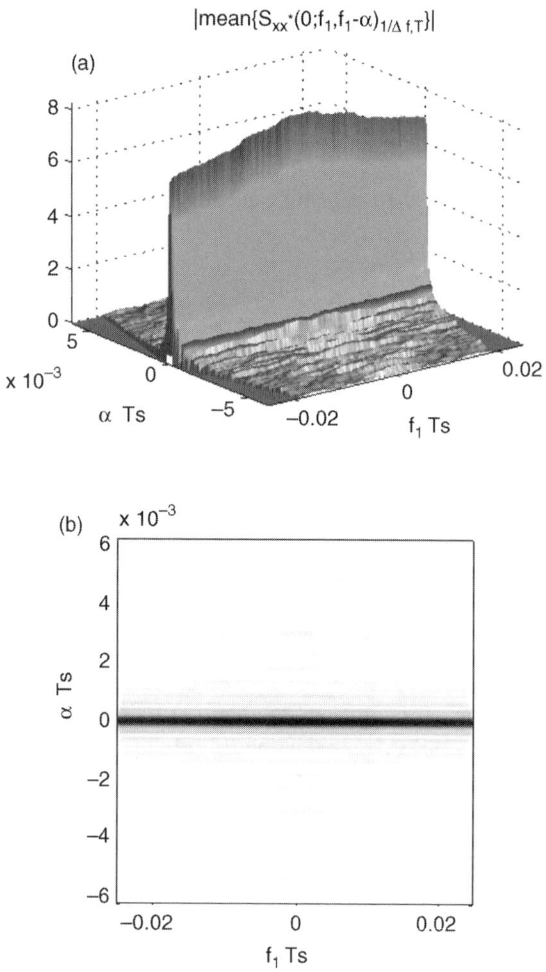

Figure 4.24 (a) Graph and (b) "checkerboard" plot, as functions of (f_1, α), of the magnitude of the sample mean of the time-smoothed periodogram of $x(t)$, for a data-record length $N = 2^{11}$. *Source:* (Napolitano 2003) © IEEE

spectral density ($\alpha = 0$) is evident. In Figure 4.25, (a) graph and (b) "checkerboard" plot of the sample standard deviation of $S_{xx^*}(0; f_1, f_1 - \alpha)_{1/\Delta f,T}$ are shown as functions of (f_1, α). In Figure 4.26, (a) graph and (b) "checkerboard" plot of the magnitude of the sample mean of the time-smoothed cross-periodogram $S_{yx^*}(0; f_1, f_1 - \alpha)_{1/\Delta f,T}$ are reported. It is evident that $y(t)$ and $x(t)$ are jointly SC processes and exhibit joint spectral correlation along the line $\alpha = (1 - 1/s)f_1$ (Figure 4.26(b)). In Figure 4.27, (a) graph and (b) "checkerboard" plot of the sample standard deviation of $S_{yx^*}(0; f_1, f_1 - \alpha)_{1/\Delta f,T}$ are drawn.

A second experiment with $N = 2^{13}$ is conducted, aimed at showing that, unlike the case of WSS and ACS processes, for (jointly) SC processes augmenting the data-record length does not necessarily imply a better performance of the spectral cross-correlation density function

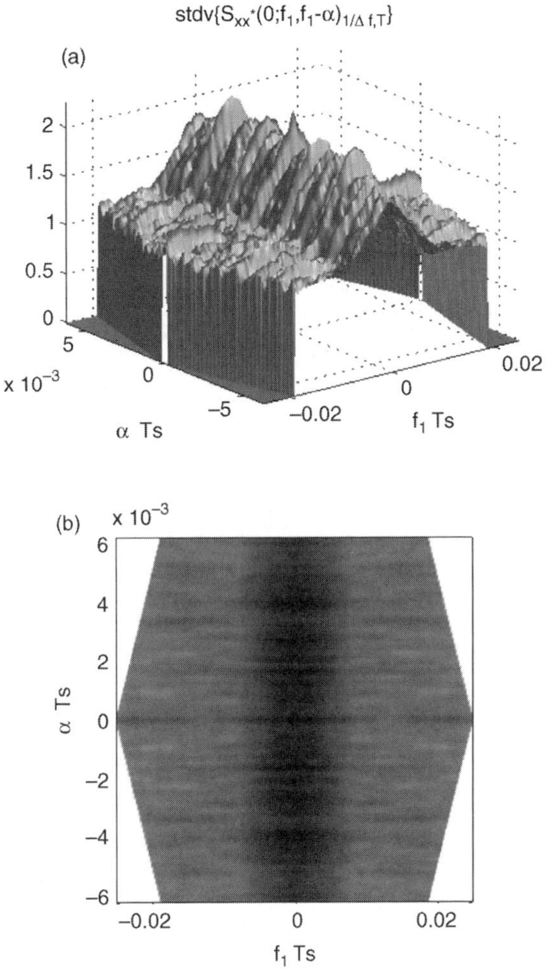

Figure 4.25 (a) Graph and (b) "checkerboard" plot, as functions of (f_1, α), of the sample standard deviation of the time-smoothed periodogram of $x(t)$, for a data-record length $N = 2^{11}$. *Source:* (Napolitano 2003) © IEEE

estimator. Figure 4.28a shows the graph of the magnitude of the sample mean and Figure 4.28b the graph of the sample standard deviation of the time-smoothed periodogram of $x(t)$. By comparing Figures 4.28a and 4.28b with Figures 4.24a and 4.25a, respectively, the result is that both bias and variance of the estimate have decreased in accordance with the well known consistency properties of the time-smoothed cross-periodogram of ACS signals. For $N = 2^{13}$, it results in $BT \simeq 1024 > 1/|1 - s|$ and, hence, also in this experiment, s cannot be approximated by 1. Moreover, $T\Delta f = 128$ so that $| - \Psi^{(n)\prime}(f_1) + 1| \simeq 0.01 \simeq 1/T\Delta f$, $\forall n \in \mathbb{I}$, that is, (4.135) is not verified. Consequently, a significant bias of the estimate is expected. In Figures 4.29a and 4.29b, the graph of the magnitude of the sample mean and the graph of the sample standard deviation of the time-smoothed cross-periodogram of $y(t)$

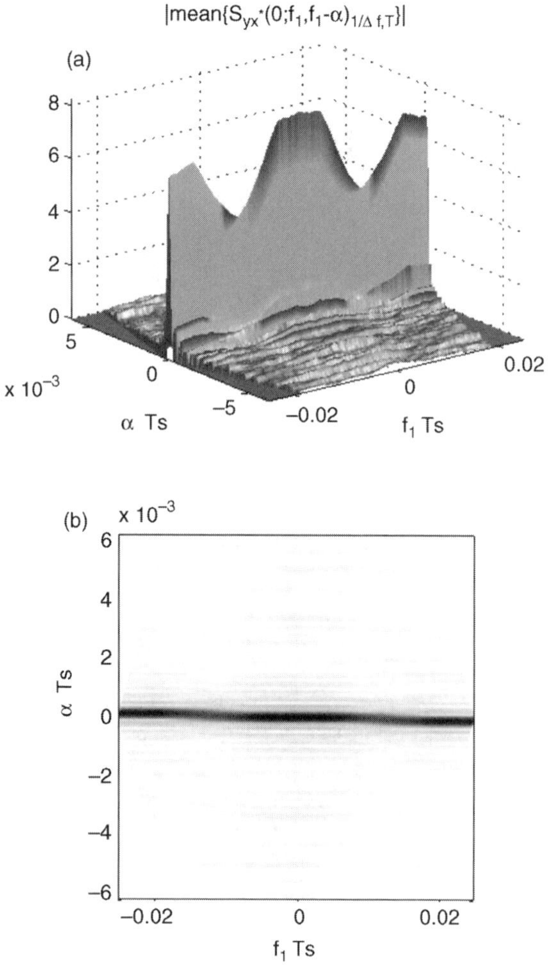

Figure 4.26 (a) Graph and (b) "checkerboard" plot, as functions of (f_1, α), of the magnitude of the sample mean of the time-smoothed cross-periodogram of $y(t)$ and $x(t)$, for a data-record length $N = 2^{11}$. *Source:* (Napolitano 2003) © IEEE

and $x(t)$ are reported. By comparing Figures 4.29a and 4.29b with Figures 4.26a and 4.27a, respectively, it is evident that, according with the theoretical results of Theorems 4.5.7 and 4.5.9 and Corollary 4.5.8, even if the variance has decreased, the bias has grown.

4.12.3 Known Support Curves

In this section, the location of one support curve of the Loève bifrequency cross-spectrum is assumed to be known. Hence, the frequency-smoothed cross-periodogram along this curve is adopted as estimator of the spectral cross-correlation density on this curve (Section 4.7).

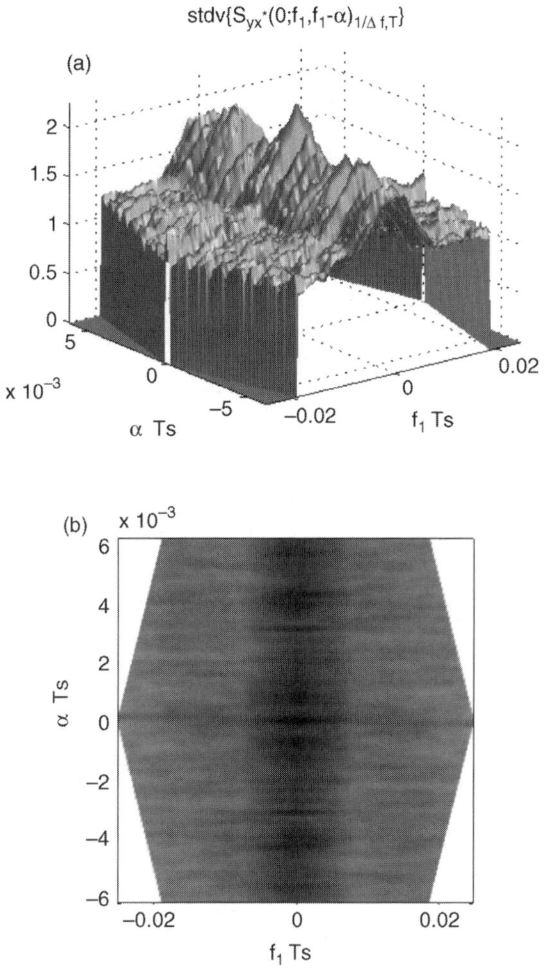

Figure 4.27 (a) Graph and (b) "checkerboard" plot, as functions of (f_1, α), of the sample standard deviation of the time-smoothed cross-periodogram of $y(t)$ and $x(t)$, for a data-record length $N = 2^{11}$. *Source:* (Napolitano 2003) © IEEE

The performance of the frequency-smoothed cross-periodogram $S^{(n)}_{yx^*}(t; f_1)_{T,\Delta f}$, defined in (4.147), along the known support curve

$$f_2 = \frac{1}{s} f_1 - \left(\alpha_p + \frac{\nu}{s}\right) \tag{4.306}$$

with $\alpha_p = \frac{1}{T_p}$ is evaluated by the sample mean and the sample standard deviation of both amplitude and phase of $S^{(n)}_{yx^{(*)}}(t; f_1)_{T,\Delta f}$. A rectangular frequency-smoothing window $A_{\Delta f}(f)$ with window-width $\Delta f = 1/(2^6 T_s)$ is used. By increasing the data-record length $T = NT_s$, both bias and standard deviation decrease, according to Theorems 4.7.5 and 4.7.7 where the

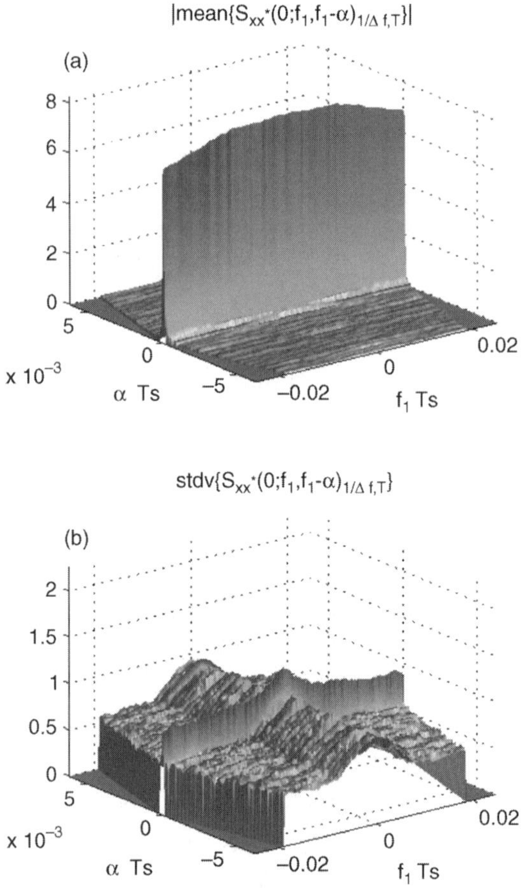

Figure 4.28 Graph of (a) the magnitude of the sample mean and (b) the sample standard deviation of the time-smoothed periodogram of $x(t)$, as function of (f_1, α), for a data-record length $N = 2^{13}$. *Source:* (Napolitano 2003) © IEEE

mean-square consistency of the estimator is proved (see Figure 4.30 for $N = 2^7$, Figure 4.31 for $N = 2^9$, and Figure 4.32 for $N = 2^{11}$).

Although it would be correct to represent mean plus/minus standard deviation for real and imaginary parts of the frequency-smoothed cross-periodogram, in Figures 4.30–4.32 mean plus/minus standard deviation of magnitude and phase are reported to show the accuracy of the phase estimate when the magnitude is small.

4.13 Spectral Analysis with Nonuniform Frequency Spacing

Spectral analysis with nonuniform (or unequal) frequency spacing finds applications in several fields such as frequency estimation. This problem has been investigated with reference to deterministic signals in (Oppenheim *et al.* 1971), (Oppenheim and Johnson 1972), (Braccini and Oppenheim 1974), (Makur and Mitra 2001), (Franz *et al.* 2003). The nonuniform frequency

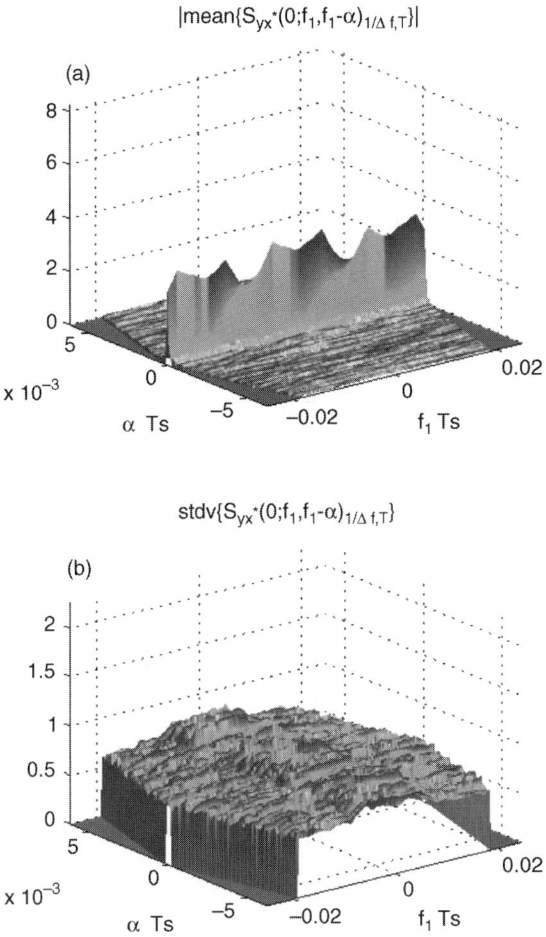

Figure 4.29 Graph of (a) the magnitude of the sample mean and (b) the sample standard deviation of the time-smoothed cross-periodogram of $y(t)$ and $x(t)$, as function of (f_1, α), for a data-record length $N = 2^{13}$. *Source:* (Napolitano 2003) © IEEE

spacing is obtained by frequency-warping techniques. For this purpose, in (Makur and Mitra 2001) and (Franz *et al.* 2003) the warped discrete Fourier transform is introduced. In this section, the problem of spectral analysis with nonuniform frequency spacing is addressed for some classes of discrete-time stochastic processes. Spectral analysis with nonuniform frequency spacing of a given process is equivalent to spectral analysis with uniform frequency spacing of a frequency-warped version of the original process. Since frequency-warping is a linear time-variant transformation, this operation modifies the nonstationarity properties of the original stochastic process under analysis. In the notable case of a WSS process, the frequency-warped process is still WSS, but jointly SC with the original process. A frequency warped ACS process is SC and jointly SC with the original ACS process.

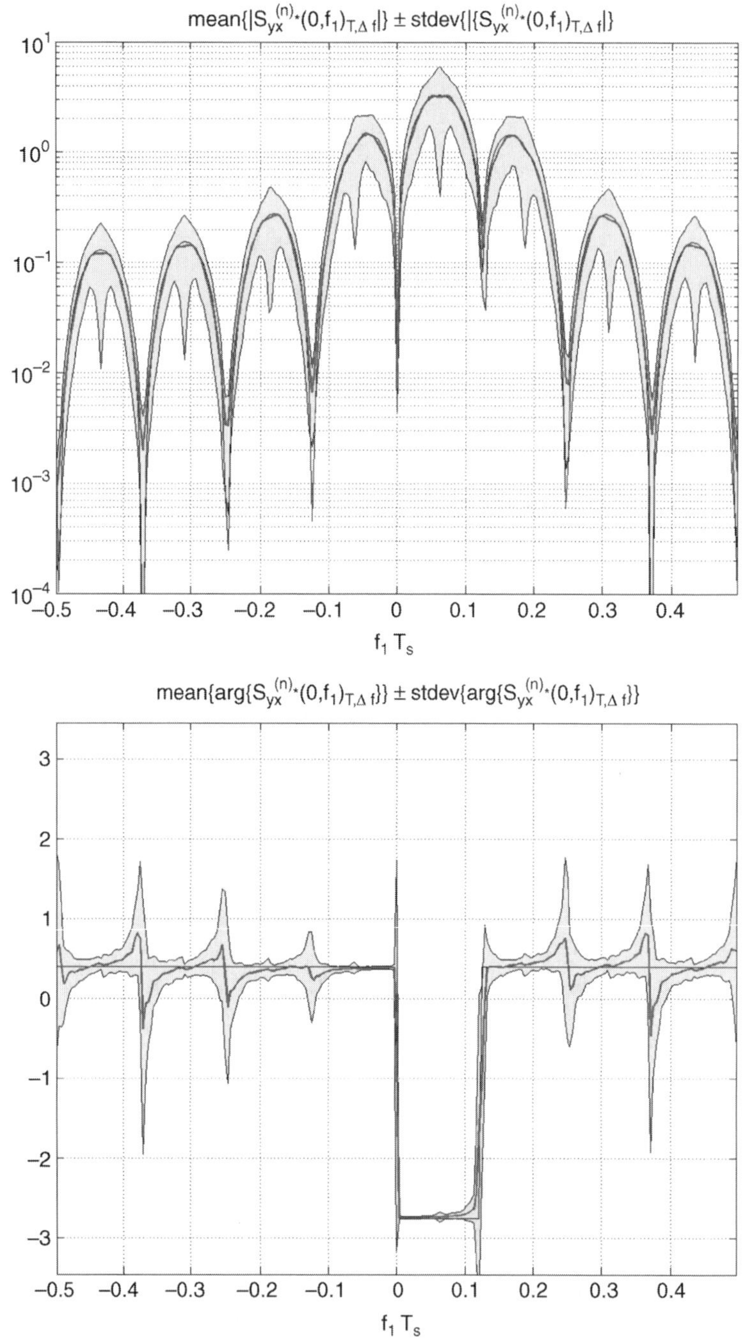

Figure 4.30 (Top) Magnitude and (bottom) phase of the frequency smoothed cross-periodogram along the known support curve $f_2 = f_1/s - (\alpha_p + \nu/s)$, with $\alpha_p = 1/T_p$, estimated by $N = 2^7$ samples. Thin line: Actual value; Thick line: Sample mean; Shaded area: Sample standard deviation

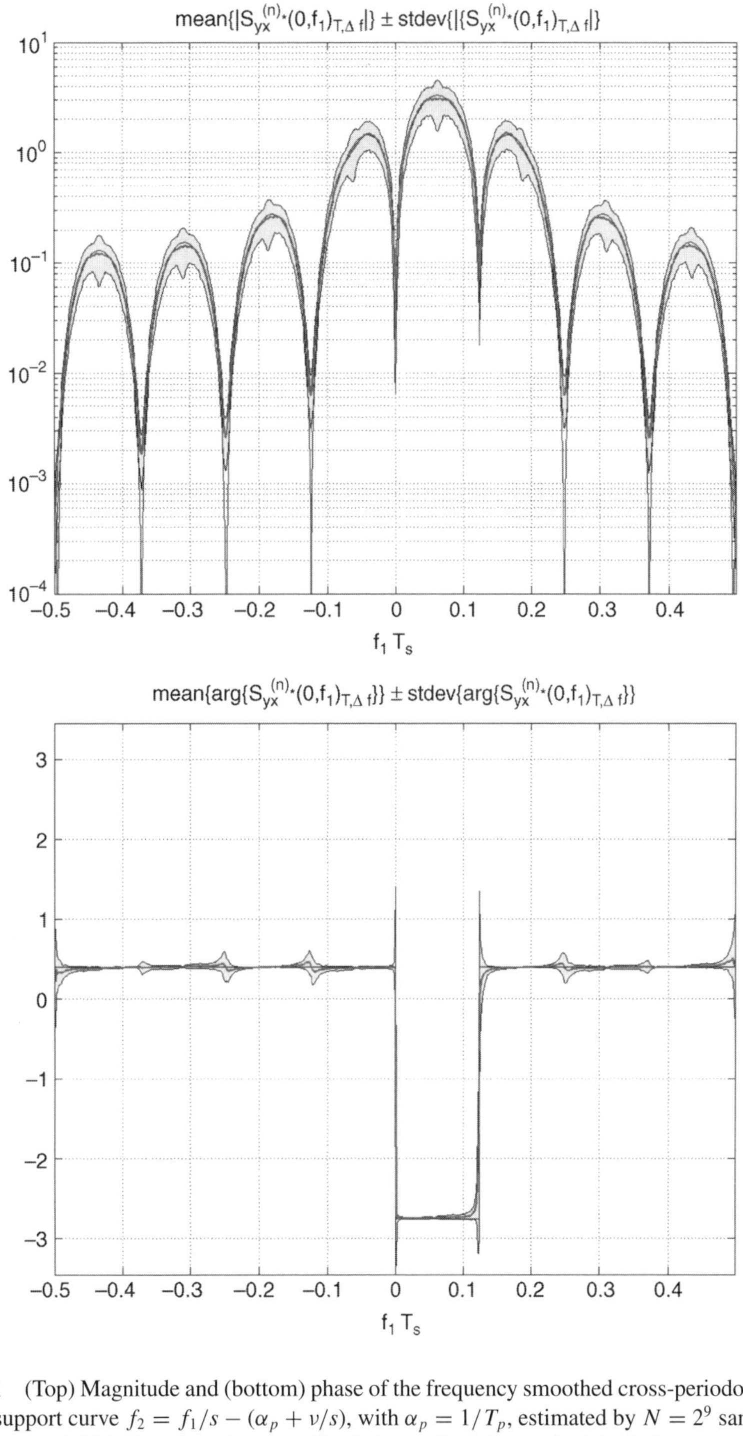

Figure 4.31 (Top) Magnitude and (bottom) phase of the frequency smoothed cross-periodogram along the known support curve $f_2 = f_1/s - (\alpha_p + \nu/s)$, with $\alpha_p = 1/T_p$, estimated by $N = 2^9$ samples. Thin line: Actual value; Thick line: Sample mean; Shaded area: Sample standard deviation

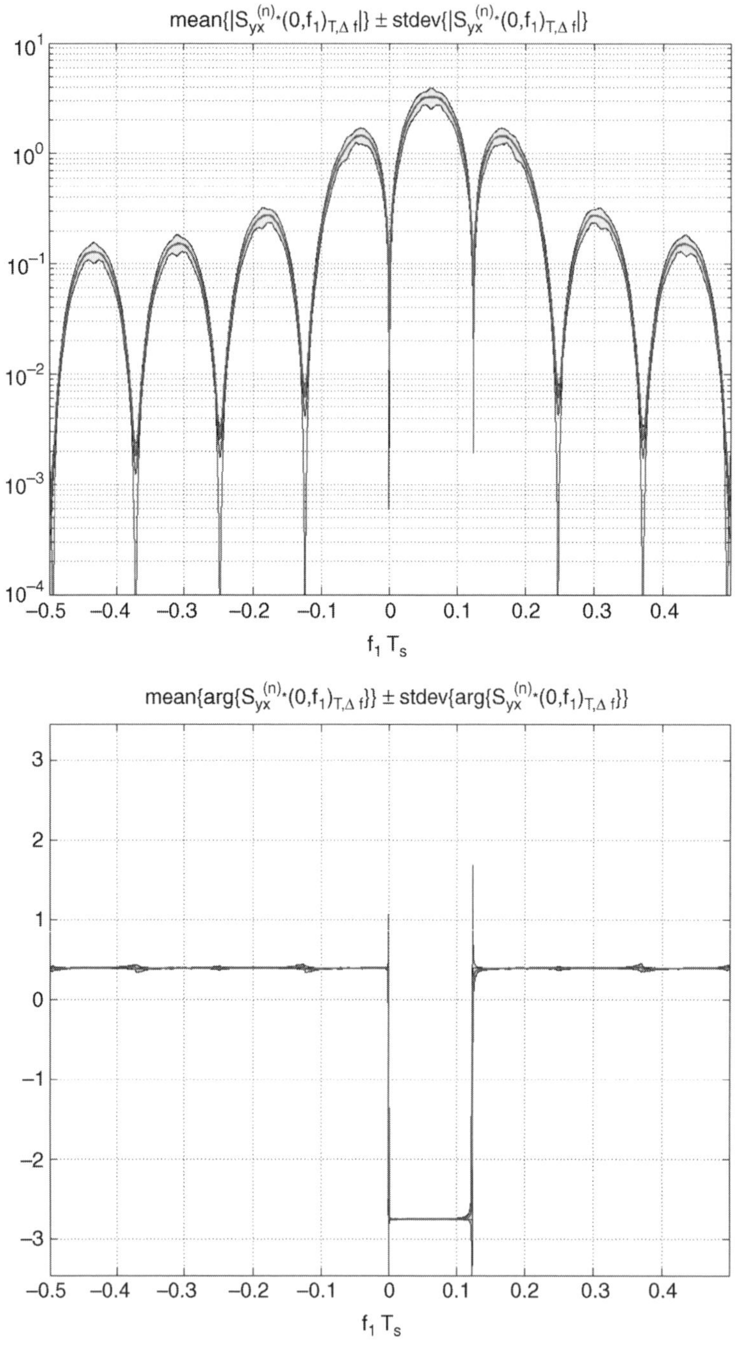

Figure 4.32 (Top) Magnitude and (bottom) phase of the frequency smoothed cross-periodogram along the known support curve $f_2 = f_1/s - (\alpha_p + \nu/s)$, with $\alpha_p = 1/T_p$, estimated by $N = 2^{11}$ samples. Thin line: Actual value; Thick line: Sample mean; Shaded area: Sample standard deviation

(Cross-)spectral analysis techniques of a discrete-time process $x(n)$ based on the Fourier transform $X(\nu)$ (4.185) and its inverse

$$x(n) = \int_{-1/2}^{1/2} X(\nu)\, e^{j2\pi\nu n}\, \mathrm{d}\nu \tag{4.307}$$

have uniform frequency spacing. That is, the discrete Fourier transform (DFT) $X(k/N)$, $k = -N/2, \ldots, N/2-1$ (N even) has frequency bins uniformly spaced in the main frequency interval $[-1/2, 1/2]$ with frequency spacing $1/N$.

Let $\psi(\nu)$ be a real-valued strictly-increasing differentiable possibly nonlinear function defined in $[-1/2, 1/2]$ and with values contained in $[-1/2, 1/2]$. The process

$$y(n) = \int_{-1/2}^{1/2} X(\psi(\nu))\, e^{j2\pi\nu n}\, \mathrm{d}\nu \tag{4.308}$$

is a *frequency-warped* version of $x(n)$. Due to the nonlinear behavior of $\psi(\cdot)$, $X(\psi(k/N))$ is a DFT of $x(n)$ with nonuniform (or unequal) frequency spacing (Oppenheim *et al.* 1971), (Oppenheim and Johnson 1972), (Braccini and Oppenheim 1974), (Makur and Mitra 2001), (Franz *et al.* 2003). That is, spectral analysis with nonuniform frequency spacing of $x(n)$ is equivalent to spectral analysis with uniform frequency spacing of a frequency-warped version of $x(n)$.

Relationship (4.308) describes a linear (time-variant) transformation of $x(n)$ into its frequency-warped version $y(n)$. In the following, the effects of this frequency warping transformation on the nonstationarity properties of $x(n)$ is analyzed with reference to the class of the ACS processes. WSS processes are considered as special case.

Let $x(n)$ be an ACS process with Loève bifrequency spectrum (4.202) with $x_1 \equiv x_2 \equiv x$ and let us denote by $\widetilde{\psi}(\nu)$ the periodic replication with period 1 of $\psi(\nu)$, that is, $\widetilde{\psi}(\nu) \triangleq \sum_{m\in\mathbb{Z}} \psi(\nu - m)$. From (4.308) and (4.202) and reasoning as in the proof of Theorem 4.8.3 (Section 5.9), it follows that

$$\begin{aligned}
\mathrm{E}\Big\{Y(\nu_1)\, Y^*(\nu_2)\Big\} &= \mathrm{E}\Big\{X\Big(\widetilde{\psi}(\nu_1)\Big)\, X^*\Big(\widetilde{\psi}(\nu_2)\Big)\Big\} \\
&= \sum_{\widetilde{\alpha}\in\widetilde{A}} \widetilde{S}_x^{\widetilde{\alpha}}\Big(\widetilde{\psi}(\nu_1)\Big)\, \widetilde{\delta}\Big(\widetilde{\psi}(\nu_2) - \widetilde{\psi}(\nu_1) + \widetilde{\alpha}\Big) \\
&= \sum_{\widetilde{\alpha}\in\widetilde{A}} \widetilde{S}_x^{\widetilde{\alpha}}\Big(\widetilde{\psi}(\nu_1)\Big)\, \Big|\widetilde{\phi}'(\widetilde{\psi}(\nu_1) - \widetilde{\alpha})\Big|\, \widetilde{\delta}\Big(\nu_2 - \widetilde{\phi}(\widetilde{\psi}(\nu_1) - \widetilde{\alpha})\Big)
\end{aligned} \tag{4.309}$$

where $\widetilde{\phi}(\cdot)$ is the periodic replication with period 1 of $\phi(\cdot)$, the inverse function of $\psi(\cdot)$, and the variable change property in the argument of the Dirac delta (Zemanian 1987, Section 1.7) is used. From (4.309) it follows that the Loève bifrequency spectrum of $y(n)$ has spectral masses concentrated on a countable set of support curves in the bifrequency plane. That is, $y(n)$ is a SC process. The Loève bifrequency cross-spectrum of $y(n)$ and $x(n)$ is

$$\begin{aligned}
\mathrm{E}\Big\{Y(\nu_1)\, X^*(\nu_2)\Big\} &= \mathrm{E}\Big\{X\Big(\widetilde{\psi}(\nu_1)\Big)\, X^*(\nu_2)\Big\} \\
&= \sum_{\widetilde{\alpha}\in\widetilde{A}} \widetilde{S}_x^{\widetilde{\alpha}}\Big(\widetilde{\psi}(\nu_1)\Big)\, \widetilde{\delta}\Big(\nu_2 - \widetilde{\psi}(\nu_1) + \widetilde{\alpha}\Big)
\end{aligned} \tag{4.310}$$

that is, $y(n)$ and $x(n)$ are jointly SC processes.

In the special case where $x(n)$ is WSS, the set $\widetilde{\mathcal{A}}$ contains the only element $\widetilde{\alpha} = 0$. Thus,

$$\mathrm{E}\left\{Y(\nu_1)\, Y^*(\nu_2)\right\} = \widetilde{S}_x^0\left(\widetilde{\psi}(\nu_1)\right) \left|\widetilde{\phi}'(\widetilde{\psi}(\nu_1))\right| \widetilde{\delta}(\nu_2 - \nu_1) \tag{4.311}$$

that is, $y(n)$ is in turn WSS accordingly with the results of (Franaszek and Liu 1967) and (Liu and Franaszek 1969). However, $x(n)$ and $y(n)$ are jointly SC with Loève bifrequency cross-spectrum

$$\mathrm{E}\left\{Y(\nu_1)\, X^*(\nu_2)\right\} = \widetilde{S}_x^0\left(\widetilde{\psi}(\nu_1)\right) \widetilde{\delta}\left(\nu_2 - \widetilde{\psi}(\nu_1)\right). \tag{4.312}$$

An illustrative example is presented to show the effects of spectral analysis with nonuniform frequency spacing on an ACS process. A sampled PAM signal $x(n)$ with raised cosine pulse with excess bandwith $\eta = 0.85$ and symbol period $T_p = 4T_s$, where T_s is the sampling period is considered. It is a discrete-time ACS process with three cycle frequencies $\widetilde{\alpha} \in \{0, \pm T_s/T_p\}$. Its frequency-warped version $y(n)$ is also considered, with frequency warping function $\psi(\nu) = B_m \tan^{-1}(\nu/B_s)$ which is typical in spectral analysis with non uniform frequency spacing (Oppenheim *et al.* 1971), (Oppenheim and Johnson 1972), (Braccini and Oppenheim 1974). In the example, $B_m = B_s = 0.2$ are assumed. In the figures, supports of impulsive functions are drawn as "checkerboard" plots with gray levels representing the magnitude of the density of the impulsive functions. In addition, for notation simplicity, the bifrequency SCD of signals a and b is denoted by $S_{ab^*}(\cdot, \cdot)$. In Figure 4.33, (top) magnitude and (bottom) "checkerboard" plot of the byfrequency spectral correlation density function of $x(n)$ are reported as functions of ν_1 and ν_2. In Figure 4.34 (top) magnitude and (bottom) "checkerboard" plot of the byfrequency spectral correlation density function of $y(n)$ are reported as functions of ν_1 and ν_2. The frequency-warping operation transforms an ACS process into a SC process. The support of the power spectral density (PSD) of the process, which is contained in the main diagonal, remains contained in the main diagonal even if the shape of the PSD is modified. This result is in accordance with the fact that frequency warping transforms WSS processes into WSS processes (Franaszek 1967), (Franaszek and Liu 1967), (Liu and Franaszek 1969).

4.14 Summary

In this chapter, the class of the spectrally correlated stochastic processes has been introduced and characterized. Processes belonging to this class have a Loève bifrequency spectrum with spectral masses concentrated on a countable set of curves in the bifrequency plane. The almost-cyclostationary processes are obtained as a special case when the separation between correlated spectral components can assume values only in a countable set. In such a case, the support curves of the Loève bifrequency spectrum are lines with unit slope. Spectrally correlated processes are an appropriate model for the output of multipath Doppler channels, when the input is a wide-band or ultra wide-band signal. Frequency-warped versions of ACS processes are SC.

The amount of spectral correlation existing between two separate spectral components of a SC process is characterized by the bifrequency spectral correlation density function which is the density of the Loève bifrequency spectrum on its support curves.

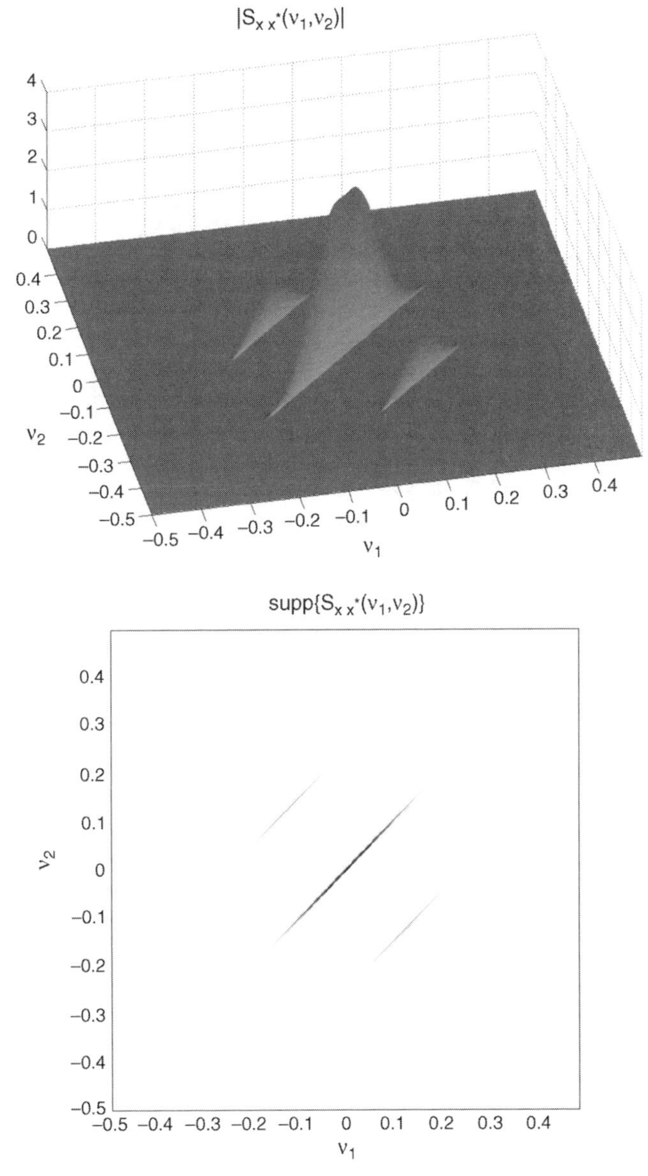

Figure 4.33 PAM signal $x(n)$ with raised cosine pulse. Bifrequency spectral cross-correlation density as a function of ν_1 and ν_2.(Top) magnitude and (bottom) support of $S_{xx*}(\nu_1, \nu_2)$

In the case of unknown support curves, the time-smoothed cross-periodogram is considered as estimator of the bifrequency spectral correlation density function. Bias and covariance are determined in the case of finite bandwidth of the spectral components and finite-time average for computing their time-correlation (Lemmas 4.5.5 and 4.5.6). The asymptotic bias (Theorem 4.5.7) and a bound for the asymptotic covariance (Theorem 4.5.9) are determined

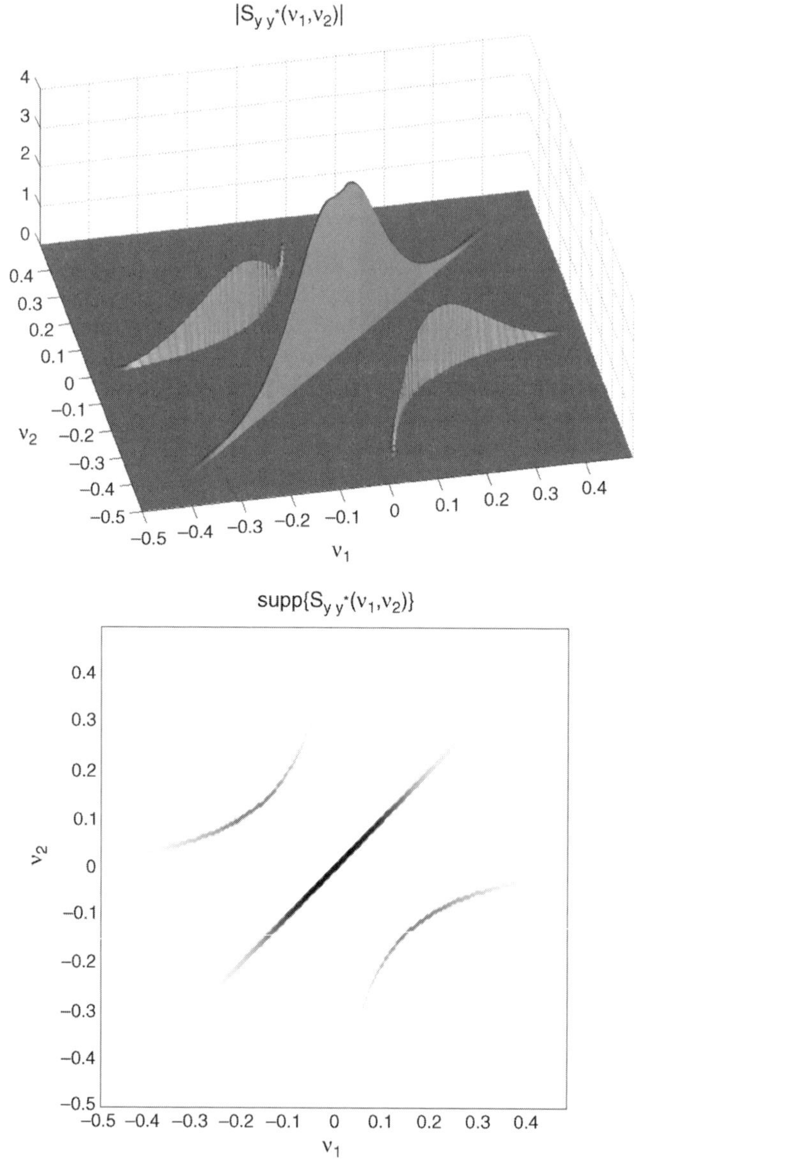

Figure 4.34 Frequency-warped version $y(n)$ of a PAM signal $x(n)$ with raised cosine pulse. Bifrequency spectral correlation density as a function of ν_1 and ν_2. (Top) magnitude and (bottom) support of $S_{yy*}(\nu_1, \nu_2)$

as the data-record length for the correlation estimate approaches infinite and the bandwidth of the spectral components approaches zero. The asymptotic covariance is of the order of the reciprocal of the smoothing product, that is, the product spectral-component-bandwidth times data-record length. Furthermore, at every point belonging to the support of the Loève bifrequency spectrum, the asymptotic expectation of the time-smoothed cross-periodogram is zero unless the slope of the curve is unit in a neighborhood of this point (as it happens for ACS processes). Consequently, in the case of SC processes that are not ACS, the variance cannot be made arbitrarily small (e.g., augmenting the data-record length) without obtaining a significant bias. Well known results of consistency of the time-smoothed cyclic periodogram of ACS processes are found as special cases of the results for SC processes.

The results on the asymptotic bias and variance of the time-smoothed cross-periodogram of SC processes imply that, in general, the spectral correlation density of SC processes can be estimated with some uncertainty. For SC processes that are not ACS, it can be estimated with some degree of reliability (small bias and variance) only if the departure of the nonstationarity from the almost-cyclostationarity is not too large. Specifically, if at a given point of the bifrequency plane the slope of the support curve is far from unity, then, to avoid a significant bias, the smoothing product cannot be made too large and, consequently, the variance cannot be too small (Corollary 4.5.8). Thus, a trade-off exists between the departure of the nonstationarity from the almost-cyclostationarity and the spectral correlation estimate accuracy obtainable by single sample-path measurements. Such a trade-off is not the usual bias-variance trade-off as for ACS processes. In fact, unlike the case of ACS processes, for SC processes the estimate accuracy cannot be improved as wished by increasing the data-record length and the spectral resolution.

In the case of known support curves, the cross-periodogram frequency smoothed along a support curve and properly normalized is proposed as an estimator of the spectral correlation density along that curve. Its expected value (Theorem 4.6.3) and covariance (Theorem 4.6.4) are determined in the case of finite data-record length and spectral resolution. Asymptotic expected value, rate of convergence to zero of the bias, and asymptotic covariance are determined as the data-record length approaches infinity and the spectral resolution approaches zero. It is shown that the asymptotic expected value equals the spectral correlation density times a function of frequency and of the slope of the support curve (Theorem 4.7.5), the rate of convergence of the bias is of the order of the reciprocal of the data-record length (Theorem 4.7.6), and the asymptotic covariance is of the order of the reciprocal of the smoothing product, that is the product data-record length times the frequency resolution (Theorem 4.7.7). Thus, the properly normalized frequency-smoothed cross-periodogram is a mean-square consistent estimator of the spectral correlation density. Moreover, frequency-smoothed cross-periodograms at different frequencies are shown to be asymptotically jointly complex Normal (Theorem 4.7.11). Well-known results for ACS processes are obtained by specializing the results for SC processes.

The problem of uniformly sampling SC processes is addressed. At first, strictly band-limited SC processes are characterized in terms of support curves and spectral correlation density functions (Theorem 4.9.2 and Corollary 4.9.3). Then, the class of the discrete-time SC processes is introduced and characterized (Theorem 4.8.3). It is shown that uniformly sampling a continuous-time SC process leads to a discrete-time SC process (Theorem 4.9.4). Moreover, aliasing issues are discussed. It is shown that for a strictly band-limited SC process, sampling at twice the bandwidth leads to non overlapping replicas in the Loève bifrequency spectrum of the SC discrete-time process (Theorem 4.9.5). However, a more stringent condition on the sampling frequency needs to be satisfied in order to assure that, for the discrete-time process, the spectral correlation densities on the support curves are scaled version of those of the continuous-time process for all values of frequencies in $[-1/2, 1/2]$ (Theorem 4.9.6). For the sake of generality, jointly SC processes and cross-statistical functions are considered.

The problem of linear time-variant processing of SC processes is addressed. The class of the SC processes is shown to be closed under linear time-variant transformations that are classified as deterministic in the FOT probability approach. For discrete-time processes, multirate processing is considered in detail. The effects of expansion and decimation operations on jointly SC processes are analyzed by considering input/output relationships of the Loève bifrequency cross-spectrum for expansors and decimators. The class of the jointly SC processes is shown to be closed under linear time-variant transformations realized by using expansors and decimators. It is proved that even if the input processes are jointly ACS, in general the output processes are jointly SC. Decimation and interpolation filters are considered. Sufficient conditions on interpolation and decimation factors and reconstruction filter bandwidths are derived to suppress imaging and aliasing replicas in the Loève bifrequency cross-spectra of interpolated or decimated processes. Sufficient conditions are derived to reconstruct a frequency-scaled image- and alias-free version of the Loève bifrequency cross-spectrum of two jointly SC processes starting from their fractionally sampled versions.

Finally, the problem of the spectral analysis with nonuniform frequency spacing is addressed when nonuniform frequency spacing is obtained by frequency-warping techniques. Such techniques are shown to modify the nonstationarity properties of the original stochastic process under analysis. Specifically, they are shown to transform ACS processes into SC processes.

5

Complements and Proofs on Spectrally Correlated Processes

In this chapter, complements and proofs for the results of Chapter 4 are presented.

5.1 Proofs for Section 4.2 "Characterization of SC Stochastic Processes"

5.1.1 *Proof of Theorem 4.2.9 Second-Order Temporal Cross-Moment of Jointly SC Processes*

Substituting (4.15) into (4.24b) and using the sampling property of the Dirac delta (Zemanian 1987, Section 1.7) leads to

$$
\begin{aligned}
\mathcal{R}_{yx^{(*)}}(t_1, t_2) &\triangleq \mathrm{E}\{y(t_1)\, x^{(*)}(t_2)\} \\
&= \int_{\mathbb{R}^2} \mathcal{S}_{yx^{(*)}}(f_1, f_2)\, e^{j2\pi(f_1 t_1 + (-) f_2 t_2)}\, \mathrm{d}f_1\, \mathrm{d}f_2 \\
&= \sum_{n \in \mathbb{I}_{yx^{(*)}}} \int_{\mathbb{R}^2} S^{(n)}_{yx^{(*)}}(f_1)\, \delta\left(f_2 - \Psi^{(n)}_{yx^{(*)}}(f_1)\right) e^{j2\pi(f_1 t_1 + (-) f_2 t_2)}\, \mathrm{d}f_1\, \mathrm{d}f_2 \\
&= \sum_{n \in \mathbb{I}_{yx^{(*)}}} \int_{\mathbb{R}} S^{(n)}_{yx^{(*)}}(f_1)\, e^{j2\pi(f_1 t_1 + (-)\Psi^{(n)}_{yx^{(*)}}(f_1) t_2)}\, \mathrm{d}f_1 .
\end{aligned}
\tag{5.1}
$$

By using the convolution theorem for the Fourier transform (Champeney 1990, Chapter 6) one obtains (4.24d).

Generalizations of Cyclostationary Signal Processing: Spectral Analysis and Applications, First Edition.

5.1.2 Proof of Theorem 4.2.10 Time-Variant Cross-Spectrum of Jointly SC Processes

Starting from (5.1) with $t_1 = t + \tau$ and $t_2 = t$ substituted into, one has

$$\begin{aligned}
\mathcal{S}_{yx^{(*)}}(t, f) &\triangleq \int_{\mathbb{R}} \mathrm{E}\left\{ y(t+\tau)\, x^{(*)}(t) \right\} e^{-j2\pi f\tau} \, \mathrm{d}\tau \\
&= \int_{\mathbb{R}} \sum_{n \in \mathbb{I}_{yx^{(*)}}} \int_{\mathbb{R}} S^{(n)}_{yx^{(*)}}(f_1)\, e^{j2\pi(f_1(t+\tau)+(-)\Psi^{(n)}_{yx^{(*)}}(f_1)t)} \, \mathrm{d}f_1 \, e^{-j2\pi f\tau} \, \mathrm{d}\tau \\
&= \sum_{n \in \mathbb{I}_{yx^{(*)}}} \int_{\mathbb{R}} S^{(n)}_{yx^{(*)}}(f_1)\, e^{j2\pi(f_1 t+(-)\Psi^{(n)}_{yx^{(*)}}(f_1)t)} \int_{\mathbb{R}} e^{-j2\pi(f-f_1)\tau} \, \mathrm{d}\tau \, \mathrm{d}f_1 \\
&= \sum_{n \in \mathbb{I}_{yx^{(*)}}} \int_{\mathbb{R}} S^{(n)}_{yx^{(*)}}(f_1)\, e^{j2\pi(f_1 t+(-)\Psi^{(n)}_{yx^{(*)}}(f_1)t)} \, \delta(f - f_1) \, \mathrm{d}f_1
\end{aligned} \tag{5.2}$$

where, in the fourth equality the Fourier transform pair $e^{-j2\pi(f-f_1)\tau} \leftrightarrow \delta(f - f_1)$ is used. Equation (4.27b) immediately follows from (5.2) by using the sampling property of Dirac delta (Zemanian 1987, Section 1.7).

5.1.3 Proof of (4.92)

Let $\psi(\cdot)$ be invertible and its inverse $\varphi(\cdot)$ differentiable. It results that (Zemanian 1987, Section 1.7)

$$\delta(f_b - \psi(f_a)) = \left|\varphi'(f_b)\right| \, \delta(f_a - \varphi(f_b)). \tag{5.3}$$

Therefore

$$\delta(\psi_2(f_2) - \psi_1(f_1)) = \left|\varphi_1'(\psi_2(f_2))\right| \, \delta(f_1 - \varphi_1(\psi_2(f_2))). \tag{5.4}$$

By using (4.91), (4.89), and (5.4), one has

$$\begin{aligned}
&\mathrm{E}\left\{ Y_1(f_1)\, Y_2^{(*)}(f_2) \right\} \\
&= \sum_{\ell_1 \in \mathbb{J}_1} \sum_{\ell_2 \in \mathbb{J}_2} H_1^{(\ell_1)}(f_1)\, H_2^{(\ell_2)}(f_2)^{(*)} \, \mathrm{E}\left\{ X_1\left(\psi_1^{(\ell_1)}(f_1)\right) X_2^{(*)}\left(\psi_2^{(\ell_2)}(f_2)\right) \right\} \\
&= \sum_{\ell_1 \in \mathbb{J}_1} \sum_{\ell_2 \in \mathbb{J}_2} H_1^{(\ell_1)}(f_1)\, H_2^{(\ell_2)}(f_2)^{(*)} \\
&\quad \sum_{n \in \mathbb{I}} S_{\boldsymbol{x}}^{(n)}\left(\psi_1^{(\ell_1)}(f_1)\right) \delta\left(\psi_2^{(\ell_2)}(f_2) - \Psi_{\boldsymbol{x}}^{(n)}\left(\psi_1^{(\ell_1)}(f_1)\right)\right) \\
&= \sum_{\ell_1 \in \mathbb{J}_1} \sum_{\ell_2 \in \mathbb{J}_2} H_1^{(\ell_1)}(f_1)\, H_2^{(\ell_2)}(f_2)^{(*)} \\
&\quad \sum_{n \in \mathbb{I}} S_{\boldsymbol{x}}^{(n)}\left(\psi_1^{(\ell_1)}(f_1)\right) \left|\varphi_2^{(\ell_2)\prime}\left(\Psi_{\boldsymbol{x}}^{(n)}\left(\psi_1^{(\ell_1)}(f_1)\right)\right)\right| \\
&\quad \delta\left(f_2 - \varphi_2^{(\ell_2)}\left(\Psi_{\boldsymbol{x}}^{(n)}\left(\psi_1^{(\ell_1)}(f_1)\right)\right)\right)
\end{aligned} \tag{5.5}$$

from which (4.92) follows by using the sampling property of the Dirac delta (Zemanian 1987, Section 1.7).

5.2 Proofs for Section 4.4 "The Bifrequency Cross-Periodogram"

In this section, proofs of lemmas and theorems presented in Section 4.4 on bias and covariance of the bifrequency cross-periodogram are reported.

The following Lemma 5.2.1 allows, in the proofs of the subsequent lemmas and theorems, the free interchange of the order of limit and sum operations.

Lemma 5.2.1 (Napolitano 2003, Lemma A.1).

(a) *Under Assumption 4.4.3a (series regularity), the function series* $\sum_{n \in \mathbb{I}_{z_1 z_2}} S^{(n)}_{z_1 z_2}(f_1)$ *is uniformly convergent. Moreover, for any function sequence* $\{w^{(n)}(f_1)\}_{n \in \mathbb{I}_{z_1 z_2}}$ *of uniformly bounded functions* ($\exists M : |w^{(n)}(f_1)| \leqslant M, \forall n, \forall f_1$) *the function series* $\sum_{n \in \mathbb{I}_{z_1 z_2}} S^{(n)}_{z_1 z_2}(f_1)\, w^{(n)}(f_1)$ *is uniformly convergent.*

(b) *Under Assumption 4.4.3b (series regularity), the function series*

$$\sum_{n \in \mathbb{I}_4} P^{(n)}_{yx^{(*)}y^*x^{(*)*}}(f_1, f_2, f_3)$$

is uniformly convergent. Moreover, for any function sequence $\{w^{(n)}(f_1, f_2, f_3)\}_{n \in \mathbb{I}_4}$ *of uniformly bounded functions, the function series*

$$\sum_{n \in \mathbb{I}_4} P^{(n)}_{yx^{(*)}y^*x^{(*)*}}(f_1, f_2, f_3)\, w^{(n)}(f_1, f_2, f_3)$$

is uniformly convergent.

Proof: The uniform convergence of the function series $\sum_n S^{(n)}_{z_1 z_2}(f_1)$ is an immediate consequence of (4.97) and the Weierstrass criterium (Smirnov 1964). Moreover, since the functions $w^{(n)}(f_1)$ are uniformly bounded, accounting for (4.97), it results in $\sum_n \|S^{(n)}_{z_1 z_2}\, w^{(n)}\|_\infty \leqslant M \sum_n \|S^{(n)}_{z_1 z_2}\|_\infty \leqslant \infty$ and hence, for the Weierstrass criterium the function series $\sum_n S^{(n)}_{z_1 z_2}(f_1)\, w^{(n)}(f_1)$ turns out to be uniformly convergent. Part b) of the Lemma can be proved analogously. □

5.2.1 *Proof of Lemma 4.4.6 Expected Value of the Bifrequency Cross-Periodogram*

By substituting into (4.93) the STFTs $Y_{\frac{1}{\Delta f}}(t, f_1)$ and $X_{\frac{1}{\Delta f}}(t, f_2)$ expressed according to (4.94b), and taking the expected value, one has

$$\mathrm{E}\left\{ I_{yx^{(*)}}(t; f_1, f_2)_{\frac{1}{\Delta f}} \right\}$$
$$= \mathrm{E}\left\{ \Delta f \int_{\mathbb{R}} Y(\nu_1)\, \mathcal{B}_{\Delta f}(f_1 - \nu_1; t)\, \mathrm{d}\nu_1 \int_{\mathbb{R}} X^{(*)}(\nu_2)\, \mathcal{B}^{(*)}_{\Delta f}(f_2 - \nu_2; t)\, \mathrm{d}\nu_2 \right\}$$

$$
\begin{aligned}
&= \Delta f \int_{\mathbb{R}} \int_{\mathbb{R}} \mathrm{E}\left\{Y(\nu_1)\, X^{(*)}(\nu_2)\right\} \mathcal{B}_{\Delta f}(f_1 - \nu_1; t)\, \mathcal{B}^{(*)}_{\Delta f}(f_2 - \nu_2; t)\, \mathrm{d}\nu_1\, \mathrm{d}\nu_2 \\
&= \Delta f \int_{\mathbb{R}} \int_{\mathbb{R}} \sum_{n \in \mathbb{I}} S^{(n)}_{yx^{(*)}}(\nu_1)\, \delta\left(\nu_2 - \Psi^{(n)}_{yx^{(*)}}(\nu_1)\right) \\
&\qquad \mathcal{B}_{\Delta f}(f_1 - \nu_1; t)\, \mathcal{B}^{(*)}_{\Delta f}(f_2 - \nu_2; t)\, \mathrm{d}\nu_1\, \mathrm{d}\nu_2 \\
&= \Delta f \int_{\mathbb{R}} \sum_{n \in \mathbb{I}} S^{(n)}_{yx^{(*)}}(\nu_1)\, \mathcal{B}_{\Delta f}(f_1 - \nu_1; t)\, \mathcal{B}^{(*)}_{\Delta f}\left(f_2 - \Psi^{(n)}_{yx^{(*)}}(\nu_1); t\right) \mathrm{d}\nu_1
\end{aligned} \tag{5.6}
$$

where, in the third equality, (4.95) with $z_1 = y$ and $z_2 = x^{(*)}$ is used.

In the derivation of (5.6), the order of integral and expectation operators can be interchanged due to the Fubini and Tonelli theorem (Champeney 1990, Chapter 3). In fact, defined

$$
F^{(n)}(\nu_1) \triangleq \Delta f\, S^{(n)}_{yx^{(*)}}(\nu_1)\, \mathcal{B}_{\Delta f}(f_1 - \nu_1; t)\, \mathcal{B}^{(*)}_{\Delta f}\left(f_2 - \Psi^{(n)}_{yx^{(*)}}(\nu_1); t\right) \tag{5.7}
$$

and accounting for Assumptions 4.4.3a and 4.4.5, for the integrand function in the right-hand side of (5.6) we have

$$
\begin{aligned}
\left|\sum_{n \in \mathbb{I}} F^{(n)}(\nu_1)\right| &\leqslant \sum_{n \in \mathbb{I}} \left|F^{(n)}(\nu_1)\right| \\
&\leqslant \Delta f \sum_{n \in \mathbb{I}} \left\|S^{(n)}_{yx^{(*)}}\right\|_{\infty} \frac{1}{\Delta f} \|W_B\|_{\infty} \frac{1}{\Delta f} \left|W_B((f_1 - \nu_1)/\Delta f)\right| \in L^1(\mathbb{R}).
\end{aligned} \tag{5.8}
$$

The interchange of sum and integral operations to obtain (4.106) from (5.6) is justified even if the set $\mathbb{I}$ is not finite by using the dominated convergence theorem (Champeney 1990, Chapter 4). Specifically, by denoting with $\{\mathbb{I}_k\}$ an increasing sequence of finite subsets of $\mathbb{I}$, i.e., such that $\mathbb{I}_h \subseteq \mathbb{I}_k$ for $h < k$ and $\lim_{k\to\infty} \mathbb{I}_k \triangleq \cup_{k\in\mathbb{N}} \mathbb{I}_k = \mathbb{I}$, we have

$$
\begin{aligned}
\lim_{k\to\infty} \sum_{n \in \mathbb{I}_k} \int_{\mathbb{R}} F^{(n)}(\nu_1)\, \mathrm{d}\nu_1 &= \lim_{k\to\infty} \int_{\mathbb{R}} \sum_{n \in \mathbb{I}_k} F^{(n)}(\nu_1)\, \mathrm{d}\nu_1 \\
&= \int_{\mathbb{R}} \lim_{k\to\infty} \sum_{n \in \mathbb{I}_k} F^{(n)}(\nu_1)\, \mathrm{d}\nu_1
\end{aligned} \tag{5.9}
$$

In fact, it results that

$$
\left|\sum_{n \in \mathbb{I}_k} F^{(n)}(\nu_1)\right| \leqslant \sum_{n \in \mathbb{I}_k} \left|F^{(n)}(\nu_1)\right| \leqslant \sum_{n \in \mathbb{I}} \left|F^{(n)}(\nu_1)\right| \tag{5.10}
$$

with the right-hand side bounded by the right-hand side of (5.8). That is, the integrand function in the second term of equality (5.9) is bounded by a summable function of ν_1 not depending on k. □

5.2.2 Proof of Lemma 4.4.7 Covariance of the Bifrequency Cross-Periodogram

For zero mean stochastic processes $x(t)$ and $y(t)$, also the STFTs $X_{\frac{1}{\Delta f}}(t, f)$ and $Y_{\frac{1}{\Delta f}}(t, f)$ are zero mean. By specializing to $N = 4$ and zero-mean processes the expression of Nth-order cumulant in terms of Nth- and lower-order moments (4.39), the result is that

$$\begin{aligned}
&\operatorname{cum}\left\{Y_{\frac{1}{\Delta f}}(t_1, f_{y1}),\ X^{(*)}_{\frac{1}{\Delta f}}(t_1, f_{x1}),\ Y^*_{\frac{1}{\Delta f}}(t_2, f_{y2}),\ X^{(*)*}_{\frac{1}{\Delta f}}(t_2, f_{x2})\right\}\\
&\quad = \mathrm{E}\left\{Y_{\frac{1}{\Delta f}}(t_1, f_{y1})\, X^{(*)}_{\frac{1}{\Delta f}}(t_1, f_{x1})\, Y^*_{\frac{1}{\Delta f}}(t_2, f_{y2})\, X^{(*)*}_{\frac{1}{\Delta f}}(t_2, f_{x2})\right\}\\
&\qquad -\mathrm{E}\left\{Y_{\frac{1}{\Delta f}}(t_1, f_{y1})\, X^{(*)}_{\frac{1}{\Delta f}}(t_1, f_{x1})\right\}\mathrm{E}\left\{Y^*_{\frac{1}{\Delta f}}(t_2, f_{y2})\, X^{(*)*}_{\frac{1}{\Delta f}}(t_2, f_{x2})\right\}\\
&\qquad -\mathrm{E}\left\{Y_{\frac{1}{\Delta f}}(t_1, f_{y1})\, Y^*_{\frac{1}{\Delta f}}(t_2, f_{y2})\right\}\mathrm{E}\left\{X^{(*)}_{\frac{1}{\Delta f}}(t_1, f_{x1})\, X^{(*)*}_{\frac{1}{\Delta f}}(t_2, f_{x2})\right\}\\
&\qquad -\mathrm{E}\left\{Y_{\frac{1}{\Delta f}}(t_1, f_{y1})\, X^{(*)*}_{\frac{1}{\Delta f}}(t_2, f_{x2})\right\}\mathrm{E}\left\{X^{(*)}_{\frac{1}{\Delta f}}(t_1, f_{x1})\, Y^*_{\frac{1}{\Delta f}}(t_2, f_{y2})\right\}\\
&\quad = \operatorname{cov}\left\{Y_{\frac{1}{\Delta f}}(t_1, f_{y1})\, X^{(*)}_{\frac{1}{\Delta f}}(t_1, f_{x1})\,,\ Y_{\frac{1}{\Delta f}}(t_2, f_{y2})\, X^{(*)}_{\frac{1}{\Delta f}}(t_2, f_{x2})\right\}\\
&\qquad -\mathrm{E}\left\{Y_{\frac{1}{\Delta f}}(t_1, f_{y1})\, Y^*_{\frac{1}{\Delta f}}(t_2, f_{y2})\right\}\mathrm{E}\left\{X^{(*)}_{\frac{1}{\Delta f}}(t_1, f_{x1})\, X^{(*)*}_{\frac{1}{\Delta f}}(t_2, f_{x2})\right\}\\
&\qquad -\mathrm{E}\left\{Y_{\frac{1}{\Delta f}}(t_1, f_{y1})\, X^{(*)*}_{\frac{1}{\Delta f}}(t_2, f_{x2})\right\}\mathrm{E}\left\{X^{(*)}_{\frac{1}{\Delta f}}(t_1, f_{x1})\, Y^*_{\frac{1}{\Delta f}}(t_2, f_{y2})\right\}.
\end{aligned} \tag{5.11}$$

Thus, accounting for the STFT expression (4.94b), from (5.11) it follows that

$$\begin{aligned}
&\operatorname{cov}\left\{\Delta f\, Y_{\frac{1}{\Delta f}}(t_1, f_{y1})\, X^{(*)}_{\frac{1}{\Delta f}}(t_1, f_{x1})\,,\ \Delta f\, Y_{\frac{1}{\Delta f}}(t_2, f_{y2})\, X^{(*)}_{\frac{1}{\Delta f}}(t_2, f_{x2})\right\}\\
&\quad = (\Delta f)^2 \int_{\mathbb{R}}\int_{\mathbb{R}}\int_{\mathbb{R}}\int_{\mathbb{R}}\\
&\qquad \Big[\mathrm{E}\left\{Y(\nu_{y1})\, Y^*(\nu_{y2})\right\}\mathrm{E}\left\{X^{(*)}(\nu_{x1})\, X^{(*)*}(\nu_{x2})\right\}\\
&\qquad +\mathrm{E}\left\{Y(\nu_{y1})\, X^{(*)*}(\nu_{x2})\right\}\mathrm{E}\left\{X^{(*)}(\nu_{x1})\, Y^*(\nu_{y2})\right\}\\
&\qquad +\operatorname{cum}\left\{Y(\nu_{y1}),\ X^{(*)}(\nu_{x1}),\ Y^*(\nu_{y2}),\ X^{(*)*}(\nu_{x2})\right\}\Big]\\
&\qquad \mathcal{B}_{\Delta f}(f_{y1} - \nu_{y1}; t_1)\, \mathcal{B}^{(*)}_{\Delta f}(f_{x1} - \nu_{x1}; t_1)\\
&\qquad \mathcal{B}^*_{\Delta f}(f_{y2} - \nu_{y2}; t_2)\, \mathcal{B}^{(*)*}_{\Delta f}(f_{x2} - \nu_{x2}; t_2)\\
&\qquad \mathrm{d}\nu_{y1}\, \mathrm{d}\nu_{y2}\, \mathrm{d}\nu_{x1}\, \mathrm{d}\nu_{x2}
\end{aligned} \tag{5.12}$$

Finally, by substituting (4.95) and (4.96) into (5.12), and using the sampling property of the Dirac delta, (4.108) follows.

The interchange of expectation, sum, and integral operations can be justified as in the proof of Lemma 4.4.6.

Lemma 5.2.2 (Napolitano 2003, Lemma A.2). *Let $\Psi^{(n)}(f_1)$ be a function satisfying Assumption 4.4.4 (support-curve regularity (I)). For any function $V(\nu)$ continuous a.e. and infinitesimal at $\pm\infty$, one has*

$$\lim_{\Delta f \to 0} V\left(\frac{f_2 - \Psi^{(n)}(f_1 - \lambda_1 \Delta f)}{\Delta f}\right) = V(\lambda_1 \Psi^{(n)\prime}(f_1))\, \delta_{f_2 - \Psi^{(n)}(f_1)} \tag{5.13}$$

for any λ_1 such that $\lambda_1 \Psi^{(n)\prime}(f_1)$ is a continuity point of V.

Proof: If $f_2 \neq \Psi^{(n)}(f_1)$, then $f_2 \neq \Psi^{(n)}(f_1)$ definitively in a neighborhood of f_1 since $\Psi^{(n)}$ is derivable and, hence, continuous. Thus, one has

$$\lim_{\Delta f \to 0} \frac{|f_2 - \Psi^{(n)}(f_1 - \lambda_1 \Delta f)|}{|\Delta f|} = +\infty \tag{5.14}$$

from which it follows that

$$\lim_{\Delta f \to 0} V\left(\frac{f_2 - \Psi^{(n)}(f_1 - \lambda_1 \Delta f)}{\Delta f}\right) = 0 \tag{5.15}$$

since $V(\nu)$ is continuous a.e. and infinitesimal at $\pm\infty$. Furthermore, if $f_2 = \Psi^{(n)}(f_1)$ then, from the assumption of derivability of $\Psi^{(n)}(f_1)$, one has

$$\lim_{\Delta f \to 0} \lambda_1 \frac{f_2 - \Psi^{(n)}(f_1 - \lambda_1 \Delta f)}{\lambda_1 \Delta f} = \lambda_1 \Psi^{(n)\prime}(f_1). \tag{5.16}$$

Thus, (5.13) follows from the a.e. continuity of V. □

Lemma 5.2.3 (Napolitano 2003, Lemma A.3). *Let $\{y(t),\ t \in \mathbb{R}\}$ and $\{x(t),\ t \in \mathbb{R}\}$ be second-order harmonizable jointly SC stochastic processes with bifrequency cross-spectrum (4.15), and define the function*

$$\mathcal{J}_{yx^{(*)}}(t_1, t_2; f_1, f_2)_{\frac{1}{\Delta f}} \triangleq \Delta f \int_{\mathbb{R}} \sum_{n \in \mathbb{I}} S^{(n)}(\nu_1)\, \mathcal{B}_{\Delta f}(f_1 - \nu_1; t_1)$$
$$\mathcal{B}^{(*)}_{\Delta f}(f_2 - \Psi^{(n)}(\nu_1); t_2)\, \mathrm{d}\nu_1. \tag{5.17}$$

Under Assumptions 4.4.3a (series regularity), 4.4.4 (support-curve regularity (I)), and 4.4.5 (data-tapering window regularity), one obtains

$$\lim_{\Delta f \to 0} \mathcal{J}_{yx^{(*)}}(t_1, t_2; f_1, f_2)_{\frac{1}{\Delta f}} = \sum_{n \in \mathbb{I}} S^{(n)}(f_1)\, \mathcal{E}^{(n)}(f_1)\, \delta_{f_2 - \Psi^{(n)}(f_1)} \tag{5.18}$$

with $\mathcal{E}^{(n)}(f_1)$ defined in (4.113).

Proof: By making into (5.17) the variable change $\lambda_1 = (f_1 - \nu_1)/\Delta f$ and accounting for (4.103) and (4.107) one has

$$\begin{aligned}
&\mathcal{I}_{yx^{(*)}}(t_1, t_2; f_1, f_2)_{\frac{1}{\Delta f}} \\
&= \int_{\mathbb{R}} \sum_{n \in \mathbb{I}} S^{(n)}(f_1 - \lambda_1 \Delta f)\, W_B(\lambda_1)\, W_B^{(*)}\left(\frac{f_2 - \Psi^{(n)}(f_1 - \lambda_1 \Delta f)}{\Delta f}\right) \\
&\quad \exp\left[j2\pi\left(-\lambda_1 \Delta f t_1 + {\scriptstyle(-)}(\Psi^{(n)}(f_1 - \lambda_1 \Delta f) - f_2)t_2\right)\right] \mathrm{d}\lambda_1 .
\end{aligned} \tag{5.19}$$

Denoted by $G_{\Delta f}(\lambda_1)$ the integrand function in (5.19) and accounting for (4.97) and Assumption 4.4.5, the result is that

$$\left|G_{\Delta f}(\lambda_1)\right| \leqslant |W_B(\lambda_1)|\, \|W_B\|_\infty \sum_{n \in \mathbb{I}} \|S^{(n)}\|_\infty \in L^1(\mathbb{R}) \tag{5.20}$$

with the right-hand side independent of Δf. Thus, the dominated convergence theorem (Champeney 1990) can be applied:

$$\begin{aligned}
&\lim_{\Delta f \to 0} \mathcal{I}_{yx^{(*)}}(t_1, t_2; f_1, f_2)_{\frac{1}{\Delta f}} \\
&= \sum_{n \in \mathbb{I}} S^{(n)}(f_1)\, e^{j2\pi[{\scriptstyle(-)}(\Psi^{(n)}(f_1) - f_2)t_2]} \\
&\quad \int_{\mathbb{R}} W_B(\lambda_1) \lim_{\Delta f \to 0} W_B^{(*)}\left(\frac{f_2 - \Psi^{(n)}(f_1 - \lambda_1 \Delta f)}{\Delta f}\right) \mathrm{d}\lambda_1
\end{aligned} \tag{5.21}$$

where, in obtaining (5.21), the limit (as $\Delta f \to 0$) and the sum (over n) operations can be interchanged since W_B is bounded (Assumption 4.4.5) and hence, by Lemma 5.2.1a, the integrand function series in (5.19) is uniformly convergent. Moreover, the sum and the integral operations can be interchanged due to the same arguments used in the proof of Lemma 4.4.6. Furthermore, since $W_B \in L^1(\mathbb{R})$ and is regular at $\pm\infty$, then is infinitesimal at $\pm\infty$. Thus, by substituting (5.13) (Lemma 5.2.2) with $V = W_B$ into (5.21) for all $\lambda_1 \in \mathbb{R}$ except, possibly, the λ_1s belonging to a set with zero Lebesgue maesure, (5.18) immediately follows. □

Observe that, strictly speaking, to prove Lemmas 4.4.6, 5.2.1, and 5.2.3, Assumptions 4.4.3a and 4.4.4 need to hold only for $z_1 = y$ and $z_2 = x^{(*)}$.

5.2.3 *Proof of Theorem 4.4.8 Asymptotic Expected Value of the Bifrequency Cross-Periodogram*

It immediately follows from Lemma 5.2.3 for $t_1 = t_2 = t$.

5.2.4 *Proof of Theorem 4.4.9 Asymptotic Covariance of the Bifrequency Cross-Periodogram*

The asymptotic covariance expression (4.115) is obtained from (4.108) as $\Delta f \to 0$. In fact, the convergence, as $\Delta f \to 0$, of the terms $\mathcal{T}_1$ and $\mathcal{T}_2$ (see (4.109) and (4.110)) to $\mathcal{T}_1'$ and $\mathcal{T}_2'$ (see (4.116) and (4.117)) , respectively, can be easily proved by using Lemma 5.2.3. Moreover, the term $\mathcal{T}_3$ (see (4.111)) approaches zero as $\Delta f \to 0$. In fact, by making into (4.111) the variable changes $\lambda_{y1} = (f_{y1} - \nu_{y1})/\Delta f$, $\lambda_{x1} = (f_{x1} - \nu_{x1})/\Delta f$, $\lambda_{y2} = (f_{y2} - \nu_{y2})/\Delta f$ and accounting for (4.103) and (4.107), one has

$$\mathcal{T}_3 = \Delta f \int_{\mathbb{R}} \int_{\mathbb{R}} \int_{\mathbb{R}} \sum_{n \in \mathbb{I}_4} P^{(n)}_{yx^{(*)}y^*x^{(*)*}}(\boldsymbol{f} - \boldsymbol{\lambda}\Delta f)$$
$$W_B(\lambda_{y1})\, W_B^{(*)}(\lambda_{x1})\, W_B^*(\lambda_{y2})\, W_B^{(*)*}\left(\frac{f_{x2} - \Psi_4^{(n)}(\boldsymbol{f} - \boldsymbol{\lambda}\Delta f)}{\Delta f}\right)$$
$$\exp\left[j2\pi\left(-(-\lambda_{y2}\Delta f + {}_{(-)}(\Psi_4^{(n)}(\boldsymbol{f} - \boldsymbol{\lambda}\Delta f) - f_{x2}))t_2\right)\right]$$
$$\exp\left[j2\pi\left((-\lambda_{y1}\Delta f - {}_{(-)}\lambda_{x1}\Delta f)t_1\right)\right] \mathrm{d}\lambda_{y1}\, \mathrm{d}\lambda_{x1}\, \mathrm{d}\lambda_{y2} \tag{5.22}$$

where $\boldsymbol{f} - \boldsymbol{\lambda}\Delta f \triangleq [f_{y1} - \lambda_{y1}\Delta f, f_{x1} - \lambda_{x1}\Delta f, f_{y2} - \lambda_{y2}\Delta f]$.
Denoted by $H_{\Delta f}(\lambda_{y1}, \lambda_{x1}, \lambda_{y2})$ the integrand function in (5.22) and accounting for (4.98) and Assumption 4.4.5 the result is that

$$|H_{\Delta f}(\lambda_{y1}, \lambda_{x1}, \lambda_{y2})| \leqslant |W_B(\lambda_{y1})\, W_B(\lambda_{x1})\, W_B(\lambda_{y2})|\, \|W_B\|_\infty \sum_{n \in \mathbb{I}_4} \|P^{(n)}_{yx^{(*)}y^*x^{(*)*}}\|_\infty \in L^1(\mathbb{R}^3) \tag{5.23}$$

with the right-hand side independent of Δf. Thus,

$$\lim_{\Delta f \to 0} \mathcal{T}_3 \leqslant \lim_{\Delta f \to 0} \Delta f \int_{\mathbb{R}^3} |H_{\Delta f}(\lambda_{y1}, \lambda_{x1}, \lambda_{y2})|\, \mathrm{d}\lambda_{y1}\, \mathrm{d}\lambda_{x1}\, \mathrm{d}\lambda_{y2} = 0. \tag{5.24}$$

5.3 Proofs for Section 4.5 "Measurement of Spectral Correlation – Unknown Support Curves"

In this section, proofs of lemmas and theorems presented in Section 4.5 on bias and covariance of the time-smoothed bifrequency cross-periodogram are reported.

Accounting for the properties of the Fourier transform of a summable function (Champeney 1990) and observing that from Assumption 4.5.2, one obtains

$$\lim_{T \to \infty} T\, a_T(t) = \lim_{T \to \infty} w_a\left(\frac{t}{T}\right) = 1 \tag{5.25}$$

the following result can easily be proved.

Lemma 5.3.1 (Napolitano 2003, Lemma B.1). *Under Assumption 4.5.2 (time-smoothing window regularity) and denoting by $W_A(f)$ the Fourier transform of $w_a(t)$, one obtains that the Fourier transform $A_{\frac{1}{T}}(f)$ of the time-smoothing window $a_T(t)$ can be expressed as*

$$A_{\frac{1}{T}}(f) = W_A(fT) \tag{5.26}$$

with $W_A(f) \in L^\infty(\mathbb{R})$, continuous, and infinitesimal for $|f| \to \infty$. Moreover, in the sense of distributions, it results that

$$\lim_{T\to\infty} T\, W_A(fT) = \delta(f). \tag{5.27}$$

If $w_a(t)$ is continuous in $t = 0$, then $\int_{\mathbb{R}} W_A(f)\,\mathrm{d}f = w_a(0) = 1$. □

5.3.1 Proof of Lemma 4.5.5 Expected Value of the Time-Smoothed Bifrequency Cross-Periodogram

From (4.123) and (5.6), and accounting for (4.15) and (4.107), it follows that

$$\begin{aligned}
&\mathrm{E}\left\{S_{yx^{(*)}}(t; f_1, f_2)_{\frac{1}{\Delta f}, T}\right\} \\
&= \int_{\mathbb{R}}\int_{\mathbb{R}} \mathrm{E}\left\{Y(\nu_1)X^{(*)}(\nu_2)\right\} \Delta f \\
&\quad \mathcal{B}_{\Delta f}(f_1 - \nu_1; t)\, \mathcal{B}^{(*)}_{\Delta f}(f_2 - \nu_2; t)\, \mathrm{d}\nu_1\, \mathrm{d}\nu_2 \underset{t}{\otimes} a_T(t) \\
&= \Delta f \int_{\mathbb{R}}\int_{\mathbb{R}}\int_{\mathbb{R}} \sum_{n\in\mathbb{I}} S^{(n)}(\nu_1)\, \delta\left(\nu_2 - \Psi^{(n)}(\nu_1)\right) \\
&\quad B_{\Delta f}(f_1 - \nu_1)\, e^{-j2\pi(f_1-\nu_1)\tau}\, B^{(*)}_{\Delta f}(f_2 - \nu_2)\, e^{-(-)j2\pi(f_2-\nu_2)\tau}\, \mathrm{d}\nu_1\, \mathrm{d}\nu_2 \\
&\qquad a_T(t-\tau)\, \mathrm{d}\tau \\
&= \Delta f \sum_{n\in\mathbb{I}} \int_{\mathbb{R}} S^{(n)}(\nu_1)\, B_{\Delta f}(f_1 - \nu_1)\, B^{(*)}_{\Delta f}\left(f_2 - \Psi^{(n)}(\nu_1)\right) \\
&\quad \int_{\mathbb{R}} e^{-j2\pi[f_1-\nu_1+(-)(f_2-\Psi^{(n)}(\nu_1))]\tau} \int_{\mathbb{R}} A_{\frac{1}{T}}(f)\, e^{j2\pi f(t-\tau)}\, \mathrm{d}f\, \mathrm{d}\tau\, \mathrm{d}\nu_1 \\
&= \Delta f \sum_{n\in\mathbb{I}} \int_{\mathbb{R}} S^{(n)}(\nu_1)\, B_{\Delta f}(f_1 - \nu_1)\, B^{(*)}_{\Delta f}\left(f_2 - \Psi^{(n)}(\nu_1)\right) \\
&\quad \int_{\mathbb{R}} A_{\frac{1}{T}}(f)\, \delta\left(f - [\nu_1 - f_1 + (-)(\Psi^{(n)}(\nu_1) - f_2)]\right) e^{j2\pi ft}\, \mathrm{d}f\, \mathrm{d}\nu_1
\end{aligned} \tag{5.28}$$

from which, accounting for the sampling property of the Dirac delta, (4.127) immediately follows. In the derivation of (5.28), the formal relationship (Zemanian 1987)

$$\int_{\mathbb{R}} e^{-j2\pi(f-f_0)\tau}\, \mathrm{d}\tau = \delta(f - f_0) \tag{5.29}$$

to be intended in the sense of distributions is used.

In the proof, the interchange of expectation, integral, and sum operations can be justified as in the proof of Lemma 4.4.6, accounting for the fact that $a_T(t) \in L^1(\mathbb{R})$ (Assumption 4.5.2).

5.3.2 Proof of Lemma 4.5.6 Covariance of the Time-Smoothed Bifrequency Cross-Periodogram

It results that (multilinearity property of cumulants (Mendel 1991))

$$\begin{aligned}
&\operatorname{cov}\left\{S_{yx^{(*)}}(t_1; f_{y1}, f_{x1})_{\frac{1}{\Delta f},T}\,,\; S_{yx^{(*)}}(t_2; f_{y2}, f_{x2})_{\frac{1}{\Delta f},T}\right\} \\
&= \operatorname{cov}\left\{I_{yx^{(*)}}(t_1; f_{y1}, f_{x1})_{\frac{1}{\Delta f}} \underset{t_1}{\otimes} a_T(t_1)\,,\; I_{yx^{(*)}}(t_2; f_{y2}, f_{x2})_{\frac{1}{\Delta f}} \underset{t_2}{\otimes} a_T(t_2)\right\} \\
&= \operatorname{cov}\left\{I_{yx^{(*)}}(t_1; f_{y1}, f_{x1})_{\frac{1}{\Delta f}}\,,\; I_{yx^{(*)}}(t_2; f_{y2}, f_{x2})_{\frac{1}{\Delta f}}\right\} \underset{t_1}{\otimes} a_T(t_1) \underset{t_2}{\otimes} a_T^*(t_2)
\end{aligned} \tag{5.30}$$

where (4.123) has been accounted for, and the interchange of integral and expectation operations can be justified by the Fubini and Tonelli theorem (Champeney 1990). In fact, by using (4.108) into the right-hand side of (5.30) one has

$$\begin{aligned}
&\operatorname{cov}\left\{I_{yx^{(*)}}(t_1; f_{y1}, f_{x1})_{\frac{1}{\Delta f}}\,,\; I_{yx^{(*)}}(t_2; f_{y2}, f_{x2})_{\frac{1}{\Delta f}}\right\} \underset{t_1}{\otimes} a_T(t_1) \underset{t_2}{\otimes} a_T^*(t_2) \\
&\quad = \mathcal{D}_1 + \mathcal{D}_2 + \mathcal{D}_3
\end{aligned} \tag{5.31}$$

where

$$\begin{aligned}
\mathcal{D}_1 \triangleq &\int_{\mathbb{R}}\int_{\mathbb{R}}\int_{\mathbb{R}}\int_{\mathbb{R}} (\Delta f)^2 \sum_{n'\in\mathbb{I}'} \sum_{n''\in\mathbb{I}''} S^{(n')}_{yy^*}(\nu_{y1})\, S^{(n'')}_{x^{(*)}x^{(*)*}}(\nu_{x1}) \\
&\mathcal{B}_{\Delta f}(f_{y1} - \nu_{y1}; \tau_1)\, \mathcal{B}^*_{\Delta f}(f_{y2} - \Psi^{(n')}_{yy^*}(\nu_{y1}); \tau_2) \\
&\mathcal{B}^{(*)}_{\Delta f}(f_{x1} - \nu_{x1}; \tau_1) \mathcal{B}^{(*)*}_{\Delta f}(f_{x2} - \Psi^{(n'')}_{x^{(*)}x^{(*)*}}(\nu_{x1}); \tau_2) \\
&a_T(t_1 - \tau_1)\, a_T^*(t_2 - \tau_2)\, \mathrm{d}\nu_{y1}\, \mathrm{d}\nu_{x1}\, \mathrm{d}\tau_1\, \mathrm{d}\tau_2
\end{aligned} \tag{5.32}$$

$$\begin{aligned}
\mathcal{D}_2 \triangleq &\int_{\mathbb{R}}\int_{\mathbb{R}}\int_{\mathbb{R}}\int_{\mathbb{R}} (\Delta f)^2 \sum_{n'''\in\mathbb{I}'''} \sum_{n^{iv}\in\mathbb{I}^{iv}} S^{(n''')}_{yx^{(*)*}}(\nu_{y1})\, S^{(n^{iv})}_{x^{(*)}y^*}(\nu_{x1}) \\
&\mathcal{B}_{\Delta f}(f_{y1} - \nu_{y1}; \tau_1)\, \mathcal{B}^{(*)*}_{\Delta f}(f_{x2} - \Psi^{(n''')}_{yx^{(*)*}}(\nu_{y1}); \tau_2) \\
&\mathcal{B}^{(*)}_{\Delta f}(f_{x1} - \nu_{x1}; \tau_1) \mathcal{B}^*_{\Delta f}(f_{y2} - \Psi^{(n^{iv})}_{x^{(*)}y^*}(\nu_{x1}); \tau_2) \\
&a_T(t_1 - \tau_1)\, a_T^*(t_2 - \tau_2)\, \mathrm{d}\nu_{y1}\, \mathrm{d}\nu_{x1}\, \mathrm{d}\tau_1\, \mathrm{d}\tau_2
\end{aligned} \tag{5.33}$$

$$\begin{aligned}
\mathcal{D}_3 \triangleq &\int_{\mathbb{R}}\int_{\mathbb{R}}\int_{\mathbb{R}}\int_{\mathbb{R}}\int_{\mathbb{R}} (\Delta f)^2 \sum_{n\in\mathbb{I}_4} P^{(n)}_{yx^{(*)}y^*x^{(*)*}}(\nu_{y1}, \nu_{x1}, \nu_{y2}) \\
&\mathcal{B}_{\Delta f}(f_{y1} - \nu_{y1}; \tau_1)\, \mathcal{B}^{(*)}_{\Delta f}(f_{x1} - \nu_{x1}; \tau_1) \\
&\mathcal{B}^*_{\Delta f}(f_{y2} - \nu_{y2}; \tau_2)\, \mathcal{B}^{(*)*}_{\Delta f}(f_{x2} - \Psi^{(n)}_4(\nu_{y1}, \nu_{x1}, \nu_{y2}); \tau_2) \\
&a_T(t_1 - \tau_1)\, a_T^*(t_2 - \tau_2)\, \mathrm{d}\nu_{y1}\, \mathrm{d}\nu_{x1}\, \mathrm{d}\nu_{y2}\, \mathrm{d}\tau_1\, \mathrm{d}\tau_2.
\end{aligned} \tag{5.34}$$

Denoted by $D_1(\nu_{y1}, \nu_{x1}, \tau_1, \tau_2)$ the integrand function in $\mathcal{D}_1$ and accounting for Assumptions 4.4.3a, 4.4.5, and 4.5.2, the result is that

$$\begin{aligned}
&\left|D_1(\nu_{y1}, \nu_{x1}, \tau_1, \tau_2)\right| \\
&\quad \leqslant \frac{1}{(\Delta f)^2} \left| W_B\left(\frac{f_{y1}-\nu_{y1}}{\Delta f}\right) W_B\left(\frac{f_{x1}-\nu_{x1}}{\Delta f}\right)\right| \\
&\quad \|W_B\|_\infty^2 \sum_{n' \in \mathbb{I}'} \|S_{yy^*}^{(n')}\|_\infty \sum_{n'' \in \mathbb{I}''} \|S_{x^{(*)}x^{(*)*}}^{(n'')}\|_\infty \\
&\quad |a_T(t_1-\tau_1)\, a_T(t_2-\tau_2)| \ \in L^1(\mathbb{R}^4).
\end{aligned} \tag{5.35}$$

Analogously, it can be shown that the integrand function in $\mathcal{D}_2$ is bounded by a summable function. Furthermore, denoted by $D_3(\nu_{y1}, \nu_{x1}, \nu_{y2}, \tau_1, \tau_2)$ the integrand function in $\mathcal{D}_3$ and accounting for Assumptions 4.4.3b, 4.4.5, and 4.5.2, the result is that

$$\begin{aligned}
&\left|D_3(\nu_{y1}, \nu_{x1}, \nu_{y2}, \tau_1, \tau_2)\right| \\
&\quad \leqslant \frac{1}{(\Delta f)^2} \left| W_B\left(\frac{f_{y1}-\nu_{y1}}{\Delta f}\right) W_B\left(\frac{f_{x1}-\nu_{x1}}{\Delta f}\right) W_B\left(\frac{f_{y2}-\nu_{y2}}{\Delta f}\right)\right| \\
&\quad \|W_B\|_\infty \sum_{n \in \mathbb{I}_4} \|P_{yx^{(*)}y^*x^{(*)*}}^{(n)}\|_\infty \ |a_T(t_1-\tau_1)\, a_T(t_2-\tau_2)| \ \in L^1(\mathbb{R}^5).
\end{aligned} \tag{5.36}$$

By reasoning as for the derivation of (5.28), terms $\mathcal{D}_1$, $\mathcal{D}_2$, and $\mathcal{D}_3$ can be shown to be equivalent to $\mathcal{T}_1''$, $\mathcal{T}_2''$, and $\mathcal{T}_3''$, respectively. Then, (4.128) easily follows from (5.30) and (5.31).

Lemma 5.3.2 *Let $g(f)$ be a real-valued function of the real variable f, defined in the neighborhood of $f = 0$. It results that*

$$\lim_{\Delta f \to 0} \delta_{g(\Delta f)} = \bar{\delta}_{g(\Delta f)} \triangleq \begin{cases} 1 & \text{if } g(\Delta f) = 0 \text{ in a neighborhood of } \Delta f = 0 \\ 0 & \text{otherwise} \end{cases} \tag{5.37}$$

where δ_f is the Kronecker delta, that is, $\delta_f = 1$ if $f = 0$ and $\delta_f = 0$ if $f \neq 0$.

Proof: Let

$$h(\Delta f) \triangleq \delta_{g(\Delta f)} = \begin{cases} 1 & g(\Delta f) = 0 \\ 0 & g(\Delta f) \neq 0 \end{cases} \tag{5.38}$$

We have

$$\lim_{\Delta f \to 0} h(\Delta f) = 1 \tag{5.39}$$

if and only if

$$\forall \epsilon > 0 \quad \exists \delta_\epsilon > 0 \ : \ 0 < |\Delta f| < \delta_\epsilon \Rightarrow |h(\Delta f) - 1| < \epsilon. \tag{5.40}$$

Since the function δ_u is discontinuous in $u = 0$, (5.40) is equivalent to

$$h(\Delta f) = 1 \quad \text{for } 0 < |\Delta f| < \delta_\epsilon \tag{5.41}$$

that is, in a neighborhood of $\Delta f = 0$. Thus, the limit in (5.39) is 1 if and only if $h(\Delta f) = 1$ in a neighborhood of $\Delta f = 0$, that is, if and only if $g(\Delta f) = 0$ in a neighborhood of $\Delta f = 0$.

Note that $g(\Delta f)$ can be possibly discontinuous or even not defined in $\Delta f = 0$. That is, we do not require $g(0) = 0$. □

5.3.3 *Proof of Theorem 4.5.7 Asymptotic Expected Value of the Time-Smoothed Bifrequency Cross-Periodogram*

By substituting (4.103) and (5.26) into the expression (4.127) of the expected value of the time-smoothed cross-periodogram, and making the variable change $\lambda_1 = (f_1 - \nu_1)/\Delta f$ one has

$$\begin{aligned}
&\mathrm{E}\left\{S_{yx^{(*)}}(t; f_1, f_2)_{\frac{1}{\Delta f},T}\right\} \\
&\quad = \int_{\mathbb{R}} \sum_{n \in \mathbb{I}} S^{(n)}(f_1 - \lambda_1 \Delta f)\, W_B(\lambda_1)\, W_B^{(*)}\left(\frac{f_2 - \Psi^{(n)}(f_1 - \lambda_1 \Delta f)}{\Delta f}\right) \\
&\qquad W_A(T(-\lambda_1 \Delta f +_{(-)} (\Psi^{(n)}(f_1 - \lambda_1 \Delta f) - f_2))) \\
&\qquad \exp\left[j2\pi\left(-\lambda_1 \Delta f +_{(-)} (\Psi^{(n)}(f_1 - \lambda_1 \Delta f) - f_2)\right) t\right] \mathrm{d}\lambda_1.
\end{aligned} \tag{5.42}$$

Denoted by $J_{\Delta f,T}(\lambda_1)$ the integrand function in (5.42) and accounting for (4.97), Assumption 4.4.5 and Lemma 5.3.1, the result is that

$$|J_{\Delta f,T}(\lambda_1)| \leqslant |W_B(\lambda_1)|\, \|W_B\|_\infty\, \|W_A\|_\infty \sum_{n \in \mathbb{I}} \|S^{(n)}\|_\infty \in L^1(\mathbb{R})$$

with the right-hand side independent of T and Δf. Thus, the dominated convergence theorem (Champeney 1990) can be applied:

$$\begin{aligned}
&\lim_{\Delta f \to 0} \lim_{T \to \infty} \mathrm{E}\left\{S_{yx^{(*)}}(t; f_1, f_2)_{\frac{1}{\Delta f},T}\right\} \\
&\quad = \sum_{n \in \mathbb{I}} S^{(n)}(f_1)\, \delta_{f_2 - \Psi^{(n)}(f_1)} \int_{\mathbb{R}} W_B(\lambda_1)\, W_B^{(*)}\left(\lambda_1 \Psi^{(n)\prime}(f_1)\right) \\
&\qquad \lim_{\Delta f \to 0} \lim_{T \to \infty} W_A\left(-T\Delta f \lambda_1 \left(1 +_{(-)} \frac{f_2 - \Psi^{(n)}(f_1 - \lambda_1 \Delta f)}{\lambda_1 \Delta f}\right)\right) \mathrm{d}\lambda_1
\end{aligned} \tag{5.43}$$

where Lemmas 5.2.1a and 5.2.2 have been accounted for, and the interchange of limit, sum, and integral operations can be justified as in the proof of Lemma 4.4.6.

Since W_A is infinitesimal at $\pm\infty$, for any finite Δf the limit for $T \to \infty$ is zero unless the argument of the inner large parenthesis in the last line of (5.43) is zero. Therefore, accounting for the continuity of $W_A(\nu)$, it results in

$$\begin{aligned}
&\lim_{\Delta f \to 0} \lim_{T \to \infty} W_A \left(-T\Delta f \lambda_1 \left(1 +_{(-)} \frac{f_2 - \Psi^{(n)}(f_1 - \lambda_1 \Delta f)}{\lambda_1 \Delta f} \right) \right) \\
&\quad = W_A(0) \lim_{\Delta f \to 0} \delta_{1 +_{(-)} (f_2 - \Psi^{(n)}(f_1 - \lambda_1 \Delta f))/(\lambda_1 \Delta f)} \\
&\quad = W_A(0)\, \delta_{f_2 - \Psi^{(n)}(f_1)}\, \bar{\delta}_{1 +_{(-)} \Psi^{(n)\prime}(f_1)}
\end{aligned} \tag{5.44}$$

where $\bar{\delta}_{g(f)}$ is defined in (5.37). To obtain the last equality, observe that, accounting for Lemma 5.3.2, when $\Delta f \to 0$ the argument of the Kronecker delta in the second line is zero only if $f_2 = \Psi^{(n)}(f_1)$ and $\Psi^{(n)\prime}(f_1) = -_{(-)}1$ in a neighborhood of f_1.

Finally, by substituting (5.44) into (5.43), (4.132) is obtained.

5.3.4 Proof of Corollary 4.5.8

At first let us observe that, for Assumption 4.5.3 there is no cluster of support curves and, consequently, for T sufficiently large, it is finite the set $\mathbb{I}_0 \subseteq \mathbb{I}$ of indices n such that, at the fixed point (f_1, f_2), it results $|f_2 - \Psi^{(n)}(f_1)| \ll T^{-1}$. In addition, there exists $T_0 > 0$ such that for $T > T_0$ the number of elements of $\mathbb{I}_0$ does not change with T.

By considering the Taylor series expansion for $\Psi^{(n)}(f_1)$ with second-order Lagrange residual term one has

$$\Psi^{(n)}(f_1 - \lambda_1 \Delta f) = \Psi^{(n)}(f_1) - \Psi^{(n)\prime}(f_1)\, \lambda_1 \Delta f + \frac{1}{2} \Psi^{(n)\prime\prime}(\bar{f}_1)\, (\lambda_1 \Delta f)^2 \tag{5.45}$$

where $\bar{f}_1 = f_1 - \xi \lambda_1 \Delta f$, $\xi \in [0, 1]$. By substituting (5.45) into

$$h^{(n)}(\lambda_1) \triangleq W_A(T(-\lambda_1 \Delta f +_{(-)} (\Psi^{(n)}(f_1 - \lambda_1 \Delta f) - f_2))) \tag{5.46}$$

it results in

$$\begin{aligned}
h^{(n)}(\lambda_1) = W_A \bigg(&{}_{(-)}T(\Psi^{(n)}(f_1) - f_2) - T\Delta f \lambda_1 (1 +_{(-)} \Psi^{(n)\prime}(f_1)) \\
&+_{(-)} \frac{1}{2} T (\lambda_1 \Delta f)^2 \Psi^{(n)\prime\prime}(\bar{f}_1) \bigg).
\end{aligned} \tag{5.47}$$

In Theorem 4.5.7, the asymptotic biasedness of the time-smoothed cross-periodogram is proved for SC processes that are not ACS. This is consequence of the fact that, when $\Psi^{(n)\prime}(f_1) \neq -_{(-)}1$ in a neighborhood of f_1, the function $h^{(n)}(\lambda_1)$ in (5.42) approaches zero, as $T \to \infty$ and $\Delta f \to 0$ with $T\Delta f \to \infty$ (see (5.44)). However, if at the point (f_1, f_2) it results that

$$\left| 1 +_{(-)} \Psi^{(n)\prime}(f_1) \right| \ll \frac{1}{T\Delta f} \tag{5.48}$$

and

$$\left|\Psi^{(n)\prime\prime}(\bar{f}_1)\right| \ll \frac{1}{T(\Delta f)^2} \tag{5.49}$$

for all $n \in \mathbb{I}_0$ i.e., for all n such that

$$|f_2 - \Psi^{(n)}(f_1)| \ll \frac{1}{T} \tag{5.50}$$

then in substituting (5.47) into (5.42) we can put

$$h^{(n)}(\lambda_1) \simeq \begin{cases} W_A(0) & n \in \mathbb{I}_0 \\ 0 & n \in \mathbb{I} - \mathbb{I}_0. \end{cases} \tag{5.51}$$

In fact, under Assumption 4.4.5 $W_B(\lambda_1)$ has approximate bandwidth equal to 1, that is, it is significantly different from zero only for $\lambda_1 = \mathcal{O}(1)$ (independently of T and Δf) and $W_A(f)$ is infinitesimal for $|f| \to \infty$. Thus, by substituting (5.51) into (5.42), for Δf sufficiently small, one obtains (4.136).

Note that (5.48) and (5.49) can be both verified if the second-order derivative of $\Psi^{(n)}$ is bounded (Assumption 4.5.4) and T and Δf are such that $T\Delta f \gg 1$ and $T(\Delta f)^2 \ll 1$. If the time-smoothed cross-periodogram is evaluated (with fixed T and Δf) for $(f_1, f_2) \in \mathbb{D}$, with $\mathbb{D}$ denoting the analysis range, then uniform boundedness in $\mathbb{D}$ of second-order derivatives $\Psi^{(n)\prime\prime}$ is necessary (Assumption 4.5.4).

5.3.5 Proof of Theorem 4.5.9 Asymptotic Covariance of the Time-Smoothed Bifrequency Cross-Periodogram

By substituting (4.103) and (5.26) into (4.129) and making the variable changes $\lambda_{y1} = (f_{y1} - \nu_{y1})/\Delta f$ and $\lambda_{x1} = (f_{x1} - \nu_{x1})/\Delta f$, one obtains

$$\begin{aligned}
T\Delta f\, \mathcal{T}_1'' = & \int_{\mathbb{R}} \int_{\mathbb{R}} \sum_{n' \in \mathbb{I}'} \sum_{n'' \in \mathbb{I}''} S^{(n')}_{yy^*}(f_{y1} - \lambda_{y1}\Delta f)\, S^{(n'')}_{x^{(*)}x^{(*)*}}(f_{x1} - \lambda_{x1}\Delta f) \\
& W_B(\lambda_{y1})\, W_B^*\left(\frac{f_{y2} - \Psi^{(n')}_{yy^*}(f_{y1} - \lambda_{y1}\Delta f)}{\Delta f}\right) \\
& W_B^{(*)}(\lambda_{x1})\, W_B^{(*)*}\left(\frac{f_{x2} - \Psi^{(n'')}_{x^{(*)}x^{(*)*}}(f_{x1} - \lambda_{x1}\Delta f)}{\Delta f}\right) \\
& W_A^*(T(\Psi^{(n')}_{yy^*}(f_{y1} - \lambda_{y1}\Delta f) - f_{y2} +_{(-)}(\Psi^{(n'')}_{x^{(*)}x^{(*)*}}(f_{x1} - \lambda_{x1}\Delta f) - f_{x2}))) \\
& T\Delta f\, W_A(T\Delta f\,(-\lambda_{y1} -_{(-)}\lambda_{x1})) \\
& \exp\left[-j2\pi\left((\Psi^{(n')}_{yy^*}(f_{y1} - \lambda_{y1}\Delta f) - f_{y2}\right.\right. \\
& \left.\left. +_{(-)}(\Psi^{(n'')}_{x^{(*)}x^{(*)*}}(f_{x1} - \lambda_{x1}\Delta f) - f_{x2}))t_2\right)\right] \\
& \exp\left[-j2\pi\left((\lambda_{y1}\Delta f +_{(-)}\lambda_{x1}\Delta f)t_1\right)\right] \mathrm{d}\lambda_{y1}\, \mathrm{d}\lambda_{x1}.
\end{aligned} \tag{5.52}$$

Thus, accounting for Assumption 4.4.3a and Lemma 5.3.1, one has

$$T\Delta f\,|\mathcal{T}_1''| \leqslant \|W_A\|_\infty \sum_{n'\in\mathbb{I}'}\sum_{n''\in\mathbb{I}''} \|S_{yy^*}^{(n')}\|_\infty \, \|S_{x^{(*)}x^{(*)*}}^{(n'')}\|_\infty$$

$$\int_{\mathbb{R}}\int_{\mathbb{R}} \left| W_B(\lambda_{y1})\, W_B^*\left(\frac{f_{y2}-\Psi_{yy^*}^{(n')}(f_{y1}-\lambda_{y1}\Delta f)}{\Delta f}\right)\right.$$

$$\left. W_B^{(*)}(\lambda_{x1})\, W_B^{(*)*}\left(\frac{f_{x2}-\Psi_{x^{(*)}x^{(*)*}}^{(n'')}(f_{x1}-\lambda_{x1}\Delta f)}{\Delta f}\right)\right|$$

$$T\Delta f\,\left|W_A(T\Delta f\,(-\lambda_{y1} - {\scriptstyle(-)}\lambda_{x1}))\right| \,\mathrm{d}\lambda_{x1}\,\mathrm{d}\lambda_{y1}. \tag{5.53}$$

From (5.27) and for $W_A(\lambda) \in L^1(\mathbb{R})$, it follows that, in the sense of distributions,

$$\lim_{Z\to\infty} Z|W_A(Z\lambda)| = \delta(\lambda). \tag{5.54}$$

Therefore, for $T \to \infty$ and $\Delta f \to 0$ with $T\Delta f \to \infty$, accounting for (5.13), (5.54), and the sampling property of Dirac delta, one obtains that $T\Delta f\,|\mathcal{T}_1''| \leqslant \mathcal{T}_1'''$ with $\mathcal{T}_1'''$ given by (4.138).

Analogously, it can be shown that

$$\lim_{\Delta f\to 0}\lim_{T\to\infty} T\Delta f\,\left|\mathcal{T}_2''\right| \leqslant \mathcal{T}_2''' \tag{5.55}$$

with $\mathcal{T}_2'''$ given by (4.139).

Furthermore, by substituting (4.103) and (5.26) into (4.131) and making the variable changes $\lambda_{y1} = (f_{y1} - \nu_{y1})/\Delta f$, $\lambda_{x1} = (f_{x1} - \nu_{x1})/\Delta f$, and $\lambda_{y2} = (f_{y2} - \nu_{y2})/\Delta f$, it results in

$$T\Delta f\,\mathcal{T}_3'' = \Delta f \int_{\mathbb{R}}\int_{\mathbb{R}}\int_{\mathbb{R}} \sum_{n\in\mathbb{I}_4} P_{yx^{(*)}y^*x^{(*)*}}^{(n)}(\boldsymbol{f} - \boldsymbol{\lambda}\Delta f)$$

$$W_B(\lambda_{y1})\, W_B^{(*)}(\lambda_{x1})\, W_B^*(\lambda_{y2})\, W_B^{(*)*}\left(\frac{f_{x2}-\Psi_4^{(n)}(\boldsymbol{f}-\boldsymbol{\lambda}\Delta f)}{\Delta f}\right)$$

$$T\Delta f\, W_A(-T\Delta f\,(\lambda_{y1} + {\scriptstyle(-)}\lambda_{x1}))$$

$$W_A^*(T(-\lambda_{y2}\Delta f + {\scriptstyle(-)}(\Psi_4^{(n)}(\boldsymbol{f}-\boldsymbol{\lambda}\Delta f) - f_{x2})))$$

$$\exp\left[j2\pi\left(-(\lambda_{y1}\Delta f + {\scriptstyle(-)}\lambda_{x1}\Delta f)t_1\right)\right]$$

$$\exp\left[j2\pi\left(-(-\lambda_{y2}\Delta f + {\scriptstyle(-)}(\Psi_4^{(n)}(\boldsymbol{f}-\boldsymbol{\lambda}\Delta f) - f_{x2}))t_2\right)\right] \,\mathrm{d}\lambda_{y1}\,\mathrm{d}\lambda_{x1}\,\mathrm{d}\lambda_{y2} \tag{5.56}$$

where $\boldsymbol{f} - \boldsymbol{\lambda}\Delta f \triangleq [f_{y1} - \lambda_{y1}\Delta f, f_{x1} - \lambda_{x1}\Delta f, f_{y2} - \lambda_{y2}\Delta f]$. Thus, accounting for Assumptions 4.4.3b and 4.4.5 and Lemma 5.3.1, one has

$$T\Delta f \left|\mathcal{T}_3''\right| \leqslant \Delta f \, \|W_A\|_\infty \, \|W_B\|_\infty \sum_{n \in \mathbb{I}_4} \|P^{(n)}_{yx^{(*)}y^*x^{(*)*}}\|_\infty$$
$$\int_{\mathbb{R}} \int_{\mathbb{R}} \int_{\mathbb{R}} \left| W_B(\lambda_{y1}) \, W_B^{(*)}(\lambda_{x1}) \, W_B^*(\lambda_{y2}) \right|$$
$$T\Delta f \left| W_A(-T\Delta f \, (\lambda_{y1} +_{(-)} \lambda_{x1})) \right| \, \mathrm{d}\lambda_{y1} \, \mathrm{d}\lambda_{x1} \, \mathrm{d}\lambda_{y2}. \tag{5.57}$$

Accounting for (5.54), one obtains

$$\lim_{T \to \infty} T\Delta f \left| W_A(-T\Delta f \, (\lambda_{y1} +_{(-)} \lambda_{x1})) \right| = \Delta f \, \delta(\Delta f \, (\lambda_{y1} +_{(-)} \lambda_{x1}))$$
$$= \delta(\lambda_{y1} +_{(-)} \lambda_{x1}). \tag{5.58}$$

Therefore, for $T \to \infty$ and $\Delta f \to 0$ with $T\Delta f \to \infty$,

$$\lim_{\Delta f \to 0} \lim_{T \to \infty} T\Delta f \left|\mathcal{T}_3''\right| \leqslant \left(\lim_{\Delta f \to 0} \Delta f \right) \|W_A\|_\infty \, \|W_B\|_\infty \sum_{n \in \mathbb{I}_4} \|P^{(n)}_{yx^{(*)}y^*x^{(*)*}}\|_\infty$$
$$\int_{\mathbb{R}} \int_{\mathbb{R}} \left| W_B(\lambda_{y1}) \, W_B^{(*)}(-_{(-)}\lambda_{y1}) \, W_B^*(\lambda_{y2}) \right| \, \mathrm{d}\lambda_{y1} \, \mathrm{d}\lambda_{y2}$$
$$= 0 \tag{5.59}$$

where the last equality holds under assumption (4.98).

As a final remark, note that the order of the two limits as $\Delta f \to 0$ and $T \to \infty$ in Theorems 4.5.7 and 4.5.9 cannot be interchanged. In fact, in the proofs of both Theorems 4.5.7 and 4.5.9 the double limit as $\Delta f \to 0$ and $T \to \infty$ is evaluated with $T\Delta f \to \infty$ by taking Δf finite and fixed, making $T \to \infty$ and after making $\Delta f \to 0$. Analogous results can be obtained if $\Delta f = \mathcal{O}(T^{-a})$ with $0 < a < 1$. However, the more restrictive condition $1/2 < a < 1$ is necessary to assure $T(\Delta f)^2 \to 0$ as requested in Corollary 4.5.8 to guarantee (5.49).

5.4 Proofs for Section 4.6 "The Frequency-Smoothed Cross-Periodogram"

In this section, proofs of lemmas and theorems presented in Section 4.6 on bias and covariance of the frequency-smoothed cross-periodogram are reported.

5.4.1 *Proof of Theorem 4.6.3 Expected Value of the Frequency-Smoothed Cross-Periodogram*

By taking the expected value of the frequency-smoothed cross-periodogram (4.147) we have

$$
\begin{aligned}
&\mathrm{E}\left\{ S^{(n)}_{yx^{(*)}}(t; f_1)_{T,\Delta f} \right\} \\
&= \mathrm{E}\left\{ \left[I_{yx^{(*)}}(t; f_1, f_2)_T \Big|_{f_2=\Psi^{(n)}_{yx^{(*)}}(f_1)} \right] \underset{f_1}{\otimes} A_{\Delta f}(f_1) \right\} \\
&= \mathrm{E}\left\{ I_{yx^{(*)}}(t; f_1, f_2)_T \Big|_{f_2=\Psi^{(n)}_{yx^{(*)}}(f_1)} \right\} \underset{f_1}{\otimes} A_{\Delta f}(f_1) \\
&= \frac{1}{T} \int_{\mathbb{R}} \sum_{m \in \mathbb{I}} \int_{\mathbb{R}} S^{(m)}_{yx^{(*)}}(\nu_1)\, \mathcal{B}_{\frac{1}{T}}(\lambda_1 - \nu_1; t) \\
&\qquad \mathcal{B}^{(*)}_{\frac{1}{T}}\left(\Psi^{(n)}_{yx^{(*)}}(\lambda_1) - \Psi^{(m)}_{yx^{(*)}}(\nu_1); t \right) \mathrm{d}\nu_1\, A_{\Delta f}(f_1 - \lambda_1)\, \mathrm{d}\lambda_1
\end{aligned}
\tag{5.60}
$$

from which (4.150) immediately follows.

In the third equality (4.106) is used. In the second equality, the interchange of expectation and convolution operations is justified by the Fubini and Tonelli Theorem (Champeney 1990, Chapter 3). In fact, defined

$$
\begin{aligned}
F^{(m)}(\lambda_1, \nu_1) \triangleq\; & \frac{1}{T}\, S^{(m)}_{yx^{(*)}}(\nu_1)\, \mathcal{B}_{\frac{1}{T}}(\lambda_1 - \nu_1; t) \\
& \mathcal{B}^{(*)}_{\frac{1}{T}}\left(\Psi^{(n)}_{yx^{(*)}}(\lambda_1) - \Psi^{(m)}_{yx^{(*)}}(\nu_1); t \right) A_{\Delta f}(f_1 - \lambda_1)
\end{aligned}
\tag{5.61}
$$

and accounting for Assumptions 4.4.3a, 4.4.5, and 4.6.2, for the integrand function in (5.60) we have

$$
\begin{aligned}
\left| \sum_{m \in \mathbb{I}} F^{(m)}(\lambda_1, \nu_1) \right| &\leqslant \sum_{m \in \mathbb{I}} \left| F^{(m)}(\lambda_1, \nu_1) \right| \\
&\leqslant \frac{1}{T} \sum_{m \in \mathbb{I}} \left\| S^{(m)}_{yx^{(*)}} \right\|_{\infty} T\, \|W_B\|_{\infty} \\
&\qquad T\, |W_B((\lambda_1 - \nu_1)T)|\, \frac{1}{\Delta f} \left| W_A\left(\frac{f_1 - \lambda_1}{\Delta f} \right) \right| \in L^1(\mathbb{R}^2).
\end{aligned}
\tag{5.62}
$$

The interchange of sum and integral operations to obtain (4.150) from (5.60) is justified even if the set $\mathbb{I}$ is not finite by using the dominated convergence theorem

(Champeney 1990, Chapter 4). Specifically, by denoting with $\{\mathbb{I}_k\}$ an increasing sequence of finite subsets of $\mathbb{I}$ such that $\lim_{k\to\infty} \mathbb{I}_k = \mathbb{I}$, we have

$$\begin{aligned}
\lim_{k\to\infty} \sum_{m\in\mathbb{I}_k} \int_{\mathbb{R}} \int_{\mathbb{R}} F^{(m)}(\lambda_1, \nu_1)\, \mathrm{d}\lambda_1\, \mathrm{d}\nu_1 &= \lim_{k\to\infty} \int_{\mathbb{R}} \int_{\mathbb{R}} \sum_{m\in\mathbb{I}_k} F^{(m)}(\lambda_1, \nu_1)\, \mathrm{d}\lambda_1\, \mathrm{d}\nu_1 \\
&= \int_{\mathbb{R}} \int_{\mathbb{R}} \lim_{k\to\infty} \sum_{m\in\mathbb{I}_k} F^{(m)}(\lambda_1, \nu_1)\, \mathrm{d}\lambda_1\, \mathrm{d}\nu_1 \qquad (5.63)
\end{aligned}$$

In fact, it results that

$$\left|\sum_{m\in\mathbb{I}_k} F^{(m)}(\lambda_1, \nu_1)\right| \leqslant \sum_{m\in\mathbb{I}_k} \left|F^{(m)}(\lambda_1, \nu_1)\right| \leqslant \sum_{m\in\mathbb{I}} \left|F^{(m)}(\lambda_1, \nu_1)\right| \qquad (5.64)$$

with the right-hand side bounded by the right-hand side of (5.62). That is, the integrand function in the second term of equality (5.63) is bounded by a summable function of (λ_1, ν_1) not depending on k. □

5.4.2 *Proof of Theorem 4.6.4 Covariance of the Frequency-Smoothed Cross-Periodogram*

By setting

$$f_{y1} = f_1, \qquad f_{x1} = \Psi^{(n_1)}_{yx^{(*)}}(f_1), \qquad f_{y2} = f_2, \qquad f_{x2} = \Psi^{(n_2)}_{yx^{(*)}}(f_2)$$

into (4.108) (with $\Delta f = 1/T$) we have (multilinearity property of cumulants (Mendel 1991))

$$\begin{aligned}
&\operatorname{cov}\left\{ S^{(n_1)}_{yx^{(*)}}(t_1; f_1)_{T,\Delta f},\ S^{(n_2)}_{yx^{(*)}}(t_2; f_2)_{T,\Delta f} \right\} \\
&= \operatorname{cov}\left\{ I_{yx^{(*)}}\left(t_1; f_1, \Psi^{(n_1)}_{yx^{(*)}}(f_1)\right)_T \underset{f_1}{\otimes} A_{\Delta f}(f_1)\,, \right. \\
&\qquad \left. I_{yx^{(*)}}\left(t_2; f_2, \Psi^{(n_2)}_{yx^{(*)}}(f_2)\right)_T \underset{f_2}{\otimes} A_{\Delta f}(f_2) \right\} \\
&= \operatorname{cov}\left\{ I_{yx^{(*)}}\left(t_1; f_1, \Psi^{(n_1)}_{yx^{(*)}}(f_1)\right)_T, I_{yx^{(*)}}\left(t_2; f_2, \Psi^{(n_2)}_{yx^{(*)}}(f_2)\right)_T \right\} \\
&\qquad \underset{f_1}{\otimes} A_{\Delta f}(f_1) \underset{f_2}{\otimes} A^{*}_{\Delta f}(f_2) \\
&= \mathcal{T}_1'' + \mathcal{T}_2'' + \mathcal{T}_3'' \qquad (5.65)
\end{aligned}$$

where $\mathcal{T}_1''$, $\mathcal{T}_2''$, and $\mathcal{T}_3''$ are defined in (4.153), (4.154), and (4.155), respectively.

The interchange of $\operatorname{cov}\{\cdot,\cdot\}$ and convolution operations $\underset{f_1}{\otimes}$ and $\underset{f_2}{\otimes}$ can be justified by the Fubini and Tonelli theorem. In fact, defined

$$\begin{aligned}
&\mathcal{D}^{(n',n'')}(\nu_{y1},\nu_{x1},\lambda_1,\lambda_2)\\
&\quad\triangleq \frac{1}{T^2}\, S^{(n')}_{yy^*}(\nu_{y1})\, S^{(n'')}_{x^{(*)}x^{(*)*}}(\nu_{x1})\\
&\quad\mathcal{B}_{\frac{1}{T}}(\lambda_1-\nu_{y1};t_1)\,\mathcal{B}^*_{\frac{1}{T}}\left(\lambda_2-\Psi^{(n')}_{yy^*}(\nu_{y1});t_2\right)\\
&\quad\mathcal{B}^{(*)}_{\frac{1}{T}}\left(\Psi^{(n_1)}_{yx^{(*)}}(\lambda_1)-\nu_{x1};t_1\right)\,\mathcal{B}^{(*)*}_{\frac{1}{T}}\left(\Psi^{(n_2)}_{yx^{(*)}}(\lambda_2)-\Psi^{(n'')}_{x^{(*)}x^{(*)*}}(\nu_{x1});t_2\right)\\
&\quad A_{\Delta f}(f_1-\lambda_1)\,A^*_{\Delta f}(f_2-\lambda_2)
\end{aligned}\tag{5.66}$$

and accounting for Assumptions 4.4.3a, 4.4.5, and 4.6.2, for the integrand function in (4.153) one obtains

$$\begin{aligned}
&\left|\sum_{n'\in\mathbb{I}'}\sum_{n''\in\mathbb{I}''}\mathcal{D}^{(n',n'')}(\nu_{y1},\nu_{x1},\lambda_1,\lambda_2)\right|\\
&\quad\leqslant \sum_{n'\in\mathbb{I}'}\sum_{n''\in\mathbb{I}''}\left|\mathcal{D}^{(n',n'')}(\nu_{y1},\nu_{x1},\lambda_1,\lambda_2)\right|\\
&\quad\leqslant \frac{1}{T^2}\sum_{n'\in\mathbb{I}'}\left\|S^{(n')}_{yy^*}\right\|_\infty \sum_{n''\in\mathbb{I}''}\left\|S^{(n'')}_{x^{(*)}x^{(*)*}}\right\|_\infty\\
&\qquad T\,W_B((\lambda_1-\nu_{y1})T)|\;T\,\|W_B\|_\infty\\
&\qquad T\left|W_B\left(\Psi^{(n_1)}_{yx^{(*)}}(\lambda_1)-\nu_{x1}\right)\right|\;T\,\|W_B\|_\infty\\
&\qquad \frac{1}{\Delta f}\left|W_A\left(\frac{f_1-\lambda_1}{\Delta f}\right)\right|\frac{1}{\Delta f}\left|W_A\left(\frac{f_2-\lambda_2}{\Delta f}\right)\right|\in L^1(\mathbb{R}^4).
\end{aligned}\tag{5.67}$$

An analogous result can be found for the integrand function in (4.154). Furthermore, defined the function

$$\begin{aligned}
&\mathcal{D}^{(n)}(\nu_{y1},\nu_{x1},\nu_{y2},\lambda_1,\lambda_2)\\
&\quad\triangleq \frac{1}{T^2}\,P^{(n)}_{yx^{(*)}y^*x^{(*)*}}(\nu_{y1},\nu_{x1},\nu_{y2})\\
&\quad\mathcal{B}^{(*)*}_{\frac{1}{T}}\left(\Psi^{(n_2)}_{yx^{(*)}}(\lambda_2)-\Psi^{(n)}_{yx^{(*)}y^*x^{(*)*}}(\nu_{y1},\nu_{x1},\nu_{y2});t_2\right)\\
&\quad\mathcal{B}_{\frac{1}{T}}(\lambda_1-\nu_{y1};t_1)\,\mathcal{B}^{(*)}_{\frac{1}{T}}\left(\Psi^{(n_1)}_{yx^{(*)}}(\lambda_1)-\nu_{x1};t_1\right)\,\mathcal{B}^*_{\frac{1}{T}}(\lambda_2-\nu_{y2};t_2)\\
&\quad A_{\Delta f}(f_1-\lambda_1)\,A^*_{\Delta f}(f_2-\lambda_2)
\end{aligned}\tag{5.68}$$

and accounting for Assumptions 4.4.3b, 4.4.5, and 4.6.2, for the integrand function in (4.155) one obtains

$$\begin{aligned}
&\left|\sum_{n\in\mathbb{I}_4} \mathcal{D}^{(n)}(\nu_{y1}, \nu_{x1}, \nu_{y2}, \lambda_1, \lambda_2)\right| \\
&\quad \leqslant \sum_{n\in\mathbb{I}_4} \left|\mathcal{D}^{(n)}(\nu_{y1}, \nu_{x1}, \nu_{y2}, \lambda_1, \lambda_2)\right| \\
&\quad \leqslant \frac{1}{T^2} \sum_{n\in\mathbb{I}_4} \left\| P^{(n)}_{yx^{(*)}y^*x^{(*)*}} \right\|_\infty T\, \|W_B\|_\infty \\
&\qquad T\,|W_B((\lambda_1 - \nu_{y1})T)|\, T \left| W_B\left((\Psi^{(n_1)}_{yx^{(*)}}(\lambda_1) - \nu_{x1})T\right)\right| T\,|W_B((\lambda_2 - \nu_{y2})T)| \\
&\qquad \frac{1}{\Delta f}\left|W_A\left(\frac{f_1 - \lambda_1}{\Delta f}\right)\right| \frac{1}{\Delta f}\left|W_A\left(\frac{f_2 - \lambda_2}{\Delta f}\right)\right| \in L^1(\mathbb{R}^5).
\end{aligned} \tag{5.69}$$

Finally, note that by using the dominated convergence theorem as in the proof of Theorem 4.6.3, it can be shown that in (4.153)–(4.155) the order of integral and sum operations can be interchanged. □

5.5 Proofs for Section 4.7.1 "Mean-Square Consistency of the Frequency-Smoothed Cross-Periodogram"

In this section, proofs of lemmas and theorems presented in Section 4.7.1 on the mean-square consistency of the frequency-smoothed cross-periodogram are reported.

Lemma 5.5.1 *Let $W(f)$ be a.e. continuous and regular as $|f| \to \infty$, $W(f) \in L^1(\mathbb{R}) \cap L^\infty(\mathbb{R})$ and $\int_{\mathbb{R}} W(f)\,\mathrm{d}f = 1$ (that is, W can be either W_A satisfying Assumption 4.4.5 or W_B satisfying Assumption 4.6.2). We have the following results.*

(a) *Let $\{\Psi^{(n)}(\lambda)\}$ be a set of a.e. derivable functions such that, for $n \neq m$, $\Psi^{(n)}(\lambda) = \Psi^{(m)}(\lambda)$ at most in a set of zero Lebesgue measure in $\mathbb{R}$. It results that*

$$\lim_{T\to\infty} W\left(\left(\Psi^{(n)}(\lambda) - \Psi^{(m)}(\lambda - \nu/T)\right)T\right) = W\left(\nu\Psi^{(n)\prime}(\lambda)\right)\,\delta_{n-m} \tag{5.70}$$

for almost all λ.

Proof: For $n = m$, the left-hand side of (5.70) can be written as

$$\lim_{T\to\infty} W\left(\nu \frac{\Psi^{(n)}(\lambda) - \Psi^{(n)}(\lambda - \nu/T)}{\nu/T}\right) = W\left(\nu\Psi^{(n)\prime}(\lambda)\right) \tag{5.71}$$

for almost all λ, provided that $\nu \neq 0$, where the a.e. continuity of $W(f)$ is accounted for.

For $n \neq m$, the left-hand side of (5.70) can be written as

$$\lim_{T\to\infty} W\left(\nu \frac{\Psi^{(n)}(\lambda) - \Psi^{(m)}(\lambda - \nu/T)}{\nu/T}\right) = W(\infty) = 0 \tag{5.72}$$

for $\nu \neq 0$ and almost all λ, where we used the fact that, for $n \neq m$, $\Psi^{(n)}(\lambda) = \Psi^{(m)}(\lambda)$ at most in a set of zero Lebesgue measure, and $W(f)$ is a.e. continuous and regular as $|f| \to \infty$.

It can be easily verified that (5.70) holds for almost all λ also for $\nu = 0$.

For those values of λ belonging to the set of zero Lebesgue measure such that $\Psi^{(n)}(\lambda) = \Psi^{(m)}(\lambda)$ for $m \in I_0$ (with I_0 containing n and depending on λ), one obtains

$$\lim_{T\to\infty} W\Big(\big(\Psi^{(n)}(\lambda) - \Psi^{(m)}(\lambda - \nu/T)\big)T\Big) = W\Big(\nu\Psi^{(m)\prime}(\lambda)\Big). \tag{5.73}$$

These values, however, give no contribution if the function in the lhs of (5.70) is integrated w.r.t. λ.

(b) *Let $\Psi(\lambda)$ be a.e. derivable. It results that*

$$\lim_{T\to\infty} W\big(\big(\lambda_2 - \Psi(\lambda_1 - \nu/T)\big)T\big) = W\Big(\nu\Psi'(\lambda_1)\Big)\,\delta_{\lambda_2 - \Psi(\lambda_1)} \tag{5.74}$$

for almost all λ_1 and λ_2.

Proof: It is similar to that of item *(a)* (see also Lemma 5.2.2).

(c) *Let $\Psi_a(\lambda)$, $\Psi_b(\lambda)$, and $\Psi_c(\lambda)$ be a.e. derivable. It results that*

$$\begin{aligned}\lim_{T\to\infty} & W\left(\{\Psi_a(\Psi_b(\lambda_1 - \nu_1/T) + \lambda_2/T) - \Psi_c(\Psi_d(\lambda_1) - \nu_2/T)\}\, T\right) \\ &= W\Big(\Psi_a'(\Psi_b(\lambda_1))\,[-\Psi_b'(\lambda_1)\,\nu_1 + \lambda_2] + \Psi_c'(\Psi_d(\lambda_1))\,\nu_2\Big) \\ &\quad \delta_{\Psi_a(\Psi_b(\lambda_1)) - \Psi_c(\Psi_d(\lambda_1))}\end{aligned} \tag{5.75}$$

for almost all λ_1 and λ_2.

Proof: By considering the Taylor series expansion for the functions Ψ_b, Ψ_a, and Ψ_c, we have a.e.

$$\begin{aligned}&\Psi_b(\lambda_1 - \nu_1/T) \\ &\quad = \Psi_b(\lambda_1) - \Psi_b'(\lambda_1)\,\nu_1/T + \mathcal{O}(1/T^2)\end{aligned} \tag{5.76}$$

$$\begin{aligned}&\Psi_a\Big(\Psi_b(\lambda_1 - \nu_1/T) + \lambda_2/T\Big) \\ &\quad = \Psi_a\Big(\Psi_b(\lambda_1) - \Psi_b'(\lambda_1)\,\nu_1/T + \mathcal{O}(1/T^2) + \lambda_2/T\Big) \\ &\quad = \Psi_a\Big(\Psi_b(\lambda_1)\Big) + \Psi_a'\Big(\Psi_b(\lambda_1)\Big)\Big[-\Psi_b'(\lambda_1)\,\nu_1/T + \mathcal{O}(1/T^2) + \lambda_2/T\Big] \\ &\quad\quad + \mathcal{O}(1/T^2)\end{aligned} \tag{5.77}$$

$$\begin{aligned}&\Psi_c\Big(\Psi_d(\lambda_1) - \nu_2/T\Big) \\ &\quad = \Psi_c\Big(\Psi_d(\lambda_1) - \nu_2/T\Big) \\ &\quad = \Psi_c\Big(\Psi_d(\lambda_1)\Big) - \Psi_c'\Big(\Psi_d(\lambda_1)\Big)\,\nu_2/T + \mathcal{O}(1/T^2)\end{aligned} \tag{5.78}$$

respectively. Thus,

$$
\begin{aligned}
& W\left(\left\{\Psi_a(\Psi_b(\lambda_1 - \nu_1/T) + \lambda_2/T) - \Psi_c(\Psi_d(\lambda_1) - \nu_2/T)\right\} T\right) \\
&= W\left(\frac{\Psi_a(\Psi_b(\lambda_1)) - \Psi_c(\Psi_d(\lambda_1))}{1/T}\right. \\
&\quad + \frac{\Psi_a'(\Psi_b(\lambda_1))\,[-\Psi_b'(\lambda_1)\,\nu_1/T + \lambda_2/T] + \Psi_c'(\Psi_d(\lambda_1))\,\nu_2/T}{1/T} \\
&\quad \left. + \frac{\mathcal{O}(1/T^2)}{1/T}\right)
\end{aligned}
\tag{5.79}
$$

from which (5.75) easily follows since $W(\infty) = 0$.

(d) *Let $\Psi_a(\lambda)$, $\Psi_b(\lambda)$, and $\Psi_d(\lambda)$ be a.e. derivable. It results*

$$
\begin{aligned}
& \lim_{T\to\infty} W\left(\left\{\Psi_a(\Psi_b(\Psi_c(\lambda_1) - \nu_1/T) + \lambda_2/T) - \Psi_d(\lambda_1 - \nu_2/T)\right\} T\right) \\
&= W\Big(\Psi_a'(\Psi_b(\Psi_c(\lambda_1)))\,\Big[-\Psi_b'(\Psi_c(\lambda_1))\,\nu_1 + \lambda_2\Big] + \Psi_d'(\lambda_1)\,\nu_2\Big) \\
&\quad \delta_{\Psi_a(\Psi_b(\Psi_c(\lambda_1))) - \Psi_d(\lambda_1)}
\end{aligned}
\tag{5.80}
$$

for almost all λ_1 and λ_2.

Proof: By considering the Taylor series expansion for the functions Ψ_b, Ψ_a, and Ψ_d, we have a.e.

$$
\begin{aligned}
& \Psi_b\Big(\Psi_c(\lambda_1) - \nu_1/T\Big) \\
&= \Psi_b\Big(\Psi_c(\lambda_1)\Big) - \Psi_b'\Big(\Psi_c(\lambda_1)\Big)\,\nu_1/T + \mathcal{O}(1/T^2)
\end{aligned}
\tag{5.81}
$$

$$
\begin{aligned}
& \Psi_a\Big(\Psi_b(\Psi_c(\lambda_1) - \nu_1/T) + \lambda_2/T\Big) \\
&= \Psi_a\Big(\Psi_b(\Psi_c(\lambda_1)) - \Psi_b'(\Psi_c(\lambda_1))\,\nu_1/T + \mathcal{O}(1/T^2) + \lambda_2/T\Big) \\
&= \Psi_a\Big(\Psi_b(\Psi_c(\lambda_1))\Big) + \Psi_a'\Big(\Psi_b(\Psi_c(\lambda_1))\Big) \\
&\quad \cdot\Big[-\Psi_b'(\Psi_c(\lambda_1))\,\nu_1/T + \mathcal{O}(1/T^2) + \lambda_2/T\Big] + \mathcal{O}(1/T^2)
\end{aligned}
\tag{5.82}
$$

$$
\begin{aligned}
& \Psi_d(\lambda_1 - \nu_2/T) \\
&= \Psi_d(\lambda_1) - \Psi_d'(\lambda_1)\,\nu_2/T + \mathcal{O}(1/T^2)
\end{aligned}
\tag{5.83}
$$

respectively. Thus,

$$
\begin{aligned}
& W\left(\left\{\Psi_a(\Psi_b(\Psi_c(\lambda_1) - \nu_1/T) + \lambda_2/T) - \Psi_d(\lambda_1 - \nu_2/T)\right\} T\right) \\
&= W\left(\frac{\Psi_a(\Psi_b(\Psi_c(\lambda_1))) - \Psi_d(\lambda_1)}{1/T}\right. \\
&\quad + \frac{\Psi_a'(\Psi_b(\Psi_c(\lambda_1)))\,[-\Psi_b'(\Psi_c(\lambda_1))\,\nu_1/T + \lambda_2/T] + \Psi_d'(\lambda_1)\,\nu_2/T}{1/T} \\
&\quad \left. + \frac{\mathcal{O}(1/T^2)}{1/T}\right)
\end{aligned}
\tag{5.84}
$$

from which (5.80) easily follows since $W(\infty) = 0$.

(e) *Let $\Psi_4(\lambda_1, \lambda_2, \lambda_3)$ have a.e. all the partial derivatives. It results*

$$\lim_{T\to\infty} W\Big(\Big\{\Psi_a(\lambda_2) - \Psi_4(\lambda_1 - \nu_1/T, \Psi_b(\lambda_1) - \nu_2/T, \lambda_2 - \nu_3/T)\Big\}T\Big)$$
$$= W\Big(\nabla\Psi_4(\lambda_1, \Psi_b(\lambda_1), \lambda_2)^{\mathrm{T}}[\nu_1, \nu_2, \nu_3]\Big)\, \delta_{\Psi_a(\lambda_2)-\Psi_4(\lambda_1,\Psi_b(\lambda_1),\lambda_2)} \tag{5.85}$$

for almost all λ_1 and λ_2, where ∇ is the gradient operator and superscript T denotes transpose.

Proof: By considering the Taylor series expansion of the function Ψ_4, one has a.e.

$$\Psi_4(\lambda_1 - \nu_1/T, \Psi_b(\lambda_1) - \nu_2/T, \lambda_2 - \nu_3/T)$$
$$= \Psi_4(\lambda_1, \Psi_b(\lambda_1), \lambda_2)$$
$$-\nabla\Psi_4(\lambda_1, \Psi_b(\lambda_1), \lambda_2)^{\mathrm{T}}[\nu_1/T, \nu_2/T, \nu_3/T] + \mathcal{O}(1/T^2) \tag{5.86}$$

Thus,

$$W\Big(\Big\{\Psi_a(\lambda_2) - \Psi_4(\lambda_1 - \nu_1/T, \Psi_b(\lambda_1) - \nu_2/T, \lambda_2 - \nu_3/T)\Big\}T\Big)$$
$$= W\left(\frac{\Psi_a(\lambda_2) - \Psi_4(\lambda_1, \Psi_b(\lambda_1), \lambda_2)}{1/T}\right.$$
$$\left. + \frac{1}{T}\nabla\Psi_4(\lambda_1, \Psi_b(\lambda_1), \lambda_2)^{\mathrm{T}}[\nu_1/T, \nu_2/T, \nu_3/T] + \frac{\mathcal{O}(1/T^2)}{1/T}\right) \tag{5.87}$$

from which (5.85) easily follows since $W(\infty) = 0$. □

5.5.1 Proof of Theorem 4.7.5 Asymptotic Expected Value of the Frequency-Smoothed Cross-Periodogram

By substituting (4.103) (with $\Delta f = 1/T$) and (4.148) into (4.150), we get

$$\mathrm{E}\Big\{S^{(n)}_{yx^{(*)}}(t; f)_{T,\Delta f}\Big\}$$
$$= \frac{1}{T}\sum_{m\in\mathbb{I}}\int_{\mathbb{R}}\int_{\mathbb{R}} S^{(m)}_{yx^{(*)}}(\nu)\, T\, W_B(\{\lambda-\nu\}T)\, \exp[-j2\pi\{\cdots\}t]$$
$$T\, W_B^{(*)}\Big(\Big\{\Psi^{(n)}_{yx^{(*)}}(\lambda) - \Psi^{(m)}_{yx^{(*)}}(\nu)\Big\}\, T\Big)\, \exp\Big[-(-)j2\pi\{\cdots\}t\Big]$$
$$\frac{1}{\Delta f} W_A\left(\frac{f-\lambda}{\Delta f}\right) \mathrm{d}\nu\, \mathrm{d}\lambda$$
$$= \sum_{m\in\mathbb{I}}\int_{\mathbb{R}}\int_{\mathbb{R}} S^{(m)}_{yx^{(*)}}(\lambda - \nu'/T)\, W_B(\nu')\, \exp[-j2\pi(\nu'/T)t]$$
$$W_B^{(*)}\Big(\Big\{\Psi^{(n)}_{yx^{(*)}}(\lambda) - \Psi^{(m)}_{yx^{(*)}}(\lambda - \nu'/T)\Big\}\, T\Big)\, \exp\Big[-(-)j2\pi\{\cdots\}t\Big]$$
$$\frac{1}{\Delta f} W_A\left(\frac{f-\lambda}{\Delta f}\right) \mathrm{d}\lambda\, \mathrm{d}\nu' \tag{5.88}$$

where, in the second equality, the variable change $\nu' = (\lambda - \nu)T$ is made. Moreover, here and in the following, for the sake of notation simplicity, we write

$$W_B(g(\lambda)T)\exp[-j2\pi g(\lambda)t] = W_B(\{g(\lambda)\}T)\exp[-j2\pi\{\cdots\}t].$$

Thus,

$$\begin{aligned}
&\lim_{\Delta f \to 0}\lim_{T\to\infty} \mathrm{E}\left\{ S^{(n)}_{yx^{(*)}}(t;f)_{T,\Delta f}\right\} \\
&= \lim_{\Delta f \to 0}\lim_{T\to\infty} \sum_{m\in\mathbb{I}} \int_{\mathbb{R}}\int_{\mathbb{R}} S^{(m)}_{yx^{(*)}}(\lambda - \nu'/T)\, W_B(\nu')\, \exp[-j2\pi(\nu'/T)t] \\
&\quad W_B^{(*)}\left(\left\{\Psi^{(n)}_{yx^{(*)}}(\lambda) - \Psi^{(m)}_{yx^{(*)}}(\lambda - \nu'/T)\right\} T\right) \exp\left[-_{(-)}j2\pi\{\cdots\}t\right] \\
&\quad \frac{1}{\Delta f} W_A\left(\frac{f-\lambda}{\Delta f}\right) \mathrm{d}\lambda\, \mathrm{d}\nu' \\
&= \lim_{\Delta f\to 0} \sum_{m\in\mathbb{I}} \int_{\mathbb{R}}\int_{\mathbb{R}} S^{(m)}_{yx^{(*)}}(\lambda)\, W_B(\nu') \\
&\quad W_B^{(*)}\left(\nu'\, \Psi^{(n)}_{yx^{(*)}}{}'(\lambda)\right)\, \delta_{n-m}\, \frac{1}{\Delta f} W_A\left(\frac{f-\lambda}{\Delta f}\right) \mathrm{d}\lambda\, \mathrm{d}\nu' \\
&= \lim_{\Delta f\to 0} \int_{\mathbb{R}}\int_{\mathbb{R}} S^{(n)}_{yx^{(*)}}(\lambda)\, W_B(\nu')\, W_B^{(*)}\left(\nu'\, \Psi^{(n)}_{yx^{(*)}}{}'(\lambda)\right) \frac{1}{\Delta f} W_A\left(\frac{f-\lambda}{\Delta f}\right) \mathrm{d}\lambda\, \mathrm{d}\nu' \\
&= \int_{\mathbb{R}}\int_{\mathbb{R}} S^{(n)}_{yx^{(*)}}(\lambda)\, W_B(\nu')\, W_B^{(*)}\left(\nu'\, \Psi^{(n)}_{yx^{(*)}}{}'(\lambda)\right) \delta(f-\lambda)\, \mathrm{d}\lambda\, \mathrm{d}\nu' \\
&= S^{(n)}_{yx^{(*)}}(f) \int_{\mathbb{R}} W_B(\nu')\, W_B^{(*)}\left(\nu'\, \Psi^{(n)}_{yx^{(*)}}{}'(f)\right) \mathrm{d}\nu'.
\end{aligned} \tag{5.89}$$

where, in the derivation of the second equality, Lemma 5.5.1a (with $W = W_B$) is used. Hence, (4.158) is proved.

In (5.89), in the second equality, the limit as $T \to \infty$ can be interchanged with the sum since the series of functions of T is uniformly convergent by the Weierstrass criterium (Smirnov 1964). In fact, let

$$\begin{aligned}
F^{(m)}(T,\Delta f;\lambda,\nu') \triangleq\; & S^{(m)}_{yx^{(*)}}(\lambda - \nu'/T)\, W_B(\nu')\, \exp[-j2\pi(\nu'/T)t] \\
& W_B^{(*)}\left(\left\{\Psi^{(n)}_{yx^{(*)}}(\lambda) - \Psi^{(m)}_{yx^{(*)}}(\lambda - \nu'/T)\right\} T\right) \exp\left[-_{(-)}j2\pi\{\cdots\}t\right] \\
& \frac{1}{\Delta f} W_A\left(\frac{f-\lambda}{\Delta f}\right)
\end{aligned} \tag{5.90}$$

one has

$$\begin{aligned}
&\sum_{m\in\mathbb{I}} \left\| \int_{\mathbb{R}}\int_{\mathbb{R}} F^{(m)}(T,\Delta f;\lambda,\nu')\, \mathrm{d}\lambda\, \mathrm{d}\nu' \right\|_{\infty} \\
&\quad \leqslant \sum_{m\in\mathbb{I}} \int_{\mathbb{R}}\int_{\mathbb{R}} \left| S^{(m)}_{yx^{(*)}}(\lambda - \nu'/T)\, W_B(\nu')\right.
\end{aligned}$$

$$W_B^{(*)}\left(\left\{\Psi^{(n)}_{yx^{(*)}}(\lambda)-\Psi^{(m)}_{yx^{(*)}}(\lambda-\nu'/T)\right\}T\right)\frac{1}{\Delta f}W_A\left(\frac{f-\lambda}{\Delta f}\right)\Bigg|\,\mathrm{d}\lambda\,\mathrm{d}\nu'$$

$$\leqslant\sum_{m\in\mathbb{I}}\left\|S^{(m)}_{yx^{(*)}}\right\|_\infty\ \|W_B\|_\infty\int_{\mathbb{R}}|W_B(\nu')|\,\mathrm{d}\nu'\int_{\mathbb{R}}\left|\frac{1}{\Delta f}W_A\left(\frac{f-\lambda}{\Delta f}\right)\right|\mathrm{d}\lambda \tag{5.91}$$

which is finite (and independent of T) due to Assumptions 4.4.3a, 4.4.5, and 4.6.2. In (5.91), the first L^∞-norm is for functions of T and the others for functions of λ and ν', respectively. Furthermore, the limit as $T\to\infty$ can be interchanged with the integral operation by the dominated convergence theorem (Champeney 1990, Chapter 4). In fact,

$$\left|F^{(m)}(T,\Delta f;\lambda,\nu')\right|\leqslant\left\|S^{(m)}_{yx^{(*)}}\right\|_\infty\ \|W_B\|_\infty\,|W_B(\nu')|\left|\frac{1}{\Delta f}W_A\left(\frac{f-\lambda}{\Delta f}\right)\right|\in L^1(\mathbb{R}^2) \tag{5.92}$$

with the right-hand side independent of T. In (5.89), in the fourth equality (4.149) is accounted for and in the fifth equality the sampling property of Dirac delta is exploited. Furthermore, in the derivation of (5.89), the a.e. continuity of the functions $S^{(n)}_{yx^{(*)}}(f)$ and $\Psi^{(n)}_{yx^{(*)}}(f)$ (Assumptions 4.4.3a and 4.4.4) is used.

Finally, note that for those values of λ such that there exist $m\in I_0(\lambda)$ such that $\Psi^{(n)}_{yx^{(*)}}(\lambda)=\Psi^{(m)}_{yx^{(*)}}(\lambda)$, the Kronecker delta δ_{n-m} in the second equality in (5.89) should be substituted by $\sum_{k\in I_0(\lambda)}\delta_{m-k}$. However, since these values of λ belong to a set with zero Lebesgue measure (see the remark following Assumption 4.4.2) and the functions $S^{(n)}_{yx^{(*)}}(f)$ are not impulsive (Assumption 4.4.3a), then these values give zero contribution to the integral.

5.5.2 *Proof of Theorem 4.7.6 Rate of Convergence of the Bias of the Frequency-Smoothed Cross-Periodogram*

For the sake of notation simplicity, let us put

$$S^{(n)}(f)\equiv S^{(n)}_{yx^{(*)}}(f)\qquad \Psi^{(n)}(f)\equiv\Psi^{(n)}_{yx^{(*)}}(f).$$

From (4.150) with (4.151) substituted into we have

$$\begin{aligned}
\text{bias}&=\mathrm{E}\left\{S^{(n)}_{yx^{(*)}}(t;f)_{T,\Delta f}\right\}-S^{(n)}(f)\,\mathcal{E}^{(n)}(f)\\
&=\sum_{m\in\mathbb{I}}\int_{\mathbb{R}}S^{(m)}(\nu)\,\frac{1}{T}\int_{\mathbb{R}}\mathcal{B}_{\frac{1}{T}}(\lambda-\nu;t)\,\mathcal{B}^{(*)}_{\frac{1}{T}}\left(\Psi^{(n)}(\lambda)-\Psi^{(m)}(\nu);t\right)\\
&\quad A_{\Delta f}(f-\lambda)\,\mathrm{d}\lambda\,\mathrm{d}\nu-S^{(n)}(f)\,\mathcal{E}^{(n)}(f)\\
&=\sum_{m\in\mathbb{I}}\int_{\mathbb{R}}S^{(m)}(\nu)\,\frac{1}{T}\int_{\mathbb{R}}T\,W_B(\{\lambda-\nu\}T)\,e^{-j2\pi\{\cdots\}t}\\
&\quad T\,W_B^{(*)}\left(\left\{\Psi^{(n)}(\lambda)-\Psi^{(m)}(\nu)\right\}T\right)\,e^{-(-)j2\pi\{\cdots\}t}\,\frac{1}{\Delta f}\,W_A\left(\frac{f-\lambda}{\Delta f}\right)\mathrm{d}\lambda\,\mathrm{d}\nu\\
&\quad -S^{(n)}(f)\int_{\mathbb{R}}W_B(\xi)\,W_B^{(*)}\left(\xi\,\Psi^{(n)\prime}(f)\right)\,\mathrm{d}\xi
\end{aligned}$$

$$
\begin{aligned}
&= \sum_{m\in\mathbb{I}} \int_{\mathbb{R}} \frac{1}{\Delta f} W_A\Big(\frac{f-\lambda}{\Delta f}\Big) \int_{\mathbb{R}} S^{(m)}\Big(\lambda - \frac{\nu'}{T}\Big) W_B(\nu')\, e^{-j2\pi(\nu'/T)t} \\
&\qquad W_B^{(*)}\Big(\Big\{\Psi^{(n)}(\lambda) - \Psi^{(m)}\Big(\lambda - \frac{\nu'}{T}\Big)\Big\}T\Big)\, e^{-(-)j2\pi\{\cdots\}t}\, \mathrm{d}\nu'\, \mathrm{d}\lambda \\
&\qquad - S^{(n)}(f) \int_{\mathbb{R}} W_B(\nu')\, W_B^{(*)}\Big(\nu'\, \Psi^{(n)\prime}(f)\Big)\, \mathrm{d}\nu' \\
&= \int_{\mathbb{R}} W_A(\lambda') \int_{\mathbb{R}} S^{(n)}\Big(f - \lambda'\Delta f - \frac{\nu'}{T}\Big) W_B(\nu')\, e^{-j2\pi(\nu'/T)t} \\
&\qquad W_B^{(*)}\Big(\Big\{\Psi^{(n)}(f-\lambda'\Delta f) - \Psi^{(n)}\Big(f - \lambda'\Delta f - \frac{\nu'}{T}\Big)\Big\}T\Big)\, e^{-(-)j2\pi\{\cdots\}t}\, \mathrm{d}\nu'\, \mathrm{d}\lambda' \\
&\qquad - S^{(n)}(f) \int_{\mathbb{R}} W_B(\nu')\, W_B^{(*)}\Big(\nu'\, \Psi^{(n)\prime}(f)\Big)\, \mathrm{d}\nu' \\
&\qquad + \Bigg[\sum_{m\in\mathbb{J}_{n,f}} + \sum_{\substack{m\notin\mathbb{J}_{n,f} \\ m\neq n}}\Bigg] \int_{\mathbb{R}} W_A(\lambda') \int_{\mathbb{R}} S^{(m)}\Big(f - \lambda'\Delta f - \frac{\nu'}{T}\Big) W_B(\nu')\, e^{-j2\pi(\nu'/T)t} \\
&\qquad W_B^{(*)}\Big(\Big\{\Psi^{(n)}(f-\lambda'\Delta f) - \Psi^{(m)}\Big(f - \lambda'\Delta f - \frac{\nu'}{T}\Big)\Big\}T\Big)\, e^{-(-)j2\pi\{\cdots\}t}\, \mathrm{d}\nu'\, \mathrm{d}\lambda'
\end{aligned}
\tag{5.93}
$$

where, in the third equality (4.103) and (4.107) (both with $\Delta f = 1/T$), (4.148) and (4.159) are used; in the fourth equality the order of integrals in $\mathrm{d}\lambda\, \mathrm{d}\nu$ is interchanged and then the variable change $\nu' = (\lambda - \nu)T$ is made in the inner integral in $\mathrm{d}\nu$ (with λ fixed); in the fifth equality the variable change $\lambda' = (f - \lambda)/\Delta f$ is made in the integral in $\mathrm{d}\lambda$ and definition (4.156) of the set $\mathbb{J}_{n,f}$ is accounted for.

In (5.93), the interchange of the order of integrals in $\mathrm{d}\nu'$ and $\mathrm{d}\lambda'$ can be justified by the Fubini and Tonelli theorem (Champeney 1990, Chapter 3). In fact, let

$$
\begin{aligned}
F(\nu', \lambda') &\triangleq \sum_{m\in\mathbb{I}} W_A(\lambda')\, S^{(m)}\Big(f - \lambda'\Delta f - \frac{\nu'}{T}\Big) W_B(\nu')\, e^{-j2\pi(\nu'/T)t} \\
&\qquad W_B^{(*)}\Big(\Big\{\Psi^{(n)}(f-\lambda'\Delta f) - \Psi^{(m)}\Big(f - \lambda'\Delta f - \frac{\nu'}{T}\Big)\Big\}T\Big)\, e^{-(-)j2\pi\{\cdots\}t}
\end{aligned}
\tag{5.94}
$$

be the integrand function of the two-dimensional integral in (5.93). One has

$$
|F(\nu', \lambda')| \leqslant \|W_B\|_\infty \sum_{m\in\mathbb{I}} \|S^{(m)}\|_\infty |W_A(\lambda')|\, |W_B(\nu')| \in L^1(\mathbb{R}^2) \tag{5.95}
$$

where Assumptions 4.4.3a (series regularity), 4.4.5 (data-tapering window regularity), and 4.6.2 (frequency-smoothing window regularity) are accounted for.

In the following, for notation simplicity, λ and ν will be used in place of λ' and ν'.

Let us consider the following Taylor series expansions with Lagrange first- or second-order residual term. These expression, holding a.e., will be substituted into integrals. Consequently, the contribution of those points where expansions do not hold is zero.

- It results that a.e.

$$\Psi^{(n)}\left(f - \frac{\nu}{T}\right) = \Psi^{(n)}(f) - \Psi^{(n)\prime}(f)\frac{\nu}{T} + \frac{1}{2}\Psi^{(n)\prime\prime}(\bar{f})\left(\frac{\nu}{T}\right)^2$$
$$= \Psi^{(n)}(f) - \Psi^{(n)\prime}(f)\frac{\nu}{T} + \mathcal{O}_\nu(1)\left(\frac{\nu}{T}\right)^2 \tag{5.96}$$

where $\bar{f} \in (f_a, f_b)$ with $f_a \triangleq \min\{f, f - \nu/T\}$, $f_b \triangleq \max\{f, f - \nu/T\}$, the assumption of bounded second-order derivative $\Psi^{(n)\prime\prime}$ is used (Assumptions 4.4.4 and 4.5.4), and $\mathcal{O}_\nu(\cdot)$ denotes "big oh" Landau symbol depending on ν, that is, $g(\nu, T) = \mathcal{O}_\nu(1/T^a)$ means $g(\nu, T) \leqslant K_\nu/T^a$ as $T \to \infty$ with K_ν depending on ν. Since the second-order derivative $\Psi^{(n)\prime\prime}$ is assumed to be uniformly bounded, then there exists K such that $K_\nu \leqslant K, \forall \nu$. Thus, when necessary, $\mathcal{O}_\nu(1)$ will be upper bounded by $\mathcal{O}(1)$.
- Using (5.96) one has a.e.

$$\Psi^{(n)}(f - \lambda\Delta f) - \Psi^{(n)}\left(f - \lambda\Delta f - \frac{\nu}{T}\right)$$
$$= \Psi^{(n)}(f) - \Psi^{(n)\prime}(f)\lambda\Delta f + \mathcal{O}_\lambda(1)\,(\Delta f)^2\lambda^2$$
$$-\left[\Psi^{(n)}(f) - \Psi^{(n)\prime}(f)\left(\lambda\Delta f + \frac{\nu}{T}\right) + \mathcal{O}_{\nu,\lambda}(1)\left(\lambda\Delta f + \frac{\nu}{T}\right)^2\right]$$
$$= \Psi^{(n)\prime}(f)\,\frac{\nu}{T} + \mathcal{O}_{\nu,\lambda}(1)\,(\Delta f)^2\lambda^2 + \mathcal{O}_{\nu,\lambda}(1)\,\Delta f\,\frac{1}{T}\,\lambda\nu + \mathcal{O}_{\nu,\lambda}(1)\left(\frac{\nu}{T}\right)^2 \tag{5.97}$$

where $\bar{f} \in (f_a, f_b)$ with $f_a \triangleq \min\{f, f - \lambda\Delta f - \nu/T\}$, $f_b \triangleq \max\{f, f - \lambda\Delta f - \nu/T\}$ and $\mathcal{O}_{\nu,\lambda}(\cdot)$ denotes "big oh" Landau symbol depending on ν and λ.
- Using (5.97) and considering the Taylor series expansion for $W_B^{(*)}(\cdot)$, one obtains a.e.

$$W_B^{(*)}\left(\left\{\Psi^{(n)}(f - \lambda\Delta f) - \Psi^{(n)}\left(f - \lambda\Delta f - \frac{\nu}{T}\right)\right\}T\right)$$
$$= W_B^{(*)}\left(\Psi^{(n)\prime}(f)\,\nu + \mathcal{O}_{\nu,\lambda}(1)\,T\,(\Delta f)^2\lambda^2 + \mathcal{O}_{\nu,\lambda}(1)\,\Delta f\,\lambda\nu + \mathcal{O}_{\nu,\lambda}(1)\,\frac{\nu^2}{T}\right)$$
$$= W_B^{(*)}\left(\nu\Psi^{(n)\prime}(f)\right) + W_B^{(*)\prime}(\bar{f})\left[\mathcal{O}_{\nu,\lambda}(1)\,T\,(\Delta f)^2\lambda^2\right.$$
$$\left.+\mathcal{O}_{\nu,\lambda}(1)\,\Delta f\,\lambda\nu + \mathcal{O}_{\nu,\lambda}(1)\,\frac{\nu^2}{T}\right]$$
$$= W_B^{(*)}\left(\nu\Psi^{(n)\prime}(f)\right) + \left[\mathcal{O}_{\nu,\lambda}(1)\,T\,(\Delta f)^2\lambda^2\right.$$
$$\left.+\mathcal{O}_{\nu,\lambda}(1)\,\Delta f\,\lambda\nu + \mathcal{O}_{\nu,\lambda}(1)\,\frac{\nu^2}{T}\right] \tag{5.98}$$

provided that $T \to \infty$ and $\Delta f \to 0$ with $T(\Delta f)^2 \to 0$. In the second equality $\bar{f}$ is appropriately chosen, and in the third equality the uniform boundedness of $W_B^{(*)\prime}$ (Assumption 4.7.3) is exploited.
- For $m \neq n$ ($m \in \mathbb{J}_{n,f}$ or $m \notin \mathbb{J}_{n,f}$), accounting for (5.96) one obtains a.e.

$$\Psi^{(n)}(f - \lambda\Delta f) - \Psi^{(m)}\left(f - \lambda\Delta f - \frac{\nu}{T}\right)$$
$$= \Psi^{(n)}(f) - \Psi^{(n)\prime}(f)\lambda\Delta f + \mathcal{O}_\lambda(1)\,(\Delta f)^2\lambda^2$$

$$
-\Big[\Psi^{(m)}(f) - \Psi^{(m)\prime}(f)\left(\lambda\Delta f + \frac{\nu}{T}\right)
+\mathcal{O}_{\nu,\lambda}(1)\,(\Delta f)^2\lambda^2 + \mathcal{O}_{\nu,\lambda}(1)\,\Delta f\,\frac{1}{T}\,\lambda\nu + \mathcal{O}_{\nu,\lambda}(1)\left(\frac{\nu}{T}\right)^2\Big] \tag{5.99}
$$

- For $m \in \mathbb{J}_{n,f}$, according to (4.156), it results that $m \neq n$ and $\Psi^{(m)}(f) = \Psi^{(n)}(f)$. Using (5.99) we have a.e.

$$
\begin{aligned}
&W_B^{(*)}\left(\left\{\Psi^{(n)}(f-\lambda\Delta f) - \Psi^{(m)}\left(f-\lambda\Delta f - \frac{\nu}{T}\right)\right\}T\right) \\
&= W_B^{(*)}\Big(-\left[\Psi^{(n)\prime}(f) - \Psi^{(m)\prime}(f)\right](\lambda T\Delta f) + \Psi^{(m)\prime}(f)\nu \\
&\quad +\mathcal{O}_{\nu,\lambda}(1)\,T\,(\Delta f)^2\lambda^2 + \mathcal{O}_{\nu,\lambda}(1)\,\Delta f\,\lambda\nu + \mathcal{O}_{\nu,\lambda}(1)\,\frac{\nu^2}{T}\Big) \\
&= \begin{cases} W_B^{(*)}\left(\Psi^{(m)\prime}(f)\nu\right) & \text{if } \lambda = 0 \\ \mathcal{O}\left(\dfrac{1}{(h'_{n,f}|\lambda|T\Delta f)^\gamma}\right) & \text{if } \lambda \neq 0 \end{cases}
\end{aligned} \tag{5.100}
$$

provided that $T \to \infty$ and $\Delta f \to 0$ with $T\Delta f \to \infty$ and $T(\Delta f)^2$ bounded, and γ is defined in Assumption 4.7.2. In (5.100), it is assumed that

$$
h'_{n,f} \triangleq \inf_{m\in\mathbb{J}_{n,f}} \left|\Psi^{(m)}_{yx^{(*)}}{}'(f) - \Psi^{(n)}_{yx^{(*)}}{}'(f)\right| > 0. \tag{5.101}
$$

That is, in the intersection points, curves have different slopes. Such a condition can be relaxed by assuming that there exists a derivative order p such that in the intersection points curves have equal derivatives up to order $p-1$ and different pth-order derivatives.

The term in (5.100) cannot be easily managed in the bias expression. For this reason, the rate of convergence of bias is derived for all values of f such that the set $\mathbb{J}_{n,f}$ is empty. That is, in all points f where two or more different curves do not intercept.

- For $m \notin \mathbb{J}_{n,f}$, $m \neq n$, according to (4.156), it results that $\Psi^{(m)}(f) \neq \Psi^{(n)}(f)$. Therefore, using (5.99) we have a.e.

$$
\begin{aligned}
&W_B^{(*)}\left(\left\{\Psi^{(n)}(f-\lambda\Delta f) - \Psi^{(m)}\left(f-\lambda\Delta f - \frac{\nu}{T}\right)\right\}T\right) \\
&= W_B^{(*)}\Big(\left[\Psi^{(n)}(f) - \Psi^{(m)}(f)\right]T - \left[\Psi^{(n)\prime}(f) - \Psi^{(m)\prime}(f)\right](\lambda T\Delta f) \\
&\quad +\Psi^{(m)\prime}(f)\nu + \mathcal{O}_{\nu,\lambda}(1)\,T\,(\Delta f)^2\lambda^2 + \mathcal{O}_{\nu,\lambda}(1)\,\Delta f\,\lambda\nu + \mathcal{O}_{\nu,\lambda}(1)\,\frac{\nu^2}{T}\Big) \\
&= \mathcal{O}\left(\frac{1}{(h_{n,f}T)^\gamma}\right)
\end{aligned} \tag{5.102}
$$

provided that $T \to \infty$ and $\Delta f \to 0$ with $T\Delta f \to \infty$ (more slowly than T), and $T(\Delta f)^2$ bounded, where $h_{n,f}$ is defined in (4.157) and γ is defined in Assumption 4.7.2.

- It results that a.e.

$$S^{(n)}\Big(f - \lambda\Delta f - \frac{\nu}{T}\Big) = S^{(n)}(f) - S^{(n)\prime}(\bar{f})\Big(\lambda\Delta f + \frac{\nu}{T}\Big)$$
$$= S^{(n)}(f) + \mathcal{O}_{\nu,\lambda}(1)\,\Big(\lambda\Delta f + \frac{\nu}{T}\Big) \tag{5.103}$$

with appropriate $\bar{f}$, provided that the first-order derivative $S^{(n)\prime\prime}(f)$ is bounded (Assumption 4.7.4).

By substituting (5.98) and (5.102)–(5.103) into (5.93), the bias for every f such that $\mathbb{J}_{n,f}$ is empty can be expressed as sum of two terms $\mathcal{B}_1$ and $\mathcal{B}_2$ with the following asymptotic behaviors.

$$\begin{aligned}
\mathcal{B}_1 &\triangleq \int_{\mathbb{R}} W_A(\lambda) \int_{\mathbb{R}} S^{(n)}\Big(f - \lambda\Delta f - \frac{\nu}{T}\Big)\, W_B(\nu)\, e^{-j2\pi(\nu/T)t} \\
&\quad W_B^{(*)}\Big(\Big\{\Psi^{(n)}(f-\lambda\Delta f) - \Psi^{(n)}\Big(f - \lambda\Delta f - \frac{\nu}{T}\Big)\Big\}T\Big)\, e^{-(-)j2\pi\{\cdots\}t}\, \mathrm{d}\nu\, \mathrm{d}\lambda \\
&\quad - S^{(n)}(f) \int_{\mathbb{R}} W_B(\nu)\, W_B^{(*)}\Big(\nu\, \Psi^{(n)\prime}(f)\Big)\, \mathrm{d}\nu \\
&= \int_{\mathbb{R}} W_A(\lambda) \int_{\mathbb{R}} \Big[\Big(S^{(n)}\Big(f - \lambda\Delta f - \frac{\nu}{T}\Big) - S^{(n)}(f)\Big) + S^{(n)}(f)\Big] \\
&\quad W_B(\nu) \Big[\Big(e^{-j2\pi(\nu/T)t} - 1\Big) + 1\Big] \\
&\quad \Big[\Big(W_B^{(*)}\Big(\Big\{\Psi^{(n)}(f-\lambda\Delta f) - \Psi^{(n)}\Big(f - \lambda\Delta f - \frac{\nu}{T}\Big)\Big\}T\Big) - W_B^{(*)}\Big(\nu\Psi^{(n)\prime}(f)\Big)\Big) \\
&\quad + W_B^{(*)}\Big(\nu\Psi^{(n)\prime}(f)\Big)\Big] \Big[\Big(e^{-(-)j2\pi\{\cdots\}t} - 1\Big) + 1\Big]\, \mathrm{d}\nu\, \mathrm{d}\lambda \\
&\quad - S^{(n)}(f) \int_{\mathbb{R}} W_B(\nu)\, W_B^{(*)}\Big(\nu\, \Psi^{(n)\prime}(f)\Big)\, \mathrm{d}\nu \\
&= \int_{\mathbb{R}} W_A(\lambda) \int_{\mathbb{R}} \Big[S^{(n)}\Big(f - \lambda\Delta f - \frac{\nu}{T}\Big) - S^{(n)}(f)\Big]\, W_B(\nu)\, e^{-j2\pi(\nu/T)t} \\
&\quad W_B^{(*)}\Big(\Big\{\Psi^{(n)}(f-\lambda\Delta f) - \Psi^{(n)}\Big(f - \lambda\Delta f - \frac{\nu}{T}\Big)\Big\}T\Big)\, e^{-(-)j2\pi\{\cdots\}t}\, \mathrm{d}\nu\, \mathrm{d}\lambda \\
&\quad + \int_{\mathbb{R}} W_A(\lambda) \int_{\mathbb{R}} S^{(n)}\Big(f - \lambda\Delta f - \frac{\nu}{T}\Big)\, W_B(\nu) \Big[e^{-j2\pi(\nu/T)t} - 1\Big] \\
&\quad W_B^{(*)}\Big(\Big\{\Psi^{(n)}(f-\lambda\Delta f) - \Psi^{(n)}\Big(f - \lambda\Delta f - \frac{\nu}{T}\Big)\Big\}T\Big)\, e^{-(-)j2\pi\{\cdots\}t}\, \mathrm{d}\nu\, \mathrm{d}\lambda \\
&\quad + \int_{\mathbb{R}} W_A(\lambda) \int_{\mathbb{R}} S^{(n)}\Big(f - \lambda\Delta f - \frac{\nu}{T}\Big)\, W_B(\nu)\, e^{-j2\pi(\nu/T)t} \\
&\quad \Big[W_B^{(*)}\Big(\Big\{\Psi^{(n)}(f-\lambda\Delta f) - \Psi^{(n)}\Big(f - \lambda\Delta f - \frac{\nu}{T}\Big)\Big\}T\Big) \\
&\quad - W_B^{(*)}\Big(\nu\Psi^{(n)\prime}(f)\Big)\Big]\, e^{-(-)j2\pi\{\cdots\}t}\, \mathrm{d}\nu\, \mathrm{d}\lambda \\
&\quad + \int_{\mathbb{R}} W_A(\lambda) \int_{\mathbb{R}} S^{(n)}\Big(f - \lambda\Delta f - \frac{\nu}{T}\Big)\, W_B(\nu)\, e^{-j2\pi(\nu/T)t}
\end{aligned}$$

$$W_B^{(*)}\left(\left\{\Psi^{(n)}(f-\lambda\Delta f)-\Psi^{(n)}\left(f-\lambda\Delta f-\frac{\nu}{T}\right)\right\}T\right)$$
$$\left[e^{-(-)j2\pi\{\cdots\}t}-1\right]\mathrm{d}\nu\,\mathrm{d}\lambda \tag{5.104}$$

where, in the third equality the fact that $\int_{\mathbb{R}} W_A(f)\,\mathrm{d}f = 1$ and the identity

$$\begin{aligned}(a_1+a_2)(b_1+b_2)(c_1+c_2)(d_1+d_2)-a_2b_2c_2d_2\\ = a_1(b_1+b_2)(c_1+c_2)(d_1+d_2)+(a_1+a_2)\,b_1\,(c_1+c_2)(d_1+d_2)\\ +(a_1+a_2)(b_1+b_2)\,c_1\,(d_1+d_2)+(a_1+a_2)(b_1+b_2)(c_1+c_2)\,d_1\end{aligned}$$

are used.

Furthermore, accounting for the bound

$$\left|e^{j\theta}-1\right|\leqslant\theta \qquad \forall\theta\in\mathbb{R} \tag{5.105}$$

which is easily obtained since

$$\left|e^{j\theta}-1\right|^2=(\cos\theta-1)^2+\sin^2\theta=2(1-\cos\theta)=4\sin^2(\theta/2)\leqslant 4(\theta/2)^2 \tag{5.106}$$

and using the Taylor series expansions (5.97)–(5.103), the following upper bound is obtained.

$$\begin{aligned}|\mathcal{B}_1|\leqslant &\int_{\mathbb{R}}|W_A(\lambda)|\int_{\mathbb{R}}\mathcal{O}(1)\left(\left|\lambda\Delta f\right|+\left|\frac{\nu}{T}\right|\right)|W_B(\nu)|\,\|W_B\|_\infty\,\mathrm{d}\nu\,\mathrm{d}\lambda\\
&+\int_{\mathbb{R}}|W_A(\lambda)|\int_{\mathbb{R}}\|S^{(n)}\|_\infty\,|W_B(\nu)|\,2\pi|t|\left|\frac{\nu}{T}\right|\,\|W_B\|_\infty\,\mathrm{d}\nu\,\mathrm{d}\lambda\\
&+\int_{\mathbb{R}}|W_A(\lambda)|\int_{\mathbb{R}}\|S^{(n)}\|_\infty\,|W_B(\nu)|\\
&\left[\mathcal{O}(T(\Delta f)^2)\lambda^2+\mathcal{O}(\Delta f)|\nu\lambda|+\mathcal{O}\left(\frac{1}{T}\right)\nu^2\right]\mathrm{d}\nu\,\mathrm{d}\lambda\\
&+\int_{\mathbb{R}}|W_A(\lambda)|\int_{\mathbb{R}}\|S^{(n)}\|_\infty\,|W_B(\nu)|\,\|W_B\|_\infty\,2\pi|t|\\
&\left[\left|\Psi^{(n)\prime}(f)\right|\left|\frac{\nu}{T}\right|+\mathcal{O}((\Delta f)^2)\lambda^2+\mathcal{O}\left(\frac{\Delta f}{T}\right)|\nu\lambda|+\mathcal{O}\left(\frac{1}{T^2}\right)\nu^2\right]\mathrm{d}\nu\,\mathrm{d}\lambda\\
=\;&\mathcal{O}((\Delta f))+\mathcal{O}\left(\frac{1}{T}\right)+\mathcal{O}(T(\Delta f)^2)\end{aligned} \tag{5.107}$$

provided that $W_A(f)|f|^p\in L^1(\mathbb{R})$ and $W_B(f)|f|^p\in L^1(\mathbb{R})$, $p=1,2$, and $W_B'\in L^\infty(\mathbb{R})$ (Assumptions 4.4.5, 4.7.2, 4.6.2, and 4.7.3).

$$\mathcal{B}_2\triangleq\sum_{\substack{m\notin\mathbb{J}_{n,f}\\ m\neq n}}\int_{\mathbb{R}}W_A(\lambda)\int_{\mathbb{R}}S^{(m)}\left(f-\lambda\Delta f-\frac{\nu}{T}\right)W_B(\nu)\,e^{-j2\pi(\nu/T)t}$$
$$W_B^{(*)}\left(\left\{\Psi^{(n)}(f-\lambda\Delta f)-\Psi^{(m)}\left(f-\lambda\Delta f-\frac{\nu}{T}\right)\right\}T\right)e^{-(-)j2\pi\{\cdots\}t}\,\mathrm{d}\nu\,\mathrm{d}\lambda \tag{5.108}$$

One has

$$|\mathcal{B}_2| \leqslant \sum_{\substack{m \notin \mathbb{J}_{n,f} \\ m \neq n}} \int_{\mathbb{R}} |W_A(\lambda)| \int_{\mathbb{R}} \|S^{(m)}\|_\infty \, |W_B(\nu)| \, \mathcal{O}\left(\frac{1}{T^\gamma}\right) \mathrm{d}\nu \, \mathrm{d}\lambda = \mathcal{O}\left(\frac{1}{T^\gamma}\right) \tag{5.109}$$

Thus,

$$\text{bias} = \mathcal{B}_1 + \mathcal{B}_2 = \mathcal{O}\left(T\,(\Delta f)^2\right) + \mathcal{O}(\Delta f) + \mathcal{O}\left(\frac{1}{T}\right) + \mathcal{O}\left(\frac{1}{T^\gamma}\right) \tag{5.110}$$

from which (4.160) follows since $\gamma \geqslant 1$.

5.5.3 Proof of Theorem 4.7.7 Asymptotic Covariance of the Frequency-Smoothed Cross-Periodogram

Let us consider the term $\mathcal{T}_1''$ defined in (4.153). Accounting for (4.103), (4.107) (both with $\Delta f = 1/T$) and (4.148), it can be written as

$$\begin{aligned}
\mathcal{T}_1'' = \frac{1}{T^2} \int_{\mathbb{R}} \int_{\mathbb{R}} \sum_{n' \in \mathbb{I}'} \int_{\mathbb{R}} & S_{yy^*}^{(n')}(\nu_{y1}) \, T \, W_B(\{\lambda_1 - \nu_{y1}\}T) \exp[-j2\pi\{\cdots\}t_1] \\
& T \, W_B^*\left(\left\{\lambda_2 - \Psi_{yy^*}^{(n')}(\nu_{y1})\right\} T\right) \exp\left[j2\pi\{\cdots\}t_2\right] \mathrm{d}\nu_{y1} \\
& \sum_{n'' \in \mathbb{I}''} \int_{\mathbb{R}} S_{x^{(*)}x^{(*)*}}^{(n'')}(\nu_{x1}) \, T \, W_B^{(*)}\left(\left\{\Psi_{yx^{(*)}}^{(n_1)}(\lambda_1) - \nu_{x1}\right\} T\right) \exp\left[-_{(-)}j2\pi\{\cdots\}t_1\right] \\
& T \, W_B^{(*)*}\left(\left\{\Psi_{yx^{(*)}}^{(n_2)}(\lambda_2) - \Psi_{x^{(*)}x^{(*)*}}^{(n'')}(\nu_{x1})\right\} T\right) \exp\left[_{(-)}j2\pi\{\cdots\}t_2\right] \mathrm{d}\nu_{x1} \\
& \frac{1}{\Delta f} W_A\left(\frac{f_1 - \lambda_1}{\Delta f}\right) \frac{1}{\Delta f} W_A^*\left(\frac{f_2 - \lambda_2}{\Delta f}\right) \mathrm{d}\lambda_1 \, \mathrm{d}\lambda_2
\end{aligned} \tag{5.111}$$

Let us make the variable changes $\nu'_{y1} = (\lambda_1 - \nu_{y1})T$ and $\nu'_{x1} = \left(\Psi_{yx^{(*)}}^{(n_1)}(\lambda_1) - \nu_{x1}\right) T$ in the inner integrals in ν_{y1} and ν_{x1} (λ_1 and λ_2 fixed) and then interchange the order of integrals so that the order is (from the innermost to the outermost) $\mathrm{d}\lambda_2 \; \mathrm{d}\lambda_1 \; \mathrm{d}\nu'_{y1} \; \mathrm{d}\nu'_{x1}$. Then, let us make the variable change $\lambda'_2 = \left(\lambda_2 - \Psi_{yy^*}^{(n')}(\lambda_1 - \nu'_{y1}/T)\right) T$ in the inner integral in λ_2 (λ_1, ν'_{y1}, and ν'_{x1} fixed) and then interchange the order of the integrals in λ_1 and λ_2. Finally, let us make the variable change $\lambda'_1 = (f_1 - \lambda_1)/\Delta f$ to obtain

$$\begin{aligned}
\mathcal{T}_1'' = \sum_{n' \in \mathbb{I}'} \sum_{n'' \in \mathbb{I}''} \int_{\mathbb{R}} \int_{\mathbb{R}} & W_B(\nu'_{y1}) \exp[-j2\pi(\nu'_{y1}/T)t_1] \\
& W_B^{(*)}(\nu'_{x1}) \exp\left[-_{(-)}j2\pi(\nu'_{x1}/T)t_1\right] \int_{\mathbb{R}} W_B^*(\lambda'_2) \exp\left[j2\pi(\lambda'_2/T)t_2\right] \\
& \int_{\mathbb{R}} S_{yy^*}^{(n')}\left(f_1 - \lambda'_1 \Delta f - \nu'_{y1}/T\right) S_{x^{(*)}x^{(*)*}}^{(n'')}\left(\Psi_{yx^{(*)}}^{(n_1)}(f_1 - \lambda'_1 \Delta f) - \nu'_{x1}/T\right) \\
& W_B^{(*)*}\left(\left\{\Psi_{yx^{(*)}}^{(n_2)}\left(\Psi_{yy^*}^{(n')}(f_1 - \lambda'_1 \Delta f - \nu'_{y1}/T) + \lambda'_2/T\right)\right.\right.
\end{aligned}$$

$$- \Psi^{(n'')}_{x^{(*)}x^{(*)*}} \left(\Psi^{(n_1)}_{yx^{(*)}}(f_1 - \lambda_1' \Delta f) - \nu_{x1}'/T \right) \Big\} T \Big) \exp \left[{\scriptstyle (-)} j2\pi \{\cdots\} t_2 \right]$$

$$\frac{1}{\Delta f} W_A^* \left(\frac{f_2 - \Psi^{(n')}_{yy^*}(f_1 - \lambda_1' \Delta f - \nu_{y1}'/T) - \lambda_2'/T}{\Delta f} \right)$$

$$W_A(\lambda_1') \, \mathrm{d}\lambda_1' \frac{\mathrm{d}\lambda_2'}{T} \, \mathrm{d}\nu_{y1}' \, \mathrm{d}\nu_{x1}' \tag{5.112}$$

Thus,

$$\lim_{T \to \infty} (T\Delta f) \, \mathcal{T}_1''$$
$$= \sum_{n' \in \mathbb{I}'} \sum_{n'' \in \mathbb{I}''} \int_{\mathbb{R}} \int_{\mathbb{R}} \int_{\mathbb{R}} \int_{\mathbb{R}} S^{(n')}_{yy^*} (f_1 - \lambda_1' \Delta f) \; S^{(n'')}_{x^{(*)}x^{(*)*}} \left(\Psi^{(n_1)}_{yx^{(*)}} (f_1 - \lambda_1' \Delta f) \right)$$
$$W_B(\nu_{y1}') \, W_B^*(\lambda_2') \, W_B^{(*)}(\nu_{x1}')$$
$$W_B^{(*)*} \left(\Psi^{(n_2)\,\prime}_{yx^{(*)}} \left(\Psi^{(n')}_{yy^*}(f_1 - \lambda_1' \Delta f) \right) \left[-\Psi^{(n')\prime}_{yy^*}(f_1 - \lambda_1' \Delta f) \, \nu_{y1}' + \lambda_2' \right] \right.$$
$$\left. + \Psi^{(n'')\,\prime}_{x^{(*)}x^{(*)*}} \left(\Psi^{(n_1)}_{yx^{(*)}}(f_1 - \lambda_1' \Delta f) \right) \nu_{x1}' \right)$$
$$\delta_{\Psi^{(n_2)}_{yx^{(*)}} \left(\Psi^{(n')}_{yy^*}(f_1 - \lambda_1' \Delta f) \right) - \Psi^{(n'')}_{x^{(*)}x^{(*)*}} \left(\Psi^{(n_1)}_{yx^{(*)}}(f_1 - \lambda_1' \Delta f) \right)}$$
$$W_A(\lambda_1') \, W_A^* \left(\frac{f_2 - \Psi^{(n')}_{yy^*}(f_1 - \lambda_1' \Delta f)}{\Delta f} \right) \mathrm{d}\lambda_1' \, \mathrm{d}\lambda_2' \, \mathrm{d}\nu_{y1}' \, \mathrm{d}\nu_{x1}' \tag{5.113}$$

where, in the derivation of (5.113), Lemma 5.5.1c (with $W = W_B$, $\Psi_a = \Psi^{(n_2)}_{yx^{(*)}}$, $\Psi_b = \Psi^{(n')}_{yy^*}$, $\Psi_c = \Psi^{(n'')}_{x^{(*)}x^{(*)*}}$, and $\Psi_d = \Psi^{(n_1)}_{yx^{(*)}}$) is used. Hence,

$$\lim_{\Delta f \to 0} \lim_{T \to \infty} (T\Delta f) \, \mathcal{T}_1''$$
$$= \sum_{n' \in \mathbb{I}'} \sum_{n'' \in \mathbb{I}''} S^{(n')}_{yy^*}(f_1) \, S^{(n'')}_{x^{(*)}x^{(*)*}} \left(\Psi^{(n_1)}_{yx^{(*)}}(f_1) \right)$$
$$\int_{\mathbb{R}} \int_{\mathbb{R}} \int_{\mathbb{R}} W_B^{(*)*} \left(\Psi^{(n_2)\,\prime}_{yx^{(*)}} \left(\Psi^{(n')}_{yy^*}(f_1) \right) \left[-\Psi^{(n')\prime}_{yy^*}(f_1) \, \nu_{y1}' + \lambda_2' \right] \right.$$
$$\left. + \Psi^{(n'')\,\prime}_{x^{(*)}x^{(*)*}} \left(\Psi^{(n_1)}_{yx^{(*)}}(f_1) \right) \nu_{x1}' \right)$$
$$W_B(\nu_{y1}') \, W_B^{(*)}(\nu_{x1}') \, W_B^*(\lambda_2') \, \mathrm{d}\nu_{y1}' \, \mathrm{d}\nu_{x1}' \, \mathrm{d}\lambda_2'$$
$$\bar{\delta}_{\Psi^{(n_2)}_{yx^{(*)}} \left(\Psi^{(n')}_{yy^*}(f_1) \right) - \Psi^{(n'')}_{x^{(*)}x^{(*)*}} \left(\Psi^{(n_1)}_{yx^{(*)}}(f_1) \right)}$$
$$\int_{\mathbb{R}} W_A(\lambda_1') \, W_A^* \left(\lambda_1' \Psi^{(n')\prime}_{yy^*}(f_1) \right) \mathrm{d}\lambda_1' \; \delta_{f_2 - \Psi^{(n')}_{yy^*}(f_1)} \tag{5.114}$$

where Lemmas 5.5.1b (with $W = W_A$) and 5.3.2 are used. From (5.114), (4.162) immediately follows.

The interchange of the order of integrals can be justified by the Fubini and Tonelli theorem (Champeney 1990, Chapter 3). In fact, let $F^{(n',n'')}(T, \Delta f; \lambda_1', \lambda_2', \nu_{y1}', \nu_{x1}')$ be $(T\Delta f)$ times the integrand function of the four-dimensional integral in (5.112). One has

$$\left|F^{(n',n'')}(T, \Delta f; \lambda_1', \lambda_2', \nu_{y1}', \nu_{x1}')\right| \leqslant \left\|S_{yy^*}^{(n')}\right\|_\infty \left\|S_{x^{(*)}x^{(*)*}}^{(n'')}\right\|_\infty$$
$$|W_B(\nu_{y1}')|\,|W_B(\nu_{x1}')|\,|W_B(\lambda_2')|\,\|W_B\|_\infty$$
$$\|W_A\|_\infty\,|W_A(\lambda_1')| \in L^1(\mathbb{R}^4) \tag{5.115}$$

(with the right-hand side independent of T and Δf), where Assumptions 4.4.3a, 4.4.5, and 4.6.2 have been accounted for. In (5.113), the interchange of the order of limit (as $T \to \infty$) and sum (over n' and n'') operations is justified since the series of functions of T

$$\sum_{n' \in \mathbb{I}'} \sum_{n'' \in \mathbb{I}''} \int_{\mathbb{R}} \int_{\mathbb{R}} \int_{\mathbb{R}} \int_{\mathbb{R}} F^{(n',n'')}(T, \Delta f; \lambda_1', \lambda_2', \nu_{y1}', \nu_{x1}')\, \mathrm{d}\lambda_1'\, \mathrm{d}\lambda_2'\, \mathrm{d}\nu_{y1}'\, \mathrm{d}\nu_{x1}'$$

is uniformly convergent. In fact,

$$\sum_{n' \in \mathbb{I}'} \sum_{n'' \in \mathbb{I}''} \left\| \int_{\mathbb{R}} \int_{\mathbb{R}} \int_{\mathbb{R}} \int_{\mathbb{R}} F^{(n',n'')}(T, \Delta f; \lambda_1', \lambda_2', \nu_{y1}', \nu_{x1}')\, \mathrm{d}\lambda_1'\, \mathrm{d}\lambda_2'\, \mathrm{d}\nu_{y1}'\, \mathrm{d}\nu_{x1}' \right\|_\infty$$
$$\leqslant \sum_{n' \in \mathbb{I}'} \left\|S_{yy^*}^{(n')}\right\|_\infty \sum_{n'' \in \mathbb{I}''} \left\|S_{x^{(*)}x^{(*)*}}^{(n'')}\right\|_\infty \|W_B\|_\infty$$
$$\int_{\mathbb{R}} |W_B(\nu_{y1}')|\, \mathrm{d}\nu_{y1}' \int_{\mathbb{R}} |W_B(\nu_{x1}')|\, \mathrm{d}\nu_{x1}' \int_{\mathbb{R}} |W_B(\lambda_2')|\, \mathrm{d}\lambda_2'$$
$$\|W_A\|_\infty \int_{\mathbb{R}} |W_A(\lambda_1')|\, \mathrm{d}\lambda_1' \tag{5.116}$$

where the first L^∞-norm is for functions of T. The right-hand side of (5.116) is bounded due to Assumptions 4.4.3a, 4.4.5, and 4.6.2. Hence, the Weierstrass criterium (Smirnov 1964) can be applied. Moreover, from (5.115) it follows that $(T\Delta f)$ times the integrand function in (5.112) is bounded by a summable function not depending on T. Thus, the dominated convergence theorem (Champeney 1990, Chapter 4) can be applied in (5.113) to interchange the order of limit and integral operations. As regards the derivation of (5.114), observe that $F^{(n',n'')}(\infty, \Delta f; \lambda_1', \lambda_2', \nu_{y1}', \nu_{x1}')$ is the integrand function of the four-dimensional integral in (5.113). In (5.114), the order of the limit (as $\Delta f \to 0$) and sum (over n' and n'') operations can be interchanged since the series of functions of Δf

$$\sum_{n' \in \mathbb{I}'} \sum_{n'' \in \mathbb{I}''} \int_{\mathbb{R}} \int_{\mathbb{R}} \int_{\mathbb{R}} \int_{\mathbb{R}} F^{(n',n'')}(\infty, \Delta f; \lambda_1', \lambda_2', \nu_{y1}', \nu_{x1}')\, \mathrm{d}\lambda_1'\, \mathrm{d}\lambda_2'\, \mathrm{d}\nu_{y1}'\, \mathrm{d}\nu_{x1}'$$

is uniformly convergent. In fact, for

$$\sum_{n' \in \mathbb{I}'} \sum_{n'' \in \mathbb{I}''} \left\| \int_{\mathbb{R}} \int_{\mathbb{R}} \int_{\mathbb{R}} \int_{\mathbb{R}} F^{(n',n'')}(\infty, \Delta f; \lambda_1', \lambda_2', \nu_{y1}', \nu_{x1}')\, \mathrm{d}\lambda_1'\, \mathrm{d}\lambda_2'\, \mathrm{d}\nu_{y1}'\, \mathrm{d}\nu_{x1}' \right\|_\infty$$

where the L^∞-norm is for functions of Δf, the same bound as in (5.116) can be obtained and, hence, the Weierstrass criterium can be applied. Moreover, the function $|F^{(n',n'')}(\infty, \Delta f; \lambda_1', \lambda_2', \nu_{y1}', \nu_{x1}')|$ is bounded by the function in the right-hand side of (5.115) which is summable and independent of Δf. Therefore, the dominated convergence theorem can be applied and the order of limit (as $\Delta f \to 0$) and integral operations can be interchanged. Furthermore, in the derivation of (5.114), the a.e. continuity of the functions $S^{(n)}_{z_1 z_2}(f)$, $\Psi^{(n)}_{z_1 z_2}(f)$, and $\Psi^{(n)}_{z_1 z_2}{}'(f)$, for $z_1, z_2 \in \{x, x^*, y, y^*\}$, (Assumptions 4.4.3a and 4.4.4) are used.

Analogously, by considering the term $\mathcal{T}_2''$ defined in (4.154) and using Lemma 5.5.1d (with $W = W_B$, $\Psi_a = \Psi^{(n_2)}_{yx^{(*)}}$, $\Psi_b = \Psi^{(n^{iv})}_{x^{(*)}y^*}$, $\Psi_c = \Psi^{(n_1)}_{yx^{(*)}}$, and $\Psi_d = \Psi^{(n''')}_{yx^{(*)*}}$), it can be shown that

$$
\begin{aligned}
&\lim_{\Delta f \to 0} \lim_{T \to \infty} (T \Delta f)\, \mathcal{T}_2'' \\
&\quad = \sum_{n''' \in \mathbb{I}'''} \sum_{n^{iv} \in \mathbb{I}^{iv}} S^{(n''')}_{yx^{(*)*}}(f_1)\, S^{(n^{iv})}_{x^{(*)}y^*}\left(\Psi^{(n_1)}_{yx^{(*)}}(f_1)\right) \\
&\qquad \int_{\mathbb{R}} \int_{\mathbb{R}} \int_{\mathbb{R}} W_B^{(*)*}\left(\Psi^{(n_2)}_{yx^{(*)}}{}'\left(\Psi^{(n^{iv})}_{x^{(*)}y^*}\left(\Psi^{(n_1)}_{yx^{(*)}}(f_1)\right)\right)\right. \\
&\qquad \left.\left[-\Psi^{(n^{iv})}_{x^{(*)}y^*}{}'\left(\Psi^{(n_1)}_{yx^{(*)}}(f_1)\right)\, \nu'_{x1} + \lambda'_2\right] + \Psi^{(n''')}_{yx^{(*)*}}{}'(f_1)\, \nu'_{y1}\right) \\
&\qquad W_B(\nu'_{y1})\, W_B^{(*)}(\nu'_{x1})\, W_B^*(\lambda'_2)\, \mathrm{d}\nu'_{y1}\, \mathrm{d}\nu'_{x1}\, \mathrm{d}\lambda'_2 \\
&\qquad \bar{\delta}_{\Psi^{(n_2)}_{yx^{(*)}}\left(\Psi^{(n^{iv})}_{x^{(*)}y^*}\left(\Psi^{(n_1)}_{yx^{(*)}}(f_1)\right)\right) - \Psi^{(n''')}_{yx^{(*)*}}(f_1)} \\
&\qquad \int_{\mathbb{R}} W_A(\lambda'_1)\, W_A^*\left(\lambda'_1\left[\Psi^{(n^{iv})}_{x^{(*)}y^*}\left(\Psi^{(n_1)}_{yx^{(*)}}(f_1)\right)\right]'\right) \mathrm{d}\lambda'_1 \\
&\qquad \delta_{f_2 - \Psi^{(n^{iv})}_{x^{(*)}y^*}\left(\Psi^{(n_1)}_{yx^{(*)}}(f_1)\right)}
\end{aligned}
\tag{5.117}
$$

from which (4.163) immediately follows.

Let us now consider the term $\mathcal{T}_3''$ defined in (4.155). Accounting for (4.103), (4.107) (both with $\Delta f = 1/T$), and (4.148), it can be written as

$$
\begin{aligned}
\mathcal{T}_3'' = \frac{1}{T^2} &\int_{\mathbb{R}} \int_{\mathbb{R}} \sum_{n \in \mathbb{I}_4} \int_{\mathbb{R}} \int_{\mathbb{R}} \int_{\mathbb{R}} P^{(n)}_{yx^{(*)}y^*x^{(*)*}}(\nu_{y1}, \nu_{x1}, \nu_{y2}) \\
&T\, W_B^{(*)*}\left(\left\{\Psi^{(n_2)}_{yx^{(*)}}(\lambda_2) - \Psi^{(n)}_{yx^{(*)}y^*x^{(*)*}}(\nu_{y1}, \nu_{x1}, \nu_{y2})\right\} T\right) \exp\left[{\scriptstyle(-)} j2\pi\{\cdots\}t_2\right] \\
&T\, W_B(\{\lambda_1 - \nu_{y1}\}T) \exp\left[-j2\pi\{\cdots\}t_1\right] \\
&T\, W_B^{(*)}\left(\left\{\Psi^{(n_1)}_{yx^{(*)}}(\lambda_1) - \nu_{x1}\right\} T\right) \exp\left[-{\scriptstyle(-)} j2\pi\{\cdots\}t_1\right] \\
&T\, W_B^*(\{\lambda_2 - \nu_{y2}\}T) \exp\left[j2\pi\{\cdots\}t_2\right] \mathrm{d}\nu_{y1}\, \mathrm{d}\nu_{x1}\, \mathrm{d}\nu_{y2} \\
&\frac{1}{\Delta f} W_A\left(\frac{f_1 - \lambda_1}{\Delta f}\right) \frac{1}{\Delta f} W_A^*\left(\frac{f_2 - \lambda_2}{\Delta f}\right) \mathrm{d}\lambda_1\, \mathrm{d}\lambda_2.
\end{aligned}
\tag{5.118}
$$

Let us make the variable changes $\nu'_{y1} = (\lambda_1 - \nu_{y1})T$, $\nu'_{x1} = \left(\Psi^{(n_1)}_{yx^{(*)}}(\lambda_1) - \nu_{x1}\right) T$, and $\nu'_{y2} = (\lambda_2 - \nu_{y2})T$ in the inner integrals in ν_{y1}, ν_{x1}, and ν_{y2} (λ_1 and λ_2 fixed). Then, let us make the variable changes $\lambda'_1 = (f_1 - \lambda_1)/\Delta f$ and $\lambda'_2 = (f_2 - \lambda_2)/\Delta f$ to obtain

$$
\mathcal{T}_3'' = \frac{1}{T} \sum_{n \in \mathbb{I}_4} \int_{\mathbb{R}} \int_{\mathbb{R}} \int_{\mathbb{R}} \int_{\mathbb{R}} \int_{\mathbb{R}} P^{(n)}_{yx^{(*)}y^*x^{(*)*}}(f_1 - \lambda'_1 \Delta f - \nu'_{y1}/T,
$$

$$
\begin{aligned}
&\Psi^{(n_1)}_{yx^{(*)}}(f_1 - \lambda'_1 \Delta f) - \nu'_{x1}/T, f_2 - \lambda'_2 \Delta f - \nu'_{y2}/T) \\
&W_B^{(*)*}\left(\left\{\Psi^{(n_2)}_{yx^{(*)}}(f_2 - \lambda'_2 \Delta f) \; - \Psi^{(n)}_{yx^{(*)}y^*x^{(*)*}}(f_1 - \lambda'_1 \Delta f - \nu'_{y1}/T,\right.\right. \\
&\left.\left.\Psi^{(n_1)}_{yx^{(*)}}(f_1 - \lambda'_1 \Delta f) - \nu'_{x1}/T, f_2 - \lambda'_2 \Delta f - \nu'_{y2}/T)\right\} T\right) \; \exp\left[{\scriptstyle (-)} j2\pi\{\cdots\}t_2\right] \\
&W_B(\nu'_{y1}) \; \exp\left[-j2\pi(\nu'_{y1}/T)t_1\right] \; W_B^{(*)}(\nu'_{x1}) \; \exp\left[-{\scriptstyle (-)} j2\pi(\nu'_{x1}/T)t_1\right] \\
&W_B^*(\nu'_{y2}) \; \exp\left[j2\pi(\nu'_{y2}/T)t_2\right] \; \mathrm{d}\nu'_{y1} \, \mathrm{d}\nu'_{x1} \, \mathrm{d}\nu'_{y2} \\
&W_A(\lambda'_1) \; W_A^*(\lambda'_2) \; \mathrm{d}\lambda'_1 \, \mathrm{d}\lambda'_2.
\end{aligned} \tag{5.119}
$$

Thus,

$$
\begin{aligned}
T\Delta f \; \left|\mathcal{T}''_3\right| \leqslant \Delta f \sum_{n \in \mathbb{I}_4} &\left\| P^{(n)}_{yx^{(*)}y^*x^{(*)*}} \right\|_\infty \; \|W_B\|_\infty \\
&\int_{\mathbb{R}} |W_B(\nu'_{y1})| \, \mathrm{d}\nu'_{y1} \int_{\mathbb{R}} |W_B(\nu'_{x1})| \, \mathrm{d}\nu'_{x1} \int_{\mathbb{R}} |W_B(\nu'_{y2})| \, \mathrm{d}\nu'_{y2} \\
&\int_{\mathbb{R}} |W_A(\lambda'_1)| \, \mathrm{d}\lambda'_1 \int_{\mathbb{R}} |W_A(\lambda'_2)| \, \mathrm{d}\lambda'_2
\end{aligned} \tag{5.120}
$$

with the rhs bounded (Assumptions 4.4.5 and 4.7.3) and independent of T. Therefore,

$$
\begin{aligned}
\lim_{\Delta f \to 0} \lim_{T \to \infty} T\Delta f \; \left|\mathcal{T}''_3\right| \leqslant \left(\lim_{\Delta f \to 0} \Delta f\right) \sum_{n \in \mathbb{I}_4} &\left\| P^{(n)}_{yx^{(*)}y^*x^{(*)*}} \right\|_\infty \; \|W_B\|_\infty \\
&\int_{\mathbb{R}} |W_B(\nu'_{y1})| \, \mathrm{d}\nu'_{y1} \int_{\mathbb{R}} |W_B(\nu'_{x1})| \, \mathrm{d}\nu'_{x1} \int_{\mathbb{R}} |W_B(\nu'_{y2})| \, \mathrm{d}\nu'_{y2} \\
&\int_{\mathbb{R}} |W_A(\lambda'_1)| \, \mathrm{d}\lambda'_1 \int_{\mathbb{R}} |W_A(\lambda'_2)| \, \mathrm{d}\lambda'_2 \\
&= 0.
\end{aligned} \tag{5.121}
$$

□

5.6 Proofs for Section 4.7.2 "Asymptotic Normality of the Frequency-Smoothed Cross-Periodogram"

In this section, proofs of lemmas and theorems presented in Section 4.7.2 on the asymptotic complex Normality of the frequency-smoothed cross-periodogram are reported.

Fact 5.6.1
Let

$$
Z_i(f_1, f_2) \triangleq \left[Y(f_1) \, X^{(*)}(f_2)\right]^{[*]_i} \qquad i = 1, \ldots, k \tag{5.122}
$$

where $[*]_i$ *represents the ith optional complex conjugation, and let us consider the* $k \times 2$ *table*

$$\begin{matrix} Y(f_{11}) & X^{(*)}(f_{21}) \\ Y(f_{12}) & X^{(*)}(f_{22}) \\ \vdots & \vdots \\ Y(f_{1k}) & X^{(*)}(f_{2k}) \end{matrix} \tag{5.123}$$

and a partition of its elements into disjoint sets $\{\nu_1, \ldots, \nu_p\}$. *The cumulant* $\text{cum}\{Z_1(f_{11}, f_{21}), \ldots, Z_k(f_{1k}, f_{2k})\}$ *can be expressed as (Leonov and Shiryaev 1959), (Brillinger 1965), (Brillinger and Rosenblatt 1967)*

$$\text{cum}\{Z_1(f_{11}, f_{21}), \ldots, Z_k(f_{1k}, f_{2k})\} = \sum_{\boldsymbol{\nu}} C_{\nu_1} \cdots C_{\nu_p} \tag{5.124}$$

where ν_m $(m = 1, \ldots, p)$ *are subsets of elements of the* $k \times 2$ *table (5.123),* C_{ν_m} *is the cumulant of the elements in* ν_m, *and the (finite) sum in (5.124) is extended over all indecomposable partitions of table (5.123), including the partition with only one element (see the discussion following (2.159) for details).*

Let

$$V_i(f_{ij}) \triangleq \begin{cases} Y(f_{1j}) & i = 1 \\ X^{(*)}(f_{2j}) & i = 2 \end{cases} \qquad j = 1, \ldots, k \tag{5.125}$$

The elements of the table (5.123) can be identified by the pair of indices (i, j), *where* $i \in \{1, 2\}$ *is the column index and* $j \in \{1, \ldots, k\}$ *is the row index.*

From Assumption 4.7.8 it follows that

$$\begin{aligned} C_{\nu_m} &\triangleq \text{cum}\{V_i(f_{ij}),\ (i, j) \in \nu_m\} \\ &= \sum_{\ell_m} P_{\nu_m}^{(\ell_m)}(\boldsymbol{f}'_{\nu_m})\, \delta\left(f_{i_m j_m} - \Psi_{\nu_m}^{(\ell_m)}(\boldsymbol{f}'_{\nu_m})\right) \end{aligned} \tag{5.126}$$

where $\boldsymbol{f}_{\nu_m}$ *is the vector with elements* f_{ij}, $(i, j) \in \nu_m$, *the pair* (i_m, j_m) *is one element (e.g., the last one) of the set* ν_m, *and* $\boldsymbol{f}'_{\nu_m}$ *is the vector containing the same elements as* $\boldsymbol{f}_{\nu_m}$ *except* (i_m, j_m). *Thus, it results that* $\boldsymbol{f}_{\nu_m} = [\boldsymbol{f}'_{\nu_m}, f_{i_m j_m}]$.

In the sequel, the following notation will be used:

$$\boldsymbol{f}_{\nu_m} = \{f_{ij}\}_{(i,j)\in\nu_m} \qquad \boldsymbol{f}'_{\nu_m} = \{f_{ij}\}_{(i,j)\in\nu'_m} \qquad \nu_m = \nu'_m \cup \{(i_m, j_m)\} \tag{5.127}$$

with elements of ν_m *ordered from left to right and from top to bottom.*

5.6.1 Proof of Lemma 4.7.10 Cumulants of Frequency-Smoothed Cross-Periodograms

Let us omit, for notation simplicity, the subscript $yx^{(*)}$ in $S_{yx^{(*)}}^{(n)}(f)$, $\Psi_{yx^{(*)}}^{(n)}(f)$ $S_{yx^{(*)}}^{(n)}(t, f)_{T,\Delta f}$, and $I_{yx^{(*)}}(t; f_1, f_2)_T$. Furthermore, in order to avoid heavy notation, the optional complex conjugations $[*]_i$ will be omitted. Results in the presence of these complex conjugations are similar.

By substituting the STFT expression (4.94b) and (4.107) (with $\Delta f = 1/T$) into the cross-periodogram definition we formally have

$$
\begin{aligned}
& I(t; f, \Psi^{(n)}(f))_T \\
& \quad \triangleq \frac{1}{T} Y_T(t, f)\, X_T^{(*)}(t, \Psi^{(n)}(f)) \\
& \quad = \frac{1}{T} \int_{\mathbb{R}} Y(\varphi_1)\, \mathcal{B}_{\frac{1}{T}}(f - \varphi_1; t)\, \mathrm{d}\varphi_1 \int_{\mathbb{R}} X^{(*)}(\varphi_2)\, \mathcal{B}_{\frac{1}{T}}^{(*)}\left(\Psi^{(n)}(f) - \varphi_2; t\right) \mathrm{d}\varphi_2 \\
& \quad = \frac{1}{T} \int_{\mathbb{R}} \int_{\mathbb{R}} Y(\varphi_1)\, X^{(*)}(\varphi_2)\, \mathcal{B}_{\frac{1}{T}}(f - \varphi_1; t)\, \mathcal{B}_{\frac{1}{T}}^{(*)}\left(\Psi^{(n)}(f) - \varphi_2; t\right) \mathrm{d}\varphi_1\, \mathrm{d}\varphi_2
\end{aligned}
\tag{5.128}
$$

Thus, let

$$
\psi_i(f_j) \triangleq \begin{cases} f_j & i = 1 \\ \Psi^{(n_j)}(f_j) & i = 2 \end{cases}
\tag{5.129}
$$

for the multilinearity property of cumulants (Mendel 1991) we have

$$
\begin{aligned}
& \operatorname{cum}\left\{ I(t_1; f_1, \Psi^{(n_1)}(f_1))_T, \ldots, I(t_k; f_k, \Psi^{(n_k)}(f_k))_T \right\} \\
& \quad = \frac{1}{T^k} \int_{\mathbb{R}^{2k}} \operatorname{cum}\{Z_1(\varphi_{11}, \varphi_{21}), \ldots, Z_k(\varphi_{1k}, \varphi_{2k})\} \\
& \qquad \prod_{j=1}^{k} \left[\mathcal{B}_{\frac{1}{T}}(f_j - \varphi_{1j}; t_j)\, \mathcal{B}_{\frac{1}{T}}^{(*)}\left(\Psi^{(n_j)}(f_j) - \varphi_{2j}; t_j\right) \right] \mathrm{d}\boldsymbol{\varphi}_1\, \mathrm{d}\boldsymbol{\varphi}_2 \\
& \quad = \frac{1}{T^k} \int_{\mathbb{R}^{2k}} \sum_{\boldsymbol{\nu}} \sum_{\ell_1} P_{\nu_1}^{(\ell_1)}(\boldsymbol{\varphi}'_{\nu_1})\, \delta\left(\varphi_{i_1 j_1} - \Psi_{\nu_1}^{(\ell_1)}(\boldsymbol{\varphi}'_{\nu_1})\right) \cdots \\
& \qquad \sum_{\ell_p} P_{\nu_p}^{(\ell_p)}(\boldsymbol{\varphi}'_{\nu_p})\, \delta\left(\varphi_{i_p j_p} - \Psi_{\nu_p}^{(\ell_p)}(\boldsymbol{\varphi}'_{\nu_p})\right) \\
& \qquad \prod_{j=1}^{k} \left[\mathcal{B}_{\frac{1}{T}}(f_j - \varphi_{1j}; t_j)\, \mathcal{B}_{\frac{1}{T}}^{(*)}\left(\Psi^{(n_j)}(f_j) - \varphi_{2j}; t_j\right) \right] \mathrm{d}\boldsymbol{\varphi}_1\, \mathrm{d}\boldsymbol{\varphi}_2 \\
& \quad = \sum_{\boldsymbol{\nu}} \sum_{\ell_1} \cdots \sum_{\ell_p} \frac{1}{T^k} \int_{\mathbb{R}^{2k-p}} P_{\nu_1}^{(\ell_1)}(\boldsymbol{\varphi}'_{\nu_1}) \cdots P_{\nu_p}^{(\ell_p)}(\boldsymbol{\varphi}'_{\nu_p}) \\
& \qquad \prod_{(i,j)\in\nu'_1} \cdots \prod_{(i,j)\in\nu'_p} \mathcal{B}_{\frac{1}{T}}^{(*)_i}\left(\psi_i(f_j) - \varphi_{ij}; t_j\right) \\
& \qquad \prod_{h=1}^{p} \mathcal{B}_{\frac{1}{T}}^{(*)_{i_h}}\left(\psi_{i_h}(f_{j_h}) - \Psi_{\nu_h}^{(\ell_h)}(\boldsymbol{\varphi}'_{\nu_h}); t_{j_h}\right) \mathrm{d}\boldsymbol{\varphi}'_{\nu_1} \cdots \mathrm{d}\boldsymbol{\varphi}'_{\nu_p}
\end{aligned}
\tag{5.130}
$$

where, $\boldsymbol{\varphi}_1 \triangleq [\varphi_{11}, \ldots, \varphi_{1k}]$, $\boldsymbol{\varphi}_2 \triangleq [\varphi_{21}, \ldots, \varphi_{2k}]$, $\boldsymbol{\varphi}'_{\nu_m} \triangleq \{\varphi_{ij}\}_{(i,j)\in\nu'_m}$. In the second equality (5.124) and (5.126) are accounted for; in the third equality the sampling property of Dirac delta is used and here and in the following $(*)_i = (*)$ if $i = 1$ and $(*)_i = (*)*$ if $i = 2$.

Using again the multilinearity property of cumulants (Mendel 1991), the result is that

$$
\begin{aligned}
&\operatorname{cum}\left\{S^{(n_1)}(t_1, f_1)_{T,\Delta f}, \ldots, S^{(n_k)}(t_k, f_k)_{T,\Delta f}\right\} \\
&= \operatorname{cum}\left\{ I(t_1; f_1, \Psi^{(n_1)}(f_1))_T \underset{f_1}{\otimes} A_{\Delta f}(f_1), \ldots, \right. \\
&\qquad \left. I(t_k; f_k, \Psi^{(n_k)}(f_k))_T \underset{f_k}{\otimes} A_{\Delta f}(f_k) \right\} \\
&= \operatorname{cum}\left\{ I(t_1; f_1, \Psi^{(n_1)}(f_1))_T, \ldots, I(t_k; f_k, \Psi^{(n_k)}(f_k))_T \right\} \\
&\qquad \underset{f_1}{\otimes} A_{\Delta f}(f_1) \cdots \underset{f_k}{\otimes} A_{\Delta f}(f_k) \\
&= \sum_{\boldsymbol{v}} \sum_{\ell_1} \cdots \sum_{\ell_p} \frac{1}{T^k} \int_{\mathbb{R}^k} \int_{\mathbb{R}^{2k-p}} \prod_{h=1}^{p} P_{v_h}^{(\ell_h)}(\boldsymbol{\varphi}'_{v_h}) \\
&\qquad \prod_{(i,j)\in v'_1} \cdots \prod_{(i,j)\in v'_p} T\, W_B^{(*)_i}\left(\left\{\psi_i(\lambda_j) - \varphi_{ij}\right\} T\right) e^{-(-)_i j2\pi\{\cdots\} t_j} \\
&\qquad \prod_{h=1}^{p} T\, W_B^{(*)_{i_h}}\left(\left\{\psi_{i_h}(\lambda_{j_h}) - \Psi_{v_h}^{(\ell_h)}(\boldsymbol{\varphi}'_{v_h})\right\} T\right) e^{-(-)_{i_h} j2\pi\{\cdots\} t_{j_h}} \\
&\qquad \mathrm{d}\boldsymbol{\varphi}'_{v_1} \cdots \mathrm{d}\boldsymbol{\varphi}'_{v_p} \prod_{j=1}^{k} \frac{1}{\Delta f} W_A\left(\frac{f_j - \lambda_j}{\Delta f}\right) \mathrm{d}\boldsymbol{\lambda} \\
&= \sum_{\boldsymbol{v}} \sum_{\ell_1} \cdots \sum_{\ell_p} \frac{1}{T^k} \int_{\mathbb{R}^k} \int_{\mathbb{R}^{2k-p}} \prod_{h=1}^{p} P_{v_h}^{(\ell_h)}\left(\left\{\psi_i(\lambda_j) - \frac{\bar{\varphi}_{ij}}{T}\right\}_{(i,j)\in v'_h}\right) \\
&\qquad \prod_{(i,j)\in v'_1} \cdots \prod_{(i,j)\in v'_p} W_B^{(*)_i}(\bar{\varphi}_{ij})\, e^{-(-)_i j2\pi(\bar{\varphi}_{ij}/T) t_j} \\
&\qquad \prod_{h=1}^{p} T\, W_B^{(*)_{i_h}}\left(\left\{\psi_{i_h}(\lambda_{j_h}) - \Psi_{v_h}^{(\ell_h)}\left(\left\{\psi_i(\lambda_j) - \frac{\bar{\varphi}_{ij}}{T}\right\}_{(i,j)\in v'_h}\right)\right\} T\right) e^{-(-)_{i_h} j2\pi\{\cdots\} t_{j_h}} \\
&\qquad \prod_{h=1}^{p} \prod_{(i,j)\in v'_h} \mathrm{d}\bar{\varphi}_{ij} \prod_{j=1}^{k} \frac{1}{\Delta f} W_A\left(\frac{f_j - \lambda_j}{\Delta f}\right) \mathrm{d}\boldsymbol{\lambda}
\end{aligned} \tag{5.131}
$$

where, in the third equality Assumptions 4.4.5 (data-tapering window regularity) (with $\Delta f = 1/T$) and 4.6.2 (frequency-smoothing window regularity) are accounted for; and in the fourth equality the variable changes

$$\bar{\varphi}_{ij} = [\psi_i(\lambda_j) - \varphi_{ij}]T \qquad (i, j) \in v'_h,\ h = 1, \ldots, p$$

are made in the inner integrals in $\mathrm{d}\boldsymbol{\varphi}'_{v_1} \cdots \mathrm{d}\boldsymbol{\varphi}'_{v_p}$ (with $\boldsymbol{\lambda}$ fixed) so that, according to notation introduced in (5.127), the result is that

$$\boldsymbol{\varphi}'_{v_h} = \{\varphi_{ij}\}_{(i,j)\in v'_h} = \{\psi_i(\lambda_j) - \bar{\varphi}_{ij}/T\}_{(i,j)\in v'_h}.$$

Starting from the bounds derived in (5.133) and (5.138), the interchange of the order of cum, expectation, and integral operations can be justified by the Fubini and Tonelli theorem (Champeney 1990, Chapter 3). Furthermore, the same bounds justify the interchange of infinite sums and integral operations by the Weierstrass criterion (Smirnov 1964).

Let us consider the only term $\mathcal{T}_{\nu_1}$ corresponding to the only partition with $p = 1$ in (5.131). The set ν_1 is coincident with the whole rectangular $k \times 2$ table and $\nu_1' = \nu_1 - \{(i_1, j_1)\}$ with $(i_1, j_1) = (2, k)$.

$$
\begin{aligned}
\mathcal{T}_{\nu_1} &= \sum_{\ell_1} \frac{1}{T^k} \int_{\mathbb{R}^k} \int_{\mathbb{R}^{2k-1}} P_{\nu_1}^{(\ell_1)}\Big(\Big\{\psi_i(\lambda_j) - \frac{\bar{\varphi}_{ij}}{T}\Big\}_{(i,j)\in\nu_1'}\Big) \\
&\quad \prod_{(i,j)\in\nu_1'} W_B^{(*)_i}(\bar{\varphi}_{ij})\, e^{-(-)_i j2\pi(\bar{\varphi}_{ij}/T)t_j} \\
&\quad T\, W_B^{(*)_2}\Big(\Big\{\psi_2(\lambda_k) - \Psi_{\nu_1}^{(\ell_1)}\Big(\Big\{\psi_i(\lambda_j) - \frac{\bar{\varphi}_{ij}}{T}\Big\}_{(i,j)\in\nu_1'}\Big)\Big\}T\Big)\, e^{-(-)_2 j2\pi\{\cdots\}t_k} \\
&\quad \prod_{(i,j)\in\nu_1'} \mathrm{d}\bar{\varphi}_{ij} \prod_{j=1}^{k} \frac{1}{\Delta f} W_A\Big(\frac{f_j - \lambda_j}{\Delta f}\Big)\, \mathrm{d}\boldsymbol{\lambda} \\
&= \sum_{\ell_1} \frac{1}{T^k} \int_{\mathbb{R}^k} \int_{\mathbb{R}^{2k-1}} P_{\nu_1}^{(\ell_1)}\Big(\Big\{\psi_i(f_j - \bar{\lambda}_j\Delta f) - \frac{\bar{\varphi}_{ij}}{T}\Big\}_{(i,j)\in\nu_1'}\Big) \\
&\quad \prod_{(i,j)\in\nu_1'} W_B^{(*)_i}(\bar{\varphi}_{ij})\, e^{-(-)_i j2\pi(\bar{\varphi}_{ij}/T)t_j} \\
&\quad T\, W_B^{(*)_2}\Big(\Big\{\psi_2(f_k - \bar{\lambda}_k\Delta f) - \Psi_{\nu_1}^{(\ell_1)}\Big(\Big\{\psi_i(f_j - \bar{\lambda}_j\Delta f) - \frac{\bar{\varphi}_{ij}}{T}\Big\}_{(i,j)\in\nu_1'}\Big)\Big\}T\Big) \\
&\quad e^{-(-)_2 j2\pi\{\cdots\}t_k} \prod_{(i,j)\in\nu_1'} \mathrm{d}\bar{\varphi}_{ij} \prod_{j=1}^{k} W_A(\bar{\lambda}_j)\, \mathrm{d}\bar{\boldsymbol{\lambda}}
\end{aligned}
\tag{5.132}
$$

where $\psi_2(\lambda_k) = \Psi^{(n_k)}(\lambda_k)$ and in the second equality the variable change

$$
\bar{\lambda}_j = \frac{f_j - \lambda_j}{\Delta f} \qquad j = 1, \ldots, k
$$

(so that $\lambda_j = f_j - \bar{\lambda}_j \Delta f$) is made.

Accounting for Assumptions 4.4.5 (data-tapering window regularity), 4.6.2 (frequency-smoothing window regularity), 4.7.8 (spectral cumulants), it results in

$$
\begin{aligned}
\left|\mathcal{T}_{\nu_1}\right| &\leqslant \frac{1}{T^{k-1}} \sum_{\ell_1} \left\|P_{\nu_1}^{(\ell_1)}\right\|_\infty \|W_B\|_\infty \int_{\mathbb{R}^{2k-1}} \Big[\prod_{(i,j)\in\nu_1'} |W_B(\bar{\varphi}_{ij})|\Big] \prod_{(i,j)\in\nu_1'} \mathrm{d}\bar{\varphi}_{ij} \\
&\quad \int_{\mathbb{R}^k} \prod_{j=1}^{k} |W_A(\bar{\lambda}_j)|\, \mathrm{d}\bar{\boldsymbol{\lambda}}.
\end{aligned}
\tag{5.133}
$$

Therefore,

$$\lim_{T\to\infty} (T\Delta f)^{k/2} \left|\mathcal{T}_{\nu_1}\right| \leqslant \lim_{T\to\infty} \frac{(\Delta f)^{k/2}}{T^{k/2-1}} \mathcal{O}(1)$$
$$= \begin{cases} \Delta f \ \mathcal{O}(1) & \text{if } k=2 \\ 0 & \text{if } k \geqslant 3 \end{cases} \tag{5.134}$$

and, hence,

$$\lim_{\Delta f\to 0}\lim_{T\to\infty} (T\Delta f)^{k/2} \left|\mathcal{T}_{\nu_1}\right| = 0 \qquad k \geqslant 2. \tag{5.135}$$

The $k \times 2$ table (5.123) has two columns and partitions are indecomposable (see comments following Assumption 2.4.15). Thus, for $p \geqslant 2$ each set ν_m hooks with one or two other sets of the partition. Specifically, two sets (those containing at least one element of the first row or at least one element of the last row) hook with one set and the remaining $p-2$ sets hook with two sets. Consequently, for $p \geqslant 2$, each set ν_m has at least one line containing only one pair of indices. Let us choose as pair (i_m, j_m) in Fact 5.6.1 the pair of indices corresponding to the lower of these lines of ν_m containing only one pair. Thus j_m is a row index not shared with any other pair $(i, j) \in \nu'_m$, where ν'_m is obtained by ν_m by removing the element (i_m, j_m). The case is different for the only partition containing only one element ($p = 1$). In such a case, $j_m = k$ is a row index shared by $(i_m, j_m) = (2, k)$ with the element $(1, k)$.

Let us consider now the generic term $\mathcal{T}_{\boldsymbol{\nu}}$ with $p > 1$ in (5.131).

Since $p \geqslant 2$ and sets ν_h hook, we can select (i_h, j_h) so that no other pair $(i, j) \in \nu'_h$ has $j = j_h$. Moreover, there are $p-1$ distinct indices j_h and $j_p = j_{p-1}$. Let us interchange the order of integrals

$$\int_{\mathbb{R}^k} (\cdot)\, \mathrm{d}\boldsymbol{\lambda} \quad \text{and} \quad \int_{\mathbb{R}^{2k-p}} (\cdot) \prod_{h=1}^{p} \prod_{(i,j)\in\nu'_h} \mathrm{d}\bar{\varphi}_{ij}$$

and let us make the variable changes

$$\bar{\lambda}_j = \frac{f_j - \lambda_j}{\Delta f} \qquad j = 1, \ldots, k, \quad j \neq j_1, \ldots, j_{p-1}.$$

Then, for $h = 1, \ldots, p-1$, order the integrals so that the innermost is that in $\mathrm{d}\lambda_{j_h}$ and make the variable change

$$\bar{\lambda}_{j_h} = \left\{ \psi_{i_h}(\lambda_{j_h}) - \Psi_{\nu_h}^{(\ell_h)}\left(\left\{ \psi_i(\lambda_j) - \frac{\bar{\varphi}_{ij}}{T} \right\}_{(i,j)\in\nu'_h} \right) \right\} T \qquad h = 1, \ldots, p-1$$

($\bar{\varphi}_{ij}$ fixed). Since no other pair $(i, j) \in \nu'_h$ has $j = j_h$, then

$$\mathrm{d}\bar{\lambda}_{j_h} = T\, \psi'_{i_h}(\lambda_{j_h})\, \mathrm{d}\lambda_{j_h} \qquad h = 1, \ldots, p-1. \tag{5.136}$$

Thus,

$$\mathcal{T}_{\boldsymbol{\nu}} = \sum_{\ell_1} \cdots \sum_{\ell_p} \frac{1}{T^k} \int_{\mathbb{R}^{2k-p}} \int_{\mathbb{R}^k} \prod_{h=1}^{p} P_{\nu_h}^{(\ell_h)}\left(\left\{\psi_i(f_j - \bar{\lambda}_j \Delta f) - \frac{\bar{\varphi}_{ij}}{T}\right\}_{(i,j)\in\nu'_h}\right)$$
$$\prod_{(i,j)\in\nu'_1} \cdots \prod_{(i,j)\in\nu'_p} W_B^{(*)_i}(\bar{\varphi}_{ij})\, e^{-(-)_i j2\pi(\bar{\varphi}_{ij}/T)t_j}$$
$$\prod_{h=1}^{p-1} W_B^{(*)_{i_h}}(\bar{\lambda}_{j_h})\, e^{-(-)_{i_h} j2\pi(\bar{\lambda}_{j_h}/T)t_{j_h}}$$
$$T\, W_B^{(*)_{i_p}}\left(\left\{\psi_{i_p}(f_{j_p} - \bar{\lambda}_{j_p}\Delta f) - \Psi_{\nu_p}^{(\ell_p)}\left(\left\{\psi_i(f_j - \bar{\lambda}_j \Delta f) - \frac{\bar{\varphi}_{ij}}{T}\right\}_{(i,j)\in\nu'_p}\right)\right\}T\right)$$
$$e^{-(-)_{i_p} j2\pi\{\cdots\}t_{j_p}}$$
$$\prod_{h=1}^{p-1} \frac{1}{\Delta f} W_A\left(\frac{f_{j_h} - U_h}{\Delta f}\right) \frac{1}{\left|\psi'_{i_h}(U_h)\right|}$$
$$W_A(\bar{\lambda}_{j_p}) \prod_{\substack{j=1 \\ j\neq j_1,\ldots,j_p}}^{k} W_A(\bar{\lambda}_j) \prod_{j=1}^{k} \mathrm{d}\bar{\lambda}_j \prod_{h=1}^{p} \prod_{(i,j)\in\nu'_h} \mathrm{d}\bar{\varphi}_{ij} \tag{5.137}$$

where

$$U_h \triangleq \psi_{i_h}^{-1}(\bar{\lambda}_{j_h}/T + \Psi_{\nu_h}^{(\ell_h)}(\{\psi_i(f_j - \bar{\lambda}_j\Delta f) - \bar{\varphi}_{ij}/T\}_{(i,j)\in\nu'_h})).$$

Let $\phi_{i_h}(\cdot)$ denote the inverse function of $\psi_{i_h}(\cdot)$. It follows that $\phi'_{i_h}(\cdot) = 1/\psi'_{i_h}(\cdot)$ is bounded due to Assumption 4.7.9. Thus,

$$|\mathcal{T}_{\boldsymbol{\nu}}| \leqslant \frac{1}{T^{k-1}(\Delta f)^{p-1}} \sum_{\ell_1} \left\|P_{\nu_1}^{(\ell_1)}\right\|_\infty \cdots \sum_{\ell_p} \left\|P_{\nu_p}^{(\ell_p)}\right\|_\infty$$
$$\|W_B\|_\infty \int_{\mathbb{R}^{2k-p}} \left[\prod_{h=1}^{p} \prod_{(i,j)\in\nu'_h} \left|W_B(\bar{\varphi}_{ij})\right|\right] \prod_{h=1}^{p} \prod_{(i,j)\in\nu'_h} \mathrm{d}\bar{\varphi}_{ij} \prod_{h=1}^{p-1} \|\phi'_{i_h}\|_\infty$$
$$\|W_A\|_\infty^{p-1} \int_{\mathbb{R}^k} \left[\prod_{h=1}^{p-1} \left|W_B(\bar{\lambda}_{j_h})\right|\right] \left|W_A(\bar{\lambda}_{j_p})\right| \prod_{\substack{j=1 \\ j\neq j_1,\ldots,j_p}}^{k} \left|W_A(\bar{\lambda}_j)\right| \mathrm{d}\bar{\boldsymbol{\lambda}} \tag{5.138}$$

That is,

$$|\mathcal{T}_{\boldsymbol{\nu}}| \leqslant \frac{1}{T^{k-1}(\Delta f)^{p-1}}\, \mathcal{O}_{\boldsymbol{\nu}}(1) \tag{5.139}$$

with $2 \leqslant p \leqslant 2k$. Hence

$$(T\Delta f)^{k/2}\,|\mathcal{T}_{\boldsymbol{\nu}}| \leqslant \frac{(\Delta f)^{k/2-p+1}}{T^{k/2-1}}\;\mathcal{O}_{\boldsymbol{\nu}}(1). \tag{5.140}$$

For $k = 2$ from (5.140), we have

$$\lim_{\Delta f\to 0}\lim_{T\to\infty}(T\Delta f)\,|\mathcal{T}_{\boldsymbol{\nu}}| \leqslant \lim_{\Delta f\to 0}(\Delta f)^{2-p}\,\mathcal{O}_{\boldsymbol{\nu}}(1) = \begin{cases} \mathcal{O}_{\boldsymbol{\nu}}(1) & p=2 \\ \infty & p=3 \\ \infty & p=4 \end{cases} \tag{5.141}$$

The above limit is bounded for all partitions $\boldsymbol{\nu}$ with $k = 2$ only if the processes are zero mean. In fact, in such a case terms with $p = 3$ and $p = 4$ are identically zero (only terms with $p = 1$ and $p = 2$ are nonzero, see (5.11)). For $k = 2$ and zero-mean processes a tighter bound is provided by Theorems 4.7.7 and 5.8.3, for which

$$\lim_{\Delta f\to 0}\lim_{T\to\infty}(T\Delta f)\mathcal{T}_{\boldsymbol{\nu}} = \mathcal{O}_{\boldsymbol{\nu}}(1) \tag{5.142}$$

For $k \geqslant 3$ from (5.140) we have

$$\lim_{\Delta f\to 0}\lim_{T\to\infty}(T\Delta f)^{k/2}\,|\mathcal{T}_{\boldsymbol{\nu}}| \leqslant \lim_{\Delta f\to 0}\lim_{T\to\infty}\frac{(\Delta f)^{k/2-p+1}}{T^{k/2-1}}\;\mathcal{O}_{\boldsymbol{\nu}}(1) = 0 \tag{5.143}$$

provided that the order of the two limits is not interchanged. In fact, for Δf finite, the limit as $T \to \infty$ is zero for $k \geqslant 3$. In contrast, for T finite, since $2 \leqslant p \leqslant 2k$, the limit as $\Delta f \to 0$ is infinite for those partitions $\boldsymbol{\nu}$ with $p \geqslant k/2 + 1$.

The interchange of the order of integrals in the expression of $\mathcal{T}_{\boldsymbol{\nu}}$ to obtain (5.137) is justified by the Fubini and Tonelli theorem (Champeney 1990, Chapter 3). In fact, let $F^{(\ell_1,\ldots,\ell_p)}(\bar{\boldsymbol{\lambda}}, \bar{\boldsymbol{\varphi}}; T, \Delta f)$ be $T^{k-1}(\Delta f)^{p-1}$ times the integrand function in the expression of $\mathcal{T}_{\boldsymbol{\nu}}$ in (5.137). One has

$$\begin{aligned}
&\left|F^{(\ell_1,\ldots,\ell_p)}(\bar{\boldsymbol{\lambda}}, \bar{\boldsymbol{\varphi}}; T, \Delta f)\right| \\
&\leqslant \prod_{h=1}^{p}\left\|P_{\nu_h}^{(\ell_h)}\right\|_\infty \left[\prod_{h=1}^{p}\prod_{(i,j)\in\nu_h'}|W_B(\bar{\varphi}_{ij})|\right]\left[\prod_{h=1}^{p-1}\left|W_B(\bar{\lambda}_{j_h})\right|\right]\|W_B\|_\infty \\
&\quad \|W_A\|_\infty^{p-1}\left|W_A(\bar{\lambda}_{j_p})\right| \prod_{\substack{j=1\\ j\neq j_1,\ldots,j_p}}^{k}\left|W_A(\bar{\lambda}_j)\right|
\end{aligned} \tag{5.144}$$

with the rhs belonging to $L^1(\mathbb{R}^{2k-p} \times \mathbb{R}^k)$ due to Assumptions 4.4.5 (data-tapering window regularity) and 4.6.2 (frequency-smoothing window regularity).

5.6.2 Proof of Theorem 4.7.11 Asymptotic Joint Complex Normality of the Frequency-Smoothed Cross-Periodograms

From Theorem 4.7.6 it follows that

$$\sqrt{T\Delta f}\ \text{bias} = \mathcal{O}\Big(\frac{(\Delta f)^{1/2}}{T^{1/2}}\Big) + \mathcal{O}(T^{1/2}(\Delta f)^{3/2}) + \mathcal{O}\left(T^{3/2}(\Delta f)^{5/2}\right). \tag{5.145}$$

The rate of decay to zero of Δf should be such that $\sqrt{T\Delta f}$ bias approaches zero as $T \to \infty$ and $\Delta f \to 0$ with $T\Delta f \to \infty$. Let us assume that

$$\Delta f \equiv \Delta f_T = \frac{1}{T^a} \tag{5.146}$$

with $0 < a < 1$. Then, as $T \to \infty$, $\Delta f_T = T^{-a} \to 0$ with $T\Delta f_T = T^{1-a} \to \infty$. In addition, we have

$$\sqrt{T\Delta f_T}\ \text{bias} = \mathcal{O}(T^{-(a+1)/2}) + \mathcal{O}(T^{1/2}T^{-3a/2}) + \mathcal{O}\left(T^{3/2}T^{-5a/2}\right). \tag{5.147}$$

Consequently, $\sqrt{T\Delta f_T}$ bias approaches zero as $T \to \infty$ provided that

$$\begin{cases} a > 1/3 \\ a > 3/5. \end{cases} \tag{5.148}$$

Therefore, the condition on a is

$$3/5 < a < 1 \tag{5.149}$$

From Theorem 4.7.7 it follows that the asymptotic covariance

$$\begin{aligned} &\lim_{\Delta f \to 0} \lim_{T \to \infty} \operatorname{cov}\left\{\sqrt{T\Delta f}\, S^{(n_1)}_{yx^{(*)}}(t_1; f_1)_{T,\Delta f},\ \sqrt{T\Delta f}\, S^{(n_2)}_{yx^{(*)}}(t_2; f_2)_{T,\Delta f}\right\} \\ &\quad = \lim_{\Delta f \to 0} \lim_{T \to \infty} \operatorname{cov}\left\{\sqrt{T\Delta f}\left[S^{(n_1)}_{yx^{(*)}}(t_1; f_1)_{T,\Delta f} - \mathcal{E}^{(n_1)}(f_1)\, S^{(n_1)}_{yx^{(*)}}(f_1)\right],\right. \\ &\qquad \left.\sqrt{T\Delta f}\left[S^{(n_2)}_{yx^{(*)}}(t_2; f_2)_{T,\Delta f} - \mathcal{E}^{(n_2)}(f_2)\, S^{(n_2)}_{yx^{(*)}}(f_2)\right]\right\} \end{aligned} \tag{5.150}$$

is finite. Analogously, from Theorem 5.8.3 it follows that the asymptotic conjugate covariance is finite. Moreover, from Lemma 4.7.10 with $k \geqslant 3$ we have

$$\begin{aligned} &\lim_{\Delta f \to 0} \lim_{T \to \infty} (T\Delta f)^{k/2} \operatorname{cum}\left\{S^{(n_1)}_{yx^{(*)}}(t_1, f_1)^{[*]_1}_{T,\Delta f}, \ldots, S^{(n_k)}_{yx^{(*)}}(t_k, f_k)^{[*]_k}_{T,\Delta f}\right\} \\ &= \lim_{\Delta f \to 0} \lim_{T \to \infty} \operatorname{cum}\left\{\sqrt{T\Delta f}\, S^{(n_1)}_{yx^{(*)}}(t_1, f_1)^{[*]_1}_{T,\Delta f}, \ldots, \sqrt{T\Delta f}\, S^{(n_k)}_{yx^{(*)}}(t_k, f_k)^{[*]_k}_{T,\Delta f}\right\} = 0 \end{aligned} \tag{5.151}$$

Since the value of the cumulant does not change by adding a constant to each of the random variables (Brillinger 1981, Theorem 2.3.1), we also have

$$\lim_{\Delta f \to 0} \lim_{T \to \infty} \operatorname{cum} \Bigg\{ \sqrt{T\Delta f} \left[S^{(n_1)}_{yx^{(*)}}(t_1, f_1)_{T,\Delta f} - \mathcal{E}^{(n_1)}(f_1)\, S^{(n_1)}_{yx^{(*)}}(f_1) \right]^{[*]_1}, \ldots,$$
$$\sqrt{T\Delta f} \left[S^{(n_k)}_{yx^{(*)}}(t_k, f_k)_{T,\Delta f} - \mathcal{E}^{(n_k)}(f_k)\, S^{(n_k)}_{yx^{(*)}}(f_k) \right]^{[*]_k} \Bigg\} = 0. \tag{5.152}$$

That is, for every fixed n_i, f_i, t_i, the random variables

$$\sqrt{T\Delta f} \left[S^{(n_i)}_{yx^{(*)}}(t_i, f_i)_{T,\Delta f} - \mathcal{E}^{(n_i)}(f)\, S^{(n_i)}_{yx^{(*)}}(f_i) \right]$$

$i = 1, \ldots, k$, are asymptotically ($T \to \infty$ and $\Delta f \to 0$ with $T\Delta f \to \infty$) zero-mean jointly complex Normal (Section 1.4.2).

□

5.7 Alternative Bounds

In this section, some bounds derived in Sections 5.2–5.6 for the proofs of theorems and lemmas of Sections 4.4–4.7.2 are suitably modified for the case where Assumption 4.7.13 is made instead of Assumption 4.4.3. Specifically,

(5.8) should be replaced by

$$\left| \sum_{n \in \mathbb{I}} F^{(n)}(\nu_1) \right| \leqslant \sum_{n \in \mathbb{I}} \left| F^{(n)}(\nu_1) \right| \leqslant \Delta f \sum_{n \in \mathbb{I}} \left| S^{(n)}_{yx^{(*)}}(\nu_1) \right| \frac{1}{(\Delta f)^2} \, \|W_B\|^2_\infty \in L^1(\mathbb{R}) \tag{5.153}$$

(5.62) should be replaced by

$$\left| \sum_{m \in \mathbb{I}} F^{(m)}(\lambda_1, \nu_1) \right| \leqslant \sum_{m \in \mathbb{I}} \left| F^{(m)}(\lambda_1, \nu_1) \right|$$
$$\leqslant \frac{1}{T} \sum_{m \in \mathbb{I}} \left| S^{(m)}_{yx^{(*)}}(\nu_1) \right| T^2 \, \|W_B\|^2_\infty \, \frac{1}{\Delta f} \left| W_A \left(\frac{f_1 - \lambda_1}{\Delta f} \right) \right| \in L^1(\mathbb{R}^2) \tag{5.154}$$

(5.67) should be replaced by

$$\left| \sum_{n' \in \mathbb{I}'} \sum_{n'' \in \mathbb{I}''} \mathcal{D}^{(n',n'')}(\nu_{y1}, \nu_{x1}, \lambda_1, \lambda_2) \right|$$
$$\leqslant \sum_{n' \in \mathbb{I}'} \sum_{n'' \in \mathbb{I}''} \left| \mathcal{D}^{(n',n'')}(\nu_{y1}, \nu_{x1}, \lambda_1, \lambda_2) \right|$$
$$\leqslant \frac{1}{T^2} \sum_{n' \in \mathbb{I}'} \left| S^{(n')}_{yy^*}(\nu_{y1}) \right| \sum_{n'' \in \mathbb{I}''} \left| S^{(n'')}_{x^{(*)}x^{(*)*}}(\nu_{x1}) \right| T^4 \, \|W_B\|^4_\infty$$
$$\frac{1}{\Delta f} \left| W_A \left(\frac{f_1 - \lambda_1}{\Delta f} \right) \right| \frac{1}{\Delta f} \left| W_A \left(\frac{f_2 - \lambda_2}{\Delta f} \right) \right| \in L^1(\mathbb{R}^4) \tag{5.155}$$

(5.69) should be replaced by

$$\left|\sum_{n\in\mathbb{I}_4} \mathcal{D}^{(n)}(\nu_{y1},\nu_{x1},\nu_{y2},\lambda_1,\lambda_2)\right|$$
$$\leqslant \sum_{n\in\mathbb{I}_4}\left|\mathcal{D}^{(n)}(\nu_{y1},\nu_{x1},\nu_{y2},\lambda_1,\lambda_2)\right|$$
$$\leqslant \frac{1}{T^2}\sum_{n\in\mathbb{I}_4}\left|P^{(n)}_{yx^{(*)}y^*x^{(*)*}}(\nu_{y1},\nu_{x1},\nu_{y2})\right| T^4 \,\|W_B\|^4_\infty$$
$$\frac{1}{\Delta f}\left|W_A\left(\frac{f_1-\lambda_1}{\Delta f}\right)\right| \frac{1}{\Delta f}\left|W_A\left(\frac{f_2-\lambda_2}{\Delta f}\right)\right| \in L^1(\mathbb{R}^5) \tag{5.156}$$

5.8 Conjugate Covariance

For complex-valued processes, both covariance and conjugate covariance are needed for a complete second-order characterization (Picinbono 1996), (Picinbono and Bondon 1997), (Schreier and Scharf 2003a), (Schreier and Scharf 2003b).

In this section, results analogous to those stated in Lemma 4.4.7 and Theorems 4.6.4, and 4.7.7 are presented for the conjugate covariance of the bifrequency cross-periodogram and the frequency-smoothed cross periodogram.

Lemma 5.8.1 Conjugate Covariance of the Bifrequency Cross-Periodogram. *Let* $\{y(t),\ t\in\mathbb{R}\}$ *and* $\{x(t),\ t\in\mathbb{R}\}$ *be second-order harmonizable zero-mean singularly and jointly SC stochastic processes with bifrequency spectra and cross-spectra (4.95). Under Assumptions 4.4.2, 4.4.3, and 4.4.5 (with* $\Delta f = 1/T$*), the conjugate covariance of the bifrequency cross-periodogram (4.93) is given by*

$$\operatorname{cov}\left\{I_{yx^{(*)}}(t_1; f_{y1}, f_{x1})_T\,,\ I^*_{yx^{(*)}}(t_2; f_{y2}, f_{x2})_T\right\} = \bar{\mathcal{T}}_1 + \bar{\mathcal{T}}_2 + \bar{\mathcal{T}}_3 \tag{5.157}$$

with

$$\bar{\mathcal{T}}_1 \triangleq \frac{1}{T^2}\sum_{n'\in\mathbb{I}'}\int_{\mathbb{R}} S^{(n')}_{yy}(\nu_{y1})\,\mathcal{B}_{\frac{1}{T}}(f_{y1}-\nu_{y1};t_1)$$
$$\mathcal{B}_{\frac{1}{T}}\left(f_{y2}-\Psi^{(n')}_{yy}(\nu_{y1});t_2\right)\,\mathrm{d}\nu_{y1}$$
$$\sum_{n''\in\mathbb{I}''}\int_{\mathbb{R}} S^{(n'')}_{x^{(*)}x^{(*)}}(\nu_{x1})\,\mathcal{B}^{(*)}_{\frac{1}{T}}(f_{x1}-\nu_{x1};t_1)$$
$$\mathcal{B}^{(*)}_{\frac{1}{T}}\left(f_{x2}-\Psi^{(n'')}_{x^{(*)}x^{(*)}}(\nu_{x1});t_2\right)\,\mathrm{d}\nu_{x1} \tag{5.158}$$
$$\bar{\mathcal{T}}_2 \triangleq \frac{1}{T^2}\sum_{n'''\in\mathbb{I}'''}\int_{\mathbb{R}} S^{(n''')}_{yx^{(*)}}(\nu_{y1})\,\mathcal{B}_{\frac{1}{T}}(f_{y1}-\nu_{y1};t_1)$$
$$\mathcal{B}^{(*)}_{\frac{1}{T}}\left(f_{x2}-\Psi^{(n''')}_{yx^{(*)}}(\nu_{y1});t_2\right)\,\mathrm{d}\nu_{y1}$$

$$\sum_{n^{iv}\in\mathbb{I}^{iv}} \int_{\mathbb{R}} S^{(n^{iv})}_{x^{(*)}y}(\nu_{x1})\, \mathcal{B}^{(*)}_{\frac{1}{T}}(f_{x1}-\nu_{x1};t_1)$$
$$\mathcal{B}_{\frac{1}{T}}\left(f_{y2}-\Psi^{(n^{iv})}_{x^{(*)}y}(\nu_{x1});t_2\right)\, \mathrm{d}\nu_{x1} \tag{5.159}$$
$$\bar{\mathcal{T}}_3 \triangleq \frac{1}{T^2}\sum_{n\in\mathbb{I}_4}\int_{\mathbb{R}}\int_{\mathbb{R}}\int_{\mathbb{R}} P^{(n)}_{yx^{(*)}yx^{(*)}}(\nu_{y1},\nu_{x1},\nu_{y2})$$
$$\mathcal{B}^{(*)}_{\frac{1}{T}}\left(f_{x2}-\Psi^{(n)}_{yx^{(*)}yx^{(*)}}(\nu_{y1},\nu_{x1},\nu_{y2});t_2\right)$$
$$\mathcal{B}_{\frac{1}{T}}(f_{y1}-\nu_{y1};t_1)\,\mathcal{B}^{(*)}_{\frac{1}{T}}(f_{x1}-\nu_{x1};t_1)$$
$$\mathcal{B}_{\frac{1}{T}}(f_{y2}-\nu_{y2};t_2)\,\mathrm{d}\nu_{y1}\,\mathrm{d}\nu_{x1}\,\mathrm{d}\nu_{y2} \tag{5.160}$$

where, for notation simplicity, $\mathbb{I}'=\mathbb{I}_{yy}$, $\mathbb{I}''=\mathbb{I}_{x^{(*)}x^{(*)}}$, $\mathbb{I}'''=\mathbb{I}_{yx^{(*)}}$, *and* $\mathbb{I}^{iv}=\mathbb{I}_{x^{(*)}y}$. □

Theorem 5.8.2 Conjugate Covariance of the Frequency-Smoothed Cross-Periodogram. *Let* $\{y(t),\ t\in\mathbb{R}\}$ *and* $\{x(t),\ t\in\mathbb{R}\}$ *be second-order harmonizable zero-mean singularly and jointly SC stochastic processes with bifrequency spectra and cross-spectra (4.95). Under Assumptions 4.4.2, 4.4.3, 4.4.5 (with* $\Delta f = 1/T$*), and 4.6.2, the conjugate covariance of the frequency-smoothed cross-periodogram (4.147) is given by*

$$\operatorname{cov}\left\{S^{(n_1)}_{yx^{(*)}}(t_1;f_1)_{T,\Delta f},\ S^{(n_2)\,*}_{yx^{(*)}}(t_2;f_2)_{T,\Delta f}\right\} = \bar{\mathcal{T}}''_1+\bar{\mathcal{T}}''_2+\bar{\mathcal{T}}''_3 \tag{5.161}$$

where

$$\bar{\mathcal{T}}''_1 \triangleq \frac{1}{T^2}\int_{\mathbb{R}}\int_{\mathbb{R}}\sum_{n'\in\mathbb{I}'}\int_{\mathbb{R}} S^{(n')}_{yy}(\nu_{y1})\,\mathcal{B}_{\frac{1}{T}}(\lambda_1-\nu_{y1};t_1)$$
$$\mathcal{B}_{\frac{1}{T}}\left(\lambda_2-\Psi^{(n')}_{yy}(\nu_{y1});t_2\right)\,\mathrm{d}\nu_{y1}$$
$$\sum_{n''\in\mathbb{I}''}\int_{\mathbb{R}} S^{(n'')}_{x^{(*)}x^{(*)}}(\nu_{x1})\,\mathcal{B}^{(*)}_{\frac{1}{T}}\left(\Psi^{(n_1)}_{yx^{(*)}}(\lambda_1)-\nu_{x1};t_1\right)$$
$$\mathcal{B}^{(*)}_{\frac{1}{T}}\left(\Psi^{(n_2)}_{yx^{(*)}}(\lambda_2)-\Psi^{(n'')}_{x^{(*)}x^{(*)}}(\nu_{x1});t_2\right)\,\mathrm{d}\nu_{x1}$$
$$A_{\Delta f}(f_1-\lambda_1)\,A_{\Delta f}(f_2-\lambda_2)\,\mathrm{d}\lambda_1\,\mathrm{d}\lambda_2 \tag{5.162}$$
$$\bar{\mathcal{T}}''_2 \triangleq \frac{1}{T^2}\int_{\mathbb{R}}\int_{\mathbb{R}}\sum_{n'''\in\mathbb{I}'''}\int_{\mathbb{R}} S^{(n''')}_{yx^{(*)}}(\nu_{y1})\,\mathcal{B}_{\frac{1}{T}}(\lambda_1-\nu_{y1};t_1)$$
$$\mathcal{B}^{(*)}_{\frac{1}{T}}\left(\Psi^{(n_2)}_{yx^{(*)}}(\lambda_2)-\Psi^{(n''')}_{yx^{(*)}}(\nu_{y1});t_2\right)\,\mathrm{d}\nu_{y1}$$
$$\sum_{n^{iv}\in\mathbb{I}^{iv}}\int_{\mathbb{R}} S^{(n^{iv})}_{x^{(*)}y}(\nu_{x1})\,\mathcal{B}^{(*)}_{\frac{1}{T}}\left(\Psi^{(n_1)}_{yx^{(*)}}(\lambda_1)-\nu_{x1};t_1\right)$$
$$\mathcal{B}_{\frac{1}{T}}\left(\lambda_2-\Psi^{(n^{iv})}_{x^{(*)}y}(\nu_{x1});t_2\right)\,\mathrm{d}\nu_{x1}$$
$$A_{\Delta f}(f_1-\lambda_1)\,A_{\Delta f}(f_2-\lambda_2)\,\mathrm{d}\lambda_1\,\mathrm{d}\lambda_2 \tag{5.163}$$

$$\bar{\mathcal{T}}_3'' \triangleq \frac{1}{T^2} \int_{\mathbb{R}} \int_{\mathbb{R}} \sum_{n \in \mathbb{I}_4} \int_{\mathbb{R}} \int_{\mathbb{R}} \int_{\mathbb{R}} P^{(n)}_{yx^{(*)}yx^{(*)}}(\nu_{y1}, \nu_{x1}, \nu_{y2})$$
$$\mathcal{B}^{(*)}_{\frac{1}{T}} \left(\Psi^{(n_2)}_{yx^{(*)}}(\lambda_2) - \Psi^{(n)}_{yx^{(*)}yx^{(*)}}(\nu_{y1}, \nu_{x1}, \nu_{y2}); t_2 \right)$$
$$\mathcal{B}_{\frac{1}{T}}(\lambda_1 - \nu_{y1}; t_1)\, \mathcal{B}^{(*)}_{\frac{1}{T}} \left(\Psi^{(n_1)}_{yx^{(*)}}(\lambda_1) - \nu_{x1}; t_1 \right)$$
$$\mathcal{B}_{\frac{1}{T}}(\lambda_2 - \nu_{y2}; t_2)\, \mathrm{d}\nu_{y1}\, \mathrm{d}\nu_{x1}\, \mathrm{d}\nu_{y2}$$
$$A_{\Delta f}(f_1 - \lambda_1)\, A_{\Delta f}(f_2 - \lambda_2)\, \mathrm{d}\lambda_1\, \mathrm{d}\lambda_2 \tag{5.164}$$

□

Theorem 5.8.3 Asymptotic Conjugate Covariance of the Frequency-Smoothed Cross-Periodogram. *Let $\{y(t),\ t \in \mathbb{R}\}$ and $\{x(t),\ t \in \mathbb{R}\}$ be second-order harmonizable zero-mean singularly and jointly SC stochastic processes with bifrequency spectra and cross-spectra (4.95). Under Assumptions 4.4.2, 4.4.3, 4.4.5 (with $\Delta f = 1/T$), and 4.6.2, the asymptotic ($T \to \infty$, $\Delta f \to 0$, with $T\Delta f \to \infty$) conjugate covariance of the frequency-smoothed cross-periodogram (4.147) is given by*

$$\lim_{\Delta f \to 0} \lim_{T \to \infty} (T\Delta f) \operatorname{cov}\left\{ S^{(n_1)}_{yx^{(*)}}(t_1; f_1)_{T,\Delta f},\ S^{(n_2)\,*}_{yx^{(*)}}(t_2; f_2)_{T,\Delta f} \right\} = \bar{\mathcal{T}}_1''' + \bar{\mathcal{T}}_2''' \tag{5.165}$$

where

$$\bar{\mathcal{T}}_1''' \triangleq \sum_{n' \in \mathbb{I}'} \sum_{n'' \in \mathbb{I}''} S^{(n')}_{yy}(f_1)\, S^{(n'')}_{x^{(*)}x^{(*)}}\left(\Psi^{(n_1)}_{yx^{(*)}}(f_1) \right)$$
$$\mathcal{J}^{(n',n'')}_1(f_1)\, \bar{\delta}_{\Psi^{(n_2)}_{yx^{(*)}}\left(\Psi^{(n')}_{yy^*}(f_1) \right) - \Psi^{(n'')}_{x^{(*)}x^{(*)}}\left(\Psi^{(n_1)}_{yx^{(*)}}(f_1) \right)}$$
$$\mathcal{I}^{(n')}_1(f_1)\, \delta_{f_2 - \Psi^{(n')}_{yy}(f_1)} \tag{5.166}$$
$$\bar{\mathcal{T}}_2''' \triangleq \sum_{n''' \in \mathbb{I}'''} \sum_{n^{iv} \in \mathbb{I}^{iv}} S^{(n''')}_{yx^{(*)}}(f_1)\, S^{(n^{iv})}_{x^{(*)}y}\left(\Psi^{(n_1)}_{yx^{(*)}}(f_1) \right)$$
$$\mathcal{J}^{(n''',n^{iv})}_2(f_1)\, \bar{\delta}_{\Psi^{(n_2)}_{yx^{(*)}}\left(\Psi^{(n^{iv})}_{x^{(*)}y^*}\left(\Psi^{(n_1)}_{yx^{(*)}}(f_1) \right) \right) - \Psi^{(n''')}_{yx^{(*)}}(f_1)}$$
$$\mathcal{I}^{(n^{iv})}_2(f_1)\, \delta_{f_2 - \Psi^{(n^{iv})}_{x^{(*)}y}\left(\Psi^{(n_1)}_{yx^{(*)}}(f_1) \right)} \tag{5.167}$$

with

$$\mathcal{J}^{(n',n'')}_1(f_1) \triangleq \int_{\mathbb{R}} \int_{\mathbb{R}} \int_{\mathbb{R}} W^{(*)}_B \left(\Psi^{(n_2)\,\prime}_{yx^{(*)}} \left(\Psi^{(n')}_{yy}(f_1) \right) \left[-\Psi^{(n')\prime}_{yy}(f_1)\, \nu_{y1} + \lambda_2 \right] \right.$$
$$\left. + \Psi^{(n'')\,\prime}_{x^{(*)}x^{(*)}} \left(\Psi^{(n_1)}_{yx^{(*)}}(f_1) \right) \nu_{x1} \right)$$
$$W_B(\nu_{y1})\, W^{(*)}_B(\nu_{x1})\, W_B(\lambda_2)\, \mathrm{d}\nu_{y1}\, \mathrm{d}\nu_{x1}\, \mathrm{d}\lambda_2 \tag{5.168}$$
$$\mathcal{I}^{(n')}_1(f_1) \triangleq \int_{\mathbb{R}} W_A(\lambda_1)\, W_A\left(\lambda_1 \Psi^{(n')\prime}_{yy}(f_1) \right) \mathrm{d}\lambda_1 \tag{5.169}$$

$$\mathcal{J}_2^{(n''',n^{iv})}(f_1) \triangleq \int_{\mathbb{R}}\int_{\mathbb{R}}\int_{\mathbb{R}} W_B^{(*)}\left(\Psi_{yx^{(*)}}^{(n_2)\,\prime}\left(\Psi_{x^{(*)}y}^{(n^{iv})}\left(\Psi_{yx^{(*)}}^{(n_1)}(f_1)\right)\right)\right.$$
$$\left[-\Psi_{x^{(*)}y}^{(n^{iv})\,\prime}\left(\Psi_{yx^{(*)}}^{(n_1)}(f_1)\right)\,\nu_{x1}+\lambda_2\right]+\left.\Psi_{yx^{(*)}}^{(n''')\prime}(f_1)\,\nu_{y1}\right)$$
$$W_B(\nu_{y1})\,W_B^{(*)}(\nu_{x1})\,W_B(\lambda_2)\,\mathrm{d}\nu_{y1}\,\mathrm{d}\nu_{x1}\,\mathrm{d}\lambda_2 \tag{5.170}$$

$$\mathfrak{J}_2^{(n^{iv})}(f_1) \triangleq \int_{\mathbb{R}} W_A(\lambda_1)\,W_A\left(\lambda_1\left[\Psi_{x^{(*)}y}^{(n^{iv})}\left(\Psi_{yx^{(*)}}^{(n_1)}(f_1)\right)\right]'\right)\mathrm{d}\lambda_1 \tag{5.171}$$

□

5.9 Proofs for Section 4.8 "Discrete-Time SC Processes"

5.9.1 *Proof of Theorem 4.8.3 Characterization of Discrete-Time Spectrally Correlated Processes*

At first, let us observe that, analogously to the case of continuous-time signals, the functions in the right-hand side of (4.188) in general are not unambiguously determined. By opportunely selecting the support of the functions $\widetilde{S}_{\boldsymbol{x}}^{(k)}(\nu_1)$, the corresponding functions $\widetilde{\Psi}_{\boldsymbol{x}}^{(k)}(\nu_1)$ can be chosen to be locally invertible in intervals $[p-1/2, p+1/2)$, with p integer, that is, their restrictions to these intervals are invertible. In addition, since $\widetilde{\Psi}_{\boldsymbol{x}}^{(k)}(\nu_1)$ are in the argument of a periodic delta train with period 1, they can always be chosen with values in $[-1/2, 1/2)$.

Every periodic function $\widetilde{\Psi}_{\boldsymbol{x}}^{(k)}(\nu)$ can be expressed as the periodic replication, with period 1, of a $L^1(\mathbb{R})$ or $L^2(\mathbb{R})$ generator function $\Psi_{\boldsymbol{x}}^{(k)}(\nu)$:

$$\widetilde{\Psi}_{\boldsymbol{x}}^{(k)}(\nu)=\sum_{p\in\mathbb{Z}}\Psi_{\boldsymbol{x}}^{(k)}(\nu-p). \tag{5.172}$$

The generator, in general, is not unambiguously determined and can have support of width larger than 1. However, there exists a (unique) generator with compact support contained in $[-1/2, 1/2)$, that is such that

$$\Psi_{\boldsymbol{x}}^{(k)}(\nu)=\widetilde{\Psi}_{\boldsymbol{x}}^{(k)}(\nu) \qquad \nu\in[-1/2, 1/2). \tag{5.173}$$

With this choice for the generator, the following useful expression holds for the periodic delta train in (4.188):

$$\widetilde{\delta}\left(\nu_2-\widetilde{\Psi}_{\boldsymbol{x}}^{(k)}(\nu_1)\right)=\sum_{p_2\in\mathbb{Z}}\delta\left(\nu_2-p_2-\sum_{p_1\in\mathbb{Z}}\Psi_{\boldsymbol{x}}^{(k)}(\nu_1-p_1)\right)$$
$$=\sum_{p_2\in\mathbb{Z}}\sum_{p_1\in\mathbb{Z}}\delta\left(\nu_2-p_2-\Psi_{\boldsymbol{x}}^{(k)}(\nu_1-p_1)\right). \tag{5.174}$$

Due to the local invertibility of functions $\widetilde{\Psi}_{\boldsymbol{x}}^{(k)}(\cdot)$, each function $\Psi_{\boldsymbol{x}}^{(k)}(\cdot)$ has compact support contained in $[-1/2, 1/2)$ where is invertible, and has values in $[-1/2, 1/2)$. Let us denote by $\Phi_{\boldsymbol{x}}^{(k)}(\cdot)$ its inverse function. $\Phi_{\boldsymbol{x}}^{(k)}(\cdot)$ has in turn compact support contained in $[-1/2, 1/2)$ and values in $[-1/2, 1/2)$. Therefore, for every $(p_1, p_2)\in\mathbb{Z}^2$ there exists only one pair $(\nu_1, \nu_2)\in[p_1-1/2, p_1+1/2)\times[p_2-1/2, p_2+1/2)$ such that $\nu_2-p_2=\Psi_{\boldsymbol{x}}^{(k)}(\nu_1-p_1)$

or, equivalently $\nu_1 - p_1 = \Phi_x^{(k)}(\nu_2 - p_2)$. Consequently, a variable change in the argument of the Dirac delta leads to (Zemanian 1987, Section 1.7)

$$\delta\Big(\nu_2 - p_2 - \Psi_x^{(k)}(\nu_1 - p_1)\Big) \\ = \left|\Phi_x^{(k)\prime}(\nu_2 - p_2)\right| \, \delta\Big(\nu_1 - p_1 - \Phi_x^{(k)}(\nu_2 - p_2)\Big). \tag{5.175}$$

Thus,

$$\begin{aligned} &\sum_{p_2\in\mathbb{Z}} \delta\Big(\nu_2 - p_2 - \Psi_x^{(k)}(\nu_1 - p_1)\Big) \\ &\quad = \sum_{p_2\in\mathbb{Z}} \left|\Phi_x^{(k)\prime}(\nu_2 - p_2)\right| \delta\Big(\nu_1 - p_1 - \Phi_x^{(k)}(\nu_2 - p_2)\Big) \\ &\quad = \left|\sum_{p_2\in\mathbb{Z}} \Phi_x^{(k)\prime}(\nu_2 - p_2)\right| \sum_{p_2\in\mathbb{Z}} \delta\Big(\nu_1 - p_1 - \Phi_x^{(k)}(\nu_2 - p_2)\Big) \\ &\quad = \left|\sum_{p_2\in\mathbb{Z}} \Phi_x^{(k)\prime}(\nu_2 - p_2)\right| \delta\Big(\nu_1 - p_1 - \sum_{p_2\in\mathbb{Z}} \Phi_x^{(k)}(\nu_2 - p_2)\Big) \end{aligned} \tag{5.176}$$

where the second and the third equalities are consequence of the finite support of $\Phi_x^{(k)}(\cdot)$ (and $\Phi_x^{(k)\prime}(\cdot)$). By substituting (5.176) into (5.174) we have

$$\begin{aligned} &\widetilde{\delta}\Big(\nu_2 - \widetilde{\Psi}_x^{(k)}(\nu_1)\Big) \\ &\quad = \sum_{p_1\in\mathbb{Z}} \sum_{p_2\in\mathbb{Z}} \delta\Big(\nu_2 - p_2 - \Psi_x^{(k)}(\nu_1 - p_1)\Big) \\ &\quad = \sum_{p_1\in\mathbb{Z}} \left|\sum_{p_2\in\mathbb{Z}} \Phi_x^{(k)\prime}(\nu_2 - p_2)\right| \delta\Big(\nu_1 - p_1 - \sum_{p_2\in\mathbb{Z}} \Phi_x^{(k)}(\nu_2 - p_2)\Big) \\ &\quad = \left|\widetilde{\Phi}_x^{(k)\prime}(\nu_2)\right| \sum_{p_1\in\mathbb{Z}} \delta\Big(\nu_1 - p_1 - \widetilde{\Phi}_x^{(k)}(\nu_2)\Big) \\ &\quad = \left|\widetilde{\Phi}_x^{(k)\prime}(\nu_2)\right| \widetilde{\delta}\Big(\nu_1 - \widetilde{\Phi}_x^{(k)}(\nu_2)\Big) \end{aligned} \tag{5.177}$$

where

$$\widetilde{\Phi}_x^{(k)}(\nu) \triangleq \sum_{p\in\mathbb{Z}} \Phi_x^{(k)}(\nu - p). \tag{5.178}$$

By using (5.177) into (4.190a), we have that (4.190b) and (4.191b) immediately follow. The proof of (4.191a) is similar. □

5.10 Proofs for Section 4.9 "Sampling of SC Processes"

5.10.1 Proof of Theorem 4.9.2 Strictly Band-Limited Spectrally Correlated Processes

The strictly band-limitedness condition (4.204) considered in the frequency domain implies that

$$\mathrm{E}\left\{H^{(B_1)}(f_1)X_{a,1}(f_1)\,H^{(B_2)}(f_2)X^{(*)}_{a,2}(f_2)\right\} = \mathrm{E}\left\{X_{a,1}(f_1)\,X^{(*)}_{a,2}(f_2)\right\} \tag{5.179}$$

where $H^{(B_i)}(f_i) = \mathrm{rect}(f_i/2B_i)$. In the case of jointly SC processes, replacing (4.209a) into (5.179) leads to

$$H^{(B_1)}(f_1)H^{(B_2)}(f_2)\sum_{k\in\mathbb{I}_a} S^{(k)}_{\boldsymbol{x}_a}(f_1)\delta\left(f_2 - \Psi^{(k)}_{\boldsymbol{x}_a}(f_1)\right)$$
$$= \sum_{k\in\mathbb{I}_a} S^{(k)}_{\boldsymbol{x}_a}(f_1)\delta\left(f_2 - \Psi^{(k)}_{\boldsymbol{x}_a}(f_1)\right). \tag{5.180}$$

Let

$$\mathbb{J}_1 \triangleq \left\{f_1 \in \mathbb{R} : \Psi^{(k)}_{\boldsymbol{x}_a}(f_1) = \Psi^{(k')}_{\boldsymbol{x}_a}(f_1) \text{ for some } k \neq k',\ k, k' \in \mathbb{I}_a\right\} \tag{5.181}$$

$$\mathbb{J}_2 \triangleq \bigcup_{f_1\in\mathbb{J}_1}\bigcup_{k\in\mathbb{I}_a}\left\{f_2 \in \mathbb{R} : f_2 = \Psi^{(k)}_{\boldsymbol{x}_a}(f_1)\right\}. \tag{5.182}$$

Without lack of generality, the functions $S^{(k)}_{\boldsymbol{x}_a}(f_1)$ and $G^{(k)}_{\boldsymbol{x}_a}(f_2)$ in (4.209a) and (4.209b) can be chosen such that both $\mathbb{J}_1$ and $\mathbb{J}_2$ are at most countable (Definition 4.2.4 and Theorem 4.2.7). Let us assume that the number of possible accumulation points (cluster points) of $\mathbb{J}_1$ and $\mathbb{J}_2$ is at most finite. Thus, from (5.180) it necessarily results that $\forall k \in \mathbb{I}_a$

$$f_1 \notin [-B_1, B_1] \cap \bar{\mathbb{J}}_1 \ \text{ or } \ \Psi^{(k)}_{\boldsymbol{x}_a}(f_1) \notin [-B_2, B_2] \cap \bar{\mathbb{J}}_2 \quad \Rightarrow \quad S^{(k)}_{\boldsymbol{x}_a}(f_1) = 0 \tag{5.183}$$

where $\bar{\mathbb{J}}_i \triangleq \mathbb{R} - \mathrm{cl}(\mathbb{J}_i)$ $(i = 1, 2)$, with cl denoting set closure, represents the set of points f_i where there is no curve intersection (or accumulation points of such f_i). Consequently, condition in (4.210a) holds almost everywhere (a.e.). However, since $\mathbb{J}_1$ is at most countable and with at most a finite number of accumulation points, the cross-correlation (4.24b) (with $[y, x^{(*)}] = [x_{a,1}, x^{(*)}_{a,2}]$) is not modified by assuming $S^{(k)}_{\boldsymbol{x}_a}(f_1) = 0$ also for f_1 such that $|f_1| > B_1$ and $f_1 \in \mathrm{cl}(\mathbb{J}_1)$. Therefore, we can assume that (4.210a) holds everywhere using suitable modifications of $S^{(k)}_{\boldsymbol{x}_a}(f_1)$. Furthermore, from (5.180) it follows that for strictly band-limited SC processes the functions $\Psi^{(k)}_{\boldsymbol{x}_a}(f_1)$ are undetermined for $f_1 \notin [-B_1, B_1]$. Thus, $\Psi^{(k)}_{\boldsymbol{x}_a}(f_1) = 0$ for $f_1 \notin [-B_1, B_1]$ can be assumed in order to have these functions with compact support. The

proof of (4.210b) is similar and, analogously, $\Phi^{(k)}_{\boldsymbol{x}_a}(f_2) = 0$ can be assumed for $f_2 \notin [-B_2, B_2]$.
The proof of the converse is straightforward.

5.10.2 Proof of Theorem 4.9.4 Loève Bifrequency Spectrum of Sampled Jointly Spectrally Correlated Processes

The Fourier transform (4.185) of the sequence $x_i(n)$ is linked to the Fourier transform (4.5) of the continuous-time signal $x_{a,i}(t)$ by the relationship (Lathi 2002, Section 9.5)

$$X_i(\nu) = \frac{1}{T_{si}} \sum_{p \in \mathbb{Z}} X_{a,i}((\nu - p) f_{si}). \tag{5.184}$$

Thus, accounting for (4.209a), one obtains

$$\begin{aligned}
&\mathrm{E}\left\{X_1(\nu_1)\, X_2^{(*)}(\nu_2)\right\} \\
&= \mathrm{E}\left\{\frac{1}{T_{s1}} \sum_{p_1 \in \mathbb{Z}} X_{a,1}((\nu_1 - p_1) f_{s1})\, \frac{1}{T_{s2}} \sum_{p_2 \in \mathbb{Z}} X_{a,2}^{(*)}((\nu_2 - p_2) f_{s2})\right\} \\
&= \frac{1}{T_{s1} T_{s2}} \sum_{p_1 \in \mathbb{Z}} \sum_{p_2 \in \mathbb{Z}} \mathrm{E}\left\{X_{a,1}((\nu_1 - p_1) f_{s1})\, X_{a,2}^{(*)}((\nu_2 - p_2) f_{s2})\right\} \\
&\quad \frac{1}{T_{s1} T_{s2}} \sum_{p_1 \in \mathbb{Z}} \sum_{p_2 \in \mathbb{Z}} \sum_{k \in \mathbb{I}_a} S^{(k)}_{\boldsymbol{x}_a}((\nu_1 - p_1) f_{s1}) \\
&\quad \delta\Big((\nu_2 - p_2) f_{s2} - \Psi^{(k)}_{\boldsymbol{x}_a}((\nu_1 - p_1) f_{s1})\Big)
\end{aligned} \tag{5.185}$$

from which (4.213) follows by a scale change in the argument of the Dirac delta (Zemanian 1987, Section 1.7).

5.10.3 Proof of Theorem 4.9.5 Loève Bifrequency Spectrum of Sampled Strictly Band-Limited Jointly SC Processes

For strictly band-limited processes $x_{a,i}(t)$, according to Corollary 4.9.3, the support of the replica with $p_1 = p_2 = 0$ in (4.213) is such that

$$\begin{aligned}
&\operatorname{supp}\left\{ S^{(k)}_{\boldsymbol{x}_a}(\nu_1 f_{s1})\, \delta\Big(\nu_2 - \Psi^{(k)}_{\boldsymbol{x}_a}(\nu_1 f_{s1}) T_{s2}\Big)\right\} \\
&\subseteq \Big\{(\nu_1, \nu_2) \in \mathbb{R} \times \mathbb{R} \,:\, |\nu_1| \leqslant B_1/f_{s1}, \\
&\qquad |\nu_2| \leqslant B_2/f_{s2},\ \nu_2 = \Psi^{(k)}_{\boldsymbol{x}_a}(\nu_1 f_{s1}) T_{s2}\Big\}.
\end{aligned} \tag{5.186}$$

Since the functions $\Psi^{(k)}_{\boldsymbol{x}_a}(\cdot)$ have compact support contained in $[-B_1, B_1]$, are invertible, and have values in $[-B_2, B_2]$ (Theorem 4.9.2), under the assumption $f_{si} \geqslant 2B_i$, the result is that for every $k \in \mathbb{I}_a$ and $(p_1, p_2) \in \mathbb{Z}^2$ there exists only one pair $(\nu_1, \nu_2) \in [p_1 - 1/2, p_1 + 1/2) \times [p_2 - 1/2, p_2 + 1/2)$ such that $(\nu_2 - p_2) f_{s2} = \Psi^{(k)}_{\boldsymbol{x}_a}((\nu_1 - p_1) f_{s1})$ in the argument of the Dirac delta in the right-hand side of (5.185). Consequently, (5.185) can be written as

$$
\begin{aligned}
&\mathrm{E}\Big\{X_1(\nu_1)\, X_2^{(*)}(\nu_2)\Big\} \\
&\quad = \sum_{k\in\mathbb{I}_a}\Big[\frac{1}{T_{s1}}\sum_{p_1\in\mathbb{Z}} S_{x_a}^{(k)}((\nu_1-p_1)f_{s1})\Big] \\
&\qquad \frac{1}{T_{s2}}\sum_{p_2\in\mathbb{Z}}\sum_{p_1\in\mathbb{Z}}\delta\Big((\nu_2-p_2)f_{s2}-\Psi_{x_a}^{(k)}((\nu_1-p_1)f_{s1})\Big) \\
&\quad = \sum_{k\in\mathbb{I}_a}\Big[\frac{1}{T_{s1}}\sum_{p_1\in\mathbb{Z}} S_{x_a}^{(k)}((\nu_1-p_1)f_{s1})\Big] \\
&\qquad \sum_{p_2\in\mathbb{Z}}\delta\Big(\nu_2-p_2-\sum_{p_1\in\mathbb{Z}}\Psi_{x_a}^{(k)}((\nu_1-p_1)f_{s1})T_{s2}\Big)
\end{aligned}
\tag{5.187}
$$

from which (4.190a) with (4.220a) and (4.221a) immediately follow. In addition, since replicas in (5.187) are separated by 1 in both variables ν_1 and ν_2 and the functions (of ν_1) $S_{x_a}^{(k)}(\nu_1 f_{s1})$ and $\Psi_{x_a}^{(k)}(\nu_1 f_{s1})$ have compact support contained in $[-B_1/f_{s1}, B_1/f_{s1}]$, condition $f_{s1} \geqslant 2B_1$ assures that replicas in (4.220a) and (4.221a) do not overlap.

The proof of (4.190b) with (4.220b) and (4.221b) is similar.

5.10.4 *Proof of Theorem 4.9.6*

Let us assume that $f_{si} \geqslant 2B_i$, $(i = 1, 2)$. Thus, according to Theorem 4.9.5, replicas in (5.187) do not overlap. However, mappings (4.222), (4.223) do not necessarily hold $\forall \nu_1 \in [-1/2, 1/2]$.

From $f_{s1} \geqslant 2B_1$ it immediately follows that $\forall \nu_1 \in [-1/2, 1/2]$

$$
|\nu_1| \leqslant \frac{1}{2} \leqslant 1 - B_1/f_{s1}. \tag{5.188}
$$

Due to condition (4.226) for $\Psi_{x_a}^{(k)}(\cdot)$, it results that $\forall \nu_1 \in [-1/2, 1/2]$

$$
|\Psi_{x_a}^{(k)}(\nu_1 f_{s1})|T_{s2} \leqslant 1 - B_2/f_{s2}. \tag{5.189}
$$

Then in (5.187), for $\nu_1 \in [-1/2, 1/2]$ the support curve $\nu_2 = \Psi_{x_a}^{(k)}(\nu_1 f_{s1})T_{s2}$ of the replica with $p_1 = p_2 = 0$ can overlap support curves of other replicas ($p_1 \neq 0$ and/or $p_2 \neq 0$) only if on these other curves the spectral correlation density is zero. In fact, $1 - B_1/f_{s1}$ $[1 - B_2/f_{s2}]$ is the smallest $|\nu_1|$ $[|\nu_2|]$ such that replicas with $p_1 = \pm 1$ $[p_2 = \pm 1]$ can have nonzero spectral correlation density. Thus conditions (4.222), (4.223) hold $\forall \nu_1 \in [-1/2, 1/2]$. (5.188) and (5.189) holding $\forall k \in \mathbb{I}_a$ prove sufficiency of condition (4.226).

Analogous considerations lead to prove sufficiency of condition (4.227).

5.11 Proofs for Section 4.10 "Multirate Processing of Discrete-Time Jointly SC Processes"

In this section, proofs of results of Section 4.10 are reported.

For future reference, let us consider the discrete-time sample train with sampling period M, its discrete Fourier series (DFS), and its Fourier transform:

$$\delta_{n \bmod M} \triangleq \sum_{\ell=-\infty}^{+\infty} \delta_{n-\ell M} = \frac{1}{M} \sum_{p=0}^{M-1} e^{j2\pi(p/M)n} \overset{\mathcal{F}}{\longleftrightarrow} \frac{1}{M} \sum_{p=0}^{M-1} \tilde{\delta}\left(\nu - \frac{p}{M}\right) \tag{5.190}$$

where δ_m is the Kronecker delta, that is, $\delta_m = 1$ if $m = 0$ and $\delta_m = 0$ if $m \neq 0$ and mod M denotes modulo operation with values in $\{0, 1, \ldots, M-1\}$. Thus, $\delta_{n \bmod M} = 1$ if $n = kM$ for some integer k and $\delta_{n \bmod M} = 0$ otherwise.

Furthermore, in the sequel, the following identities are used

$$\sum_{p=0}^{M-1} \tilde{\delta}(\nu - p/M - \psi) = \sum_{p=0}^{M-1} \sum_{h \in \mathbb{Z}} \delta(\nu - h - p/M - \psi) = \sum_{h \in \mathbb{Z}} \delta(\nu - h/M - \psi) \tag{5.191}$$

$$\sum_{p=0}^{M-1} \sum_{h \in \mathbb{Z}} \delta(\nu - p - hM - \psi) = \sum_{h \in \mathbb{Z}} \delta(\nu - h - \psi) = \tilde{\delta}(\nu - \psi). \tag{5.192}$$

In the following proofs, the bounded variation assumption (4.192) allows to use the Fubini and Tonelli theorem (Champeney 1990, Chapter 3) to interchange the order of the integrals in ν_1 and ν_2.

5.11.1 *Expansion (Section 4.10.1): Proof of (4.241), (4.242b), and (4.242c)*

From the identity

$$x_I(n) = \sum_{\ell=-\infty}^{+\infty} \delta_{\ell-n/L}\, x(\ell) \tag{5.193}$$

it follows that the impulse-response function of the LTV system that operates expansion is

$$h_I(n, \ell) = \delta_{\ell-n/L} = \delta_{\ell L-n} = \begin{cases} 1 & \ell = n/L \\ 0 & \text{otherwise} \end{cases} \tag{5.194}$$

Thus, the transmission function is

$$\begin{aligned} H_I(\nu, \lambda) &\triangleq \sum_{n \in \mathbb{Z}} \sum_{\ell \in \mathbb{Z}} h_I(n, \ell)\, e^{-j2\pi(\nu n - \lambda \ell)} \\ &= \sum_{n \in \mathbb{Z}} \sum_{\ell \in \mathbb{Z}} \delta_{\ell L-n}\, e^{-j2\pi(\nu n - \lambda \ell)} \end{aligned}$$

$$= \sum_{\ell \in \mathbb{Z}} e^{j2\pi\lambda\ell} \sum_{n \in \mathbb{Z}} \delta_{\ell L - n}\, e^{-j2\pi\nu n}$$
$$= \sum_{\ell \in \mathbb{Z}} e^{j2\pi(\lambda - \nu L)\ell}$$
$$= \sum_{k \in \mathbb{Z}} \delta(\lambda - \nu L - k) \tag{5.195}$$

where in the last equality the Poisson's sum formula (Zemanian 1987, p. 189)

$$\sum_{n=-\infty}^{+\infty} e^{-j2\pi(f/f_s)n} = f_s \sum_{k=-\infty}^{+\infty} \delta(f - kf_s) \tag{5.196}$$

is used. Equation (4.242b) follows by using the definition (4.243) of $\widetilde{\delta}_\lambda(\lambda)$, where equalities should be intended in the sense of distributions (generalized functions) (Zemanian 1987).

From (5.195), using the scaling property of the Dirac's delta $\delta(bt) = \delta(t)/|b|$ (Zemanian 1987, p. 27) we have

$$H_I(\nu, \lambda) = \sum_{k \in \mathbb{Z}} \delta(\lambda - \nu L - k)$$
$$= \frac{1}{L} \sum_{k \in \mathbb{Z}} \delta\Big(\frac{\lambda - k}{L} - \nu\Big)$$
$$= \frac{1}{L} \sum_{p=0}^{L-1} \sum_{k \in \mathbb{Z}} \delta\Big(\frac{\lambda - p}{L} - \nu - k\Big) \tag{5.197}$$

from which (4.242c) follows accounting for the definition of $\widetilde{\delta}_\nu(\nu)$.

5.11.2 Expansion (Section 4.10.1): Proof of (4.250)

The inverse double Fourier transform of both sides of (4.248c) provides the cross-correlation of $x_{I1}(n)$ and $x_{I2}(n)$ (see (4.182) and (4.187))

$$\mathrm{E}\left\{x_{I1}(n_1)\, x_{I2}^{(*)}(n_2)\right\}$$
$$= \int_{[-1/2,1/2]^2} \mathrm{E}\Big\{X_{I1}(\nu_1)\, X_{I2}^{(*)}(\nu_2)\Big\}\, e^{j2\pi[\nu_1 n_1 + (-)\nu_2 n_2]}\, \mathrm{d}\nu_1\, \mathrm{d}\nu_2$$
$$= \sum_{\widetilde{\alpha} \in \widetilde{\mathcal{A}}} \int_{-1/2}^{1/2} \widetilde{S}_{\boldsymbol{x}}^{\widetilde{\alpha}}(\nu_1 L_1)\, e^{j2\pi\nu_1 n_1}$$
$$\int_{-1/2}^{1/2} \frac{1}{L_2} \sum_{p=0}^{L_2-1} \widetilde{\delta}_{\nu_2}\Big(\nu_2 - \frac{p}{L_2} - (-)\Big(\frac{\widetilde{\alpha}}{L_2} - \frac{L_1}{L_2}\nu_1\Big)\Big)\, e^{(-)j2\pi\nu_2 n_2}\, \mathrm{d}\nu_2\, \mathrm{d}\nu_1$$
$$= \sum_{\widetilde{\alpha} \in \widetilde{\mathcal{A}}} \int_{-1/2}^{1/2} \widetilde{S}_{\boldsymbol{x}}^{\widetilde{\alpha}}(\nu_1 L_1)\, e^{j2\pi\nu_1 n_1}$$

$$\frac{1}{L_2} \sum_{p=0}^{L_2-1} e^{(-)j2\pi[p/L_2+(-)(\widetilde{\alpha}/L_2-(L_1/L_2)\nu_1)]n_2} \, d\nu_1$$

$$= \sum_{\widetilde{\alpha} \in \widetilde{A}} e^{j2\pi(\widetilde{\alpha}/L_2)n_2} \int_{-1/2}^{1/2} \widetilde{S}_{x}^{\widetilde{\alpha}}(\nu_1 L_1) \, e^{j2\pi\nu_1[n_1-(L_1/L_2)n_2]} \, d\nu_1 \, \delta_{n_2 \bmod L_2}$$

$$= \sum_{\widetilde{\alpha} \in \widetilde{A}} e^{j2\pi(\widetilde{\alpha}/L_2)n_2} \, \widetilde{\widetilde{R}}_{x}^{\alpha}\left[\frac{n_1 - (L_1/L_2)n_2}{L_1}\right] \delta_{n_2 \bmod L_2} \tag{5.198}$$

The third equality in (5.198) follows from the Fourier pair

$$e^{j2\pi\nu_0 n} \overset{\mathcal{F}}{\longleftrightarrow} \widetilde{\delta}(\nu - \nu_0) \tag{5.199}$$

and the fourth equality is due to (5.190). The fifth equality in (5.198) is a consequence of the Fourier-transform pair (4.244)–(4.245).

By substituting $n_1 = n + m$ and $n_2 = n$ into (5.198), (4.250) immediately follows.

5.11.3 Expansion (Section 4.10.1): Proof of (4.253)

From (4.250) with $L_1 = L_2 = L$ and using (5.190) it follows that

$$\begin{aligned}
\widetilde{\widetilde{R}}^{\alpha}_{x_{I1}x^{(*)}_{I2}}(m) &\triangleq \left\langle \mathrm{E}\left\{x_{I1}(n+m)\, x^{(*)}_{I2}(n)\right\} e^{-j2\pi\widetilde{\alpha}n} \right\rangle_n \\
&= \left\langle \sum_{\widetilde{\beta} \in \widetilde{A}} e^{j2\pi(\widetilde{\beta}/L)n} \, \widetilde{\widetilde{R}}_{x}^{\beta}\left[\frac{m}{L}\right] \delta_{n \bmod L} \, e^{-j2\pi\widetilde{\alpha}n} \right\rangle_n \\
&= \sum_{\widetilde{\beta} \in \widetilde{A}} \widetilde{\widetilde{R}}_{x}^{\beta}\left[\frac{m}{L}\right] \left\langle e^{j2\pi(\widetilde{\beta}/L)n} \, \frac{1}{L} \sum_{\ell=0}^{L-1} e^{j2\pi(\ell/L)n} \, e^{-j2\pi\widetilde{\alpha}n} \right\rangle_n \\
&= \frac{1}{L} \sum_{\widetilde{\beta} \in \widetilde{A}} \widetilde{\widetilde{R}}_{x}^{\beta}\left[\frac{m}{L}\right] \sum_{\ell=0}^{L-1} \delta_{[\widetilde{\alpha}-\widetilde{\beta}/L-\ell/L] \bmod 1} \\
&= \frac{1}{L} \sum_{\widetilde{\beta} \in \widetilde{A}} \widetilde{\widetilde{R}}_{x}^{\beta}\left[\frac{m}{L}\right] \sum_{\ell=0}^{L-1} \delta_{[L\widetilde{\alpha}-\widetilde{\beta}-\ell] \bmod L} \\
&= \frac{1}{L} \, \widetilde{\widetilde{R}}_{x}^{L\widetilde{\alpha}-\ell_0}\left[\frac{m}{L}\right].
\end{aligned} \tag{5.200}$$

The sixth equality of (5.200) follows observing that for every $\widetilde{\beta} \in \widetilde{A}$ there is at most one value of $\ell \in \mathbb{Z}$, say ℓ_0 (depending on $\widetilde{\alpha}$, $\widetilde{\beta}$, and L), such that $L\widetilde{\alpha} - \widetilde{\beta} - \ell_0 = 0$, provided that $L\widetilde{\alpha} \bmod 1 \in \widetilde{A}$. In fact, it results that

$$\sum_{\ell=0}^{L-1} \delta_{[L\widetilde{\alpha}-\widetilde{\beta}-\ell] \bmod L} = \sum_{\ell=0}^{L-1} \sum_{h \in \mathbb{Z}} \delta_{L\widetilde{\alpha}-\widetilde{\beta}-\ell-hL} = \sum_{\ell \in \mathbb{Z}} \delta_{L\widetilde{\alpha}-\widetilde{\beta}-\ell}. \tag{5.201}$$

Thus, (4.253) immediately follows since $\widetilde{\widetilde{R}}_{x}^{\alpha}(m)$ is periodic in $\widetilde{\alpha}$ with period 1.

5.11.4 Sampling (Section 4.10.2): Proof of (4.261)

Let be $M_1 = M_2 = M$. Each process $x_{\delta i}(n)$ is the product of the almost-cyclostationary discrete-time process $x_i(n)$ and the (real-valued) periodic train $\delta_{n \bmod M}$ with Fourier series expansion given in (5.190). Thus, by using the discrete-time counterpart of (1.143) or (Napolitano 1995, eq. (46)) (specialized to 2 single-input single-output systems, second-order, and reduced-dimension) we have

$$\begin{aligned} \widetilde{R}^{\widetilde{\alpha}}_{x_{\delta 1} x_{\delta 2}^{(*)}}(m) &= \frac{1}{M^2} \sum_{p_1=0}^{M-1} \sum_{p_2=0}^{M-1} \widetilde{R}_{\boldsymbol{x}}^{\widetilde{\alpha}-(p_1+p_2)/M}(m)\, e^{j2\pi(p_1/M)m} \\ &= \frac{1}{M^2} \sum_{p=0}^{M-1} \sum_{q=p}^{M-1+p} \widetilde{R}_{\boldsymbol{x}}^{\widetilde{\alpha}-q/M}(m)\, e^{j2\pi(p/M)m} \\ &= \frac{1}{M} \sum_{q=0}^{M-1} \widetilde{R}_{\boldsymbol{x}}^{\widetilde{\alpha}-q/M}(m)\, \frac{1}{M} \sum_{p=0}^{M-1} e^{j2\pi(p/M)m} \end{aligned} \tag{5.202}$$

from which, accounting for (5.190), one obtains (4.261).

In (5.202), in the second equality the variable changes $p = p_1$ and $q = p_1 + p_2$ are made and in the third equality the fact that $\widetilde{R}_{\boldsymbol{x}}^{\widetilde{\alpha}}(m)$ is periodic in $\widetilde{\alpha}$ with period 1 is used, so that the sum over $\{\widetilde{\alpha} - p/M, \widetilde{\alpha} - p/M - 1/M, \ldots, \widetilde{\alpha} - p/M - (M-1)/M\}$ can be equivalently extended over $\{\widetilde{\alpha}, \widetilde{\alpha} - 1/M, \ldots, \widetilde{\alpha} - (M-1)/M\}$.

5.11.5 Decimation (Section 4.10.3): Proof of (4.268), (4.269b), and (4.269c)

From the identity

$$x_D(n) = \sum_{\ell=-\infty}^{+\infty} \delta_{\ell-nM}\, x(\ell) \tag{5.203}$$

it follows that the impulse-response function of the LTV system that operates decimation is

$$h_D(n,\ell) = \delta_{\ell-nM} = \begin{cases} 1 & \ell = nM \\ 0 & \text{otherwise} \end{cases} \tag{5.204}$$

Consequently, the transmission function is

$$\begin{aligned} H_D(\nu,\lambda) &\triangleq \sum_{n\in\mathbb{Z}} \sum_{\ell\in\mathbb{Z}} h_D(n,\ell)\, e^{-j2\pi(\nu n-\lambda\ell)} \\ &= \sum_{n\in\mathbb{Z}} \sum_{\ell\in\mathbb{Z}} \delta_{\ell-nM}\, e^{-j2\pi(\nu n-\lambda\ell)} \\ &= \sum_{n\in\mathbb{Z}} e^{-j2\pi\nu n} \sum_{\ell\in\mathbb{Z}} \delta_{\ell-nM}\, e^{j2\pi\lambda\ell} \end{aligned}$$

$$= \sum_{n\in\mathbb{Z}} e^{-j2\pi(\nu-\lambda M)n}$$
$$= \sum_{k\in\mathbb{Z}} \delta(\nu - \lambda M - k) \tag{5.205}$$

where in the last equality the Poisson's sum formula (5.196) is used. Equation (4.269b) follows by the definition of $\widetilde{\delta}_\nu(\nu)$. From (5.205), using the scaling property of the Dirac delta, we have

$$\begin{aligned} H_D(\nu,\lambda) &= \sum_{k\in\mathbb{Z}} \delta(\nu - \lambda M - k) \\ &= \frac{1}{M}\sum_{k\in\mathbb{Z}} \delta\Big(\frac{\nu-k}{M} - \lambda\Big) \\ &= \frac{1}{M}\sum_{p=0}^{M-1}\sum_{k\in\mathbb{Z}} \delta\Big(\frac{\nu-p}{M} - \lambda - k\Big) \end{aligned} \tag{5.206}$$

from which (4.269c) follows accounting for the definition of $\widetilde{\delta}_\lambda(\lambda)$.

5.11.6 Decimation (Section 4.10.3): Proof of (4.277)

By using (4.275) we have

$$\begin{aligned} &\mathrm{E}\left\{x_{D1}(n_1)\, x_{D2}^{(*)}(n_2)\right\} \\ &= \int_{[-1/2,1/2]^2} \mathrm{E}\left\{X_{D1}(\nu_1)\, X_{D2}^{(*)}(\nu_2)\right\} e^{j2\pi[\nu_1 n_1 + (-)\nu_2 n_2]}\,\mathrm{d}\nu_1\,\mathrm{d}\nu_2 \\ &= \sum_{\widetilde{\alpha}\in\widetilde{A}} \int_{-1/2}^{1/2} \frac{1}{M_1}\sum_{p_1=0}^{M_1-1} \widetilde{S}_{x}^{\widetilde{\alpha}}\Big(\frac{\nu_1-p_1}{M_1}\Big)\, e^{j2\pi\nu_1 n_1} \\ &\qquad \int_{-1/2}^{1/2} \widetilde{\delta}_{\nu_2}\Big(\nu_2 - (-)\Big(\widetilde{\alpha}M_2 - (\nu_1-p_1)\frac{M_2}{M_1}\Big)\Big) e^{(-)j2\pi\nu_2 n_2}\,\mathrm{d}\nu_2\,\mathrm{d}\nu_1 \\ &= \sum_{\widetilde{\alpha}\in\widetilde{A}} e^{j2\pi\widetilde{\alpha}M_2 n_2} \int_{-1/2}^{1/2} \frac{1}{M_1}\sum_{p_1=0}^{M_1-1} \widetilde{S}_{x}^{\widetilde{\alpha}}\Big(\frac{\nu_1-p_1}{M_1}\Big) \\ &\qquad e^{-j2\pi[(\nu_1-p_1)/M_1]M_2 n_2}\, e^{j2\pi\nu_1 n_1}\,\mathrm{d}\nu_1 \\ &= \sum_{\widetilde{\alpha}\in\widetilde{A}} e^{j2\pi\widetilde{\alpha}M_2 n_2}\, \widetilde{R}_{x}^{\widetilde{\alpha}}(M_1 n_1 - M_2 n_2) \end{aligned} \tag{5.207}$$

where, in the fourth equality, the Fourier-transform pairs (see (4.270) and (4.272))

$$\widetilde{R}_{x}^{\widetilde{\alpha}}(n_1 - M_2 n_2) \overset{\mathcal{F}}{\longleftrightarrow} \widetilde{S}_{x}^{\widetilde{\alpha}}(\nu_1)\, e^{-j2\pi\nu_1 M_2 n_2} \tag{5.208}$$

$$\widetilde{R}_{x}^{\widetilde{\alpha}}(M_1 n_1 - M_2 n_2) \overset{\mathcal{F}}{\longleftrightarrow} \frac{1}{M_1}\sum_{p_1=0}^{M_1-1} \widetilde{S}_{x}^{\widetilde{\alpha}}\Big(\frac{\nu_1-p_1}{M_1}\Big)\, e^{-j2\pi[(\nu_1-p_1)/M_1]M_2 n_2} \tag{5.209}$$

are accounted for.

Equation (4.277) follows by substituting $n_1 = n + m$ and $n_2 = n$.

5.11.7 Decimation (Section 4.10.3): Proof of (4.278)

From (4.277) with $M_1 = M_2 = M$ it follows that

$$\begin{aligned}
\widetilde{R}^{\widetilde{\alpha}}_{x_{D1}x^{(*)}_{D2}}(m) &\triangleq \left\langle \mathrm{E}\left\{x_{D1}(n+m)\, x^{(*)}_{D2}(n)\right\} e^{-j2\pi\widetilde{\alpha}n} \right\rangle_n \\
&= \left\langle \sum_{\widetilde{\beta}\in\widetilde{\mathcal{A}}} e^{j2\pi\widetilde{\beta}Mn}\, \widetilde{R}^{\widetilde{\beta}}_{\boldsymbol{x}}(mM)\, e^{-j2\pi\widetilde{\alpha}n} \right\rangle_n \\
&= \sum_{\widetilde{\beta}\in\widetilde{\mathcal{A}}} \widetilde{R}^{\widetilde{\beta}}_{\boldsymbol{x}}(mM)\, \delta_{[\widetilde{\beta}M-\widetilde{\alpha}] \bmod 1} \\
&= \sum_{\widetilde{\beta}\in\widetilde{\mathcal{A}}} \widetilde{R}^{\widetilde{\beta}}_{\boldsymbol{x}}(mM) \sum_{h\in\mathbb{Z}} \delta_{\widetilde{\beta}-\widetilde{\alpha}/M+h/M}
\end{aligned} \tag{5.210}$$

In the rhs of (5.210), there are at most M (consecutive) values of h such that $\widetilde{\beta} - \widetilde{\alpha}/M + h/M = 0$ since when $\widetilde{\beta}$ ranges in $\widetilde{\mathcal{A}}$ it results that $\widetilde{\beta} \in [-1/2, 1/2)$. Therefore, (4.278) follows, accounting for the periodicity in $\widetilde{\alpha}$ with period 1 of $\widetilde{R}^{\widetilde{\alpha}}_{\boldsymbol{x}}(m)$.

5.11.8 Expansion and Decimation (Section 4.10.4): Proof of (4.286)

Accounting for (4.182) and (4.187) and using (4.285) and the sampling property of the Dirac delta, the cross-correlation function of $x_{I1}(n)$ and $x_{D2}(n)$ can be expressed as

$$\begin{aligned}
&\mathrm{E}\left\{x_{I1}(n_1)\, x^{(*)}_{D2}(n_2)\right\} \\
&= \int_{[-1/2,1/2]^2} \mathrm{E}\left\{X_{I1}(\nu_1)\, X^{(*)}_{D2}(\nu_2)\right\} e^{j2\pi[\nu_1 n_1 + (-)\nu_2 n_2]}\, \mathrm{d}\nu_1\, \mathrm{d}\nu_2 \\
&= \sum_{\widetilde{\alpha}\in\widetilde{\mathcal{A}}} \int_{-1/2}^{1/2} \widetilde{S}^{\widetilde{\alpha}}_{\boldsymbol{x}}(\nu_1 L_1)\, e^{j2\pi\nu_1 n_1} \\
&\qquad \int_{-1/2}^{1/2} \delta_{\nu_2}\Big(\nu_2 - (-)(\widetilde{\alpha}M_2 - \nu_1 L_1 M_2)\Big)\, e^{(-)j2\pi\nu_2 n_2}\, \mathrm{d}\nu_2\, \mathrm{d}\nu_1 \\
&= \sum_{\widetilde{\alpha}\in\widetilde{\mathcal{A}}} e^{j2\pi\widetilde{\alpha}M_2 n_2} \int_{-1/2}^{1/2} \widetilde{S}^{\widetilde{\alpha}}_{\boldsymbol{x}}(\nu_1 L_1)\, e^{j2\pi(n_1 - L_1 M_2 n_2)\nu_1}\, \mathrm{d}\nu_1 \\
&= \sum_{\widetilde{\alpha}\in\widetilde{\mathcal{A}}} e^{j2\pi\widetilde{\alpha}M_2 n_2}\, \widetilde{R}^{\widetilde{\alpha}}_{\boldsymbol{x}}\left[\frac{n_1 - L_1 M_2 n_2}{L_1}\right]
\end{aligned} \tag{5.211}$$

where, in the last equality, the Fourier-transform pair (4.244) and (4.245) is used. By substituting $n_1 = n + m$ and $n_2 = n$ we have (4.286).

5.11.9 Strictly Band-Limited SC Processes (Section 4.9.1): Proof of (4.287a)–(4.288b)

Let $x_1(n)$ and $x_2(n)$ be strictly band-limited processes with bandwidths B_1 and B_2, respectively. It results that

$$\mathrm{E}\left\{H_{B_1}(\nu_1)\,X_1(\nu_1)\,H_{B_2}(\nu_2)\,X_2^{(*)}(\nu_2)\right\} = \mathrm{E}\left\{X_1(\nu_1)\,X_2^{(*)}(\nu_2)\right\}. \tag{5.212}$$

If $x_1(n)$ and $x_2(n)$ are jointly SC with Loève bifrequency cross-spectrum (4.190a), then (5.212) specializes into

$$H_{B_1}(\nu_1)\,H_{B_2}(\nu_2)\sum_{k\in\mathbb{I}}\widetilde{S}_{\boldsymbol{x}}^{(k)}(\nu_1)\,\widetilde{\delta}\left(\nu_2-\widetilde{\Psi}_{\boldsymbol{x}}^{(k)}(\nu_1)\right)$$
$$=\sum_{k\in\mathbb{I}}\widetilde{S}_{\boldsymbol{x}}^{(k)}(\nu_1)\,\widetilde{\delta}\left(\nu_2-\widetilde{\Psi}_{\boldsymbol{x}}^{(k)}(\nu_1)\right). \tag{5.213}$$

Let us define the sets

$$\mathbb{J}_1^{(m)} \triangleq \Big\{\nu_1\in[m-1/2,m+1/2)\ :$$
$$\widetilde{\Psi}_{\boldsymbol{x}}^{(k)}(\nu_1)=\widetilde{\Psi}_{\boldsymbol{x}}^{(k')}(\nu_1)\ \text{for some}\ k\neq k',\ \ k,k'\in\mathbb{I}\Big\} \tag{5.214}$$

$$\mathbb{J}_2 \triangleq \bigcup_{\nu_1\in\mathbb{J}_1^{(0)}}\ \bigcup_{k\in\mathbb{I}}\left\{\nu_2\in[-1/2,1/2)\ :\ \nu_2=\widetilde{\Psi}_{\boldsymbol{x}}^{(k)}(\nu_1)\right\} \tag{5.215}$$

Analogously to the continuous-time case, without lack of generality the functions $\widetilde{S}_{\boldsymbol{x}}^{(k)}(\nu_1)$ and $\widetilde{G}_{\boldsymbol{x}}^{(k)}(\nu_2)$ in (4.190a) and (4.190b) can be chosen with supports such that the sets $\mathbb{J}_1^{(m)}$ and $\mathbb{J}_2$ are at most countable (Definition 4.2.4 and Theorem 4.2.7). In addition, let us assume that the number of cluster points (accumulation points) of $\mathbb{J}_1^{(m)}$ and $\mathbb{J}_2$ is at most finite. Thus, from (5.213) it follows that $\forall k\in\mathbb{I}$

$$\nu_1\notin\bigcup_{m\in\mathbb{Z}}[m-B_1,m+B_1)\cap\bar{\mathbb{J}}_1^{(m)}\quad\text{or}\quad\widetilde{\Psi}_{\boldsymbol{x}}^{(k)}(\nu_1)\notin[-B_2,B_2]\cap\bar{\mathbb{J}}_2$$
$$\Rightarrow\quad\widetilde{S}_{\boldsymbol{x}}^{(k)}(\nu_1)=0 \tag{5.216}$$

where $\bar{\mathbb{J}}_1^{(m)}\triangleq[m-1/2,m+1/2)-\mathrm{cl}\{\mathbb{J}_1^{(m)}\}$ and $\bar{\mathbb{J}}_2\triangleq[-1/2,1/2)-\mathrm{cl}\{\mathbb{J}_2\}$, with $\mathrm{cl}\{\cdot\}$ denoting set closure. From (5.216) it follows that

$$\widetilde{S}_{\boldsymbol{x}}^{(k)}(\nu_1)=0\ \text{ if }\ |\nu_1 \bmod 1|>B_1\ \text{ or }\ \left|\widetilde{\Psi}_{\boldsymbol{x}}^{(k)}(\nu_1)\right|>B_2 \tag{5.217}$$

almost everywhere for $\nu_1\in[-1/2,1/2)$. However, since $\mathbb{J}_1^{(m)}$ is at most countable, the cross-correlation (4.198a) is not modified assuming that (5.217) holds everywhere. Thus, (4.287a) and (4.287b) immediately follow.

Analogously, starting from (4.190b), the band-limitedness condition (5.212) leads to

$$\widetilde{G}_{\boldsymbol{x}}^{(k)}(\nu_2)=0\ \text{ if }\ |\nu_2 \bmod 1|>B_2\ \text{ or }\ \left|\widetilde{\Phi}_{\boldsymbol{x}}^{(k)}(\nu_2)\right|>B_1 \tag{5.218}$$

from which (4.288a) and (4.288b) follow.

5.11.10 Interpolation Filters (Section 4.10.6): Proof of the Image-Free Sufficient Condition (4.290)

Accounting for (4.246a), (4.287a), and (4.287b), the support of each term in the sum over $k \in \mathbb{I}$ in (4.289) is given by

$$\begin{aligned}
&\operatorname{supp}\left\{ \widetilde{S}_x^{(k)}(\nu_1 L_1)\, \widetilde{\delta}\left(\nu_2 L_2 - \widetilde{\Psi}_x^{(k)}(\nu_1 L_1)\right)\right\} \\
&\subseteq \left\{(\nu_1, \nu_2) \in \mathbb{R}\times\mathbb{R} \,:\, |(\nu_1 L_1) \bmod 1| \leqslant B_1,\ \left|\widetilde{\Psi}_x^{(k)}(\nu_1 L_1)\right| \leqslant B_2,\right. \\
&\qquad \left.\nu_2 L_2 = \widetilde{\Psi}_x^{(k)}(\nu_1 L_1) \bmod 1\right\} \\
&= \left\{(\nu_1, \nu_2) \in \mathbb{R}\times\mathbb{R} \,:\, |(\nu_1 L_1) \bmod 1| \leqslant B_1,\ |(\nu_2 L_2) \bmod 1| \leqslant B_2,\right. \\
&\qquad \left.\nu_2 L_2 = \widetilde{\Psi}_x^{(k)}(\nu_1 L_1) \bmod 1\right\} \\
&= \left\{(\nu_1, \nu_2) \in \mathbb{R}\times\mathbb{R} \,:\, |\nu_1 \bmod 1/L_1| \leqslant B_1/L_1,\ |\nu_2 \bmod 1/L_2| \leqslant B_2/L_2,\right. \\
&\qquad \left.\nu_2 = \frac{1}{L_2}\widetilde{\Psi}_x^{(k)}(\nu_1 L_1) \bmod 1/L_2\right\} \\
&= \bigcup_{\ell_1\in\mathbb{Z}} \bigcup_{\ell_2\in\mathbb{Z}} \Gamma_{\ell_1\ell_2}^{(k)}
\end{aligned} \tag{5.219}$$

where

$$\begin{aligned}
\Gamma_{\ell_1\ell_2}^{(k)} \triangleq &\left\{(\nu_1, \nu_2) \in \mathbb{R}\times\mathbb{R} \,:\, |\nu_1 - \ell_1/L_1| \leqslant B_1/L_1,\ |\nu_2 - \ell_2/L_2| \leqslant B_2/L_2,\right. \\
&\quad \left.\nu_2 - \ell_2/L_2 = \frac{1}{L_2}\widetilde{\Psi}_x^{(k)}(\nu_1 L_1)\right\}.
\end{aligned} \tag{5.220}$$

Let $H_{W_i}(\nu) \triangleq \sum_{m\in\mathbb{Z}} \operatorname{rect}((\nu - m)/(2W_i))$ be the Fourier transform of $h_{W_i}(n)$. It results that

$$\operatorname{supp}\left\{ H_{W_1}(\nu_1)\, H_{W_2}(\nu_2)\right\} = \bigcup_{m_1\in\mathbb{Z}} \bigcup_{m_2\in\mathbb{Z}} \Delta_{m_1 m_2} \tag{5.221}$$

where

$$\Delta_{m_1 m_2} \triangleq \left\{(\nu_1, \nu_2) \in \mathbb{R}\times\mathbb{R} \,:\, |\nu_1 - m_1| \leqslant W_1,\ |\nu_2 - m_2| \leqslant W_2,\right\}. \tag{5.222}$$

The following conditions are sufficient to assure that the Loève bifrequency cross-spectrum $\mathrm{E}\{Y_1(\nu_1)\, Y_2^{(*)}(\nu_2)\}$ in (4.289) is a frequency-scaled image-free version of $\mathrm{E}\{X_1(\nu_1)\, X_2^{(*)}(\nu_2)\}$:

1. $B_i \leqslant 1/2$, $(i = 1, 2)$. So images do not overlap. That is, for every k and k' in $\mathbb{I}$ (possibly $k = k'$), one has

$$\Gamma_{\ell_1\ell_2}^{(k)} \cap \Gamma_{\ell_1'\ell_2'}^{(k')} = \emptyset \ \text{ for } \ (\ell_1, \ell_2) \neq (\ell_1', \ell_2'). \tag{5.223}$$

2. $W_i \leqslant 1/2$, $(i = 1, 2)$. So the filter passbands do not overlap. That is,

$$\Delta_{m_1 m_2} \cap \Delta_{m_1' m_2'} = \emptyset \text{ for } (m_1, m_2) \neq (m_1', m_2'). \tag{5.224}$$

3. $B_i/L_i \leqslant W_i$, $(i = 1, 2)$. So the low-pass filters $H_{W_1}(\nu_1)$ and $H_{W_2}(\nu_2)$ capture the main images $((\ell_1, \ell_2) = (m_1 L_1, m_2 L_2), m_1, m_2 \in \mathbb{Z}$ in (5.220)), i.e., those centered in $(\nu_1, \nu_2) = (m_1, m_2) \in \mathbb{Z}^2$. That is, $\forall k \in \mathbb{I}$ one has

$$\Gamma^{(k)}_{m_1 L_1,\, m_2 L_2} \subseteq \Delta_{m_1 m_2}. \tag{5.225}$$

4. $W_i \leqslant 1/(2L_i)$, $(i = 1, 2)$. So the low-pass filters $H_{W_1}(\nu_1)$ and $H_{W_2}(\nu_2)$ do not capture images different from the main ones. That is, $\forall k \in \mathbb{I}$ one has

$$\Gamma^{(k)}_{\ell_1 \ell_2} \cap \Delta_{m_1 m_2} = \emptyset \text{ for } (\ell_1, \ell_2) \neq (m_1 L_1, m_2 L_2). \tag{5.226}$$

The sufficient conditions in items 1–4 can be summarized into the sufficient condition (4.290).

5.11.11 Interpolation Filters (Section 4.10.6): Proof of the Sufficient Conditions (4.291) and (4.292)

Let us assume that (4.290) holds. Thus the Loève bifrequency cross-spectrum $\mathrm{E}\{Y_1(\nu_1)\, Y_2^{(*)}(\nu_2)\}$ in (4.289) is a frequency-scaled image-free version of $\mathrm{E}\{X_1(\nu_1)\, X_2^{(*)}(\nu_2)\}$ with $\mathrm{E}\{X_{I1}(\nu_1)\, X_{I2}^{(*)}(\nu_2)\}$ given by (4.246a)–(4.246c). In addition, form (4.290) and $B_1 \leqslant 1/2$ it follows that

$$|\nu_1| \leqslant \frac{1}{2L_1} \leqslant \frac{1}{L_1} - \frac{B_1}{L_1} \qquad \forall \nu_1 \in [-1/2L_1, 1/2L_1]. \tag{5.227}$$

For a fixed $\widetilde{\Psi}^{(k)}_x(\cdot)$, due to condition (4.291), it results that

$$\frac{1}{L_2} \left| \widetilde{\Psi}^{(k)}_x(\nu_1 L_1) \right| \leqslant \frac{1}{L_2} - \frac{B_2}{L_2} \qquad \forall \nu_1 \in [-1/2L_1, 1/2L_1]. \tag{5.228}$$

If both conditions (5.227) and (5.228) are satisfied, then in expression (4.246b) of the Loève bifrequency cross-spectrum $\mathrm{E}\{X_{I1}(\nu_1)\, X_{I2}^{(*)}(\nu_2)\}$, for $\nu_1 \in [-1/(2L_1), 1/(2L_1)]$ the support curve $\nu_2 = \widetilde{\Psi}^{(k)}_x(\nu_1 L_1)/L_2$ of the replica with $h = 0$ can intersect other support curves with the same k $(h \neq 0)$ only if on these other curves the spectral correlation density is zero. In fact, $1/L_1 - B_1/L_1$ is the smallest $|\nu_1|$ in correspondence of which the image around $\nu_1 = \pm 1/L_1$ of the replica centered in $\nu_2 = 0$ ($h = 0$ in (4.246b)) can have nonzero spectral correlation density. Moreover, $1/L_2 - B_2/L_2$ is the smallest $|\nu_2|$ in correspondence of which the image around $\nu_1 = 0$ of the replica centered in $\nu_2 = \pm 1/L_2$ ($h = \pm 1$ in (4.246b)) can have nonzero spectral correlation density. Thus, the density of Loève bifrequency cross-spectrum (4.289) along the considered support curve is a frequency-scaled image-free version of the corresponding density of Loève bifrequency cross-spectrum $\mathrm{E}\{X_1(\nu_1)\, X_2^{(*)}(\nu_2)\}\ \forall \nu_1 \in [-1/(2L_1), 1/(2L_1)]$.

Condition (5.228) holding for every $k \in \mathbb{I}$ jointly with $B_2 \leqslant 1/2$ gives

$$1 \geqslant \sup_{k\in\mathbb{I}} \sup_{\nu_1\in[-1/2,1/2]} \left|\widetilde{\Psi}_{\boldsymbol{x}}^{(k)}(\nu_1)\right| + B_2 \tag{5.229}$$

$$1 \geqslant 2B_2 \tag{5.230}$$

which prove sufficiency of (4.291).

The proof of sufficiency of (4.292) is similar.

5.11.12 Decimation Filters (Section 4.10.7): Proof of the Aliasing-Free Sufficient Condition (4.293)

The Loève bifrequency cross-spectrum of $x_{D1}(n)$ and $x_{D2}(n)$ is given by (4.273a)–(4.273c). Accounting for (4.287a) and (4.287b), the support of each term in the sums over k, p_1, and p_2 in (4.273a) is given by

$$\begin{aligned}
&\operatorname{supp}\left\{ \widetilde{S}_{\boldsymbol{x}}^{(k)}\left(\frac{\nu_1 - p_1}{M_1}\right) \widetilde{\delta}\left(\frac{\nu_2 - p_2}{M_2} - \widetilde{\Psi}_{\boldsymbol{x}}^{(k)}\left(\frac{\nu_1 - p_1}{M_1}\right)\right)\right\} \\
&\quad \subseteq \left\{(\nu_1, \nu_2) \in \mathbb{R}\times\mathbb{R} : \left|\frac{\nu_1 - p_1}{M_1} \bmod 1\right| \leqslant B_1, \left|\widetilde{\Psi}_{\boldsymbol{x}}^{(k)}\left(\frac{\nu_1 - p_1}{M_1}\right)\right| \leqslant B_2,\right. \\
&\qquad \left.\frac{\nu_2 - p_2}{M_2} = \widetilde{\Psi}_{\boldsymbol{x}}^{(k)}\left(\frac{\nu_1 - p_1}{M_1}\right) \bmod 1\right\} \\
&\quad = \left\{(\nu_1, \nu_2) \in \mathbb{R}\times\mathbb{R} : \left|\frac{\nu_1 - p_1}{M_1} \bmod 1\right| \leqslant B_1, \left|\frac{\nu_2 - p_2}{M_2} \bmod 1\right| \leqslant B_2,\right. \\
&\qquad \left.\frac{\nu_2 - p_2}{M_2} = \widetilde{\Psi}_{\boldsymbol{x}}^{(k)}\left(\frac{\nu_1 - p_1}{M_1}\right) \bmod 1\right\} \\
&\quad = \left\{(\nu_1, \nu_2) \in \mathbb{R}\times\mathbb{R} : \left|(\nu_1 - p_1) \bmod M_1\right| \leqslant M_1 B_1,\right. \\
&\qquad \left|(\nu_2 - p_2) \bmod M_2\right| \leqslant M_2 B_2, \\
&\qquad \left.\nu_2 - p_2 = M_2 \widetilde{\Psi}_{\boldsymbol{x}}^{(k)}\left(\frac{\nu_1 - p_1}{M_1}\right) \bmod M_2\right\} \\
&\quad = \bigcup_{q_1\in\mathbb{Z}} \bigcup_{q_2\in\mathbb{Z}} \Gamma_{p_1 p_2 q_1 q_2}^{(k)}
\end{aligned} \tag{5.231}$$

where

$$\begin{aligned}
\Gamma_{p_1 p_2 q_1 q_2}^{(k)} \triangleq &\left\{(\nu_1, \nu_2) \in \mathbb{R}\times\mathbb{R} : \left|\nu_1 - p_1 - q_1 M_1\right| \leqslant M_1 B_1,\right. \\
&\left|\nu_2 - p_2 - q_2 M_2\right| \leqslant M_2 B_2, \\
&\left.\nu_2 - p_2 - q_2 M_2 = M_2 \widetilde{\Psi}_{\boldsymbol{x}}^{(k)}\left(\frac{\nu_1 - p_1}{M_1}\right)\right\}
\end{aligned} \tag{5.232}$$

with $p_i \in \{0, 1, \ldots, M_i - 1\}$ and $q_i \in \mathbb{Z}$, $(i = 1, 2)$.

If $M_i B_i \leqslant 1/2$, $(i = 1, 2)$, then

$$\Gamma^{(k)}_{p_1 p_2 q_1 q_2} \cap \Gamma^{(k')}_{p'_1 p'_2 q'_1 q'_2} = \emptyset \text{ for } (p_1, p_2, q_1, q_2) \neq (p'_1, p'_2, q'_1, q'_2) \tag{5.233}$$

for every k and k' in $\mathbb{I}$ (possibly $k = k'$). Therefore, (4.293) is a sufficient condition to assure that the Loève bifrequency cross-spectrum $\mathrm{E}\{X_{D1}(\nu_1)\, X^{(*)}_{D2}(\nu_2)\}$ is a frequency-scaled alias-free version of $\mathrm{E}\{X_1(\nu_1)\, X^{(*)}_2(\nu_2)\}$.

5.11.13 Decimation Filters (Section 4.10.7): Proof of the Sufficient Conditions (4.294) and (4.295)

Let us assume that condition (4.293) holds. Thus, the Loève bifrequency cross-spectrum $\mathrm{E}\{X_{D1}(\nu_1)\, X^{(*)}_{D2}(\nu_2)\}$ in (4.273a)–(4.273c) is a frequency-scaled alias-free version of $\mathrm{E}\{X_1(\nu_1)\, X^{(*)}_2(\nu_2)\}$. From (4.293) it immediately follows that

$$|\nu_1| \leqslant \frac{1}{2} \leqslant 1 - M_1 B_1 \qquad \forall \nu_1 \in [-1/2, 1/2]. \tag{5.234}$$

For a fixed $\widetilde{\Psi}^{(k)}_{x}(\cdot)$, due to condition (4.294) one obtains

$$M_2 \left| \widetilde{\Psi}^{(k)}_{x}(\nu_1/M_1) \right| \leqslant 1 - M_2 B_2 \qquad \forall \nu_1 [-1/2, 1/2]. \tag{5.235}$$

If conditions (5.234) and (5.235) are both verified, then in expression (4.273b) of the Loève bifrequency cross-spectrum $\mathrm{E}\{X_{D1}(\nu_1)\, X^{(*)}_{D2}(\nu_2)\}$, for $\nu_1 \in [-1/2, 1/2]$ the support curve $\nu_2 = M_2 \widetilde{\Psi}^{(k)}_{x}(\nu_1/M_1)$ of the replica with $h = 0$, $p_1 = 0$ can intersect other support curves with the same k ($h \neq 0$, $p_1 \neq 0$) only if on these other curves the spectral correlation density is zero. In fact, $1 - M_1 B_1$ is the smallest $|\nu_1|$ in correspondence of which replicas with $p_1 = \pm 1$ can have nonzero spectral correlation density and $1 - M_2 B_2$ is the smallest $|\nu_2|$ in correspondence of which replicas with $h = \pm 1$ can have nonzero spectral correlation density. Thus, the density of Loève bifrequency cross-spectrum (4.273b) along the considered support curve is a frequency-scaled alias-free version of the corresponding density of Loève bifrequency cross-spectrum $\mathrm{E}\{X_1(\nu_1)\, X^{(*)}_2(\nu_2)\}$ for every $\nu_1 \in [-1/2, 1/2]$.

Condition (5.235) holding for every $k \in \mathbb{I}$ jointly with $B_2 \leqslant 1/(2M_2)$ prove sufficiency of (4.294).

The proof of sufficiency of (4.295) is similar.

5.11.14 Fractional Sampling Rate Converters (Section 4.10.8): Proof of (4.296)

It results that

$$\mathrm{E}\Big\{Y_{D1}(\nu_1)\, Y^{(*)}_{D2}(\nu_2)\Big\}$$
$$= \frac{1}{M_1 M_2} \sum_{p_1=0}^{M_1-1} \sum_{p_2=0}^{M_2-1} \mathrm{E}\Big\{Y_1\Big(\frac{\nu_1 - p_1}{M_1}\Big)\, Y^{(*)}_2\Big(\frac{\nu_2 - p_2}{M_2}\Big)\Big\}$$

$$= \frac{1}{M_1 M_2} \sum_{p_1=0}^{M_1-1} \sum_{p_2=0}^{M_2-1} H_{W_1}\left(\frac{\nu_1 - p_1}{M_1}\right) H_{W_2}\left(\frac{\nu_2 - p_2}{M_2}\right)$$

$$\mathrm{E}\left\{X_{I1}\left(\frac{\nu_1 - p_1}{M_1}\right) X_{I2}^{(*)}\left(\frac{\nu_2 - p_2}{M_2}\right)\right\}$$

$$= \frac{1}{M_1 M_2} \sum_{p_1=0}^{M_1-1} \sum_{p_2=0}^{M_2-1} H_{W_1}\left(\frac{\nu_1 - p_1}{M_1}\right) H_{W_2}\left(\frac{\nu_2 - p_2}{M_2}\right)$$

$$\mathrm{E}\left\{X_1\left((\nu_1 - p_1)\frac{L_1}{M_1}\right) X_2^{(*)}\left((\nu_2 - p_2)\frac{L_2}{M_2}\right)\right\} \tag{5.236}$$

where, in the first and third equality, (4.272) and (4.245) are accounted for, respectively. By using (4.190a) into (5.236), (4.296) immediately follows.

6

Functional Approach for Signal Analysis

In this chapter, the functional approach for signal analysis (Gardner 1987d, 1991c), (Gardner and Brown 1991), (Gardner 1994), (Leśkow and Napolitano 2006) is addressed. This approach is an alternative to the classical one based on stochastic processes (Section 1.1) since signals are modeled as single functions of time rather than realizations (sample paths) of stochastic processes. Statistical functions are defined through infinite-time averages of single functions of time rather than as ensemble averages of an ensemble of realizations.

6.1 Introduction

An approach for signal analysis which is alternative to classical one based on stochastic processes is the *fraction-of-time (FOT)* probability framework or *functional approach* (Gardner 1987d, 1991c), (Gardner and Brown 1991), (Gardner 1994), (Leśkow and Napolitano 2006). In such an approach, signals are modeled as single functions of time (time series) rather than sample paths of stochastic processes. This is a more appropriate model when an ensemble of realizations does not exist and, consequently, the stochastic process turns out to be artificially introduced just to create a mathematical model. Common pitfalls that can arise from the adoption of a non appropriate stochastic process model are described in (Gardner 1987d, 1991c), (Gardner and Brown 1991), (Gardner 1994), (Izzo and Napolitano 2002a).

In the FOT approach, starting from the concept of relative measurability of sets and functions (Kac and Steinhaus 1938), (Leśkow and Napolitano 2006), for a single function of time (or time series) a distribution function is constructed and its corresponding expected value is shown to be the infinite-time average. All familiar probabilistic parameters and functions, such as variance, moments, and cumulants, are built starting from the single time series at hand $\{x(t), t \in \mathbb{R}\}$ in terms of infinite-time averages. Time series for which this model does not lead to trivial results should be persistent or finite power, that is, the time-averaged power

$$P_x \triangleq \lim_{T\to\infty} \frac{1}{T} \int_{-T/2}^{T/2} |x(t)|^2 \, \mathrm{d}t \tag{6.1}$$

Generalizations of Cyclostationary Signal Processing: Spectral Analysis and Applications, First Edition.

must exist and be finite. Such a necessary, but not sufficient, condition, with lim replaced by lim sup is also required in Wiener work on generalized harmonic analysis (Wiener 1930), where autocorrelation function and power spectrum are defined for single functions of time.

In the functional approach, the relative measure plays the role played by the probability measure in the stochastic approach. The relative measure, however, does not posses the sigma-additivity property and is not continuous. Such a result constitutes a strong motivation to adopt the functional approach since it enlightens a deep difference between properties of stochastic processes and properties of functions. In other words, the stochastic process model for a single realization at hand should be used carefully, since properties of the stochastic process could not correspond to analogous properties of the function of time at hand.

A time-variant probabilistic model which is based on a single time series is introduced in (Gardner 1987d), (Gardner and Brown 1991), (Gardner 1994), by showing that the almost-periodic component extraction operator is an expectation operator. Starting from this expectation operator, almost-periodically time-variant distributions, moments, and cumulants are defined. Moreover, concepts such as statistical independence, stationarity, and nonstationarity can be introduced. Therefore, this model can be used to statistically characterize ACS and GACS signals. In contrast, at the moment, no single-function-based probabilistic model exists for SC signals.

A rigorous link between the time-average-based and the stochastic-process frameworks in the stationary case is established in (Wold 1948), where an isometric isomorphism (Wold isomorphism) between a stationary ergodic stochastic process and the Hilbert space generated by a single sample path is singled out. The Wold isomorphism is extended to cyclostationary signals in (Gardner 1987d), (Gardner and Brown 1991), (Hurd and Koski 2004).

6.2 Relative Measurability

6.2.1 Relative Measure of Sets

Given a set $A \in \mathcal{B}_{\mathbb{R}}$, $\mathcal{B}_{\mathbb{R}}$ being the σ-field of the Borel subsets and μ the Lebesgue measure on the real line $\mathbb{R}$, the *relative measure* of A is defined as (Kac and Steinhaus 1938), (Leśkow and Napolitano 2006)

$$\mu_R(A) \triangleq \lim_{T\to\infty} \frac{1}{T} \mu(A \cap [t_0 - T/2, t_0 + T/2]) \tag{6.2}$$

provided that the limit exists. If the limit in (6.2) exists, it is independent of t_0 and the set A is said to be relatively measurable. From definition (6.2) it follows that the relative measure of a set is the Lebesgue measure of the set normalized to that of the whole real line. Thus, sets with finite Lebesgue measure have zero relative measure, and only sets with infinite Lebesgue measure can have nonzero relative measure. In (Leśkow and Napolitano 2006), the following properties of the relative measure are proved: The class of RM sets is not closed under union and intersection; The relative measure μ_R is additive, but not σ-additive; If A is a RM set, then $\bar{A} \triangleq \mathbb{Z} - A$ is RM. Moreover, the lack of σ-additivity does not allow to prove the continuity of μ_R. In (Leśkow and Napolitano 2006) it is also shown that non RM sets are "not so rare or sophisticated" as non-Lebesgue-measurable sets. In fact, non RM sets can be easily constructed and visualized.

The lack of σ-additivity *proved* for the relative measure μ_R, in contrast with the σ-additivity *assumed* for the probability P in the classical stochastic approach, constitutes one of the motivations of using the functional approach in signal analysis. In fact, due to such a difference between μ_R and P, results holding for stochastic processes do not necessarily have a counterpart in terms of functions of time representing sample paths of these stochastic processes.

6.2.2 *Relatively Measurable Functions*

Let $\{x(t),\ t \in \mathbb{R}\}$ be a Lebesgue measurable real-valued function (or signal). The function $x(t)$ is said to be *relatively measurable* if and only if the set $\{t \in \mathbb{R} : x(t) \leqslant \xi\}$ is RM for every $\xi \in \mathbb{R} - \Xi_0$, where Ξ_0 is an at most countable set of points. Each RM function $x(t)$ generates a function

$$\begin{aligned} F_x(\xi) &\triangleq \mu_R(\{t \in \mathbb{R} : x(t) \leqslant \xi\}) \\ &= \lim_{T\to\infty} \frac{1}{T}\mu(\{t \in [t_0 - T/2, t_0 + T/2] : x(t) \leqslant \xi\}) \\ &= \lim_{T\to\infty} \frac{1}{T}\int_{t_0-T/2}^{t_0+T/2} \mathbf{1}_{\{x(t)\leqslant\xi\}}\, \mathrm{d}t \end{aligned} \tag{6.3}$$

in all points ξ where the limit exists. In (6.3),

$$\mathbf{1}_{\{t\,:\,x(t)\leqslant\xi\}} \triangleq \begin{cases} 1, & t : x(t) \leqslant \xi, \\ 0, & t : x(t) > \xi \end{cases} \tag{6.4}$$

is the indicator of the set $\{t \in \mathbb{R} : x(t) \leqslant \xi\}$ (note the difference with definition (1.2)). The function $F_x(\xi)$ defined in (6.3) has values in $[0, 1]$ and is non decreasing. Thus, it has all the properties of a distribution function, except the right-continuity in the discontinuity points. Furthermore, as for every bounded nondecreasing function, the set of discontinuity points is at most countable. The function $F_x(\xi)$ represents the *fraction-of-time probability* that the signal $x(t)$ is below the threshold ξ (Gardner 1987d) and hence is named fraction-of-time distribution function.

Since non relatively measurable sets can be easily constructed (unlike non Lebesgue measurable sets), the lack of relative measurability of a function is not a rare property as the lack of Lebesgue measurability.

The function $F_x(\xi)$ allows to define all familiar probabilistic parameters such as mean, variance, moments, and cumulants. Furthermore, if $x(t)$ is a RM, not necessarily bounded, function and $g(\xi)$ satisfies appropriate regularity conditions, then the following *fundamental theorem of expectation* can be proved (Leśkow and Napolitano 2006, Theorem 3.2):

$$\lim_{T\to\infty} \frac{1}{T}\int_{-T/2}^{T/2} g(x(t))\, \mathrm{d}t = \int_{\mathbb{R}} g(\xi)\, \mathrm{d}F_x(\xi) \tag{6.5}$$

where the second integral is in the Riemann-Stiltjes sense. That is, the infinite-time average is the expectation operator of the distribution (6.3)

$$\int_{\mathbb{R}} \xi\, \mathrm{d}F_x(\xi) = \lim_{T\to\infty} \frac{1}{T}\int_{-T/2}^{T/2} x(t)\, \mathrm{d}t \equiv \langle x(t)\rangle_t\ . \tag{6.6}$$

The set of RM functions can be shown to be not closed under addition and multiplication operations. Consequently, RM functions do not constitute a vector space. On the contrary, the linear combination of two stochastic processes is still a stochastic process, provided that the two sample spaces are *assumed* to be jointly measurable, which is an implicitly made assumption.

6.2.3 Jointly Relatively Measurable Functions

In (Leśkow and Napolitano 2006), the concept of joint relative measurability between functions is introduced. It is shown that operations on jointly RM functions can lead to a RM function. Therefore, for such a function, a probabilistic model based on time averages can be constructed. The joint relative measurability is an analytical property of functions and, hence, easier to be verified than the analogous property in the stochastic process framework, that is, the joint measurability of sample spaces. The latter property, in fact, cannot be easily verified in applications since, generally, the sample spaces are not specified.

Two Lebesgue measurable functions $x(t)$ and $y(t)$ are said to be *jointly RM* (Leśkow and Napolitano 2006) if the limit

$$\begin{aligned} F_{yx}(\xi_1, \xi_2) &\triangleq \mu_R(\{t \in \mathbb{R} : y(t) \leqslant \xi_1\} \cap \{t \in \mathbb{R} : x(t) \leqslant \xi_2\}) \\ &= \lim_{T\to\infty} \frac{1}{T} \mu(\{t \in [-T/2, T/2] : y(t) \leqslant \xi_1, \ x(t) \leqslant \xi_2\}) \\ &= \lim_{T\to\infty} \frac{1}{T} \int_{-T/2}^{T/2} \mathbf{1}_{\{y(t)\leqslant\xi_1\}} \, \mathbf{1}_{\{x(t)\leqslant\xi_2\}} \, \mathrm{d}t \end{aligned} \tag{6.7}$$

exists for all $(\xi_1, \xi_2) \in \mathbb{R}^2 - \Xi_0$, where Ξ_0 is an at most countable set of lines of $\mathbb{R}^2$. The function F_{yx} has all the properties of a bivariate joint distribution function except the right continuity in the discontinuity points.

Let $x(t)$ and $y(t)$ be jointly RM functions. Then both $x(t)$ and $y(t)$ are RM. Moreover, the sum $x(t) + y(t)$ and the product $x(t)\, y(t)$ are RM, provided that at least one of the functions is bounded. In contrast, if $y(t)$ is not RM, then $y(t)$ is not jointly RM with any RM $x(t)$ (Leśkow and Napolitano 2006), (Leśkow and Napolitano 2007).

If $x(t)$ is RM, then the lag product $x(t)\, x(t+\tau)$ is not necessarily RM $\forall \tau \in \mathbb{R}$. However, if $x(t)$ and its shifted version $x(t+\tau)$ are jointly RM, then the lag product $x(t)\, x(t+\tau)$ is RM.

The notion of joint relative measurability can be extended to the multidimensional case. A finite collection of Lebesgue measurable functions $x_1(t), \ldots, x_n(t)$ is jointly RM if the limit

$$\begin{aligned} F_{x_1\cdots x_n}(\xi_1, \ldots, \xi_n) &\triangleq \mu_R(\{t \in \mathbb{R} : x_1(t) \leqslant \xi_1\} \cap \cdots \cap \{t \in \mathbb{R} : x_n(t) \leqslant \xi_n\}) \\ &= \lim_{T\to\infty} \frac{1}{T} \mu(\{t \in [-T/2, T/2] : x_1(t) \leqslant \xi_1, \ldots, x_n(t) \leqslant \xi_n\}) \\ &= \lim_{T\to\infty} \frac{1}{T} \int_{-T/2}^{T/2} \mathbf{1}_{\{x_1(t)\leqslant\xi_1\}} \cdots \mathbf{1}_{\{x_n(t)\leqslant\xi_n\}} \, \mathrm{d}t \end{aligned} \tag{6.8}$$

exists for all $(\xi_1, \ldots, \xi_n) \in \mathbb{R}^n - \Xi_0$, where Ξ_0 is at most a countable set of $(n-1)$-dimensional hyperplanes of $\mathbb{R}^n$.

The function $F_{x_1\cdots x_n}$ has all the properties of a nth-order joint distribution function, except the right continuity property with respect to each of the $\xi_1, \ldots, \xi_n$ variables in the discontinuity points.

Let $x_1(t+\tau_1), \ldots, x_{n-1}(t+\tau_{n-1}), x_n(t)$ be not necessarily bounded functions jointly RM for any $\tau_1, \tau_2, \ldots, \tau_{n-1}$ and let $g(\xi_1, \ldots, \xi_n)$ a bounded function satisfying appropriate regularity conditions. Then, for any $t_0 \in \mathbb{R}$ the following *fundamental theorem of expectation for the multivariate case* holds (Leśkow and Napolitano 2006, Theorem 4.5)

$$\lim_{T\to\infty} \frac{1}{T} \int_{t_0-T/2}^{t_0+T/2} g\left(x_1(t+\tau_1), \ldots, x_{n-1}(t+\tau_{n-1}), x_n(t)\right) \mathrm{d}t$$
$$= \int_{\mathbb{R}^n} g(\xi_1, \ldots, \xi_n) \, \mathrm{d}F_{x_1\cdots x_n}(\xi_1, \ldots, \xi_n; \tau_1, \ldots, \tau_{n-1}) \tag{6.9}$$

where the first integral is in the Lebesgue sense and the second is in the Riemann-Stieltjes sense and

$$F_{x_1\cdots x_n}(\xi_1, \ldots, \xi_{n-1}, \xi_n; \tau_1, \ldots, \tau_{n-1})$$
$$\triangleq \mu_R\left(\{t \in \mathbb{R} : x_1(t+\tau_1) \leqslant \xi_1, \ldots, x_{n-1}(t+\tau_{n-1}) \leqslant \xi_{n-1}, x_n(t) \leqslant \xi_n\}\right) . \tag{6.10}$$

Consequently, if $x_1(t+\tau_1), \ldots, x_{n-1}(t+\tau_{n-1}), x_n(t)$ are bounded jointly RM functions for any $\tau_1, \tau_2, \ldots, \tau_{n-1}$, then their temporal cross moment can be expressed as

$$\lim_{T\to\infty} \frac{1}{T} \int_{t_0-T/2}^{t_0+T/2} \prod_{i=1}^{n-1} x_i(t+\tau_i)\, x_n(t) \, \mathrm{d}t$$
$$= \int_{\mathbb{R}^n} \prod_{i=1}^{n} \xi_i \, \mathrm{d}F_{x_1\cdots x_n}(\xi_1, \ldots, \xi_{n-1}, \xi_n; \tau_1, \ldots, \tau_{n-1}) . \tag{6.11}$$

In the special case of $n = 2$, if $x(t)$ and $y(t)$ are bounded functions and and $y(t+\tau)$ and $x(t)$ are jointly RM for any τ, then the cross-correlation function of x and y is given by

$$R_{yx}(\tau) \triangleq \lim_{T\to\infty} \frac{1}{T} \int_{t_0-T/2}^{t_0+T/2} y(t+\tau)\, x(t) \, \mathrm{d}t \tag{6.12a}$$
$$= \int_{\mathbb{R}^2} \xi_1 \xi_2 \, \mathrm{d}F_{yx}(\xi_1, \xi_2; \tau) \tag{6.12b}$$

where

$$F_{yx}(\xi_1, \xi_2; \tau) \triangleq \mu_R\left(\{t \in \mathbb{R} : y(t+\tau) \leqslant \xi_1, x(t) \leqslant \xi_2\}\right) . \tag{6.13}$$

That is, the cross-correlation function (6.12a) is the expected value corresponding to the distribution (6.13).

As a consequence of the lack of σ-additivity of the relative measure μ_R, the corresponding expectation operator, the infinite-time average, is linear, but not σ-linear. The infinite-time average of the linear combination of a finite number of jointly RM functions (with at least one of them bounded) is equal to the linear combination of the time averages. This is not always true if we have a countable infinity of functions of time. For example, the periodic function $\cos^2(t)$ has infinite-time average equal to 1/2. However, $\cos^2(t) = \sum_{k=-\infty}^{+\infty} \cos^2(t)\mathbf{1}_{[k,k+1)}(t)$ and the time average of $\cos^2(t)\mathbf{1}_{[k,k+1)}(t)$ is zero. This result is different form the corresponding one in the stochastic approach where the expectation operator is σ-linear, provided that the underlying infinite series of random variables is absolutely convergent (Kolmogorov 1933).

Finally, note that the case of complex-valued signals can be treated with obvious changes by considering the joint characterization of real and imaginary parts (which leads to a doubling of the order of the distributions).

6.2.4 Conditional Relative Measurability and Independence

The definition of independence between two signals in the functional approach is given starting from the definition of conditional relative measurability. Then the result that the joint distribution function of two independent signals factorizes into the product of the two marginal distributions is obtained as a theorem and an intuitive concept of independence is shown to correspond to such a mathematical property.

Let A and B be Lebesgue measurable sets and $\{B_n\}$ be an arbitrary increasing sequence of Lebesgue measurable subsets of B with $0 < \mu(B_n) < \infty$, such that $m < n \Rightarrow B_m \subseteq B_n$; $\lim_n B_n \triangleq \cup_{n \in \mathbb{N}} B_n = B$; and $0 < \lim_n \mu(B_n)/n < \infty$. The *conditional relative measure* $\mu_R(\cdot|B)$ of the set A given B is defined as (Leśkow and Napolitano 2006)

$$\mu_R(A|B) \triangleq \lim_n \frac{\mu(A \cap B_n)}{\mu(B_n)} \tag{6.14}$$

provided that the limit exists. Note that in definition (6.14) the two sets A and B can be such that neither of the two is RM, nor $A \cap B$ is RM, but $\mu_R(A|B)$ exists (Leśkow and Napolitano 2006, Example 5.1).

Let the sets A and B be such that $\mu_R(A|B)$ exists and A is RM. The sets A and B are said to be *independent* if

$$\mu_R(A|B) = \mu_R(A)\,. \tag{6.15}$$

Let $x(t)$ be a RM function and $A \triangleq \{t \in \mathbb{R} \,:\, x(t) \leqslant \xi_1\}$. Let $y(t)$ be a Lebesgue measurable function and $B \triangleq \{t \in \mathbb{R} \,:\, y(t) \leqslant \xi_2\}$. Assume also that $\forall(\xi_1, \xi_2) \in \mathbb{R}^2 - \Xi_0$, where Ξ_0 is at most a countable set of lines, $\mu_R(A|B)$ exists. The *signals $x(t)$ and $y(t)$ are called independent* if (6.15) holds $\forall(\xi_1, \xi_2) \in \mathbb{R}^2 - \Xi_0$.

Assume that $x(t)$ and $y(t)$ are jointly RM. The signals $x(t)$ and $y(t)$ are independent, if and only if, $\forall(\xi_1, \xi_2) \in \mathbb{R}^2$ except at most a countable set of lines, it results that (Leśkow and Napolitano 2006, Theorem 5.2)

$$F_{xy}(\xi_1, \xi_2) = F_x(\xi_1)\, F_y(\xi_2) \tag{6.16}$$

where $F_{xy}(\xi_1, \xi_2)$ is the joint distribution function of $x(t)$ and $y(t)$ in the sense of definition (6.7) and $F_x(\xi_1)$ and $F_y(\xi_2)$ are the distribution functions of $x(t)$ and $y(t)$, respectively, in the sense of definition (6.3).

The definition of independence of two signals $x(t)$ and $y(t)$ based on (6.15) leads to the following intuitive interpretation of the concept of independence. If $x(t)$ and $y(t)$ are independent, then defined the sets $A \triangleq \{t \in \mathbb{R} \,:\, x(t) \leqslant \xi_1\}$ and $B \triangleq \{t \in \mathbb{R} \,:\, y(t) \leqslant \xi_2\}$, we have $\mu_R(A|B) = \mu_R(A)$. That is, the normalization in (6.14) of the measure of the set A, constructed from $x(t)$, made by subsets B_n of the set B, constructed from $y(t)$, gives rise to the same result obtained considering the normalization $B_n = [-n/2, n/2]$. In other words, the function $y(t)$ from which the normalizing sets B_n are constructed, has no influence on the relative measure $\mu_R(A|B)$. Therefore, according with the intuitive concept of independence, the two functions or signals $x(t)$ and $y(t)$ have no link each other. Note that such an intuitive interpretation of independence has no counterpart in the stochastic process approach where independence of processes is *defined* as the factorization of the joint distribution function into the product of the marginal ones (Kolmogorov 1933), (Doob 1953), (Billingsley 1968).

6.2.5 Examples

Signals of interest that can be characterized in the functional approach are almost-periodic functions and pseudorandom functions. The latter are appropriate models for realizations of several digital communications signals. They are shown to be RM and characterized in (Leśkow and Napolitano 2006).

In (Leśkow and Napolitano 2006), it is shown that if $x(t)$ is a uniformly almost-periodic function (Section 1.2.1) then the limit (6.3) does not necessarily exist at the discontinuity points of $F_x(\xi)$. The function of t, $\mathbf{1}_{\{x(t)\leqslant\xi\}}$ is discontinuous in t for any ξ such that the limit (6.3) exists. Moreover, it is shown that for any ξ such that the limit (6.3) exists, the function of t, $\mathbf{1}_{\{x(t)\leqslant\xi\}}$ is a W^p-AP function (Definition 1.2.4). Analogously, it can be shown that the function of t, $\mathbf{1}_{\{x(t)\leqslant\xi\}}$ is a B^p-AP function. In (Leśkow and Napolitano 2006) uniformly AP functions are proved to be RM and jointly RM. Moreover they are proved to be equal in distribution to the asymptotically AP functions which are obtained by adding $L^1_{\mathrm{loc}}(\mathbb{R})$ terms to the uniformly AP functions.

In (Kac 1959, p. 52), it is shown that the functions $x(t) = \cos(\lambda_1 t)$ and $y(t) = \cos(\lambda_2 t)$, with λ_1 and λ_2 incommensurate, are independent.

Examples of non-RM functions are provided in (Leśkow and Napolitano 2006) and (Leśkow and Napolitano 2007). These functions exhibit statistical functions defined in terms of time averages that are not convergent. Thus they can be suitably exploited to design secure communications systems where an unauthorized user cannot discover the modulation format by second- and higher-order (cyclic) spectral analysis of the transmitted signal.

6.3 Almost-Periodically Time-Variant Model

In the probabilistic framework built by distribution functions like that in (6.10) and (6.13), the infinite-time average plays the role of the expectation operator. Consequently, statistical functions defined in terms of this expectation operator do not depend on t (see e.g., the nth-order moment (6.11) and the cross-correlation function (6.12a)). That is, such a model corresponds to a stationary description of the signal.

6.3.1 Almost-Periodic Component Extraction Operator

A time-variant probabilistic model which is based on a single time series is proposed in (Gardner 1987d), (Gardner and Brown 1991), (Gardner 1994), and referred to as *fraction-of-time probability for time series that exhibit cyclostationarity*. It is developed in (Gardner and Spooner 1994), (Spooner and Gardner 1994), (Izzo and Napolitano 1998a, 2002a). Such an approach is based on the decomposition of time series, or functions of time series, into a (possibly zero) almost-periodic component and a residual term. Several kinds of decompositions are possible according to the results presented in Sections 1.2.3, 1.2.4, and 1.2.5. In this section, the decomposition of Hartman and Ryll-Nardzewski (Section 1.2.5) will be considered. In fact, this is the only decomposition which is compatible with a finite-power residual term.

Let $z(t) \in H^1_{\text{ap}} \cap \mathcal{M}^p$, $p > 1$, where H^1_{ap} is the set of AP functions in the sense of Hartman (Section 1.2.5) and $\mathcal{M}^p$ is the Marcinkiewicz space (Section 1.2.2). According to Theorem 1.2.17, the following decomposition holds

$$z(t) \triangleq z_{\text{Bap}}(t) + z_{\text{r}}(t) \tag{6.17}$$

where $z_{\text{Bap}}(t)$ is a B^p-AP function in the sense of Definition 1.2.4 and $z_{\text{r}}(t)$ is a residual term not containing any finite-strength additive sinewave component, that is,

$$\lim_{T\to\infty} \frac{1}{T} \int_{-T/2}^{T/2} z_{\text{r}}(t)\, e^{-j2\pi\alpha t}\, \mathrm{d}t = 0 \qquad \forall \alpha \in \mathbb{R}\,. \tag{6.18}$$

Note that $z_{\text{r}}(t)$ can be a finite-power signal.

According to Theorem 1.2.5, $z_{\text{Bap}}(t)$ is the limit in B^p of a sequence of trigonometric polynomials in t. That is, denoting by A_n an increasing sequence of countable sets such that $m < n \Rightarrow A_m \subseteq A_n$ and $\lim_n A_n \triangleq \cup_{n\in\mathbb{N}} A_n = A$, it results that

$$\lim_n \left[\limsup_{T\to\infty} \frac{1}{T} \int_{-T/2}^{T/2} \left| z_{\text{Bap}}(t) - \sum_{\alpha\in A_n} z_\alpha\, e^{j2\pi\alpha t} \right|^p \mathrm{d}t \right]^{1/p} = 0 \tag{6.19}$$

shortly

$$z_{\text{Bap}}(t) = \sum_{\alpha\in A} z_\alpha\, e^{j2\pi\alpha t} \quad \text{in } B^p \tag{6.20}$$

where z_α are the complex-valued Fourier coefficients

$$z_\alpha \triangleq \lim_{T\to\infty} \frac{1}{T} \int_{-T/2}^{T/2} z(t)\, e^{-j2\pi\alpha t}\, \mathrm{d}t \tag{6.21}$$

and the real-valued frequencies α range in the countable set A which contains possibly incommensurate elements. In particular, $z_{\text{Bap}}(t)$ can be AP in the sense of Bohr (Bohr 1933, paragraphs 84–92) or, equivalently, uniformly almost periodic in the sense of Besicovitch (Besicovitch 1932, Chapter 1), (Corduneanu 1989, Section 1.2).

With reference to decomposition (6.17), the *almost-periodic component extraction operator* is defined as the operator that extracts the almost-periodic component of its argument (Gardner and Brown 1991):

$$\mathrm{E}^{\{\alpha\}}\left\{z(t)\right\} \triangleq z_{\text{Bap}}(t)\,. \tag{6.22}$$

If the set A needs to be evidenced, then the symbol $\mathrm{E}^{\{A\}}\{\cdot\}$ instead of $\mathrm{E}^{\{\alpha\}}\{\cdot\}$ is adopted. Moreover, if the variable (e.g., t) with respect to which the almost-periodic component is taken needs to be evidenced, then the notations $\mathrm{E}_t^{\{A\}}\{\cdot\}$ or $\mathrm{E}_t^{\{\alpha\}}\{\cdot\}$ are used.

6.3.2 Second-Order Statistical Characterization

In this section, the second-order characterization of time series that exhibit (generalized) almost-cyclostationarity is discussed in detail. Similar arguments hold for first- and higher-order characterizations.

Let $x(t)$ be a real-valued continuous-time time series. The function $z(t) \triangleq \mathbf{1}_{\{t\,:\,x(t+\tau)\leqslant\xi_1\}}\,\mathbf{1}_{\{t\,:\,x(t)\leqslant\xi_2\}}$ is bounded by the constant 1 which is a finite-power signal. Thus, the almost-periodic component of $z(t)$, if it exists, for every $(\xi_1, \xi_2, \tau) \in \mathbb{R}^3$ has a set of frequencies $\Gamma_{\xi_1,\xi_2,\tau}$ which is at most countable. If also the set $\Gamma_\tau \triangleq \cup_{(\xi_1,\xi_2)\in\mathbb{R}^2}\Gamma_{\xi_1,\xi_2,\tau}$ is countable for every $\tau \in \mathbb{R}$, then the time series is said to be *second-order generalized almost-cyclostationary in the strict sense* (Izzo and Napolitano 1998a), (Izzo and Napolitano 2005). In the special case where also $\Gamma = \cup_{\tau\in\mathbb{R}}\Gamma_\tau$ is countable, then the time series is said to be *second-order almost-cyclostationary in the strict sense*. Since $z(t)$ is a discontinuous function of t, accordingly with the results of Sections 6.2.5 and 6.3.1, almost-periodicity is considered in the B^p sense. Thus, for fixed $\tau \in \mathbb{R}$, the AP component of $z(t)$ can be expressed by the (generalized) Fourier series

$$\mathrm{E}^{\{\alpha\}}\left\{\mathbf{1}_{\{t\,:\,x(t+\tau)\leqslant\xi_1\}}\mathbf{1}_{\{t\,:\,x(t)\leqslant\xi_2\}}\right\} = \sum_{\gamma\in\Gamma_\tau} F_x^\gamma(\xi_1, \xi_2; \tau)\, e^{j2\pi\gamma t} \tag{6.23}$$

where equality should be intended in the sense of (6.19), (6.20).

In (Gardner and Brown 1991), in the special case of ACS signals, it is shown that the function of ξ_1 and ξ_2

$$F_x(\xi_1, \xi_2; t, \tau) \triangleq \mathrm{E}^{\{\alpha\}}\left\{\mathbf{1}_{\{t\,:\,x(t+\tau)\leqslant\xi_1\}}\,\mathbf{1}_{\{t\,:\,x(t)\leqslant\xi_2\}}\right\} \tag{6.24}$$

is a valid second-order joint cumulative distribution function for every fixed t and τ, except for the right-continuity property (in the discontinuity points) with respect to ξ_1 and ξ_2. With similar arguments the same result can be obtained for GACS time series.

By definition, the function $F_x(\xi_1, \xi_2; t, \tau)$ is almost periodic in t. Therefore, it provides a generalized almost-cyclostationary (at second-order and in the strict-sense) probabilistic model for the time series $x(t)$ and the almost-periodic component extraction operator $\mathrm{E}^{\{\alpha\}}\{\cdot\}$ is the expectation operator. Moreover, the function

$$\mathcal{R}_x(t, \tau) \triangleq \mathrm{E}^{\{\alpha\}}\{x(t+\tau)\,x(t)\} \tag{6.25}$$

is a valid autocorrelation function and can be characterized by (Gardner and Brown 1991), (Izzo and Napolitano 1998a)

$$\begin{aligned}\mathcal{R}_x(t, \tau) &= \int_{\mathbb{R}^2} \xi_1\,\xi_2\,\mathrm{d}F_x(\xi_1, \xi_2; t, \tau)\\ &= \sum_{\alpha\in A_\tau} R_x(\alpha, \tau)\, e^{j2\pi\alpha t}\end{aligned} \tag{6.26}$$

where, for each fixed τ, $A_\tau \subseteq \Gamma_\tau$ is a countable set and

$$R_x(\alpha, \tau) \triangleq \lim_{T\to\infty} \frac{1}{T}\int_{-T/2}^{T/2} x(t+\tau)\,x(t)\,e^{-j2\pi\alpha t}\,\mathrm{d}t \tag{6.27a}$$

$$= \int_{\mathbb{R}^2} \xi_1\,\xi_2\,\mathrm{d}F_x^\alpha(\xi_1, \xi_2; \tau) \tag{6.27b}$$

is the (nonstochastic) *cyclic autocorrelation function* at cycle frequency α. If $A \triangleq \cup_{\tau\in\mathbb{R}}A_\tau$ is also countable, then $A \subseteq \Gamma$ and the time series is said to be ACS.

Under mild regularity conditions on the function $g(\xi_1, \xi_2)$, the following *Gardner Theorem of Expectation* holds (Gardner and Brown 1991)

$$\int_{\mathbb{R}^2} g(\xi_1, \xi_2)\, \mathrm{d}F_x(\xi_1, \xi_2; t, \tau) = \mathrm{E}^{\{\alpha\}}\left\{g(x(t+\tau), x(t))\right\} . \tag{6.28}$$

In the fraction-of-time probability for time series that exhibit cyclostationarity, the almost-periodic functions play the role of the deterministic signals. In fact, if $x(t)$ is a B^p-AP function, it is coincident with its almost-periodic component which is extracted by the almost-periodic component extraction operator:

$$\mathrm{E}^{\{\alpha\}}\left\{x(t)\right\} = x(t) . \tag{6.29}$$

In addition, for a uniformly AP function $x(t)$ the indicator of the set $\{t : x(t) \leqslant \xi\}$ is B^p-AP in t so that

$$\mathrm{E}^{\{\alpha\}}\left\{\mathbf{1}_{\{t : x(t) \leqslant \xi\}}\right\} = \mathbf{1}_{\{t : x(t) \leqslant \xi\}} \tag{6.30}$$

where equality in (6.30) should be intended in the sense of (6.19), (6.20). Signals which are not almost periodic functions are the random signals. Relations (6.29) and (6.30) are the nonstochastic counterparts of analogous relations for deterministic processes in the stochastic approach. For further results on higher-order statistics, see (Izzo and Napolitano 1998a), (Izzo and Napolitano 2002a), (Izzo and Napolitano 2005).

In (Leśkow and Napolitano 2006, Example 6.1) it is shown that even if $x(t)$ is RM, $x(t)\cos(2\pi\alpha t + \phi)$ could be not RM for some α and ϕ. For these values of α and ϕ, $x(t)$ and $\cos(2\pi\alpha t + \phi)$ are not jointly RM and the cyclic autocorrelation (6.27a) does not exist.

According to (Leśkow and Napolitano 2006, Corollary 4.2), a sufficient condition for the existence of the cyclic autocorrelation (6.27a) is that the three signals $x(t)$, $x(t+\tau)$, and $\cos(2\pi\alpha t + \phi)$ are jointly RM for every τ, α, and ϕ, that is

$$\begin{aligned} &\mu_R\Big(\{t \in \mathbb{R} \,:\, x(t) \leqslant \xi_1,\ x(t+\tau) \leqslant \xi_2,\ \cos(2\pi\alpha t + \phi) \leqslant \xi_3\}\Big) \quad \text{exists} \\ &\forall\, (\xi_1, \xi_2, \xi_3) \in \mathbb{R}^3 - \Xi_0, \quad \forall\, (\tau, \alpha, \phi) \in \mathbb{R}^2 \times [0, 2\pi) \end{aligned} \tag{6.31}$$

where Ξ_0 is an at most countable set of planes in $\mathbb{R}^3$. An alternative sufficient condition for the existence of limit in (6.27a) is that the lag-product $x(t)x(t+\tau)$ and $\cos(2\pi\alpha t + \phi)$ are jointly RM for every τ, α, and ϕ, that is

$$\begin{aligned} &\mu_R\Big(\{t \in \mathbb{R} \,:\, x(t)\,x(t+\tau) \leqslant \xi_1,\ \cos(2\pi\alpha t + \phi) \leqslant \xi_2\}\Big) \quad \text{exists} \\ &\forall\, (\xi_1, \xi_2) \in \mathbb{R}^2 - \Xi_0, \quad \forall\, (\tau, \alpha, \phi) \in \mathbb{R}^2 \times [0, 2\pi) \end{aligned} \tag{6.32}$$

where Ξ_0 is an at most countable set of lines in $\mathbb{R}^2$.

By solving inequality $\cos(2\pi\alpha t + \phi) \leqslant \xi_3$ with respect to t it can be shown that condition (6.31) is equivalent to

$$\begin{aligned} &\mu_R\Big(\{t \in \mathbb{R} \,:\, x(t) \leqslant \xi_1,\ x(t+\tau) \leqslant \xi_2 \mid S_\alpha\}\Big) \quad \text{exists} \\ &\forall\, (\xi_1, \xi_2) \in \mathbb{R}^2 - \Xi_0, \quad \forall\, (\tau, \alpha) \in \mathbb{R}^2, \alpha \neq 0 \end{aligned} \tag{6.33}$$

where $\mu_R(\cdot|S_\alpha)$ is the conditional relative measure (defined according to (6.14)) conditioned to the set S_α, S_α is any set of the form

$$S_\alpha = \bigcup_{k\in\mathbb{Z}} \left[a + \frac{k}{\alpha}, b + \frac{k}{\alpha}\right) \tag{6.34}$$

and Ξ_0 is an at most countable set of lines in $\mathbb{R}^2$.

6.3.3 Spectral Line Regeneration

The time series decomposition (6.17) explains a phenomenon called spectral line regeneration, that is, the origin of finite-strength additive sinewave components present in the second-order lag-product of a GACS signal but not present in the signal.

Let us consider decomposition (6.17) for the signal $x(t)$ in the special case where the AP component is uniformly AP in the sense of Besicovitch (Definition 1.2.1)

$$x(t) = x_{\text{uap}}(t) + x_r(t) \tag{6.35}$$

where the residual term $x_r(t)$ does not contain any finite-strength additive sinewave component and has possibly finite power. The second-order lag-product waveform is expressed as

$$\begin{aligned} x(t+\tau)\,x(t) &= [x_{\text{uap}}(t+\tau) + x_r(t+\tau)][x_{\text{uap}}(t) + x_r(t)] \\ &= x_{\text{uap}}(t+\tau)\,x_{\text{uap}}(t) + x_r(t+\tau)\,x_{\text{uap}}(t) \\ &\quad + x_{\text{uap}}(t+\tau)\,x_r(t) + x_r(t+\tau)\,x_r(t) \end{aligned} \tag{6.36}$$

where

a) $x_{\text{uap}}(t+\tau)\,x_{\text{uap}}(t)$ is a uniformly AP function since the product of uniformly AP functions is a uniformly AP function (Besicovitch 1932);
b) $x_r(t+\tau)\,x_{\text{uap}}(t)$ and $x_{\text{uap}}(t+\tau)\,x_r(t)$ do not contain any additive finite-strength sinewave component;
c) $x_r(t+\tau)\,x_r(t)$ can contain additive finite-strength sinewave components.

The (possible) almost-periodic component contained in $x_r(t+\tau)\,x_r(t)$ is the second-order pure almost-periodic component contained in $x(t)$ (Gardner and Spooner 1994). That is, it represents the finite-strength additive sinewave components present in the second-order lag-product $x(t+\tau)\,x(t)$ that are not generated by beats of (first-order) finite-strength sinewaves present in the time series (that is, term $x_{\text{uap}}(t+\tau)\,x_{\text{uap}}(t)$). In particular, if $x(t)$ does not contain any additive finite-strength sinewave component, that is, $x_{\text{uap}}(t) \equiv 0$ in (6.35), then the possible AP component in the lag product waveform $x(t+\tau)\,x(t)$ is only due to the product of the residual term $x_r(t)$ and its time-shifted version $x_r(t+\tau)$. That is, periodicities that are hidden at first order are regenerated in the second-order lag-product waveform. If the second-order lag-product waveform $x(t+\tau)\,x(t)$ is uniformly continuous and bounded, possibly possessing an uniformly AP component, according to Theorem 1.2.20, this uniformly AP component is given by the product of $x(t+\tau) \equiv x_r(t+\tau) \in R_0$ and $x(t) \equiv x_r(t) \in R_0$.

Accordingly with (6.25), (6.26), and (6.36), the second-order lag-product waveform $x(t + \tau)x(t)$ can be decomposed into the sum of its almost-periodic component and a residual term

$$\begin{aligned} x(t+\tau)\,x(t) &= \mathrm{E}^{\{\alpha\}}\left\{x(t+\tau)\,x(t)\right\} + \ell_x(t,\tau) \\ &= \sum_{\alpha\in A_\tau} R_x(\alpha,\tau)\,e^{j2\pi\alpha t} + \ell_x(t,\tau) \end{aligned} \tag{6.37}$$

where $\ell_x(t, \tau)$ does not contain any finite-strength additive sinewave component (see (6.18)).

In the special case of ACS time series ($A \triangleq \cup_{\tau\in\mathbb{R}} A_\tau$ countable), any homogeneous quadratic time-invariant (QTI) transformation of the lag-product waveform has the form

$$\begin{aligned} y(t) &\triangleq \int_{\mathbb{R}} k(\tau)\,x(t+\tau)\,x(t)\,\mathrm{d}\tau \\ &= \sum_{\alpha\in A} \int_{\mathbb{R}} k(\tau)\,R_x(\alpha,\tau)\,\mathrm{d}\tau\,e^{j2\pi\alpha t} + \int_{\mathbb{R}} k(\tau)\,\ell_x(t,\tau)\,\mathrm{d}\tau \end{aligned} \tag{6.38}$$

where $k(\tau)$ is the kernel of the QTI transformation. That is, finite-strength additive sinewave components can be regenerated by homogeneous QTI transformations (Gardner 1987d), (Gardner and Brown 1991), (Gardner 1994). In contrast, for stationary time series (A containing the only element $\alpha = 0$), no spectral line at nonzero frequency can be regenerated in the lag-product or by QTI transformations.

More generally, a time series $x(t)$ is said to exhibit *higher-order cyclostationarity* if finite-strength additive sinewave components can be regenerated by homogeneous nonlinear time-invariant transformations of $x(t)$ of order greater than two (Gardner and Spooner 1994), (Spooner and Gardner 1994). In such a case, almost-periodically time-variant higher-order moment and cumulant functions can be defined by the almost-periodic component extraction operator (Gardner and Spooner 1994), (Spooner and Gardner 1994), (Izzo and Napolitano 1998a). For communications ACS signals, cycle frequencies of second- and higher-order statistical functions are related to parameters such as sinewave carrier frequency, pulse rate, symbol rate, frame rate, sampling frequency. Therefore, spectral line regeneration by second- or higher-order time-invariant transformations leads to signals suitable for synchronization purposes (Gardner 1987d).

6.3.4 Spectral Correlation

Let us define the (nonstochastic) *cyclic spectrum* at cycle frequency α

$$S_x^\alpha(f) \triangleq \lim_{\Delta f\to 0}\lim_{T\to\infty} \frac{1}{T}\int_{-T/2}^{T/2} \Delta f\; X_{1/\Delta f}(t,f)\,X^*_{1/\Delta f}(t,f-\alpha)\,\mathrm{d}t \tag{6.39}$$

where the order of the two limits cannot be reversed and $X_{1/\Delta f}(t, f)$ is the short-time Fourier transform (STFT) defined according to

$$X_Z(t,f) \triangleq \int_{t-Z/2}^{t+Z/2} x(s)\,e^{-j2\pi f s}\,\mathrm{d}s\,. \tag{6.40}$$

The cyclic spectrum $S_x^{\alpha}(f)$ is also called the *spectral correlation density function*. It represents the time-averaged correlation (with zero lag) of two spectral components at frequencies f and $f-\alpha$, as the bandwidth approaches zero.

In (Gardner 1985, 1987d), it is shown that for ACS time series, $\forall\alpha \in A$,

$$S_x^{\alpha}(f) = \int_{\mathbb{R}} R_x(\alpha, \tau)\, e^{-j2\pi f\tau}\, \mathrm{d}\tau \tag{6.41}$$

which is referred to as the *Gardner Relation* (also called the *Cyclic Wiener-Khinchin Relation*). Therefore, for an ACS time series, correlation exists between spectral components that are separated by amounts equal to the cycle frequencies. That is, for ACS time series, the spectral correlation property ($S_x^{\alpha}(f) \not\equiv 0$ for $\alpha \neq 0$) is equivalent to the property of regeneration of spectral lines with nonzero frequency by homogeneous QTI transformations. In contrast, for WSS time series, the set A contains the only element $\alpha = 0$, no correlation exists between distinct spectral components, an no spectral lines (with nonzero frequency) can be regenerated by homogeneous QTI transformations.

For $\alpha = 0$, the cyclic spectrum reduces to the power spectrum or spectral density function $S_x^0(f)$ and (6.41) reduces to the *Wiener-Khinchin Relation*.

By reasoning as in the stochastic process framework (Sections 2.2.2 and 2.2.3), it can be shown that the cyclic autocorrelation function of GACS time series is a discontinuous function of the lag parameter τ and a spectral characterization by the Fourier transform of the cyclic autocorrelation function cannot be made in terms of ordinary functions (Izzo and Napolitano 1998a), (Izzo and Napolitano 2002a), (Izzo and Napolitano 2005). In particular, GACS time series that are not ACS do not exhibit the spectral correlation property.

6.3.5 *Statistical Function Estimators*

Estimators of the FOT probabilistic parameters of a time series $\{x(u), u \in \mathbb{R}\}$ based on the observation $\{x(u), u \in [t - T/2, t + T/2]\}$ are obtained by considering finite-time averages over $[t - T/2, t + T/2]$ of the same quantities involved in the infinite-time averages. The performance of the estimator is expressed in terms of FOT bias and variance which are defined by using as expected value the almost-periodic component extraction operator acting on the time variable t viz., the central point of the finite-length time-series segment adopted for the estimation (Gardner 1991c), (Gardner and Brown 1991), (Gardner *et al.* 2006). Thus, unlike the stochastic process framework where the variance accounts for fluctuations of the estimates over the ensemble of sample paths, in the FOT probability framework the variance accounts for the fluctuations of the estimates in the time parameter t, that is, the location on the real axis of the time segment adopted for the estimation. Mean-square consistency of the estimator means that both FOT bias and variance of the estimator approach zero as the length T of the observation interval approaches infinity. In such a case, estimates obtained by using different time segments asymptotically do not depend on the central point of the segment.

By following the guidelines in (Gardner 1987d), (Brown 1987), let us consider the convergence of time series in the temporal mean-square sense (t.m.s.s.). Given a time series $z(t)$ (such as a lag product of another time series with a fixed value of the lag parameter), we define

$$z_{\beta}(t)_T \triangleq \frac{1}{T}\int_{t-T/2}^{t+T/2} z(u)\, e^{-j2\pi\beta u}\, \mathrm{d}u \tag{6.42}$$

to be an estimator of the parameter

$$z_\beta \triangleq \lim_{T\to\infty} \frac{1}{T} \int_{-T/2}^{T/2} z(u)\, e^{-j2\pi\beta u}\, du \tag{6.43}$$

and we assume that

$$\lim_{T\to\infty} z_\beta(t)_T = z_\beta \qquad \text{(t.m.s.s.)}\,, \qquad \forall\beta \in \mathbb{R}\,; \tag{6.44}$$

that is,

$$\lim_{T\to\infty} \left\langle \left| z_\beta(t)_T - z_\beta \right|^2 \right\rangle_t = 0\,, \quad \forall\beta \in \mathbb{R}\,. \tag{6.45}$$

This is a more stringent condition w.r.t. the convergence in B^2 (Section 1.2.2). It can be shown that, if the time series $z(t)$ has finite-average-power (i.e., $\langle |z(t)|^2 \rangle_t < \infty$), then the set $B \triangleq \{\beta \in \mathbb{R} : z_\beta \neq 0\}$ is countable, the series $\sum_{\beta\in B} |z_\beta|^2$ is summable (Brown 1987) and, accounting for (6.44), it follows that

$$\lim_{T\to\infty} \sum_{\beta\in B} z_\beta(t)_T\, e^{j2\pi\beta t} = \sum_{\beta\in B} z_\beta\, e^{j2\pi\beta t} \quad \text{(t.m.s.s.)}\,. \tag{6.46}$$

The magnitude and phase of z_β are the amplitude and phase of the finite-strength additive complex sinewave with frequency β contained in the time series $z(t)$. Moreover, the right-hand side in (6.46) is just the almost-periodic component contained in the time series $z(t)$.

The function $z_\beta(t)_T$ is an estimator of z_β based on the observation $\{z(u),\ u \in [t - T/2, t + T/2]\}$. It is worthwhile emphasizing that, in the FOT probability framework, probabilistic functions are defined in terms of the almost-periodic component extraction operation, which plays the same role as that played by the statistical expectation operation in the stochastic process framework (Section 6.3.2) (Gardner 1987d, 1994). Therefore,

$$\begin{aligned} \text{bias}\left\{z_\beta(t)_T\right\} &\triangleq \mathrm{E}^{\{\alpha\}}\left\{z_\beta(t)_T\right\} - z_\beta \\ &\simeq \left\langle z_\beta(t)_T \right\rangle_t - z_\beta \end{aligned} \tag{6.47}$$

$$\begin{aligned} \text{var}\left\{z_\beta(t)_T\right\} &\triangleq \mathrm{E}^{\{\alpha\}}\left\{\left| z_\beta(t)_T - \mathrm{E}^{\{\alpha\}}\left\{z_\beta(t)_T\right\}\right|^2\right\} \\ &\simeq \left\langle \left| z_\beta(t)_T - \left\langle z_\beta(t)_T \right\rangle_t \right|^2 \right\rangle_t \end{aligned} \tag{6.48}$$

where $\mathrm{E}^{\{\alpha\}}\{\cdot\}$ extracts the almost-periodic component of its argument. Approximations in (6.47) and (6.48) become exact equalities in the limit as $T \to \infty$. Thus, unlike the stochastic process framework where the variance accounts for fluctuations of the estimates over the ensemble of sample paths, in the FOT probability framework the variance accounts for the fluctuations of the estimates in the time parameter t, viz., the central point of the finite-length time series segment adopted for the estimation. Therefore, the assumption that the estimator asymptotically approaches the true value (the infinite-time average) in the t.m.s.s. leads to the result that the estimator is mean-square consistent in the FOT probability sense. In fact, from (6.45), (6.47), and (6.48), it follows that

$$\left\langle \left| z_\beta(t)_T - z_\beta \right|^2 \right\rangle_t \simeq \mathrm{var}\left\{ z_\beta(t)_T \right\} + \left| \mathrm{bias}\left\{ z_\beta(t)_T \right\} \right|^2 \tag{6.49}$$

and this approximation becomes exact as $T \to \infty$. In such a case, estimates obtained by using different time segments asymptotically do not depend on the central point of the segment.

6.3.6 *Sampling, Aliasing, and Cyclic Leakage*

In this section, the problems of aliasing and cyclic leakage in the discrete-time estimator of the (conjugate) cyclic autocorrelation function of a continuous-time GACS signal are addressed.

Let $x(n)$ be the discrete-time signal obtained by uniformly sampling with period $T_s = 1/f_s$ the complex-valued continuous-time GACS signal $x_a(t)$. That is,

$$x(n) = x_a(t)|_{t=nT_s} \quad n \in \mathbb{Z}\,. \tag{6.50}$$

According to the results of Section 6.3.3, the second-order lag product of the continuous-time signal $x_a(t)$ can be decomposed into its almost-periodic component and a residual term not containing any finite-strength additive sinewave component

$$x_a(t+\tau)\, x_a^{(*)}(t) = \sum_{\beta \in A_\tau} R_{\boldsymbol{x}_a}(\beta, \tau)\, e^{j2\pi\beta t} + \ell_{\boldsymbol{x}_a}(t, \tau) \tag{6.51}$$

where $\boldsymbol{x}_a = [x_a\, x_a^{(*)}]$, $A_\tau = A_{\boldsymbol{x}_a,\tau}$ is the set of second-order (conjugate) cycle frequencies of $x_a(t)$ at lag τ, and

$$\left\langle \ell_{\boldsymbol{x}_a}(t, \tau)\, e^{-j2\pi\beta t} \right\rangle_t = 0 \qquad \forall \beta \in \mathbb{R}\,. \tag{6.52}$$

It results that

$$\begin{aligned} x(n+m)\, x^{(*)}(n) &\triangleq x_a(t+\tau)\, x_a^{(*)}(t)|_{t=nT_s, \tau=mT_s} \\ &= \sum_{\beta \in A_{mT_s}} R_{\boldsymbol{x}_a}(\beta, mT_s)\, e^{j2\pi\beta n T_s} + \ell_{\boldsymbol{x}_a}(nT_s, mT_s) \end{aligned} \tag{6.53}$$

where the discrete-time signal $\ell_{\boldsymbol{x}_a}(nT_s, mT_s)$ does not contain any discrete-time finite-strength additive sinewave component (see (6.91) and (Izzo and Napolitano 2003), (Izzo and Napolitano 2005)).

The discrete-time estimate of the (conjugate) cyclic autocorrelation function is

$$\begin{aligned} \widetilde{\widetilde{R}}{}_{\boldsymbol{x}}^{\widetilde{\alpha}}(m)_N &\triangleq \frac{1}{N} \sum_{n=0}^{N-1} x(n+m)\, x^{(*)}(n)\, e^{-j2\pi\widetilde{\alpha}n} \\ &= \frac{1}{N} \sum_{n=0}^{N-1} \left[\sum_{\beta \in A_{mT_s}} R_{\boldsymbol{x}_a}(\beta, mT_s)\, e^{j2\pi\beta n T_s} + \ell_{\boldsymbol{x}_a}(nT_s, mT_s) \right] e^{-j2\pi\widetilde{\alpha}n} \\ &= \sum_{\beta \in A_{mT_s}} R_{\boldsymbol{x}_a}(\beta, mT_s) \left[\frac{1}{N} \sum_{n=0}^{N-1} e^{j2\pi(\beta T_s - \widetilde{\alpha})n} \right] \end{aligned}$$

$$
\begin{aligned}
&+\frac{1}{N}\sum_{n=0}^{N-1} \ell_{x_a}(nT_s, mT_s)\, e^{-j2\pi\widetilde{\alpha}n} \\
&= \sum_{\beta\in A_{mT_s}} R_{x_a}(\beta, mT_s)\, D_{\frac{1}{N}}(\beta T_s - \widetilde{\alpha}) \\
&+\frac{1}{N}\sum_{n=0}^{N-1} \ell_{x_a}(nT_s, mT_s)\, e^{-j2\pi\widetilde{\alpha}n}
\end{aligned}
\tag{6.54}
$$

where

$$
\begin{aligned}
D_{\frac{1}{N}}(\widetilde{\gamma}) &\triangleq \frac{1}{N}\sum_{n=0}^{N-1} e^{j2\pi\widetilde{\gamma}n} = \frac{1}{N}\frac{e^{j2\pi\widetilde{\gamma}N}-1}{e^{j2\pi\widetilde{\gamma}}-1} \\
&= \frac{1}{N}\, e^{j\pi\widetilde{\gamma}(N-1)}\, \frac{\sin[\pi\widetilde{\gamma}N]}{\sin[\pi\widetilde{\gamma}]}
\end{aligned}
\tag{6.55}
$$

is $(1/N)$ times the Dirichlet kernel.

In the first sum in the rhs of (6.54), terms with $\beta T_s \bmod 1 \neq \widetilde{\alpha}$ describe the cyclic leakage of cycle frequencies different from $\alpha = \widetilde{\alpha} f_s \bmod f_s$ on the estimate. Their effect becomes negligible as the number of samples N approaches infinite. Terms with $\beta T_s \bmod 1 = \widetilde{\alpha}$ and $\beta \neq \widetilde{\alpha} f_s$ describe the effect of aliasing due to uniform sampling. Their effect becomes negligible when the sampling period T_s approaches zero. The second sum in the rhs of (6.54) describes the randomness (in the FOT sense) of the measure. It becomes negligible as N approaches infinite.

By using the limit

$$
\lim_{N\to\infty} D_{\frac{1}{N}}(\widetilde{\gamma}) = \delta_{\widetilde{\gamma} \bmod 1} = \begin{cases} 1 & \widetilde{\gamma}\in\mathbb{Z} \\ 0 & \text{otherwise} \end{cases}
\tag{6.56}
$$

where δ_γ is the Kronecker delta (that is, $\delta_\gamma = 1$ if $\gamma = 0$ and $\delta_\gamma = 0$ if $\gamma \neq 0$), and accounting for the fact that (see (6.91) and (Izzo and Napolitano 2003), (Izzo and Napolitano 2005))

$$
\lim_{N\to\infty}\frac{1}{N}\sum_{n=0}^{N-1} \ell_{x_a}(nT_s, mT_s)\, e^{-j2\pi\widetilde{\alpha}n} = 0 \qquad \forall\widetilde{\alpha}\in\mathbb{R}
\tag{6.57}
$$

we get the FOT counterpart of (1.174) (extended to GACS time series), that is

$$
\begin{aligned}
\widetilde{\widetilde{R}}{}_{x}^{\alpha}(m) &= \lim_{N\to\infty} \widetilde{\widetilde{R}}{}_{x}^{\alpha}(m)_N \\
&= \sum_{\beta\in A_{mT_s}} R_{x_a}(\beta, mT_s)\, \delta_{(\beta T_s - \widetilde{\alpha}) \bmod 1} \\
&= \sum_{\beta\in A_{mT_s}} R_{x_a}(\beta, mT_s) \sum_{p\in\mathbb{Z}} \delta_{\beta T_s - \widetilde{\alpha} + p} \\
&= \sum_{p\in\mathbb{Z}} R_{x_a}((\widetilde{\alpha} - p) f_s, mT_s)
\end{aligned}
\tag{6.58}
$$

which is formally identical to (1.174). In the sum in (6.58) give nonzero contribution all values of p such that $(\widetilde{\alpha} - p) f_s \in A_{mT_s}$.

6.3.7 FOT-Deterministic Systems

In Section 6.3.1, it is shown that almost-periodic functions are the deterministic signals in the functional approach, when an almost-periodically time-variant model is adopted, that is, when the expectation operator is coincident with the almost-periodic component extraction operator. Therefore, within this signal model, a *FOT-deterministic system* is defined as a possibly complex (and not necessarily linear) system that for every deterministic (i.e., almost periodic) input signal delivers a deterministic output signal (Izzo and Napolitano 2002a), (Izzo and Napolitano 2005). Systems that are not FOT deterministic are referred to as *FOT-random systems* in (Izzo and Napolitano 2002a), (Izzo and Napolitano 2005). Therefore, for the input time series

$$x(t) = e^{j2\pi\lambda t} \tag{6.59}$$

a FOT-deterministic system delivers the output almost-periodic time series

$$y_\lambda(t) = \sum_{\eta \in E_\lambda} G'_\eta \, e^{j2\pi\eta t} \tag{6.60}$$

where, for each fixed λ, $E_\lambda \triangleq \{\eta_1(\lambda), \ldots, \eta_n(\lambda), \ldots\}$ is the countable set of the output frequencies (depending on the input frequency λ) and the G'_η are complex coefficients.

The set of points $(\eta, \lambda) \in E_\lambda \times \mathbb{R}$ such that $G'_\eta \neq 0$ is constituted by a set of not necessarily continuous curves defined by the implicit equations

$$\Phi_{\sigma'}(\eta, \lambda) = 0, \qquad \sigma' \in \Omega' \tag{6.61}$$

where Ω' is a countable set. Thus, it results that

$$\begin{aligned} &\operatorname{cl}\left\{(\eta, \lambda) \in E_\lambda \times \mathbb{R} \,:\, G'_\eta \neq 0\right\} \\ &\quad = \operatorname{cl} \bigcup_{\sigma' \in \Omega'} \left\{(\eta, \lambda) \in \mathbb{R} \times \mathbb{R} \,:\, \Phi_{\sigma'}(\eta, \lambda) = 0,\; G'_\eta \neq 0\right\} \\ &\quad = \operatorname{cl} \bigcup_{\sigma \in \Omega} \left\{(\eta, \lambda) \in \mathbb{R} \times D_{\varphi_\sigma} \,:\, \eta = \varphi_\sigma(\lambda)\right\} \end{aligned} \tag{6.62}$$

where cl denotes closure, Ω is a countable set and, in the last equality, it is assumed that each curve described by the implicit equation $\Phi_{\sigma'}(\eta, \lambda) = 0$ can be decomposed into a countable set of curves, each described by the explicit equation $\eta = \varphi_\sigma(\lambda)$, where $\varphi_\sigma(\cdot)$ can always be chosen such that $\varphi_{\sigma_1}(\lambda) \neq \varphi_{\sigma_2}(\lambda)$ for $\sigma_1 \neq \sigma_2$, and D_{φ_σ} is the domain of $\varphi_\sigma(\cdot)$. Therefore, defined the functions

$$G_\sigma(\lambda) \triangleq \begin{cases} G'_\eta \Big|_{\eta = \varphi_\sigma(\lambda)}, & \lambda \in D_{\varphi_\sigma} \\ 0, & \text{elsewhere} \end{cases} \tag{6.63}$$

the output almost-periodic time series (6.60) can be written as

$$y_\lambda(t) = \sum_{\sigma \in \Omega} G_\sigma(\lambda) e^{j2\pi\varphi_\sigma(\lambda)t} . \tag{6.64}$$

A mild regularity assumption on the system is that a small change in the input frequency λ gives rise to small changes in the output frequencies $\eta_n(\lambda) \in E_\lambda$. Thus, the $\eta_n(\lambda)$ are continuous functions (not necessarily invertible) of the input frequency λ, that is, the set E_λ is continuous with respect to λ. In such a case, the functions $\Phi_{\sigma'}(\eta, \lambda)$ and, hence, the functions $\varphi_\sigma(\lambda)$ are continuous in their domain.

For a given system, the functions $\varphi_\sigma(\cdot)$ and $G_\sigma(\cdot)$ are not univocally determined, since, in general, for each curve described by an implicit equation, several decompositions into curves described by explicit equations are possible by properly choosing the supports of the corresponding $G_\sigma(\cdot)$ functions. Moreover, if more functions $\varphi_{\sigma_1}(\lambda), \ldots, \varphi_{\sigma_K}(\lambda)$ are defined in K (not necessarily coincident) neighborhoods of the same point λ_0, all have the same limit, say φ_0, for $\lambda \to \lambda_0$, and only one is defined in λ_0, then it is convenient to assume all the functions defined in λ_0 with $\varphi_{\sigma_1}(\lambda_0) = \cdots = \varphi_{\sigma_K}(\lambda_0) = \varphi_0$ and, consequently, to define

$$G_{\sigma_i}(\lambda_0) \triangleq \lim_{\lambda \to \lambda_0} G_{\sigma_i}(\lambda), \quad i = 1, \ldots, K \tag{6.65}$$

where, for each i, the limit is made with λ ranging in the neighborhood of λ_0 where the function $\varphi_{\sigma_i}(\lambda)$ is defined.

Finally, let us observe that the FOT deterministic systems are those called "stationary" in (Claasen and Mecklenbräuker 1982).

6.3.8 *FOT-Deterministic Linear Systems*

For a LTV system the input/output relationship is

$$y(t) = \int_{\mathbb{R}} h(t, u) x(u) \, du \tag{6.66}$$

where $h(t, u)$ is the system impulse-response function (Section 1.1.7). By Fourier transforming both sides of (6.66), one obtains the input/output relationship in the frequency domain

$$\begin{aligned} Y(f) &\triangleq \int_{\mathbb{R}} y(t) e^{-j2\pi f t} \, dt \\ &= \int_{\mathbb{R}} H(f, \lambda) X(\lambda) \, d\lambda \end{aligned} \tag{6.67}$$

where the transmission function $H(f, \lambda)$ (Claasen and Mecklenbräuker 1982) is the double Fourier transform of the impulse-response function:

$$H(f, \lambda) \triangleq \int_{\mathbb{R}^2} h(t, u) e^{-j2\pi(ft - \lambda u)} \, dt \, du . \tag{6.68}$$

In (6.67), (6.68), and the following, Fourier transforms are assumed to exist in the sense of distributions (generalized functions) (Zemanian 1987).

For a FOT-deterministic LTV system we have

$$x(t) = e^{j2\pi\lambda t} \quad \Rightarrow \quad y_\lambda(t) = \int_{\mathbb{R}} h(t,u)\, e^{j2\pi\lambda u}\, \mathrm{d}u\,. \tag{6.69}$$

The Fourier transform of $y_\lambda(t)$ is

$$\begin{aligned} Y_\lambda(f) &= \int_{\mathbb{R}} e^{-j2\pi ft} \int_{\mathbb{R}} h(t,u)\, e^{j2\pi\lambda u}\, \mathrm{d}u\, \mathrm{d}t \\ &= H(f,\lambda) \end{aligned} \tag{6.70}$$

where (6.68) and the Fubini and Tonelli theorem (Champeney 1990, Chapter 3) are used to interchange the order of integrals under the summability assumption

$$\int_{\mathbb{R}^2} |h(t,u)|\, \mathrm{d}t\, \mathrm{d}u < \infty\,. \tag{6.71}$$

Therefore, the transmission function (6.68) represents the Fourier transform of the output $y_\lambda(t)$ corresponding to the input (6.59). Consequently, for FOT-deterministic linear systems the transmission function can be obtained by Fourier transforming the right-hand side of (6.64):

$$\begin{aligned} H(f,\lambda) &= \sum_{\sigma\in\Omega} G_\sigma(\lambda) \int_{\mathbb{R}} e^{j2\pi\varphi_\sigma(\lambda)t}\, e^{-j2\pi ft}\, \mathrm{d}t \\ &= \sum_{\sigma\in\Omega} G_\sigma(\lambda)\, \delta(f-\varphi_\sigma(\lambda)) \qquad (6.72\text{a}) \\ &= \sum_{\sigma\in\Omega} H_\sigma(f)\, \delta(\lambda-\psi_\sigma(f)) \qquad (6.72\text{b}) \end{aligned}$$

In the derivation of (6.72b), the functions $\varphi_\sigma(\cdot)$ are assumed to be invertible and differentiable, with inverse functions $\psi_\sigma(\cdot)$ also differentiable and referred to as *frequency mapping functions*. Moreover, the functions $G_\sigma(\cdot)$ and $H_\sigma(\cdot)$ in (6.72a) and (6.72b), accounting for the fact that $\delta(f-\varphi_\sigma(\lambda)) = |\psi'_\sigma(f)|\, \delta(\lambda-\psi_\sigma(f))$ and $\delta(\lambda-\psi_\sigma(f)) = |\varphi'_\sigma(\lambda)|\, \delta(f-\varphi_\sigma(\lambda))$ (Zemanian 1987, Section 1.7), are linked by the relationships

$$H_\sigma(f) = |\psi'_\sigma(f)|\, G_\sigma(\psi_\sigma(f)) \tag{6.73}$$

$$G_\sigma(\lambda) = |\varphi'_\sigma(\lambda)|\, H_\sigma(\varphi_\sigma(\lambda)) \tag{6.74}$$

with $\psi'_\sigma(\cdot)$ and $\varphi'_\sigma(\cdot)$ denoting the derivative of $\psi_\sigma(\cdot)$ and $\varphi_\sigma(\cdot)$, respectively.

Accounting for (6.68), from (6.72a) and (6.72b) it follows that the impulse-response function of FOT deterministic linear systems can be expressed as

$$\begin{aligned} h(t,u) &= \int_{\mathbb{R}^2} H(f,\lambda)\, e^{j2\pi(ft-\lambda u)}\, \mathrm{d}f\, \mathrm{d}\lambda \\ &= \sum_{\sigma\in\Omega} \int_{\mathbb{R}} G_\sigma(\lambda)\, e^{j2\pi\varphi_\sigma(\lambda)t} e^{-j2\pi\lambda u}\, \mathrm{d}\lambda \qquad (6.75\text{a}) \\ &= \sum_{\sigma\in\Omega} \int_{\mathbb{R}} H_\sigma(f)\, e^{-j2\pi\psi_\sigma(f)u} e^{j2\pi ft}\, \mathrm{d}f\,. \qquad (6.75\text{b}) \end{aligned}$$

By substituting (6.72a) and (6.72b) into (6.67), one obtains the input/output relationship for FOT deterministic linear systems in the frequency domain:

$$Y(f) = \sum_{\sigma \in \Omega} \int_{\mathbb{R}} G_\sigma(\lambda)\, \delta(f - \varphi_\sigma(\lambda))\, X(\lambda)\, d\lambda \tag{6.76a}$$

$$= \sum_{\sigma \in \Omega} H_\sigma(f) X(\psi_\sigma(f)) \tag{6.76b}$$

from which, by inverse Fourier transforming, it follows that

$$y(t) = \sum_{\sigma \in \Omega} \int_{\mathbb{R}} G_\sigma(\lambda)\, X(\lambda)\, e^{j2\pi\varphi_\sigma(\lambda)t}\, d\lambda \tag{6.77a}$$

$$= \sum_{\sigma \in \Omega} h_\sigma(t) \otimes x_{\psi_\sigma}(t) \tag{6.77b}$$

where $\otimes$ denotes convolution and

$$x_{\psi_\sigma}(t) \triangleq \int_{\mathbb{R}} X(\psi_\sigma(f))\, e^{j2\pi f t}\, df\,. \tag{6.78}$$

In other words, the output of FOT-deterministic LTV systems is constituted by frequency warped and then LTI filtered versions of the input.

In Section 4.3.1, it is shown that LAPTV systems and systems performing a time-scale change are special cases of FOT-deterministic systems.

Finally, note that in (Claasen and Mecklenbräuker 1982), the class of LTV systems called here FOT-deterministic are analyzed considering a concept of stationarity with reference to a single time series.

6.4 Nonstationarity Classification in the Functional Approach

In the classical approach for statistical signal analysis, signals are modeled as stochastic processes, that is, ensemble of sample paths or realizations. In such a framework, nonstationarity is the property that statistical functions defined by ensemble averages depend on the time parameter t.

In the functional approach, once the almost-periodically time-variant model (characterized at second-order by (6.24)–(6.27b)) is adopted, the classification of the kind of second-order nonstationarity in the wide sense for a time series can be made on the basis of the elements contained in the set A_τ in (6.26). If A_τ contains incommensurate cycle frequencies α, then the time series $x(t)$ is said to be *wide-sense generalized almost cyclostationary*. If $A \triangleq \cup_{\tau \in \mathbb{R}} A_\tau$ is countable, then the time series is said to be *wide-sense almost cyclostationary*. In the special case where $A \equiv \{k/T_0\}_{k \in \mathbb{Z}}$, the time series is said to be *wide-sense cyclostationary*. If the set A contains only the element $\alpha = 0$, then the time series is said to be *wide-sense stationary*. In this classification, WSS time series are a subclass of the cyclostationary time series which are a subclass of the ACS time series which in turn are a subclass of the GACS time series. A similar classification is made in the strict sense in Section 6.3.2 with reference to the sets Γ_τ and Γ.

In the functional approach a different classification can be made on the basis of the existence or not of the stationary and almost-cyclostationary probabilistic models. Specifically, time series for which a probabilistic model based on almost-periodic distributions like (6.24) exists are a subclass of time series for which a probabilistic model based on stationary distributions like (6.13) (with $y \equiv x$) exists. In fact, the existence of (6.24) requires more stringent conditions on the time series $x(t)$ with respect to those for the existence of (6.13) (with $y \equiv x$) (Section 6.3.2), (Leśkow and Napolitano 2006, Ex. 6.1). Therefore, from this point of view, almost-cyclostationary time series are a subclass of the stationary time series.

The existence of two possible classifications of nonstationarity of time series in the functional approach constitutes a significative difference with respect to the classical stochastic approach. This difference is consequence of the difference existing between the expectation operators in the two approaches. In the stochastic approach the expectation operator makes an average on the variable ranging in the sample space and produces statistical functions of a stochastic process (possibly) depending on the time variable t (see (1.1)–(1.5)). The kind of time variability with respect to t characterizes the kind of nonstationarity of the process. In contrast, in the functional approach, the expectation operator (the infinite-time average or the almost-periodic component extraction operator) acts on the time variable t and is *a priori* chosen with the probabilistic model (see (6.3)–(6.13) for the stationary model and (6.24)–(6.26) for the (generalized) almost-cyclostationary model). The choice of the model implicitly imposes conditions on the admissible time series $x(t)$, i.e., the conditions for which the statistical functions of the considered model exist.

Nonstationary signal analysis in the functional approach should be carefully handled. In (Gardner 1994), possible pitfalls associated with the possibility of choosing non unique statistical models (e.g., stationary and almost cyclostationary) are discussed. In addition, it should be noted that not every kind of nonstationarity (i.e., time variability of the statistical functions) that can be modeled in the stochastic approach can also be modeled in the functional approach. This is consequence of the fact that statistical functions, and in particular autocorrelation functions, of finite-power nonstationary stochastic processes can be reliably estimated starting from a single realization only if the kind of nonstationarity is of known form or of almost-periodic type (Gardner 1988a). For example, as shown in Chapter 4, a consistent estimator of the spectral correlation density of SC processes is obtained only when the location in the bifrequency plane of the support curves of the Loève bifrequency spectrum is known (Napolitano 2003), (Napolitano 2007c).

6.5 Proofs of FOT Counterparts of Some Results on ACS and GACS Signals

In this section, proofs of some results presented in Section 1.3 in the stochastic approach are carried out in the functional approach. Moreover, when possible, results are derived in the more general case of GACS signals.

6.5.1 *Proof of FOT Counterpart of (1.114)*

By substituting (1.112) into (1.113) we have

$$\begin{aligned}
&\mathrm{E}^{\{\alpha\}}\{y_1(t+\tau)\, y_2^{(*)}(t)\} \\
&= \mathrm{E}^{\{\alpha\}}\left\{\int_{\mathbb{R}} h_1(t+\tau, u_1)\, x_1(u_1)\, \mathrm{d}u_1 \int_{\mathbb{R}} h_2^{(*)}(t, u_2)\, x_2^{(*)}(u_2)\, \mathrm{d}u_2\right\}
\end{aligned}$$

$$= \mathrm{E}^{\{\alpha\}} \left\{ \int_{\mathbb{R}} \sum_{\sigma_1 \in J_1} h_{\sigma_1}(t+\tau-u_1)\, e^{j2\pi\sigma_1 u_1}\, x_1(u_1)\, \mathrm{d}u_1 \right.$$

$$\left. \int_{\mathbb{R}} \sum_{\sigma_2 \in J_2} h_{\sigma_2}^{(*)}(t-u_2)\, e^{(-)j2\pi\sigma_2 u_2}\, x_2^{(*)}(u_2)\, \mathrm{d}u_2 \right\}$$

$$= \mathrm{E}^{\{\alpha\}} \left\{ \int_{\mathbb{R}} \sum_{\sigma_1 \in J_1} h_{\sigma_1}(s_1+\tau)\, e^{j2\pi\sigma_1 (t-s_1)}\, x_1(t-s_1)\, \mathrm{d}s_1 \right.$$

$$\left. \int_{\mathbb{R}} \sum_{\sigma_2 \in J_2} h_{\sigma_2}^{(*)}(s_2)\, e^{(-)j2\pi\sigma_2 (t-s_2)}\, x_2^{(*)}(t-s_2)\, \mathrm{d}s_2 \right\} \tag{6.79}$$

where, in the last equality, the variable changes $s_1 = t - u_1$ and $s_2 = t - u_2$ are made. Thus by assuming that the order of sum, integral, and almost-periodic component extraction operations can be interchanged, and observing that almost-periodic functions can be led outside the almost-periodic component extraction operator, we get

$$\mathrm{E}^{\{\alpha\}}\{y_1(t+\tau)\, y_2^{(*)}(t)\}$$

$$= \sum_{\sigma_1 \in J_1} \sum_{\sigma_2 \in J_2} \int_{\mathbb{R}} \int_{\mathbb{R}} h_{\sigma_1}(s_1+\tau)\, h_{\sigma_2}^{(*)}(s_2)\, e^{-j2\pi\sigma_1 s_1}\, e^{-(-)j2\pi\sigma_2 s_2}$$

$$\mathrm{E}^{\{\alpha\}} \left\{ x_1(t-s_1)\, x_2^{(*)}(t-s_2) \right\} \mathrm{d}s_1\, \mathrm{d}s_2\, e^{j2\pi(\sigma_1 + (-)\sigma_2)t}$$

$$= \sum_{\alpha \in A_{12}} \sum_{\sigma_1 \in J_1} \sum_{\sigma_2 \in J_2} \int_{\mathbb{R}} \int_{\mathbb{R}} R^{\alpha}_{x_1 x_2^{(*)}}(s_2 - s_1)\, h_{\sigma_1}(s_1+\tau)\, h_{\sigma_2}^{(*)}(s_2)$$

$$e^{-j2\pi\sigma_1 s_1}\, e^{-(-)j2\pi\sigma_2 s_2}\, e^{-j2\pi\alpha s_2}\, \mathrm{d}s_1\, \mathrm{d}s_2\, e^{j2\pi(\alpha+\sigma_1+(-)\sigma_2)t}$$

$$= \sum_{\alpha \in A_{12}} \sum_{\sigma_1 \in J_1} \sum_{\sigma_2 \in J_2} \int_{\mathbb{R}} R^{\alpha}_{x_1 x_2^{(*)}}(u)\, e^{j2\pi\sigma_1 u}$$

$$\int_{\mathbb{R}} h_{\sigma_1}(s_2-u+\tau)\, h_{\sigma_2}^{(*)}(s_2)\, e^{-j2\pi(\alpha+\sigma_1+(-)\sigma_2)s_2}\, \mathrm{d}s_2\, \mathrm{d}u\, e^{j2\pi(\alpha+\sigma_1+(-)\sigma_2)t} \tag{6.80}$$

where, in the last equality, the variable change $s_2 - s_1 = u$ is made and, according to (1.95) (with the replacements $t - s_1 \curvearrowright t + \tau$ and $t - s_2 \curvearrowright t$), we put

$$\mathrm{E}^{\{\alpha\}} \left\{ x_1(t-s_1)\, x_2^{(*)}(t-s_2) \right\} = \sum_{\alpha \in A_{12}} R^{\alpha}_{x_1 x_2^{(*)}}(s_2 - s_1)\, e^{j2\pi\alpha(t-s_2)}\, . \tag{6.81}$$

By substituting (1.115) into (6.80), we get the FOT counterpart of (1.114) which is formally the same as (1.114).

6.5.2 Proof of FOT Counterpart of (1.116)

By taking the Fourier coefficient at frequency β of the almost periodic-function in the FOT counterpart of (1.114), we have

$$R^{\beta}_{y_1 y_2^{(*)}}(\tau) \triangleq \left\langle y_1(t+\tau)\, y_2^{(*)}(t)\, e^{-j2\pi\beta t} \right\rangle_t$$
$$= \sum_{\alpha\in A_{12}} \sum_{\sigma_1\in J_1} \sum_{\sigma_2\in J_2} \left[R^{\alpha}_{x_1 x_2^{(*)}}(\tau)\, e^{j2\pi\sigma_1\tau} \right] \underset{\tau}{\otimes} r^{\alpha+\sigma_1+(-)\sigma_2}_{\sigma_1\sigma_2(*)}(\tau)$$
$$\left\langle e^{j2\pi(\alpha+\sigma_1+(-)\sigma_2)t}\, e^{-j2\pi\beta t} \right\rangle_t . \tag{6.82}$$

Since

$$\left\langle e^{j2\pi(\alpha+\sigma_1+(-)\sigma_2)t}\, e^{-j2\pi\beta t} \right\rangle_t = \delta_{\beta-(\alpha+\sigma_1+(-)\sigma_2)} \tag{6.83}$$

the FOT counterpart of (1.116), which is formally identical to (1.116), easily follows.

6.5.3 Proof of FOT Counterparts of (1.135) and (1.136)

By using the FOT counterparts of (1.131) and (1.132), which are formally identical to (1.131) and (1.132), we get

$$R^{\alpha}_{y_1 y_2^{(*)}}(\tau) \triangleq \left\langle y_1(t+\tau)\, y_2^{(*)}(t)\, e^{-j2\pi\alpha t} \right\rangle_t$$
$$= \left\langle \mathrm{E}^{\{\alpha\}} \left\{ y_1(t+\tau)\, y_2^{(*)}(t) \right\} e^{-j2\pi\alpha t} \right\rangle_t$$
$$= \left\langle \mathrm{E}^{\{\alpha\}} \left\{ c_1(t+\tau)\, c_2^{(*)}(t) \right\} \mathrm{E}^{\{\alpha\}} \left\{ x_1(t+\tau)\, x_2^{(*)}(t) \right\} e^{-j2\pi\alpha t} \right\rangle_t$$
$$= \left\langle \sum_{\alpha_c\in A_{c_1 c_2^{(*)}}} R^{\alpha_c}_{c_1 c_2^{(*)}}(\tau)\, e^{j2\pi\alpha_c t} \sum_{\alpha_x\in A_{x_1 x_2^{(*)}}} R^{\alpha_x}_{x_1 x_2^{(*)}}(\tau)\, e^{j2\pi\alpha_x t}\, e^{-j2\pi\alpha t} \right\rangle_t$$
$$= \sum_{\alpha_x\in A_{x_1 x_2^{(*)}}} \sum_{\alpha_c\in A_{c_1 c_2^{(*)}}} R^{\alpha_x}_{x_1 x_2^{(*)}}(\tau)\, R^{\alpha_c}_{c_1 c_2^{(*)}}(\tau)$$
$$\left\langle e^{j2\pi\alpha_x t}\, e^{j2\pi\alpha_c t}\, e^{-j2\pi\alpha t} \right\rangle_t \tag{6.84}$$

from which the FOT counterpart of (1.135), formally identical to (1.135), follows observing that

$$\left\langle e^{j2\pi\alpha_x t}\, e^{j2\pi\alpha_c t}\, e^{-j2\pi\alpha t} \right\rangle_t = \delta_{\alpha-\alpha_x-\alpha_c} . \tag{6.85}$$

The FOT counterpart of (1.136) easily follows by Fourier transforming both sides of the FOT counterpart of (1.135).

6.5.4 *Proof of the FOT Counterpart of (1.173) for GACS Signals*

Let $x(n) = x_a(t)|_{t=nT_s}$, $n \in \mathbb{Z}$, be the discrete-time signal obtained by uniformly sampling the complex-valued continuous-time signal $x_a(t)$ with sampling period $T_s = 1/f_s$. The FOT counterpart of (1.173) is

$$\widetilde{\mathrm{E}^{\{\alpha\}}}\{x(n+m)\, x^{(*)}(n)\} = \mathrm{E}^{\{\alpha\}}\{x_a(t+\tau)\, x_a(t)\}\Big|_{t=nT_s,\tau=mT_s} \tag{6.86}$$

where $\mathrm{E}^{\{\alpha\}}\{\cdot\}$ is the *continuous-time* almost-periodic component extraction operator (see (6.22)) and $\widetilde{\mathrm{E}^{\{\alpha\}}}\{\cdot\}$ is the *discrete-time* almost-periodic component extraction operator which is defined similarly to its continuous-time counterpart.

Proof: The second-order lag product $x_a(t+\tau)\, x_a^{(*)}(t)$ can be decomposed into the sum of its almost-periodic component and a residual term $\ell_{\boldsymbol{x}_a}(t,\tau)$ not containing any finite-strength additive sinewave component (Section 6.3.3). For a GACS signal, the result is that

$$\begin{aligned} x_a(t+\tau)\, x_a^{(*)}(t) &= \mathrm{E}^{\{\alpha\}}\left\{x_a(t+\tau)\, x_a^{(*)}(t)\right\} + \ell_{\boldsymbol{x}_a}(t,\tau) \\ &= \sum_{\alpha\in A_\tau} R_{\boldsymbol{x}_a}(\alpha,\tau)\, e^{j2\pi\alpha t} + \ell_{\boldsymbol{x}_a}(t,\tau) \end{aligned} \tag{6.87}$$

where $\boldsymbol{x}_a = [x_a\; x_a^{(*)}]$, $A_\tau \triangleq A_{\boldsymbol{x}_a,\tau}$, and

$$\left\langle \ell_{\boldsymbol{x}_a}(t,\tau)\, e^{-j2\pi\alpha t}\right\rangle_t \equiv 0 \quad \forall\alpha\in\mathbb{R}\,. \tag{6.88}$$

For any $0 < \epsilon < T_s$ we formally have

$$\begin{aligned} &\frac{1}{2N+1}\sum_{n=-N}^{N} \ell_{\boldsymbol{x}_a}(t,\tau)\Big|_{t=nT_s,\tau=mT_s} e^{-j2\pi\tilde{\alpha}n} \\ &= \frac{1}{2N+1}\int_{-NT_s-\epsilon}^{NT_s+\epsilon} \ell_{\boldsymbol{x}_a}(t,mT_s) \sum_{n=-\infty}^{+\infty}\delta(t-nT_s)\, e^{-j2\pi\tilde{\alpha}t/T_s}\,\mathrm{d}t \\ &= \sum_{p=-\infty}^{+\infty} \frac{1}{(2N+1)T_s}\int_{-NT_s-\epsilon}^{NT_s+\epsilon} \ell_{\boldsymbol{x}_a}(t,mT_s)\, e^{-j2\pi(\tilde{\alpha}-p)t/T_s}\,\mathrm{d}t \end{aligned} \tag{6.89}$$

where, in the second equality, the Poisson sum formula

$$\sum_{n=-\infty}^{+\infty}\delta(t-nT_s) = \frac{1}{T_s}\sum_{p=-\infty}^{+\infty} e^{j2\pi pt/T_s} \tag{6.90}$$

is accounted for. Taking the limit for $N\to\infty$ in (6.89) and accounting for (6.88) one has

$$\lim_{N\to\infty}\frac{1}{2N+1}\sum_{n=-N}^{N} \ell_{\boldsymbol{x}_a}(t,\tau)\Big|_{t=nT_s,\tau=mT_s} e^{-j2\pi\tilde{\alpha}n} = 0\,. \tag{6.91}$$

That is, if the continuous-time residual term $\ell_{x_a}(t, \tau)$ does not contain any (continuous-time) finite-strength additive sinewave component (see (6.88)), then also the sampled residual term $\ell_{x_a}(nT_s, mT_s)$ does not contain any discrete-time finite-strength additive sinewave component.

Thus,

$$
\begin{aligned}
&\mathrm{E}^{\{\widetilde{\alpha}\}}\{x(n+m)\, x^{(*)}(n)\} \\
&\quad = \mathrm{E}^{\{\widetilde{\alpha}\}}\left\{ x_a(t+\tau)\, x_a^{(*)}(t)\Big|_{t=nT_s, \tau=mT_s} \right\} \\
&\quad = \mathrm{E}^{\{\widetilde{\alpha}\}}\left\{ \mathrm{E}^{\{\alpha\}}\{x_a(t+\tau)\, x_a^{(*)}(t)\}\Big|_{t=nT_s, \tau=mT_s} \right\} \\
&\qquad + \mathrm{E}^{\{\widetilde{\alpha}\}}\left\{ \ell_{x_a}(t, \tau)\Big|_{t=nT_s, \tau=mT_s} \right\} \\
&\quad = \mathrm{E}^{\{\widetilde{\alpha}\}}\left\{ \sum_{\alpha \in A_\tau} R_{x_a}(\alpha, \tau)\, e^{j2\pi\alpha t}\Bigg|_{t=nT_s, \tau=mT_s} \right\} \\
&\quad = \sum_{\alpha \in A_\tau} R_{x_a}(\alpha, \tau)\, e^{j2\pi\alpha t}\Bigg|_{t=nT_s, \tau=mT_s}
\end{aligned} \tag{6.92}
$$

where in the third equality (6.91) is accounted for and in the last equality the fact that the right-hand side of (6.92) is a discrete-time almost periodic function is exploited.

Equation (6.86) is then proved observing that the right-hand side of (6.92) is coincident with the right-hand side of (6.86).

Note that, as in the stochastic approach, uniformly sampling a continuous-time GACS signal leads to a discrete-time ACS signal (Theorem 2.5.2).

7

Applications to Mobile Communications and Radar/Sonar

In this chapter, the (non-relativistic) Doppler effect due to motion between transmitter and receiver and/or surrounding scatterers is described by a LTV system introducing time-varying delays. Since a non-relativistic analysis is carried out, time dilation existing between clocks in relative motion is neglected (Gray and Addison 2003). In the case of constant relative radial speed v between transmitter and receiver, according to Lorentz transformations, the time-dilation factor $\gamma = (1 - (v/c)^2)^{-1/2} \simeq 1 + (v/c)^2/2$ is approximated to 1 where c is the medium propagation speed. Moreover, aberration effect is neglected. Applications to mobile communications and radar/sonar problems are considered when the transmitted signal is ACS. Conditions on the transmitted signal and channel parameters are derived under which the received signal can be modeled as GACS or SC.

7.1 Physical Model for the Wireless Channel

The propagation problem should be afforded by solving the Maxwell's equation with the boundary conditions on buildings, ground, vehicles, sea, etc. Boundary conditions should be specified at least with accuracy of the order of the wavelength λ, where $\lambda f = c$, f being the transmitted frequency and $c \simeq 3 \cdot 10^8 \mathrm{m\,s}^{-1}$ the propagation speed. In the case of mobile cellular communications, $f \simeq 1$ GHz $= 10^9$ Hz so that $\lambda \simeq 3 \cdot 10^{-1}$ m. Since describing the boundary conditions with such an accuracy is a formidable problem, a simplified approach is used to afford the propagation problem in the wireless channel (Gallagher 2008).

7.1.1 Assumptions on the Propagation Channel

In the sequel, the following assumptions on the propagation channel are made.

The transmitter (TX) is located in the origin of the (x, y, z) axis system (TX reference system). The receiver (RX) is located in $\boldsymbol{P} \equiv (r, \theta, \psi)$ (spherical coordinates with respect to TX reference system) where the electric and magnetic fields are $\boldsymbol{e}(\boldsymbol{P}, t)$ and $\boldsymbol{h}(\boldsymbol{P}, t)$, respectively. Free-space propagation and far field ($r \gg \lambda$) conditions are satisfied. Thus $\boldsymbol{e}$ is orthogonal to

Generalizations of Cyclostationary Signal Processing: Spectral Analysis and Applications, First Edition.

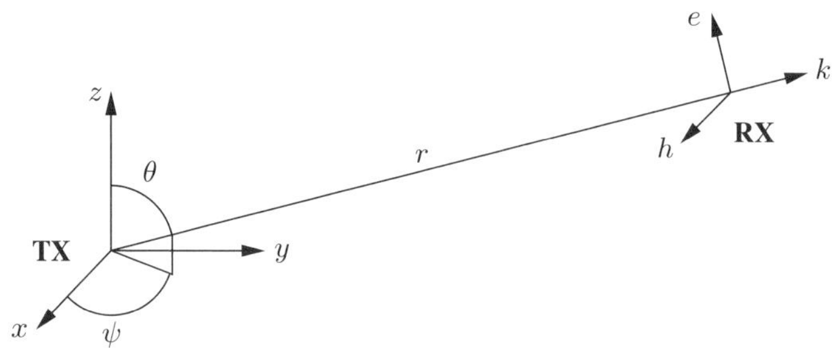

Figure 7.1 Plane wave

$\boldsymbol{h}$, both $\boldsymbol{e}$ and $\boldsymbol{h}$ are orthogonal to the propagation vector $\boldsymbol{k}$, and $\|\boldsymbol{e}(\boldsymbol{P},t)\|$ is proportional to $\|\boldsymbol{h}(\boldsymbol{P},t)\|$ (Figure 7.1). A spherical coordinate system (r', θ', ψ') with origin in the RX is also defined. Furthermore, in the case of both moving TX and RX, also a fixed reference system (r_0, θ_0, ψ_0) will be considered.

The radiation pattern of the TX antenna $\alpha_T(\theta, \psi, f_T)$ in the TX reference system is in general depending on the transmitted frequency f_T. The radiation pattern of the RX antenna $\alpha_R(\theta', \psi', f_R)$ in the RX reference system is in general depending on the received frequency f_R. The presence of RX antenna only causes local modification in the electric and magnetic fields and does not affect the *propagation delay* r/c and the *far-field attenuation factor* $1/r$.

7.1.2 Stationary TX, Stationary RX

Let us consider the propagation channel between a stationary TX and a stationary RX. For a complex sinewave transmitted signal at the TX antenna clamps $x_T(t) = e^{j2\pi ft}$ (TX frequency $f_T = f$) the component of the electric field along the direction of $\boldsymbol{e}(\boldsymbol{P},t)$ in the RX location $\boldsymbol{P}$ is

$$e(\boldsymbol{P}, t; f) = \alpha_T(\theta, \psi, f) \frac{e^{j2\pi f(t-r/c)}}{r} \tag{7.1}$$

where $\alpha_T(\theta, \psi, f)$ describes the TX radiation pattern. The received frequency is coincident with the transmitted frequency: $f_R = f_T = f$. The signal at the (possibly loaded) RX antenna clamps is given by

$$x_R(t; f) = \alpha(\theta, \psi, \theta', \psi', f) \frac{e^{j2\pi f(t-r/c)}}{r} \tag{7.2}$$

where

$$\alpha(\theta, \psi, \theta', \psi', f) \triangleq \alpha_T(\theta, \psi, f)\, \alpha_R(\theta', \psi', f) \tag{7.3}$$

describes both TX and RX radiation patterns. More generally, α possibly includes also antenna-polarization and input-impedance mismatchs. Thus, the result is that

$$x_R(t; f) = H(f)\, x_T(t) = H(f)\, e^{j2\pi ft} \tag{7.4}$$

with

$$H(f) = \alpha(\theta, \psi, \theta', \psi', f)\frac{e^{-j2\pi fr/c}}{r}. \tag{7.5}$$

That is, the overall propagation channel is LTI with harmonic-response function $H(f)$.

Let us consider now a generic (non sinusoidal) transmitted signal $x(t)$ with Fourier transform $X(f)$, that is,

$$x(t) = \int_{\mathbb{R}} X(f)\, e^{j2\pi ft}\, \mathrm{d}f. \tag{7.6}$$

Since Maxwell'equations are linear, for the superposition principle and accounting for (7.5), the signal at the RX antenna clamps is given by

$$y(t) = \int_{\mathbb{R}} X(f)\, H(f)\, e^{j2\pi ft}\, \mathrm{d}f \tag{7.7a}$$

$$= \int_{\mathbb{R}} X(f)\, \frac{\alpha(\theta, \psi, \theta', \psi', f)}{r}\, e^{j2\pi f(t-r/c)}\, \mathrm{d}f \tag{7.7b}$$

$$= x(t) \otimes a(t) \otimes \delta(t - r/c) \tag{7.7c}$$

where $a(t)$, depending on $\theta, \psi, \theta', \psi'$, is the inverse Fourier transform of $\alpha(\theta, \psi, \theta', \psi', f)/r$ and describes the linear distortion on the transmitted signal introduced by the TX and RX (frequency-dependent) radiation patterns.

Assumption 7.1.1 Wide-Band TX and RX Antennas (Stationary TX and RX). *The overall TX and RX radiation pattern can be assumed to be constant with respect to frequency in the frequency interval where the spectral components of the transmitted (and hence also of the received) signal are relevant. That is, for some f_0 it results that*

$$\alpha(\theta, \psi, \theta', \psi', f) \simeq \alpha(\theta, \psi, \theta', \psi', f_0) = \alpha_0(\theta, \psi, \theta', \psi') \quad \forall f\ :\ X(f) \neq 0. \tag{7.8}$$

Under the wide-band antenna assumption (Assumption 7.1.1), (7.7a)–(7.7c) specialize to

$$y(t) = \int_{\mathbb{R}} X(f)\, H(f)\, e^{j2\pi ft}\, \mathrm{d}f \tag{7.9a}$$

$$= \frac{\alpha_0(\theta, \psi, \theta', \psi')}{r} \int_{\mathbb{R}} X(f)\, e^{j2\pi f(t-r/c)}\, \mathrm{d}f \tag{7.9b}$$

$$= \frac{\alpha_0(\theta, \psi, \theta', \psi')}{r}\, x(t - r/c) \tag{7.9c}$$

That is, the propagation LTI channel introduces constant attenuation and delay:

$$h(t) = \frac{\alpha_0(\theta, \psi, \theta', \psi')}{r}\, \delta(t - r/c) \overset{\mathfrak{F}}{\longleftrightarrow} H(f) = \frac{\alpha_0(\theta, \psi, \theta', \psi')}{r}\, e^{-j2\pi fr/c}. \tag{7.10}$$

7.1.3 Moving TX, Moving RX

Let both source (or transmitter (TX)) and observer (or receiver (RX)) be moving and let $\boldsymbol{P}_{\rm T}(u)$ and $\boldsymbol{P}_{\rm R}(t)$ be the three-dimensional vectors of TX and RX positions, respectively, with respect to a fixed reference system (r_0, θ_0, ψ_0), where t is time measured at RX and u is time measured at TX.

Let t be the time at the receiver and $D(t)$ the (time-varying) delay experienced by a transmitted wavefront before being received at time t. In other words, a wavefront transmitted at $u = t - D(t)$ is received at time t. Consequently, the delay $D(t)$ is due to the time-varying distance between TX located in $\boldsymbol{P}_{\rm T}(t - D(t))$ and RX located in $\boldsymbol{P}_{\rm R}(t)$ (Kapoulitsas 1981; Neipp *et al.* 2003):

$$c\, D(t) = \|\boldsymbol{P}_{\rm R}(t) - \boldsymbol{P}_{\rm T}(t - D(t))\| \triangleq \varrho(t, D(t)) \tag{7.11}$$

where c is the medium propagation speed and $\|\boldsymbol{P}\|$ is the norm of the vector $\boldsymbol{P}$. In the considered fixed reference system (r_0, θ_0, ψ_0), TX reference system (r, θ, ψ), and RX reference system (r', θ', ψ') we have

$$\boldsymbol{P}_{\rm T}(t - D(t)) = (r_0(t - D(t)), \theta_0(t - D(t)), \psi_0(t - D(t))) \tag{7.12}$$

$$\boldsymbol{P}_{\rm R}(t) = (r_0(t), \theta_0(t), \psi_0(t)) \tag{7.13}$$

$$\alpha_{\rm T} = \alpha_{\rm T}(\theta(t - D(t)), \psi(t - D(t)), f_{\rm T}) \tag{7.14}$$

$$\alpha_{\rm R} = \alpha_{\rm R}(\theta'(t), \psi'(t), f_{\rm R}). \tag{7.15}$$

The function $\varrho(t, D(t))$ is the time-varying distance covered by the wavefront transmitted at time $t - D(t)$ in $\boldsymbol{P}_{\rm T}(t - D(t))$ and received at time t in $\boldsymbol{P}_{\rm R}(t)$. It should not be confused with the time-varying distance between TX and RX at time t, that is, the distance between $\boldsymbol{P}_{\rm T}(t)$ and $\boldsymbol{P}_{\rm R}(t)$, which is

$$R(t) \triangleq \|\boldsymbol{P}_{\rm R}(t) - \boldsymbol{P}_{\rm T}(t)\|. \tag{7.16}$$

Note that the time-varying distance $\varrho(t, D(t))$ (not $R(t)$) is responsible of the Doppler effect in the received signal. In fact, the time-varying delay $D(t)$ is a consequence of the propagation from $\boldsymbol{P}_{\rm T}(t - D(t))$ to $\boldsymbol{P}_{\rm R}(t)$ (see (7.11)) and not from $\boldsymbol{P}_{\rm T}(t)$ to $\boldsymbol{P}_{\rm R}(t)$. Even if $R(t)$ is constant, we can still have Doppler effect if $\varrho(t, D(t))$, and hence $D(t)$, are time-varying. For example, if both TX and RX move along two parallel lines with the same nonuniform motion law, $R(t)$ is constant while $\varrho(t, D(t))$ is time-varying giving rise to Doppler effect. However, as shown in Sections 7.1.4 and 7.1.5, if either TX or RX is stationary, then $\varrho(t, D(t))$ can be easily linked to $R(t)$.

If the transmitted signal is a sinewave, the received signal is in turn a sinewave only if the propagation channel can be modeled as a (linear) FOT-deterministic time-variant system (Section 6.3.8 and (Izzo and Napolitano 2002a)). In all other cases, the received signal contains spectral components spread on a band depending on the time-varying nature of the propagation channel.

For a complex sinewave transmitted signal $x_{\rm T}(t) = e^{j2\pi f_{\rm T} t}$ the component of the electric field along the direction of $\boldsymbol{e}(\boldsymbol{P}_{\rm R}(t), t)$ in the RX location $\boldsymbol{P}_{\rm R}(t)$ at time t is due to transmission at time $t - D(t)$ and is given by

$$e(\boldsymbol{P}_{\rm R}(t), t; f_{\rm T}) = \alpha_{\rm T}(\theta(t - D(t)), \psi(t - D(t)), f_{\rm T}) \, \frac{e^{j2\pi f_{\rm T}(t - D(t))}}{\varrho(t, D(t))}. \tag{7.17}$$

In general, this function is not sinusoidal. Its Fourier transform

$$E(\boldsymbol{P}_{\mathrm{R}}(\cdot), f_{\mathrm{R}}; f_{\mathrm{T}}) = \int_{\mathbb{R}} e(\boldsymbol{P}_{\mathrm{R}}(t), t; f_{\mathrm{T}})\, e^{-j2\pi f_{\mathrm{R}} t}\, \mathrm{d}t \tag{7.18}$$

depends on the RX position $\boldsymbol{P}_{\mathrm{R}}(t)$ for $t \in (-\infty, +\infty)$. Each infinitesimal spectral component

$$E(\boldsymbol{P}_{\mathrm{R}}(\cdot), f_{\mathrm{R}}; f_{\mathrm{T}})\, e^{j2\pi f_{\mathrm{R}} t}\, \mathrm{d}f_{\mathrm{R}}$$

of the received electrical field gives rise to an infinitesimal voltage at the antenna clamps that depends on the time-varying RX radiation pattern

$$\alpha_{\mathrm{R}}(\theta'(t), \psi'(t); f_{\mathrm{R}})\, E(\boldsymbol{P}_{\mathrm{R}}(\cdot), f_{\mathrm{R}}; f_{\mathrm{T}})\, e^{j2\pi f_{\mathrm{R}} t}\, \mathrm{d}f_{\mathrm{R}}.$$

Since Maxwell's equations are linear, the overall received signal $x_{\mathrm{R}}(t; f_{\mathrm{T}})$ corresponding to the transmitted complex sinewave $x_{\mathrm{T}}(t) = e^{j2\pi f_{\mathrm{T}} t}$ is given by the superposition integral

$$x_{\mathrm{R}}(t; f_{\mathrm{T}}) = \int_{\mathbb{R}} \alpha_{\mathrm{R}}(\theta'(t), \psi'(t), f_{\mathrm{R}})\, E(\boldsymbol{P}_{\mathrm{R}}(\cdot), f_{\mathrm{R}}; f_{\mathrm{T}})\, e^{j2\pi f_{\mathrm{R}} t}\, \mathrm{d}f_{\mathrm{R}}. \tag{7.19}$$

Let us consider now a generic (non sinusoidal) transmitted signal $x(t)$ with Fourier transform $X(f)$, that is,

$$x(t) = \int_{\mathbb{R}} X(f_{\mathrm{T}})\, e^{j2\pi f_{\mathrm{T}} t}\, \mathrm{d}f_{\mathrm{T}}. \tag{7.20}$$

Since Maxwell's equations are linear, for the superposition principle the signal at the RX antenna clamps is given by

$$\begin{aligned} y(t) &= \int_{\mathbb{R}} X(f_{\mathrm{T}})\, x_{\mathrm{R}}(t; f_{\mathrm{T}})\, \mathrm{d}f_{\mathrm{T}} \\ &= \int_{\mathbb{R}} \int_{\mathbb{R}} x(u)\, e^{-j2\pi f_{\mathrm{T}} u}\, \mathrm{d}u\, x_{\mathrm{R}}(t; f_{\mathrm{T}})\, \mathrm{d}f_{\mathrm{T}} \\ &= \int_{\mathbb{R}} x(u) \int_{\mathbb{R}} x_{\mathrm{R}}(t; f_{\mathrm{T}})\, e^{-j2\pi f_{\mathrm{T}} u}\, \mathrm{d}f_{\mathrm{T}}\, \mathrm{d}u. \end{aligned} \tag{7.21}$$

That is, the propagation channel is a LTV system (Section 1.1.7) with impulse-response function

$$\begin{aligned} h(t, u) &= \int_{\mathbb{R}} x_{\mathrm{R}}(t; f_{\mathrm{T}})\, e^{-j2\pi f_{\mathrm{T}} u}\, \mathrm{d}f_{\mathrm{T}} \\ &= \int_{\mathbb{R}} \int_{\mathbb{R}} \alpha_{\mathrm{R}}(\theta'(t), \psi'(t), f_{\mathrm{R}})\, E(\boldsymbol{P}_{\mathrm{R}}(\cdot), f_{\mathrm{R}}; f_{\mathrm{T}})\, e^{j2\pi f_{\mathrm{R}} t}\, \mathrm{d}f_{\mathrm{R}}\, e^{-j2\pi f_{\mathrm{T}} u}\, \mathrm{d}f_{\mathrm{T}} \\ &= \int_{\mathbb{R}} \int_{\mathbb{R}} \int_{\mathbb{R}} \alpha_{\mathrm{R}}(\theta'(t), \psi'(t), f_{\mathrm{R}})\, \alpha_{\mathrm{T}}(\theta(\xi - D(\xi)), \psi(\xi - D(\xi)), f_{\mathrm{T}}) \\ &\quad \frac{e^{j2\pi f_{\mathrm{T}}(\xi - D(\xi))}}{\varrho(\xi, D(\xi))}\, e^{-j2\pi f_{\mathrm{R}} \xi}\, e^{-j2\pi f_{\mathrm{T}} u}\, e^{j2\pi f_{\mathrm{R}} t}\, \mathrm{d}\xi\, \mathrm{d}f_{\mathrm{R}}\, \mathrm{d}f_{\mathrm{T}} \end{aligned} \tag{7.22}$$

where, in the second and third equality, (7.17)–(7.19) are accounted for.

Assumption 7.1.2 Wide-Band TX and RX Antennas (Moving TX and RX). *The overall TX and RX radiation pattern can be assumed to be constant with respect to frequency in the* *frequency interval where the spectral components of both the transmitted and received signals are relevant. That is,*

$$\alpha_{\mathrm{T}}(\theta, \psi, f_{\mathrm{T}})\, \alpha_{\mathrm{R}}(\theta', \psi', f_{\mathrm{R}}) \simeq \alpha_{\mathrm{T}}(\theta, \psi)\, \alpha_{\mathrm{R}}(\theta', \psi')$$
$$\forall f_{\mathrm{T}} \,:\, X(f_{\mathrm{T}}) \neq 0 \quad \forall f_{\mathrm{R}} \,:\, Y(f_{\mathrm{R}}) \neq 0 \tag{7.23}$$

where $Y(f_{\mathrm{R}})$ is the Fourier transform of $y(t)$.

Under the wide-band antenna assumption (Assumption 7.1.2), the impulse-response function (7.22) of the propagation channel specializes into

$$\begin{aligned}
h(t, u) &= \alpha_{\mathrm{R}}(\theta'(t), \psi'(t)) \int_{\mathbb{R}} \int_{\mathbb{R}} \alpha_{\mathrm{T}}(\theta(\xi - D(\xi)), \psi(\xi - D(\xi))) \\
&\quad \frac{e^{j2\pi f_{\mathrm{T}}(\xi - D(\xi))}}{\varrho(\xi, D(\xi))} \int_{\mathbb{R}} e^{j2\pi f_{\mathrm{R}}(t-\xi)}\, \mathrm{d}f_{\mathrm{R}} e^{-j2\pi f_{\mathrm{T}} u}\, \mathrm{d}\xi\, \mathrm{d}f_{\mathrm{T}} \\
&= \alpha_{\mathrm{R}}(\theta'(t), \psi'(t)) \int_{\mathbb{R}} \int_{\mathbb{R}} \alpha_{\mathrm{T}}(\theta(\xi - D(\xi)), \psi(\xi - D(\xi))) \\
&\quad \frac{e^{j2\pi f_{\mathrm{T}}(\xi - D(\xi))}}{\varrho(\xi, D(\xi))}\, \delta(t - \xi) e^{-j2\pi f_{\mathrm{T}} u}\, \mathrm{d}\xi\, \mathrm{d}f_{\mathrm{T}} \\
&= \frac{\alpha_{\mathrm{R}}(\theta'(t), \psi'(t))\, \alpha_{\mathrm{T}}(\theta(t - D(t)), \psi(t - D(t)))}{\varrho(t, D(t))} \\
&\quad \int_{\mathbb{R}} e^{j2\pi f_{\mathrm{T}}(t - D(t))} e^{-j2\pi f_{\mathrm{T}} u}\, \mathrm{d}f_{\mathrm{T}} \\
&= A(t)\, \delta(u - t + D(t))
\end{aligned} \tag{7.24}$$

where

$$A(t) \triangleq \frac{\alpha_{\mathrm{R}}(\theta'(t), \psi'(t))\, \alpha_{\mathrm{T}}(\theta(t - D(t)), \psi(t - D(t)))}{\varrho(t, D(t))} \tag{7.25}$$

is the time-varying complex gain. Therefore, in the case of wide-band antennas the received signal is given by

$$\begin{aligned}
y(t) &= \int_{\mathbb{R}} h(t, u)\, x(u)\, \mathrm{d}u \\
&= \int_{\mathbb{R}} A(t)\, \delta(u - t + D(t))\, x(u)\, \mathrm{d}u \\
&= A(t)\, x(t - D(t)).
\end{aligned} \tag{7.26}$$

Let us consider the transmitted signal

$$x(t) = \mathrm{Re}\left\{ \widetilde{x}(t)\, e^{j2\pi f_c t} \right\} \tag{7.27}$$

where $\mathrm{Re}\{\cdot\}$ denotes real part, f_c is the carrier frequency, and $\widetilde{x}(t)$ is a complex signal which is coincident with the complex envelope provided that $\widetilde{x}(t)\, e^{j2\pi f_c t}$ is an analytic signal, that is, the Fourier transform of $\widetilde{x}(t)\, e^{j2\pi f_c t}$ is zero for negative frequencies (Gardner 1987b; Izzo and

Napolitano 1997). According to (7.26), the received signal is given by

$$y(t) = \text{Re}\left\{A(t)\,\widetilde{x}(t - D(t))\,e^{j2\pi f_c(t-D(t))}\right\}$$
$$\triangleq \text{Re}\left\{\widetilde{y}(t)\,e^{j2\pi f_c t}\right\} \tag{7.28}$$

where

$$\widetilde{y}(t) = A(t)\,e^{-j2\pi f_c D(t)}\,\widetilde{x}(t - D(t)). \tag{7.29}$$

That is, the complex signal $\widetilde{y}(t)$ can be obtained by a LTV transformation of the complex signal $\widetilde{x}(t)$. The impulse-response function of the LTV system is

$$\widetilde{h}(t, u) = A(t)\,e^{-j2\pi f_c D(t)}\,\delta(u - t + D(t)). \tag{7.30}$$

In a first approximation, the dependence on t in the gain $A(t)$ can be neglected (Quinn and Hannan 2001, pp. 21–22) (see Assumption 7.3.1 and comments in Section 7.3.2). Thus, the impulse–response function of the LTV system can be written as

$$\widetilde{h}(t, u) = A\,\delta(u - t + D(t))\,e^{-j2\pi f_c D(t)}. \tag{7.31}$$

In Section 7.3, it is shown that in the special case of constant relative radial speed between TX and RX, the complex signal $\widetilde{y}(t)$ is obtained from $\widetilde{x}(t)$ by the LTV transformation with impulse-response function

$$\widetilde{h}(t, u) = a\,\delta(u - st + d_0)\,e^{j2\pi\nu t}. \tag{7.32}$$

which is obtained as a special case of (7.31). If $s \simeq 1$, then the LTV system can be approximated as linear periodically time-variant. If $s \simeq 1$ and $\nu \simeq 0$ then the LTV system can be approximated as LTI.

Note that (7.32) is not the complex envelope of a real impulse-response function. In fact, even in the special case of LTI channel introducing just the complex gain a ($s = 1$, $d_0 = 0$, and $\nu = 0$), the complex envelope of the impulse-response function cannot obtained from (7.32) since it is given by

$$\widetilde{h}(\tau) = a\left[\delta(\tau) + j\frac{1}{\pi\tau}\right]e^{-j2\pi f_c \tau} \tag{7.33}$$

where f_c is the carrier frequency of the real signal. On this subject, see also (Aiken 1967; Yoo 2008).

7.1.4 Stationary TX, Moving RX

Let be TX stationary and RX moving. That is, $\boldsymbol{P}_{\mathrm{T}}(u) = \boldsymbol{P}_0\ \forall u \in \mathbb{R}$. Accordingly with (7.11) and (7.16) we have

$$R(t) = \|\boldsymbol{P}_{\mathrm{R}}(t) - \boldsymbol{P}_0\| = \varrho(t, D(t)) \tag{7.34}$$

and $\varrho(t, D(t))$ is independent of $D(t)$. Equation (7.11) specializes into

$$c\,D(t) = R(t) \tag{7.35}$$

from which it follows that

$$\dot{D}(t) = \frac{\dot{R}(t)}{c} \tag{7.36}$$

where $\dot{D}(t)$ denotes the first-order derivative of $D(t)$ and $\dot{R}(t)$ is the relative radial speed between $\boldsymbol{P}_{\mathrm{T}}(t)$ and $\boldsymbol{P}_{\mathrm{R}}(t)$.

7.1.5 Moving TX, Stationary RX

Let be TX moving and RX stationary. That is, $\boldsymbol{P}_{\mathrm{R}}(t) = \boldsymbol{P}_0 \; \forall t \in \mathbb{R}$. Accordingly with (7.11) and (7.16) we have

$$R(t - D(t)) = \|\boldsymbol{P}_0 - \boldsymbol{P}_{\mathrm{T}}(t - D(t))\| = \varrho(t, D(t)) \tag{7.37}$$

and (7.11) specializes into

$$c\, D(t) = R(t - D(t)) \tag{7.38}$$

from which it follows that

$$c\; \dot{D}(t) = (1 - \dot{D}(t))\, \dot{R}(t - D(t)) \tag{7.39}$$

and, hence,

$$\dot{D}(t) = \frac{\dot{R}(t - D(t))}{c + \dot{R}(t - D(t))}. \tag{7.40}$$

7.1.6 Reflection on Point Scatterer

Let us consider a transmitted signal reflected on a point scatterer and then received. This is a model for radar/sonar systems where the scatterer is the target or for communications systems where the scatterer is a reflecting object in surroundings of transmitter or receiver. In radar systems also the clutter contribution should be accounted for and in some communications systems also the line-of-sight path should be considered. Such contributions are not considered here for the sake of simplicity and can be easily added in more realistic models. The equivalent system model is described in Figure 7.2, where

- $x(t)$ is the transmitted signal;
- $h_1(t, u)$ is the impulse-response function of the propagation channel between transmitter and scatterer, which also includes the transmitter radiation pattern and the scatterer receiving radiation pattern;

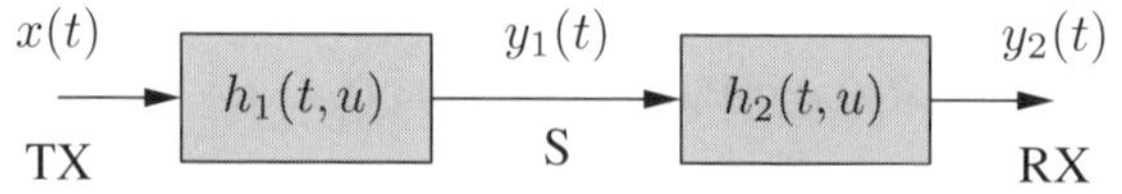

Figure 7.2 Reflection on point scatterer: transmitter TX, scatterer S, receiver RX

- $y_1(t)$ is the signal received by the scatterer and retransmitted;
- $h_2(t, u)$ is the impulse-response function of the propagation channel between scatterer and receiver, which also includes the scatterer transmitting radiation pattern and the receiver radiation pattern;
- $y_2(t)$ is the received signal.

Let $\boldsymbol{P}_{\mathrm{T}}$, $\boldsymbol{P}_{\mathrm{S}}$, and $\boldsymbol{P}_{\mathrm{R}}$ be the three-dimensional vectors of positions of possibly moving transmitter (TX), scatterer (S), and receiver (RX), respectively. The scatterer is characterized by the radiation pattern $\alpha_{\mathrm{S}}(\theta_{\mathrm{S}}, \psi_{\mathrm{S}})$ to be considered in the receiving $(\theta_{\mathrm{SR}}, \psi_{\mathrm{SR}})$ and retransmitting $(\theta_{\mathrm{ST}}, \psi_{\mathrm{ST}})$ directions, with spherical coordinates taken in a reference system centered in the scatterer. Under the wide-band antenna assumption (Assumption 7.1.2), accounting for (7.11), (7.24), and (7.25), the LTV system describing the TX-S channel is

$$h_1(t, u) = A_1(t)\, \delta(u - t + D_1(t)) \tag{7.41}$$

where $D_1(t)$ is the time-varying delay such that

$$c\, D_1(t) = \|\boldsymbol{P}_{\mathrm{S}}(t) - \boldsymbol{P}_{\mathrm{T}}(t - D_1(t))\| \tag{7.42}$$

and

$$A_1(t) \triangleq \frac{\alpha_{\mathrm{S}}(\theta_{\mathrm{SR}}(t), \psi_{\mathrm{SR}}(t))\, \alpha_{\mathrm{T}}(\theta(t - D_1(t)), \psi(t - D_1(t)))}{\|\boldsymbol{P}_{\mathrm{S}}(t) - \boldsymbol{P}_{\mathrm{T}}(t - D_1(t))\|} \tag{7.43}$$

is the time-varying complex gain introduced by the TX-S channel. Furthermore, the LTV system describing the S-RX channel is

$$h_2(t, u) = A_2(t)\, \delta(u - t + D_2(t)) \tag{7.44}$$

where $D_2(t)$ is the time-varying delay such that

$$c\, D_2(t) = \|\boldsymbol{P}_{\mathrm{R}}(t) - \boldsymbol{P}_{\mathrm{S}}(t - D_2(t))\| \tag{7.45}$$

and

$$A_2(t) \triangleq \frac{\alpha_{\mathrm{R}}(\theta'(t), \psi'(t))\, \alpha_{\mathrm{S}}(\theta_{\mathrm{ST}}(t - D_2(t)), \psi_{\mathrm{ST}}(t - D_2(t)))}{\|\boldsymbol{P}_{\mathrm{R}}(t) - \boldsymbol{P}_{\mathrm{S}}(t - D_2(t))\|} \tag{7.46}$$

is the time-varying complex gain introduced by the S-RX channel.

Let u, t_s, and t, time measured at transmitter, scatterer, and receiver, respectively. From (7.41) it follows that a wavefront received at S at time t_s is transmitted from TX at time

$$u = t_s - D_1(t_s). \tag{7.47}$$

From (7.44) it follows that a wavefront received at RX at time t is transmitted from S at time

$$t_s = t - D_2(t). \tag{7.48}$$

According to (7.26), the signal received from the scatterer and retransmitted is

$$\begin{aligned} y_1(t_s) &= \int_{\mathbb{R}} h_1(t_s, u)\, x(u)\, \mathrm{d}u \\ &= A_1(t_s)\, x(t_s - D_1(t_s)) \end{aligned} \tag{7.49}$$

and the signal at the receiver is

$$\begin{aligned} y_2(t) &= \int_{\mathbb{R}} h_2(t, t_s)\, y_1(t_s)\, \mathrm{d}t_s \\ &= A_2(t)\, y_1(t - D_2(t)) \\ &= \underbrace{A_2(t)\, A_1(t - D_2(t))}_{A(t)}\, x(t - \underbrace{[D_2(t) + D_1(t - D_2(t))]}_{D(t)}). \end{aligned} \tag{7.50}$$

Thus, the overall LTV system introduces a time-varying delay

$$D(t) = D_2(t) + D_1(t - D_2(t)) \tag{7.51}$$

which, accordingly with (7.47) and (7.48), is the overall delay experienced by a wavefront transmitted at

$$u = t_s - D_1(t_s) = t - D_2(t) - D_1(t - D_2(t)) = t - D(t). \tag{7.52}$$

Moreover, the overall attenuation is

$$A(t) = A_2(t)\, A_1(t - D_2(t)). \tag{7.53}$$

The radar cross-section (RCS) of the scatterer is

$$\begin{aligned} \mathrm{RCS} = {} & \alpha_{\mathrm{S}}(\theta_{\mathrm{SR}}(t - D_2(t)), \psi_{\mathrm{SR}}(t - D_2(t))) \\ & \alpha_{\mathrm{S}}(\theta_{\mathrm{ST}}(t - D_2(t)), \psi_{\mathrm{ST}}(t - D_2(t))). \end{aligned} \tag{7.54}$$

7.1.7 Stationary TX, Reflection on Point Moving Scatterer, Stationary RX (Stationary Bistatic Radar)

In a stationary bistatic radar system, both TX and RX are stationary and a reflection occurs on a moving target S located in $\boldsymbol{P}_{\mathrm{S}}$ (Figure 7.3). This case is obtained by specializing that in Section 7.1.6.

Let t_s be the time at the moving scatterer S and $D_1(t_s)$ the delay experienced by the transmitted wavefront before being received by the scatterer S at time t_s. Let $R_1(t_s)$ be the time-varying distance between the stationary TX and the moving scatterer S (first path). Define, analogously to (7.34),

$$R_1(t_s) \triangleq \|\boldsymbol{P}_{\mathrm{S}}(t_s) - \boldsymbol{P}_{\mathrm{T}}\|. \tag{7.55}$$

According to (7.35) we have

$$c\, D_1(t_s) = R_1(t_s). \tag{7.56}$$

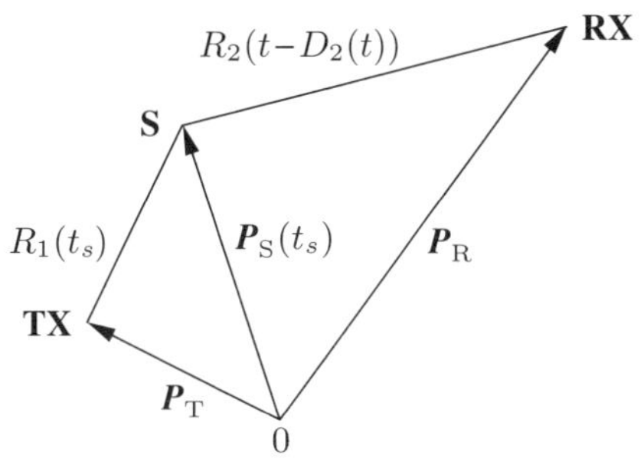

Figure 7.3 TX stationary, RX stationary, reflection on a moving scatterer S (stationary bistatic radar)

Now, let t be the time at the stationary RX and $D_2(t)$ the delay experienced by the wavefront transmitted at time t_s by the moving scatterer S before being received at time t. Due to (7.48), $t_s = t - D_2(t)$. Let $R_2(t)$ be the time-varying distance between the moving scatterer S and the stationary RX (second path). Define, analogously to (7.37),

$$R_2(t - D_2(t)) \triangleq \|\boldsymbol{P}_{\mathrm{R}} - \boldsymbol{P}_{\mathrm{S}}(t - D_2(t))\|. \tag{7.57}$$

According to (7.38) we have

$$c\, D_2(t) = R_2\,(t - D_2(t))\,. \tag{7.58}$$

The overall delay $D(t)$ experienced by a wavefront transmitted at $t - D(t)$ by TX and received at time t by RX is given by (7.51) which specializes into

$$\begin{aligned} D(t) &\triangleq D_1(t_s) + D_2(t) \\ &= D_1(t - D_2(t)) + D_2(t) \\ &= \frac{R_1(t - D_2(t))}{c} + \frac{R_2(t - D_2(t))}{c}. \end{aligned} \tag{7.59}$$

7.1.8 (Stationary) Monostatic Radar

By considering the special case of (7.59) when $\boldsymbol{P}_{\mathrm{T}} \equiv \boldsymbol{P}_{\mathrm{R}}$, we obtain the time-varying delay for the monostatic radar. In such a case, setting $R_1(t) = R_2(t) = R(t)$ and $D_1(t) = D_2(t) = D(t)/2$ in (7.59) leads to

$$c\, D(t) = 2R\left(t - \frac{D(t)}{2}\right). \tag{7.60}$$

Equation (7.60) can be interpreted as follows. Let t be the time at the receiver and $D(t)$ the delay experienced by the transmitted wavefront. The wavefront transmitted at $t - D(t)$ is reflected by the target at $t - D(t)/2$. Consequently, the delay $D(t)$ is due to the round-trip corresponding to the distance of the target from the monostatic stationary radar antenna at the time of reflection, that is at $t - D(t)/2$.

A more detailed analysis can be carried out accounting for boundary conditions for the electric and magnetic fields with relativistic correction on the target (Cooper 1980). In the

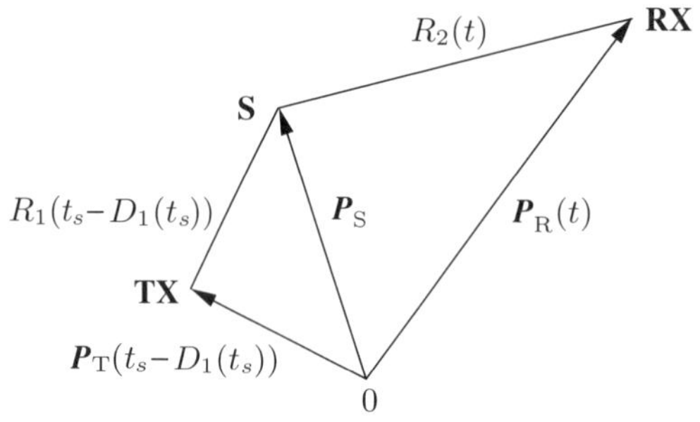

Figure 7.4 TX moving, RX moving, reflection on a stationary scatterer S

special case of a plane wave scattered by a plane surface moving along the same direction of the propagation vector, the scattered waveform is amplitude modulated by a term proportional to $(1 + \dot{R}(t)/c)/(1 - \dot{R}(t)/c)$ (Cooper 1980).

7.1.9 Moving TX, Reflection on a Stationary Scatterer, Moving RX

Let both TX and RX be moving, with a reflection occurring on a stationary scatterer S located in $\boldsymbol{P}_{\rm S}$ (Figure 7.4). This case is obtained by specializing that considered in Section 7.1.6 when both TX and RX are moving and the scatterer is stationary.

Let t_s be the time at the stationary scatterer S and $D_1(t_s)$ the delay experienced by the transmitted wavefront before being received by the scatterer S at time t_s. Let $R_1(t_s)$ be the time-varying distance between the moving TX and the stationary scatterer S (first path). Define, analogously to (7.37),

$$R_1(t_s - D_1(t_s)) \triangleq \|\boldsymbol{P}_{\rm S} - \boldsymbol{P}_{\rm T}(t_s - D_1(t_s))\|. \tag{7.61}$$

According to (7.38) we have

$$c\, D_1(t_s) = R_1\,(t_s - D_1(t_s))\,. \tag{7.62}$$

Now, let t be the time at RX and $D_2(t)$ the delay experienced by the wavefront transmitted at time t_s by the scatterer S before being received at time t. Due to (7.48), $t_s = t - D_2(t)$. Let $R_2(t)$ be the time-varying distance between the stationary scatterer S and the moving RX (second path). Define, analogously to (7.34),

$$R_2(t) \triangleq \|\boldsymbol{P}_{\rm R}(t) - \boldsymbol{P}_{\rm S}\|. \tag{7.63}$$

According to (7.35) we have

$$c\, D_2(t) = R_2\,(t)\,. \tag{7.64}$$

The overall delay $D(t)$ experienced by a wavefront received by RX at time t is given by (7.51)

$$D(t) \triangleq D_1(t_s) + D_2(t) = D_1(t - D_2(t)) + D_2(t). \tag{7.65}$$

Therefore,

$$\begin{aligned} c\,D(t) &\triangleq c\,[D_1(t_s) + D_2(t)] \\ &= R_1\,(t_s - D_1(t_s)) + R_2(t) \\ &= R_1\,(t - D_2(t) - D_1(t - D_2(t))) + R_2(t). \end{aligned} \tag{7.66}$$

Due to (7.52), this results agrees with that in (Sadowsky and Kafedziski 1998).

The example considered in this section includes, as a special case, a synthetic aperture radar (SAR) observing a stationary scene. For a SAR, both TX and RX antennas move along the same direction and stay at the same distance provided that the speed of the aircraft is constant.

7.2 Constant Velocity Vector

In this section, the case of constant velocity vector is analyzed. The relative radial speed is not necessarily constant.

7.2.1 Stationary TX, Moving RX

By specializing (7.34) with

$$\boldsymbol{P}_{\mathrm{R}}(t) = v\,(t - t_0)\,\boldsymbol{i}_1 \qquad\qquad \boldsymbol{P}_0 = R_0\,\boldsymbol{i}_2 \tag{7.67}$$

with $\boldsymbol{i}_1 \perp \boldsymbol{i}_2$ (without lack of generality) and $\|\boldsymbol{i}_1\| = \|\boldsymbol{i}_2\| = 1$, we get (Figure 7.5)

$$c\,D(t) = R(t) = \sqrt{R_0^2 + v^2(t - t_0)^2}. \tag{7.68}$$

where the first equality is due to (7.35). RX is moving with constant velocity vector $v\boldsymbol{i}_1$. The (non-constant) radial speed is

$$\overset{\bullet}{R}(t) = \frac{\mathrm{d}}{\mathrm{d}t}\left[R_0^2 + v^2(t - t_0)^2\right]^{1/2} = \frac{1}{2}\left[R_0^2 + v^2(t - t_0)^2\right]^{-1/2} v^2\,2(t - t_0) \tag{7.69}$$

and it results that

$$\lim_{t \to +\infty} \overset{\bullet}{R}(t) = |v|. \tag{7.70}$$

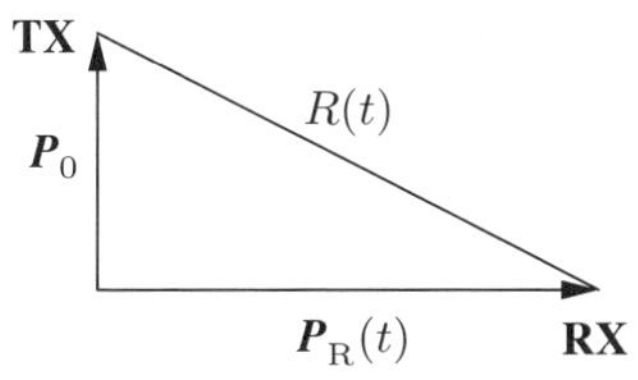

Figure 7.5 Constant velocity vector: TX stationary, RX moving

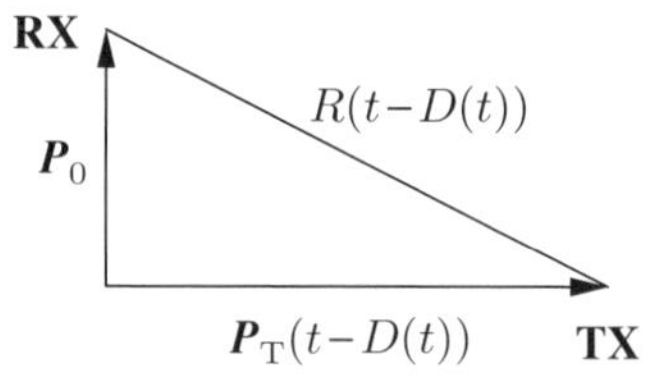

Figure 7.6 Constant velocity vector: TX moving, RX stationary

7.2.2 *Moving TX, Stationary RX*

By specializing (7.37) with

$$\boldsymbol{P}_{\mathrm{T}}(t) = v\,(t - t_0)\,\boldsymbol{i}_1 \qquad\qquad \boldsymbol{P}_0 = R_0\,\boldsymbol{i}_2 \tag{7.71}$$

with $\boldsymbol{i}_1 \perp \boldsymbol{i}_2$ (without lack of generality) and $\|\boldsymbol{i}_1\| = \|\boldsymbol{i}_2\| = 1$, we get (Figure 7.6)

$$R(t - D(t)) = \sqrt{R_0^2 + v^2(t - D(t) - t_0)^2}. \tag{7.72}$$

From (7.38) we have

$$\begin{aligned} c^2\,D^2(t) &= R^2(t - D(t)) \\ &= R_0^2 + v^2(t - D(t) - t_0)^2 \\ &= R_0^2 + v^2\left[(t - t_0)^2 - 2\,(t - t_0)\,D(t) + D^2(t)\right]. \end{aligned} \tag{7.73}$$

That is,

$$\underbrace{(c^2 - v^2)}_{A}\;D^2(t) + \underbrace{2\,v^2\,(t - t_0)}_{B}\;D(t) + \underbrace{[-R_0^2 - v^2\,(t - t_0)^2]}_{C} = 0. \tag{7.74}$$

Hence

$$\begin{aligned} B^2 - 4AC &= 4\,v^4\,(t - t_0)^2 + 4\,(c^2 - v^2)\,(R_0^2 + v^2\,(t - t_0)^2) \\ &= 4\,(c^2 - v^2)R_0^2 + 4\,c^2\,v^2\,(t - t_0)^2 \end{aligned} \tag{7.75}$$

with $B^2 - 4AC \geqslant 0\;\; \forall t \in \mathbb{R}$ since $c > |v|$. Thus,

$$\begin{aligned} D(t) &= \frac{-B \pm \sqrt{B^2 - 4AC}}{2A} \\ &= \frac{-v^2\,(t - t_0) \pm \sqrt{(c^2 - v^2)R_0^2 + c^2\,v^2\,(t - t_0)^2}}{c^2 - v^2} \end{aligned} \tag{7.76}$$

where the solution with "+" sign provides a positive delay $D(t)$. This result is in accordance with (Quinn and Hannan 2001).

7.3 Constant Relative Radial Speed

Let $\xi(t)$ be the projection of the vector $\boldsymbol{P}_{\mathrm{R}}(t) - \boldsymbol{P}_{\mathrm{T}}(t - D(t))$ along the propagation direction $\boldsymbol{i}_k$ (oriented from $\boldsymbol{P}_{\mathrm{T}}(t - D(t))$ to $\boldsymbol{P}_{\mathrm{R}}(t)$), that is

$$\xi(t) \triangleq [\boldsymbol{P}_{\mathrm{R}}(t) - \boldsymbol{P}_{\mathrm{T}}(t - D(t))] \cdot \boldsymbol{i}_k. \tag{7.77}$$

The delay $D(t)$ experienced by the wavefront transmitted at time $t - D(t)$ in $\boldsymbol{P}_{\mathrm{T}}(t - D(t))$ and received at time t in $\boldsymbol{P}_{\mathrm{R}}(t)$ depends on the distance $\|\boldsymbol{P}_{\mathrm{R}}(t) - \boldsymbol{P}_{\mathrm{T}}(t - D(t))\|$. Accordingly with (7.11) we have

$$cD(t) = \|\boldsymbol{P}_{\mathrm{R}}(t) - \boldsymbol{P}_{\mathrm{T}}(t - D(t))\| = \varrho(t, D(t)) = |\xi(t)|. \tag{7.78}$$

As observed in Section 7.1.3, $|\xi(t)|$ is the time-varying distance responsible of the Doppler effect, viz, the time-varying delay $D(t)$, and should not be confused with

$$R(t) = \|\boldsymbol{P}_{\mathrm{R}}(t) - \boldsymbol{P}_{\mathrm{T}}(t)\|. \tag{7.79}$$

Let us assume that the relative radial speed between $\boldsymbol{P}_{\mathrm{T}}(t - D(t))$ and $\boldsymbol{P}_{\mathrm{R}}(t)$ is constant within the observation interval. Thus, the projection of the vector $\boldsymbol{P}_{\mathrm{R}}(t) - \boldsymbol{P}_{\mathrm{T}}(t - D(t))$ along the propagation direction $\boldsymbol{i}_k$ is a linear function of t. That is,

$$\xi(t) \triangleq [\boldsymbol{P}_{\mathrm{R}}(t) - \boldsymbol{P}_{\mathrm{T}}(t - D(t))] \cdot \boldsymbol{i}_k = \xi_0 + v_\xi\,(t - t_0). \tag{7.80}$$

Furthermore, let us assume that, within the observation interval, RX does not collapse on TX. Thus, $\xi(t)$ does not change sign and it is always $\varrho(t, D(t)) > 0$. Under such an assumption, $\varrho(t, D(t)) = \xi(t)$ or $\varrho(t, D(t)) = -\xi(t)$ is a linear function of time and the angles in TX and RX radiation patterns are constant with time:

$$(\theta(t), \psi(t)) = (\theta, \psi) \qquad \text{constant with respect to } t \tag{7.81}$$

$$(\theta'(t), \psi'(t)) = (\theta', \psi') \qquad \text{constant with respect to } t \tag{7.82}$$

7.3.1 *Moving TX, Moving RX*

The expression of $D(t)$ depends on the sign of $\xi(t)$. Specifically,

1) For all $t \in \mathbb{R}$ such that $\xi(t) \geqslant 0$ from (7.78) it follows that

$$cD(t) = \xi(t). \tag{7.83}$$

2) For all $t \in \mathbb{R}$ such that $\xi(t) < 0$ from (7.78) it follows that

$$cD(t) = -\xi(t). \tag{7.84}$$

Therefore, from (7.80) we have (upper sign for $\xi(t) \geqslant 0$ and lower sign for $\xi(t) < 0$)

$$D(t) = \pm\frac{1}{c}\xi(t) = \pm\left(\frac{\xi_0}{c} + \frac{v_\xi}{c}\,(t - t_0)\right) \tag{7.85}$$

and

$$\begin{aligned} t - D(t) &= t \mp \frac{\xi_0}{c} \mp \frac{v_\xi}{c}(t - t_0) = \left(1 \mp \frac{v_\xi}{c}\right)t - \left(\pm\frac{\xi_0}{c} \mp \frac{v_\xi}{c}t_0\right) \\ &= st - d_0 \end{aligned} \tag{7.86}$$

where

$$s \triangleq \begin{cases} 1 - v_\xi/c & \forall t \,:\, \xi(t) \geqslant 0 \\ 1 + v_\xi/c & \forall t \,:\, \xi(t) < 0 \end{cases} \tag{7.87}$$

$$d_0 \triangleq \begin{cases} (\xi_0 - v_\xi t_0)/c & \forall t \,:\, \xi(t) \geqslant 0 \\ -(\xi_0 - v_\xi t_0)/c & \forall t \,:\, \xi(t) < 0 \end{cases} \tag{7.88}$$

For a complex sinewave transmitted signal $x_T(t) = e^{j2\pi f_T t}$ the component of the electric field along the direction of $\boldsymbol{e}(\boldsymbol{P}_R(t), t)$ in the RX location $\boldsymbol{P}_R(t)$ at time t is

$$\begin{aligned} e(\boldsymbol{P}_R(t), t) &= \alpha_T(\theta, \psi, f_T)\, \frac{e^{j2\pi f_T(t - D(t))}}{|\xi(t)|} \\ &= \alpha_T(\theta, \psi, f_T)\, \frac{e^{j2\pi f_T(st - d_0)}}{|\xi(t)|} \end{aligned} \tag{7.89}$$

where s is given in (7.87).

Assumption 7.3.1 *It results that* $|v_\xi\,(t - t_0)| \ll |\xi_0| \quad \forall t \in [t_0, t_0 + T]$ *(observation interval). That is,*

$$|v_\xi| T \ll |\xi_0|. \tag{7.90}$$

Thus, since $(1 + x)^\alpha \simeq 1 + \alpha x$ *for* $|x| \ll 1$, *the following approximation holds for the attenuation factor:*

$$\begin{aligned} \frac{1}{|\xi_0 + v_\xi\,(t - t_0)|} &= \frac{1}{|\xi_0|} \left| \frac{1}{1 + v_\xi\,(t - t_0)/\xi_0} \right| \\ &\simeq \frac{1}{|\xi_0|} \left| 1 - \frac{v_\xi\,(t - t_0)}{\xi_0} \right| \qquad \forall t \in [t_0, t_0 + T]. \end{aligned} \tag{7.91}$$

Note that a similar approximation cannot be made in the argument of the complex exponential $e^{j2\pi f_T(t - D(t))} = e^{j2\pi f_T(t \mp \xi(t)/c)}$ *since phases must be considered modulo* 2π. □

That is, the attenuation factor equals a constant term $1/|\xi_0|$ plus an amplitude modulating factor proportional to $t - t_0$. The amplitude-modulation operator t, in the frequency domain, corresponds to the derivative operator $\mathrm{d}/\mathrm{d}f$. Therefore, when $|v_\xi| T \ll |\xi_0|$, the approximation $(\xi_0 + v_\xi\,(t - t_0))^{-1} \simeq 1/\xi_0, \forall t \in [t_0, t_0 + T]$ consists in neglecting a small term which, in the frequency domain, is proportional to a T-finite approximation of $\delta^{(1)}(f - sf_T)$, the first-order derivative of the Dirac delta centered in sf_T (Zemanian 1987, Section 1.3).

Under Assumption 7.3.1 and making the approximation $1/|\xi(t)| \simeq 1/|\xi_0|$ into the expression of $e(\boldsymbol{P}_{\mathrm{R}}(t), t)$, the electric field at RX is a complex sinewave with frequency $f_{\mathrm{R}} = s f_{\mathrm{T}}$. Thus, the signal at the RX antenna clamps can be expressed in the following equivalent forms

$$x_{\mathrm{R}}(t; f_{\mathrm{T}}) = \alpha_{\mathrm{R}}(\theta', \psi', s f_{\mathrm{T}})\, \alpha_{\mathrm{T}}(\theta, \psi, f_{\mathrm{T}})\, \frac{e^{j2\pi f_{\mathrm{T}}(st-d_0)}}{|\xi_0|}$$

$$= B(f_{\mathrm{T}})\, e^{j2\pi f_{\mathrm{T}}\,(st-d_0)} \tag{7.92a}$$

$$= G(f_{\mathrm{T}})\, e^{j2\pi s f_{\mathrm{T}} t} \tag{7.92b}$$

where

$$G(f_{\mathrm{T}}) \triangleq B(f_{\mathrm{T}})\, e^{-j2\pi f_{\mathrm{T}} d_0} \triangleq \frac{\alpha_{\mathrm{R}}(\theta', \psi', s f_{\mathrm{T}})\, \alpha_{\mathrm{T}}(\theta, \psi, f_{\mathrm{T}})}{|\xi_0|}\, e^{-j2\pi f_{\mathrm{T}} d_0}. \tag{7.93}$$

The transmitted frequency f_{T} is transformed into the received frequency

$$f_{\mathrm{R}} = s f_{\mathrm{T}} = (1 \mp v_{\xi}/c)\, f_{\mathrm{T}} = f_{\mathrm{T}} + f_{\mathrm{D}} \tag{7.94}$$

where

$$f_{\mathrm{D}} \triangleq \mp \frac{v_{\xi}}{c}\, f_{\mathrm{T}} \tag{7.95}$$

is referred to as *Doppler shift* introduced by the time-varying channel.

Let us consider a generic transmitted signal $x(t)$ with Fourier transform $X(f)$. Since Maxwell's equations are linear, starting from (7.92a) the signal at the RX antenna clamps can be expressed as

$$y(t) = \int_{\mathbb{R}} X(f_{\mathrm{T}})\, x_{\mathrm{R}}(t; f_{\mathrm{T}})\, \mathrm{d} f_{\mathrm{T}}$$

$$= \int_{\mathbb{R}} X(f_{\mathrm{T}})\, B(f_{\mathrm{T}})\, e^{j2\pi f_{\mathrm{T}}\,(st-d_0)}\, \mathrm{d} f_{\mathrm{T}}$$

$$= \Big[x(u) \underset{u}{\otimes} b(u) \Big]_{u=st-d_0} \tag{7.96}$$

where $b(t)$ is the inverse Fourier transform of $B(f_{\mathrm{T}})$. Equivalently, from (7.92b), we have

$$y(t) = \int_{\mathbb{R}} X(\lambda)\, G(\lambda)\, e^{j2\pi s \lambda t}\, \mathrm{d}\lambda \tag{7.97}$$

which compared with (4.82a) shows that the Doppler channel is a FOT-deterministic LTV system with the set Ω containing only one element σ and characterized by

$$G_{\sigma}(\lambda) \equiv G(\lambda) = B(\lambda)\, e^{-j2\pi\lambda d_0} \tag{7.98}$$

$$\varphi_{\sigma}(\lambda) = s\lambda. \tag{7.99}$$

Moreover, by taking the Fourier transform of both sides of (7.97), we have

$$Y(f) = \int_{\mathbb{R}} y(t)\, e^{-j2\pi f t}\, \mathrm{d}t$$

$$= \int_{\mathbb{R}} X(\lambda)\, G(\lambda) \underbrace{\int_{\mathbb{R}} e^{-j2\pi(f-s\lambda)t}\, dt}_{\delta(f-\lambda s)=\delta(\lambda-f/s)/|s|} d\lambda$$

$$= X\left(\frac{f}{s}\right) \frac{1}{|s|}\, G\left(\frac{f}{s}\right). \tag{7.100}$$

Thus, accordingly with (4.81b), the FOT-LTV system can also be characterized by the functions

$$H_\sigma(f) = \frac{1}{|s|}\, G\left(\frac{f}{s}\right) \tag{7.101}$$

$$\psi_\sigma(f) = \frac{f}{s}. \tag{7.102}$$

Under the wide-band TX and RX antenna assumption (Assumption 7.1.2), (7.96) specializes into

$$y(t) = b\, x(st - d_0) \tag{7.103}$$

where

$$b \triangleq \frac{\alpha_{\mathrm{R}}(\theta', \psi')\, \alpha_{\mathrm{T}}(\theta, \psi)}{|\xi_0|}. \tag{7.104}$$

Thus, by considering the transmitted signal

$$x(t) = \mathrm{Re}\left\{\widetilde{x}(t)\, e^{j2\pi f_c t}\right\} \tag{7.105}$$

where $\mathrm{Re}\{\cdot\}$ denotes real part and f_c is the carrier frequency, the result is that

$$\begin{aligned} y(t) &= \mathrm{Re}\left\{b\, \widetilde{x}(st - d_0)\, e^{j2\pi f_c(st-d_0)}\right\} \\ &= \mathrm{Re}\left\{b\, e^{-j2\pi f_c d_0}\, \widetilde{x}(st - d_0)\, e^{j2\pi(s-1)f_c t}\, e^{j2\pi f_c t}\right\} \\ &= \mathrm{Re}\left\{\widetilde{y}(t)\, e^{j2\pi f_c t}\right\} \end{aligned} \tag{7.106}$$

where

$$\widetilde{y}(t) \triangleq a\, \widetilde{x}(st - d_0)\, e^{j2\pi \nu t} \tag{7.107}$$

with complex gain

$$a \triangleq b\, e^{-j2\pi f_c d_0} \tag{7.108}$$

and frequency shift

$$\nu \triangleq (s-1)\, f_c = \mp \frac{\upsilon_\xi}{c}\, f_c. \tag{7.109}$$

7.3.2 Stationary TX, Moving RX

In the case of stationary TX and moving RX, (7.11) specializes into (7.34) and (7.35). Assuming that during the observation interval RX does not collapse on TX, we have

$$cD(t) = \|\boldsymbol{P}_{\mathrm{R}}(t) - \boldsymbol{P}_0\| = \varrho(t, D(t)) = |\xi(t)| \tag{7.110}$$

$$cD(t) = R(t). \tag{7.111}$$

From (7.80), (7.110), and (7.111), it follows

$$|\xi(t)| = R(t) = R_0 + v\,(t - t_0) \qquad \forall t\ :\ R_0 + v\,(t - t_0) > 0 \tag{7.112}$$

and comparing (7.80) and (7.112) it results in $v = v_\xi$ for $\xi \geqslant 0$ and $v = -v_\xi$ for $\xi < 0$. In addition, (7.111) and (7.112) lead to

$$D(t) = \frac{R_0}{c} + \frac{v}{c}(t - t_0). \tag{7.113}$$

The same expression can be obtained observing that, accounting for (7.112), (7.36) specializes into

$$\dot{D}\,(t) = \frac{v}{c}. \tag{7.114}$$

The time-varying delay $D(t)$ depends linearly on t: $D(t) = d_0 + d_1 t$, with $d_0 \triangleq (R_0 - vt_0)/c$ and $d_1 = v/c \triangleq 1 - s$. Furthermore

$$\begin{aligned} t - D(t) &= t - \frac{R_0}{c} - \frac{v}{c}(t - t_0) = \frac{c - v}{c}t - \left(\frac{R_0}{c} - \frac{v}{c}t_0\right) \\ &= st - d_0 \end{aligned} \tag{7.115}$$

where

$$s = 1 - \frac{v}{c} \tag{7.116}$$

is the time-scale factor due to the time-varying delay.

For a complex sinewave transmitted signal $x_T(t) = e^{j2\pi f_{\mathrm{T}} t}$ the component of the electric field along the direction of $\boldsymbol{e}(\boldsymbol{P}_{\mathrm{R}}(t), t)$ in the RX location $\boldsymbol{P}_{\mathrm{R}}(t)$ at time t is

$$e(\boldsymbol{P}_{\mathrm{R}}(t), t) = \alpha_{\mathrm{T}}(\theta, \psi, f_{\mathrm{T}})\,\frac{e^{j2\pi f_{\mathrm{T}}(st - d_0)}}{R_0 + v\,(t - t_0)} \tag{7.117}$$

where (7.17), (7.34), and (7.81) are accounted for. Thus, the electric field at RX is a sinewave with frequency $f_{\mathrm{R}} = sf_{\mathrm{T}} = (1 - v/c) f_{\mathrm{T}}$ amplitude-modulated by $(R_0 + v\,(t - t_0))^{-1}$.

Under Assumption 7.3.1, $|v|T \ll R_0$, and

$$\begin{aligned} \frac{1}{R_0 + v\,(t - t_0)} &= \frac{1}{R_0}\,\frac{1}{1 + v\,(t - t_0)/R_0} \\ &\simeq \frac{1}{R_0}\left(1 - \frac{v\,(t - t_0)}{R_0}\right) \qquad \forall t \in [t_0, t_0 + T]. \end{aligned} \tag{7.118}$$

Under Assumption 7.3.1, and making the approximation $(R_0 + v\,(t - t_0))^{-1} \simeq 1/R_0$, the electric field at RX is a sinewave with frequency $f_R = s f_T$. Thus, the signal at the RX antenna clamps is

$$x_R(t; f_T) = G(f_T)\, e^{j2\pi s f_T t} \tag{7.119}$$

where

$$G(f_T) \triangleq \frac{\alpha_R(\theta', \psi', s f_T)\, \alpha_T(\theta, \psi, f_T)}{R_0}\, e^{-j2\pi f_T d_0}. \tag{7.120}$$

The transmitted frequency f_T is transformed into the received frequency

$$f_R = s f_T = (1 - v/c) f_T = f_T + f_D \tag{7.121}$$

where

$$f_D \triangleq -\frac{v}{c} f_T \tag{7.122}$$

is the *Doppler shift* introduced by the time-varying channel. For $s \neq 1$ the propagation channel is a FOT-deterministic LTV channel (Section 6.3.8), (Izzo and Napolitano 2002a).

For a transmitted frequency $f_T > 0$, accounting for (7.112), we have:

- $\boldsymbol{P}_R(t)$ moves away from TX $\Rightarrow\ v > 0\ \Rightarrow\ f_D = -(v/c) f_T < 0\ \Rightarrow\ f_R < f_T$
- $\boldsymbol{P}_R(t)$ approaches TX $\Rightarrow\ v < 0\ \Rightarrow\ f_D = -(v/c) f_T > 0\ \Rightarrow\ f_R > f_T$

Let us now consider a generic transmitted signal $x(t)$ with Fourier transform $X(f)$. Since Maxwell's equations are linear, by reasoning as for (7.96) the signal at the RX antenna clamps can be expressed as

$$y(t) = \Big[x(u) \underset{u}{\otimes} b(u)\Big]_{u = st - d_0} \tag{7.123}$$

where $b(t)$ is the inverse Fourier transform of

$$B(f_T) \triangleq \frac{\alpha_R(\theta', \psi', s f_T)\, \alpha_T(\theta, \psi, f_T)}{R_0} \tag{7.124}$$

which describes the filtering effect of the TX and RX antenna radiation patterns. Similarly to the case of moving TX and moving RX, the system with input/output relationship (7.123) can be expressed in terms of functions $G_\sigma(\lambda)$, $\varphi_\sigma(\lambda)$, $H_\sigma(f)$, $\psi_\sigma(f)$ that characterize FOT-deterministic LTV systems.

Under the wide-band TX and RX antenna assumption (Assumption 7.1.2), (7.123) specializes into

$$y(t) = b\, x(st - d_0) \tag{7.125}$$

where

$$b \triangleq \frac{\alpha_R(\theta', \psi')\, \alpha_T(\theta, \psi)}{R_0}. \tag{7.126}$$

Thus, by considering the transmitted signal

$$x(t) = \mathrm{Re}\left\{\widetilde{x}(t)\, e^{j2\pi f_c t}\right\} \tag{7.127}$$

where Re{·} denotes real part and f_c is the carrier frequency, and reasoning as for (7.107), the result is that

$$y(t) = \mathrm{Re}\left\{\widetilde{y}(t)\, e^{j2\pi f_c t}\right\} \tag{7.128}$$

where

$$\widetilde{y}(t) \triangleq a\, \widetilde{x}(st - d_0)\, e^{j2\pi \nu t} \tag{7.129}$$

with complex gain

$$a \triangleq b\, e^{-j2\pi f_c d_0} \tag{7.130}$$

and frequency shift

$$\nu \triangleq (s-1)\, f_c = -\frac{v}{c}\, f_c = -d_1\, f_c. \tag{7.131}$$

7.3.3 Moving TX, Stationary RX

In the case of moving TX and stationary RX, (7.11) specializes into (7.37) and (7.38). Assuming that during the observation interval TX does not collapse on RX, we have

$$cD(t) = \|\boldsymbol{P}_0 - \boldsymbol{P}_T(t - D(t))\| = \varrho(t, D(t)) = |\xi(t)| \tag{7.132}$$

$$cD(t) = R(t - D(t)). \tag{7.133}$$

From (7.80), (7.132), and (7.133), it follows

$$D(t) = \pm\frac{\xi(t)}{c} = \pm\left[\frac{\xi_0}{c} + \frac{v_\xi}{c}\,(t - t_0)\right] \tag{7.134}$$

where the upper sign is for $\xi(t) \geqslant 0$ and the lower for $\xi(t) < 0$. By substituting this expression of $D(t)$ into (7.133) leads to

$$\pm(\xi_0 + v_\xi\,(t - t_0)) = R\left(t \mp \left[\frac{\xi_0}{c} + \frac{v_\xi}{c}\,(t - t_0)\right]\right) \tag{7.135}$$

which implies that $R(t)$ must be a linear function of t, say

$$R(t) = R_0 + v\,(t - t_0) \qquad \forall t \,:\, R_0 + v\,(t - t_0) > 0 \tag{7.136}$$

where the strict inequality is consequence of the assumptions that within the observation interval TX does not collapse on RX. Substituting (7.136) into (7.135) we have

$$\pm\xi_0 \pm v_\xi (t - t_0) = R_0 + v \left(t \mp \left[\frac{\xi_0}{c} + \frac{v_\xi}{c} (t - t_0) \right] - t_0 \right). \tag{7.137}$$

Since (7.137) holds for all t such that $R(t - D(t)) > 0$, it necessarily results

$$R_0 = \pm\xi_0 \left(1 + \frac{v}{c} \right) \tag{7.138}$$

$$v (t - t_0) = \pm v_\xi \left(1 + \frac{v}{c} \right) (t - t_0). \tag{7.139}$$

Therefore,

$$D(t) = \pm\frac{\xi_0}{c} \pm \frac{v_\xi}{c} (t - t_0) = \frac{R_0}{c + v} + \frac{v}{c + v} (t - t_0). \tag{7.140}$$

That is, the time-varying delay $D(t)$ depends linearly on t: $D(t) = d_0 + d_1 t$, with $d_0 \triangleq (R_0 - v t_0)/(c + v)$ and $d_1 = v/(c + v) \triangleq 1 - s$. In addition, the time-varying distance $R(t)$ between TX and RX is such that

$$\begin{aligned} R(t - D(t)) &= R_0 + v (t - D(t) - t_0) \\ &= R_0 + v \left(t - \frac{R_0}{c + v} - \frac{v}{c + v} (t - t_0) - t_0 \right) \\ &= \frac{c}{c + v} \Big(R_0 + v (t - t_0) \Big). \end{aligned} \tag{7.141}$$

Furthermore

$$\begin{aligned} t - D(t) &= t - \frac{R_0}{c + v} - \frac{v}{c + v} (t - t_0) = \frac{c}{c + v} t - \left(\frac{R_0}{c + v} - \frac{v}{c + v} t_0 \right) \\ &= st - d_0 \end{aligned} \tag{7.142}$$

where

$$s \triangleq \frac{c}{c + v} \tag{7.143}$$

is the time-scale factor due to the time-varying delay.

Finally, note that the expression of $D(t)$ in the rhs of (7.140) can be also obtained observing that, accounting for (7.136), (7.40) specializes into

$$\dot{D}(t) = \frac{v}{c + v}. \tag{7.144}$$

For a complex sinewave transmitted signal $x_T(t) = e^{j2\pi f_T t}$ the component of the electric field along the direction of $\boldsymbol{e}(\boldsymbol{P}_R(t), t)$ in the RX location $\boldsymbol{P}_R(t)$ at time t is

$$e(\boldsymbol{P}_R(t), t) = \alpha_T(\theta, \psi, f_T) \frac{e^{j2\pi f_T (st - d_0)}}{[R_0 + v (t - t_0)] c/(c + v)}. \tag{7.145}$$

Thus, the electric field at RX is a sinewave with frequency $f_R = sf_T = (c/(c+v))f_T$ amplitude-modulated by $[[R_0 + v(t - t_0)]c/(c+v)]^{-1}$.

Under Assumption 7.3.1, that is $|vT| \ll R_0$, and making the approximation $[[R_0 + v(t - t_0)]c/(c+v)]^{-1} \simeq (c+v)/(cR_0)$, $t \in (t_0, t_0 + T)$, the electric field at RX is a sinewave with frequency $f_R = sf_T$. Thus, the signal at the RX antenna clamps is

$$x_R(t; f_T) = G(f_T)\, e^{j2\pi s f_T t} \tag{7.146}$$

where

$$G(f_T) \triangleq \frac{\alpha_R(\theta', \psi', sf_T)\, \alpha_T(\theta, \psi, f_T)}{cR_0/(c+v)}\, e^{-j2\pi f_T d_0}. \tag{7.147}$$

The transmitted frequency f_T is transformed into the received frequency

$$f_R = s\, f_T = \frac{c}{c+v}\, f_T = f_T + f_D \tag{7.148}$$

where

$$f_D \triangleq -\frac{v}{c+v}\, f_T \tag{7.149}$$

is referred to as *Doppler shift* introduced by the time-varying channel. For $s \neq 1$ the propagation channel is a FOT-deterministic LTV channel (Section 4.3.1), (Izzo and Napolitano 2002a).

For a transmitted frequency $f_T > 0$, if $\boldsymbol{P}_R(t)$ moves away from TX or approaches TX we have the same kind of order relationships between f_R and f_T as in the case of stationary TX and moving RX (Section 7.3.2).

Let us now consider a generic transmitted signal $x(t)$ with Fourier transform $X(f)$. Since Maxwell's equations are linear, by reasoning as for (7.96) the signal at the RX antenna clamps can be expressed as

$$y(t) = \Big[x(u) \underset{u}{\otimes} b(u)\Big]_{u = st - d_0} \tag{7.150}$$

where $b(t)$ is the inverse Fourier transform of

$$B(f_T) \triangleq \frac{\alpha_R(\theta', \psi', sf_T)\, \alpha_T(\theta, \psi, f_T)}{cR_0/(c+v)} \tag{7.151}$$

which describes the filtering effect of the TX and RX antenna radiation patterns. Similarly to the case of moving TX and moving RX, the system with input/output relationship (7.150) can be expressed in terms of functions $G_\sigma(\lambda)$, $\varphi_\sigma(\lambda)$, $H_\sigma(f)$, $\psi_\sigma(f)$ that characterize FOT-deterministic LTV systems.

Under the wide-band TX and RX antenna assumption (Assumption 7.1.2), (7.150) specializes into

$$y(t) = b\, x(st - d_0) \tag{7.152}$$

where

$$b \triangleq \frac{\alpha_{\mathrm{R}}(\theta', \psi')\, \alpha_{\mathrm{T}}(\theta, \psi)}{c R_0/(c+v)}. \tag{7.153}$$

Thus, by considering the transmitted signal

$$x(t) = \mathrm{Re}\left\{\widetilde{x}(t)\, e^{j2\pi f_c t}\right\} \tag{7.154}$$

where Re{·} denotes real part and f_c is the carrier frequency, and reasoning as for (7.107), the result is that

$$y(t) = \mathrm{Re}\left\{\widetilde{y}(t)\, e^{j2\pi f_c t}\right\} \tag{7.155}$$

where

$$\widetilde{y}(t) \triangleq a\, \widetilde{x}(st - d_0)\, e^{j2\pi \nu t} \tag{7.156}$$

with complex gain

$$a \triangleq b\, e^{-j2\pi f_c d_0} \tag{7.157}$$

and frequency shift

$$\nu \triangleq (s-1)\, f_c = -\frac{v}{c+v}\, f_c = -d_1 f_c. \tag{7.158}$$

Remark 7.3.2 *The results of Section 7.1 are non-relativistic. For $|\dot{R}(t)/c| \ll 1$, the asymmetry in the results for stationary TX–moving RX and moving TX–stationary RX in the case of constant relative radial speed disappears.*

In fact, for stationary TX– moving RX, from (7.116) and (7.131), one obtains

$$s = 1 - \frac{v}{c} \tag{7.159}$$

$$\nu = (s-1)\, f_c = -\frac{v}{c}\, f_c. \tag{7.160}$$

Furthermore, for moving TX–stationary RX, from (7.143) and (7.158) one has

$$s = \frac{c}{c+v} = \frac{1}{1+v/c} \simeq 1 - \frac{v}{c} \tag{7.161}$$

$$\nu = (s-1)\, f_c \simeq -\frac{v}{c}\, f_c \tag{7.162}$$

where the approximate equalities hold for $|v/c| \ll 1$ and the error is of the order of $(v/c)^2$.

7.3.4 Stationary TX, Reflection on a Moving Scatterer, Stationary RX (Stationary Bistatic Radar)

In this section, the combined Doppler effect in the case of stationary TX, reflection on a moving scatterer, and stationary RX is analyzed. This scenario corresponds to a stationary bistatic radar in the presence of a moving target. The propagation model is that described in Section 7.1.7 (see also Section 7.1.6 for notation).

Assumption 7.3.3 *Both relative radial speeds of target with respect to TX (Radar 1) and RX (Radar 2) can be assumed to be constant within the observation interval.*

Up link: From stationary TX (Radar 1) to moving RX (Target or Scatterer).

The propagation channel is characterized by

$$h_1(t,u) = a_1\,\delta(u - t + D_1(t)) \tag{7.163}$$

$$D_1(t) = (1 - s_1)t + d_{01} \tag{7.164}$$

$$s_1 = 1 - v_1/c \tag{7.165}$$

$$R_1(t) = R_{01} + v_1\,(t - t_0) \tag{7.166}$$

$$cD_1(t) = R_1(t). \tag{7.167}$$

Let $x_T(t) = e^{j2\pi f_T t}$ be the complex sinewave transmitted by radar 1. Under Assumptions 7.3.1 and 7.3.3, and according to (7.119), $1/R_1(t) \simeq 1/R_{01}$ and the signal received from the target is

$$x_S(t; f_T) = G_1(f_T)\,e^{j2\pi s_1 f_T t} \tag{7.168}$$

where (for $t_0 = 0$)

$$G_1(f_T) \triangleq \frac{\alpha_S(\theta_{SR}, \psi_{SR}, s_1 f_T)\,\alpha_T(\theta, \psi, f_T)}{R_{01}}\,e^{-j2\pi f_T R_{01}/c}. \tag{7.169}$$

Down link: From moving TX (Target or Scatterer) to stationary RX (Radar 2).

The propagation channel is characterized by

$$h_2(t,u) = a_2\,\delta(u - t + D_2(t)) \tag{7.170}$$

$$D_2(t) = (1 - s_2)t + d_{02} \tag{7.171}$$

$$s_2 = c/(c + v_2) \tag{7.172}$$

$$R_2(t) = R_{02} + v_2\,(t - t_0) \tag{7.173}$$

$$cD_2(t) = R_2(t - D_2(t)). \tag{7.174}$$

The complex sinewave transmitted by the target is $x_S(t; f_T) = G_1(f_T)\,e^{j2\pi s_1 f_T t}$. Under Assumptions 7.3.1 and 7.3.3, and according to (7.146), $1/R_2(t) \simeq 1/R_{02}$ and the signal at the RX antenna clamps is

$$x_R(t; f_T) = G_2(f_T)\,G_1(f_T)\,e^{j2\pi s_2 s_1 f_T t} \tag{7.175}$$

where (for $t_0 = 0$)

$$G_2(f_\mathrm{T}) \triangleq \frac{\alpha_\mathrm{R}(\theta', \psi', s_2 s_1 f_\mathrm{T})\, \alpha_\mathrm{S}(\theta_\mathrm{ST}, \psi_\mathrm{ST}, s_1 f_\mathrm{T})}{c R_{02}/(c + v_2)}\, e^{-j2\pi s_1 f_\mathrm{T} R_{02}/(c+v_2)} \tag{7.176}$$

The overall time-scale factor is

$$s = s_2 s_1 = \frac{c}{c + v_2}\, \frac{c - v_1}{c} = \frac{c - v_1}{c + v_2}. \tag{7.177}$$

The transmitted frequency f_T is transformed into the received frequency

$$f_\mathrm{R} = s_2\, s_1\, f_\mathrm{T} = \frac{c - v_1}{c + v_2}\, f_\mathrm{T} = \left(1 - \frac{v_1 + v_2}{c + v_2}\right) f_\mathrm{T} = f_\mathrm{T} + f_\mathrm{D} \tag{7.178}$$

where

$$f_\mathrm{D} \triangleq -\frac{v_1 + v_2}{c + v_2}\, f_\mathrm{T} = (s - 1)\, f_\mathrm{T} \tag{7.179}$$

is the *Doppler shift* introduced by the overall (TX-S-RX) time-varying channel.

Let us consider a generic transmitted signal $x(t)$ with Fourier transform $X(f)$. Since Maxwell's equations are linear, the signal at the RX antenna clamps is

$$\begin{aligned} y(t) &= \int_{\mathbb{R}} X(f_\mathrm{T})\, x_\mathrm{R}(t; f_\mathrm{T})\, \mathrm{d} f_\mathrm{T} \\ &= \int_{\mathbb{R}} X(f_\mathrm{T})\, G_2(f_\mathrm{T})\, G_1(f_\mathrm{T})\, e^{j2\pi s_2 s_1 f_\mathrm{T} t}\, \mathrm{d} f_\mathrm{T} \end{aligned} \tag{7.180}$$

Under Assumption 7.1.2 on TX and RX radar antennas and modeling the target as a RX and TX antenna satisfying Assumption 7.1.2, (7.169) and (7.176) (for $t_0 = 0$) reduce to

$$G_1(f_\mathrm{T}) = \alpha_1\, e^{-j2\pi f_\mathrm{T} R_{01}/c} \tag{7.181}$$

$$G_2(f_\mathrm{T}) = \alpha_2\, e^{-j2\pi s_1 f_\mathrm{T} R_{02}/(c+v_2)} \tag{7.182}$$

with $\alpha_1 \triangleq \alpha_\mathrm{S}(\theta_\mathrm{SR}, \psi_\mathrm{SR})\, \alpha_\mathrm{T}(\theta, \psi)/R_{01}$, $\alpha_2 \triangleq \alpha_\mathrm{R}(\theta', \psi')\, \alpha_\mathrm{S}(\theta_\mathrm{ST}, \psi_\mathrm{ST})/(c R_{02}/(c + v_2))$, and (7.180) specializes into

$$\begin{aligned} y(t) &= \int_{\mathbb{R}} X(f_\mathrm{T})\, \alpha_2\, e^{-j2\pi f_\mathrm{T} s_1 R_{02}/(c+v_2)}\, \alpha_1\, e^{-j2\pi f_\mathrm{T} R_{01}/c}\, e^{j2\pi s_2 s_1 f_\mathrm{T} t}\, \mathrm{d} f_\mathrm{T} \\ &= \alpha_1 \alpha_2 \int_{\mathbb{R}} X(f_\mathrm{T})\, e^{j2\pi f_\mathrm{T}[s_1 s_2 t - R_{01}/c - s_1 R_{02}/(c+v_2)]}\, \mathrm{d} f_\mathrm{T} \\ &= a\, x(st - d) \end{aligned} \tag{7.183}$$

with $a \triangleq \alpha_1 \alpha_2$, s given by (7.177), and

$$d \triangleq \frac{R_{01}}{c} + s_1\, \frac{R_{02}}{c + v_2} = \frac{1}{c}(R_{01} + s R_{02}). \tag{7.184}$$

7.3.5 (Stationary) Monostatic Radar

For a stationary monostatic radar TX and RX antennas are coincident. Thus, the analysis is made by specializing the results for the bistatic radar to the case $R_1(t) = R_2(t)$. It follows that $v_1 = v_2 \triangleq v$, $R_{01} = R_{02} = R_0$, and

$$s = \frac{c - v}{c + v} \tag{7.185}$$

$$f_{\mathrm{D}} = -\frac{2v}{c + v} f_{\mathrm{T}} = (s - 1) f_{\mathrm{T}} \tag{7.186}$$

$$d = \frac{1}{c}(R_0 + sR_0) = \frac{2R_0}{c + v}. \tag{7.187}$$

7.3.6 Moving TX, Reflection on a Stationary Scatterer, Moving RX

In this section, the combined Doppler effect in the case of moving TX, reflection on a stationary scatterer S, and moving RX is analyzed. If TX and RX move along the same direction and stay always at the same distance, this scenario corresponds to a SAR illuminating a stationary scene. The propagation model is that described in Section 7.1.9.

If both the relative radial speeds between TX and S and between S and RX are constant within the observation interval, then

$$R_i(t) = R_{0i} + v_i\,(t - t_0) \quad i = 1, 2, \qquad \forall t\ :\ R_{0i} + v_i\,(t - t_0) > 0. \tag{7.188}$$

In the first path (moving TX, stationary receiver S), for all values of t_s such that

$$R_1(t_s - D_1(t_s)) = R_{01} + v_1\,(t_s - D_1(t_s) - t_0) > 0$$

that is, since $t_s = t - D_2(t)$, for all values of t such that

$$R_{01} + v_1\,(t - D_2(t) - D_1(t - D_2(t)) - t_0) > 0$$

accounting for (7.140), we have

$$D_1(t_s) = \frac{R_{01}}{c + v_1} + \frac{v_1}{c + v_1}\,(t_s - t_0). \tag{7.189}$$

In the second path (stationary transmitter S, moving RX) for all values of t such that

$$R_2(t) = R_{02} + v_2\,(t - t_0) > 0$$

accounting for (7.113), we have

$$D_2(t) = \frac{R_{02}}{c} + \frac{v_2}{c}\,(t - t_0). \tag{7.190}$$

Thus, the expression of the overall delay $D(t)$ in (7.65) specializes into

$$\begin{aligned}
D(t) &\triangleq D_1(t_s) + D_2(t) \\
&= D_1(t - D_2(t)) + D_2(t) \\
&= \frac{R_{01}}{c+v_1} + \frac{v_1}{c+v_1}(t - D_2(t) - t_0) + D_2(t) \\
&= \frac{R_{01}}{c+v_1} + \frac{v_1}{c+v_1}\left(t - \frac{R_{02}}{c} - \frac{v_2}{c}(t-t_0) - t_0\right) + \frac{R_{02}}{c} + \frac{v_2}{c}(t-t_0) \\
&= \frac{R_{01}+R_{02}}{c+v_1} + \frac{v_1+v_2}{c+v_1}(t-t_0).
\end{aligned} \tag{7.191}$$

7.3.7 *Non-synchronized TX and RX oscillators*

The received signal models (7.129) and (7.156) can also account for possible frequency and phase mismatch, ν_{osc} and φ_{osc}, respectively, of the local (RX) oscillator with respect to the transmitter oscillator. In such a case, it results that

$$\widetilde{y}(t) = a\, e^{j\varphi_{\text{osc}}}\, \widetilde{x}(st - d_0)\, e^{j2\pi(\nu+\nu_{\text{osc}})t} \tag{7.192}$$

and the overall frequency shift $\nu + \nu_{\text{osc}} = (s-1)f_c + \nu_{\text{osc}}$ is not proportional to f_c only through the factor $(s-1)$.

7.4 Constant Relative Radial Acceleration

According to notation introduced in Section 7.3, we have

$$\xi(t) = [\boldsymbol{P}_{\text{R}}(t) - \boldsymbol{P}_{\text{T}}(t - D(t))] \cdot \boldsymbol{i}_k \tag{7.193}$$

$$cD(t) = \|\boldsymbol{P}_{\text{R}}(t) - \boldsymbol{P}_{\text{T}}(t - D(t))\| = |\xi(t)| \tag{7.194}$$

$$R(t) = \|\boldsymbol{P}_{\text{R}}(t) - \boldsymbol{P}_{\text{T}}(t)\|\,. \tag{7.195}$$

Let us assume that the relative radial acceleration between $\boldsymbol{P}_{\text{T}}(t - D(t))$ and $\boldsymbol{P}_{\text{R}}(t)$ is constant within the observation interval, that is,

$$\ddot{\xi}(t) = a_\xi \qquad \forall t \in (t_0, t_0 + T) \tag{7.196}$$

then

$$\dot{\xi}(t) = v_\xi + a_\xi\,(t - t_0) \tag{7.197}$$

$$\xi(t) = \xi_0 + v_\xi\,(t-t_0) + \frac{1}{2}\,a_\xi\,(t-t_0)^2 \tag{7.198}$$

where $v_\xi = \dot{\xi}(t_0)$ and $\xi_0 = \xi(t_0)$.

Note that motion with constant acceleration is not compatible with the special relativity theory. In fact, from (7.197) it follows that

$$\lim_{t\to+\infty} \left|\dot{\xi}(t)\right| = +\infty > c. \tag{7.199}$$

Consequently, such a motion model can be used provided that the length T of the observation interval is not too large.

Let us assume that TX and RX do not collapse within the observation interval. Then, from (7.194) and (7.198) we have

$$D(t) = \pm \frac{\xi(t)}{c} = \pm \frac{1}{c} \left[\xi_0 + v_\xi\,(t - t_0) + \frac{1}{2}\,a_\xi\,(t - t_0)^2 \right] \tag{7.200}$$

where the upper sign is for $\xi(t) \geqslant 0$ and the lower is for $\xi(t) < 0$. That is, $D(t)$ depends quadratically on t:

$$D(t) = d_0 + d_1 t + d_2 t^2 \tag{7.201}$$

with $d_0 = \pm(\xi_0 - v_\xi t_0 + a_\xi t_0^2/2)/c$, $d_1 = \pm(v_\xi - a_\xi t_0)/c$, $d_2 = \pm a_\xi/(2c)$.

7.4.1 Stationary TX, Moving RX

In the case of stationary TX and moving RX, (7.11) specializes into (7.34) and (7.35). Assuming that during the observation interval RX does not collapse on TX, we have

$$cD(t) = \|\boldsymbol{P}_{\mathrm{R}}(t) - \boldsymbol{P}_0\| = |\xi(t)| \tag{7.202}$$

$$cD(t) = R(t). \tag{7.203}$$

From (7.198), (7.202), and (7.203), it follows

$$\begin{aligned} |\xi(t)| = R(t) = R_0 + v\,(t - t_0) + \frac{1}{2}\,a\,(t - t_0)^2 \\ \forall t\ :\ R_0 + v\,(t - t_0) + \frac{1}{2}\,a\,(t - t_0)^2 > 0 \end{aligned} \tag{7.204}$$

and comparing (7.198) and (7.204) it results in $v = v_\xi$, $a = a_\xi$ for $\xi(t) \geqslant 0$ and $v = -v_\xi$, $a = -a_\xi$ for $\xi(t) < 0$. In addition, (7.203) and (7.204) lead to

$$D(t) = \frac{R_0}{c} + \frac{v}{c}\,(t - t_0) + \frac{a}{2c}\,(t - t_0)^2. \tag{7.205}$$

That is, the time-varying delay depends quadratically on t and in (7.201) one has $d_0 = (R_0 - vt_0 + at_0^2/2)/c$, $d_1 = (v - at_0)/c$, $d_2 = a/(2c)$.

7.4.2 Moving TX, Stationary RX

In the case of moving TX and stationary RX, (7.11) specializes into (7.37) and (7.38). Assuming that during the observation interval TX does not collapse on RX, we have

$$cD(t) = \|\boldsymbol{P}_0 - \boldsymbol{P}_{\mathrm{T}}(t - D(t))\| = |\xi(t)| \tag{7.206}$$

$$cD(t) = R(t - D(t)). \tag{7.207}$$

From (7.198) and (7.206) it follows

$$D(t) = \pm\frac{\xi(t)}{c} = \pm\left[\frac{\xi_0}{c} + \frac{v_\xi}{c}(t-t_0) + \frac{a_\xi}{2c}(t-t_0)^2\right] \tag{7.208}$$

where the upper sign is for $\xi(t) \geqslant 0$ and the lower for $\xi(t) < 0$. As in the case of stationary TX and moving RX, $D(t)$ depends quadratically on t. The time behavior of $R(t)$ cannot be easily derived using (7.207) with (7.208) substituted into.

7.5 Transmitted Signal: Narrow-Band Condition

Models (7.129) and (7.156) are derived in (Van Trees 1971) for the case of relative motion between transmitter and receiver when the relative radial speed v can be considered constant within the observation interval $(t_0, t_0 + T)$. Moreover, in (Van Trees 1971) it is shown that the time–scale factor s can be assumed unity, provided that the condition

$$BT \ll \left|1 + \frac{c}{v}\right| \tag{7.209}$$

is fulfilled, where B is the bandwidth of $\widetilde{x}(t)$ and T the length of the observation interval. Even if condition (7.209) involves both the signal bandwidth and observation interval, it is generally referred to as *narrow-band condition*.

In the following, the narrow-band condition is derived for a deterministic signal with Fourier transform approaching zero sufficiently fast. The special case of a strictly band-limited signal is considered. Then, the case of a stochastic process with power spectrum approaching zero sufficiently fast is addressed.

Theorem 7.5.1 Narrow-Band Condition – Deterministic Signals. *Let $x(t)$ be a differentiable deterministic signal with Fourier transform $X(f)$ such that $X(f) = \mathcal{O}(|f|^{-\gamma})$ with $\gamma > 2$ for $|f| > B$ (that is, $|X(f)| \leqslant K|f|^{-\gamma}$ for $|f| > B$). It results that*

$$x(t+\Delta t) = x(t) + \mathcal{O}(B|\Delta t|) + \mathcal{O}(B^{-\gamma+2}|\Delta t|) \tag{7.210}$$

Proof: Let us consider the first-order Taylor series expansion with Lagrange residual term of a delayed version of the signal $x(t)$

$$\begin{aligned} x(t+\Delta t) &= x(t) + \dot{x}(\bar{t})\Delta t \\ &= x(t) + \int_{-\infty}^{+\infty} j2\pi f\, X(f)\, e^{j2\pi f\bar{t}}\, \mathrm{d}f\; \Delta t \end{aligned} \tag{7.211}$$

where $\bar{t} \in [\min(t, t+\Delta t), \max(t, t+\Delta t)]$. One has

$$\begin{aligned} &\left|\int_{-\infty}^{+\infty} j2\pi f\, X(f)\, e^{j2\pi f\bar{t}}\, \mathrm{d}f\right| \\ &\quad = \left|\int_{-B}^{B} j2\pi f\, X(f)\, e^{j2\pi f\bar{t}}\, \mathrm{d}f + \int_{|f|>B} j2\pi f\, X(f)\, e^{j2\pi f\bar{t}}\, \mathrm{d}f\right| \\ &\quad \leqslant 2\pi B \int_{-B}^{B} |X(f)|\, \mathrm{d}f + 2\pi \int_{|f|>B} |f|\, K|f|^{-\gamma}\, \mathrm{d}f \end{aligned}$$

$$
\begin{aligned}
&= 2\pi B \int_{-B}^{B} |X(f)| \, \mathrm{d}f + 2\pi 2K \int_{B}^{+\infty} f^{-\gamma+1} \, \mathrm{d}f \\
&= 2\pi B \int_{-B}^{B} |X(f)| \, \mathrm{d}f + 4K\pi \frac{1}{-\gamma+2} \left[f^{-\gamma+2} \right]_{B}^{+\infty} \\
&= 2\pi B \int_{-B}^{B} |X(f)| \, \mathrm{d}f + \frac{4K\pi}{\gamma-2} B^{-\gamma+2}
\end{aligned} \tag{7.212}
$$

from which (7.210) follows. Condition $\gamma > 2$ assures the existence of all the involved integrals.

In the special case of strictly band-limited signal, the term with the integral on $|f| > B$ is identically zero so that the term $\mathcal{O}(B^{-\gamma+2}|\Delta t|)$ is absent in (7.210). □

From (7.210) it follows that when $B|\Delta t| \ll 1$ (and, hence, a fortiori $B^{-\gamma+2}|\Delta t| \ll 1$ if $B >$ 1 Hz and $\gamma > 2$), that is when the time shift $|\Delta t|$ is much smaller that the reciprocal of the bandwidth B of the signal, then the time shift can be neglected in the argument of x. Similar results can be found in (Swick 1969).

The case of stochastic processes can be addressed by the following result.

Theorem 7.5.2 Narrow-Band Condition – Stochastic Processes. *Let $x(t)$ be a finite-power stochastic process with a.e. differentiable time-averaged autocorrelation function $R_{xx^*}(\tau)$ (Section 1.1.2) and with power spectrum $S_{xx^*}(f)$ such that $S_{xx^*}(f) = \mathcal{O}(|f|^{-\gamma})$ with $\gamma > 2$ for $|f| > B$ (that is, $S_{xx^*}(f) \leqslant K|f|^{-\gamma}$ for $|f| > B$). For all values of Δt such that $R_{xx^*}(\tau)$ is differentiable in $(0, |\Delta t|)$, one obtains*

$$
\left\langle \mathrm{E}\left\{ |x(t+\Delta t) - x(t)|^2 \right\} \right\rangle_t = \mathcal{O}(B|\Delta t|) + \mathcal{O}(B^{-\gamma+2}|\Delta t|) \tag{7.213}
$$

Proof: It results that

$$
\begin{aligned}
&\left\langle \mathrm{E}\left\{ |x(t+\Delta t) - x(t)|^2 \right\} \right\rangle_t \\
&\quad = \left\langle \mathrm{E}\left\{ |x(t+\Delta t)|^2 \right\} \right\rangle_t + \left\langle \mathrm{E}\left\{ |x(t)|^2 \right\} \right\rangle_t - 2\mathrm{Re}\left[\left\langle \mathrm{E}\left\{ x(t+\Delta t)\, x(t)^* \right\} \right\rangle_t \right] \\
&\quad = 2\mathrm{Re}\left[R_{xx^*}(0) - R_{xx^*}(\Delta t) \right]
\end{aligned} \tag{7.214}
$$

where the second equality holds since the time-average does not depend on time shifts of the signal and $R_{xx^*}(0)$ is real and nonnegative.

Let us consider the first-order Taylor series expansion with Lagrange residual term of the time-averaged autocorrelation function around the origin

$$
R_{xx^*}(\Delta t) = R_{xx^*}(0) + \dot{R}_{xx^*}(\bar{\tau})\Delta t \tag{7.215}
$$

where $\dot{R}_{xx^*}(\tau)$ denotes the first-order derivative of $R_{xx^*}(\tau)$ and $\bar{\tau} \in (0, \Delta t)$ if $\Delta t > 0$, $\bar{\tau} \in (\Delta t, 0)$ if $\Delta t < 0$. One has

$$
\begin{aligned}
&|R_{xx^*}(0) - R_{xx^*}(\Delta t)| \\
&\quad = \left| \dot{R}_{xx^*}(\bar{\tau}) \right| |\Delta t|
\end{aligned}
$$

$$
\begin{aligned}
&= \left| \int_{-\infty}^{+\infty} j2\pi f\, S_{xx^*}(f)\, e^{j2\pi f \bar{\tau}}\, \mathrm{d}f \right| \, |\Delta t| \\
&= |\Delta t| \left| \int_{-B}^{B} j2\pi f\, S_{xx^*}(f)\, e^{j2\pi f \bar{\tau}}\, \mathrm{d}f + \int_{|f|>B} j2\pi f\, S_{xx^*}(f)\, e^{j2\pi f \bar{\tau}}\, \mathrm{d}f \right| \\
&\leqslant 2\pi B |\Delta t| \int_{-B}^{B} S_{xx^*}(f)\, \mathrm{d}f + 2\pi |\Delta t| \int_{|f|>B} |f|\, K |f|^{-\gamma}\, \mathrm{d}f \\
&= 2\pi B |\Delta t| \int_{-B}^{B} S_{xx^*}(f)\, \mathrm{d}f + \frac{4K\pi}{\gamma - 2} B^{-\gamma+2} |\Delta t|.
\end{aligned}
\tag{7.216}
$$

From (7.214) and (7.216), condition (7.213) immediately follows.

Analogously to the case of deterministic signals, in the special case of strictly band-limited stochastic process, the term with the integral on $|f| > B$ is identically zero so that the term $\mathcal{O}(B^{-\gamma+2}|\Delta t|)$ is absent in (7.213). □

7.5.1 *Constant Relative Radial Speed*

Theorems 7.5.1 and 7.5.2 can be applied in the special case of the linearly time-varying delay introduced by the relative motion between transmitter and receiver with constant relative radial speed. Specifically, from (7.115) (stationary TX, moving RX) and (7.142) (moving TX, stationary RX) we have

$$
t - D(t) = st - d_0 = t - [(1-s)t + d_0]. \tag{7.217}
$$

Thus, for $t \in (t_0, t_0 + T)$ the maximum variation of $D(t)$ is

$$
\begin{aligned}
\Delta t_{\max} &= |D(t_0 + T) - D(t_0)| = |(1-s)(t_0 + T) + d_0 - (1-s)t_0 - d_0| \\
&= |1 - s| T.
\end{aligned}
\tag{7.218}
$$

Let us assume that

$$
\Delta t_{\max} = |1 - s| T \ll \frac{1}{B}. \tag{7.219}
$$

According to Theorems 7.5.1 and 7.5.2, time variations in intervals of width much smaller than $1/B$ can be neglected. Thus, for $t \in (t_0, t_0 + T)$

$$
\widetilde{x}(st - d_0) = \widetilde{x}(t - (1-s)t - d_0) \simeq \widetilde{x}(t - d_0) \tag{7.220}
$$

and

$$
y(t) \simeq \mathrm{Re}\left\{ a\, \widetilde{x}(t - d_0)\, e^{j2\pi \nu t}\, e^{j2\pi f_c t} \right\}. \tag{7.221}
$$

That is, the Doppler channel for the complex signals can be modeled as linear periodically time-variant.

In the case of stationary TX and moving RX, accounting for (7.116), $1 - s = v/c$ and condition (7.219) can be written as

$$
BT \ll \frac{1}{|1 - s|} = \left| \frac{c}{v} \right|. \tag{7.222}
$$

Moreover, in the case of moving TX and stationary RX, accounting for (7.143), $1 - s = \upsilon/(c + \upsilon)$ and condition (7.219) can be written as

$$BT \ll \frac{1}{|1 - s|} = \left|1 + \frac{c}{\upsilon}\right| \tag{7.223}$$

which is coincident with (7.209).

Note that when condition (7.219) is satisfied, then $s \simeq 1$ can be assumed in the argument of $\widetilde{x}(\cdot)$. Such a condition, however, does not allow to assume $s \simeq 1$ (and, hence, $\nu \simeq 0$) in the argument of the complex exponential $e^{j2\pi\nu t} = e^{j2\pi(s-1)f_c t}$. If condition

$$|\nu T| \ll 1 \tag{7.224}$$

holds, then

$$e^{j2\pi\nu t} \simeq e^{j2\pi\nu t_0} \quad \forall t \in (t_0, t_0 + T). \tag{7.225}$$

Thus, if both conditions (7.219) and (7.224) are satisfied, then

$$y(t) \simeq \operatorname{Re}\left\{a_1\, \widetilde{x}(t - d_0)\, e^{j2\pi f_c t}\right\} \tag{7.226}$$

where $a_1 \triangleq a\, e^{j2\pi\nu t_0}$, and the channel model for the complex signals is LTI.

In the case of stationary TX and moving RX, accounting for (7.131), condition (7.224) can be written as

$$|\nu T| = |(s - 1) f_c T| = \left|\frac{\upsilon}{c} f_c T\right| \ll 1 \tag{7.227}$$

that is

$$f_c T \ll \left|\frac{c}{\upsilon}\right|. \tag{7.228}$$

In the case of moving TX and stationary RX, accounting for (7.158), condition (7.224) can be written as

$$|\nu T| = |(s - 1) f_c T| = \left|\frac{\upsilon}{c + \upsilon} f_c T\right| \ll 1 \tag{7.229}$$

that is

$$f_c T \ll \left|1 + \frac{c}{\upsilon}\right|. \tag{7.230}$$

Conditions (7.228) and (7.230) are more stringent then (7.222) and (7.223), respectively, since $f_c \gg B$.

7.5.1.1 Example: DSSS Signal

Let us consider a direct-sequence spread-spectrum (DSSS) signal (4.60) used in code-division multiple-access (CDMA) systems with full duty-cycle rectangular chip pulse, chip period T_c, number of chips per bit N_c, and bit period $T_b = N_c T_c$. By assuming a bandwidth $B \simeq 1/T_c$

for the DSSS signal, one has

$$B \simeq 1/T_c = r_b N_c \tag{7.231}$$

$$T = N_b T_b = N_b / r_b \tag{7.232}$$

where $r_b = 1/T_b = 1/(N_c T_c)$ is the bit rate and N_b the number of processed bits.

If $v = 50\,\text{km h}^{-1} = 14\,\text{m s}^{-1}$, $c = 3 \cdot 10^8 \text{ms}^{-1}$, and $N_c = 31$, then, in the case of stationary TX and moving RX, $1 - s \simeq 0.5 \cdot 10^{-7}$ and the narrow band condition (7.222) is

$$BT \ll \frac{1}{|1-s|} = \left|\frac{c}{v}\right| \quad \Leftrightarrow \quad N_b N_c \ll \left|\frac{c}{v}\right| \quad \Leftrightarrow \quad N_b \ll 7 \cdot 10^5 \text{ bits.} \tag{7.233}$$

In order to model the channel as LTI, also condition $e^{j2\pi\nu t} \simeq 1$ must hold within the observation interval $(0, T)$. Such a condition is generally assumed to be verified when the maximum phase term introduced by the complex exponential within the observation interval is less than $\Delta\theta = \pi/4$. If $f_c = 2$ GHz, in case of stationary TX and moving RX, accounting for (7.131) the magnitude of the Doppler shift is

$$|\nu| = f_c \left|\frac{v}{c}\right| \simeq 100 \text{ Hz} \tag{7.234}$$

and, hence, assuming $r_b = 128 \text{ kb s}^{-1}$ we have

$$2\pi|\nu|T < \Delta\theta = \pi/4 \quad \Leftrightarrow \quad 2\pi f_c \left|\frac{v}{c}\right| \frac{N_b}{r_b} < \pi/4 \quad \Leftrightarrow \quad N_b < 172 \text{ bits} \tag{7.235}$$

The coherence time τ_c is the time necessary for a wavefront to cross a $\lambda/2$ distance, that is, $\tau_c = \lambda/(2|\nu|) = c/(2f_c|v|)$ (Sklar 1997, p. 145, eq. (21)). According to such a definition, the channel can be considered stationary within observation intervals less than $\tau_c/4$. That is, the condition to be fulfilled in order to model the channel as LTI is

$$T = N_b/r_b < \tau_c/4 \quad \Leftrightarrow \quad N_b < \frac{r_b c}{8 f_c |v|}$$

which is coincident with (7.235).

7.5.1.2 Example: Sonar Systems

In sonar systems, the narrow band conditions (7.222) or (7.223) involve values of $|1-s|$ significantly bigger than those for the case of electromagnetic propagation. If the medium is air, then $c = 340 \text{ ms}^{-1}$. Therefore, if $v = 100 \text{ km h}^{-1} = 28 \text{ m s}^{-1}$ then $1 - s \simeq 8 \cdot 10^{-2}$. If the medium is water, then $c = 1480 \text{ ms}^{-1}$. Thus, if $v = 100 \text{ km h}^{-1}$ then $1 - s \simeq 2 \cdot 10^{-2}$.

7.5.1.3 Example: Satellite and Space Communications

In the Cassini-Huygens mission, the Cassini probe speed with respect to Huygens in some circumstances is $v = 5.5 \cdot 10^3 \text{ m s}^{-1}$ (Oberg 2004). Therefore, for $c = 3 \cdot 10^8 \text{ ms}^{-1}$ one has $1 - s \simeq 2 \cdot 10^{-5}$. In (Oberg 2004), it is shown that, for synchronization purposes, the time-scale factor s cannot be assumed to be unity in the argument of the received complex envelope signal.

Satellites of the Global Position System (GPS) move with a speed $v_0 \simeq 3.9$ km s^{-1}. The radial speed w.r.t. a stationary observer on the Earth surface is $v = v_0 \cos(E + \phi)$ where E and ϕ are the elevation angle and the Earth central angle, respectively. For $E + \phi = \pi/2 - \pi/16$, one has $|1 - s| = |v/(v + c)| \simeq 2.5 \cdot 10^{-6}$.

7.5.2 *Constant Relative Radial Acceleration*

Let us consider the received signal (7.28) (with $A(t) = b$ not depending on t):

$$\begin{aligned} y(t) &= \operatorname{Re}\left\{\widetilde{y}(t)\, e^{j2\pi f_c t}\right\} \\ &= \operatorname{Re}\left\{b\, \widetilde{x}(t - D(t))\, e^{j2\pi f_c (t - D(t))}\right\}. \end{aligned} \tag{7.236}$$

with (see (7.201))

$$D(t) = d_0 + d_1 t + d_2 t^2 \tag{7.237}$$

where $d_0 = \pm(\xi_0 - v_\xi t_0 + a_\xi t_0^2/2)/c$, $d_1 = \pm(v_\xi - a_\xi t_0)/c$, $d_2 = \pm a_\xi/(2c)$.

The time-varying part of the delay $D(t)$ can be neglected if

$$\widetilde{x}(t - D(t)) \simeq \widetilde{x}(t - d_0) \qquad t \in (t_0, t_0 + T). \tag{7.238}$$

Time variations in intervals of width much smaller than $1/B$, where B is the bandwidth of $\widetilde{x}(t)$, can be neglected (Theorems 7.5.1 and 7.5.2). Thus, a sufficient condition to assure (7.238) is

$$\Delta t_{\max} \ll 1/B \tag{7.239}$$

where $\Delta t_{\max}$ is the maximum delay variation for $t \in (t_0, t_0 + T)$. If condition (7.239) is satisfied, from (7.236) it follows that

$$\begin{aligned} \widetilde{y}(t) &= b\, \widetilde{x}(t - D(t))\, e^{-j2\pi f_c D(t)} \\ &= b\, \widetilde{x}(t - d_0 - d_1 t - d_2 t^2)\, e^{-j2\pi f_c [d_0 + d_1 t + d_2 t^2]} \\ &\simeq b\, \widetilde{x}(t - d_0)\, e^{-j2\pi f_c [d_0 + d_1 t + d_2 t^2]} \\ &= \bar{b}\, \widetilde{x}(t - d_0)\, e^{j2\pi \nu t}\, e^{j\pi \gamma t^2} \end{aligned} \tag{7.240}$$

where,

$$\bar{b} \triangleq b\, e^{-j2\pi f_c d_0} \qquad \text{complex gain} \tag{7.241a}$$

$$\nu \triangleq -f_c d_1 = -f_c \frac{v}{c} \qquad \text{frequency shift} \tag{7.241b}$$

$$\gamma \triangleq -2 f_c d_2 = -f_c \frac{a}{c} \qquad \text{chirp rate} \tag{7.241c}$$

and the second equality in (7.241b) and (7.241c) is obtained for $t_0 = 0$ in the case of stationary TX and moving RX (Section 7.4.1).

Assuming that for $t \in (t_0, t_0 + T)$ the function $D(t)$ is monotone, then its maximum variation is

$$\Delta t_{\max} = |D(t_0 + T) - D(t_0)|. \tag{7.242}$$

From (7.200) we have

$$\begin{aligned} |D(t_0 + T) - D(t_0)| &= \left| \frac{1}{c} \left(\xi_0 + v_\xi T + \frac{1}{2} a_\xi T^2 - \xi_0 \right) \right| \\ &\leqslant \left| \frac{v_\xi}{c} \right| T + \left| \frac{a_\xi}{2c} \right| T^2 . \end{aligned} \tag{7.243}$$

Therefore, (7.239) is verified if

$$\left| \frac{v_\xi}{c} \right| T + \left| \frac{a_\xi}{2c} \right| T^2 \ll \frac{1}{B}. \tag{7.244}$$

Since both quantities involved in the left-hand side of (7.244) are positive, a necessary and sufficient condition assuring (7.244) holds is

$$\begin{cases} \left| \dfrac{v_\xi}{c} \right| T \ll \dfrac{1}{B} \\ \left| \dfrac{a_\xi}{2c} \right| T^2 \ll \dfrac{1}{B} \end{cases} \tag{7.245}$$

that is,

$$BT \ll \left| \frac{c}{v_\xi} \right| \tag{7.246a}$$

$$BT^2 \ll \left| \frac{2c}{a_\xi} \right| \tag{7.246b}$$

Condition (7.246a) is the same as condition (7.222) derived for constant relative radial speed, stationary TX and moving RX. Therefore, (7.246a) and (7.246b) are referred to as *narrow-band conditions* for the case of constant relative radial acceleration.

In the case of stationary TX an moving RX, $|a_\xi| = |a|$ and $|v_\xi| = |v|$. Accounting for (7.241b) and (7.241c), for $t_0 = 0$ conditions (7.246a) and (7.246b) can be written as

$$BT \ll \left| \frac{c}{v_\xi} \right| = \left| \frac{f_c}{\nu} \right| = \left| \frac{1}{d_1} \right| \tag{7.247a}$$

$$BT^2 \ll \left| \frac{2c}{a_\xi} \right| = \left| \frac{2 f_c}{\gamma} \right| = \left| \frac{1}{d_2} \right| . \tag{7.247b}$$

7.5.2.1 Example: DSSS Signal

Let us consider a DSSS signal used in CDMA systems, with chip period T_c, number of chips per bit N_c, and bit period $T_b = N_c T_c$. It results that $B \simeq 1/T_c = r_b N_c$ and $T = N_b T_b = N_b / r_b$, where $r_b = 1/T_b = 1/(N_c T_c)$ is the bit rate and N_b the number of processed bits.

The narrow band conditions (7.246a) and (7.246b) specialize into

$$BT \ll |c/v_\xi| \quad \Leftrightarrow \quad N_b N_c \ll |c/v_\xi| \tag{7.248a}$$

$$BT^2 \ll |2c/a_\xi| \quad \Leftrightarrow \quad (N_b N_c)^2 T_c \ll |2c/a_\xi| \tag{7.248b}$$

respectively.

7.6 Multipath Doppler Channel

7.6.1 Constant Relative Radial Speeds – Discrete Scatterers

In the case of multiple reflections on point scatterers, propagation can be described by the multipath Doppler channel (Izzo and Napolitano 2002b), (Napolitano 2003), (Izzo and Napolitano 2005), that is a LTV system with impulse-response function

$$h(t, u) = \sum_{k=1}^{K} a_k \, \delta(u - s_k t + d_k) \, e^{j2\pi \nu_k t} \tag{7.249}$$

where K is the number of the channel paths.

By double Fourier transforming the right-hand side of (7.249), one obtains the transmission function (see (1.44) and (6.68))

$$H(f, \lambda) = \sum_{k=1}^{K} a_k \, e^{-j2\pi\lambda d_k} \, \delta(f - \nu_k - \lambda s_k) \tag{7.250}$$

from which, accounting for (4.77a), it follows that the multipath Doppler channel is a FOT deterministic LTV system with $\Omega = \{1, \dots, K\}$,

$$\varphi_k(\lambda) = s_k \lambda + \nu_k \, , \qquad k = 1, \dots, K \tag{7.251}$$

$$G_k(\lambda) = a_k \, e^{-j2\pi\lambda d_k} \, , \qquad k = 1, \dots, K. \tag{7.252}$$

Equivalently, accounting for (4.78), we get

$$\psi_k(f) = \frac{1}{s_k}(f - \nu_k) \, , \qquad k = 1, \dots, K \tag{7.253}$$

$$H_k(f) = \frac{a_k}{|s_k|} \, e^{-j2\pi(f-\nu_k)d_k/s_k} \, , \qquad k = 1, \dots, K. \tag{7.254}$$

The output $y(t)$ corresponding to the input signal $x(t)$ is obtained by substituting (7.249) into (1.41):

$$y(t) = \sum_{k=1}^{K} a_k \, x(s_k t - d_k) \, e^{j2\pi\nu_k t}. \tag{7.255}$$

7.6.2 Continuous Scatterer

In the case of distributed moving reflector, assuming that each volume element $(\boldsymbol{\xi}, \boldsymbol{\xi} + \mathrm{d}\boldsymbol{\xi})$ has constant relative radial speed with respect to transmitter and receiver, the input/output relationship for the propagation channel is

$$y(t) = \int_{\mathbb{R}^3} a(\boldsymbol{\xi}) \, x(s(\boldsymbol{\xi})t - d(\boldsymbol{\xi})) \, e^{j2\pi\nu(\boldsymbol{\xi})t} \, \mathrm{d}\boldsymbol{\xi} \tag{7.256}$$

which corresponds to a LTV system with impulse-response function

$$h(t, u) = \int_{\mathbb{R}^3} a(\boldsymbol{\xi})\, \delta(u - s(\boldsymbol{\xi})t + d(\boldsymbol{\xi}))\, e^{j2\pi\nu(\boldsymbol{\xi})t}\, \mathrm{d}\boldsymbol{\xi}. \tag{7.257}$$

The system with impulse-response function (7.249) is deterministic in the FOT probability sense (Section 6.3.8) (Izzo and Napolitano 2002a,b) whereas the system (7.257) is random in the FOT probability sense. Similar models are considered in (Middleton 1967; Sadowsky and Kafedziski 1998).

If the narrow-band condition (7.219) can be assumed to be verified for the speed of each volume element, then the input/output relation can be expressed in terms of the *(narrow-band) spreading function* $\mathrm{SF}(\tau, \nu)$ (Bello 1963):

$$y(t) = \int_{\mathbb{R}^2} \mathrm{SF}(\tau, \nu)\, x(t - \tau)\, e^{j2\pi\nu t}\, \mathrm{d}\tau\, \mathrm{d}\nu. \tag{7.258}$$

Analogously, if the narrow-band condition is not satisfied, the input/output relation can be expressed in terms of the *wide-band spreading function* $\mathrm{WSF}(\tau, s)$ (Weiss 1994):

$$y(t) = \int_{\mathbb{R}^2} \mathrm{WSF}(\tau, s)\, x(st - \tau)\, \mathrm{d}\tau\, \mathrm{d}s. \tag{7.259}$$

If a frequency shift $\nu = (s - 1)f_c$ is present in the propagation model as in (7.129) and (7.156), then (7.259) modifies into

$$y(t) = \int_{\mathbb{R}^2} \mathrm{WSF}_1(\tau, s)\, x(st - \tau)\, e^{j2\pi(s-1)f_c t}\, \mathrm{d}\tau\, \mathrm{d}s. \tag{7.260}$$

7.7 Spectral Analysis of Doppler-Stretched Signals – Constant Radial Speed

In this section, the spectral analysis of the Doppler stretched signal received in the case of constant relative radial speed between TX and RX is addressed. The second-order statistical characterization is made in both time and frequency domains and both continuous- and discrete-time cases are considered. Sampling theorems are proved to obtain input/output relationships for discrete-time sampled signals which are formally analogous to their continuous-time counterparts. Proofs are reported in Section 7.10.

Let us consider the continuous-time Doppler-stretched signal (Section 7.3):

$$y_a(t) = A e^{j\varphi}\, x_a(st - d_a)\, e^{j2\pi f_a t}. \tag{7.261}$$

It is a time-scaled, delayed, and frequency-shifted version of the transmitted signal $x_a(t)$. Its Fourier transform is given by (Section 7.10)

$$Y_a(f) = A e^{j\varphi}\, \frac{1}{|s|} X_a\left(\frac{f - f_a}{s}\right) e^{-j2\pi(f - f_a)d_a/s} \tag{7.262}$$

7.7.1 *Second-Order Statistics (Continuous-Time)*

Let $x_a(t)$ exhibit (conjugate) cyclostationarity. Then, also $y_a(t)$ exhibits (conjugate) cyclostationarity and its cyclic statistics are linked to those of $x_a(t)$ as follows.

7.7.1.1 Cyclic Autocorrelation Function and Cyclic Spectrum

The cyclic autocorrelation function and the cyclic spectrum of $y_a(t)$ are given by (Section 7.10)

$$R^{\alpha}_{y_a y_a^*}(\tau) \triangleq \left\langle \mathrm{E}\left\{ y_a(t+\tau)\, y_a^*(t) \right\} e^{-j2\pi\alpha t} \right\rangle_t$$

$$= A^2 e^{j2\pi f_a \tau}\, e^{-j2\pi\alpha d_a/s}\, R^{\alpha/s}_{x_a x_a^*}(s\tau) \tag{7.263}$$

$$\overset{\mathcal{F}}{\longleftrightarrow} S^{\alpha}_{y_a y_a^*}(f) = A^2 e^{-j2\pi\alpha d_a/s} \frac{1}{|s|} S^{\alpha/s}_{x_a x_a^*}\left(\frac{f-f_a}{s}\right) \tag{7.264}$$

Equivalently,

$$R^{s\alpha}_{y_a y_a^*}(\tau) = A^2 e^{j2\pi f_a \tau}\, e^{-j2\pi\alpha d_a}\, R^{\alpha}_{x_a x_a^*}(s\tau) \tag{7.265}$$

$$\overset{\mathcal{F}}{\longleftrightarrow} S^{s\alpha}_{y_a y_a^*}(f) = A^2 e^{-j2\pi\alpha d_a} \frac{1}{|s|} S^{\alpha}_{x_a x_a^*}\left(\frac{f-f_a}{s}\right) \tag{7.266}$$

Thus, if $x_a(t)$ exhibits cyclostationarity at cycle frequency α_x, then $y_a(t)$ exhibits cyclostationarity at cycle frequency $\alpha = s\alpha_x$.

7.7.1.2 Conjugate Cyclic Autocorrelation Function and Conjugate Cyclic Spectrum

The conjugate cyclic autocorrelation function and the conjugate cyclic spectrum of $y_a(t)$ are given by (Section 7.10)

$$R^{\beta}_{y_a y_a}(\tau) \triangleq \left\langle \mathrm{E}\left\{ y_a(t+\tau)\, y_a(t) \right\} e^{-j2\pi\beta t} \right\rangle_t$$

$$= A^2 e^{j2\varphi}\, e^{j2\pi f_a \tau}\, e^{-j2\pi(\beta-2f_a)d_a/s}\, R^{(\beta-2f_a)/s}_{x_a x_a}(s\tau) \tag{7.267}$$

$$\overset{\mathcal{F}}{\longleftrightarrow} S^{\beta}_{y_a y_a}(f) = A^2 e^{j2\varphi}\, e^{-j2\pi(\beta-2f_a)d_a/s} \frac{1}{|s|} S^{(\beta-2f_a)/s}_{x_a x_a}\left(\frac{f-f_a}{s}\right) \tag{7.268}$$

Equivalently,

$$R^{s\beta+2f_a}_{y_a y_a}(\tau) = A^2 e^{j2\varphi}\, e^{j2\pi f_a \tau}\, e^{-j2\pi\beta d_a}\, R^{\beta}_{x_a x_a}(s\tau) \tag{7.269}$$

$$\overset{\mathcal{F}}{\longleftrightarrow} S^{s\beta+2f_a}_{y_a y_a}(f) = A^2 e^{j2\varphi}\, e^{-j2\pi\beta d_a} \frac{1}{|s|} S^{\beta}_{x_a x_a}\left(\frac{f-f_a}{s}\right) \tag{7.270}$$

Thus, if $x_a(t)$ exhibits conjugate cyclostationarity at conjugate cycle frequency β_x, then $y_a(t)$ exhibits conjugate cyclostationarity at conjugate cycle frequency $\beta = s\beta_x + 2f_a$.

Let $x_a(t)$ be ACS. Its (conjugate) autocorrelation function (1.88) and Loève bifrequency spectrum (1.94) are reported here for convenience:

$$\mathrm{E}\{x_a(t_1)\, x_a^{(*)}(t_2)\} = \sum_{\alpha_n \in A_{x_a x_a^{(*)}}} R^{\alpha_n}_{x_a x_a^{(*)}}(t_1 - t_2)\, e^{j2\pi\alpha_n t_2} \tag{7.271}$$

$$\mathrm{E}\left\{X_a(f_1)\, X_a^{(*)}(f_2)\right\} = \sum_{\alpha_n \in A_{x_a x_a^{(*)}}} S^{\alpha_n}_{x_a x_a^{(*)}}(f_1)\, \delta(f_2 - {}_{(-)}(\alpha_n - f_1)). \tag{7.272}$$

In the following it is shown that for $s \neq 1$, $x_a(t)$ and $y_a(t)$ are singularly but not jointly ACS. Specifically, $x_a(t)$ and $y_a(t)$ are jointly SC.

7.7.1.3 Loève Bifrequency Spectrum – ACS Input Signal

The Loève bifrequency spectrum of $y_a(t)$ is given by (Section 7.10)

$$\begin{aligned}
&\mathrm{E}\left\{Y_a(f_1)\, Y_a^{(*)}(f_2)\right\} \\
&\quad = A^2\, e^{j\kappa\varphi}\, \frac{1}{|s|}\, e^{-j2\pi(f_1 - f_a)d_a/s}\, e^{-(-)j2\pi(f_2 - f_a)d_a/s} \\
&\qquad \sum_{\alpha_n \in A_{x_a x_a^{(*)}}} S^{\alpha_n}_{x_a x_a^{(*)}}\left(\frac{f_1 - f_a}{s}\right) \delta\left(f_2 + {}_{(-)}f_1 - [{}_{(-)}s\alpha_n + \kappa f_a]\right)
\end{aligned} \tag{7.273}$$

where $\kappa \triangleq (1 + {}_{(-)}1)$, that is, $\kappa = 0$ if ${}_{(-)} = -$ and $\kappa = 2$ if ${}_{(-)} = +$. From (7.273), it follows that the Loève bifrequency spectrum of $y_a(t)$ has support contained in lines with slopes ± 1. That is, the Doppler-stretched signal $y_a(t)$ is ACS. Moreover, its (conjugate) cyclic spectra can be related to those of the transmitted signal $x_a(t)$ by (7.264), (7.266) (7.268), and (7.270). Therefore, due to the Doppler channel, cycle frequencies of x_a are scaled by s and conjugate cycle frequencies of x_a are scaled by s and shifted by $2f_a$.

Accounting for (1.178a) and (1.178b), the support of the generic term of the sum over α in (7.389) (which is equivalent to (7.273)) is such that

$$\begin{aligned}
&\mathrm{supp}\left\{S^{\alpha_n}_{x_a x_a^{(*)}}\left(\frac{f_1 - f_a}{s}\right)\delta\left(\frac{f_2 - f_a}{s} - {}_{(-)}\left(\alpha_n - \frac{f_1 - f_a}{s}\right)\right)\right\} \\
&\subseteq \left\{(f_1, f_2) \in \mathbb{R}\times\mathbb{R} : \left|\frac{f_1 - f_a}{s}\right| \leqslant B, \right. \\
&\qquad \left. \left|\frac{f_1 - f_a}{s} - \alpha_n\right| \leqslant B,\ \frac{f_2 - f_a}{s} = {}_{(-)}\left(\alpha_n - \frac{f_1 - f_a}{s}\right)\right\} \\
&= \left\{(f_1, f_2) \in \mathbb{R}\times\mathbb{R} : \left|\frac{f_1 - f_a}{s}\right| \leqslant B, \right. \\
&\qquad \left. \left|\frac{f_2 - f_a}{s}\right| \leqslant B,\ f_2 = {}_{(-)}s\alpha_n + \kappa f_a - {}_{(-)}f_1\right\} \\
&\subseteq \left\{(f_1, f_2) \in \mathbb{R}\times\mathbb{R} : |f_1| - |f_a| \leqslant B|s|, \right.
\end{aligned}$$

$$
\begin{aligned}
&|f_2| - |f_a| \leqslant B|s|,\ f_2 = {}_{(-)}s\alpha_n + \kappa f_a - {}_{(-)}f_1 \Big\} \\
&= \Big\{ (f_1, f_2) \in \mathbb{R} \times \mathbb{R} \ : \ |f_1| \leqslant |f_a| + B|s|, \\
&|f_2| \leqslant |f_a| + B|s|,\ f_2 = {}_{(-)}s\alpha_n + \kappa f_a - {}_{(-)}f_1 \Big\}
\end{aligned}
\tag{7.274}
$$

Note that the ACS nature of $y_a(t)$ is not an obvious result. In fact, relation (7.261) describes a transformation of the input signal $x_a(t)$ into the output signal $y_a(t)$ which is linear time-variant but not almost-periodically time-variant. Only in this last case it is known that ACS signals are transformed into ACS signals (Section 1.3.3), (Napolitano 1995), (Gardner *et al.* 2006). Transformation described by (7.261) is a special case of deterministic transformation in the FOT probability sense (Section 4.3.1), (Izzo and Napolitano 2002a, 2005).

7.7.1.4 (Conjugate) Autocorrelation Function – ACS Input Signal

In Section 7.10, it is shown that for an input ACS signal the (conjugate) autocorrelation function of the signal $y_a(t)$ in (7.261) is given by

$$
\begin{aligned}
&\mathrm{E}\Big\{ y_a(t+\tau)\, y_a^{(*)}(t) \Big\} \\
&= A^2\, e^{j\kappa\varphi} \sum_{\alpha_n \in A_{x_a x_a^{(*)}}} R^{\alpha_n}_{x_a x_a^{(*)}}(s\tau)\, e^{j2\pi f_a \tau}\, e^{-j2\pi\alpha_n d_a}\, e^{j2\pi(s\alpha_n + \kappa f_a)t}
\end{aligned}
\tag{7.275}
$$

which agrees with (7.265) and (7.269) derived in the more general case of $x_a(t)$ that exhibits cyclostationarity.

7.7.1.5 Loève Bifrequency Cross-Spectrum – ACS Input Signal

The Loève bifrequency cross-spectrum of $y_a(t)$ and $x_a(t)$ is given by (Section 7.10)

$$
\begin{aligned}
&\mathrm{E}\Big\{ Y_a(f_1)\, X_a^{(*)}(f_2) \Big\} \\
&= A\, e^{j\varphi}\, \frac{1}{|s|}\, e^{-j2\pi(f_1 - f_a)d_a/s} \\
&\quad \sum_{\alpha_n \in A_{x_a x_a^{(*)}}} S^{\alpha_n}_{x_a x_a^{(*)}}\left(\frac{f_1 - f_a}{s}\right) \delta\left(f_2 - {}_{(-)}\left(\alpha_n - \frac{f_1 - f_a}{s}\right)\right)
\end{aligned}
\tag{7.276}
$$

Thus, when $s \neq 1$, the Loève bifrequency cross-spectrum has support contained in lines with slopes different from ± 1. That is, even if both signals $x_a(t)$ and $y_a(t)$ are (singularly) ACS, they are not jointly ACS. The joint nonstationary behavior of $x_a(t)$ and $y_a(t)$ can be described resorting to the class of the SC signals. In fact, jointly SC signals have Loève bifrequency cross-spectrum with spectral masses concentrated on a countable set of curves in the bifrequency plane. From (7.276) it follows that $y_a(t)$ and $x_a(t)$ are jointly SC with support curves constituted by lines with slopes $\pm 1/s$. Specifically, with reference to the notation

introduced in Chapter 4, we have

$$\mathrm{E}\left\{Y_a(f_1)\,X_a^{(*)}(f_2)\right\} = \sum_{n\in\mathbb{I}} S^{(n)}_{y_a x_a^{(*)}}(f_1)\,\delta\left(f_2 - \Psi^{(n)}_{y_a x_a^{(*)}}(f_1)\right) \tag{7.277}$$

with

$$n \in \mathbb{I} \Leftrightarrow \alpha_n \in A_{x_a x_a^{(*)}} \qquad \text{(one-to-one correspondence)} \tag{7.278}$$

$$S^{(n)}_{y_a x_a^{(*)}}(f_1) \triangleq A\,e^{j\varphi}\,\frac{1}{|s|}\,e^{-j2\pi(f_1-f_a)d_a/s}\,S^{\alpha_n}_{x_a x_a^{(*)}}\left(\frac{f_1-f_a}{s}\right) \tag{7.279}$$

$$\Psi^{(n)}_{y_a x_a^{(*)}}(f_1) \triangleq {}_{(-)}(\alpha_n - (f_1-f_a)/s)\,. \tag{7.280}$$

Assuming the transmitted signal $x_a(t)$ strictly band-limited with bandwidth B, and accounting for (1.178a) and (1.178b), the support of the generic term of the sum over α in (7.276) is such that

$$\begin{aligned}
&\operatorname{supp}\left\{S^{\alpha_n}_{x_a x_a^{(*)}}\left(\frac{f_1-f_a}{s}\right)\delta\left(f_2 - {}_{(-)}\left(\alpha_n - \frac{f_1-f_a}{s}\right)\right)\right\}\\
&\subseteq \left\{(f_1,f_2)\in\mathbb{R}\times\mathbb{R} : \left|\frac{f_1-f_a}{s}\right| \leqslant B,\right.\\
&\qquad \left.\left|\frac{f_1-f_a}{s} - \alpha_n\right| \leqslant B,\ f_2 = {}_{(-)}\left(\alpha_n - \frac{f_1-f_a}{s}\right)\right\}\\
&= \left\{(f_1,f_2)\in\mathbb{R}\times\mathbb{R} : \left|\frac{f_1-f_a}{s}\right| \leqslant B,\right.\\
&\qquad \left.|f_2| \leqslant B,\ f_2 = {}_{(-)}\left(\alpha_n - \frac{f_1-f_a}{s}\right)\right\}\\
&\subseteq \left\{(f_1,f_2)\in\mathbb{R}\times\mathbb{R} : |f_1| \leqslant |f_a| + B|s|,\right.\\
&\qquad \left.|f_2| \leqslant B,\ f_2 = {}_{(-)}\left(\alpha_n - \frac{f_1-f_a}{s}\right)\right\}.
\end{aligned} \tag{7.281}$$

7.7.1.6 Cross-Correlation Function – ACS Input Signal

By using (7.261) and (7.271) leads to

$$\begin{aligned}
&\mathrm{E}\left\{y_a(t_1)\,x_a^{(*)}(t_2)\right\}\\
&= A\,e^{j\varphi}\,e^{j2\pi f_a t_1}\,\mathrm{E}\left\{x_a(st_1 - d_a)\,x_a^{(*)}(t_2)\right\}\\
&= A\,e^{j\varphi}\,e^{j2\pi f_a t_1} \sum_{\alpha_n\in A_{x_a x_a^{(*)}}} R^{\alpha_n}_{x_a x_a^{(*)}}\left(st_1 - t_2 - d_a\right) e^{j2\pi\alpha_n t_2}.
\end{aligned} \tag{7.282}$$

Moreover, by the variable change $t_1 = t + \tau$, $t_2 = t$ we have

$$\mathrm{E}\left\{ y_a(t+\tau)\, x_a^{(*)}(t) \right\} = A\, e^{j\varphi}\, e^{j2\pi f_a \tau} \sum_{\alpha_n \in A_{x_a x_a^{(*)}}} R^{\alpha_n}_{x_a x_a^{(*)}}\Big((s-1)t + s\tau - d_a \Big)\, e^{j2\pi(\alpha_n + f_a)t}. \tag{7.283}$$

Thus, under the mild assumption that $R^{\alpha_n}_{x_a x_a^{(*)}}(\tau) \in L^1(\mathbb{R})$, $\forall \alpha_n$, the function of t in the rhs of (7.283) can contain finite-strength additive sinewave components if and only if $s = 1$, i.e., if there is no dependence on t in the argument of $R^{\alpha_n}_{x_a x_a^{(*)}}(\cdot)$. That is, according to the frequency-domain result, for $s \neq 1$ the cross-correlation function $\mathrm{E}\{y_a(t+\tau)\, x_a^{(*)}(t)\}$ is not an almost-periodic function of t, that is, $y_a(t)$ and $x_a(t)$ are not jointly ACS.

7.7.2 Multipath Doppler Channel

In Section 7.6.1, it is shown that the output signal $y(t)$ of a multipath Doppler channel introducing complex gain, time-scale factor, delay, and frequency shift for each path, for input signal $x(t)$ is

$$y(t) = \sum_{k=1}^{K} a_k\, x(s_k t - d_k)\, e^{j2\pi \nu_k t}. \tag{7.284}$$

7.7.2.1 Output Statistics

From (7.284), it follows that the (conjugate) autocorrelation function of $y(t)$ is

$$\mathrm{E}\left\{ y(t+\tau)\, y^{(*)}(t) \right\} = \sum_{k_1=1}^{K} \sum_{k_2=1}^{K} a_{k_1}\, a_{k_2}^{(*)}\, e^{j2\pi \nu_{k_1}(t+\tau)}\, e^{(-)j2\pi \nu_{k_2} t}\ \mathrm{E}\left\{ x(s_{k_1}(t+\tau) - d_{k_1})\, x^{(*)}(s_{k_2} t - d_{k_2}) \right\} \tag{7.285}$$

In Section 7.10 it is shown that for an ACS input signal $x(t)$, under the mild assumption of (conjugate) cyclic autocorrelation function summability (1.89), if $s_{k_1} \neq s_{k_2}$ for $k_1 \neq k_2$, then the almost-periodic component of the (conjugate) autocorrelation function is given by

$$\mathrm{E}^{\{\alpha\}}\left\{ \mathrm{E}\left\{ y(t+\tau)\, y^{(*)}(t) \right\} \right\} = \sum_{k=1}^{K} a_k\, a_k^{(*)}\, e^{j2\pi \nu_k [(t+\tau)+(-)t]}\ \mathrm{E}^{\{\alpha\}}\left\{ \mathrm{E}\left\{ x(s_k(t+\tau) - d_k)\, x^{(*)}(s_k t - d_k) \right\} \right\}$$

$$= \sum_{k=1}^{K} a_k\, a_k^{(*)}\, e^{j2\pi[(1+(-)1)\nu_k]t}\, e^{j2\pi\nu_k\tau} \sum_{\alpha\in A_{xx^{(*)}}} R^{\alpha}_{xx^{(*)}}(s_k\tau)\, e^{j2\pi\alpha(s_k t - d_k)} \tag{7.286}$$

In addition, the (conjugate) cyclic autocorrelation function of the output signal $y(t)$ is

$$R^{\beta}_{yy^{(*)}}(\tau) = \sum_{k=1}^{K} a_k\, a_k^{(*)}\, e^{j2\pi\nu_k\tau}\, R^{[\beta-(1+(-)1)\nu_k]/s_k}_{xx^{(*)}}(s_k\tau)\, e^{-j2\pi[\beta-(1+(-)1)\nu_k]d_k/s_k} \tag{7.287}$$

In contrast, if the length T of the observation interval is such that the narrow-band condition (see (7.222) and (7.223))

$$BT \ll \frac{1}{|1 - s_k|} \tag{7.288}$$

holds for all paths, then $s_{k_1} \simeq 1$ and $s_{k_2} \simeq 1$ can be assumed in (7.285) and the multipath Doppler channel is LAPTV. Thus, it transforms ACS input signals into ACS output signals. Consequently, for an ACS input signal $x(t)$, the (conjugate) autocorrelation function of the output signal $y(t)$ is coincident with its almost-periodic component and, accounting for (7.392) with $s_{k_1} = s_{k_2} = 1$, the result is

$$\begin{aligned}
&\mathrm{E}^{\{\alpha\}}\left\{\mathrm{E}\left\{y(t+\tau)\, y^{(*)}(t)\right\}\right\} \\
&= \sum_{k_1=1}^{K}\sum_{k_2=1}^{K} a_{k_1}\, a_{k_2}^{(*)}\, e^{j2\pi\nu_{k_1}(t+\tau)}\, e^{(-)j2\pi\nu_{k_2}t}\ \mathrm{E}\left\{x(t+\tau-d_{k_1})\, x^{(*)}(t-d_{k_2})\right\} \\
&= \sum_{k_1=1}^{K}\sum_{k_2=1}^{K} a_{k_1}\, a_{k_2}^{(*)}\, e^{j2\pi\nu_{k_1}(t+\tau)}\, e^{(-)j2\pi\nu_{k_2}t} \sum_{\alpha\in A_{xx^{(*)}}} R^{\alpha}_{xx^{(*)}}(\tau - d_{k_1} + d_{k_2})\, e^{j2\pi\alpha(t-d_{k_2})}
\end{aligned} \tag{7.289}$$

and

$$R^{\beta}_{yy^{(*)}}(\tau) = \sum_{k_1=1}^{K}\sum_{k_2=1}^{K} a_{k_1}\, a_{k_2}^{(*)}\, e^{j2\pi\nu_{k_1}\tau}\, R^{[\beta-\nu_{k_1}+(-)\nu_{k_2}]}_{xx^{(*)}}(\tau - d_{k_1} + d_{k_2})\, e^{-j2\pi[\beta-\nu_{k_1}+(-)\nu_{k_2}]d_{k_2}} \tag{7.290}$$

To corroborate the result that a different behavior of the cyclic autocorrelation function is predicted by (7.287) or (7.290) depending on the data-record length adopted for the cyclic correlogram estimate, the following experiment is carried out. A binary PAM signal with bit

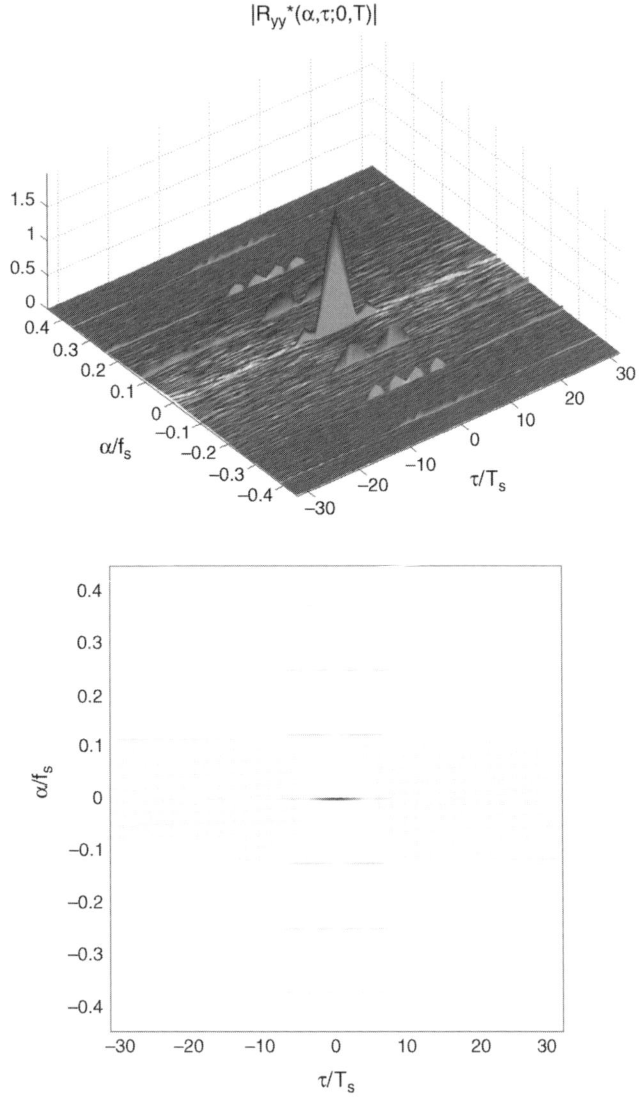

Figure 7.7 Output signal of a multipath Doppler channel excited by a PAM input signal. (Top) graph and (bottom) "checkerboard" plot of the magnitude of the cyclic correlogram as a function of αT_s and τ/T_s, estimated by a data-record length of 2^{15} samples (narrow-band condition (7.288) not satisfied)

rate $1/8T_s$ and full-duty-cycle rectangular pulse excites a multipath Doppler channel characterized by $a_1 = 1$, $d_1 = 0$, $\nu_1 = 0.015/Ts$, $s_1 = 1.001$, $a_2 = 0.9$, $d_2 = 15T_s$, $\nu_2 = -0.1/Ts$, and $s_2 = 0.999$. In Figures 7.7 and 7.8, (top) graph and (bottom) "checkerboard" plot of the magnitude of the cyclic correlogram as a function of αT_s and τ/T_s, estimated by a data-record length of 2^{15} samples and 2^{10} samples, respectively, are reported. For $T = 2^{15}T_s$ one has $BT \simeq 4096$ and $1/|1 - s_k| = 1000$, $k = 1, 2$. Thus, the narrow-band condition (7.288) is not

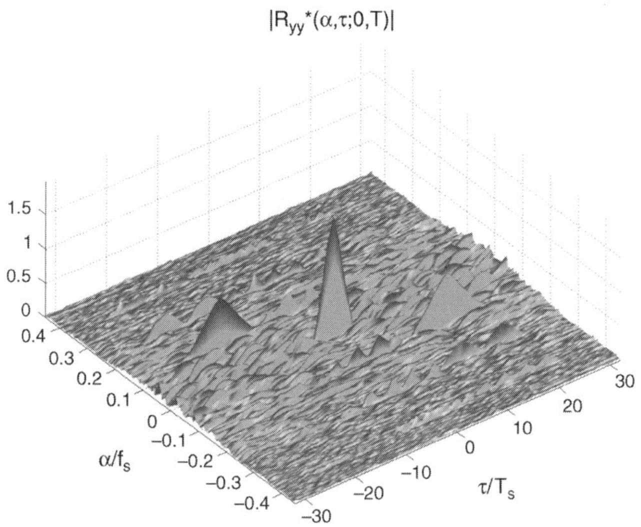

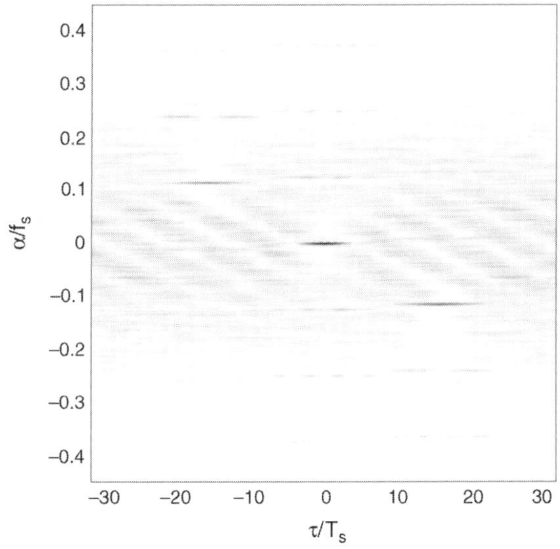

Figure 7.8 Output signal of a multipath Doppler channel excited by a PAM input signal. (Top) graph and (bottom) "checkerboard" plot of the magnitude of the cyclic correlogram as a function of αT_s and τ / T_s, estimated by a data-record length of 2^{10} samples (narrow-band condition (7.288) satisfied)

satisfied and (7.287) describes the cyclic autocorrelation function of the output (Figure 7.7). In contrast, for $T = 2^{10} T_s$ one has $BT \simeq 128$ and $1/|1 - s_k| = 1000$, $k = 1, 2$, and the narrow-band condition (7.288) holds. Therefore, (7.290) describes the cyclic autocorrelation function of the output. Also terms with $k_1 \neq k_2$ give nonzero contribution (Figure 7.8).

Briefly, both the above analysis and previous experiment show that if the product bandwidth data-record length is sufficiently small, then the multipath Doppler channel can be modeled as

a LAPTV system since the effects of the time scaling factors can be ignored in the argument of the complex envelope of the received signal. In contrast, if the data-record length is increased (e.g., to obtain high noise immunity in cyclostationarity-based algorithms), the time-scaling factors must be taken into account, and the system cannot be modeled as LAPTV but, rather, as FOT deterministic.

Equations (7.285), (7.286), (7.287), (7.289), and (7.290) can also be written in the functional approach by substituting the statistical expectation operator with the almost-periodic component extraction operator $\mathrm{E}^{\{\alpha\}}\{\cdot\}$ (Izzo and Napolitano 2002b), (Izzo and Napolitano 2005) and observing that $\mathrm{E}^{\{\alpha\}}\{\mathrm{E}^{\{\alpha\}}\{\cdot\}\} \equiv \mathrm{E}^{\{\alpha\}}\{\cdot\}$. In such a case, however, equalities must be intended in the temporal mean-square sense (Section 6.3.5, (Gardner *et al.* 2006, Section 4.2), (Izzo and Napolitano 2005, Appendix A)). Alternatively, almost periodicity should be considered in some of the generalized senses addressed in Sections 1.2.2, 1.2.3, 1.2.4, 1.2.5 (see also (Besicovitch 1932, Chapter 2), (Bohr 1933, paragraphs 94–102), (Corduneanu 1989, Chapter VI)).

Accounting for (7.262), we have the following expression for the Loève bifrequency spectrum of $y(t)$

$$\mathrm{E}\left\{Y(f_1)\,Y^{(*)}(f_2)\right\} = \sum_{k_1=1}^{K}\sum_{k_2=1}^{K} \frac{a_{k_1} a_{k_2}^{(*)}}{|s_{k_1}|\,|s_{k_2}|}\, e^{-j2\pi(f_1-\nu_{k_1})d_{k_1}/s_{k_1}}\, e^{-(-)j2\pi(f_2-\nu_{k_2})d_{k_2}/s_{k_2}}\, \mathrm{E}\left\{X\left(\frac{f_1-\nu_{k_1}}{s_{k_1}}\right) X^{(*)}\left(\frac{f_2-\nu_{k_2}}{s_{k_2}}\right)\right\}. \tag{7.291}$$

which, in the special case of ACS input signal, accounting for (7.273), reduces to

$$\mathrm{E}\left\{Y(f_1)\,Y^{(*)}(f_2)\right\} = \sum_{k_1=1}^{K}\sum_{k_2=1}^{K}\sum_{\alpha\in A_{xx^{(*)}}} \frac{a_{k_1} a_{k_2}^{(*)}}{|s_{k_1}|}\, e^{-j2\pi(f_1-\nu_{k_1})(d_{k_1}-d_{k_2})/s_{k_1}}\, e^{-j2\pi\alpha d_{k_2}}\, S^{\alpha}_{xx^{(*)}}\left(\frac{f_1-\nu_{k_1}}{s_{k_1}}\right) \delta\left(f_2-\nu_{k_2}-(-)\left(s_{k_2}\alpha-\frac{s_{k_2}}{s_{k_1}}(f_1-\nu_{k_1})\right)\right). \tag{7.292}$$

Equation (7.292) shows that for an input ACS signal, the Loève bifrequency spectrum of the output signal has support contained in lines with slopes s_{k_2}/s_{k_1}. Thus, the output signal is SC provided that at least two time-scale factors are different.

7.7.2.2 Input/Output Cross-Statistics

Accounting for (7.276), the Loève bifrequency cross-spectrum of $y(t)$ and $x(t)$ is given by

$$\mathrm{E}\left\{Y(f_1)\,X^{(*)}(f_2)\right\} = \sum_{k=1}^{K} \frac{a_k}{|s_k|}\, e^{-j2\pi(f_1-\nu_k)d_k/s_k}\, \mathrm{E}\left\{X\left(\frac{f_1-\nu_k}{s_k}\right) X^{(*)}(f_2)\right\}$$

$$= \sum_{k=1}^{K} \frac{a_k}{|s_k|} e^{-j2\pi(f_1-\nu_k)d_k/s_k}$$
$$\sum_{\beta \in A_{xx^{(*)}}} S^{\beta}_{xx^{(*)}}\left(\frac{f_1-\nu_k}{s_k}\right) \delta\left(f_2 -_{(-)} \left(\beta - \frac{f_1-\nu_k}{s_k}\right)\right) \tag{7.293}$$

That is, the input and output signals of the multipath Doppler channel are jointly SC signals.

7.7.3 Doppler-Stretched Signal (Discrete-Time)

In this section, the equivalent discrete-time Doppler channel obtained by uniformly sampling input and output signals of a continuous-time Doppler channel is considered. Input/output relationships in both time and frequency domains are derived. Moreover, a sampling theorem is proved such that the input/output relationship in the frequency domain is formally equivalent to its continuous-time counterpart.

In the case of constant relative radial speed between TX and RX (Section 7.3) the continuous-time Doppler-stretched received signal and its Fourier transform are given by (7.261) and (7.262), respectively, reported here for convenience

$$y_a(t) = A e^{j\varphi}\, x_a(st - d_a)\, e^{j2\pi f_a t} \tag{7.294}$$
$$\overset{\mathfrak{F}}{\longleftrightarrow} Y_a(f) = A e^{j\varphi} \frac{1}{|s|} X_a\left(\frac{f - f_a}{s}\right) e^{-j2\pi(f-f_a)d_a/s}. \tag{7.295}$$

Let us consider the uniformly sampled discrete-time signal

$$y(n) \triangleq y_a(t)|_{t=nT_s} = A e^{j\varphi}\, x_a(snT_s - d_a)\, e^{j2\pi f_a n T_s} \tag{7.296}$$

where $T_s = 1/f_s$ is the sampling period and let us define

$$x(n) \triangleq x_a(t)|_{t=nT_s} \tag{7.297}$$
$$\bar{\nu} \triangleq f_a T_s \tag{7.298}$$
$$d \triangleq d_a/T_s \tag{7.299}$$

where $\bar{\nu}$ is the normalized frequency shift and d is the (possibly non-integer) normalized delay.

Accounting for the relationship (5.184) between the Fourier transform of a continuous-time signal and that of its uniformly sampled version

$$X(\nu) = \frac{1}{T_s} \sum_{k=-\infty}^{+\infty} X_a((\nu - k) f_s) \tag{7.300}$$

the Fourier transform of the discrete-time signal $y(n)$ is given by

$$Y(\nu) = \frac{1}{T_s} \sum_{k=-\infty}^{+\infty} Y_a((\nu - k) f_s)$$

$$= \frac{1}{T_s} \sum_{k=-\infty}^{+\infty} A e^{j\varphi} \frac{1}{|s|} X_a\left(\frac{(\nu-k)f_s - f_a}{s}\right) e^{-j2\pi((\nu-k)f_s - f_a)d_a/s}$$

$$= A e^{j\varphi} e^{-j2\pi(\nu - f_a/f_s)d_a f_s/s} \frac{1}{|s|} \frac{1}{T_s} \sum_{k=-\infty}^{+\infty} X_a\left(\frac{(\nu-k)f_s - f_a}{s}\right) e^{j2\pi k f_s d_a/s}$$

$$= A e^{j\varphi} e^{-j2\pi(\nu-\bar{\nu})d/s} \frac{1}{|s|} \frac{1}{T_s} \sum_{k=-\infty}^{+\infty} X_a\left(\frac{(\nu-\bar{\nu})f_s}{s} - k\frac{f_s}{s}\right) e^{j2\pi k d/s} \tag{7.301}$$

The following remarks are in order.

1. Replicas of functions of ν are separated by 1. With the mapping $f = \nu f_s$, replicas in (7.301) are separated by f_s.
2. Accounting for (7.300), one has

$$X\left(\frac{\nu-\bar{\nu}}{s}\right) = \frac{1}{T_s} \sum_{k=-\infty}^{+\infty} X_a\left(\frac{\nu-\bar{\nu}}{s} f_s - k f_s\right). \tag{7.302}$$

   Since $X(\nu)$ is periodic with period 1, it follows that $X((\nu-\bar{\nu})/s)$ is periodic with period s. Therefore, $X((\nu-\bar{\nu})/s)$ is not the Fourier transform of a sequence.
3. $e^{-j2\pi(\nu-\bar{\nu})d/s}$ is periodic with period 1 only if $d/s \in \mathbb{Z}$.

From the above remarks we find that a discrete-time counterpart of (7.295) formally analogous to (7.295) cannot be straightforwardly derived from (7.301).

Let $x_a(t)$ be strictly band-limited, that is, $X_a(f) = 0$ for $|f| > B$. Thus,

$$\frac{1}{|s|} X_a\left(\frac{f - f_a}{s}\right) = 0 \qquad \left|\frac{f - f_a}{s}\right| > B. \tag{7.303}$$

With the mapping $f = \nu f_s$, replicas in (7.301) are centered in multiples of f_s. Hence, a sufficient condition to avoid replicas' overlap in (7.301) is $f_a + |s|B \leqslant f_a + f_s - |s|B$ (Figure 7.9), that is

$$f_s \geqslant 2B|s|. \tag{7.304}$$

Note that the frequency shift f_a has no influence in this condition. A sufficient condition to obtain that *only* the kth replica has support contained in the interval

$$I_k \triangleq \left[k f_s - \frac{f_s}{2},\ k f_s + \frac{f_s}{2}\right) \tag{7.305}$$

is (Figure 7.9)

$$\frac{f_s}{2} \geqslant |f_a| + B|s|. \tag{7.306}$$

In such a case, the frequency shift f_a has influence and condition (7.306) is more restrictive than condition (7.304) that only avoids replicas' overlap.

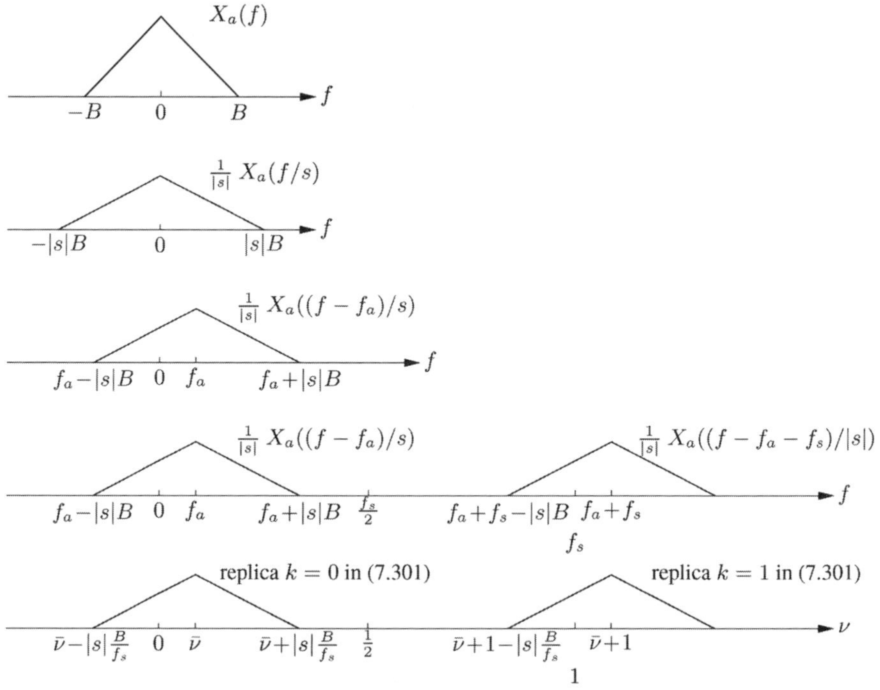

Figure 7.9 Sampled stretched spectra ($f_a > 0$, $|s| > 1$, $s \gtrless 0$)

Conditions (7.304) and (7.306) can be written in terms of frequency ν accounting for the mapping $\nu = f/f_s$. Specifically (see Figure 7.9), condition (7.304) of replicas' non-overlap becomes

$$\frac{1}{2} \geqslant \frac{B}{f_s}|s| \tag{7.307}$$

and condition (7.306) assuring that only the kth replica falls in the interval I_k becomes

$$\frac{1}{2} \geqslant |\bar{\nu}| + \frac{B}{f_s}|s|. \tag{7.308}$$

Assumption 7.7.1 No Aliasing for the Discrete-Time Doppler Stretched Signal

1. $x_a(t)$ strictly band-limited, i.e.,

$$X_a(f) = 0 \qquad |f| > B \tag{7.309}$$

2. Only the kth replica in (7.301) has support contained in the interval I_k defined in (7.305). A sufficient condition for this is (7.306), that is

$$f_s \geqslant 2(|f_a| + B|s|). \tag{7.310}$$

Theorem 7.7.2 Sampling Theorem for Doppler-Stretched Signals. *Under Assumption 7.7.1, it results that*

$$Y(\nu) = Ae^{j\varphi}\, e^{-j2\pi(\nu-\bar{\nu})d/s}\, \frac{1}{|s|}\, X\left(\frac{\nu-\bar{\nu}}{s}\right) \qquad |\nu| \leqslant \frac{1}{2} \tag{7.311}$$

which is formally analogous to its continuous-time counterpart (7.295).

Proof: By comparing replicas for $k = 0$ in (7.301) and (7.302), we obtain (7.311), provided that only replica for $k = 0$ in (7.301) is contained in the interval $[-1/2,1/2)$ (that is, I_0/f_s, where I_0 is defined according to (7.305)). □

In the special case $s = 1$, (7.301) becomes

$$Y(\nu) = Ae^{j\varphi}\, e^{-j2\pi(\nu-\bar{\nu})d}\, \frac{1}{T_s} \sum_{k=-\infty}^{+\infty} X_a\big((\nu-\bar{\nu}-k)f_s\big)\, e^{j2\pi kd} \tag{7.312}$$

from which it follows

$$Y(\nu) = Ae^{j\varphi}\, e^{-j2\pi(\nu-\bar{\nu})d}\, X(\nu-\bar{\nu}) \qquad |\nu| \leqslant \frac{1}{2}. \tag{7.313}$$

Equality in (7.313) holds $\forall \nu \in \mathbb{R}$ only if $d \in \mathbb{Z}$.

7.7.4 Simulation of Discrete-Time Doppler-Stretched Signals

In this section, an interpolation procedure is outlined to generate samples of a discrete-time Doppler-stretched signal. The method is extended to the generation of samples of time-warped version of a given signal with generic time-warping function. Then, its frequency-domain counterpart is considered to generate a frequency-warped version of a given signal.

Let $x_a(t)$ be strictly bandlimited, i.e., $X_a(f) = 0 \quad |f| > B$ and let $f_s \geqslant 2B$ (sufficient condition assuring the replicas in (7.300) do not overlap). The following *Shannon interpolation formula* holds

$$x_a(t) = \sum_{k\in\mathbb{Z}} x(k)\, \text{sinc}\left(\frac{t}{T_s} - k\right) \qquad \forall t \in \mathbb{R} \tag{7.314}$$

where $x(k) = x_a(t)|_{t=kT_s}$. In addition, if $f_s \geqslant 2B|s|$ we also have

$$x_a(st - d_a) = \sum_{k\in\mathbb{Z}} x(k)\, \text{sinc}\left(\frac{st - d_a}{T_s} - k\right) \qquad \forall t \in \mathbb{R}. \tag{7.315}$$

The more stringent condition (7.310) has to be verified to have that Theorem 7.7.2 holds. Thus, from (7.261) it follows that samples of the sequence $y(n)$ can be generated by the following

interpolation formula (see also (Chan and Ho 2005))

$$\begin{aligned} y(n) &\triangleq y_a(t)|_{t=nT_s} \\ &= A\, e^{j\varphi}\, x_a(snT_s - d_a)\, e^{j2\pi f_a n T_s} \\ &= A\, e^{j\varphi}\, e^{j2\pi f_a n T_s} \sum_{k\in\mathbb{Z}} x(k) \operatorname{sinc}\left(\frac{snT_s - d_a}{T_s} - k\right) \\ &= A\, e^{j\varphi}\, e^{j2\pi \bar{\nu} n} \sum_{k\in\mathbb{Z}} x(k) \operatorname{sinc}(sn - d - k)\,. \end{aligned} \tag{7.316}$$

Analogously, a time-warped version of $x_a(t)$ say

$$y_a(t) = x_a(\psi(t)) \tag{7.317}$$

where $\psi(t)$ is the real-valued time-warping function, can be expressed as

$$y_a(t) = \sum_{k\in\mathbb{Z}} x(k) \operatorname{sinc}\left(\frac{\psi(t)}{T_s} - k\right) \tag{7.318}$$

provided that $f_s \geqslant 2B'$, with B' the monolateral bandwidth of $x_a(\psi(t))$. Consequently, samples of $y_a(t)$ can be generated by the interpolation formula

$$y(n) \triangleq y_a(t)|_{t=nT_s} = \sum_{k\in\mathbb{Z}} x(k) \operatorname{sinc}\left(\frac{\psi(nT_s)}{T_s} - k\right). \tag{7.319}$$

In particular, this relation can be used with $\psi(t) = t - D(t)$ in the case of constant relative radial acceleration with $D(t)$ given by (7.201).

The problem of the bandlimitedness of time-warped versions of bandlimited signals is addressed in (Cochran and Clark 1990), (Xia and Zhang 1992), (Clark and Cochran 1993), (Azizi and Cochran 1999), (Azizi *et al.* 2002). In particular, in (Xia and Zhang 1992) it is shown that if $x_a(t)$ is strictly band-limited, within a large class of time-warping functions, the time-warped signal $y_a(t) = x_a(\psi(t))$ is strictly band-limited if and only if $\psi(t)$ is affine, i.e., $\psi(t) = st - d$. Consequently, for a generic time-warping function $\psi(t)$, even if $x_a(t)$ is strictly band-limited, in general $y_a(t)$ can only be approximatively band-limited.

A similar interpolation procedure can be adopted to generate samples of a frequency-warped version of $x(n)$. Let $X(k/N)$ be the DFT of the finite N-length sequence $x(n)$ defined for $n \in \{0, 1, \ldots, N-1\}$ (see (3.215)). The Fourier transform of $x(n)$ can be expressed by the interpolation formula

$$X(\nu) = \sum_{k=0}^{N-1} X\left(\frac{k}{N}\right) \frac{1}{N} D_{\frac{1}{N}}\left(\nu - \frac{k}{N}\right) \tag{7.320}$$

where $D_{\frac{1}{N}}(\nu)$ is the Dirichlet kernel defined in (3.216b). Thus, for a generic real-valued frequency-warping function $\Psi(\nu)$ (with $-1/2 \leqslant \Psi(\nu) \leqslant 1/2$) we have that the

frequency-warped version of $x(n)$ has Fourier transform

$$X(\Psi(\nu)) = \sum_{k=0}^{N-1} X\left(\frac{k}{N}\right) \frac{1}{N} D_{\frac{1}{N}}\left(\Psi(\nu) - \frac{k}{N}\right) \tag{7.321}$$

and DFT

$$Y_h = X\left(\Psi\left(\frac{h}{N}\right)\right) \qquad h = 0, 1, \ldots, N-1 \tag{7.322}$$

The samples $y(n)$ of the frequency-warped version of $x(n)$ are obtained by inverse discrete Fourier transform (IDFT) of the sequence Y_k

$$y(n) = \text{IDFT}[Y_h] = \frac{1}{N} \sum_{h=0}^{N-1} Y_h \, e^{j2\pi(h/N)n}. \tag{7.323}$$

7.7.5 *Second-Order Statistics (Discrete-Time)*

In this section, sampling theorems are proved for the statistical functions characterizing the uniformly sampled Doppler-stretched received signal and jointly characterizing the sampled transmitted and received signals. In addition, sufficient conditions are derived to avoid aliasing in the relationships linking spectral functions of sampled transmitted and received signals in order to make these relationships formally analogous to their continuous-time counterparts (7.264), (7.268), (7.273), and (7.276) which are obtained starting from the continuous-time physical model (7.261)–(7.262) for the propagation channel. This formal analogy allows, for example, straightforward discrete-time implementations of the Doppler-channel parameter estimation algorithm proposed in continuous-time in (De Angelis *et al.* 2005), and the range-Doppler estimation algorithms proposed in (Napolitano and Doğançay 2011).

Analogous results could be obtained by specializing results of Section 4.9.2 to the case of support curves which are lines. In this section, however, by exploiting the linearity of the support curves, less stringent constraints on the sampling frequency f_s are obtained with respect to those that can be obtained specializing the results of Section 4.9.2.

In the following, for convenience, equations (7.264) and (7.268) will be expressed by the unique equation

$$S^{\alpha}_{y_a y_a^{(*)}}(f_1) = A^2 \, e^{j\kappa\varphi} \, e^{-j2\pi(\alpha-\kappa f_a)d_a/s} \, \frac{1}{|s|} \, S^{(\alpha-\kappa f_a)/s}_{x_a x_a^{(*)}}\left(\frac{f_1 - f_a}{s}\right). \tag{7.324}$$

where $\kappa \triangleq (1 + {\scriptstyle(-)}1)$, that is, $\kappa = 0$ if ${\scriptstyle(-)} = -$ and $\kappa = 2$ if ${\scriptstyle(-)} = +$.

Accounting for the Fourier transform (7.262) of the Doppler-stretched signal, the Loève bifrequency spectrum (7.272) of ACS signals, and the Fourier transform (7.300) of sampled signals, the following aliasing formula for the Loève bifrequency spectrum of $y(n)$ is obtained

when $x_a(t)$ is an ACS signal:

$$\begin{aligned}\mathrm{E}\left\{Y(\nu_1)\, Y^{(*)}(\nu_2)\right\} &= \frac{1}{T_s}\, A^2\, e^{j\kappa\varphi}\, \frac{1}{|s|}\, e^{-j2\pi[\nu_1 +_{(-)} \nu_2 - \kappa\bar{\nu}]d/s} \\ &\sum_{n_1\in\mathbb{Z}}\sum_{n_2\in\mathbb{Z}}\sum_{\alpha\in A_{x_a x_a^{(*)}}} S^{\alpha}_{x_a x_a^{(*)}}\left(\frac{(\nu_1 - n_1 - \bar{\nu}) f_s}{s}\right) e^{j2\pi(n_1 +_{(-)} n_2)d/s} \\ &\delta\left(\nu_2 - n_2 - \bar{\nu} -_{(-)}\left(s\frac{\alpha}{f_s} - (\nu_1 - n_1 - \bar{\nu})\right)\right). \end{aligned} \tag{7.325}$$

The discrete-time signal $y(n)$ is ACS since its Loève bifrequency spectrum has support contained in lines with slope ± 1 in the bifrequency plane (ν_1, ν_2). In addition, if $x_a(t)$ is band-limited with bandwidth B, accounting for (1.178b), the support of the replica with $n_1 = n_2 = 0$ in the aliasing formula (7.325) is a frequency scaled version of that in (7.274) and is reported here for future reference

$$\begin{aligned}&\operatorname{supp}\left\{S^{\alpha}_{x_a x_a^{(*)}}\left(\frac{\nu_1 - \bar{\nu}}{s} f_s\right)\delta\left(\frac{\nu_2 - \bar{\nu}}{s} -_{(-)}\left(\frac{\alpha}{f_s} - \frac{\nu_1 - \bar{\nu}}{s}\right)\right)\right\} \\ &\subseteq \left\{(\nu_1, \nu_2) \in \mathbb{R}\times\mathbb{R} \,:\, |\nu_1 - \bar{\nu}| \leqslant \frac{B}{f_s}|s|,\ |\nu_2 - \bar{\nu}| \leqslant \frac{B}{f_s}|s|,\right. \\ &\qquad \left.\frac{\alpha}{f_s} = \frac{\nu_1 - \kappa\bar{\nu}}{s} +_{(-)} \frac{\nu_2}{s}\right\} \qquad\qquad (7.326a) \\ &\subseteq \left\{(\nu_1, \nu_2) \in \mathbb{R}\times\mathbb{R} \,:\, |\nu_1| \leqslant |\bar{\nu}| + \frac{B}{f_s}|s|,\ |\nu_2| \leqslant |\bar{\nu}| + \frac{B}{f_s}|s|,\right. \\ &\qquad \left.\frac{\alpha}{f_s} = \frac{\nu_1 - \kappa\bar{\nu}}{s} +_{(-)} \frac{\nu_2}{s}\right\}. \qquad\qquad (7.326b)\end{aligned}$$

Starting from (7.325)–(7.326b), the following results can be proved, which provide sufficient conditions to avoid aliasing in the Loève bifrequency spectrum of $y(n)$ and its density.

Theorem 7.7.3 Doppler-Stretched Signal: Sampling Theorem for Loève Bifrequency Spectrum (Napolitano 2010b, Theorem 2). *Let $x_a(t)$ be ACS and band-limited with bandwidth B. For $f_s \geqslant 2B|s|$ replicas in the aliasing formula (7.325) do not overlap. Furthermore, for $f_s \geqslant 2(B|s| + |f_a|)$ one obtains*

$$\begin{aligned}\mathrm{E}\left\{Y(\nu_1)\, Y^{(*)}(\nu_2)\right\} &= \frac{1}{T_s}\, A^2\, e^{j\kappa\varphi}\, \frac{1}{|s|}\, e^{-j2\pi(\nu_1 +_{(-)} \nu_2 - \kappa\bar{\nu})d/s} \\ &\sum_{\alpha\in A_{x_a x_a^{(*)}}} S^{\alpha}_{x_a x_a^{(*)}}\left(\frac{\nu_1 - \bar{\nu}}{s} f_s\right)\delta\left(\nu_2 - \bar{\nu} -_{(-)}\left(\frac{\alpha}{f_s}s - (\nu_1 - \bar{\nu})\right)\right) \\ &|\nu_1| \leqslant \frac{1}{2},\ |\nu_2| \leqslant \frac{1}{2}\end{aligned} \tag{7.327}$$

which is coincident with its continuous-time counterpart (7.273) but for the amplitude scale factor $1/T_s$ and the scaling factor f_s in both frequency variables.

Proof: See Section 7.10. □

An application of Theorem 7.7.3 is illustrated in Section 7.7.6 (Figures 7.10, 7.11, and 7.12).

Theorem 7.7.4 Doppler Stretched Signal: Sampling Theorem for the Density of Loève Bifrequency Spectrum (Napolitano 2010b, Theorem 3). *Let $x_a(t)$ be ACS and band-limited with bandwidth B. For $f_s \geqslant 2(2B|s| + |f_a|)$, the density of Loève bifrequency spectrum of $y(n)$ is given by*

$$\widetilde{S}^{(k)}_{yy^{(*)}}(\nu_1) = \frac{1}{|s|} A^2\, e^{j\kappa\varphi}\, e^{-j2\pi(\alpha/f_s)d/s}\, \frac{1}{T_s} S^{\alpha}_{x_a x_a^{(*)}}\left(\frac{\nu_1 - \bar{\nu}}{s} f_s\right)$$

$$|\nu_1| \leqslant 1/2, \quad \nu_2 = {(-)}\left(\frac{\alpha}{f_s}s - \nu_1 + \kappa\bar{\nu}\right) \qquad \forall \alpha \tag{7.328}$$

where k is an index in one-to-one correspondence with $\alpha \in A_{x_a x_a^{(*)}}$. In (7.328), for $(-) = -$ one has $\nu_1 - \nu_2 = \alpha s/f_s$ and for $(-) = +$ one has $\nu_1 + \nu_2 = \alpha s/f_s + 2\bar{\nu}$.

Proof: See Section 7.10. □

The sufficient condition on the sampling frequency in Theorem 7.7.4 can be made more stringent by requiring that not only the mapping between spectral frequencies $\nu_1 = f_1/f_s$ holds for $\nu_1 \in [-1/2, 1/2]$ but also that the mapping between cycle frequencies $\widetilde{\alpha} = \alpha/f_s$ holds for $\widetilde{\alpha} \in [-1/2, 1/2]$. This new condition assures the formal analogy between formulas in terms of cyclic spectra for continuous- and discrete-time cases. It can be obtained starting from the aliasing formula for the discrete-time (conjugate) cyclic spectrum of $y(n)$ which can be derived accounting for (7.324) and (1.175):

$$\begin{aligned}\widetilde{S}^{\widetilde{\alpha}}_{yy^{(*)}}(\nu) &= \frac{1}{T_s} \sum_{p\in\mathbb{Z}} \sum_{q\in\mathbb{Z}} S^{\alpha - pf_s}_{y_a y_a^{(*)}}(f - qf_s)\Big|_{f=\nu f_s, \alpha=\widetilde{\alpha} f_s} \\ &= A^2\, e^{j\kappa\varphi}\, e^{-j2\pi(\widetilde{\alpha}-\kappa\bar{\nu})d/s}\, \frac{1}{|s|}\frac{1}{T_s} \\ &\quad \sum_{p\in\mathbb{Z}} \sum_{q\in\mathbb{Z}} S^{[(\widetilde{\alpha}-\kappa\bar{\nu}-p)f_s]/s}_{x_a x_a^{(*)}}\left(\frac{\nu - \bar{\nu}}{s} f_s - q\frac{f_s}{s}\right) e^{j2\pi pd/s}.\end{aligned} \tag{7.329}$$

It is worth observing that the formal analogy between cyclic spectra of the discrete-time sampled signals and their continuous-time counterparts is not straightforward and should be carefully stated, as clarified by the following consideration. By replacing $\widetilde{\alpha}$ with $(\widetilde{\alpha} - \kappa\bar{\nu})/s$ and ν with $(\nu - \bar{\nu})/s$ into the aliasing formula for cyclic spectra (1.175), it follows that

$$\widetilde{S}^{(\widetilde{\alpha}-\kappa\bar{\nu})/s}_{xx^{(*)}}\left(\frac{\nu - \bar{\nu}}{s}\right) = \frac{1}{T_s} \sum_{p\in\mathbb{Z}} \sum_{q\in\mathbb{Z}} S^{[(\widetilde{\alpha}-\kappa\bar{\nu})/s - p]f_s}_{x_a x_a^{(*)}}\left(\frac{\nu - \bar{\nu}}{s} f_s - qf_s\right). \tag{7.330}$$

Even if the function in the left-hand-side (lhs) of (7.330) can be seen as the discrete-time counterpart of the cyclic spectrum in the right-hand-sides (rhs) of (7.324), both functions in the rhs and lhs of (7.330) are periodic in ν and $\widetilde{\alpha}$ with period s and, hence, are not (conjugate) cyclic spectra of discrete-time signals. The discrete-time counterpart of (7.324) is obtained by the following result.

Theorem 7.7.5 Doppler Stretched Signal: Sampling Theorem for the (Conjugate) Cyclic Spectrum (Napolitano 2010b, Theorem 4). *Let $x_a(t)$ be ACS and band-limited with bandwidth B. For $f_s \geqslant 4(B|s| + |f_a|)$ one has*

$$\widetilde{\widetilde{S}}^{\alpha}_{yy^{(*)}}(\nu) = A^2\, e^{j\kappa\varphi}\, e^{-j2\pi(\widetilde{\alpha}-\kappa\bar{\nu})d/s}\, \frac{1}{|s|}\, \widetilde{S}^{(\widetilde{\alpha}-\kappa\bar{\nu})/s}_{xx^{(*)}}\left(\frac{\nu-\bar{\nu}}{s}\right)$$
$$|\nu| \leqslant \frac{1}{4},\ |\widetilde{\alpha}| \leqslant \frac{1}{2} \tag{7.331}$$

which is formally analogous to its continuous-time counterpart (7.324).

Proof: See Section 7.10. □

An application of Theorems 7.7.4 and 7.7.5 is illustrated in Section 7.7.6 (Figure 7.13).

The spectral cross-characterization of $y(n)$ and $x(n)$ requires to prove sampling theorems in the case of signals that are not jointly ACS but, rather, jointly SC. Therefore, results for auto-statistics cannot be extended to cross-statistics.

Accounting for the Fourier transform (7.262) of the Doppler-stretched signal, the Loève bifrequency spectrum (7.272) of ACS signals, and the Fourier transform (7.300) of sampled signals, the following aliasing formula for the Loève bifrequency cross-spectrum of $y(n)$ and $x(n)$ can be derived when $x_a(t)$ is ACS:

$$\mathrm{E}\left\{Y(\nu_1)\, X^{(*)}(\nu_2)\right\} = \frac{1}{T_s}\, A\, e^{j\varphi}\, \frac{1}{|s|}\, e^{-j2\pi(\nu_1-\bar{\nu})d/s}$$
$$\sum_{n_1\in\mathbb{Z}} \sum_{n_2\in\mathbb{Z}} \sum_{\alpha\in A_{x_a x_a^{(*)}}} S^{\alpha}_{x_a x_a^{(*)}}\left(\frac{\nu_1-n_1-\bar{\nu}}{s} f_s\right)$$
$$\delta\left((\nu_2-n_2) - _{(-)}\left(\alpha/f_s - \frac{\nu_1-\bar{\nu}-n_1}{s}\right)\right) e^{j2\pi n_1 d/s}. \tag{7.332}$$

From (7.332), it follows that the discrete-time signals $x(n)$ and $y(n)$ are jointly SC with the Loève bifrequency cross-spectrum having support contained in lines with slope $\pm 1/s$ in the bifrequency plane (ν_1, ν_2).

If $x_a(t)$ is band-limited with bandwidth B, accounting for (1.178b), the support of the replica with $n_1 = n_2 = 0$ and fixed α in the aliasing formula (7.332) is such that

$$\mathrm{supp}\left\{S^{\alpha}_{x_a x_a^{(*)}}\left(\frac{\nu_1-\bar{\nu}}{s} f_s\right)\delta\left(\nu_2 - _{(-)}\left(\alpha/f_s - \frac{\nu_1-\bar{\nu}}{s}\right)\right)\right\}$$
$$\subseteq \left\{(\nu_1,\nu_2)\in\mathbb{R}\times\mathbb{R} : |\nu_1-\bar{\nu}| \leqslant \frac{B}{f_s}|s|,\ |\nu_2| \leqslant \frac{B}{f_s},\ \frac{\alpha}{f_s} = \left(\frac{\nu_1-\bar{\nu}}{s} + _{(-)}\nu_2\right)\right\} \tag{7.333a}$$
$$\subseteq \left\{(\nu_1,\nu_2)\in\mathbb{R}\times\mathbb{R} : |\nu_1|-|\bar{\nu}| \leqslant \frac{B}{f_s}|s|,\ |\nu_2| \leqslant \frac{B}{f_s},\ \frac{\alpha}{f_s} = \left(\frac{\nu_1-\bar{\nu}}{s} + _{(-)}\nu_2\right)\right\}. \tag{7.333b}$$

Equations (7.333a) and (7.333b) allow to establish the following results.

Theorem 7.7.6 Sampling Theorem for Loève Bifrequency Cross-Spectrum (Napolitano 2010b, Theorem 5). *Let $x_a(t)$ be ACS and band-limited with bandwidth B. For*

$$f_s \geqslant \max\{2(B|s| + |f_a|), 2B\} \tag{7.334}$$

one obtains

$$\begin{aligned} \mathrm{E}\left\{Y(\nu_1)\, X^{(*)}(\nu_2)\right\} &= \frac{1}{T_s}\, A\, e^{j\varphi}\, \frac{1}{|s|}\, e^{-j2\pi(\nu_1-\bar{\nu})d/s} \\ &\quad \sum_{\alpha \in A_{x_a x_a^{(*)}}} S^{\alpha}_{x_a x_a^{(*)}}\left(\frac{\nu_1 - \bar{\nu}}{s} f_s\right) \delta\left(\nu_2 - (-)\left(\frac{\alpha}{f_s} - \frac{\nu_1 - \bar{\nu}}{s}\right)\right) \\ &\quad |\nu_1| \leqslant \frac{1}{2},\ |\nu_2| \leqslant \frac{1}{2} \end{aligned} \tag{7.335}$$

which is formally analogous to its continuous-time counterpart (7.276).

Proof: See Section 7.10. □

An application of Theorem 7.7.6 is illustrated in Section 7.7.6 (Figures 7.14, 7.15, and 7.16).

Theorem 7.7.7 Sampling Theorem for the Density of Loève Bifrequency Cross-Spectrum (Napolitano 2010b, Theorem 6). *Let $x_a(t)$ be ACS and band-limited with bandwidth B. For*

$$f_s \geqslant \max\{2(B|s| + |f_a|), 4B\} \tag{7.336}$$

one obtains that the density of the Loève bifrequency cross-spectrum (7.332) along the support line $\nu_2 = {}_{(-)}(\alpha/f_s - (\nu_1 - \bar{\nu})/s)$ is given by

$$\widetilde{S}^{(k)}_{yx^{(*)}}(\nu_1) = \frac{1}{|s|} A\, e^{j\varphi}\, e^{-j2\pi(\nu_1-\bar{\nu})d/s}\, \frac{1}{T_s} S^{\alpha}_{x_a x_a^{(*)}}\left(\frac{\nu_1 - \bar{\nu}}{s} f_s\right) \qquad |\nu_1| \leqslant 1/2 \tag{7.337}$$

where k is an index in one-to-one correspondence with $\alpha \in A_{x_a x_a^{(*)}}$.

Proof: See Section 7.10. □

An application of Theorem 7.7.7 is illustrated in Section 7.7.6 (Figure 7.17).

By inverse Fourier transforming both sides of (7.324) gives the following expression for the (conjugate) cyclic autocorrelation function of the Doppler-stretched signal $y_a(t)$ (see also (7.263) and (7.267))

$$R^{\alpha}_{y_a y_a^{(*)}}(\tau) = A^2\, e^{j\kappa\varphi}\, e^{j2\pi f_a \tau}\, e^{-j2\pi(\alpha-\kappa f_a)d_a/s}\, R^{(\alpha-\kappa f_a)/s}_{x_a x_a^{(*)}}(s\tau). \tag{7.338}$$

Then, accounting for (1.174) the following aliasing formula for the (conjugate) cyclic autocorrelation function of $y(n)$ is obtained

$$\begin{aligned}\widetilde{\widetilde{R}}^{\alpha}_{yy^{(*)}}(m) &= \sum_{p\in\mathbb{Z}} R^{(\widetilde{\alpha}-p)f_s}_{y_a y_a^{(*)}}(mT_s) \\ &= A^2\, e^{j\kappa\varphi}\, e^{j2\pi\bar{\nu}m}\, e^{-j2\pi[\widetilde{\alpha}-\kappa\bar{\nu}]d/s} \\ &\quad \sum_{p\in\mathbb{Z}} R^{(\widetilde{\alpha}-\kappa\bar{\nu}-p)f_s}_{x_{as} x_{as}^{(*)}}(mT_s)\, e^{j2\pi pd/s}.\end{aligned} \tag{7.339}$$

In (7.339), $x_{as}(t) \triangleq x_a(st)$ and, hence, according to (7.338) (with $y_a(t) = x_a(st)$, $A = 1, \varphi = 0$, $f_a = 0$, and $d_a = 0$) one has $R^{\alpha}_{x_{as} x_{as}^{(*)}}(\tau) = R^{\alpha/s}_{x_a x_a^{(*)}}(s\tau)$.

From (7.339), the following result can be derived.

Theorem 7.7.8 Sampling Theorem for the (Conjugate) Cyclic Autocorrelation Function (Napolitano 2010b, Theorem 7). *Let $x_a(t)$ be ACS and band-limited with bandwidth B. For*

$$f_s \geqslant 2B|s| \tag{7.340}$$

one obtains

$$\widetilde{\widetilde{R}}^{\alpha}_{yy^{(*)}}(m) = A^2\, e^{j\kappa\varphi}\, e^{j2\pi\bar{\nu}m}\, e^{-j2\pi[\widetilde{\alpha}-\kappa\bar{\nu}]d/s}\, R^{(\widetilde{\alpha}-\kappa\bar{\nu})f_s/s}_{x_a x_a^{(*)}}(smT_s) \tag{7.341a}$$

$$= A^2\, e^{j\kappa\varphi}\, e^{j2\pi\bar{\nu}m}\, e^{-j2\pi[\widetilde{\alpha}-\kappa\bar{\nu}]d/s}\, \widetilde{\widetilde{R}}^{\alpha-\kappa\bar{\nu}}_{x_s x_s^{(*)}}(m) \qquad |\widetilde{\alpha}| \leqslant 1/2 \tag{7.341b}$$

with $x_s(n) \triangleq x_a(st)|_{t=nT_s}$. Equation (7.341b) is the unaliased sampled version of (7.338).

Proof: See Section 7.10. □

An expression for $\widetilde{\widetilde{R}}^{\alpha}_{yy^{(*)}}(m)$ alternative to (7.341a) and (7.341b) can be obtained starting from (7.331). In fact, for $f_s \geqslant 4(B|s| + |f_a|)$ the (conjugate) cyclic autocorrelation function of $y(n)$ can be expressed as

$$\begin{aligned}\widetilde{\widetilde{R}}^{\alpha}_{yy^{(*)}}(m) &= \int_{-1/2}^{1/2} \widetilde{\widetilde{S}}^{\alpha}_{yy^{(*)}}(\nu)\, e^{j2\pi\nu m}\, \mathrm{d}\nu \\ &= A^2\, e^{j\kappa\varphi}\, e^{-j2\pi(\widetilde{\alpha}-\kappa\bar{\nu})d/s}\, \frac{1}{|s|} \int_{-1/4}^{1/4} \widetilde{\widetilde{S}}^{(\widetilde{\alpha}-\kappa\bar{\nu})/s}_{xx^{(*)}}\left(\frac{\nu-\bar{\nu}}{s}\right) e^{j2\pi\nu m}\, \mathrm{d}\nu \\ &\quad |\widetilde{\alpha}| \leqslant \frac{1}{2}.\end{aligned} \tag{7.342}$$

Finally, the cross-characterization of $y(n)$ and $x(n)$ is obtained by setting $t = nT_s$ and $\tau = mT_s$ into (7.283)

$$\begin{aligned}
\mathrm{E}\left\{ y(n+m)\, x^{(*)}(n) \right\} \\
&= A\, e^{j\varphi}\, e^{j2\pi\bar{\nu} m} \sum_{\alpha \in A_{x_a x_a^{(*)}}} R^{\alpha}_{x_a x_a^{(*)}}\Big((s-1)nT_s + smT_s - dT_s \Big)\, e^{j2\pi(\alpha + \bar{\nu} f_s) nT_s} \\
&= A\, e^{j\varphi}\, e^{j2\pi\bar{\nu}(n+m)} \sum_{\widetilde{\alpha} \in \widetilde{A}} e^{j2\pi\widetilde{\alpha} n} \sum_{p \in \mathbb{Z}} R^{(\widetilde{\alpha} - p) f_s}_{x_a x_a^{(*)}}\Big((sm + (s-1)n - d) T_s \Big) \qquad (7.343)
\end{aligned}$$

where $\widetilde{A}$ is defined according to (1.171) and (1.172) and $(sm + (s-1)n - d)$ in general is non-integer.

Analogously to the continuous-time case, the cross-correlation $\mathrm{E}\left\{ y(n+m)\, x^{(*)}(n) \right\}$ does not contain any finite-strength additive sinewave component if $s \neq 1$.

7.7.6 *Illustrative Examples*

In this section, examples are presented aimed at illustrating the theoretical results of Section 7.7.5. These results, which are formulated in terms of Loève bifrequency spectra, can be equivalently expressed in terms of bifrequency spectral cross-correlation densities (4.19) by replacing Dirac with Kronecker deltas. For example, (7.276) becomes

$$\begin{aligned}
\bar{S}_{y_a x_a^*}(f_1, f_2) &= A\, e^{j\varphi}\, \frac{1}{|s|}\, e^{-j2\pi(f_1 - f_a)d_a/s} \\
&\quad \sum_{\alpha \in A_{x_a x_a^{(*)}}} S^{\alpha}_{x_a x_a^{(*)}}\left(\frac{f_1 - f_a}{s} \right) \delta_{[f_2 - (-)(\alpha - (f_1 - f_a)/s)]} \qquad (7.344)
\end{aligned}$$

where δ_γ denotes Kronecker delta, that is, $\delta_\gamma = 1$ for $\gamma = 0$ and $\delta_\gamma = 0$ for $\gamma \neq 0$.

In the experiment, a PAM signal $x_a(t)$ is considered with stationary white modulating sequence, raised cosine pulse with excess bandwidth η, and symbol period T_p. It is second-order cyclostationary with period T_p and strictly bandlimited with bandwidth $B = (1+\eta)/(2T_p)$ which can be slightly overestimated by $B' = 1/T_p$. Thus, it has three cycle frequencies $\alpha_h = h/T_p$, $h \in \{0, \pm 1\}$ (Gardner *et al.* 2006). Furthermore, $\eta = 0.85$ and $T_p = 4/f_r$ are assumed, where f_r is a fixed reference frequency. The Doppler-stretched signal $y_a(t)$ in (7.261) has $A = 1$, $\varphi = 0$, $d_a = 0$, $f_a = 0.075 f_r$, and $s = 0.75$.

In the figures, supports of impulsive functions are drawn as "checkerboard" plots with gray levels representing the magnitude of the density of the impulsive functions. In addition, for notation simplicity, the bifrequency SCD of signals a and b is denoted by $S_{ab^*}(\cdot,\cdot)$. The bifrequency spectral correlation density $\bar{S}_{x_a x_a^*}(f_1, f_2)$ is the same as that in Figure 4.6. In Figure 7.10, (top) magnitude and (bottom) "checkerboard" plot of the bifrequency spectral correlation density $\bar{S}_{y_a y_a^*}(f_1, f_2)$ are reported as functions of f_1/f_r and f_2/f_r. In Figures 7.11 and 7.12, the magnitude of the bifrequency spectral correlation density $\bar{S}_{yy^*}(\nu_1, \nu_2)$ of the discrete-time signal $y(n) = y_a(t)|_{t=nT_s}$, is represented as a function of ν_1 and ν_2 for two values of the sampling frequency. Specifically, in Figure 7.11 $f_s = 2|s|B'$ and in Figure 7.12 $f_s = 4|s|B'$

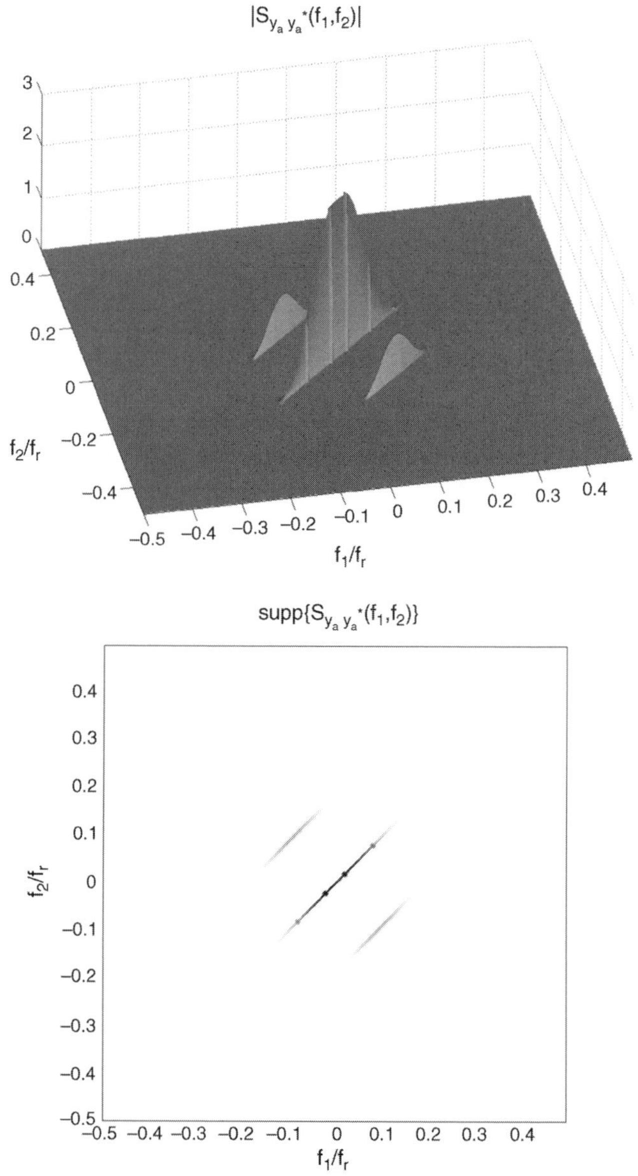

Figure 7.10 (Continuous-time) Doppler-stretched signal $y_a(t) = x_a(st)\, e^{j2\pi f_a t}$, where $x_a(t)$ is a PAM signal with raised cosine pulse. (Top) magnitude and (bottom) support of the bifrequency spectral correlation density of $y_a(t)$, as a function of f_1/f_r and f_2/f_r. *Source:* (Napolitano 2010b).

are considered. Since the support lines have unit slope, both continuous- and discrete-time signals are ACS. According to Theorem 7.7.3, $f_s = 2|s|B' \geqslant 2|s|B$ is a sufficient condition to assure non-overlapping replicas in the Loève bifrequency spectrum of $y_a(t)$. However, only the more stringent condition $f_s \geqslant 2(B|s| + |f_a|)$ can assure that (7.327) holds. For the considered

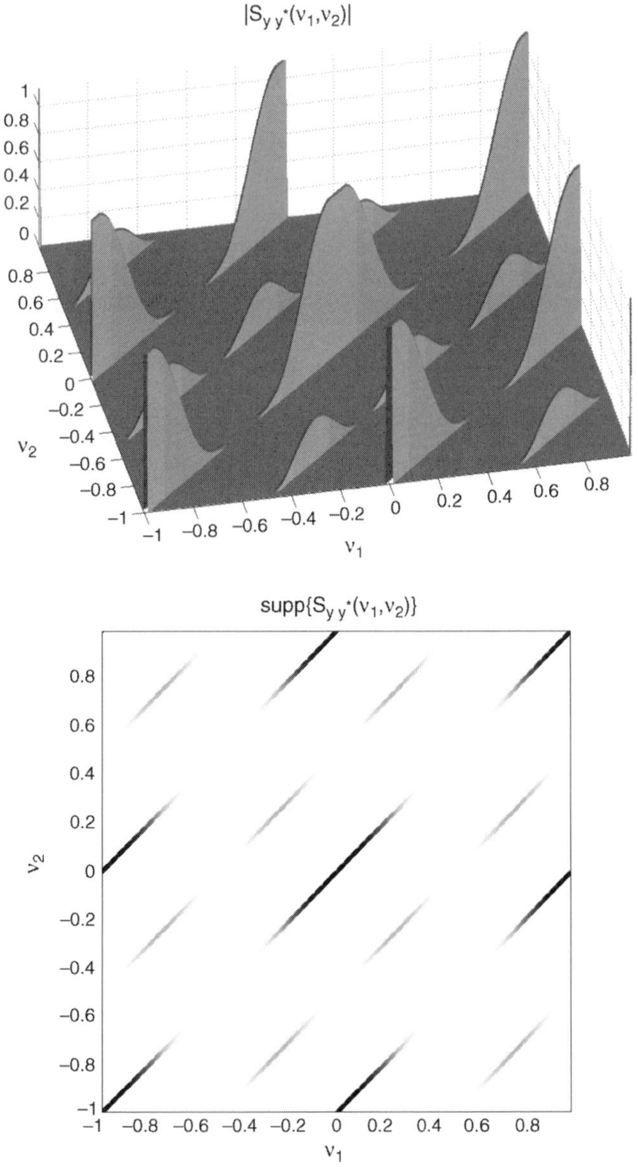

Figure 7.11 (Continuous-time) Doppler-stretched signal $y_a(t) = x_a(st)\, e^{j2\pi f_a t}$, where $x_a(t)$ is a PAM signal with raised cosine pulse. Bifrequency spectral correlation density of the discrete-time signal $y(n) = y_a(t)|_{t=nT_s}$, as a function of ν_1 and ν_2. (Top) magnitude and (bottom) support for $f_s = 2|s|B'$. *Source:* (Napolitano 2010b) Copyright of Elsevier

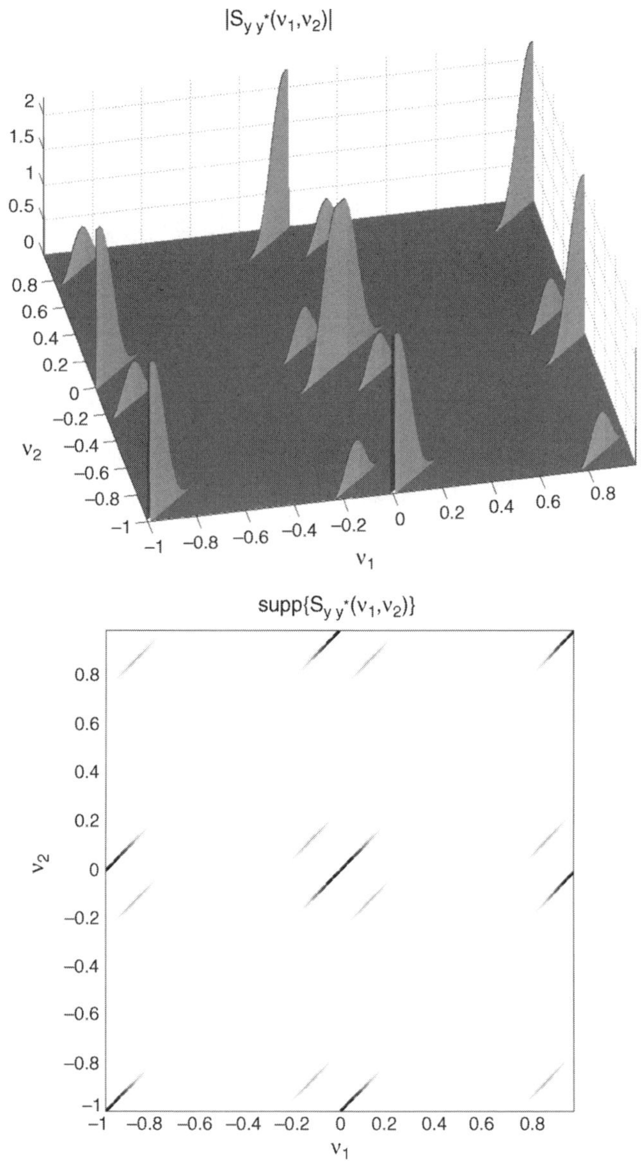

Figure 7.12 (Continuous-time) Doppler-stretched signal $y_a(t) = x_a(st)\, e^{j2\pi f_a t}$, where $x_a(t)$ is a PAM signal with raised cosine pulse. Bifrequency spectral correlation density of the discrete-time signal $y(n) = y_a(t)|_{t=nT_s}$, as a function of ν_1 and ν_2. (Top) magnitude and (bottom) support for $f_s = 4|s|B'$. *Source:* (Napolitano 2010b) Copyright of Elsevier.

numerical values, $f_s = 4|s|B'$ is such that the sufficient conditions on f_s of Theorems 7.7.4 and 7.7.5 are both satisfied.

In Figure 7.13, the slice of the magnitude of the spectral correlation density of the continuous-time signal $y_a(t)$ along the support line $f_2 = f_1 - s/T_p$ (thin line) and of the rescaled ($\nu_1 =$

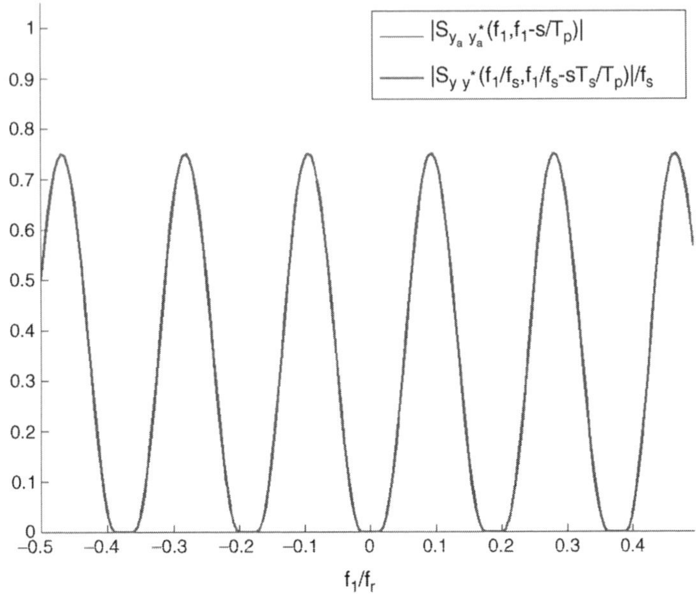

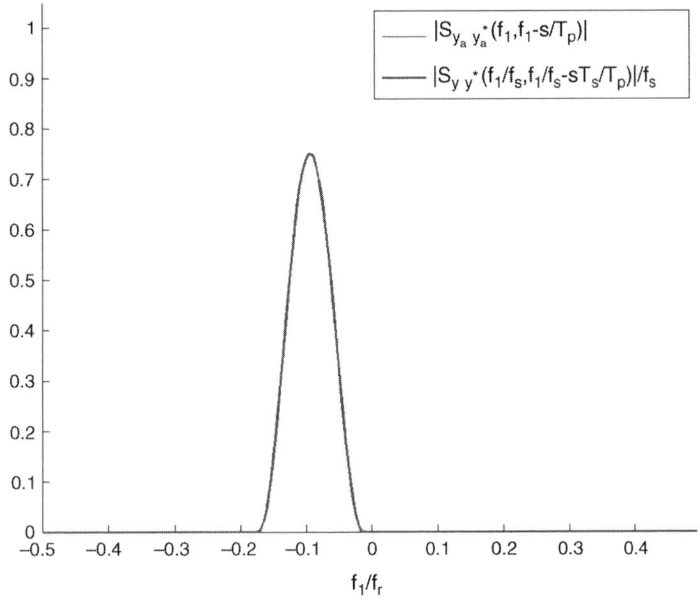

Figure 7.13 (Continuous-time) Doppler-stretched signal $y_a(t) = x_a(st)\, e^{j2\pi f_a t}$, where $x_a(t)$ is a PAM signal with raised cosine pulse. Slice of the magnitude of the spectral correlation density of the continuous-time signal $y_a(t)$ along the support line $f_2 = f_1 - s/T_p$ (thin line) and of the rescaled ($\nu_1 = f_1/f_s$) spectral correlation density of the discrete-time signal $y(n) = y_a(t)|_{t=nT_s}$, along $\nu_2 = \nu_1 - s(T_s/T_p)$ (thick line) as a function of f_1/f_r. (Top) $f_s = 2|s|B'$ $[f_s/2 = 0.1875\, f_r]$ and (bottom) $f_s = 4|s|B'$ $[f_s/2 = f_r/2]$. *Source:* (Napolitano 2010b) Copyright of Elsevier

f_1/f_s) spectral correlation density of the discrete-time signal $y(n) = y_a(t)|_{t=nT_s}$ along $\nu_2 = \nu_1 - s(T_s/T_p)$ (thick line) as a function of f_1/f_r are reported (top) for $f_s = 2|s|B'$ and (bottom) for $f_s = 4|s|B'$. According with the results of Theorems 7.7.4 and 7.7.5, only in case $f_s = 4|s|B'$ there are no aliasing replicas and the spectral correlation density of $y_a(t)$ (coincident with the cyclic spectrum) is equal to that rescaled of $y(n)$ considered in the main frequency domain.

Let us consider now the bifrequency spectral cross-correlation density $\bar{S}_{y_a x_a^*}(f_1, f_2)$ between the jointly SC signals $y_a(t)$ and $x_a(t)$. Accounting for (7.344) we have

$$\bar{S}_{y_a x_a^*}(f_1, f_2) = \frac{1}{|s|} \sum_{h=-1}^{1} S_{x_a x_a^*}^{\alpha_h}\left(\frac{f_1 - f_a}{s}\right) \delta_{f_2-(f_1-f_a)/s+\alpha_h} \tag{7.345}$$

where $S_{x_a x_a^*}^{\alpha_h}(f_1)$ denote the cyclic spectra of the PAM signal $x_a(t)$.

In Figure 7.14, (top) magnitude and (bottom) "checkerboard" plot of the spectral cross-correlation density (7.345) of the continuous-time signals $y_a(t)$ and $x_a(t)$ are reported as functions of f_1/f_r and f_2/f_r.

In Figures 7.15 and 7.16, the bifrequency spectral cross-correlation density $\bar{S}_{yx^*}(\nu_1, \nu_2)$ of the discrete-time signals $y(n) \triangleq y_a(t)|_{t=nT_s}$ and $x(n) = x_a(t)|_{t=nT_s}$ is represented as a function of ν_1 and ν_2 for two values of the sampling frequency. Specifically, in Figure 7.15 $f_s = 2B'$ and in Figure 7.16 $f_s = 4B'$ are considered. Since $B' \geqslant B \geqslant B|s| + |f_a|$, due to Theorem 7.7.6, condition $f_s = 2B'$ assures non-overlapping replicas in the Loève bifrequency cross-spectrum, and, hence, in its bifrequency density $\bar{S}_{yx^*}(\nu_1, \nu_2)$. However, such a condition does not assure, for each support line, the mapping in the whole principal frequency domain between spectral densities of continuous- and discrete-time signals. In fact, in Figure 7.17, the slice of the magnitude of the spectral cross-correlation density of the continuous-time signals $y_a(t)$ and $x_a(t)$ along the support line $f_2 = (f_1 - f_a)/s - 1/T_p$ (thin line) and of the rescaled ($\nu_1 = f_1/f_s$) spectral cross-correlation density of the discrete-time signals $y(n) = y_a(t)|_{t=nT_s}$ and $x(n) = x_a(t)|_{t=nT_s}$ along $\nu_2 = (\nu_1 - \bar{\nu})/s - (T_s/T_p)$ (thick line) as a function of f_1/f_r is reported (top) for $f_s = 2B'$ and (bottom) for $f_s = 4B'$. According to Theorem 7.7.7 and due to the numerical values of s and f_a, only condition $f_s = 4B'$ assures the lack of aliasing replicas in the considered cross-spectral density function. In such a case, the spectral cross-correlation density of the continuous-time signals and that rescaled of the discrete-time signals are coincident in the principal frequency domain.

7.7.7 Concluding Remarks

The Loève bifrequency spectrum is used as framework to describe in the spectral domain the nonstationarity of a Doppler-stretched signal when the transmitted signal is ACS. Even if the Doppler channel is linear not almost-periodically time-variant, the received signal is still ACS. The transmitted and received signals are jointly SC with spectral masses in the bifrequency plane concentrated on lines with slope equal to the reciprocal of the time-scale factor introduced by the Doppler channel. Sampling theorems are proved for spectral statistical functions characterizing the Doppler-stretched received signal and jointly characterizing the transmitted and received signals. Sufficient conditions are found in terms of sampling frequency f_s, transmitted signal bandwidth B, frequency-shift f_a, and time-scale factor s. For the ACS Doppler-stretched signal, $f_s \geqslant 2B|s|$ and $f_s \geqslant 2(B|s| + |f_a|)$ assure, in the Loève

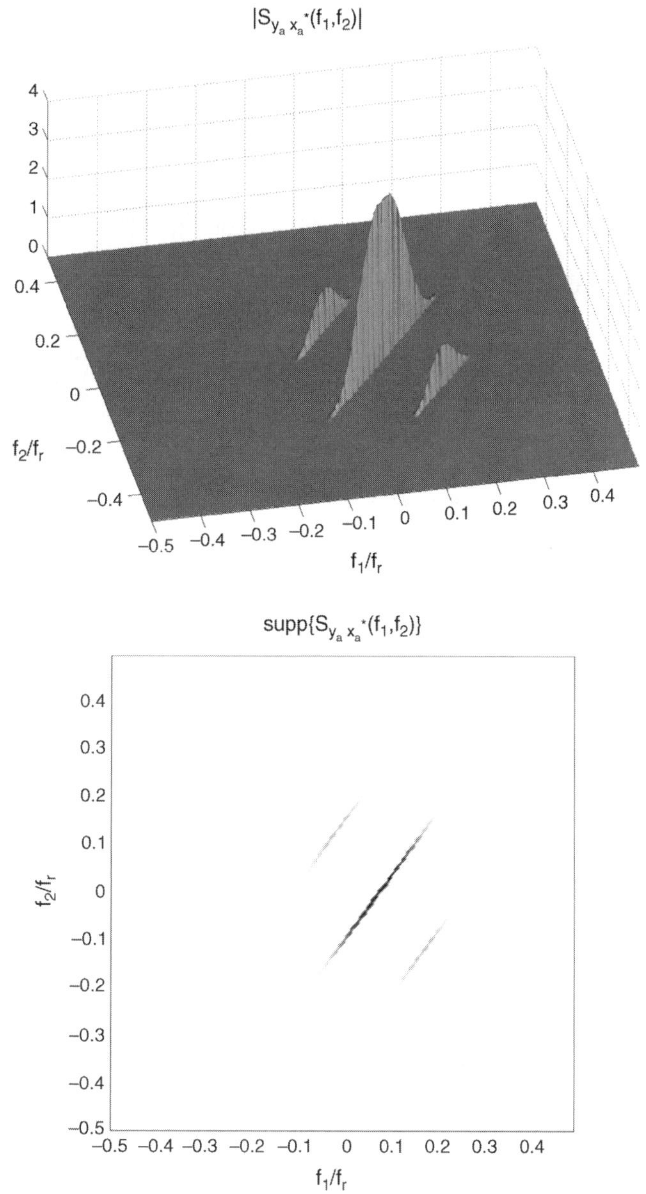

Figure 7.14 (Continuous-time) PAM signal $x_a(t)$ with raised cosine pulse. (Top) magnitude and (bottom) support of the bifrequency spectral cross-correlation density of $y_a(t) = x_a(st)\, e^{j2\pi f_a t}$ and $x_a(t)$, as a function of f_1/f_r and f_2/f_r. *Source:* (Napolitano 2010b) Copyright of Elsevier

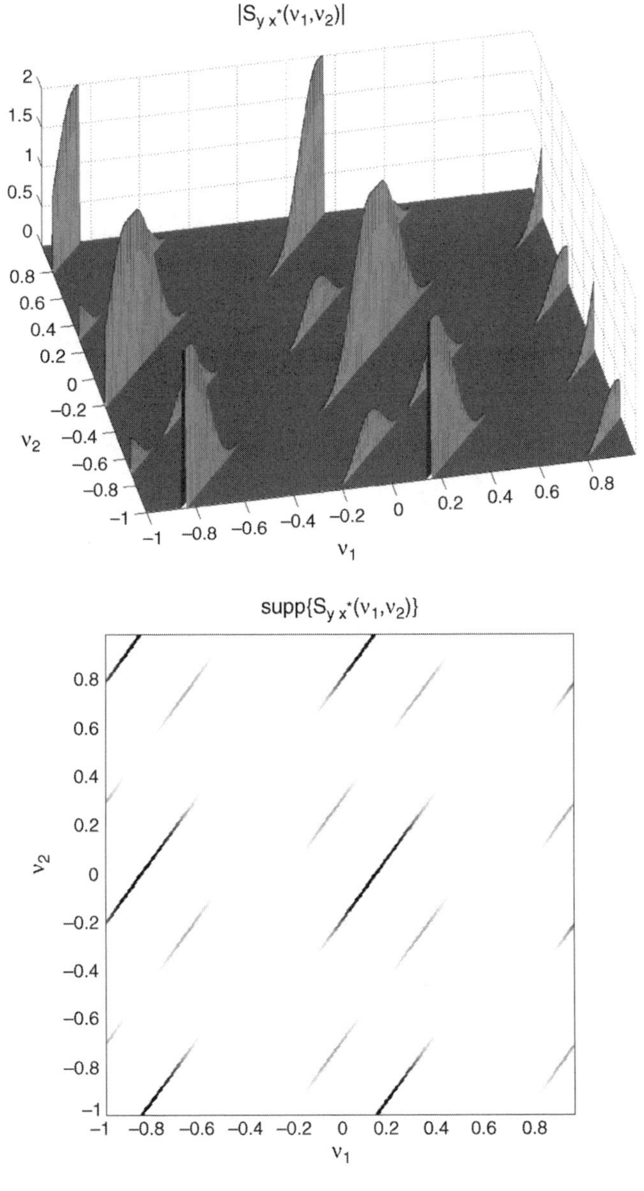

Figure 7.15 (Continuous-time) PAM signal $x_a(t)$ with raised cosine pulse and its Doppler-stretched version $y_a(t) = x_a(st)\, e^{j2\pi f_a t}$. Bifrequency spectral cross-correlation density of the discrete-time signals $y(n) \triangleq y_a(t)|_{t=nT_s}$ and $x(n) = x_a(t)|_{t=nT_s}$, as a function of ν_1 and ν_2. (Top) magnitude and (bottom) support for $f_s = 2B'$. *Source:* (Napolitano 2010b) Copyright of Elsevier

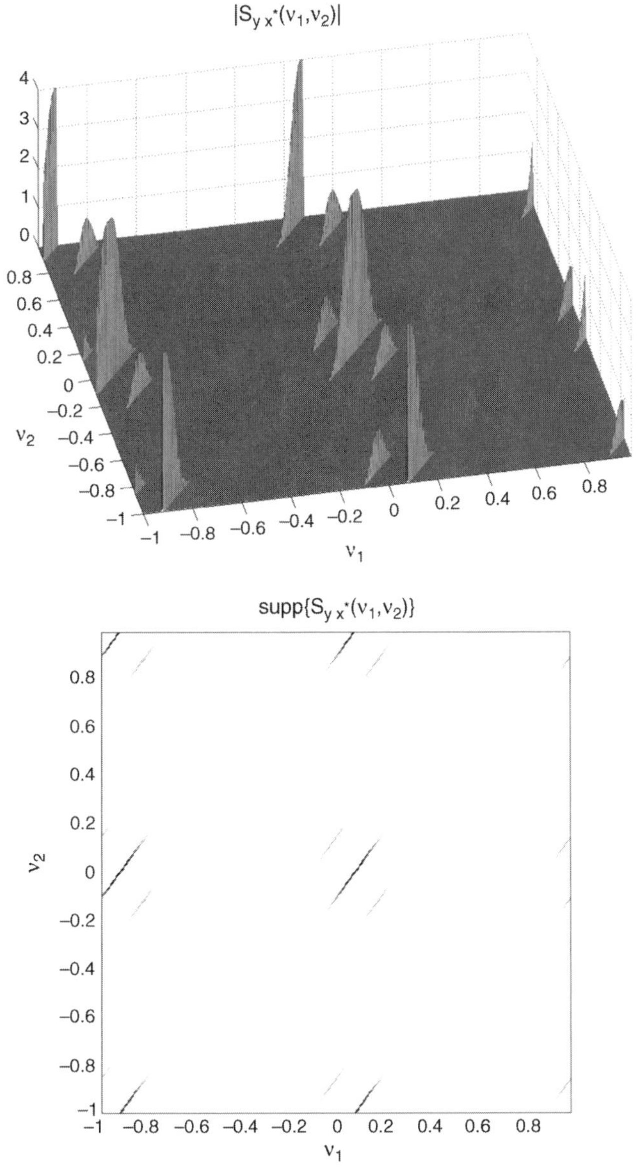

Figure 7.16 (Continuous-time) PAM signal $x_a(t)$ with raised cosine pulse and its Doppler-stretched version $y_a(t) = x_a(st)\, e^{j2\pi f_a t}$. Bifrequency spectral cross-correlation density of the discrete-time signals $y(n) \triangleq y_a(t)|_{t=nT_s}$ and $x(n) = x_a(t)|_{t=nT_s}$, as a function of ν_1 and ν_2. (Top) magnitude and (bottom) support for $f_s = 4B'$. *Source:* (Napolitano 2010b) Copyright of Elsevier

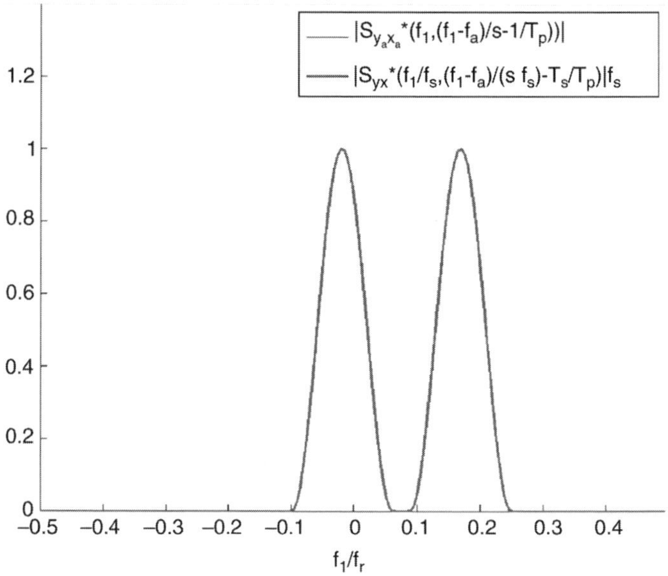

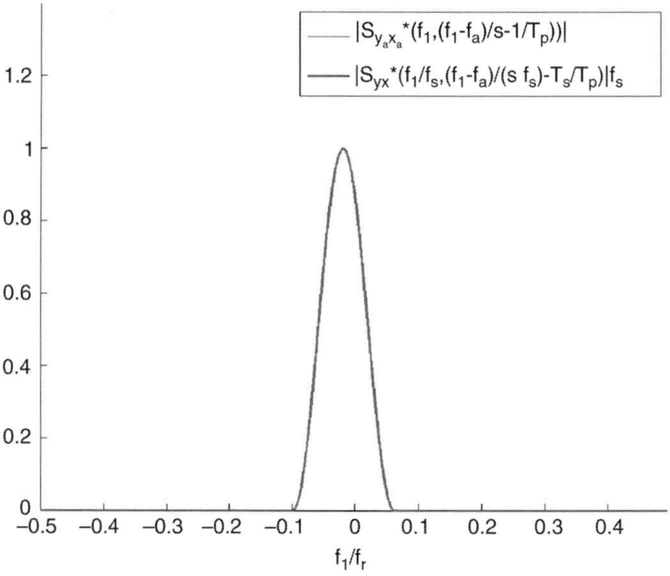

Figure 7.17 (Continuous-time) PAM signal $x_a(t)$ with raised cosine pulse and its Doppler-stretched version $y_a(t) = x_a(st)\, e^{j2\pi f_a t}$. Slice of the magnitude of the spectral cross-correlation density of the continuous-time signals $y_a(t)$ and $x_a(t)$ along the support line $f_2 = (f_1 - f_a)/s - 1/T_p$ (thin line) and of the rescaled ($\nu_1 = f_1/f_s$) spectral cross-correlation density of the discrete-time signals $y(n) = y_a(t)|_{t=nT_s}$ and $x(n) = x_a(t)|_{t=nT_s}$, along $\nu_2 = (\nu_1 - \bar{\nu})/s - (T_s/T_p)$ (thick line) as a function of f_1/f_r. (top) $f_s = 2B'$ $[f_s/2 = 0.25\, f_r]$ and (bottom) $f_s = 4B'$ $[f_s/2 = f_r/2]$. *Source:* (Napolitano 2010b) Copyright of Elsevier

bifrequency spectrum of the sampled signal, no overlap of replicas and main replica in the principal frequency domain, respectively (Theorem 7.7.3); $f_s \geqslant 2(2B|s| + |f_a|)$ guarantees the mapping between spectral densities of the continuous- and discrete-time signals in the whole principal frequency domain (Theorem 7.7.4); $f_s \geqslant 4(B|s| + |f_a|)$ assures that the main replica of the (conjugate) cyclic spectrum is in the principal spectral and cycle frequency domain (Theorem 7.7.5). For the jointly SC transmitted and received Doppler-stretched signals, $f_s \geqslant \max\{2B|s|, 2B\}$ and $f_s \geqslant \max\{2(B|s| + |f_a|), 2B\}$ assure, in the Loève bifrequency cross-spectrum, no overlap of replicas and main replica in the principal frequency domain, respectively (Theorem 7.7.6); $f_s \geqslant \max\{2(B|s| + |f_a|), 4B\}$ provides the mapping between cross-spectral densities of the continuous- and discrete-time signals in the whole principal frequency domain (Theorem 7.7.7). Condition $f_s \geqslant 4B|s|$ is sufficient to avoid aliasing in the cycle-frequency domain for the (conjugate) cyclic autocorrelation function (Theorem 7.7.8).

It is worth observing that in communications and radar applications the several obtained bounds on f_s can be reduced to the two bounds $f_s \geqslant 2B$ and $f_s \geqslant 4B$ since $|s| \simeq 1$ and $|f_a| \ll B$ can be generally assumed. Such conditions, however, do not hold in sonar and acoustic aircraft applications (Ferguson 1999; Weiss 1994). The derived lower bounds for the sampling frequency are such that relationships in continuous-time linking spectral statistical functions of transmitted ACS and received Doppler-stretched signals are formally analogous to their discrete-time counterparts. Thus, signal-parameter estimation procedures described in continuous-time in (De Angelis *et al.* 2005) and (Napolitano and Doğançay 2011) can be straightforwardly implemented in discrete-time by substituting cyclic spectra of continuous-time signals with cyclic spectra of their sampled versions and integrals with sums.

7.8 Spectral Analysis of Doppler-Stretched Signals – Constant Relative Radial Acceleration

In the case of constant relative radial acceleration, under the narrow-band conditions (7.246a) and (7.246b), the complex received signal is given by the chirp modulated signal (7.240) (with $\widetilde{x}$ and $\widetilde{y}$ replaced by x and y, respectively, for notation simplicity):

$$y(t) = \bar{b}\, x(t - d_0)\, e^{j2\pi\nu t}\, e^{j\pi\gamma t^2} \tag{7.346}$$

Accounting for the Fourier transform pairs

$$x(t - d_0)\, e^{j2\pi\nu t} \overset{\mathfrak{F}}{\longleftrightarrow} X(f - \nu)\, e^{-j2\pi(f-\nu)d_0} \tag{7.347}$$

and (Champeney 1990, p. 81)

$$e^{jat^2} \overset{\mathfrak{F}}{\longleftrightarrow} \left(\frac{\pi}{a}\right)^{1/2} e^{j\pi/4}\, e^{-j\pi^2 f^2/a} \tag{7.348}$$

with $a = \pi\gamma$, we have that the Fourier transform of a delayed, frequency shifted, and chirp modulated signal (7.346) can be expressed as

$$\begin{aligned} Y(f) &= \left[X(f-\nu)\, e^{-j2\pi(f-\nu)d_0}\right] \underset{f}{\otimes} \mathcal{F}\left[e^{j\pi\gamma t^2}\right] \\ &= \left[X(f-\nu)\, e^{-j2\pi(f-\nu)d_0}\right] \underset{f}{\otimes} \frac{e^{j\pi/4}}{\sqrt{\gamma}}\, e^{-j\pi f^2/\gamma} \\ &= \int_{\mathbb{R}} X(\lambda-\nu)\, e^{-j2\pi(\lambda-\nu)d_0}\, \frac{e^{j\pi/4}}{\sqrt{\gamma}}\, e^{-j\pi(f-\lambda)^2/\gamma}\, d\lambda \\ &= \mathcal{O}_A\left[X(\lambda-\nu)\, e^{-j2\pi(\lambda-\nu)d_0}\right] \end{aligned} \tag{7.349}$$

where $\mathcal{O}_A[\cdot]$ denotes the Linear Canonical Transform (LCT) operator (Ozaktas *et al.* 2001) with parameter matrix

$$\boldsymbol{A} = \begin{bmatrix} 1 & -\dfrac{\gamma}{2\pi} \\ 0 & 1 \end{bmatrix}. \tag{7.350}$$

which is coincident with a Fresnel transform with $\lambda z = -\gamma$.

In the case of finite-power stochastic processes, the Fourier transform should be intended in the sense of generalized functions (Section 1.1.2), (Gelfand and Vilenkin 1964; Henniger 1970).

7.8.1 *Second-Order Statistics (Continuous-Time)*

7.8.1.1 Narrow-Band Model

In the case of constant relative radial acceleration, under the narrow-band condition (7.246a) and (7.246b), the complex received signal is given by (7.346). It results that

$$\begin{aligned} \mathrm{E}\left\{y(t+\tau)\, y^{(*)}(t)\right\} &= \bar{b}\, \bar{b}^{(*)}\, \mathrm{E}\left\{x(t+\tau-d_0)\, x^{(*)}(t-d_0)\right\} \\ &\quad e^{j2\pi\nu(t+\tau)}\, e^{j\pi\gamma(t^2+2t\tau+\tau^2)}\, e^{(-)j2\pi\nu t}\, e^{(-)j\pi\gamma t^2} \end{aligned} \tag{7.351}$$

If $(*)$ is present in (7.351), one obtains the *autocorrelation function*

$$\mathrm{E}\left\{y(t+\tau)\, y^{*}(t)\right\} = |\bar{b}|^2\, \mathrm{E}\left\{x(t+\tau-d_0)\, x^{*}(t-d_0)\right\}\, e^{j2\pi\nu\tau}\, e^{j\pi\gamma\tau^2}\, e^{j2\pi\gamma\tau t}. \tag{7.352}$$

If $x(t)$ is ACS, that is, with autocorrelation function

$$\mathrm{E}\left\{x(t+\tau)\, x^{*}(t)\right\} = \sum_{n\in\mathbb{I}} R_{xx^*}^{\alpha_n}(\tau)\, e^{j2\pi\alpha_n t} \tag{7.353}$$

where $\mathbb{I}$ is a countable set and $\{\alpha_n\}_{n\in\mathbb{I}} = A_{xx^*}$ is the set of cycle frequencies, (7.352) specializes into

$$\mathrm{E}\Big\{y(t+\tau)\, y^*(t)\Big\} = |\bar{b}|^2 \sum_{n\in\mathbb{I}} R^{\alpha_n}_{xx^*}(\tau)\, e^{j2\pi\nu\tau}\, e^{j\pi\gamma\tau^2}\, e^{-j2\pi\alpha_n d_0}\, e^{j2\pi[\alpha_n+\gamma\tau]t}. \tag{7.354}$$

Thus, $y(t)$ is GACS and its autocorrelation function can be expressed as

$$\mathrm{E}\Big\{y(t+\tau)\, y^*(t)\Big\} = \sum_{n\in\mathbb{I}} R^{(n)}_{yy^*}(\tau)\, e^{j2\pi\eta_n(\tau)t} \tag{7.355}$$

with lag dependent cycle frequencies

$$\eta_n(\tau) = \alpha_n + \gamma\tau \quad n \in \mathbb{I} \tag{7.356}$$

and generalized cyclic autocorrelation functions

$$R^{(n)}_{yy^*}(\tau) = |\bar{b}|^2\, R^{\alpha_n}_{xx^*}(\tau)\, e^{j2\pi\nu\tau}\, e^{j\pi\gamma\tau^2}\, e^{-j2\pi\alpha_n d_0} \quad n \in \mathbb{I} \tag{7.357}$$

The functions $\eta_n(\tau)$ are parallel lines with slope γ in the (α, τ) plane. Thus, they do not intersect each other.

If $(*)$ is absent in (7.351), one obtains the *conjugate autocorrelation function*

$$\begin{aligned}\mathrm{E}&\Big\{y(t+\tau)\, y(t)\Big\}\\ &= \bar{b}^2\, \mathrm{E}\Big\{x(t+\tau-d_0)\, x(t-d_0)\Big\}\, e^{j2\pi\nu\tau}\, e^{j\pi\gamma\tau^2}\, e^{j2\pi[2\nu+\gamma\tau]t}\, e^{j2\pi\gamma t^2}\end{aligned} \tag{7.358}$$

If $x(t)$ is ACS, that is, with conjugate autocorrelation function

$$\mathrm{E}\Big\{x(t+\tau)\, x(t)\Big\} = \sum_{n\in\mathbb{I}'} R^{\beta_n}_{xx}(\tau)\, e^{j2\pi\beta_n t} \tag{7.359}$$

where $\mathbb{I}'$ is a countable set and $\{\beta_n\}_{n\in\mathbb{I}'} = A_{xx}$ is the set of conjugate cycle frequencies, (7.358) specializes into

$$\begin{aligned}\mathrm{E}&\Big\{y(t+\tau)\, y(t)\Big\}\\ &= \bar{b}^2 \sum_{n\in\mathbb{I}'} R^{\beta_n}_{xx}(\tau)\, e^{-j2\pi\beta_n d_0}\, e^{j2\pi\nu\tau}\, e^{j\pi\gamma\tau^2}\, e^{j2\pi[2\nu+\beta_n+\gamma\tau]t}\, e^{j2\pi\gamma t^2}.\end{aligned} \tag{7.360}$$

Due to the presence of the term $e^{j2\pi\gamma t^2}$, the conjugate autocorrelation function is not periodic or almost-periodic in t. Therefore, with reference to the conjugate autocorrelation function, the continuous-time signal $y(t)$ is not ACS or GACS. Note that in discrete-time the chirp signal $e^{j2\pi\widetilde{\gamma}n^2}$ is periodic when $\widetilde{\gamma} \in \mathbb{Q}$ and is not AP if $\widetilde{\gamma} \notin \mathbb{Q}$.

Accounting for the fact that for $b \neq 0$ both integrals

$$\int_0^{+\infty} \sin(at+bt^2)\, \mathrm{d}t \quad \text{and} \quad \int_0^{+\infty} \cos(at+bt^2)\, \mathrm{d}t$$

are finite (Champeney 1990, p. 81) one obtains

$$\left\langle e^{j2\pi(at+bt^2)} \right\rangle_t = \delta_a \, \delta_b. \tag{7.361}$$

Thus,

$$\begin{aligned}
R^{\varrho}_{yy}(\tau) &\triangleq \left\langle \mathrm{E}\left\{ y(t+\tau)\, y(t) \right\} e^{-j2\pi\varrho t} \right\rangle_t \\
&= \bar{b}^2 \sum_{n\in\mathbb{I}'} R^{\beta_n}_{xx}(\tau)\, e^{-j2\pi\beta_n d_0}\, e^{j2\pi\nu\tau}\, e^{j\pi\gamma\tau^2} \left\langle e^{j2\pi[2\nu+\beta_n+\gamma\tau-\varrho]t}\, e^{j2\pi\gamma t^2} \right\rangle_t \\
&= \bar{b}^2 \sum_{n\in\mathbb{I}'} R^{\beta_n}_{xx}(\tau)\, e^{-j2\pi\beta_n d_0}\, e^{j2\pi\nu\tau}\, e^{j\pi\gamma\tau^2}\, \delta_{[2\nu+\beta_n+\gamma\tau-\varrho]}\, \delta_\gamma \\
&= \bar{b}^2\, R^{\varrho-2\nu}_{xx}(\tau)\, e^{-j2\pi\beta_n d_0}\, e^{j2\pi\nu\tau}\, \delta_\gamma
\end{aligned} \tag{7.362}$$

Therefore, the conjugate autocorrelation function can contain a finite-strength additive sinewave component at frequency ϱ only if $\gamma = 0$ and $x(t)$ has nonzero conjugate cyclic autocorrelation function at conjugate cycle frequency $\beta_n = \varrho - 2\nu$ for some $\beta_n \in A_{xx}$.

Finally, note that the above model for the received signal depends on the length of the observation interval T since it is derived under the narrow-band conditions (7.246a) and (7.246b), that is, $BT \ll |c/v_\xi|$ and $BT^2 \ll |2c/a_\xi|$. In order to model the received signal as GACS, however, the further condition $|\gamma|\tau_{\text{corr}} > 1/T$ must be satisfied in order to appreciate the variability of the lag-dependent cycle frequencies with cycle-frequency resolution $\Delta\alpha = 1/T$, where τ_{corr} is the maximum value of $|\tau|$ such that $R_{xx^*}(0, \tau)$ is significantly different form zero (Section 2.7.6). For $t_0 = 0$ conditions (7.246a) and (7.246b) can be written as (7.247a) and (7.247b), respectively, that is $BT \ll |f_c/\nu|$ and $BT^2 \ll |2f_c/\gamma|$. Therefore, we must have

$$BT^2 \ll \left| \frac{2f_c}{\gamma} \right| < 2f_c\, T\tau_{\text{corr}} \tag{7.363}$$

that is

$$BT \ll 2f_c\tau_{\text{corr}}. \tag{7.364}$$

7.8.1.2 Wide-Band Model

In the case of constant relative radial acceleration the delay is quadratically time varying (see (7.201))

$$D(t) = d_0 + d_1 t + d_2 t^2 \tag{7.365}$$

According to (7.236) (with $\widetilde{x}$ and $\widetilde{y}$ replaced by x and y, respectively, for notation simplicity) the complex received signal is given by

$$\begin{aligned}
y(t) &= b\, x(t - D(t))\, e^{-j2\pi f_c D(t)} \\
&= b\, x(t - d_0 - d_1 t - d_2 t^2)\, e^{-j2\pi f_c [d_0 + d_1 t + d_2 t^2]}.
\end{aligned} \tag{7.366}$$

Thus,

$$\begin{aligned}
&\mathrm{E}\Big\{y(t+\tau)\, y^{(*)}(t)\Big\} \\
&\quad = \mathrm{E}\Big\{x(t+\tau-D(t+\tau))\, x^{(*)}(t-D(t))\Big\}\, e^{-j2\pi f_c[D(t+\tau)+(-)D(t)]} \\
&\quad = \mathrm{E}\Big\{x\Big((1-d_1)(t+\tau)-d_0-d_2(t^2+2t\tau+\tau^2)\Big)\, x^{(*)}\Big((1-d_1)t-d_0-d_2t^2\Big)\Big\} \\
&\qquad e^{-j2\pi f_c[d_0+d_1(t+\tau)+d_2(t^2+2t\tau+\tau^2)+(-)(d_0+d_1t+d_2t^2)]} \\
&\quad = \mathrm{E}\Big\{x\Big(\underbrace{(1-d_1)t-d_0-d_2t^2}_{t'}+\underbrace{(1-d_1)\tau-2d_2t\tau-d_2\tau^2}_{\tau'}\Big) \\
&\qquad x^{(*)}\Big(\underbrace{(1-d_1)t-d_0-d_2t^2}_{t'}\Big)\Big\} \\
&\qquad e^{-j2\pi f_c[(d_0+(-)d_0)+d_1\tau+d_2\tau^2+(d_1+(-)d_1)t+(d_2+(-)d_2)t^2+2d_2\tau t]}.
\end{aligned} \tag{7.367}$$

If $(*)$ is present and $x(t)$ is WSS, then the autocorrelation function of $y(t)$ is

$$\begin{aligned}
&\mathrm{E}\Big\{y(t+\tau)\, y^{*}(t)\Big\} \\
&\quad = R^0_{xx^*}(\tau')\, e^{-j2\pi f_c[d_1\tau+d_2\tau^2]}\, e^{-j2\pi f_c 2d_2\tau t} \\
&\quad = R^0_{xx^*}\Big((1-d_1)\tau-2d_2t\tau-d_2\tau^2\Big)\, e^{-j2\pi f_c[d_1\tau+d_2\tau^2]}\, e^{-j2\pi f_c 2d_2\tau t}
\end{aligned} \tag{7.368}$$

with $R^0_{xx^*}(\tau) \in L^1(\mathbb{R})$ or $L^2(\mathbb{R})$. Thus,

$$\begin{aligned}
R^\alpha_{yy^*}(\tau) &\triangleq \Big\langle \mathrm{E}\Big\{y(t+\tau)\, y^*(t)\Big\}\, e^{-j2\pi\alpha t}\Big\rangle_t \\
&= e^{-j2\pi f_c[d_1\tau+d_2\tau^2]}\Big\langle R^0_{xx^*}\Big((1-d_1)\tau-2d_2t\tau-d_2\tau^2\Big)\, e^{-j2\pi f_c 2d_2\tau t}\, e^{-j2\pi\alpha t}\Big\rangle_t \\
&= R^0_{xx^*}(0)\,\delta_\tau\,\delta_{[\alpha+2f_cd_2\tau]} \\
&= R^0_{xx^*}(0)\,\delta_\tau\,\delta_\alpha.
\end{aligned} \tag{7.369}$$

This result is due to the presence of t in the argument of $R^0_{xx^*}(\cdot) \in L^1(\mathbb{R})$ or $L^2(\mathbb{R})$. The infinite time average can be nonzero only if $\tau = 0$. A similar result is obtained if $x(t)$ is ACS.

In (Izzo and Napolitano 2002b, 2005, pp. 200–201), it is shown that cyclic higher-order statistics (moments and cumulants) can have supports contained in lines.

An experiment is conducted to illustrate the behavior of the cyclic autocorrelation function of the output of the Doppler channel existing between transmitter and receiver with constant relative radial acceleration. The transmitted signal $x(t)$ is a binary PAM signal with pulse period $T_p = 8T_s$ and raised cosine pulse with excess bandwidth $\eta = 0.25$. The channel introduces a quadratically time-variant delay (see (7.365)) with $d_0 = 20T_s$, $d_1 = -0.1$, and $d_2 = -0.00375/T_s$. For an observation interval of 2^{10} samples, the narrow-band conditions (7.247a) and (7.247b) are not satisfied. In fact, since $B = (1+\eta)/(2T_p)$, we have $BT = 80$ and $|1/d_1| = 10$, $BT^2 = 81920T_s$ and $|1/d_2| \simeq 266.67T_s$.

In Figure 7.18, (top) graph and (bottom) "checkerboard" plot of the magnitude of the cyclic correlogram of the output signal $y(t)$, is reported as a function of αT_s and τ/T_s. Accordingly

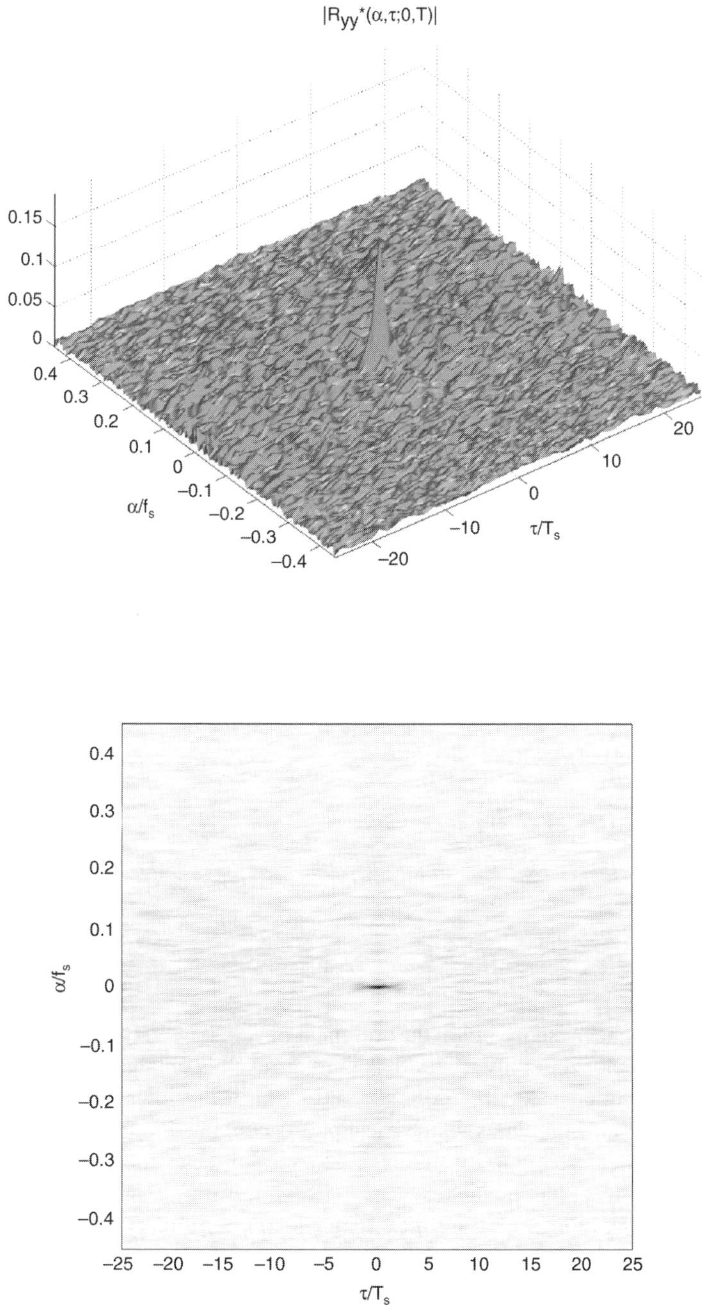

Figure 7.18 Output signal of the Doppler channel between transmitter and receiver with constant relative acceleration under the wide-band condition. (Top) graph and (bottom) "checkerboard" plot of the magnitude of the cyclic correlogram as a function of αT_s and τ / T_s.

with (7.369), the cyclic correlogram is significantly different from zero only in a neighborhood of $(\alpha, \tau) = (0, 0)$.

7.9 Other Models of Time-Varying Delays

In this section, models of time-varying delays which are not linearly or quadratically time variant are briefly considered.

7.9.1 *Taylor Series Expansion of Range and Delay*

The time-varying range $R(t)$ can be expanded in Taylor series, around t_0, with Lagrange residual term:

$$R(t) = R(t_0) + \dot{R}(t)\Big|_{t=t_0} (t - t_0) + \frac{1}{2} \ddot{R}(t)\Big|_{t=t_0} (t - t_0)^2 + \cdots + \frac{1}{(n-1)!} R^{(n-1)}(t)\Big|_{t=t_0} (t - t_0)^{n-1} + \frac{1}{n!} R^{(n)}(t)\Big|_{t=t'_0} (t - t_0)^n \qquad (7.370)$$

where $R^{(n)}(\cdot)$ denotes the nth-order derivative of $R(\cdot)$, and $t'_0 \in (t_0, t)$ if $t > t_0$ and $t'_0 \in (t, t_0)$ if $t < t_0$. This expression can be substituted into (7.38) to get an n-order algebric equation in $D(t)$. One of the roots of this equation is the time-varying delay $D(t)$ for the given $R(t)$ (see Section 7.4 for the case $n = 2$ and $t'_0 = t_0$).

Alternatively, by following the approach in (Kelly 1961) and (Kelly and Wishner 1965), the time-varying delay $D(t)$ can be expanded in Taylor series, around t_0, with Lagrange residual term:

$$D(t) = D(t_0) + \dot{D}(t)\Big|_{t=t_0} (t - t_0) + \frac{1}{2} \ddot{D}(t)\Big|_{t=t_0} (t - t_0)^2 + \cdots + \frac{1}{(n-1)!} D^{(n-1)}(t)\Big|_{t=t_0} (t - t_0)^{n-1} + \frac{1}{n!} D^{(n)}(t)\Big|_{t=t'_0} (t - t_0)^n \qquad (7.371)$$

where $t'_0 \in (t_0, t)$ if $t > t_0$ and $t'_0 \in (t, t_0)$ if $t < t_0$.

This approach, when $x(t) \in L^2(\mathbb{R})$ is embedded in additive white Gaussian noise (AWGN), allows to address the joint detection-estimation problem by the generalized ambiguity function (Kelly 1961; Kelly and Wishner 1965).

7.9.2 *Periodically Time-Variant Delay*

A rotating reflecting object gives rise to a periodically time-variant delay in the received signal. If the object is also moving with constant relative radial speed, then the delay contains also a linearly time-variant term (Chen *et al.* 2006):

$$D(t) = d_0 + d_1 t + d_m \cos(2\pi f_m t). \qquad (7.372)$$

The effect on the received signal is called micro-Doppler. In (Chen *et al.* 2006), with reference to the complex-envelope received-signal

$$\widetilde{y}(t) = A\, e^{j2\pi f_c (t - D(t))} \qquad (7.373)$$

the "Doppler frequency" is defined as

$$f_D(t) \triangleq \frac{\mathrm{d}}{\mathrm{d}t} D(t) \tag{7.374}$$

and the micro-Doppler effect is characterized by time-frequency analysis.

7.9.3 Periodically Time-Variant Carrier Frequency

Let $x_0(t)$ be the complex-envelope of a transmitted signal and let

$$x(t) = a\, x_0(t - d_0)\, e^{j2\pi f_c[s+\Delta \cos(2\pi f_m t+\phi)]t} \tag{7.375}$$

the complex signal reflected under the narrow-band condition by an oscillating scatterer whose effect is to produce a sinusoidally time-varying time-scale factor. The same mathematical model is obtained if the carrier frequency is not constant but is sinusoidally time-varying due to imperfection of the transmitting oscillator.

The (conjugate) autocorrelation function of $x(t)$ is

$$\begin{aligned} \mathrm{E}\{x(t+\tau)\, x^{(*)}(t)\} &= a\, a^{(*)}\, \mathrm{E}\{x_0(t+\tau-d_0)\, x_0^{(*)}(t-d_0)\} \\ &\quad e^{j2\pi f_c s\tau}\, e^{j2\pi f_c s[1+(-)1]t}\, z(t+\tau)\, z^{(*)}(t) \end{aligned} \tag{7.376}$$

where

$$z(t) \triangleq e^{j2\pi f_c \Delta \cos(2\pi f_m t+\phi)\, t}. \tag{7.377}$$

Let $a \triangleq 2\pi f_c \Delta$ and $\omega_m \triangleq 2\pi f_m$. By using the expansion of a complex exponential in terms of Bessel functions (NIST 2010, eqs. 10.12.1-3) (with $z = au$ and $\theta = 2\pi f_m t + \phi$) we have

$$\xi(u, t) \triangleq e^{jau \cos(\omega_m t+\phi)} = \sum_{n=-\infty}^{+\infty} j^n\, J_n(au)\, e^{jn(\omega_m t+\phi)} \tag{7.378}$$

where $J_n(z)$ is the Bessel function of the first kind of order n (NIST 2010, eqs. 10.2.2, 10.9.1, 10.9.2) . The function of two variables $\xi(u, t)$ can be evaluated along the diagonal $u = t$ leading to

$$z(t) = e^{ja \cos(\omega_m t+\phi)t} = \sum_{n=-\infty}^{+\infty} j^n\, J_n(at)\, e^{jn(\omega_m t+\phi)}. \tag{7.379}$$

The second-order lag product of $z(t)$ is given by

$$\begin{aligned} z(t+\tau)\, z^{(*)}(t) &= e^{ja\ \cos(\omega_m(t+\tau)+\phi)\,(t+\tau)}\, e^{(-)ja\ \cos(\omega_m t+\phi)\, t} \\ &= \underbrace{e^{ja\ \cos(\omega_m(t+\tau)+\phi)\,\tau}}_{r_1(t)}\ \underbrace{e^{ja[\cos(\omega_m(t+\tau)+\phi)+(-)\cos(\omega_m t+\phi)]\, t}}_{r_2(t)}. \end{aligned} \tag{7.380}$$

The function $r_1(t)$ is periodic in t. From (7.378) with $u = \tau$ we have

$$r_1(t) = \sum_{n=-\infty}^{+\infty} j^n\, J_n(a\tau)\, e^{jn(\omega_m(t+\tau)+\phi)}. \tag{7.381}$$

Due to the function

$$r_2(t) \triangleq e^{ja[\cos(\omega_m(t+\tau)+\phi)+(-)\cos(\omega_m t+\phi)]\,t}$$
$$= e^{ja[\cos(\omega_m \tau)\cos(\omega_m t+\phi)-\sin(\omega_m \tau)\sin(\omega_m t+\phi)+(-)\cos(\omega_m t+\phi)]\,t} \qquad (7.382)$$

the signal $x(t)$ turns out to be GACS with nonlinear lag-dependent cycle frequencies with complicate analytical expression.

A simulation experiment is carried out to estimate the cyclic correlogram of $x(t)$. In the experiment, $x_0(t)$ is a PAM signal with raised cosine pulse with excess bandwidth $\eta = 0.25$, stationary white binary modulating sequence, and symbol period $T_p = 64T_s$, where T_s is the sampling period. Furthermore, $f_c = 0.125/T_s$, $s = 0.99$, $f_m = 0.0233/T_s$, $\Delta = 0.005$, and $\phi = 0$.

In Figure 7.19, (top) graph and (bottom) "checkerboard" plot of the magnitude of the cyclic correlogram of $x(t)$, estimated by 2^{13} samples as a function of αT_s and τ/T_s, are reported.

7.10 Proofs

7.10.1 Proof of (7.262)

Fourier transform of a time-scaled, delayed, and frequency-shifted signal:

$$x_a(st) \overset{\mathcal{F}}{\longleftrightarrow} \frac{1}{|s|} X_a\left(\frac{f}{s}\right) \qquad (7.383)$$

$$x_a(s(t - d_a/s)) \overset{\mathcal{F}}{\longleftrightarrow} \frac{1}{|s|} X_a\left(\frac{f}{s}\right) e^{-j2\pi f d_a/s} \qquad (7.384)$$

$$y_a(t) = Ae^{j\varphi}\, x_a(st - d_a)\, e^{j2\pi f_a t}$$
$$\overset{\mathcal{F}}{\longleftrightarrow} Y_a(f) = Ae^{j\varphi} \frac{1}{|s|} X_a\left(\frac{f - f_a}{s}\right) e^{-j2\pi(f - f_a)d_a/s} \qquad (7.385)$$

7.10.2 Proof of (7.263) and (7.264)

By using (7.261), the result is that

$$R^{\alpha}_{y_a y_a^*}(\tau) \triangleq \left\langle \mathrm{E}\left\{y_a(t+\tau)\, y_a^*(t)\right\} e^{-j2\pi\alpha t} \right\rangle_t$$
$$= \left\langle \mathrm{E}\left\{ Ae^{j\varphi}\, x_a(s(t+\tau) - d_a) e^{j2\pi f_a(t+\tau)}\, Ae^{-j\varphi}\, x_a^*(st - d_a) e^{-j2\pi f_a t} \right\} e^{-j2\pi\alpha t} \right\rangle_t$$
$$= A^2 e^{j2\pi f_a \tau} \left\langle \mathrm{E}\left\{x_a(st + s\tau - d_a)\, x_a^*(st - d_a)\right\} e^{-j2\pi\alpha t} \right\rangle_t$$
$$= A^2 e^{j2\pi f_a \tau} \left\langle \mathrm{E}\left\{x_a(t' + s\tau)\, x_a^*(t')\right\} e^{-j2\pi\alpha(t'+d_a)/s} \right\rangle_{t'}$$
$$= A^2 e^{j2\pi f_a \tau}\, e^{-j2\pi\alpha d_a/s}\, R^{\alpha/s}_{x_a x_a^*}(s\tau) \qquad (7.386)$$

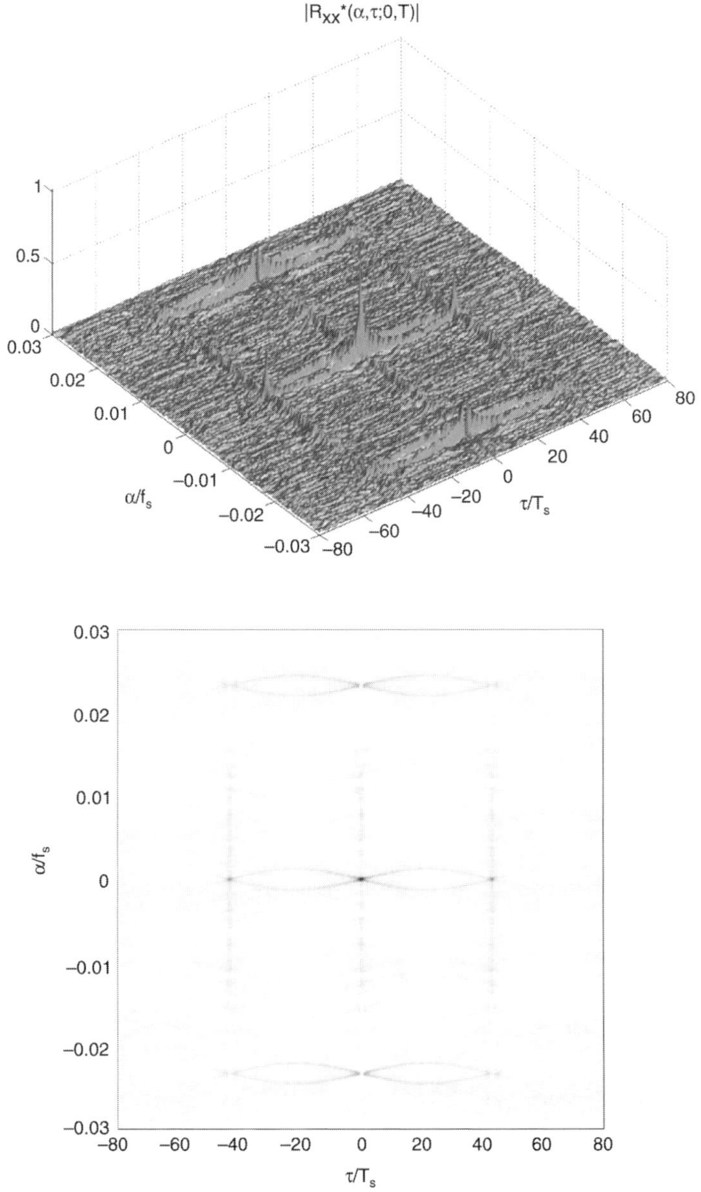

Figure 7.19 Complex PAM signal with sinusoidally-varying carrier frequency. (Top) graph and (bottom) "checkerboard" plot of the magnitude of the cyclic correlogram as a function of αT_s and τ/T_s

where $\langle\cdot\rangle_t \triangleq \lim_{T\to\infty} \frac{1}{T}\int_{-T/2}^{T/2}(\cdot)\,\mathrm{d}t$ is the continuous-time infinite average with respect to t, and, in the third equality, the variable change $t' = st - d_a$ is made, so that

$$\frac{1}{T}\int_{-T/2}^{T/2} \mathrm{E}\left\{x_a(st+s\tau-d_a)\, x_a^*(st-d_a)\right\}\, e^{-j2\pi\alpha t}\,\mathrm{d}t$$

$$= \frac{1}{|s|T}\int_{-|s|T/2-d_a}^{|s|T/2-d_a} \mathrm{E}\left\{x_a(t'+s\tau)\, x_a^*(t')\right\}\, e^{-j2\pi\alpha(t'+d_a)/s}\,\mathrm{d}t'. \tag{7.387}$$

This proof can be equivalently carried out in the FOT approach by simply removing the expectation operator $\mathrm{E}\{\cdot\}$ in (7.386) and (7.387).

(7.264) is obtained by Fourier transforming both sides of (7.263).

7.10.3 Proof of (7.267) and (7.268)

By reasoning as for (7.386), we have

$$\begin{aligned}
R^{\beta}_{y_a y_a}(\tau) &\triangleq \left\langle \mathrm{E}\{y_a(t+\tau)\, y_a(t)\}\, e^{-j2\pi\beta t}\right\rangle_t \\
&= \left\langle \mathrm{E}\left\{A e^{j\varphi}\, x_a(s(t+\tau)-d_a) e^{j2\pi f_a(t+\tau)}\, A e^{j\varphi}\, x_a(st-d_a) e^{j2\pi f_a t}\right\}\, e^{-j2\pi\beta t}\right\rangle_t \\
&= A^2 e^{j2\varphi}\, e^{j2\pi f_a\tau}\left\langle \mathrm{E}\{x_a(st+s\tau-d_a)\, x_a(st-d_a)\}\, e^{-j2\pi(\beta-2f_a)t}\right\rangle_t \\
&= A^2 e^{j2\varphi}\, e^{j2\pi f_a\tau}\left\langle \mathrm{E}\left\{x_a(t'+s\tau)\, x_a(t')\right\}\, e^{-j2\pi(\beta-2f_a)(t'+d_a)/s}\right\rangle_{t'} \\
&= A^2 e^{j2\varphi}\, e^{j2\pi f_a\tau}\, e^{-j2\pi(\beta-2f_a)d_a/s}\, R^{(\beta-2f_a)/s}_{x_a x_a}(s\tau)
\end{aligned} \tag{7.388}$$

(7.268) is obtained by Fourier transforming both sides of (7.267).

7.10.4 Proof of (7.273)

Accounting for (7.262) and (7.272), one obtains

$$\begin{aligned}
&\mathrm{E}\left\{Y_a(f_1)\, Y_a^{(*)}(f_2)\right\} \\
&\quad= \mathrm{E}\left\{A\, e^{j\varphi}\, \frac{1}{|s|}\, X_a\left(\frac{f_1-f_a}{s}\right)\, e^{-j2\pi(f_1-f_a)d_a/s}\right. \\
&\qquad \left. A\, e^{(-)j\varphi}\, \frac{1}{|s|}\, X_a^{(*)}\left(\frac{f_2-f_a}{s}\right)\, e^{-(-)j2\pi(f_2-f_a)d_a/s}\right\} \\
&\quad= A^2\, e^{j\varphi[1+(-)1]}\, \frac{1}{|s|^2}\, e^{-j2\pi(f_1-f_a)d_a/s}\, e^{-(-)j2\pi(f_2-f_a)d_a/s} \\
&\qquad \mathrm{E}\left\{X_a\left(\frac{f_1-f_a}{s}\right)\, X_a^{(*)}\left(\frac{f_2-f_a}{s}\right)\right\}
\end{aligned}$$

$$= A^2 \, e^{j\varphi[1+(-)1]} \, \frac{1}{|s|^2} \, e^{-j2\pi(f_1-f_a)d_a/s} \, e^{-(-)j2\pi(f_2-f_a)d_a/s}$$
$$\sum_{\alpha_n \in A_{x_a x_a^{(*)}}} S^{\alpha_n}_{x_a x_a^{(*)}}\left(\frac{f_1 - f_a}{s}\right) \delta\left(\frac{f_2 - f_a}{s} -_{(-)} \left(\alpha_n - \frac{f_1 - f_a}{s}\right)\right) \tag{7.389}$$

from which (7.273) follows by a variable change in the argument of the Dirac delta (Zemanian 1987, Section 1.7).

7.10.5 *Proof of (7.275)*

Using (7.261), we have

$$\mathrm{E}\left\{y_a(t+\tau)\, y_a^{(*)}(t)\right\}$$
$$= A^2 \, e^{j\kappa\varphi} \, \mathrm{E}\left\{x_a(s(t+\tau) - d_a)\, x_a^{(*)}(st - d_a)\right\} \, e^{j2\pi f_a(t+\tau)} \, e^{(-)j2\pi f_a t}$$
$$= A^2 \, e^{j\kappa\varphi} \left[\sum_{\alpha_n \in A_{x_a x_a^{(*)}}} R^{\alpha_n}_{x_a x_a^{(*)}}(s\tau)\, e^{j2\pi\alpha_n(st-d_a)}\right] e^{j2\pi\kappa f_a t} \, e^{j2\pi f_a \tau} \tag{7.390}$$

from which (7.275) immediately follows. The same result is obtained by taking the double inverse Fourier transform (see (1.15)) of both sides of (7.273) and then making the variable changes $t_1 = t + \tau$, $t_2 = t$.

7.10.6 *Proof of (7.276)*

Accounting for (7.262), one obtains

$$\mathrm{E}\left\{Y_a(f_1)\, X_a^{(*)}(f_2)\right\}$$
$$= \mathrm{E}\left\{A \, e^{j\varphi} \, \frac{1}{|s|} \, X_a\left(\frac{f_1 - f_a}{s}\right) e^{-j2\pi(f_1-f_a)d_a/s} \, X_a^{(*)}(f_2)\right\}$$
$$= A \, e^{j\varphi} \, \frac{1}{|s|} \, e^{-j2\pi(f_1-f_a)d_a/s} \, \mathrm{E}\left\{X_a\left(\frac{f_1 - f_a}{s}\right) X_a^{(*)}(f_2)\right\} \tag{7.391}$$

from which (7.276) immediately follows using (7.272).

7.10.7 *Proof of (7.286) and (7.287)*

Accounting for (7.284) and (1.88), we have

$$\mathrm{E}\{y(t+\tau)\, y(t)\}$$
$$= \mathrm{E}\left\{\sum_{k_1=1}^{K} a_{k_1}\, x(s_{k_1}(t+\tau) - d_{k_1})\, e^{j2\pi\nu_{k_1}(t+\tau)}\right.$$
$$\left.\sum_{k_2=1}^{K} a_{k_2}^{(*)}\, x^{(*)}(s_{k_2}t - d_{k_2})\, e^{(-)j2\pi\nu_{k_2}t}\right\}$$

$$
\begin{aligned}
&= \sum_{k_1=1}^{K} \sum_{k_2=1}^{K} a_{k_1}\, a_{k_2}^{(*)}\, e^{j2\pi[\nu_{k_1} + (-)\nu_{k_2}]t}\, e^{j2\pi\nu_{k_1}\tau} \\
&\quad \mathrm{E}\left\{ x(s_{k_1}(t+\tau) - d_{k_1})\, x^{(*)}(s_{k_2}t - d_{k_2}) \right\} \\
&= \sum_{k_1=1}^{K} \sum_{k_2=1}^{K} a_{k_1}\, a_{k_2}^{(*)}\, e^{j2\pi[\nu_{k_1} + (-)\nu_{k_2}]t}\, e^{j2\pi\nu_{k_1}\tau} \\
&\quad \sum_{\alpha \in A_{xx^{(*)}}} R^{\alpha}_{xx^{(*)}}((s_{k_1} - s_{k_2})t + s_{k_1}\tau - d_{k_1} + d_{k_2})\, e^{j2\pi\alpha(s_{k_2}t - d_{k_2})}
\end{aligned}
\tag{7.392}
$$

where the last equality is obtained by putting $t_1 = s_{k_1}(t+\tau) - d_{k_1}$ and $t_2 = s_{k_2}t - d_{k_2}$ into (1.88).

Under assumption (1.89), $R^{\alpha}_{xx^{(*)}}(\tau) \in L^1(\mathbb{R})$. Thus, if $s_{k_1} \neq s_{k_2}$ for $k_1 \neq k_2$, then each term in the sum over α in (7.392) is a finite-strength additive sinewave component of the cyclic autocorrelation function only when $k_1 = k_2 = k$ so that there is no dependence on t in the argument of $R^{\alpha}_{xx^{(*)}}(\cdot)$. Thus we have the first and second equality in (7.286).

By taking the Fourier coefficient at frequency β in (7.286) we have

$$
\begin{aligned}
&\left\langle \mathrm{E}\{y(t+\tau)\, y(t)\}\, e^{-j2\pi\beta t} \right\rangle_t \\
&\quad = \sum_{k=1}^{K} a_k\, a_k^{(*)}\, e^{j2\pi\nu_k\tau} \sum_{\alpha \in A_{xx^{(*)}}} R^{\alpha}_{xx^{(*)}}(s_k\tau)\, e^{-j2\pi\alpha d_k} \\
&\quad \left\langle e^{j2\pi[(1+(-)1)\nu_k + \alpha s_k - \beta]t} \right\rangle_t
\end{aligned}
\tag{7.393}
$$

from which (7.287) easily follows observing that the last time average on t equals to the Kronecker delta $\delta_{\alpha-[\beta-(1+(-)1)\nu_k]/s_k}$.

Note that (7.287) can be obtained as special case of (Izzo and Napolitano 2002b, eq. (90)) for $N = 2$ with only the second optional complex conjugation present, by using (Izzo and Napolitano 1998a, eq. (36)) that expresses reduced-dimension (generalized) cyclic temporal moment functions in terms of (generalized) cyclic temporal moment functions.

7.10.8 *Proof of Theorem 7.7.3 Doppler-Stretched Signal: Sampling Theorem for Loève Bifrequency Spectrum*

Replicas in (7.325) are separated by 1 in both ν_1 and ν_2 variables. Thus, from (7.326a) it follows that $f_s \geqslant 2B|s|$ is a sufficient condition such that replicas do not overlap. Note that, even if replicas do not overlap, the mapping $\nu_i = f_i/f_s, i = 1, 2$, does not link (7.273) and (7.325) for $\nu_i \in [-1/2, 1/2]$. A sufficient condition to assure such a mapping or, equivalently, that only replica (n_1, n_2) in (7.325) lies in the square $(\nu_1, \nu_2) \in (n_1 - 1/2, n_1 + 1/2) \times (n_2 - 1/2, n_2 + 1/2)$, according to (7.326b) is $|\bar{\nu}| + B|s|/f_s \leqslant 1/2$, that is, $f_s \geqslant 2(B|s| + |f_a|)$, where (7.298) is accounted for.

7.10.9 *Proof of Theorem 7.7.4 Doppler Stretched Signal: Sampling Theorem for the Density of Loève Bifrequency Spectrum*

According to (7.326a), if $f_s \geqslant 2B|s|$, the mapping $\nu_1 = f_1/f_s$ for $\nu_1 \in [-1/2, 1/2]$ between the densities in (7.325) with $n_1 = n_2 = 0$ and those in (7.273) is assured provided that the support of replica with $n_1 = n_2 = 0$ does not intercept that of replicas with $n_1 = 0, n_2 = \pm 1$. That is, on every support line

$$\nu_2 = {}_{(-)}\left(\frac{\alpha}{f_s}s - \nu_1 + \kappa\bar{\nu}\right) \qquad \alpha \in A_{x_a x_a^{(*)}} \tag{7.394}$$

the following implication

$$\nu_1 = \bar{\nu} \pm \frac{B}{f_s}|s| \;\Rightarrow\; |\nu_2| \leqslant 1 - \left(|\bar{\nu}| + \frac{B}{f_s}|s|\right) \tag{7.395}$$

must hold, where (7.394) is obtained by rearranging the argument of the Dirac delta in (7.325). From (7.394) with $\nu_1 = \bar{\nu} \pm B|s|/f_s$ substituted into we have

$$\begin{aligned} |\nu_2| &= \left|\frac{\alpha}{f_s}s - \nu_1 + \kappa\bar{\nu}\right| = \left|\frac{\alpha}{f_s}s - \bar{\nu} \mp \frac{B}{f_s}|s| + \kappa\bar{\nu}\right| \\ &\leqslant \sup_{\alpha \in A_{x_a x_a^{(*)}}} \left|\frac{\alpha}{f_s}\right| |s| + \frac{B}{f_s}|s| + |\kappa\bar{\nu} - \bar{\nu}| = \frac{2B}{f_s}|s| + \frac{B}{f_s}|s| + \frac{|f_a|}{f_s} \end{aligned} \tag{7.396}$$

where the bound for $\sup|\alpha|$ comes form (1.176b). Thus, inequality in (7.395) is satisfied provided that $(3B/f_s)|s| + (|f_a|/f_s) \leqslant 1 - \big((|f_a|/f_s) + (B/f_s)|s|\big)$ that is, $f_s \geqslant 2(2B|s| + |f_a|)$.

These results agree with those in (Napolitano 2010b, Theorem 3) since ${}_{(-)}\kappa = \kappa$.

7.10.10 *Proof of Theorem 7.7.5 Doppler Stretched Signal: Sampling Theorem for the (Conjugate) Cyclic Spectrum*

By setting $\alpha = (\widetilde{\alpha} - \kappa\bar{\nu})f_s/s$ and $f = (\nu - \bar{\nu})f_s/s$ into the cyclic-spectrum support expression (1.176a), we have

$$\begin{aligned} &\operatorname{supp}\left\{S_{x_a x_a^{(*)}}^{(\widetilde{\alpha}-\kappa\bar{\nu})f_s/s}\left(\frac{\nu - \bar{\nu}}{s}f_s\right)\right\} \\ &\subseteq \left\{(\widetilde{\alpha}, \nu) \in \mathbb{R}\times\mathbb{R} \,:\, \left|\frac{\nu - \bar{\nu}}{s}\right| f_s \leqslant B,\ \left|\frac{\widetilde{\alpha} - \kappa\bar{\nu}}{s} - \frac{\nu - \bar{\nu}}{s}\right| f_s \leqslant B\right\} \\ &\subseteq \left\{(\widetilde{\alpha}, \nu) \in \mathbb{R}\times\mathbb{R} \,:\, |\nu - \bar{\nu}| \leqslant \frac{B}{f_s}|s|,\ |\widetilde{\alpha} - \kappa\bar{\nu}| - |\nu - \bar{\nu}| \leqslant \frac{B}{f_s}|s|\right\} \\ &\subseteq \left\{(\widetilde{\alpha}, \nu) \in \mathbb{R}\times\mathbb{R} \,:\, |\nu| \leqslant |\bar{\nu}| + \frac{B}{f_s}|s|,\ |\widetilde{\alpha}| \leqslant \kappa|\bar{\nu}| + 2\frac{B}{f_s}|s|\right\} \\ &\subseteq \left\{(\widetilde{\alpha}, \nu) \in \mathbb{R}\times\mathbb{R} \,:\, |\nu| \leqslant \frac{1}{4},\ |\widetilde{\alpha}| \leqslant \frac{1}{2}\right\} \end{aligned} \tag{7.397}$$

where the last inclusion relationship holds provided that inequality $f_s \geqslant 4(B|s| + |f_a|)$ is satisfied. In fact, accounting for (7.298), such an inequality is equivalent to $B|s|/f_s + |\bar{\nu}| \leqslant 1/4$.

Replicas in (7.329) are separated by 1 in both $\widetilde{\alpha}$ and ν domains. Consequently, condition $f_s \geqslant 4(B|s| + |f_a|)$ assures that only replica with $p = 0$ and $q = 0$ gives nonzero contribution in the principal domain $(\widetilde{\alpha}, \nu) \in [-1/2, 1/2] \times [-1/2, 1/2]$. Then, observing that only replicas with $p = q = 0$ are coincident in (7.329) and (7.330), equation (7.331) immediately follows.

7.10.11 *Proof of Theorem 7.7.6 Sampling Theorem for Loève Bifrequency Cross-Spectrum*

Replicas in the aliasing formula (7.332) are separated by 1 in both ν_1 and ν_2 variables. Thus, from the support bound (7.333a) it follows that $B|s|/f_s \leqslant 1/2$ and $B/f_s \leqslant 1/2$ that is,

$$f_s \geqslant \max\{2B|s|, 2B\} \tag{7.398}$$

is a sufficient condition such that replicas do not overlap. Moreover, from (7.333b), it follows that a sufficient condition to obtain that only the replica (n_1, n_2) lies in the square $(\nu_1, \nu_2) \in (n_1 - 1/2, n_1 + 1/2) \times (n_2 - 1/2, n_2 + 1/2)$ is $|\bar{\nu}| + B|s|/f_s \leqslant 1/2$ and $B/f_s \leqslant 1/2$. Thus, accounting for (7.298), for the replica with $n_1 = n_2 = 0$ we have (7.334).

7.10.12 *Proof of Theorem 7.7.7 Sampling Theorem for the Density of Loève Bifrequency Cross-Spectrum*

Due to (7.333a), the mapping $\nu_1 = f_1/f_{s1}$ for $\nu_1 \in [-1/2, 1/2]$ between the densities with $n_1 = n_2 = 0$ in (7.332) and those in (7.276) is assured provided that the support replica with $n_1 = n_2 = 0$ does not intercept that of the replica with $n_1 = 0$, $n_2 = \pm 1$. That is, on every support line

$$\nu_2 = {}^{(-)}\left(\frac{\alpha}{f_s} - \frac{\nu_1 - \bar{\nu}}{s}\right) \qquad \alpha \in A_{x_a x_a^{(*)}} \tag{7.399}$$

the following implication

$$\nu_1 = \bar{\nu} \pm \frac{B}{f_s}|s| \;\Rightarrow\; |\nu_2| \leqslant 1 - \frac{B}{f_s} \tag{7.400}$$

must hold. From (7.399) with $\nu_1 = \bar{\nu} \pm B|s|/f_s$ substituted into we have

$$\begin{aligned} |\nu_2| &= \left|\frac{\alpha}{f_s} - \frac{\nu_1 - \bar{\nu}}{s}\right| = \left|\frac{\alpha}{f_s} - \frac{\bar{\nu} \pm B|s|/f_s - \bar{\nu}}{s}\right| \\ &\leqslant \sup_{\alpha \in A_{x_a x_a^{(*)}}} \left|\frac{\alpha}{f_s}\right| + \frac{B}{f_s} = \frac{2B}{f_s} + \frac{B}{f_s} = \frac{3B}{f_s}. \end{aligned} \tag{7.401}$$

Thus, inequality in (7.400) is satisfied provided that $3B/f_s \leqslant 1 - B/f_s$, that is, $f_s \geqslant 4B$. Then, accounting for (7.334), we obtain (7.336).

7.10.13 *Proof of Theorem 7.7.8 Sampling Theorem for the (Conjugate) Cyclic Autocorrelation Function*

From (7.339) (with $y(n) = x_s(n)$, $A = 1$, $\varphi = 0$, $\bar{\nu} = 0$, and $d = 0$) it follows that

$$\widetilde{\widetilde{R}}^{\alpha}_{x_s x_s^{(*)}}(m) = \sum_{p \in \mathbb{Z}} R^{(\widetilde{\alpha}-p)f_s}_{x_{as} x_{as}^{(*)}}(mT_s). \tag{7.402}$$

Thus, if $d/s \in \mathbb{Z}$ or if $R^{\alpha}_{x_{as} x_{as}^{(*)}}(\tau) = 0$ for $|\alpha| > f_s/2$, then for $|\widetilde{\alpha}| \leqslant 1/2$ the sum in the rhs of (7.339) equals the rhs of (7.402) and, hence, $\widetilde{\widetilde{R}}^{\alpha}_{x_s x_s^{(*)}}(m)$. A sufficient condition to assure $R^{\alpha}_{x_{as} x_{as}^{(*)}}(\tau) = 0$ for $|\alpha| > f_s/2$ is that the bandwidth of $x_{as}(t)$ is less than $f_s/2$ (Section 1.3.9), (Napolitano 1995; Gardner *et al.* 2006). That is, $B|s| \leqslant f_s/2$, i.e., (7.340).

8

Bibliographic Notes

In this chapter, citations are classified into categories and listed in chronological order.

8.1 Almost-Periodic Functions

8.1.1 Fundamental Treatments

(Besicovitch 1932), (Bohr 1933), (Amerio and Prouse 1971), (Corduneanu 1989).

8.1.2 Generalizations

(von Neumann 1934), (Jessen and Tornehave 1945), (Eberlein 1949), (Eberlein 1956), (Kahane 1961), (Urbanik 1962), (Kahane 1962), (Zhang 1994), (Zhang 1995), (Ait Dads and Arino 1996), (Han and Hong 2007), (Phong 2007), (Zhang and Liu 2010).

8.1.3 Reviews

(Andreas *et al.* 2006), (Chérif 2011a), (Chérif 2011b).

8.2 Cyclostationary Signals

8.2.1 Fundamental Treatments

(Gladyshev 1961), (Gladyshev 1963), (Gardner 1985), (Gardner 1986c), (Gardner 1987d), (Hurd 1989b), (Dehay and Hurd 1994), (Gardner 1994), (Giannakis 1998), (Gardner *et al.* 2006), (Hurd and Miamee 2007).

8.2.2 Developments and Applications

(Ogura 1971), (Hurd 1974), (Gardner and Franks 1975), (Gardner 1978), (Boyles and Gardner 1983), (Gardner 1986a), (Gardner 1986b), (Brown 1987), (Gardner 1987a), (Gardner 1987b), (Gardner 1987c), (Gardner *et al.* 1987), (Alekseev 1988), (Gardner 1988a), (Gardner 1988b),

---

Generalizations of Cyclostationary Signal Processing: Spectral Analysis and Applications, First Edition.

(Gardner 1988c), (Gardner and Chen 1988), (Hurd 1989a), (Agee *et al.* 1990), (Gardner 1991a), (Hurd 1991), (Hurd and Gerr 1991), (Gardner and Chen 1992), (Chen and Gardner 1992), (Gardner and Spooner 1992), (Hurd and Leśkow 1992b), (Hurd and Leśkow 1992a), (Gardner 1993), (Gardner and Spooner 1993), (Sathe and Vaidyanathan 1993), (Dehay 1994), (Franks 1994), (Genossar *et al.* 1994), (Gerr and Allen 1994), (Schell 1995), (Dehay and Leśkow 1996), (Swift 1996), (Akkarakaran and Vaidyanathan 2000), (De Angelis *et al.* 2005), (Lii and Rosenblatt 2006), (Dehay 2007), (Javors'kyj *et al.* 2007), (Wahlberg and Schreier 2008), (Antoni 2009), (Spooner and Nicholls 2009), (Napolitano and Doğançay 2011).

8.2.3 *Higher-order Cyclostationarity*

(Dandawaté and Giannakis 1994), (Gardner and Spooner 1994), (Spooner and Gardner 1994), (Dandawaté and Giannakis 1995), (Napolitano 1995), (Izzo and Napolitano 1996), (Izzo and Napolitano 1997), (Izzo and Napolitano 1998b), (Sadler and Dandawaté 1998), (Flagiello *et al.* 2000), (Napolitano and Spooner 2000), (Napolitano and Spooner 2001), (Spooner and Nicholls 2009).

8.3 Generalizations of Cyclostationarity

8.3.1 *Generalized Almost-Cyclostationary Signals*

(Izzo and Napolitano 1998a), (Izzo and Napolitano 2002a), (Izzo and Napolitano 2002b), (Izzo and Napolitano 2003), (Izzo and Napolitano 2005), (Napolitano 2007a), (Napolitano 2009), (Napolitano and Tesauro 2011).

8.3.2 *Spectrally Correlated Signals*

(Allen and Hobbs 1992), (Genossar 1992), (Lii and Rosenblatt 1998), (Lii and Rosenblatt 2002), (Napolitano 2001), (Napolitano 2003), (Øigård *et al.* 2006), (Napolitano 2007b), (Napolitano 2007c), (Napolitano 2010b), (Napolitano 2010a), (Napolitano 2011).

8.4 Other Nonstationary Signals

(Silverman 1957), (Kampé de Fériet and Frenkiel 1962), (Priestley 1965), (Mandelbrot and Van Ness 1968), (Hurd 1973), (Martin and Flandrin 1985), (Hurd 1988), (Flandrin 1989), (Cohen 1989), (Gardner 1991b), (Amin 1992), (Hlawatsch and Bourdeaux-Bartels 1992), (Wornell 1993), (Cohen 1995), (Allen and Hobbs 1997), (Matz *et al.* 1997), (Wu and Lev-Ari 1997), (Hanin and Schreiber 1998), (Wang and Zhou 1998), (Flandrin 1999), (Ozaktas *et al.* 2001), (Soedjack 2002), (Hanssen and Scharf 2003), (Hopgood and Rayner 2003), (Scharf *et al.* 2005), (Dmochowski *et al.* 2009), (Flandrin *et al.* 2011).

8.5 Functional Approach and Generalized Harmonic Analysis

(Wiener 1930), (Wiener 1933), (Kac and Steinhaus 1938), (Wiener 1949), (Kac 1959), (Lee 1967), (Pfaffelhuber 1975), (Gardner 1991c), (Gardner and Brown 1991), (Benedetto 1996), (Hurd and Koski 2004), (Leśkow and Napolitano 2006), (Leśkow and Napolitano 2007), (Casinovi 2007), (Casinovi 2009).

8.6 Linear Time-Variant Processing

8.6.1 Channels

(Bello 1963), (Middleton 1967), (Crochiere and Rabiner 1981), (Vaidyanathan 1990), (Varghese and Donohue 1994), (Munk *et al.* 1995), (Sadowsky and Kafedziski 1998).

8.6.2 Doppler Effect

(Kelly 1961), (Kelly and Wishner 1965), (Rihaczek 1967), (Swick 1969), (Censor 1973), (Cooper 1980), (Kapoulitsas 1981), (Censor 1984), (Chiu 1986), (Weiss 1994), (Jin *et al.* 1995), (Ho and Chan 1998), (Ferguson 1999), (Schlotz 2002), (Gray and Addison 2003), (Neipp *et al.* 2003), (Oberg 2004), (Chan and Ho 2005).

8.6.3 Time Warping

(Cochran and Clark 1990), (Xia and Zhang 1992), (Clark and Cochran 1993), (Azizi and Cochran 1999), (Azizi *et al.* 2002).

8.6.4 Frequency Warping

(Franaszek 1967), (Franaszek and Liu 1967), (Liu and Franaszek 1969), (Rabiner *et al.* 1969), (Oppenheim *et al.* 1971), (Oppenheim and Johnson 1972), (Braccini and Oppenheim 1974), (Claasen and Mecklenbräuker 1982), (Oppenheim and Schafer 1989), (von Schroeter 1999), (Makur and Mitra 2001), (Franz *et al.* 2003).

8.7 Sampling

(Lloyd 1959), (Zakai 1965), (Gardner 1972), (Lee 1978), (Garcia *et al.* 2001), (Song *et al.* 2007).

8.8 Complex Random Variables, Signals, and Systems

(Aiken 1967), (Brandwood 1983), (van den Bos 1994), (van den Bos 1995), (Picinbono 1996), (Picinbono and Bondon 1997), (Schreier and Scharf 2003a), (Schreier and Scharf 2003b), (Politis 2005), (Yoo 2008), (Adali *et al.* 2011).

8.9 Stochastic Processes

(Kolmogorov 1933), (Cramér 1940), (Wold 1948), (Doob 1953), (Grenander and Rosenblatt 1957), (Parzen 1957), (Leonov and Shiryaev 1959), (Loève 1963), (Brillinger 1965), (Brillinger and Rosenblatt 1967), (Billingsley 1968), (Brillinger 1969), (Gikhman and Skorokhod 1969), (Prohorov and Rozanov 1989), (Van Trees 1971), (Brillinger 1974), (Rosenblatt 1974), (Cambanis and Masry 1976), (Brillinger 1981), (Thomson 1982), (Rosenblatt 1985), (Mendel 1991), (Papoulis 1991), (Scharf 1991), (Boashash *et al.* 1995), (Politis 1998), (Lahiri 2003), (Politis 2005), (Rao 2008).

8.10 Mathematics

(Morse and Transue 1956), (Kolmogorov 1960), (Gelfand and Vilenkin 1964), (Smirnov 1964), (Henniger 1970), (Kolmogorov and Fomin 1970), (Rudin 1987), (Zemanian 1987), (Champeney 1990), (Johnsonbaugh and Pfaffenberger 2002), (NIST 2010).

8.11 Signal Processing and Communications

(Sklar 1997), (Quinn and Hannan 2001), (Lathi 2002), (Gallagher 2008).

References

Adali T, Schreier PJ and Scharf LL 2011 Complex-valued signal processing: The proper way to deal with impropriety. *IEEE Transactions on Signal Processing* **59**(11), 5101–5125.

Agee BC, Schell SV and Gardner WA 1990 Spectral self-coherence restoral: A new approach to blind adaptive signal extraction using antenna arrays. *Proceedings of the IEEE* **78**, 753–767.

Aiken RT 1967 Time-variant filters and analytic signals. *IEEE Transactions on Information Theory* **IT-13**, 331–333.

Ait Dads E and Arino O 1996 Exponential dichotomy and existence of pseudo almost-periodic solutions of some differential equations. *Nonlinear Analysis, Theory, Methods & Applications* **27**(4), 369–386.

Akkarakaran S and Vaidyanathan PP 2000 Bifrequency and bispectrum maps: A new look at multirate systems with stochastic inputs. *IEEE Transactions on Signal Processing* **48**(3), 723–736.

Alekseev VG 1988 Estimating the spectral densities of a Gaussian periodically correlated stochastic process. *Problemy Peredachi Informatsii* **24**, 31–38.

Allen JC and Hobbs SL 1992 Detecting target motion by frequency-plane smoothing. *Twenty-Sixth Annual Asilomar Conference on Signals, Systems, and Computers*, Pacific Grove, CA.

Allen JC and Hobbs SL 1997 Spectral estimation of non-stationary white noise. *Journal of Franklin Institute* **334B**(1), 99–116.

Amerio L and Prouse G 1971 *Almost-Periodic Functions and Functional Equations*. Van Nostrand, New York.

Amin MG 1992 Time-frequency spectrum analysis and estimation for nonstationary random processes In *Time-Frequency Signal Analysis: Methods and Applications* (ed. Boashash B). Longman, Cheshire.

Andreas J, Bersani AM and Grande RF 2006 Hierarchy of almost-periodic function spaces. *Rendiconti di Matematica* **26**, 121–188.

Antoni J 2009 Cyclostationarity by examples. *Mechanical Systems and Signal Processing* **23**(4), 987–1036.

Azizi S and Cochran D 1999 Reproducing kernel structure and sampling on time-warped Kramer spaces. *Proceedings of IEEE International Conference on Acoustics, Speech, and Signal Processing (ICASSP 1999)*, vol. 3, pp. 1649–1652.

Azizi S, Cochran D and McDonald J 2002 Reproducing kernel structure and sampling on time-warped spaces with application to warped wavelets. *IEEE Transactions on Information Theory* **48**(3), 789–790.

Bello PA 1963 Characterization of randomly time-variant channels. *IEEE Transactions on Communications Systems* **CS-11**, 360–393.

Benedetto JJ 1996 *Harmonic Analysis and Applications*. CRC Press, New York.

Besicovitch AS 1932 *Almost Periodic Functions* Cambridge University Press, London. (New York: Dover Publications, Inc., 1954).

Billingsley P 1968 *Convergence in Probability Measures*. Wiley, New York.

Boashash B, Powers EJ and Zoubir AM 1995 *Higher-Order Statistical Signal Processing*. Longman, Australia.

Bohr H 1933 *Almost Periodic Functions* Springer, Berlin. (New York: Chelsea Publishing Company, 1947).

Boyles RA and Gardner WA 1983 Cycloergodic properties of discrete parameter non-stationary stochastic processes. *IEEE Transactions on Information Theory* **IT-29**, 105–114.

Braccini C and Oppenheim AV 1974 Unequal bandwidth spectral analysis using digital frequency warping. *IEEE Transactions on Acoustics, Speech, and Signal Processing* **ASSP-22**(4), 236–244.

Generalizations of Cyclostationary Signal Processing: Spectral Analysis and Applications, First Edition.

Brandwood DH 1983 A complex gradient operator and its application in adaptive array theory. *IEE Proceedings, Parts F and H* **130**(1), 11–16.

Brillinger DR 1965 An introduction to polyspectra. *The Annals of Mathematics* **36**(5), 1351–1374.

Brillinger DR 1969 Asymptotic properties of spectral estimates of second order. *Biometrika* **56**, 375–390.

Brillinger DR 1974 Fourier analysis of stationary processes. *Proceedings of the IEEE* **62**(12), 1628–1643.

Brillinger DR 1981 *Time Series. Data Analysis and Theory*. Holden Day, Inc., San Francisco, CA. SIAM 2001.

Brillinger DR and Rosenblatt M 1967 Asymptotic theory of estimates of *k*th-order spectra In *Spectral Analysis of Time Series* (ed. Harris B), Wiley, New York, pp. 153–188.

Brown WA 1987 On the theory of cyclostationary signals Ph.D. Dissertation Dep. Elect. Eng. Comput. Sci., Univ. California, Davis, CA.

Cambanis S and Masry E 1976 Zakai's class of bandlimited functions and processes: Its characterization and properties. *SIAM Journal of Applied Mathematics* **30**(1), 10–21.

Casinovi G 2007 l^1-norm convergence properties of correlogram spectral estimates. *IEEE Transactions on Signal Processing* **55**(9), 4354–4365.

Casinovi G 2009 Sampling and ergodic theorems for weakly almost periodic signals. *IEEE Transactions on Information Theory* **55**(4), 1883–1897.

Censor D 1973 The generalized Doppler effect and applications. *Journal of the Franklin Institute* **295**(2), 103–116.

Censor D 1984 Theory of the Doppler effect: Fact, fiction, and approximation. *Radio Science* **19**(4), 1027–1040.

Champeney DC 1990 *A Handbook of Fourier Theorems*. Cambridge University Press, New York.

Chan YT and Ho KC 2005 Joint time-scale and TDOA estimation: Analysis and fast approximation. *IEEE Transactions on Signal Processing* **53**, 2625–2634.

Chen CK and Gardner WA 1992 Signal-selective time-difference-of-arrival estimation for passive location of manmade signal sources in highly corruptive environments. Part II: Algorithms and performance. *IEEE Transactions on Signal Processing* **40**, 1185–1197.

Chen VC, Li F, Ho SS and Wechsler H 2006 Micro-Doppler effect in radar: Phenomenon, model, and simulation study. *IEEE Transactions on Aerospace and Electronic Systems* **42**(1), 2–21.

Chérif F 2011a Various types of almost periodic functions on Banach spaces: Part I. *International Mathematical Forum* **6**(19), 921–952.

Chérif F 2011b Various types of almost periodic functions on Banach spaces: Part II. *International Mathematical Forum* **6**(20), 953–985.

Chiu ST 1986 Statistical estimation of the parameters of a moving source form array data. *The Annals of Statistics* **14**(2), 559–578.

Claasen TACM and Mecklenbräuker WFG 1982 On stationary linear time-varying systems. *IEEE Transactions on Circuits Systems* pp. 169–184.

Clark J and Cochran D 1993 Time-warped bandlimited signals: Sampling, bandlimitedness, and uniqueness of representation. *Proceedings of 1993 IEEE International Symposium on Information Theory*, p. 331.

Cochran D and Clark J 1990 On the sampling and reconstruction of time-warped bandlimited signals *IEEE International Conference on Acoustics, Speech, and Signal Processing, (ICASSP 1990),*, vol. 3, pp. 1539–1541.

Cohen L 1989 Time-frequency distributions: A review. *Proceedings of the IEEE* **77**(7), 941–981.

Cohen L 1995 *Time Frequency Analysis: Theory and Applications*. Prentice Hall, Upper Saddle River, NJ.

Cooper J 1980 Scattering of electromagnetic fields by a moving boundary: The one-dimensional case. *IEEE Transactions on Antennas and Propagation* **AP-28**(6), 791–795.

Corduneanu C 1989 *Almost Periodic Functions* Chelsea Publishing Company New York.

Cramér H 1940 On the theory of stationary random processes. *Annals of Mathematics* **1**, 215–230.

Crochiere RE and Rabiner LR 1981 Interpolation and decimation of digital signals: A tutorial review. *Proceedings of the IEEE* **69**, 300–331.

Dandawaté AV and Giannakis GB 1994 Nonparametric polyspectral estimators for *k*th-order (almost) cyclostationary processes. *IEEE Transactions on Information Theory* **40**(1), 67–84.

Dandawaté AV and Giannakis GB 1995 Asymptotic theory of mixed time averages and *k*th-order cyclic-moment and cumulant statistics. *IEEE Transactions on Information Theory* **41**(1), 216–232.

De Angelis V, Izzo L, Napolitano A and Tanda M 2005 Cyclostationarity-based parameter estimation of wide-band signals in mobile communications. *IEEE Statistical Signal Processing Workshop (SSP'05)*, Bordeaux, France.

Dehay D 1994 Spectral analysis of the covariance of the almost periodically correlated processes. *Stochastic Processes and their Applications* **50**, 315–330.

Dehay D 2007 Discrete time observation of almost periodically correlated processes and jitter phenomena. *XV European Signal Processing Conference (EUSIPCO 2007)*, Poznan, Poland.

Dehay D and Hurd HL 1994 Representation and estimation for peridically and almost periodically correlated random processes In *Cyclostationarity in Communications and Signal Processing* (ed. Gardner WA) IEEE Press, New York, Chapter 6, pp. 295–328.

Dehay D and Leśkow J 1996 Functional limit theory for the spectral covariance estimator. *Journal Applied Probability* **33**, 1077–1092.

Dmochowski J, Benesty J and Affes S 2009 On spatial aliasing in microphone arrays. *IEEE Transactions on Signal Processing* **57**(4), 1383–1395.

Doob JL 1953 *Stochastic Processes*. John Wiley & Sons, Inc. New York.

Eberlein WF 1949 Abstract ergodic theorems and weak almost periodic functions. *Transactions of the American Mathematical Society* **67**, 217–240.

Eberlein WF 1956 The point spectrum of weakly almost periodic functions. *Michigan Journal of Mathematics* **3**, 137–139.

Ferguson BG 1999 Time-delay estimation techniques applied to the acoustic detection of jet aircraft transit. *The Journal of the Acoustical Society of America* **106**(1), 225–264.

Flagiello F, Izzo L and Napolitano A 2000 A computationally efficient and interference tolerant nonparametric algorithm for LTI system identification based on higher order cyclostationarity. *IEEE Transactions on Signal Processing* **48**(4), 1040 –1052.

Flandrin P 1989 On the spectrum of fractional Brownian motions. *IEEE Transactions on Information Theory* **35**(1), 197 –199.

Flandrin P 1999 *Time-Frequency/Time-Scale Analysis*. Academic Press, San Diego, CA.

Flandrin P, Napolitano A, Ozaktas HM and Thompson DJ 2011 Recent advances in theory and methods for nonstationary signal analysis (editorial of the special issue). *EURASIP Journal on Advances in Signal Processing (JASP)*. Article ID 963642.

Franaszek PA 1967 On linear systems which preserve wide sense stationarity. *SIAM Journal of Applied Mathematics* **15**, 1481–1484.

Franaszek PA and Liu B 1967 On a class of linear time-varying filters. *IEEE Transactions on Information Theory* **IT-13**, 477–481.

Franks LE 1994 Polyperiodic linear filtering In *Cyclostationarity in Communications and Signal Processing* (ed. Gardner WA) IEEE Press, New York, Chapter 4, pp. 240–266.

Franz S, Mitra SK and Doblinger G 2003 Frequency estimation using warped discrete Fourier transform. *Signal Processing* **83**, 1661–1671.

Gallagher RG 2008 *Principles of Digital Communication*. Cambridge University Press, Cambridge, UK.

Garcia FM, Lourtie IMG and Buescu J 2001 $l^2(\mathbb{R})$ nonstationary processes and the sampling theorem. *IEEE Signal Processing Letters* **8**(4), 117–119.

Gardner WA 1972 A sampling theorem for nonstationay random processes. *IEEE Transactions on Information Theory* **IT-18**, 808–809.

Gardner WA 1978 Stationarizable random processes. *IEEE Transactions on Information Theory* **IT-24**, 8–22.

Gardner WA 1985 *Introduction to Random Processes with Applications to Signals and Systems*. Macmillan, New York. (1990, 2nd Edn., McGraw-Hill, New York).

Gardner WA 1986a Measurements of spectral correlation. *IEEE Trans. on Acoustic, Speech, and Signal Processing* **ASSP-34**, 1111–1123.

Gardner WA 1986b The role of spectral correlation in design and performance analysis of synchronizers. *IEEE Transactions on Communications* **COM-34**(11), 1089–1095.

Gardner WA 1986c The spectral correlation theory of cyclostationary time series. *Signal Processing* **11**, 13–36. (Erratum: *Signal Processing*, vol. 11, p. 405).

Gardner WA 1987a Common pitfalls in the application of stationary process theory to time-sampled and modulated signals. *IEEE Transactions on Communications* **COM-35**, 529–534.

Gardner WA 1987b Rice's representation for cyclostationary processes. *IEEE Transactions on Communications* **COM-35**, 74–78.

Gardner WA 1987c Spectral correlation of modulated signals, Part I – Analog modulation. *IEEE Transactions on Communications* **COM-35**(6), 584–594.

Gardner WA 1987d *Statistical Spectral Analysis: A Nonprobabilistic Theory*. Prentice-Hall, Englewood Cliffs, NJ.

Gardner WA 1988a Correlation estimation and time-series modeling for nonstationary processes. *Signal Processing* **15**(1), 31–41.

Gardner WA 1988b Signal interception: A unifying theoretical framework for feature detection. *IEEE Transactions on Communications* **COM-36**, 897–906.

Gardner WA 1988c Simplification of MUSIC and ESPRIT by exploitation of cyclostationarity. *Proceedings of the IEEE* **76**(7), 845–847.

Gardner WA 1991a Exploitation of spectral redundancy in cyclostationary signals. *IEEE Signal Processing Magazine* **8**, 14–36.

Gardner WA 1991b On the spectral coherence of nonstationary processes. *IEEE Transactions on Signal Processing* **39**(2), 424–430.

Gardner WA 1991c Two alternative philosophies for estimation of the parameters of time-series. *IEEE Transaction on Information Theory* **37**, 216–218.

Gardner WA 1993 Cyclic Wiener filtering: Theory and method. *IEEE Transaction on Communications* **41**(1), 151–163.

Gardner WA 1994 An introduction to cyclostationary signals In *Cyclostationarity in Communications and Signal Processing* (ed. Gardner WA) IEEE Press, New York, Chapter 1, pp. 1–90.

Gardner WA and Brown WA 1991 Fraction-of-time probability for time-series that exhibit cyclostationarity. *Signal Processing* **23**, 273–292.

Gardner WA and Chen CK 1988 Interference-tolerant time-difference-of-arrival estimation for modulated signals. *IEEE Transactions on Acoustics, Speech, and Signal Processing* **ASSP-36**, 1385–1395.

Gardner WA and Chen CK 1992 Signal-selective time-difference-of-arrival estimation for passive location of man-made signal sources in highly corruptive environments. Part I: Theory and method. *IEEE Transactions on Signal Processing* **40**, 1168–1184.

Gardner WA and Franks LE 1975 Characterization of cyclostationary random signal processes. *IEEE Transactions on Information Theory* **IT-21**, 4–14.

Gardner WA and Spooner CM 1992 Signal interception: Performance advantages of cyclic feature detectors. *IEEE Transactions on Communications* **40**(1), 149–159.

Gardner WA and Spooner CM 1993 Detection and source location of weak cyclostationary signals: Simplifications of the maximum-likelihood receiver. *IEEE Transactions on Communications* **41**(6), 905–916.

Gardner WA and Spooner CM 1994 The cumulant theory of cyclostationary time-series, Part I: Foundation. *IEEE Transactions on Signal Processing* **42**, 3387–3408.

Gardner WA, Brown WA and Chen CK 1987 Spectral correlation of modulated signals, Part II – Digital modulation. *IEEE Transactions on Communications* **COM-35**(6), 595–601.

Gardner WA, Napolitano A and Paura L 2006 Cyclostationarity: Half a century of research. *Signal Processing* **86**(4), 639–697.

Gelfand IM and Vilenkin NY 1964 *Generalized Functions, Applications of Harmonic Analysis* vol. 4 Academic Press New York.

Genossar MJ 1992 Spectral characterization of nonstationary processes Ph.D. Dissertation Stanford University Stanford, CA.

Genossar MJ, Lev-Ari H and Kailath T 1994 Consistent estimation of the cyclic autocorrelation. *IEEE Transactions on Signal Processing* **42**, 595–603.

Gerr NL and Allen JC 1994 The generalized spectrum and spectral coherence of harmonizable time series. *Digital Signal Processing* **4**, 222–238.

Giannakis GB 1998 Cyclostationary signal analysis In *The Digital Signal Processing Handbook* (ed. Madisetti VK and Williams DB) CRC Press and IEEE Press, New York, Chapter 17.

Gikhman II and Skorokhod AV 1969 *Introduction to the Theory of Random Processes*. Dover, New York.

Gladyshev EG 1961 Periodically correlated random sequences. *Soviet Math. Dokl.* **2**, 385–388.

Gladyshev EG 1963 Periodically and almost periodically correlated random processes with continuous time parameter. *Theory Prob. Appl.* pp. 137–177. (in Russian).

Gray JE and Addison SR 2003 Effect of nonuniform target motion on radar backscattered waveforms. *IEE Proceedings on Radar, Sonar and Navigation* **150**(4), 262–270.

Grenander U and Rosenblatt M 1957 *Statistical Analysis of Stationary Time Series*. Wiley, New York.

Han Y and Hong J 2007 Almost periodic random sequences in probability. *Journal of Mathematical Analysis and Applications* **336**, 962–974.

Hanin LG and Schreiber BM 1998 Discrete spectrum of nonstationary stochastic processes on groups. *Journal of Theoretical Probability* **11**(4), 1111–1133.

Hanssen A and Scharf LL 2003 A theory of polyspectra for nonstationary stochastic processes. *IEEE Transactions on Signal Processing* **51**(5), 1243–1252.

Henniger J 1970 Functions of bounded mean square, and generalized Fourier-Stieltjes transforms. *Canadian Journal of Mathematics* **XXII**(5), 1016–1034.

Hlawatsch F and Bourdeaux-Bartels GF 1992 Linear and quadratic time-frequency signal representations. *IEEE Signal Processing Magazine* pp. 21–67.

Ho KC and Chan YT 1998 Optimum discrete wavelet scaling and its application to delay and Doppler estimation. *IEEE Transactions on Signal Processing* **46**(9), 2285–2290.

Hopgood JR and Rayner PJW 2003 Single channel nonstationary stochastic signal separation using linear time-varying filters. *IEEE Transactions on Signal Processing* **51**(7), 1739–1752.

Hurd HL 1973 Testing for harmonizability. *IEEE Transactions on Information Theory* **IT-19**, 316–320.

Hurd HL 1974 Periodically correlated processes with discontinuous correlation function. *Theory of Probability and its Applications* **19**, 804–808.

Hurd HL 1988 Spectral coherence of nonstationary and transient stochastic processes *IEEE Fourth Annual ASSP Workshop on Spectrum Estimation and Modeling*, Minneapolis, MN.

Hurd HL 1989a Nonparametric time series analysis for periodically correlated processes. *IEEE Transactions on Information Theory* **35**, 350–359.

Hurd HL 1989b Representation of strongly harmonizable periodically correlated processes and their covariances. *Journal of Multivariate Analysis* **29**, 53–67.

Hurd HL 1991 Correlation theory of almost periodically correlated processes. *Journal of Multivariate Analysis* **37**, 24–45.

Hurd HL and Gerr NL 1991 Graphical methods for determining the presence of periodic correlation. *Journal of Time Series Analysis* **12**, 337–350.

Hurd HL and Koski T 2004 The Wold isomorphism for cyclostationary sequences. *Signal Processing* **84**, 813–824.

Hurd HL and Leśkow J 1992a Estimation of the Fourier coefficient functions and their spectral densities for ϕ-mixing almost periodically correlated processes. *Statistics & Probability Letters* **14**, 299–306.

Hurd HL and Leśkow J 1992b Strongly consistent and asymptotically normal estimation of the covariance for almost periodically correlated processes. *Statist. & Decision* **10**, 201–225.

Hurd HL and Miamee AG 2007 *Periodically Correlated Random Sequences: Spectral Theory and Practice* John Wiley & Sons Inc., New Jersey.

Izzo L and Napolitano A 1996 Higher-order cyclostationarity properties of sampled time-series. *Signal Processing* **54**, 303–307.

Izzo L and Napolitano A 1997 Higher-order statistics for Rice's representation of cyclostationary signals. *Signal Processing* **56**, 279–292.

Izzo L and Napolitano A 1998a The higher-order theory of generalized almost-cyclostationary time-series. *IEEE Transactions on Signal Processing* **46**(11), 2975–2989.

Izzo L and Napolitano A 1998b Multirate processing of time series exhibiting higher order cyclostationarity. *IEEE Transactions on Signal Processing* **46**(2), 429–439.

Izzo L and Napolitano A 2002a Linear time-variant transformations of generalized almost-cyclostationary signals, Part I: Theory and method. *IEEE Transactions on Signal Processing* **50**(12), 2947–2961.

Izzo L and Napolitano A 2002b Linear time-variant transformations of generalized almost-cyclostationary signals, Part II: Development and applications. *IEEE Transactions on Signal Processing* **50**(12), 2962–2975.

Izzo L and Napolitano A 2003 Sampling of generalized almost-cyclostationary signals. *IEEE Transactions on Signal Processing* **51**(6), 1546–1556.

Izzo L and Napolitano A 2005 Generalized almost-cyclostationary signals In *Advances in Imaging and Electron Physics* (ed. Hawkes PW) vol. 135, Elsevier, Oxford, Chapter 3, pp. 103–223.

Javors'kyj I, Iasyev I, Zakrzewski Z and Brooks SP 2007 Coherent covariance analysis of periodically correlated random processes. *Signal Processing* **87**, 13–32.

Jessen B and Tornehave H 1945 Mean motions and zeros of almost periodic functions. *Acta Mathematica* **77**(1), 137–279.

Jin Q, Wong KM and Luo ZQ 1995 The estimation of time delay and Doppler stretch of wideband signals. *IEEE Transactions on Signal Processing* **43**, 904–916.

Johnsonbaugh R and Pfaffenberger WE 2002 *Foundations of Mathematical Analysis*. Dover Publications, Inc., Mineola, NY.

Kac M 1959 *Statistical Independence in Probability, Analysis and Number Theory*. The Mathematical Association of America, USA.

Kac M and Steinhaus H 1938 Sur les fonctions indépendantes IV. *Studia Mathematica* **7**, 1–15.

Kahane JP 1961 Sur les coefficients de Fourier-Bohr. *Studia Mathematica* **21**, 103–106.

Kahane JP 1962 Sur les fonctions presque-périodiques généralisées dont le spectre est vide. *Studia Mathematica* **21**, 231–236.

Kampé de Fériet J and Frenkiel FN 1962 Correlations and spectra for non-stationary random functions. *Mathematics of Computation* **16**(77), 1–21.

Kapoulitsas GM 1981 On the non-relativistic Doppler effect. *European Journal of Physics* **2**, 174–177.

Kelly EJ 1961 The radar measurement of range, velocity, and acceleration. *IRE Transactions on Military Electronics* **MIL-5**, 51–57.

Kelly EJ and Wishner RP 1965 Matched-filter theory for high-velocity, accelerating targets. *IEEE Transactions on Military Electronics* **MIL-9**, 56–69.

Kolmogorov AN 1933 *Foundations of the Theory of Probability*. and Chelsea, New York, 1956.

Kolmogorov AN 1960 On the $\phi^{(n)}$ classes of Fortet and Blanc-Lapierre. *Thoery of Probability and its Applications* **5**, 337.

Kolmogorov AN and Fomin SV 1970 *Introductory Real Analysis*. Prentice Hall, Englewood Cliffs, NJ. (Dover, New York, 1975).

Lahiri SN 2003 A necessary and sufficient condition for independence of discrete Fourier transforms under short- and long-range dependence. *The Annals of Statistics* **31**(2), 613–641.

Lathi BP 2002 *Linear Systems and Signals*. Oxford University Press, New York.

Lee AJ 1978 Sampling theorems for nonstationary random processes. *Transactions of the American Mathematical Society* **242**, 225–241.

Lee YW 1967 *Statistical Theory of Communication*. Wiley, New York.

Lenart L, Leśkow J and Synowiecki R 2008 Subsampling in testing autocovariance for periodically correlated time series. *Journal of Time Series Analysis* **29**(6), 995–1018.

Leonov VP and Shiryaev AN 1959 On a method of calculation of semi-invariants. *Theory of Probability and its Applications*.

Leśkow J and Napolitano A 2006 Foundations of the functional approach for signal analysis. *Signal Processing* **86**(12), 3796–3825.

Leśkow J and Napolitano A 2007 Non-relatively measurable functions for secure-communications signal design. *Signal Processing* **87**(11), 2765–2780.

Lii KS and Rosenblatt M 1998 Line spectral analysis for harmonizable processes. *Proceedings of the National Academy of Science of USA* **95**, 4800–4803.

Lii KS and Rosenblatt M 2002 Spectral analysis for harmonizable processes. *The Annals of Statistics* **30**(1), 258–297.

Lii KS and Rosenblatt M 2006 Estimation for almost periodic processes. *The Annals of Statistics* **34**(3), 1115–1139.

Liu B and Franaszek PA 1969 A class of time-varying digital filters. *IEEE Transactions on Circuit Theory* **CT-16**, 467–471.

Lloyd SP 1959 A sampling theorem for stationary (wide sense) stochastic processes. *Transactions of the American Mathematical Society* **92**(1), 1–12.

Loève M 1963 *Probability Theory*. Van Nostrand, Princeton, NJ.

Makur A and Mitra SK 2001 Warped discrete-Fourier transform: Theory and applications. *IEEE Transactions on Circuits and Systems I: Fundamental Theory and Applications* **48**(9), 1086–1093.

Mandelbrot BB and Van Ness HW 1968 Fractional Brownian motions, fractional noises and applications. *SIAM Review* **10**, 422–436.

Martin W and Flandrin P 1985 Wigner-Ville spectral analysis of nonstationary processes. *IEEE Transactions on Acoustics, Speech, and Signal Processing* (6), 1461–1470.

Matz G, Hlawatsch F and Kozek W 1997 Generalized evolutionary spectral analysis and the Weyl spectrum of nonstationary random processes. *IEEE Transactions on Signal Processing* **45**(6), 1520–1534.

Mendel JM 1991 Tutorial on higher-order statistics (spectra) in signal processing and system theory: Theoretical results and some applications. *Proceedings of the IEEE* **79**, 278–305.

Middleton D 1967 A statistical theory of reverberation and similar first-order scattered fields, Part II: Moments spectra, and special distributions. *IEEE Transactions on Information Theory* **IT-13**(3), 393–414.

Morse M and Transue W 1956 ℂ-bimeasures and their integral extensions. *Annals of Mathematics* **64**, 480–504.

Munk W, Worchester P and Wunsch C 1995 *Ocean Acoustic Tomography*. Cambridge University Press, New York.

Napolitano A 1995 Cyclic higher-order statistics: input/output relations for discrete- and continuous-time MIMO linear almost-periodically time-variant systems. *Signal Processing* **42**(2), 147–166.

Napolitano A 2001 On the spectral correlation measurement of nonstationary stochastic processes *Thirty-Fifth Annual Asilomar Conference on Signals, Systems, and Computers*, Pacific Grove, CA.

Napolitano A 2003 Uncertainty in measurements on spectrally correlated stochastic processes. *IEEE Transactions on Information Theory* **49**(9), 2172–2191.

Napolitano A 2007a Estimation of second-order cross-moments of generalized almost-cyclostationary processes. *IEEE Transactions on Information Theory* **53**(6), 2204–2228.

Napolitano A 2007b Generalized almost-cyclostationary processes and spectrally correlated processes: Two extensions of the class of the almost-cyclostationary processes *IEEE-EURASIP International Symposium on Signal Processing and its Applications (ISSPA 2007)*, Sharjah, United Arab Emirates. Invited Paper.

Napolitano A 2007c Mean-square consistent estimation of the spectral correlation density for spectrally correlated stochastic processes *Proceedings of IEEE International Conference on Acoustics, Speech, and Signal Processing (ICASSP 2007)*, Honolulu, Hawaii, USA.

Napolitano A 2009 Discrete-time estimation of second-order statistics of generalized almost-cyclostationary processes. *IEEE Transactions on Signal Processing* **57**(5), 1670–1688.

Napolitano A 2010a Interpolation and decimation of spectrally correlated stochastic processes *XVIII European Signal Processing Conference (EUSIPCO 2010)*, Aalborg, Denmark.

Napolitano A 2010b Sampling theorems for Doppler-stretched wide-band signals. *Signal Processing* **90**(7), 2276–2287.

Napolitano A 2011 Sampling of spectrally correlated processes. *IEEE Transactions on Signal Processing* **59**(2), 525–539.

Napolitano A and Doğançay K 2011 On waveform design in interference-tolerant range-Doppler estimation for wideband multistatic radars. *XIX European Signal Processing Conference (EUSIPCO 2011)*, Barcelona, Spain.

Napolitano A and Spooner CM 2000 Median-based cyclic polyspectrum estimation. *IEEE Transactions on Signal Processing* **48**(5), 1462–1466.

Napolitano A and Spooner CM 2001 Cyclic spectral analysis of continuous-phase modulated signals. *IEEE Transactions on Signal Processing* **49**(1), 30–44.

Napolitano A and Tesauro M 2011 Almost-periodic higher-order statistic estimation. *IEEE Transactions on Information Theory* **57**(1), 514–533.

Neipp C, Hernandez A, Rodes JJ, Marquez A, Belendez T and Belendez A 2003 An analysis of the classical Doppler effect. *European Journal of Physics* **24**, 497–505.

NIST 2010 In *NIST Handbook of Mathematical Functions* (ed. Oliver FWJ, Lozier DW, Boisvert RF and Clark CW) Cambridge University Press, New York.

Oberg J 2004 Titan calling. *IEEE Spectrum* **41**, 28–33.

Ogura H 1971 Spectral representation of a periodic nonstationary random process. *IEEE Transactions on Information Theory* **IT-17**, 143–149.

Øigård TA, Scharf LL and Hanssen A 2006 Spectral correlations of fractional Brownian motion. *Physical Review E.* E74 031114.

Oppenheim AV and Johnson DH 1972 Discrete representations of signals. *Proceedings of the IEEE* **60**(6), 681–691.

Oppenheim A, Johnson DH and Steiglitz K 1971 Computation of spectra with unequal resolution using the fast Fourier transform. *Proceedings of the IEEE* **59**(2), 299–301.

Oppenheim AV and Schafer RW 1989 *Discrete-Time Signal Processing*. Prentice-Hall, Englewood Cliffs, NJ.

Ozaktas H, Zalevsky Z and Kutay M 2001 *The Fractional Fourier Transform with Applications in Optics and Signal Processing*. John Wiley and Sons, New York, Series in Pure and Applied Optics.

Papoulis A 1991 *Probability, Random Variables, and Stochastic Processes*, 3rd edn. McGraw-Hill, New York.

Parzen E 1957 On consistent estimates of the spectrum of stationary time series. *The Annals of Mathematical Statistics* **28**(2), 329–348.

Pfaffelhuber E 1975 Generalized harmonic analysis for distributions. *IEEE Transactions on Information Theory* **IT-21**, 605–611.

Phong VQ 2007 A new proof and generalizations of Gearhart's theorem. *Proceedings of the American Mathematical Society* **135**(7), 2065–2072.

Picinbono B 1996 Second-order complex random vectors and Normal distributions. *IEEE Transactions on Signal Processing* **44**, 2637–2640.

Picinbono B and Bondon P 1997 Second-order statistics of complex signals. *IEEE Transactions on Signal Processing* **45**(2), 411–420.

Politis DN 1998 Computer intensive methods in statistical analysis. *IEEE Signal Processing Magazine* **15**(1), 39–55.

Politis DN 2005 Complex-valued tapers. *IEEE Signal Processing Letters* **12**(7), 512–515.

Priestley MB 1965 Evolutionary spectra and non-stationary processes. *Journal of the Royal Statistical Society. Series B (Methodological)* **27**(2), 204–237.

Prohorov YV and Rozanov YA 1989 *Probability Theory*. Springer-Verlag, New York.

Quinn BG and Hannan EJ 2001 *The Estimation and Tracking of Frequency*. Cambridge University Press, Cambridge, UK.

Rabiner LR, Shafer RW and Rader CM 1969 The chirp z-transform algorithm. *IEEE Transactions on Audio and Electroacoustics* **AU-17**(2), 86–92.

Rao MM 2008 Integral representations of second-order processes. *Nonlinear Analysis* **69**, 979–986.

Rihaczek AW 1967 Radar resolution of moving targets. *IEEE Transactions on Information Theory* **IT-13**, 51–56.

Rosenblatt M 1974 *Random Processes*. Springer-Verlag, New York.

Rosenblatt M 1985 *Stationary Sequences and Random Fields*. Birkhäuser, Boston.

Rudin W 1987 *Real and Complex Analysis*, 3rd edn. McGraw-Hill, New York.

Sadler BM and Dandawaté AV 1998 Nonparametric estimation of the cyclic cross-spectrum. *IEEE Transactions on Information Theory* **44**, 351–358.

Sadowsky JS and Kafedziski V 1998 On the correlation and scattering functions of the WSS channel for mobile communications. *IEEE Transactions on Vehicular Technology* **47**(1), 270–282.

Sathe VP and Vaidyanathan PP 1993 Effects of multirate systems on the statistical properties of random signals. *IEEE Transactions on Signal Processing* **41**, 131–146.

Scharf LL 1991 *Statistical Signal Processing. Detection, Estimation, and Time Series Analysis*. Addison-Wesley Publishing Company, New York.

Scharf LL, Schreier PJ and Hanssen A 2005 The Hilbert space geometry of the Rihaczek distribution for stochastic analytic signals. *IEEE Signal Processing Letters* **12**(4), 297–300.

Schell SV 1995 Asymptotic moments of estimated cyclic correlation matrices. *IEEE Transactions on Signal Processing* **43**(1), 173–180.

Schlotz RA 2002 Problems in modeling UWB channels *Thirty-Sixth Asilomar Conference on Signals, Systems, and Computers*, Pacific Grove, CA.

Schreier PJ and Scharf LL 2003a Second-order analysis of improper complex random vectors and processes. *IEEE Transactions on Signal Processing* **51**(3), 714–725.

Schreier PJ and Scharf LL 2003b Stochastic time-frequency analysis using the analytic signal: Why the complementary distribution matters. *IEEE Transactions on Signal Processing* **51**(12), 3071–3079.

Silverman RA 1957 Locally stationary random processes. *IRE Transactions on Information Theory* pp. 182–187.

Sklar B 1997 Rayleigh fading channels in mobile digital communication systems, Part I: Characterization. *IEEE Communications Magazine* **35**, 136–146.

Smirnov VI 1964 *A Course of Higher Mathematics* vol. I. Pergamon, Oxford, UK.

Soedjack H 2002 Consistent estimation of the bispectral density function of a harmonizable process. *Journal of Statistical Planning and Inference* **100**, 159–170.

Song Z, Sun W, Zhou X and Hou Z 2007 An average sampling theorem for bandlimited stochastic processes. *IEEE Transactions on Information Theory* **53**(12), 4798–4800.

Spooner CM and Gardner WA 1994 The cumulant theory of cyclostationary time-series, Part II: Development and applications. *IEEE Transactions on Signal Processing* **42**, 3409–3429.

Spooner CM and Nicholls RB 2009 Spectrum sensing based on spectral correlation In *Cognitive Radio Technology* (ed. Fette B), 2nd edn. Elsevier, Chapter 18.

Swick DA 1969 A review of wideband ambiguity functions. Technical Report 6994, Naval Research Laboratory.

Swift RJ 1996 Almost periodic harmonizable processes. *Georgian Mathematical Journal* **3**(3), 275–292.

Thomson DJ 1982 Spectrum estimation and harmonic analysis. *Proceedings of the IEEE* **70**, 1055–1096.

Urbanik K 1962 Fourier analysis in Marcinkiewicz spaces. *Studia Mathematica* **21**, 93–102.

Vaidyanathan PP 1990 Multirate digital filters, filter banks, polyphase networks, and applications: A tutorial review. *Proceedings of the IEEE* **78**, 56–93.

van den Bos A 1994 Complex gradient and Hessian. *IEE Proceedings on Vision, Image, and Signal Processing* **141**(6), 380–382.

van den Bos A 1995 The multivariate complex Normal distribution: A generalization. *IEEE Transactions on Information Theory* **41**, 537–539.

Van Trees HL 1971 *Detection, Estimation, and Modulation Theory. Part III*. John Wiley & Sons, Inc. New York.

Varghese T and Donohue KD 1994 Mean-scatter spacing estimates with spectral correlation. *Journal of Acoustical Society of America* **96**, 3504–3515.

von Neumann J 1934 Almost periodic functions in a group I. *Transactions of the American Mathematical Society* **36**, 445–492.

von Schroeter T 1999 Frequency warping with arbitrary allpass maps. *IEEE Signal Processing Letters* **6**(5), 116–118.

Wahlberg P and Schreier PJ 2008 Spectral relations for multidimensional complex improper stationary and (almost) cyclostationary processes. *IEEE Transactions on Information Theory* **54**(4), 1670–1682.

Wang Y and Zhou G 1998 On the use of higher-order ambiguity function for multicomponent polynomial phase signals. *Signal Processing* **65**, 283–296.

Weiss LG 1994 Wavelets and wideband correlation processing. *IEEE Signal Processing Magazine* **11**(1), 13–22.

Wiener N 1930 Generalized harmonic analysis. *Acta Mathematica* **55**, 117–258.

Wiener N 1933 *The Fourier Integral and Certain of its Applications*. Cambridge University Press, London. (Dover, New York, 1958).

Wiener N 1949 *Extrapolation, Interpolation, and Smoothing of Stationary Time Series*. MIT Press, Cambridge, MA.

Wold HOA 1948 On prediction in stationary time series. *Ann. Math. Statist.* **19**, 558–567.

Wornell GW 1993 Wavelet-based representations for the $1/f$ family of fractal processes. *Proceedings of the IEEE* **81**(10), 1428–1450.

Wu PW and Lev-Ari H 1997 Optimized estimation of moments for nonstationary signals. *IEEE Transactions on Signal Processing* **45**(5), 1210–1221.

Xia XG and Zhang Z 1992 On a conjecture on time-warped band-limited signals. *IEEE Transactions on Signal Processing* **40**(1), 252–254.

Yoo DS 2008 Equivalent complex baseband representations of linear time-variant systems and signals. *IEEE Transactions on Signal Processing* **56**(8), 3775–3778.

Zakai M 1965 Band-limited functions and the sampling theorem. *Information and Control* **8**, 143–158.

Zemanian

Zemanian AH 1987 *Distribution Theory and Transform Analysis*. Dover, New York.

Zhang C 1994 Pseudo almost periodic solutions of some differential equations. *Journal of Mathematical Analysis and Applications* **181**, 62–76.

Zhang C 1995 Pseudo almost periodic solutions of some differential equations, II. *Journal of Mathematical Analysis and Applications* **192**, 543–561.

Zhang C and Liu J 2010 Two unsolved problems on almost periodic type functions. *Applied Mathematics Letters* **23**, 1133–1136.

Index